Lecture Notes in Electrical Engineering

Volume 331

About this Series

"Lecture Notes in Electrical Engineering (LNEE)" is a book series which reports the latest research and developments in Electrical Engineering, namely:

- Communication, Networks, and Information Theory
- Computer Engineering
- Signal, Image, Speech and Information Processing
- Circuits and Systems
- Bioengineering

LNEE publishes authored monographs and contributed volumes which present cutting edge research information as well as new perspectives on classical fields, while maintaining Springer's high standards of academic excellence. Also considered for publication are lecture materials, proceedings, and other related materials of exceptionally high quality and interest. The subject matter should be original and timely, reporting the latest research and developments in all areas of electrical engineering.

The audience for the books in LNEE consists of advanced level students, researchers, and industry professionals working at the forefront of their fields. Much like Springer's other Lecture Notes series, LNEE will be distributed through Springer's print and electronic publishing channels.

More information about this series at http://www.springer.com/series/7818

James J. (Jong Hyuk) Park
Yi Pan · Han-Chieh Chao
Gangman Yi
Editors

Ubiquitous Computing Application and Wireless Sensor

UCAWSN-14

Springer

Editors
James J. (Jong Hyuk) Park
Department of Computer Science and Engineering
Seoul National University of Science and Technology
Seoul
Korea, Republic of (South Korea)

Yi Pan
Department of Computer Science
Georgia State University
Atlanta, GA
USA

Han-Chieh Chao
National Ilan University
Yilan City
Taiwan

Gangman Yi
Department of Computer Science and Engineering
Gangneung-Wonju National University
Wonju
Korea, Republic of (South Korea)

ISSN 1876-1100 ISSN 1876-1119 (electronic)
Lecture Notes in Electrical Engineering
ISBN 978-94-017-9617-0 ISBN 978-94-017-9618-7 (eBook)
DOI 10.1007/978-94-017-9618-7

Library of Congress Control Number: 2014955621

Springer Dordrecht Heidelberg New York London

Printed on acid-free paper

Springer Science+Business Media B.V. Dordrecht is part of Springer Science+Business Media (www.springer.com)

Message from the UCAWSN 2014 General Chairs

The 2nd FTRA International Conference on Ubiquitous Computing Application and Wireless Sensor Network (UCAWSN-14) is an event of the series of international scientific conferences. This conference took place in Jeju, Korea, during July 7–10, 2014. UCAWSN-14 was the most comprehensive conference focused on various aspects of Ubiquitous Computing Application and Wireless Sensor Network (UCA-WSN). The UCAWSN-14 will provided an opportunity for academic and industry professionals to discuss the latest issues and progress in the area of UCAWSN. In addition, the conference published high-quality papers that are closely related to various theories and practical applications. Furthermore, we expect that the conference and its publications will be a trigger for further related research and technology improvements in this important subject.

The papers included in the proceedings cover two tracks: Track 1–Ubiquitous Computing and Track 2–Wireless Sensor Network. Accepted and presented papers highlight new trends and challenges of Ubiquitous Computing Application and Wireless Sensor Network. The presenters showed how new research could lead to novel and innovative applications. We hope you will find these results useful and inspiring for your future research.

We would like to express our sincere thanks to the Program Chairs: Gangman Yi (Gangneung-Wonju National University, Korea), Yuh-Shyan Chen (National Taipei University, Taiwan), Xiaohong Peng (Aston University, UK), Ali Abedi (Maine University, US), Neil Y. Yen (the University of Aizu, Japan), Jen Juan Li (North Dakota State University, USA), Hongxue (Harris) Wang (Athabasca University, Canada), Lijun Zhu (ISTIC, China), and all Program Committee members and additional reviewers for their valuable efforts in the review process, which helped us to guarantee the highest quality of the selected papers for the conference. Our special thanks go to the invited speakers who kindly accepted our invitation, and helped to meet the objectives of the conference: Prof. Young-Sik Jeong (Dongguk University, Korea) and Mohammad S. Obaidat (Monmouth University,

USA), and another special thank you goes to Hak Hyun Choi (Seoul Women's University) for his efforts to make the conference successful.

We cordially thank all the authors for their valuable contributions and the other participants of this conference. The conference would not have been possible without their support. Thanks are also due to the many experts who contributed to making the event a success.

July 2014

Hwa-Young Jeong
Kyung Hee University, Korea

Han-Chieh Chao
National Ilan University, Taiwan

Yi Pan
Georgia State University, USA

Qun Jin
Waseda University, Japan

Message from the UCAWSN 2014 Program Chairs

Welcome to the 2nd FTRA International Conference on Ubiquitous Computing Application and Wireless Sensor Network (UCAWSN-14), held in Jeju during July 7–10, 2014. UCAWSN 2014 is the most comprehensive conference focused on various aspects of information technology. UCAWSN 2014 provides an opportunity for academic and industry professionals to discuss the latest issues and progress in the area of UCAWSN such as Ubiquitous and context-aware computing, context-awareness reasoning and representation, locations awareness services, architectures, protocols and algorithms of WSN, energy, management and control of WSN, etc. In addition, the conference will publish high-quality papers that are closely related to the various theories and practical applications in UCA-WSN. Furthermore, we expect that the conference and its publications will be a trigger for further related research and technology improvements in this important subject.

For UCAWSN 2014, we received many paper submissions after a rigorous peer-review process; we accepted articles of high quality for the UCAWSN 2014 proceedings, published by Springer. All submitted papers underwent blind reviews by at least three reviewers from the Technical Program Committee, which consisted of leading researchers around the globe. Without their hard work, achieving such a high-quality proceeding would not have been possible. We take this opportunity to thank them for their great support and cooperation. We would like to sincerely thank the following invited speakers who kindly accepted our invitation, and, in this way, helped to meet the objectives of the conference: Prof. Young-Sik Jeong (Dongguk University, Korea) and Mohammad S. Obaidat (Monmouth University,

USA). Finally, we would like to thank all of you for your participation in our conference, and we also thank all the authors, reviewers, and organizing committee members. Thank you and enjoy the conference!

July 2014

Gangman Yi
Gangneung-Wonju
National University, Korea

Yuh-Shyan Chen
National Taipei University, Taiwan

Xiaohong Peng
Aston University, UK

Ali Abedi
Maine University, US

Neil Y. Yen
The University of Aizu, Japan

Jen Juan Li
North Dakota State University, USA

Hongxue (Harris) Wang
Athabasca University, Canada

Lijun Zhu, ISTIC, China
UCAWSN 2014 Program Chairs

Organization

Steering Chairs
James J. Park, SeoulTech, Korea
Hamid R. Arabnia, The University of Georgia, USA

General Chairs
Hwa-Young Jeong, Kyung Hee University, Korea
Han-Chieh Chao, National Ilan University, Taiwan
Yi Pan, Georgia State University, USA
Qun Jin, Waseda University, Japan

General Vice-Chairs
Young-Sik Jeong, Dongguk University, Korea
Jason C. Hung, Overseas Chinese University, Taiwan
Hanmin Jung, KISTI, Korea
Yang Xiao, University of Alabama, USA

Program Chairs
Gangman Yi, Gangneung-Wonju National University, Korea
Yuh-Shyan Chen, National Taipei University, Taiwan
Xiaohong Peng, Aston University, UK
Ali Abedi, Maine University, US
Neil Y. Yen, the University of Aizu, Japan
Jen Juan Li, North Dakota State University, USA
Hongxue (Harris) Wang, Athabasca University, Canada
Lijun Zhu, ISTIC, China

International Advisory
Sherali Zeadally, University of the District of Columbia, USA
Naveen Chilamkurti, La Trobe University, Australia
Luis Javier Garcia Villalba, Universidad Complutense de Madrid (UCM), Spain
Mohammad S. Obaidat, Monmouth University, USA
Jianhua Ma, Hosei University, Japan

Laurence T. Yang, St. Francis Xavier University, Canada
Hai Jin, HUST, China
Weijia Jia, City U. of Hong Kong, Hong Kong
Albert Zomaya, University of Sydney, Australia
Bin Hu, Lanzhou University, China
Doo-soon Park, SoonChunHyang University, Korea

Publicity Chair

Deok-Gyu Lee, ETRI, Korea
Hak Hyun Choi, Seoul Women's University, Korea
Weiwei Fang, Beijing Jiaotong University, China
Cain Evans, Birmingham City University, UK
Chun-Cheng Lin, National Chiao Tung University, Taiwan
Antonio Coronato, ICAR, Italy
Rung-Shiang Cheng, Kunshan University, Taiwan
Haixia Zhang, Shandong University, China
Rafael Falcon, Larus Technologies, Canada
Xu Shao, Institute for Infocomm Research, Singapore
Bong-Hwa Hong, Kyunghee Cyber University, Korea

Local Arrangement Chairs

Namje Park, Jeju National University, Korea
Cheonshik Kim, Sejong University, Korea
Min Choi, Chungbuk National University, Korea
Aziz Nasridinov, Dongguk University, Korea

Contents

1 **Voronoi Diagram and Microstructure of Weldment** 1
Jungho Cho and Min Choi

2 **High-SNR Approximate Closed-Form Formulas for the Average Error Probability of *M*-ary Modulation Schemes Over Nakagami-*q* Fading Channels** 11
Hoojin Lee

3 **Performance Evaluation of IEEE 802.15.6 MAC with User Priorities for Medical Applications** 23
Li Yang, Changle Li, Yueyang Song, Xiaoming Yuan and Yanle Lei

4 **Using Dead Reckoning, GPS and Fingerprinting for Ubiquitous Positioning** 31
Hsiao-Hsien Chiu, Yi-Jiun Tang and Ming-Shi Wang

5 **Multi-core Scheduling Scheme for Wireless Sensor Nodes with NVRAM-Based Hybrid Memory** 45
Seokho Oh and Yeonseung Ryu

6 **Security Scheme for LTE Initial Attach** 53
Uijin Jang, Hyungmin Lim and Hyungjoo Kim

7 **ASiPEC: An Application Specific Instruction-Set Processor for High Performance Entropy Coding** 67
Seung-Hyun Choi, Neungsoo Park, Yong Ho Song and Seong-Won Lee

8 **An Enhanced Cooperative Spectrum Sensing Scheme Based on New Rule of Combining Evidences in Cognitive Radio** 77
Muhammad Sajjad Khan and Insoo Koo

9 **Trajectory Prediction for Using Real Data and Real Meteorological Data** 89
Yong Kyun Kim, Jong Wook Han and Hyodal Park

10 **The ADS-B Protection Method for Next-Generation Air Traffic Management System** 105
Seoung-Hyeon Lee, Jong-Wook Han and Deok-Gyu Lee

11 **Interactive Drawing Based on Hand Gesture** 115
Tae-Eun Kim

12 **Nullifying Malicious Users for Cooperative Spectrum Sensing in Cognitive Radio Networks Using Outlier Detection Methods** 123
Prakash Prasain and Dong-You Choi

13 **A Study on Electronic-Money Technology Using Near Field Communication** 133
Min-Soo Jung

14 **A Novel Android Memory Management Policy Focused on Periodic Habits of a User** 143
Jang Hyun Kim, Junghwan Sung, Sang Yun Hwang and Hyo-Joong Suh

15 **Care Record Summary Validation Tool with Social Network Service** 151
Jae Woo Sin, Joon Hyun Song, Do Yun Lee and Il Kon Kim

16 **Performance Study of an Adaptive Trickle Scheme for Wireless Sensor Networks** 163
Yen-Wen Lin and Po-Hsiang Wang

17 **A New Distributed Grid Scheme Utilizing Node-based Preprocessing Technique for Supporting k-NN Queries in Location-based Services** 175
Hyunjo Lee, Min Yoon and Jae-Woo Chang

18 A Centroid-GPS Model to Improving Positioning Accuracy for a Sensitive Location-Based System 187
Md. Rashedul Islam and Jong-Myon Kim

19 Intelligent Evaluation Models Based on Different Routing Protocols in Wireless Sensor Networks 197
Ning Cao, Russell Higgs and Gregory M.P. O'Hare

20 Implementation of Personalized Wellness Service 211
Marie Kim, Namje Park and Hyo-Chan Bang

21 SOM Clustering Method Using User's Features to Classify Profitable Customer for Recommender Service in u-Commerce 219
Young Sung Cho, Song Chul Moon and Keun Ho Ryu

22 A Performance Improvement Scheme for Discovery Service 229
Peng Liu, Ning Kong, Ye Tian, Xiaodong Lee and Baoping Yan

23 A Cross Cloud Authorization Mechanism Using NFC and RBAC Technology 239
Jun-Fu Chan, Ta-Chih Yang and Horng Twu Liaw

24 A Study on High Security Authentication Mechanism for 3G/WLAN Networks 247
Wei-Chen Wu and Horng-Twu Liaw

25 A Simple Authentication Scheme and Access Control Protocol for VANETs 259
Wei-Chen Wu and Yi-Ming Chen

26 Decision Tree Approach to Predict Lung Cancer the Data Mining Technology 273
Jui-Hung Kao, Hui-I Chen, Feipei Lai, Li-Min Hsu and Horng-Twu Liaw

27 Study of Customer Value and Supplier Dependence with the RFM Model 283
Jui-Hung Kao, Feipei Lai, Horng-Twu Liaw and Pei-hua Hsieh

28 Feature Selection for Support Vector Machines Base on Modified Artificial Fish Swarm Algorithm 297
Kuan-Cheng Lin, Sih-Yang Chen and Jason C. Hung

29 **A New Remote Desktop Approach with Mobile Devices: Design and Implementation** 305
Teng-Yao Huang, Hai-Hui Wang, Chun-Lung Peng and Hsin-Mao Huang

30 **Implementation of Low Power LED Display Controller Using Adiabatic Operation** 323
Kyung-Ryang Lee, Sung-Dae Yeo, Seung-il Cho and Seong-Kweon Kim

31 **Hand Gesture Recognition Using 8-Directional Vector Chains in Quantization Space** 333
Seongjo Lee, Sohyun Sim, Kyhyun Um, Young-Sik Jeong and Kyungeun Cho

32 **Forensic Artifacts in Network Surveillance Systems** 341
Kyung-Soo Lim, Jeong-Nye Kim and Deok-Gyu Lee

33 **Routing Protocol for Hierarchical Clustering Wireless Sensor Networks** 349
Alghanmi Ali Omar, ChunGun Yu and ChongGun Kim

34 **The Design of a New Virtualization-Based Server Cluster System Targeting for Ubiquitous IT Systems** 361
Jungmin Lim, Sihoo Song, Soojin Lee, Seokjoo Doo and Hyunsoo Yoon

35 **A Conceptualisation of a Strategy to Shiftwork Scheduling Optimisation in an Emergency Department** 377
Muhaimin Noor Azhar, Aida Bustam, Rashidi Ahmad, Rishya Manikam and Shamimi A. Halim

36 **The Evaluation of Pen Gestures in a Digital Painting Environment** 385
Chih-Hsiang Ko

37 **Disease Pattern Analysis Using Electronic Discharge Summary of EMR in HL7 for Computerized Treatment Plan of Cancer Patients in Korea** 393
Young Sung Cho, Song Chul Moon, Kwang Sun Ryu and Keun Ho Ryu

38 **Research on the Network Assisted Teaching System of College Physics Experiments** 405
Yanling Du, Hongxia Bu and Xiaonan Fang

39 **Scheduling Model and Algorithm of Vertical Replenishment** 413
Shufen Liu and Bin Li

40 **Assisted Lung Ventilation Control System as a Human Centered Application: The Project and Its Educational Impact on the Course of Embedded Systems** 421
Diego Cabezas, Alexei Vassiliev and Evgeny Pyshkin

41 **Research of SSO Based on the Fingerprint Key in Cloud Computing** 429
Yong-Sheng Zhang, Jia-Shun Zou and Yan Gao

42 **Study of Local Monitoring System Based on SMS Under the Cloud Environment** 437
Yong-sheng Zhang, Jia-shun Zou and Yan Gao

43 **Effects of Meal Size on the SDA of the Taimen** 443
Guiqiang Yang, Zhanquan Wang, Ding Yuan, Shaogang Xu and Junfeng Ma

44 **Management Information Systems** 449
Lidia Ogiela and Marek R. Ogiela

45 **Security in Management of Distributed Information** 457
Marek R. Ogiela, Lidia Ogiela and Urszula Ogiela

46 **Research of ABAC Mechanism Based on the Improved Encryption Algorithm Under Cloud Environment** 463
Long-xiang Zhang and Jia-shun Zou

47 **A Novel Design of Education Video Personalized Recommendation System Based on Collaborative Filtering Recommendation Technology** 471
Xiao Jun and Wang Min

48 **The Design of a Medical Rules Synchronization System** 481
Yi-Hsing Chang, Leng-Kang Chang Chien and Rong-Jyue Fang

49 Associative Recommendation of Learning Contents Aided by Eye-Tracking in a Social Media Enhanced Environment 493
Guangyu Piao, Xiaokang Zhou and Qun Jin

50 A Novel Strategy for Colonies Recognition and Classification in Super-Resolution Images 503
Qi Zhang, Xueqing Li and Xianlun Dong

51 Study on Intelligent Course Scheduling System 511
Teng Li and Xuqing Li

52 The Study and Research on Chinese Sports Network Marketing Strategy 519
Jiang Yong

53 Interactive Visualization of Enrollment Data Using Parallel Coordinates 529
Wei Xinglei and Li Xueqing

54 The Effect of Peer's Progress on Learning Achievement in e-Learning: A Social Facilitation Perspective 537
Po-Sheng Chiu, Ting-Ting Wu, Yueh-Ming Huang and Hong-Leok Ho

55 An Integration Framework for Clinical Decision Support Applications 543
Xiang Zheng, Yin-sheng Zhang, Zhen-zhen Huang, Zheng Jia, Hui-long Duan and Hao-min Li

56 Method of Generating Intelligent Group Animation by Fusing Motion Capture Data 553
Jie Song, Xiang-wei Zheng and Gui-juan Zhang

57 Histogram-Based Masking Technique for Retinal Fundus Images 561
Rachel M. Chong and Jeziel C. Suniel

58 Effect of Electronic Scoring System for Scenario Group Tutorial Implementation for Supporting Medical Student Studies 569
Piyapong Khumrin and Volaluck Supajatura

59 **Perceived Risk of Anti-corruption e-Learning, Email Phishing Literacy, and Anomia** 577
Juneman Abraham and Sharron

60 **A Model of Business Intelligence Systems for School: A Case Study (Perception Applications and Benefits)** 585
Adhi Nugroho Chandra and Yohannes Kurniawan

61 **A Surveillance Platform of Antimicrobial Use in Hospital Based on Defined Daily Dose** 595
Guowei Liang, Yinsheng Zhang, Haomin Li, Weihong Chen and Huilong Duan

62 **Identification of Adverse Drug Events in Chinese Clinical Narrative Text** 605
Caixia Ge, Yinsheng Zhang, Huilong Duan and Haomin Li

63 **Control System and Control Method for Automatic Adjustment of Outdoor LED Display Brightness** 613
Feng Yang, Xu-fei Qin and Lin-bo Zhai

64 **A Novel Quantitative Evaluation Metric of 3D Mesh Segmentation** 621
Xiao-peng Sun, Lu Wang, Xingyue Wang and Xiaona Zhao

65 **A Survey Analysis of Chinese Virtues Questionnaire in Medical Postgraduates** 629
Miao Yu, Wei Wang, Zhilei Hu, Huibin Ji and Wei Xing

66 **Practice and Exploration of All-in-English Teaching of Compiler Principles** 637
Xinxin Liu and Hongyun Xu

67 **An Application of Ant Colony Optimization Clustering Approach for Primary Headache Diagnosis** 643
Wu Yanping and Duan Huilong

68 **Comparisons of Iterative Closest Point Algorithms** 649
Lu Wang and Xiaopeng Sun

69 **Study on the Effectiveness of Distance Education in the Judicial Examination Training** 657
Rongxia Zhang and Weisheng Wang

70 Greedy Strategy Based Self-adaption Ant Colony Algorithm for 0/1 Knapsack Problem . . . 663
De-peng Du and Yue-ran Zu

71 Non-Chinese Students Speak: Sectional and Clinical Anatomy Learning in a Chinese Medical School . . . 671
Xu He, Fu-Xiang Liu and Aihua Pan

72 Dynamic and Efficient Search System for Digital Encyclopedia of Intangible Cultural Heritage: The Case Study of ICHPEDIA . . . 679
Jung Song Lee, Soon Cheol Park and Han Heeh Hahm

73 On the Performance of Quasi-orthogonal Space Time Block Coded Massive MIMO with up to 16 Antennas . . . 687
Khin Zar Chi Winn, Phyu Phyu Han, Kasun Bandara and Yeon-Ho Chung

74 Encrypted Data Group Authentication for Outsourced Databases . . . 695
Miyoung Jang, Ara Jo and Jae-Woo Chang

75 De-Word Classification Algorithm Based on the Electric Power of Large Data Library Retrieval . . . 707
Xiaoli Guo, Huiyu Sun, Ling Wang, Zhaoyang Qu and Wei Ding

Author Index . . . 717

Subject Index . . . 721

Chapter 1
Voronoi Diagram and Microstructure of Weldment

Jungho Cho and Min Choi

Abstract One of the famous space decomposition algorithm known as Voronoi diagram is applied to express metal's microstructure for the first time by fortuitous discovery of superficial analogy of Voronoi cell and metal's crystal grain. Areas of Voronoi cells are controlled by locations and the number of seed points. And it can be correlated to grain size of microstructure and nuclei numbers. Therefore grain coarsening and refinement of microstructure can be described by Voronoi tessellation for simple case. In addition to this, columnar crystal caused by rapid cooling rate in one direction is also described by anisotropic locations of seed points which can be observed in typical weldment in easy. Although it needs more profound research about correlation between crystal grain growth and Voronoi diagram control variables, it shows fairly reasonable feasibility of adopting Voronoi tesselation as metal's microstructure description and prediction tool.

Keywords Voronoi · Microstructure · Grain size · Weld · Columnar crystal

1.1 Introduction

Voronoi diagram, also referred as Voronoi tessellation, is famous space decomposition method in mathematics. It is simple and easy to apply. It also provides triangulation algorithm which makes triangle patches in 2D or 3D space with the given vertices to construct virtual polygon models. And the triangulation technique is known as Delaunay triangulation [1, 2]. Therefore, it was very popular in the early time of computer graphics development and still quite important as basic tool.

J. Cho
School of Mechanical Engineering, Chungbuk National University, Cheongju, Korea

M. Choi (✉)
School of Information and Communication Engineering, Chungbuk National University,
52 Naesudong-ro, Heunduk-gu, Cheongju 363-761, Korea
e-mail: mchoi@cbnu.ac.kr

J.J. (Jong Hyuk) Park et al. (eds.), *Ubiquitous Computing Application and Wireless Sensor*, Lecture Notes in Electrical Engineering 331,
DOI 10.1007/978-94-017-9618-7_1

Besides the computer graphics area, the Voronoi diagram has been applied in vast area including science, technology and art.

Except for the applications in computer graphics, another popular application of Voronoi diagram in engineering is finite element simulation field [3–7]. When it is hard to generate meshes for some circumstances or interesting nodes are not in regular arrangement, arbitrarily generated mesh by Voronoi diagram becomes an alternative solution. There were also practical usage examples of Voronoi diagram such as derivation of meaningful quantitative data from SEM pictures [8, 9] by matching the diagram to image. Trials of direct use of Voronoi diagram as a tool for expression of polycrystalline structure and its morphological analysis were also reported [10, 11] but their application was not metal's microstructure. Practical application to metal structure was reported once [12] which was a trial to express severely deformed aluminum's microstructure in simple way. And it showed size reduced grain only without simulation of various grain shapes.

Microstructure of metal is composed of multiple crystal grains. And border of the grains, in other words, grain boundary determines mechanical strength in most cases. Therefore the grain size and grain boundary structure are very important features in metal component microstructure analysis. Grain boundary structure is decided by grain growth and its process is like following. Nucleus is generated at first in solidification temperature then this nucleus grows to a crystal by attaching metal or alloying elements. In rapid cooling rate, a number of nuclei are generated in simultaneous way and they grow fast then made coarse grain boundary system while relatively slow cooling rate makes fine grain boundary with less nuclei. In this viewpoint, Voronoi diagram has morphological similarity with metal's microstructure, exactly speaking, grain boundary. Because Voronoi cell can be matched to crystal and grain boundary to Voronoi edge. In addition to this, seed points of Voronoi diagram play a same role of nuclei.

Microstructure formation of material is extremely complex problem since every element takes part in crystallization and chemical bonding process with respect to circumstance variable such as temperature and pressure. Therefore it is almost impossible to foresee the microstructure exactly. However, prediction of grain boundary structure has some degree of potential at least because relationship between crystal size, type, temperature and time is already known through equilibrium phase diagram and continuous cooling temperature diagram based on numerous experiments. Inspired by aforementioned morphological similarity to microstructure of metal and prediction possibility of grain boundary, Voronoi diagram is suggested as a microstructure prediction tool in this research. Final purpose and role will be visual presentation of grain boundary for given temperature history in the future but now its feasibility is tested by simulation of typical weldment's microstructure for the first time.

1.2 Building Voronoi Diagram

Basic concept of Voronoi tessellation is quite simple. It just needs to compute and build equidistance separating lines between given seed points in space. For example in 2D, if there are only two seed points, just build an equidistance separation line. For more seed points over 3, circumcircle and its circumcenter should be computed for every 3 points.

More concrete methodology of Voronoi diagram is explained through Fig. 1.1. There are 4 seed points in the given 2D space as shown in (a). First work to do is random selection of 3 points then compute circumcircle fitting these points. When outer 3 points are selected like figure (b), the other unselected point is inside of circumcircle. This case is not a satisfied condition in Voronoi diagram therefore nothing is proceeded further. For the cases of (c) and (d) where computed circumcircle of selected 3 points does not contain other seed point inside of it, then the circumcenters are memorized. A line connecting these 2 newly generated center points then becomes a Voronoi edge because it exactly divides a line connecting circles' intersection points into 2 pieces with same length. Like this way, (n-2) circumcenters marked as Voronoi vertices are determined for every 3 neighboring vertices for given n seed points and neighboring vertices connected to each other by Voronoi edges. And each divided subspace, i.e. polygon surrounded by Voronoi edges, is referred as Voronoi cell. Figure 1.2 shows designation of each component in Voronoi diagram

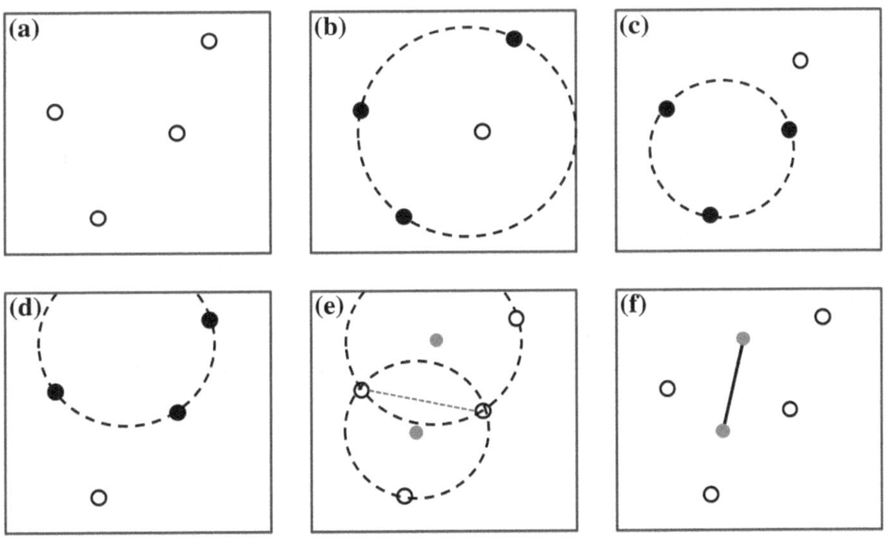

Fig. 1.1 Voronoi diagram methodology. **a** Arbitrarily located 4 seed points, **b** selection of outer 3 points and circumcircle, **c** lower 3 points and circumcircle, **d** upper 3 points and circumcircle, **e** circumcenters of circumcircles, **f** generated Voronoi edge

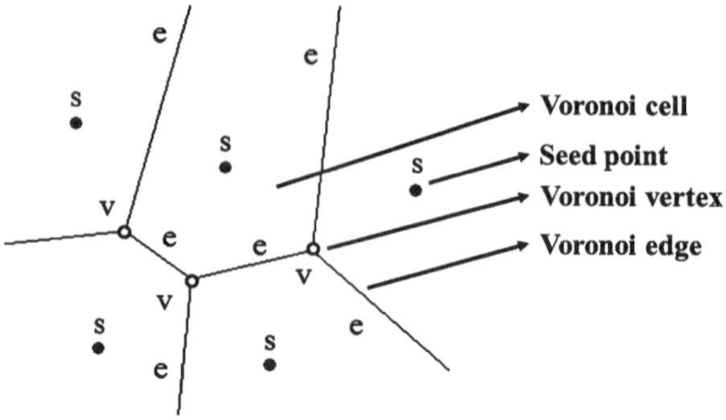

Fig. 1.2 Designation of Voronoi diagram component

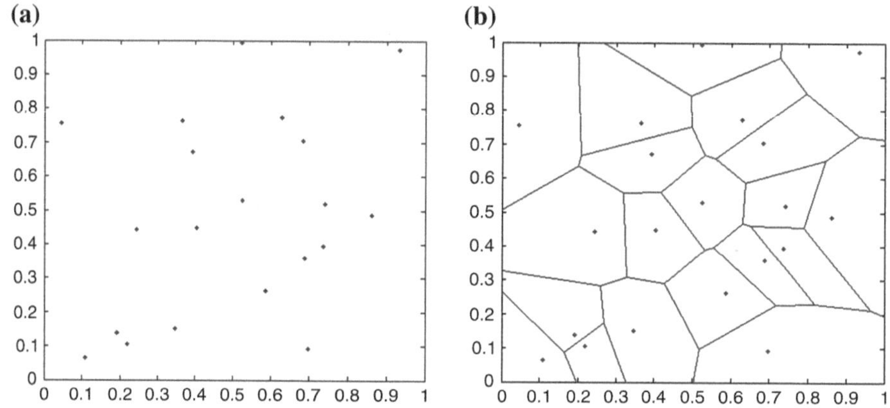

Fig. 1.3 Generated **b** Voronoi diagram for **a** randomly distributed seed point

Simple example of Voronoi diagram is shown in Fig. 1.3. Several seed points are randomly distributed in 2D space and generated Voronoi cells are shown through the figure. As mentioned in previous paragraph, the only required condition for 3 neighboring points is that there should not be any other seed point inside of the circumcircle. Three seed points satisfying the condition can consist of a triangle patch for 2D or 3D polygon construction. This technique is called as Delaunay triangulation [1, 2] which is famous in computer graphic applications.

1.3 Expression of Weldment

As already mentioned in introduction section, grain boundary observed in metal's microstructure has morphological similarity to typical Voronoi diagram. Nucleus of single metal crystal and grain boundary can be matched to seed point and Voronoi edge consecutively. Therefore if the position of seed point can be controlled according to solidification process parameter, Voronoi diagram can be used as microstructure, exactly speaking, grain boundary, prediction tool. Inspired by the morphological similarity and possibility of predicting grain boundary structure, the author gave an idea to develop Voronoi diagram as microstructure prediction tool then tested its feasibility by reproducing microstructure of weldment. The reason of selection of weldment for feasibility test target is that welding process shows various microstructure types in small domain. It has locally molten and re-solidified zone. Therefore it shows recrystallized zones with and without melting at the same time. And the weldment undergoes comparatively rapid cooling rate. So, it is a good example which contains coarse and fine grain boundary structures and anisotropically shaped crystals altogether in common domain.

Before practical Voronoi diagram simulation, formation of weldment microstructure should be understood first. Typical heat sources for welding are arc and laser, and a region directly exposed to heat source will be molten and re-solidified. This section is referred as fusion zone (FZ) and it undergoes comparatively rapid cooling rate because cumulated heat is quickly conducted to surrounding solid part. Of course its microstructure is changed by re-crystallization and therefore it has different grain boundary structure from base metal. If its temperature is high enough, metal can be re-crystallized in solid state without melting. A region surrounding FZ and receiving heat from it is also undergoes re-crystallization by this way. This re-crystallized solid region is called as heat affected zone (HAZ). Therefore weldment is categorized into three types, FZ, HAZ and base metal (BM). For most cases, HAZ shows coarsened grain boundary because it continuously receives heat from FZ therefore has relatively slow cooling rate which gives enough growth time to grains. Interesting feature of FZ is columnar crystal. It shows slab-sided shape because of directional growth of grain. When the nucleation is started at the liquid/solid boundary, grains can competitively grow only in cooling direction towards liquid side. Then it is resulted as epitaxial columnar crystals. And the crystal's longitudinal direction is, of course, perpendicular to liquid/solid boundary which is corresponding to FZ/HAZ boundary in this research.

First of all, expressibility of coarse and fine grain boundary structure is tested. Figures 1.4 and 1.5 show Voronoi diagrams corresponding to relatively fine and coarse microstructure. Different numbers of seed points are randomly distributed and Voronoi diagrams are generated in each case. It is showing successful expression of grain size distinction through Voronoi cell which was possible by controlling number of seed points in the space.

Finally, feasibility test of typical weldment expression is conducted. And the result is shown through Fig. 1.6 which is expressing cross sectional bead shape of typical laser welding. Unlike arc welding bead, laser weld has narrow and deep penetration like Fig. 1.6 therefore it is named as keyhole. 2D space is divided into three regions to express BM, HAZ and FZ. Boundary of each section is denoted by black dashed line in the figure and these quadratic curves were actually defined in the program. Simulated Voronoi diagram is showing laser keyhole bead of which BM is assumed to have fine microstructure at first. Distinctive region upon the first

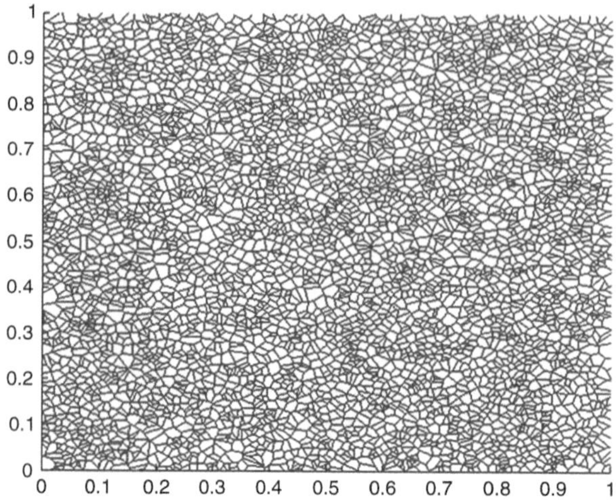

Fig. 1.4 Example of Voronoi diagram corresponding to fine microstructure of metal

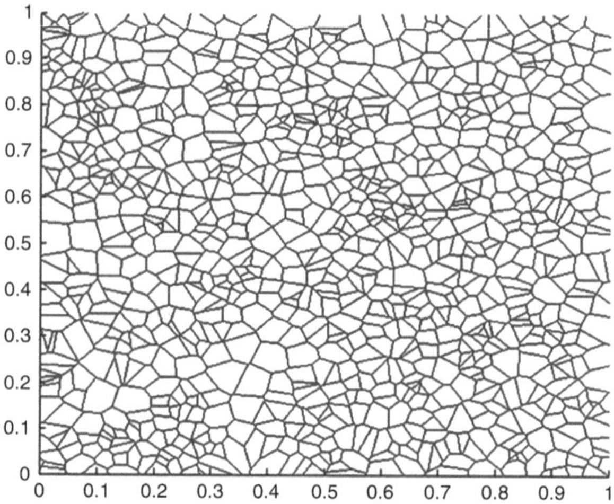

Fig. 1.5 Example of Voronoi diagram corresponding to relatively coarsened microstructure of metal

border line from the bottom in the figure is showing HAZ with relatively coarsened grain boundary structure. Randomly distributed seed point with less density zone is successfully converted to Voronoi diagram with large cells comparing to BM domain. Upon the second border, Voronoi cell is quite different from that of HAZ and BM zone. Unlike the others, generated cells in this FZ region have slab-sided shape representing columnar crystal. And it can be noticed that its longitudinal direction is perpendicular to border which is same to aforementioned feature of columnar crystal. So, the second border in the figure is depicting boundary of FZ/HAZ which is also a boundary of molten liquid and solid during the welding process. In the FZ, generated Voronoi cells have even larger cell area comparing to HAZ and BM zone because a few seed points are distributed in that area along

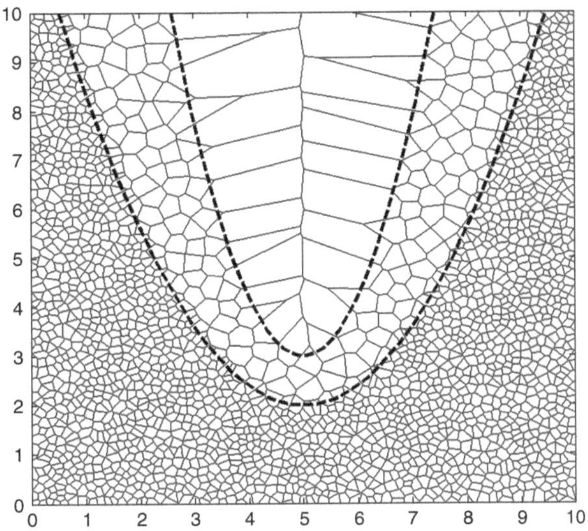

Fig. 1.6 Voronoi diagram corresponding to typical laser weld. Fine structure of base metal, coarsened HAZ and columnar crystal in FZ are expressed

FZ/HAZ border. It seems like it has somewhat exaggerated columnar crystal shape but small bead of precision laser welding shows almost same bead formation indeed.

1.4 Discussion and Conclusion

Microstructure of metal is a result of crystallization of various metallic and nonmetallic elements. Therefore it is almost impossible to predict the microstructure exactly in detail level. However, if we focus on grain boundary structure related to nucleation and grain growth only, it becomes a relatively simple problem. While permanent plastic deformation is obtained by slip mechanism inside of crystal, crack starts from grain boundary in most cases because segregated nonmetallic elements or impurities are stacked up in grain boundary. Therefore, the more complex grain boundary structure, the higher mechanical strength in general. Mechanical strength of metal component is highly correlated to grain boundary structure like this way. Therefore, if possible, it is worthy to develop a prediction technique of grain boundary formation. As mentioned in introduction section, famous space decomposition method Voronoi diagram shows similar morphological features to grain boundary of microstructure. Consequently, the Voronoi diagram is suggested as a tool for metal's microstructure prediction and its feasibility is tested by several simulations in this research.

As first trial, different size of grain boundary is expressed by Voronoi diagram and the result shows successfully subdivided structure with respect to randomly distributed seed points in 2D space. Coarse and fine microstructures are simulated and the relative difference was obtained by variation of seed point density. After basic trial, typical weldment was targeted for authentic feasible test. Selected microstructure was cross sectional bead shape of laser welding which has 3 different microstructure regions, BM (base metal zone), HAZ (heat affected zone) and FZ (fusion zone). In most cases, grain boundary structure of HAZ is coarsened comparing to fine BM. And FZ can be characterized by columnar crystals in epitaxial growth which have slap-sided shape. As shown in the figure, Voronoi diagram successfully expressed all three different microstructures according to their special features. HAZ is virtually re-constructed by larger Voronoi cells which are consistent with coarsened grain boundary. BM of fine microstructure is also visualized with denser distribution of seed points and it was clearly distinctive from HAZ. Unlike other two microstructures, FZ has different structure and its characteristic feature is also obtained and shown by expressing columnar crystals. Much less seed points are randomly distributed near virtual liquid/solid boundary to express columnar crystals and directional growth by elongated polygons as seen in the figure.

As conclusion, Voronoi diagram is proposed as a prediction tool of metal's microstructure. And its feasibility is tested by expressing coarse and fine grain boundary structure. Then, it is finally applied to expression of typical laser weldment consisted of BM, HAZ and FZ. And generated Voronoi diagram successfully showed its possibility by visualizing distinctive differences between microstructures of weldment including columnar crystals.

Generated Vonoroi diagram is not directly compared to practical cross sectional bead shape in here yet. However, it does not mean that the suggestion is less meaningful because control variable of Voronoi diagram can be adjusted in any way to fit given bead image. Important factors in succession of Voronoi diagram as a microstructure prediction tool is correlation between Voronoi diagram variables and practical grain boundary formation variables. Exactly speaking, correlation factors between seed point location and grain size, cooling direction should be defined. And it is further works of this research.

Acknowledgments This research was supported by Basic Science Research Program through the National Research Foundation of Korea (NRF) funded by the Ministry of Science, ICT and Future Planning. (Grant No. 2012R1A1A1012487)

References

1. Lee DT, Schachter BJ (1980) Two algorithms for constructing a delaunay triangulation. Int J Comput Inform Sci 9(3):219–242
2. Chew LP (1989) Constrained delaunay triangulations. Algorithmica 4:97–108
3. Cao J, Zhuang W, Wang S, Ho KC, Zhang N, Lin J, Dean TA (2009) An integrated crystal plasticity FE system for microforming simulation. J Multiscale Model 1(1):107–204
4. Zhang P, Balint D, Lin J (2011) An integrated scheme for crystal plasticity analysis: virtual grain structure generation. Comput Mater Sci 50:2854–2864

5. Zhang P, Karimpour M, Balint D, Lin J, Farrugia D (2012) A controlled poisson Voronoi tessellation for grain and cohesive boundary generation applied to crystal plasticity analysis. Comput Mater Sci 64:84–89
6. Zhang P, Karimpour M, Balint D, Lin J (2012) Three-dimensional virtual grain structure generation with grain size control. Mech Mater 55:89–101
7. Ghosh S, Nowak Z, Lee K (1997) Quantitative characterization and modeling of composite microstructures by Voronoi cells. Acta Mater 45(6):2215–2234
8. Heijman MJGW, Benes NE, ten Elshof JE, Verweij H (2002) Quantitative analysis of the microstructural homogeneity of zirconia-toughened alumina composite. Mater Res Bull 37:141–149
9. Jacobs LJM, Danen KCH, Kemmere MF, Keurentjes JTF (2007) Quantitative morphology analysis of polymers foamed with supercritical carbon dioxide using Voronoi diagrams. Comput Mater Sci 38:751–758
10. Fan Z, Wu Y, Zhao X, Lu Y (2004) Simulation of polycrystalline structure with Voronoi diagram in laguerre geometry based on random closed packing of spheres. Comput Mater Sci 29:301–308
11. Wu Y, Zhou W, Wang B, Yang F (2010) Modeling and characterization of two-phase composites by Voronoi diagram in the laguerre geometry based on random close packing of spheres. Comput Mater Sci 47:951–961
12. Jafari R, Kazeminezhad M (2011) Microstructure generation of severely deformed materials using Voronoi diagram in laguerre geometry: full algorithm. Comput Mater Sci 50:2698–2705

Chapter 2
High-SNR Approximate Closed-Form Formulas for the Average Error Probability of *M*-ary Modulation Schemes Over Nakagami-*q* Fading Channels

Hoojin Lee

Abstract In this paper, we present high signal-to-noise ratio (SNR) approximate closed-form formulas for the average error probabilities of several M-ary signals, particularly in Nakagami-q (i.e., Hoyt) fading channels. The derived formulas are much more concise than the existing exact closed-form expressions, however, showing the very tightness to error performance obtained from the exact formulas in the high SNR regime. The validity of our derivations and analyses is verified though the rigorous numerical results.

Keywords Nakagami-q fading · Hoyt · M-ary modulation schemes · Error probability

2.1 Introduction

In wireless communication and network systems, it is well-known that the systems are easily subject to severe fading, which is able to seriously degrade their performance. Thus, there is a very wide range of statistical models for the fading channels, the veracity of which in general depends on several communication environments. Specifically, Rayleigh, Nakagami-n (i.e., Rice), Nakagami-q (i.e., Hoyt), Nakagami-m statistical models are popularly adopted in the literature [1–8], which is due to the fact that these models exhibit an excellent fit to several experimental fading channel measurements. Among these statistical fading models, Nakagami-q distribution is typically used to describe the short-term signal variation observed in various wireless communication and network systems, where the in-phase and quadrature signal components of the fading channels have zero-mean

H. Lee (✉)
Department of Information and Communications Engineering,
Hansung University, Seoul 136-792, Korea
e-mail: hjlee@hansung.ac.kr

J.J. (Jong Hyuk) Park et al. (eds.), *Ubiquitous Computing Application and Wireless Sensor*, Lecture Notes in Electrical Engineering 331,
DOI 10.1007/978-94-017-9618-7_2

and arbitrary variance. For example, Nakagami-q distribution, where q denotes the fading severity parameter, is generally observed in satellite-based communication systems subject to strong ionospheric scintillation [3] and multipath wireless links with heavy shadowing [5], ranges from one-sided Gaussian distribution (cf. $q = 0.0$) to Rayleigh distribution (cf. $q = 1.0$), and thus is extensively utilized.

The bit error probability (BEP) and symbol error probability (SEP) for various types of modulation schemes have been well-known as key performance metrics of wireless communication and network systems. In particular, the average error performance of M-ary coherent, M-ary differentially coherent, and noncoherent correlated binary modulation schemes in Nakagami-q are given in integral forms in [3, 6], and the novel exact-form solutions for the integral expressions are presented in [8]. Based on the analytical results given in [8], our purpose of this paper is to derive much simpler high signal-to-noise ratio (SNR) approximate closed-form formulas for the SEP or BEP of various M-ary modulation schemes, especially in Nakagami-q fading channels. By judiciously exploiting the derived formulas, analytical and useful insights into the impact of Nakagami-q fading severity on the error performance can be easily achieved, which is obviously validated through various numerical results in this paper.

2.2 Channel Model

We consider the quasi-static frequency-nonselective Nakagami-q fading channels. Then, the corresponding probability density function (PDF) of the received SNR γ is given by [3]

$$f_\Gamma(\gamma) = \frac{(1+q^2)}{2q\bar{\gamma}} \exp\left(-\frac{(1+q^2)^2\gamma}{4q^2\bar{\gamma}}\right) I_0\left(\frac{(1-q^4)\gamma}{4q^2\bar{\gamma}}\right), \tag{2.1}$$

where $I_0(\cdot)$ is the zeroth-order modified Bessel function of the first kind and q is the Nakagami-q fading severity parameter with $0 \leq q \leq 1$, in which both extremes of the range of q are interpreted as the limits.

In addition, the moment-generating function (MGF) of the received SNR γ is defined as [1–3].

$$\mathcal{MGF}(s) = \int_0^\infty e^{-s\gamma} f_\Gamma(\gamma)d\gamma. \tag{2.2}$$

Then, the MGF of γ for Nakagami-q fading channels is expressed as [3, 8]

$$\mathcal{MGF}(s) = \left(1 + 2s\bar{\gamma} + \frac{(2sq)^2\bar{\gamma}}{(1+q^2)^2}\right)^{-\frac{1}{2}}, \tag{2.3}$$

where $\bar{\gamma}$ denotes the average received SNR.

2.3 Existing Expressions for Average Error Probability in Nakagami-q Fading Channels

2.3.1 M-ary Differentially Coherent Phase Shift-Keying (DPSK)

According to [3, 8], the average SEP of M-ary differentially coherent phase shift-keying (DPSK) signals in Nakagami-q fading channels is given by

$$P_S = \frac{2}{\pi} \int_0^{\frac{\pi}{2}\left(1-\frac{1}{M}\right)} \mathcal{MGF}\left(\varsigma \sin^2\left(\frac{\pi}{M}\right)\right) d\phi, \tag{2.4}$$

where M is the constellation size of M-ary signal and $\varsigma = \left(1+\cos\left(\frac{\pi}{M}\right) - 2\cos\left(\frac{\pi}{M}\right)\sin^2\phi\right)^{-1}$. Then, the corresponding closed-form formula is given in [8] as

$$P_S = \frac{2\cos\left(\frac{\pi}{2M}\right)}{\pi} \mathcal{MGF}(g_{DPSK}) \times \left[F_D^{(3)}\left(\frac{1}{2},\frac{1}{2},\frac{1}{2},\frac{1}{2};\frac{3}{2};z_1,z_2,z_3\right) - \frac{\cos\left(\frac{\pi}{M}\right)}{3} F_D^{(3)}\left(\frac{3}{2},\frac{1}{2},\frac{1}{2},\frac{1}{2};\frac{5}{2};z_1,z_2,z_3\right)\right], \tag{2.5}$$

where

$$\begin{aligned} g_{DPSK} &= 2\sin^2\left(\frac{\pi}{2M}\right), \\ z_1 &= \cos\left(\frac{\pi}{M}\right) \Big/ \left(1 + g_{DPSK}\left(\frac{2}{1+q^2}\right)\bar{\gamma}\right), \\ z_2 &= \cos\left(\frac{\pi}{M}\right) \Big/ \left(1 + g_{DPSK}\left(\frac{2q^2}{1+q^2}\right)\bar{\gamma}\right), \\ z_3 &= \cos^2\left(\frac{\pi}{2M}\right), \end{aligned} \tag{2.6}$$

and the fourth Lauricella hypergeometric function $F_D^{(n)}$ of n variables is defined by [9–12]

$$\begin{aligned}
&F_D^{(n)}(a, b_1, \ldots, b_n; c; x_1, \ldots, x_n) \\
&= \sum_{i_1, \cdots, i_n}^{\infty} \frac{(a)_{i_1+\cdots+i_n}}{(c)_{i_1+\cdots+i_n}} \prod_{j=1}^{n} (b_j)_{i_j} \frac{x_j^{i_j}}{i_j!}, \max\{|x_i|\}_{i=1}^{n} < 1, \\
&= \frac{1}{B(a, c-a)} \int_0^1 t^{a-1}(1-t)^{c-a-1} \prod_{i=1}^{n} (1 - x_i t)^{-b_i} dt, \\
&\Re(c) > \Re(a) > 0,
\end{aligned} \tag{2.7}$$

where $(x)_i = \Gamma(x+i)/\Gamma(x)$ is the Pochhammer symbol for $i \geq 0$ with the gamma function, $\Gamma(\cdot)$, $B(a, b) = \Gamma(a)\Gamma(b)/\Gamma(a+b)$ denotes the beta function, and $\Re(\cdot)$ stands for the real part [12].

2.3.2 Noncoherent Correlated Binary Signals

The average BEP of the noncoherent correlated binary signals in Nakagami-q fading channels can be expressed as [7, 8]

$$P_b = \frac{1}{\pi} \int_0^{\frac{\pi}{2}} \mathcal{MGF}\left(\frac{(b^2 - a^2)^2}{2(a+b)^2 - 8ab\cos^2\phi}\right) d\phi, \tag{2.8}$$

where for the signal correlation coefficient $|\rho| \leq 1$

$$\begin{aligned}
a &= \sqrt{\frac{1 - \sqrt{1 - |\rho|^2}}{2}}, \\
b &= \sqrt{\frac{1 + \sqrt{1 - |\rho|^2}}{2}}.
\end{aligned} \tag{2.9}$$

In [8], the closed-form formula of (8) is given by

$$\begin{aligned}
P_b = \frac{1}{2} \mathcal{MGF}(g_K) \Bigg[&F_1\left(\frac{1}{2}, \frac{1}{2}, \frac{1}{2}; 1; \hat{z}_1, \hat{z}_2\right) \\
&- \frac{g_Q}{2} F_1\left(\frac{3}{2}, \frac{1}{2}, \frac{1}{2}; 2; \hat{z}_1, \hat{z}_2\right) \Bigg],
\end{aligned} \tag{2.10}$$

where F_1 denotes the Appell hypergeometric function defined by [12]

$$\begin{aligned} & F_1(a, b_1, b_2; c; x_1, x_2) \\ & = \sum_{i_1, i_2}^{\infty} \frac{(a)_{i_1+i_2}(b_1)_{i_1}(b_2)_{i_1}}{(c)_{i_1+i_2} i_1! i_2!} x_1^{i_1} x_2^{i_2}, \max\{|x_i|\}_{i=1}^{2} < 1, \\ & = \frac{1}{B(a, c-a)} \int_0^1 t^{a-1}(1-t)^{c-a-1} \prod_{i=1}^{2} (1 - x_i t)^{-b_i} dt, \\ & \Re(c-a) > 0, \Re(a) > 0, \end{aligned} \tag{2.11}$$

and

$$\begin{aligned} g_Q &= \frac{4ab}{(a+b)^2}, \\ g_K &= \frac{(b-a)^2}{2}, \\ \hat{z}_1 &= \frac{g_Q}{1 + g_K\left(\frac{2}{1+q^2}\right)\bar{\gamma}}, \\ \hat{z}_2 &= \frac{g_Q}{1 + g_K\left(\frac{2q^2}{1+q^2}\right)\bar{\gamma}}. \end{aligned} \tag{2.12}$$

2.3.3 M-ary Coherent Phase Shift-Keying (PSK)

For the M-ary coherent phase shift-keying (PSK) signals, the average SEP is given as [7, 8]

$$P_S = \frac{1}{\pi} \left[\int_0^{\frac{\pi}{2}} \mathcal{MGF}\left(\frac{g_{PSK}}{\sin^2 \phi}\right) d\phi + \int_{\frac{\pi}{2}}^{\pi\left(1-\frac{1}{M}\right)} \mathcal{MGF}\left(\frac{g_{PSK}}{\sin^2 \phi}\right) d\phi \right], \tag{2.13}$$

where $g_{PSK} = \sin^2\left(\frac{\pi}{M}\right)$. Then, the corresponding closed-form formula is derived in [8] as

$$\begin{aligned} P_S = & \frac{1}{4} \mathcal{MGF}(g_{PSK}) F_1\left(\frac{1}{2}, \frac{1}{2}, \frac{1}{2}; 2; \frac{1}{1 + x_{PSK}}, \frac{1}{1 + y_{PSK}}\right) \\ & + \frac{\cos\left(\frac{\pi}{M}\right)}{\pi} \mathcal{MGF}(g_{PSK}) F_D^{(3)}\left(\frac{1}{2}, -\frac{1}{2}, \frac{1}{2}, \frac{1}{2}; \frac{3}{2}; \right. \\ & \left. \cos^2\left(\frac{\pi}{M}\right), \frac{\cos^2\left(\frac{\pi}{M}\right)}{1 + x_{PSK}}, \frac{\cos^2\left(\frac{\pi}{M}\right)}{1 + y_{PSK}}\right), \end{aligned} \tag{2.14}$$

where

$$
\begin{aligned}
x_{PSK} &= g_{PSK}\left(\frac{2}{1+q^2}\right)\bar{\gamma},\\
y_{PSK} &= g_{PSK}\left(\frac{2q^2}{1+q^2}\right)\bar{\gamma}.
\end{aligned}
\tag{2.15}
$$

2.3.4 M-ary Square Quadrature Amplitude Modulation (QAM)

The average SEP of M-ary square quadrature amplitude modulation (QAM) is expreesed as [7, 8]

$$
P_S = \frac{4\beta}{\pi}\left[\int_0^{\frac{\pi}{2}} \mathcal{MGF}\left(\frac{g_{QAM}}{\sin^2\phi}\right)d\phi - \beta\int_0^{\frac{\pi}{4}} \mathcal{MGF}\left(\frac{g_{QAM}}{\sin^2\phi}\right)d\phi\right], \tag{2.16}
$$

where

$$
\begin{aligned}
\beta &= 1 - \frac{1}{\sqrt{M}},\\
g_{QAM} &= \frac{3}{2(M-1)}.
\end{aligned}
\tag{2.17}
$$

Then, from [8], the closed-form expression is expressed by

$$
\begin{aligned}
P_S = {} & \beta\mathcal{MGF}(g_{QAM})F_1\left(\frac{1}{2},\frac{1}{2},\frac{1}{2};2;\frac{1}{1+x_{QAM}},\frac{1}{1+y_{QAM}}\right)\\
& -\frac{2\beta^2}{3\pi}\mathcal{MGF}(2g_{QAM})F_D^{(3)}\left(1,1,\frac{1}{2},\frac{1}{2};\frac{5}{2};\right.\\
& \left.\frac{1}{2},\frac{1+x_{QAM}}{1+2x_{QAM}},\frac{1+y_{QAM}}{1+2y_{QAM}}\right),
\end{aligned}
\tag{2.18}
$$

where the associated parameters are

$$
\begin{aligned}
x_{QAM} &= g_{QAM}\left(\frac{2}{1+q^2}\right)\bar{\gamma},\\
y_{QAM} &= g_{QAM}\left(\frac{2q^2}{1+q^2}\right)\bar{\gamma}.
\end{aligned}
\tag{2.19}
$$

2.4 High-SNR Approximate Expressions for Average Error Probability in Nakagami-*q* Fading Channels

2.4.1 M-ary DPSK

For the exact closed-form SEP expression of M-ary DPSK in (5), we consider the following high-SNR (i.e., $\bar{\gamma} \to \infty$) approximations as [9–12]

$$\begin{aligned} & z_1, z_2 \to 0, \\ & F_D^{(3)}\left(\frac{1}{2}, \frac{1}{2}, \frac{1}{2}, \frac{1}{2}; \frac{3}{2}; 0, 0, z_3\right) \\ & = {}_2F_1\left(\frac{1}{2}, \frac{1}{2}; \frac{3}{2}; z_3\right) = \frac{\sin^{-1}(\sqrt{z_3})}{\sqrt{z_3}}, \\ & F_D^{(3)}\left(\frac{3}{2}, \frac{1}{2}, \frac{1}{2}, \frac{1}{2}; \frac{5}{2}; 0, 0, z_3\right) \\ & = {}_2F_1\left(\frac{3}{2}, \frac{1}{2}; \frac{5}{2}; z_3\right) = \frac{3\left(\sin^{-1}(\sqrt{z_3}) - \sqrt{1-z_3}\sqrt{z_3}\right)}{2z_3^{3/2}}, \end{aligned} \tag{2.20}$$

where ${}_2F_1(a, b; c; x)$ is the Gauss hypergeometric function [12] defined by

$$\begin{aligned} & {}_2F_1(a, b; c; x) \\ & = \sum_i^{\infty} \frac{(a)_i (b)_i}{(c)_i i!} x^i, \; |x| < 1, \; \Re(c - a - b) > 0, \\ & = \frac{1}{B(a, c-a)} \int_0^1 t^{a-1} (1-t)^{c-a-1} (1-xt)^{-b} dt, \\ & \Re(c) > \Re(a) > 0, \end{aligned} \tag{2.21}$$

Then, by applying (20)–(5), we can obtain the high-SNR approximate closed-form formula as

$$\begin{aligned} P_S^{High-SNR} & = \frac{2\cos\left(\frac{\pi}{2M}\right)}{\pi} \mathcal{MGF}(g_{DPSK}) \left[\frac{\sin^{-1}(\sqrt{z_3})}{\sqrt{z_3}} \right. \\ & \left. - \cos\left(\frac{\pi}{M}\right) \frac{\left(\sin^{-1}(\sqrt{z_3}) - \sqrt{1-z_3}\sqrt{z_3}\right)}{2z_3^{3/2}}\right]. \end{aligned} \tag{2.22}$$

We note that there are no hypergeometric functions in (22), which usually entail a high computational cost, and thus the error probability evaluation can be more easily performed.

2.4.2 Noncoherent Correlated Binary Signals

In the high-SNR regime, $\bar{\gamma} \to \infty$, we have the following approximations [9–12]

$$\begin{aligned} &\hat{z}_1, \hat{z}_2 \to 0, \\ &F_1\left(\frac{1}{2}, \frac{1}{2}, \frac{1}{2}; 1; 0, 0\right) = 1, \\ &F_1\left(\frac{3}{2}, \frac{1}{2}, \frac{1}{2}; 2; 0, 0\right) = 1. \end{aligned} \tag{2.23}$$

Then, from (10), the exact expression of the average BEP for the noncoherent correlated binary signals can be well approximated at high SNRs as

$$P_b^{High-SNR} = \frac{1}{2}\mathcal{MGF}(g_K)\left[1 - \frac{g_Q}{2}\right]. \tag{2.24}$$

2.4.3 M-ary Coherent PSK

For $\bar{\gamma} \to \infty$, the high-SNR approximate SEP formula for M-ary coherent PSK can be obtained as

$$\begin{aligned} P_S^{High-SNR} &= \frac{1}{4}\mathcal{MGF}(g_{PSK}) + \frac{1}{2\pi}\mathcal{MGF}(g_{PSK}) \\ &\quad \times \left(\sin\left(\frac{\pi}{M}\right)\cos\left(\frac{\pi}{M}\right) + \sin^{-1}\left(\cos\left(\frac{\pi}{M}\right)\right)\right), \end{aligned} \tag{2.25}$$

which can be straightforwardly derived by utilizing the following high-SNR approximations as [9–12]

$$\begin{aligned} &\frac{1}{1 + x_{PSK}}, \frac{1}{1 + y_{PSK}} \to 0, \\ &F_1\left(\frac{1}{2}, \frac{1}{2}, \frac{1}{2}; 2; 0, 0\right) = 1, \\ &F_D^{(3)}\left(\frac{1}{2}, -\frac{1}{2}, \frac{1}{2}, \frac{1}{2}; \frac{3}{2}; 0, 0, z_3\right) \\ &\quad = {}_2F_1\left(\frac{1}{2}, -\frac{1}{2}; \frac{3}{2}; \cos^2\left(\frac{\pi}{M}\right)\right) \\ &\quad = \frac{\sin\left(\frac{\pi}{M}\right)\cos\left(\frac{\pi}{M}\right) - \sin^{-1}\left(\cos\left(\frac{\pi}{M}\right)\right)}{2\cos\left(\frac{\pi}{M}\right)}. \end{aligned} \tag{2.26}$$

2.4.4 *M-ary Square QAM*

By exploiting the following high-SNR approximate formulas [9–12]

$$\begin{aligned}
&\frac{1}{1+x_{QAM}}, \frac{1}{1+y_{QAM}} \to 0,\\
&F_1\left(\frac{1}{2},\frac{1}{2},\frac{1}{2};2;0,0\right) = 1,\\
&F_D^{(3)}\left(1,1,\frac{1}{2},\frac{1}{2};\frac{5}{2};\frac{1}{2},\frac{1}{2},\frac{1}{2}\right)\\
&\quad = 2_2F_1\left(\frac{1}{2},1;\frac{5}{2};-1\right) = \frac{3}{2}(\pi-2),
\end{aligned} \tag{2.27}$$

we can simplify the SEP expression for M-ary square QAM given in (18), in the high-SNR regime, as

$$P_S^{High-SNR} = \beta\mathcal{MGF}(g_{QAM}) - \beta^2\left(1-\frac{2}{\pi}\right)\mathcal{MGF}(2g_{QAM}). \tag{2.28}$$

2.5 Numerical Results

In this section, by utilizing all the preceding formulations, numerical results for the SEP and/or BEP performances of various M-ary modulation schemes under consideration are presented over Nakagami-q fading channels. That is, we demonstrate the comparisons between the SEP/BEP curves from the existing exact closed-form expressions given in [8] and those obtained from our derived high-SNR approximate closed-form formulas for several modulation types over various fading environments. Specifically, Figs. 2.1–2.4 illustrate the SEP/BEP curves for M-ary DPSK, noncoherent correlated binary signals, M-ary coherent PSK, and M-ary square QAM, respectively, for the several values of fading severity parameter, q, and modulation size, M. As can be seen from all the figures, it can be easily observe that all the error probability curves obtained from the proposed high-SNR approximations are very close to those from the conventional exact formulas in the medium and high SNR regimes, which apparently confirms that the proposed concise high-SNR approximate closed-form formulas can be adopted for the effective error probability performance evaluations of various modulation schemes in Nakagami-q fading channels.

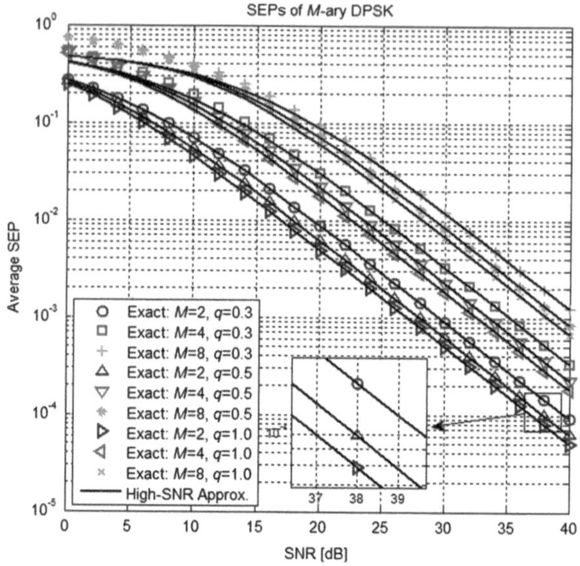

Fig. 2.1 Exact SEPs from (5) and high-SNR approximated SEPs from (22) versus SNR (i.e., $\bar{\gamma}$) for M-ary DPSK with $M = (2, 4, 8)$ (i.e., DBPSK, DQPSK, and D8PSK) over Nakagami-q fading channels with $q = (0.3, 0.5, 1.0)$

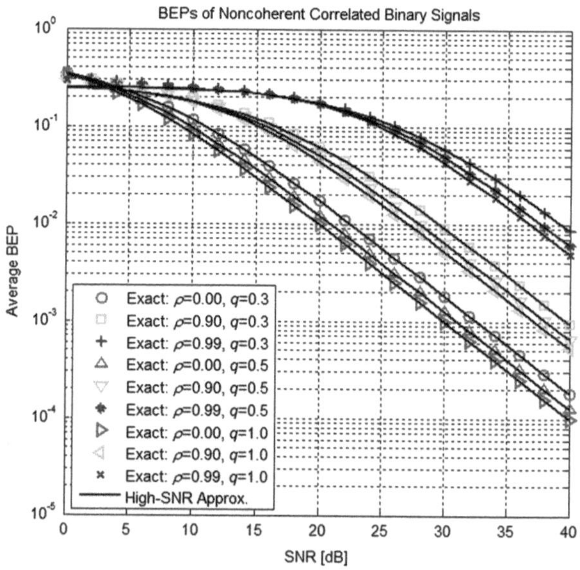

Fig. 2.2 Exact BEPs from (10) and high-SNR approximated BEPs from (24) versus SNR (i.e., $\bar{\gamma}$) for noncoherent correlated binary signals with various correlation coefficients $\rho = (0.00, 0.9, 0.99)$ over Nakagami-q fading channels with $q = (0.3, 0.5, 1.0)$

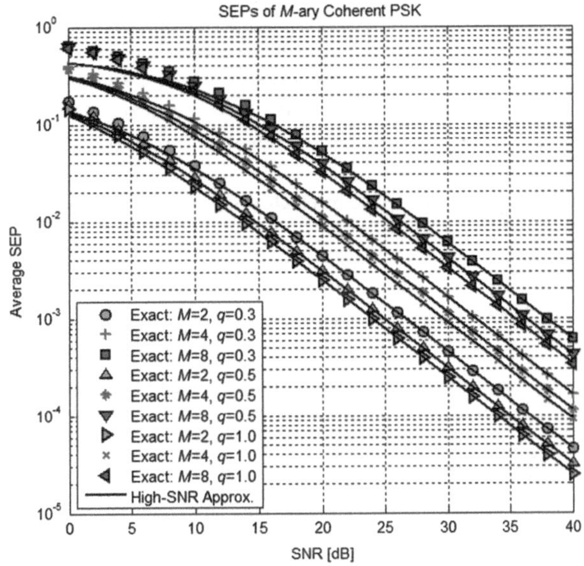

Fig. 2.3 Exact SEPs from (14) and high-SNR approximated SEPs from (25) versus SNR (i.e., $\bar{\gamma}$) for M-ary coherent PSK with $M = (2, 4, 8)$ (i.e., BPSK, QPSK, and 8-PSK) over Nakagami-q fading channels with $q = (0.3, 0.5, 1.0)$

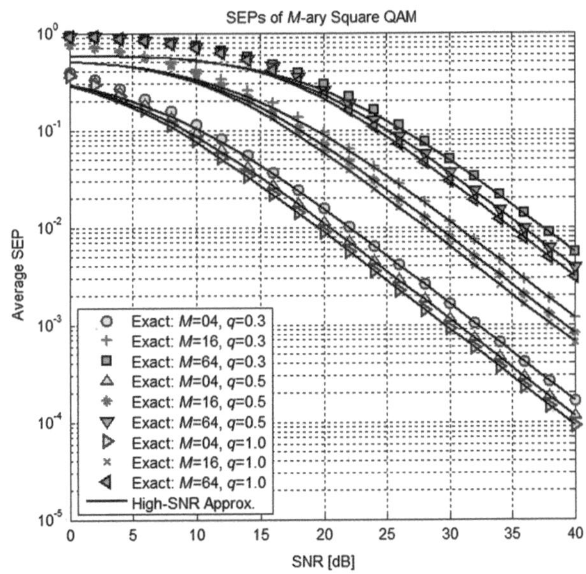

Fig. 2.4 Exact SEPs from (18) and high-SNR approximated SEPs from (28) versus SNR (i.e., $\bar{\gamma}$) for M-ary square QAM with $M = (4, 16, 64)$ (i.e., 4-QAM, 16-QAM, and 64-QAM) over Nakagami-q fading channels with $q = (0.3, 0.5, 1.0)$

2.6 Conclusions

In this paper, we have derived concise high-SNR approximate closed-form expressions of the average SEP or BEP for several M-ary modulation schemes (i.e., M-ary DPSK, noncoherent correlated binary signals, M-ary coherent PSK, and

M-ary square QAM) especially in Nakagami-q (i.e., also known as Hoyt) fading channels. Compared with the existing exact error probability formulas, the derived high-SNR approximate expressions have much simpler formulations, still demonstrating good agreement with the exact error probabilities in the high-SNR regions, which have been evidently verified through extensive numerical results.

Acknowledgments This research was financially supported by Hansung University.

References

1. Proakis JG (1995) Digital communications, 3rd edn. McGraw-Hill, New York
2. Goldsmith A (2005) Wireless communications. Cambridge University Press, New York
3. Simon MK, Alouini M-S (2005) Digital communication over fading channels, 2nd edn. Wiley, New York
4. Simon MK, Alouini M-S (1998) A unified approach to the performance analysis of digital communication over generalized fading channels. Proc IEEE 86(9):1860–1877
5. Youssef N, Wang C-X, Patzold M (2005) A study on the second order statistics of Nakagami-Hoyt mobile fading channels. IEEE Trans Veh Technol 54(4):1259–1265
6. Tellambura C, Mueller AJ, Bhargawa VK (1997) Analysis of M-ary phase-shift keying with diversity reception for land-mobile satellite channels. IEEE Trans Veh Technol 46(4):910–922
7. Annamalai A, Tellambura C (2001) Error rates for Nakagami-m fading multichannel reception of binary and M-ary signals. IEEE Trans Commun 49(1):58–68
8. Radaydeh RM (2007) Average error performance of M-ary modulation schemes in Nakagami-q (Hoyt) fading channels. IEEE Commun Lett 11(3):255–257
9. Erdelyi A (1953) Higher transcendental function. McGraw Hill, New York
10. Exton H (1976) Multiple hypergeometric functions and applications. Wiley, New York
11. Gradshteyn IS, Ryzhik IM (1994) Table of integrals, series, and products, 5th edn. Academic Press, San Diego
12. Wolfram Research The Wolfram functions site. http://functions.wolfram.com

Chapter 3
Performance Evaluation of IEEE 802.15.6 MAC with User Priorities for Medical Applications

Li Yang, Changle Li, Yueyang Song, Xiaoming Yuan and Yanle Lei

Abstract In order to satisfy the heterogeneous service requirements from different applications and the complex channel characters owing to body motions, IEEE 802.15.6 standard was established as a new solution for Wireless Body Area Networks (WBANs). In this paper, we evaluate the effect of user priorities (UPs) on the performance of IEEE 802.15.6 CSMA/CA channel access mechanism in narrow band. Simulation metrics mainly focus on the normalized throughput and average packet delay in which the traffic arrival rate and traffic distribution vary. In addition, we make a performance comparison with the non-priority CSMA/CA which concludes that the IEEE 802.15.6 with user priorities performs better in specific situation.

Keywords WBAN · IEEE 802.15.6 · CSMA/CA · User priorities

3.1 Introduction

Wireless Body Area Network (WBAN) is a kind of short-range, wireless communication networks, aiming at providing the access services of low-power, high reliability and low latency, which can be exploited in many fields that vary from the medical application (e.g. health monitoring), the consumer electronics to the personal entertainment. Therefore, the broad prospects have motivated many researches on the key technologies and standardization process of WBANs.

As a solution for Wireless Personal Area Network (WPAN), IEEE 802.15.4 [1] provides technologies to operate around human body, but due to the effect of human body on the radio channel and the diversified demands in different applications, it suffers restrictions when used in the timely and reliable situation. Therefore, IEEE 802.15.6 [2] was released as the appropriate standard for short-range, wireless communication in the vicinity of, or inside human body.

L. Yang · C. Li (✉) · Y. Song · X. Yuan · Y. Lei
State Key Laboratory of Integrated Services Networks, Xidian University, Xi'an, Shaanxi 710071, China
e-mail: clli@mail.xidian.edu.cn

J.J. (Jong Hyuk) Park et al. (eds.), *Ubiquitous Computing Application and Wireless Sensor*, Lecture Notes in Electrical Engineering 331,
DOI 10.1007/978-94-017-9618-7_3

Compared with IEEE 802.15.4, several improvements have been made by IEEE 802.15.6 to provide high quality of service (QoS) and extreme high power efficiency. IEEE 802.15.6 designs three different Physical Layers (PHYs) to adapt the broad range of possible applications and constructs a more flexible frame structure that supports multiple access modes. However, the greatest difference is that assigning different priorities based on the traffic type, which guarantees the high-priority data, for example, the emergency data, transmit timely and reliably.

Most of existing researches pay attention to the Medium Access Control (MAC) of IEEE 802.15.6. Cavallari et al. [3] draws an overview of WBAN main applications, technologies and summarizes the advantages of IEEE 802.15.6, while [4] and [5] give detailed descriptions of MAC functionalities in IEEE 802.15.6. Ullah et al. [6] presents the theoretical maximum throughput and minimum delay limits of IEEE 802.15.6 under different frequency bands and data rates. Jung et al. [7] evaluates the performance of IEEE 802.15.6 MAC in terms of energy consumption and energy efficiency.

There are also some literatures that study the effect of user priorities on the performance of IEEE 802.15.6. An et al. [8] analyzes the performance of IEEE 802.15.6 MAC where there are three user priorities (UP1, UP3 and UP5) in the whole network, which is not rigorous due to ignoring the other UPs especially the emergency medical traffic of UP6 and UP7. Li et al. [9] presents a three-dimensional Markov model to evaluate the performance of CSMA/CA scheme in saturation condition, which means there is always a packet in the queue waiting to transmit, thus it can't be used in the medical application directly. Rashwand et al. [10] investigates the effectiveness of user priorities in IEEE 802.15.6 with focusing on the influence of the access phase lengths, but the model of arrival traffic is Bernoulli arrival process, which deviates from the real medical scenarios because the primary type of traffic is periodic type traffic. Compared with the previous works, we evaluate the effect of user priorities on the performance of IEEE 802.15.6 CSMA/CA versus the traffic load and traffic distribution in narrow band. Both the unsaturated and saturated situations are taken into account, periodically arrival process is considered as the model of arrival traffic, which is more suitable for the real medical scenario, such as the continuous monitoring of the human body. In addition, we also make a comparison with the non-priority CSMA/CA.

The remainder of the paper is organized as follows: Sect. 3.2 briefly describes IEEE 802.15.6 MAC with the details of CSMA/CA. In Sect. 3.3 simulation models, parameters and results are presented. According to the simulation results, Sect. 3.4 concludes the paper and makes an outlook on future work.

3.2 Overview of IEEE 802.15.6 MAC

IEEE 802.15.6 supports three access modes. Here we focus on the beacon mode with superframes that the hub transmits a beacon at the start of each superframe to inform the information related to BAN identification, synchronization, and superframe structure.

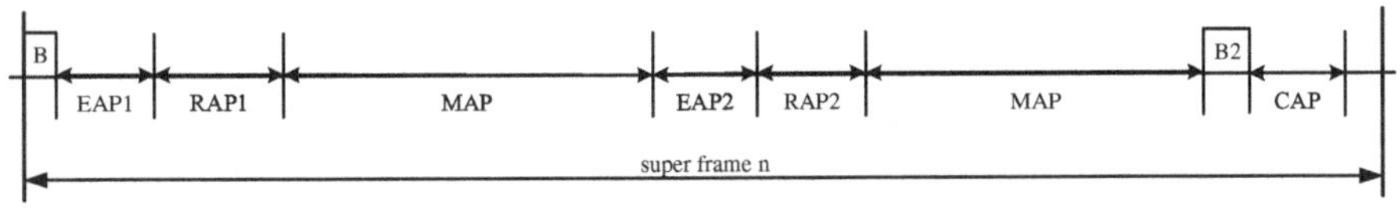

Fig. 3.1 Superframe structure of the beacon mode with superframes

Table 3.1 CW bounds and UP mapping for CSMA/CA

Priority	UP	Traffic designation	CWmin	CWmax
Lowest	0	Background (BK)	16	64
	1	Best effort (BE)	16	32
	2	Excellent effort (EE)	8	32
	3	Video (VI)	8	16
	4	Voice (VO)	4	16
	5	Medical data or network control	4	8
	6	High-priority medical data or network control	2	8
Highest	7	Emergency or medical implant event report	1	4

Figure 3.1 shows the superframe structure under beacon mode with superframes. Among the seven periods, EAP1, EAP2, RAP1, RAP2 and CAP access phases are used for the contended allocation based on CSMA/CA or slotted Aloha scheme while the MAP is composed of uplink allocation intervals, downlink allocation intervals and bilink allocation intervals.

Depending on the characteristics of different traffic flows, eight user priorities (UPs) which span from UP0 to UP7 are assigned to distinguish them. The higher priorities represent urgent traffic while the lowers are for general traffic. The pre-defined relationship between UP and contention window (CW) bounds, CWmin and CWmax, for CSMA/CA are showed in Table 3.1.

The details of CSMA/CA scheme are described as follows: When data arrives, the node firstly selects a random integer uniformly distributed over the interval [1, CW] as the back-off counter value. Then the node will perform the clear channel assessment (CCA) detection, the back-off counter subtracts by one in the unlock state. Once the back-off counter decreases to zero and the remaining time is long enough, it transmits the data immediately. If this transmission failed, the CW and back-off counter change their stage based on the corresponding rules in [2] until the packet is transmitted successfully or has reached the max retransmission times.

3.3 Performance Evaluation

In this section, we make simulation on the platform MIRAI-SF, a program written in JAVA. The Two performance metrics considered in our work are the normalized throughput and average packet delay, respectively. The normalized throughput is defined as the ratio of the amount of successful data transmission in unit time to information data rate, while the average packet delay refers the interval that a packet arrives at local MAC layer until the peer node receives the data successfully.

3.3.1 Simulation Scenario

A one-hop star network is chosen as the network topology which consists of a hub and eight nodes, as shown in Fig. 3.2. Each node generates a periodic data flow that represents a kind of UPs varied from UP0 to UP7, competes for allocation slot by CSMA/CA. Considering the medical sensing where most data transmissions are initiated by the sensor nodes, the downlink traffic is not considered in this paper.

The key parameters we used in the simulation are listed in Table 3.2.

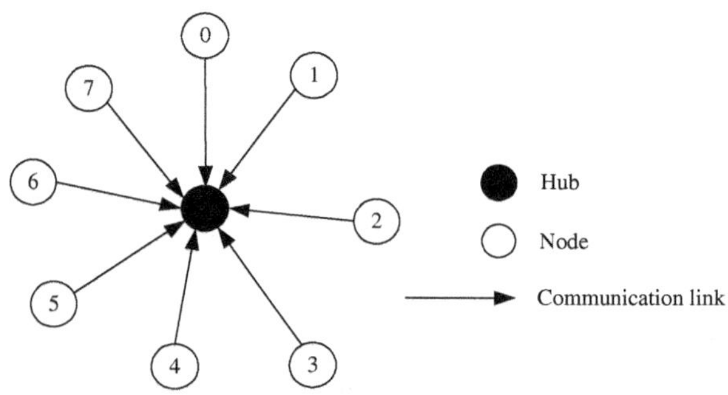

Fig. 3.2 The network topology used in the simulation

Table 3.2 Simulation parameters

IEEE 802.15.6 CSMA/CA	
Parameters	Default value
Frequency band	2,400–2,483.5 MHz
Information data rate	971.4 kbps
Payload size	255 Bytes
Symbol rate	600 ksps
MAC header	9 Bytes
PLCP header	31 bits
PLCP preamble	90 bits
ACK	193 bits
pSIFS	75 μs
CSMA/CA slot time	145 μs
Non-priority CSMA/CA	
Parameters	Default value
CWmin	3
CWmax	5

3.3.2 Simulation Results

In this part, we evaluate the performance of IEEE 802.15.6 CSMA/CA for different UPs versus the arrival rate of packets and make a comparison with non-priority CSMA/CA where all nodes have fair chances to access channel.

From Figs. 3.3 and 3.4, one can see that large difference exists on the performance of nodes with different UPs. With the arrival rate of traffic in the network increasing, the normalized throughput of UP7 and UP6 can achieve improvement constantly, whereas the other UPs increase firstly but then fall down. The average packet delay of each UP is almost constant at first, then increases rapidly, and finally flattens out, whereas the UP7 remains almost unchanged. One reason is when the arrival rate is low, each packet can access channel effectively, but after getting a certain value, fierce collisions lead the network to be saturated. The other is the different CW values for different UPs, the higher UPs access channel

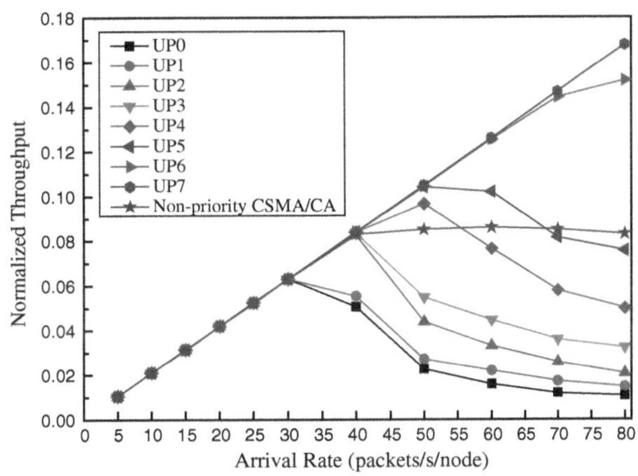

Fig. 3.3 Comparison of normalized throughput for all UPs in IEEE 802.15.6 CSMA/CA and non-priority CSMA/CA

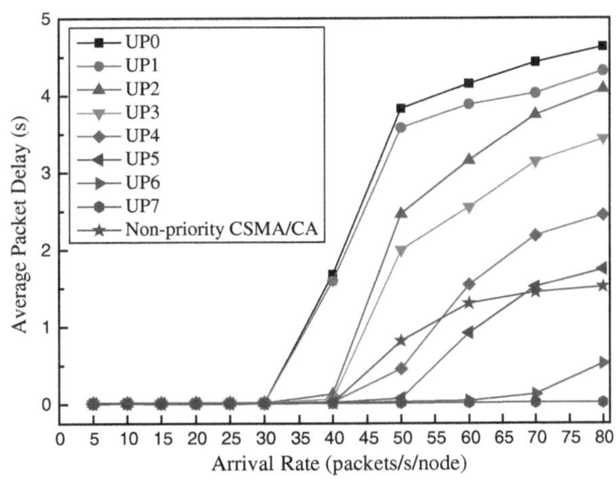

Fig. 3.4 Comparison of average packet delay of all UPs in IEEE 802.15.6 CSMA/CA and non-priority CSMA/CA

frequently as the lower CWmin and CWmax, while the lower UPs may make more backoffs before accessing channel. The higher of UP can obtain the higher maximum throughput and lower average packet delay simultaneously.

Figures 3.3 and 3.4 also show that the non-priority CSMA/CA has the approximate performance with IEEE 802.15.6 CSMA/CA when the arrival rate is below 30 packets/s. From then it performs better than IEEE 802.15.6 CSMA/CA for UP0, UP1 firstly, better than the UP2 and UP3, UP4, UP5 at the arrival rate of 40 packets/s, 55 packets/s and 70 packets/s, respectively. Nodes in the non-priority CSMA/CA access channel in fair, whereas the higher UPs occupy the channel frequently such that gives rise to the collision with other UPs in the IEEE 802.15.6 CSMA/CA. After the arrival rate has reached a relatively high value, the UP6 and UP7 still perform better than the non-priority CSMA/CA, which indicates the IEEE 802.15.6 CSMA/CA can guarantee higher reliability and lower delay for the high UPs, especially the emergency data and high-priority medical data.

In this part, we evaluate the performance of IEEE 802.15.6 CSMA/CA varies with the traffic distribution in the whole network. Here we assume only two UPs in the network simultaneously. According the Table 3.1, we treat the UP6 and UP7 as the medical traffic while the UP0 and UP2 as the non-medical traffic. Performance under three different cases-UP0&UP6, UP2&UP6, UP2&UP7, are conducted when the proportion of medical traffic to the overall traffic varies from 0 to 100 %.

In Figs. 3.5 and 3.6, the normalized throughput and average packet delay in non-priority CSMA/CA are constant because the nodes have equal chances to access channel. For IEEE 802.15.6 CSMA/CA, the two curves with same color represent a kind of scenarios where the square is for the average performance of the whole network and the triangle is only for the medical traffic. The IEEE 802.15.6 CSMA/CA and non-priority CSMA/CA offer the same throughput and approximate delay when the medical traffic occupies a few part of traffic in the network. Considering the average packet delay over 250 ms is unacceptable for general WBAN applications [11], Fig. 3.6 only shows the performance of IEEE 802.15.6 CSMA/CA in the

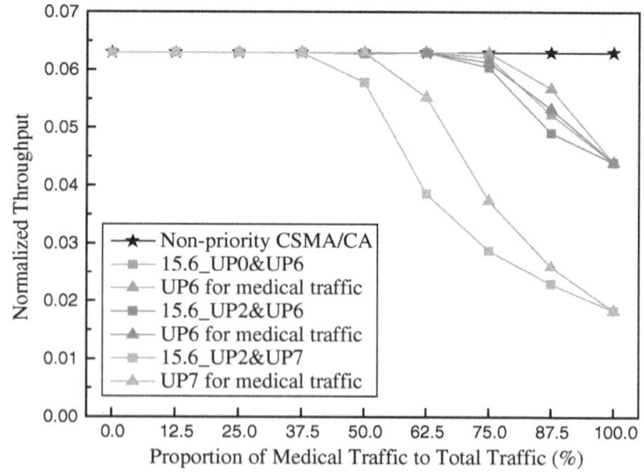

Fig. 3.5 Comparison of normalized throughput versus traffic distribution

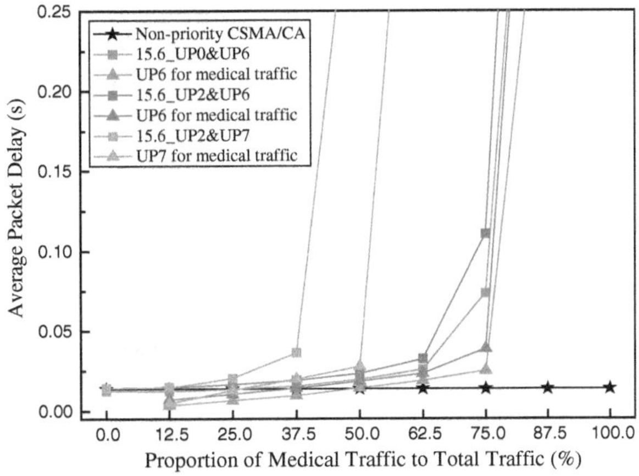

Fig. 3.6 Comparison of average packet delay versus traffic distribution

acceptable range of delay. It should be noted that the IEEE 802.15.6 CSMA/CA guarantees a lower average delay for the medical traffic when there is few medical traffic. The reduction of delay for medical traffic is up to 74.5 % in the case of UP0&UP6 when the proportion of medical traffic is 12.5 %, which indicates that employing the IEEE 802.15.6 CSMA/CA is better because the latency requirement of a medical data is much lower than a background data.

However, after the occupancy of medical traffic exceeds a certain value, the IEEE 802.15.6 CSMA/CA will suffer deterioration and perform poorer than the non-priority CSMA/CA due to the frequent collisions. For the medical traffic, the corresponding contention window is so small that increases the probability to choose same back-off counter, which always leads collisions in the first few transmission times. Even so, both the throughput and delay for medical traffic are still well above the average level.

Compared with the performance in the case of UP2&UP6, the one in UP2&UP7 is much worse because the higher UPs for medical traffic have smaller CW to contend for channel frequently, which is more selfish to complete its data transmission at the expense of hindering transmission of lower UPs and of increasing the internal competition among higher UPs. However, the one in UP0&UP6 has better performance because the lower UPs have larger CW to avoid conflicting with higher UPs, thus improves the performance of the whole network by alleviating the congestion of the network.

3.4 Conclusion

For the IEEE 802.15.6 CSMA/CA, user priorities mechanism makes a big influence on network performance, it guarantees reliable and timely service for higher UPs with sacrificing the performance of lower UPs. What's more, the performance of

IEEE 802.15.6 with user priorities is better when there is lightweight medical traffic in the network, which can obtain both higher throughput and lower delay especially for the medical traffic. But it suffers deterioration when the medical traffic holds a large proportion of total traffic due to the fierce collisions. It also concludes that employing the CSMA/CA scheme alone is not enough to meet all needs of medical traffic, thus we will also consider the other access modes in IEEE 802.15.6 such as the improvised access and scheduled access in the future.

Acknowledgments This work was supported by the National Natural Science Foundation of China under Grant No. 61271176, the National Science and Technology Major Project under Grant No. 2013ZX03005007-003, the Fundamental Research Funds for the Central Universities, and the 111 Project (B08038).

References

1. IEEE Standard for Local and metropolitan area networks, Part 15.4 (2003) Wireless medium access control (MAC) and physical layer (PHY) specifications for low-rate wireless personal area networks (WPANs). IEEE Std 802.15.4-2003
2. IEEE Standard for Local and metropolitan area networks, Part 15.6 (2012) Wireless body area networks. IEEE Std 802.15.6-2012
3. Cavallari R, Martelli F, Rosini R, Buratti C (2014) A survey on wireless body area networks: technologies and design challenges. IEEE Commun Surv Tutor 16(3):1635–1657
4. Kwak KS, Ullah S, Ullah N (2010) An overview of IEEE 802.15.6 standard. In: 3rd IEEE international symposium on applied sciences in biomedical and communication technologies (ISABEL), Rome, pp 1–6
5. Brada N, Belhaj S, Chaari L, Kamoun L (2011) Study of medium access mechanisms under IEEE 802.15.6 Standard. In: 4th joint IFIP on wireless and mobile networking conference (WMNC), Toulouse, pp 1–6
6. Ullah S, Chen M, Kwak KS (2012) Throughput and delay analysis of IEEE 802.15.6-based CSMA/CA protocol. J Med Syst 36(3):3875–3891
7. Jung BH, Akbar RU, Sung DK (2012) Throughput, energy consumption, and energy efficiency of IEEE 802.15.6 body area network (BAN) MAC protocol. In: 23rd IEEE international symposium on personal indoor and mobile radio communications (PIMRC), Sydney, pp 584–589
8. An N, Wang P, Yi C, Li Y (2013) Performance analysis of CSMA/CA based on the IEEE 802.15.6 MAC protocol. In: 15th IEEE international conference on communication technology (ICCT), China, pp 539–544
9. Li C, Geng X, Yuan J, Sun T (2013) Performance analysis of IEEE 802.15.6 MAC protocol in beacon mode with superframes. KSII Trans Internet Inf 7(5):1108–1130
10. Rashwand S, Mišic J, Mišic VB Analysis of CSMA/CA mechanism of IEEE 802.15.6 under non-saturation regime, to appear in IEEE Transactions on Parallel and Distributed Systems.
11. Cordeiro C (2007) Use cases, applications, and requirements for BANs. Doc.: 802.15-07-0564

Chapter 4
Using Dead Reckoning, GPS and Fingerprinting for Ubiquitous Positioning

Hsiao-Hsien Chiu, Yi-Jiun Tang and Ming-Shi Wang

Abstract In many Internet of Things (IoT) scenarios, the applications need to identify the location of sensors/actuators and interact with them. Thus, a basic, common and primary mechanism is needed to identify the location of things. Everybody knows that Global Positioning System (GPS) is widely accepted as a reliable, available and accurate source of positioning, able to operate across the globe. But it is unavailable for indoors, it is due to the absence of line of sight to satellites. For indoor positioning, there are many wireless technologies have been developed, includes WiFi, RFID, Bluetooth, etc. But the sensors and actuators of IoT will be deployed in anywhere and they maybe move between indoor and outdoor environment. So ubiquitous positioning systems is expected that can work in both environments. In this paper, we proposed a hybrid positioning algorithm, which combined the Complementary Extended Kalman Filter, Dead Reckoning, GPS and WiFi Fingerprinting technologies for ubiquitous positioning. Based on the algorithm, the mobile things can report its geographic coordinate no matter where they are.

Keywords Ubiquitous positioning · Dead reckoning · GPS · Fingerprinting · Complementary extended Kalman filter

4.1 Introduction

4.1.1 Scenario

As the Internet of Things is strongly rooted in the physical world, the notion of physical location is very important, especially for spatial identification.

H.-H. Chiu · Y.-J. Tang · M.-S. Wang (✉)
Department of Engineering Science, National Cheng-Kung University,
No. 1 University Road, Tainan City, Taiwan
e-mail: tomwolfcit@gmail.com

J.J. (Jong Hyuk) Park et al. (eds.), *Ubiquitous Computing Application and Wireless Sensor*, Lecture Notes in Electrical Engineering 331,
DOI 10.1007/978-94-017-9618-7_4

For example, there is a scenario about Intelligent Commuter proposed by IoT-i [1]. In this scenario, there are several location-based services (LBS) as carpool service, transit information delivery service, etc. But the LBS are not only suitable in outdoors area. Especially in the urban area, travelers must move between indoors and outdoors very often. Even, travelers may move in underground for a long time. Thus, the LBS of course shall have the capability to monitor travelers no matter where they are. Therefore, the infrastructure of IoT has to support finding things according to location [2].

4.1.2 GPS

Generally, in outdoors, Global Positioning System (GPS) [3] is the most widely used satellite-based positioning system, which offers maximum coverage. GPS capability can be added to various devices by adding GPS cards and accessories in these devices, which enable location-based services, such as navigation, tourism, etc. However, GPS is unavailable in indoors, because line-of-sight transmission between receivers and satellites [4].

4.1.3 WiFi Positioning

There are many technologies have been developed for indoor positioning system (IPS), such as infrared (IR), ultrasound, radio-frequency identification (RFID), wireless local area network (WLAN), Bluetooth, sensor networks, ultra-wideband (UWB), magnetic signals, vision analysis and audible sound [4].

In these technologies, WiFi is very popular and has been implemented in many public and private areas. WiFi-based positioning systems reuse the existed WiFi infrastructures, which is a low-cost technology of indoor positioning. The accuracy of location estimations based on the signal strength of WiFi signals is affected by various elements in indoor environments such as movement and orientation of human body, the overlapping of APs, the nearby tracked mobile devices, walls, doors, etc. But, we consider WiFi-based positioning system is still a suitable solution for indoors.

Several positioning algorithms have been developed for WiFi positioning. These fall into the broad categories of geometric techniques, statistical techniques, Fingerprinting and particle filters. Fingerprinting is also referred to in the literature as radio mapping, database correlation or pattern recognition [5]. For reviewing of these techniques, please see [6].

4.1.4 Pedestrian Dead Reckoning

In indoors, if there is no WiFi signal, the use of inertial sensors is becoming widespread for pedestrian navigation. There are two approaches, one is using accelerometer and gyroscope to estimate the step length and direction, a person's position can be determined by dead-reckoning (DR) [7, 8]. Another approach is based on full six degree of freedom (6DOF) inertial navigation. A foot-mounted 6DOF strap down inertial platform comprising triads of accelerometers and gyroscopes is used to dead reckoning via a conventional strapdown navigation algorithm [9].

4.1.5 Relative Integration Technologies

In IoT scenario, the things move between indoors and outdoors very often. Therefore, it is a basic capability to switch indoor and outdoor positioning systems. Singh et al. [10] and Cheong et al. [11] proposed strategies for switching between GPS and WiFi positioning systems.

In general, GPS system uses geographic coordinate system and WiFi system uses Cartesian coordinate system. They are two very different systems. But for IoT, all things should be placed in one universal coordinate system. Ogawa et al. [12] and Lin et al. [13] tried to establish seamless indoor/outdoor systems, but they kept the original coordinate of GPS and IPS independently. When the mobile things transfer to different environments, the ordinate system will be changed.

For extending the GPS coverage with WiFi, most researches are focus on enhancement accuracy of GPS [14, 15]. Otherwise, [16] ever tried to use WiFi to extend the coverage of GPS with virtual satellite concept. But, when the area can't receive any signal of satellites, it just can use indoor positioning system only and the indoor positioning system is independent of GPS. Reference [17] proposed a hybrid positioning system, it distinguish two kinds of Access Point (AP), Anchor Node (AN) and Unknown Node (UN). It used AN to detect the position of UN and smart phone.

There also are several commercial WiFi positioning systems already published as Skyhook, Indoor Google Maps. About Skyhook, there is a description of the system and extensive tests of it can be found in Gallagher et al. [18].

4.1.6 The Guideline of this Paper

In this paper, we proposed a hybrid positioning algorithm, which is based on Complementary Extended Kalman Filter, and integrating with Pedestrian Dead Reckoning (PDR), GPS and WiFi Fingerprinting together. In Chap. 3, we will describe the test environment and path. In Chapter 4, we will introduce the ubiquitous positioning algorithm. Then we will show the experiment results in Chap. 5. At final, we will have the conclusions and future works.

4.2 Test Environment and Path

About test environment, we selected the building, which is belong to the Department of Engineering Science of Cheng Kong University as Fig. 4.1. The building has basement, we can walk through the building and down to the basement for the every kind environments.

For testing the positioning system can switch between indoors and outdoors, we designed the test path from outside to inside and go to outside again as Fig. 4.1. For testing in a GPS-denied area, we designed the path from first floor down to basement and go up again, as Figs. 4.2 and 4.3. Then we calculated the geographic coordinate of the test points

At experiment phase, on every test point, we collected the information by smartphone as followed:

- Geographic coordinate by GPS;
- WiFi AP Mac Address;
- WiFi Radio Signal Strength (RSS);
- Direction Angle and Movement of Smartphone.

After collecting the information on each test point, we send these data to positioning server immediately. We designed a serious experiment to examine our approach as followed:

- Using GPS only;
- Using WiFi Fingerprinting only;
- Using Dead Reckoning only;
- Using Our Proposed Algorithm.

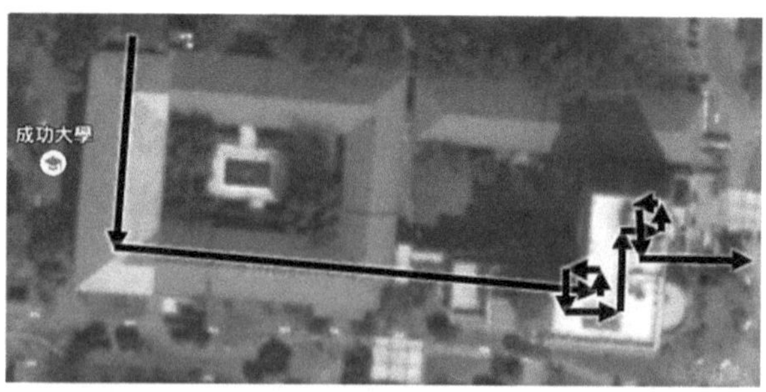

Fig. 4.1 Test path through the building

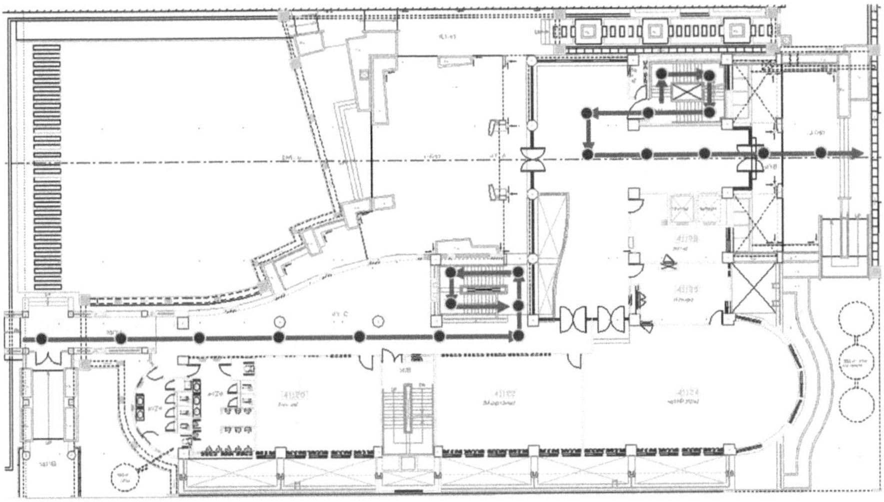

Fig. 4.2 Test points on first floor

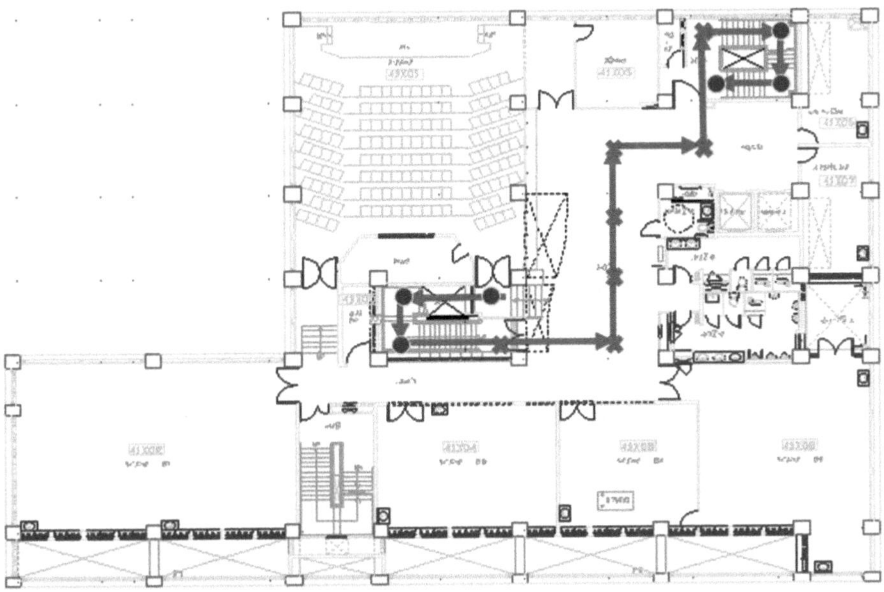

Fig. 4.3 Test points on basement

4.3 The Proposed Algorithm

4.3.1 Issue Description

For ubiquitous positioning, at first, we have to analysis how many kind environments in the world. Thus, we found out four different kind environments as the Table 4.1 shows:

Due to each kind environment has its own characteristics, we have to choose suitable positioning technologies for each environment. We abstracted three common positioning technologies, GPS, WiFi Fingerprinting and Dead Reckoning for these environments. GPS is suitable for the outdoors, WiFi Fingerprinting is suitable for the environment with sufficient WiFi signal and Dead Reckoning is suitable for any environment, especially for the Indoors without sufficient WiFi signal.

For ubiquitous positioning, it is no doubt to integrate the three positioning technologies together, but there are three key issues needed to be solved as followed:

- How to switch the different positioning systems smoothly?
- How to output unified location information?
- How to extend the coverage of positioning system in large-scale indoors where cannot receive any signals of satellites or WiFi?

4.3.2 Pedestrian Dead Reckoning

Due to pedestrian dead reckoning is the only way to positioning in all environments. We proposed to use PDR to be the base of the algorithm.

We used Gyroscope and 3D-Accelerometer sensors on smartphone to detect the direction and movement of smartphone, HTC-J. The Acceleration of Z-axis situation is shown as Fig. 4.4. We detected the step when the acceleration had oscillation and assumed the distance of every step is fixed. Then we used the direction, step counts and fixed distance to estimate the position of smartphone. Then we record the tracking path as Fig. 4.5. But the problem of the PDR is the cumulative errors.

Table 4.1 Different kind environments definition

	WiFi sufficient	WiFi insufficient
Outdoors	GPS/WiFi/DR	GPS/DR
Indoors	WiFi/DR	DR

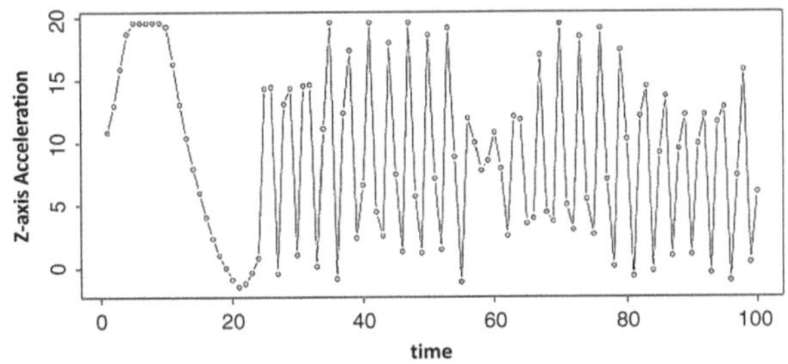

Fig. 4.4 Acceleration of Z-axis diagram

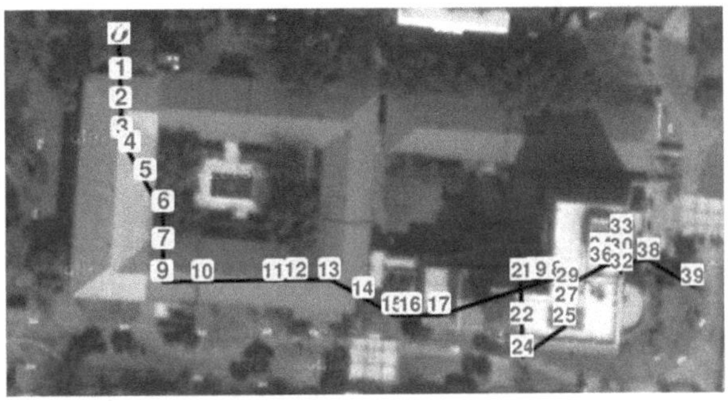

Fig. 4.5 The tracking path of pedestrian dead reckoning

4.3.3 GPS and WiFi Fingerprinting

For correcting the error of DR, we thought that need other positioning systems to help us correcting the estimation result of DR. The GPS and WiFi Fingerprinting are two candidates.

For understanding the positioning situation of GPS, we used GPS to positioning only. As Fig. 4.6, when we walked in outdoors, the accuracy of GPS is higher (Point 0–2). But, when we walked in indoors, the accuracy is lower and the GPS coordinates are offset from original path.

At the same time, we tried to only use WiFi Fingerprinting technology to positioning. At offline phase, we collect the RSS of APs at reference points and

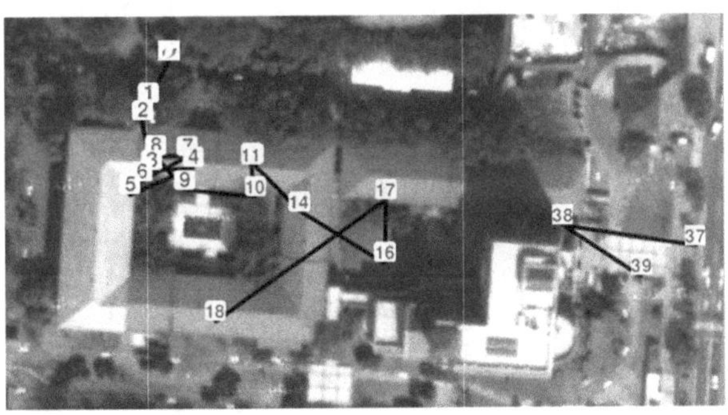

Fig. 4.6 Tracking path by GPS

Fig. 4.7 Reference points for WiFi fingerprinting

Fig. 4.8 Tracking path by WiFi fingerprinting

stored them in database as Fig. 4.7. Then at online phase, we compared the RSS of APs between target and reference points by Mahalanobis equation [19]. Using the method, we will find-out the minimal value of reference points by Mahalanobis equation to be the estimated coordinate. The tracking path with WiFi Fingerprinting is shown as Fig. 4.8.

Although the result is good in indoors. But, the problem of Fingerprinting is it needs a lot of man-power and time to collect the RSS of APs for reference points. When we setup too many reference points, the positioning performance becomes very low at online phase. On the other hand, when we setup too few reference points, it must sacrifice the accuracy of positioning.

4.3.4 Complementary Extended Kalman Filter

Due to there are cumulative errors in dead reckoning, and there is error model of GPS and WiFi Fingerprinting positioning. So we proposed to use Complementary Extended Kalman Filter (CEKF) to update the estimation of dead reckoning by GPS and WiFi Fingerprinting as Leppakoski et al. [7].

CEKF is derived from Extended Kalman Filter (EKF). EKF is the nonlinear version of the Kalman Filter (KF). In the EKF, the state transition and observation models need not be linear functions of the state but may instead be differentiable functions.

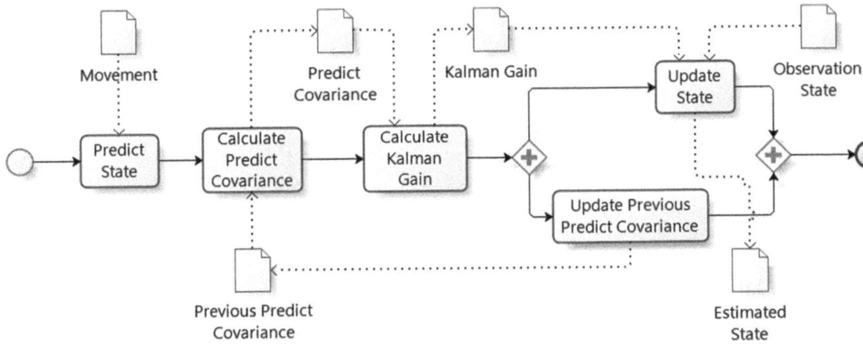

Fig. 4.9 CEKF process diagram

The CEKF Process is shown as Fig. 4.9. The first step is predicting the state by DR. The predicted state is propagated by using

$$\hat{X}_k^- = \hat{X}_{k-1} + \begin{bmatrix} \theta_k \\ \Delta S_{long} \\ \Delta S_{lati} \end{bmatrix} \tag{4.1}$$

where $\hat{X}_{k-1}$ is the posterior estimate after the measurement update using the (k-1)th measurement samples, while $\hat{X}_k^-$ is the prior estimate for kth time step. θ_k is the angle measured by the gyroscope sensor. ΔS_{long} and ΔS_{lati} are the travel differences of longitude and latitude.

The second step is to calculate the Predict Covariance:

$$P_k^- = F_k P_{k-1} F_k^T + Q_k \tag{4.2}$$

where P_{k-1} is the posterior covariance from the previous time step. F_k is the state transition matrix. Q_k is the noise of the state.

$$F_k = \begin{bmatrix} 1 & 0 & 0 \\ -\Delta S_{lati} & 1 & 0 \\ \Delta S_{long} & 0 & 1 \end{bmatrix} \tag{4.3}$$

The third step is to calculate the Kalman Gain.

$$K_k = P_k^- H^T \left(H P_k^- H^T + R\right)^{-1} \tag{4.4}$$

where $H = \begin{bmatrix} 0 & 1 & 0 \\ 0 & 0 & 1 \end{bmatrix}$ is observation model which maps the true state space into the observed space. R is the covariance of measurement noise.

The fourth step is to update the state.

$$\hat{X}_k = \hat{X}_k^- + K_k(z_k - H\hat{X}_k^-) \tag{4.5}$$

where $\hat{X}_k$ is the final estimated state. z_k is the measurement state at kth time step.

Final step is to update the previous predict covariance and provide for next step estimation.

$$P_k = (I_{3\times 3} - K_k H)P_k^- \tag{4.6}$$

4.3.5 Proposed Algorithm

For ubiquitous positioning, we proposed the algorithm integrated Dead Reckoning, GPS and WiFi Fingerprinting and Complementary Extended Kalman Filter as Fig. 4.10. The key point is the observation state z_k of CEKF, as (5). We used smartphone to collect relative information includes GPS Geo-Coordinate, WiFi RSS of APs, Direction and Movement…etc. Then we sent the information to positioning server to estimate a coordinate to be observation state of CEKF. The observation state selection principle is that while the WiFi signal is sufficient, the positioning by WiFi fingerprinting is first priority. The second choice is GPS. If GPS and WiFi both are unavailable, Dead Reckoning is the final choice. After

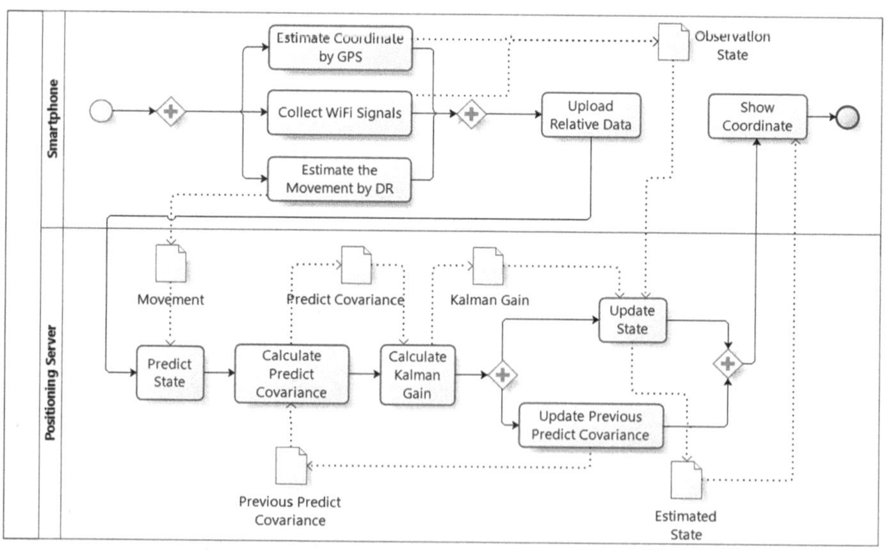

Fig. 4.10 Integrated hybrid positioning algorithm

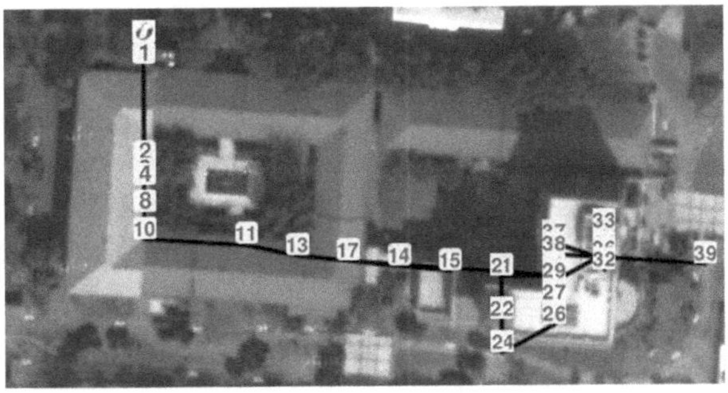

Fig. 4.11 Tracking path by proposed algorithm

decision of observation state, we put it in CEKF to calculate the estimated coordinate. Then we can get a smooth tracking path with unified coordination in different environments as Fig. 4.11.

4.4 Experimental Results and Discussions

As Fig. 4.12, path of GPS is shown by arrow icon, path of WiFi Fingerprinting is shown by dot icon, and path of Dead Reckoning is shown by diamond icon. We can clearly see that result of GPS has significant error in indoors. Although the Dead Reckoning is more accuracy than GPS, but the error is accumulated. The result of WiFi Fingerprinting is closest to test path, but it can't work without WiFi signal.

The path of our proposed positioning methodology is shown by cross icon. The result shows that it is the closest path to the whole test path. Because the proposed method merges all the advantages of each positioning methods, it can be used to positioning in outdoors and indoors, with or without WiFi signal.

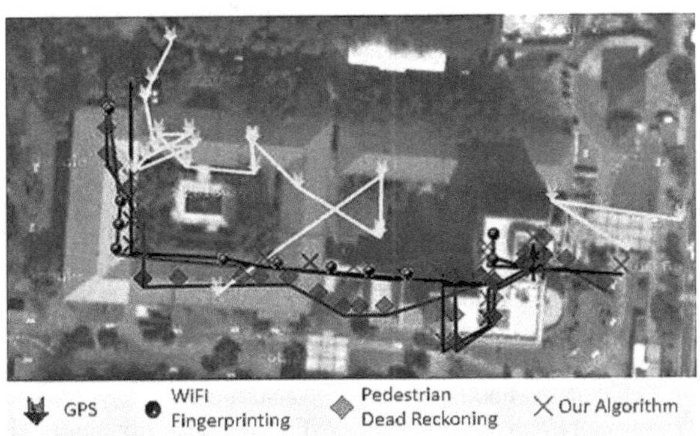

Fig. 4.12 Comparison of positioning methods

4.5 Conclusions and Future Works

GPS, WiFi Fingerprinting and Dead Reckoning have their own advantages for positioning. For ubiquitous positioning we combined them for different environments. Using our algorithm users can be located no matter where they are. Even that we can monitor their moving path by Google Maps immediately. The proposed algorithm is suitable to positioning in wider and deeper indoor environment, especially for underground.

But there still are some issues need to be solved. One is WiFi Fingerprinting, due to we use Fingerprinting technology to positioning in WiFi environment. It still needs scan the WiFi RSS for every reference points at offline phase. It is a wasting time job. In the future, we hope can toward a way to develop an easier and faster WiFi positioning method. At the same time, we also hope to improve the approach for 3-Dimension positioning. If these two issues are solved, we will make things of IoT being located more automatically and quickly. It will be helpful for delivering spatial resolution function in IoT platform.

Acknowledgments Authors are grateful to thank the Ministry of Science and Technology of Taiwan, for financial support under Grants NSC 102-2221-E-006-144.

References

1. Presser M (2011) D2.1 Initial report on IoT applications of strategic interest
2. Woysch G (2012) D1.3 Analysis of existing IoT strategic research directions and priorities
3. Hofmann-Wellenhof B, Lichtenegger H, Collins J (1997) GPS-global positioning system. Theory and practice. GPS-global positioning system. In: Hofmann-Wellenhof B, Lichtenegger H, Collins J (eds) Theory and practice, vol XXIII+. Springer, Wien (Austria), p 389. ISBN 3-211-82839-7, Price DM 86.00
4. Gu Y, Lo A, Niemegeers I (2009) A survey of indoor positioning systems for wireless personal networks. IEEE Commun Surv Tutor 11(1):13–32
5. Zandbergen PA (2009) Accuracy of iPhone locations: a comparison of assisted GPS, WiFi and cellular positioning. Trans GIS 13(s1):5–25
6. Gezici S (2008) A survey on wireless position estimation. Wirel Pers Commun 44(3):263–282
7. Leppakoski H, Collin J, Takala J (2013) Pedestrian navigation based on inertial sensors, indoor map, and WLAN signals. J Sign Process Syst Sign Image Video Technol 71 (3):287–296
8. Jin Y et al (2011) A robust dead-reckoning pedestrian tracking system with low cost sensors. In: IEEE international conference on pervasive computing and communications (PerCom)
9. Frank K et al (2009) Development and evaluation of a combined WLAN and inertial indoor pedestrian positioning system. In: ION GNSS
10. Singh R, Guainazzo M, Regazzoni CS (2004) Location determination using WLAN in conjunction with GPS network (Global Positioning System). In: 59th IEEE conference on vehicular technology, VTC 2004-Spring
11. Cheong JW et al (2009) GPS/WiFi real-time positioning device: an initial outcome. In: Gartner G, Rehrl K (eds) Location based services and telecartography II. Springer, Heidelberg, pp 439–456
12. Ogawa K et al (2011) Toward seamless indoor-outdoor applications: developing stakeholder-oriented location-based services. Geo-spat Inf Sci 14(2):109–118

13. Lin CY, Cheng WT, Wang SC (2011) An end-to-end logistics management application over heterogeneous location systems. Wirel Pers Commun 59(1):5–16
14. Gallagher T et al (2009) Wi-Fi+ GPS for urban canyon positioning. In: Symposium on GPS/GNSS (IGNSS2009)
15. NUR K et al (2013) Integration of GPS with a WiFi high accuracy ranging functionality. Geospat Inf Sci 16:1–14 (ahead-of-print)
16. Cacopardi S et al (2010) WiFi assisted GPS for extended location services personal satellite services. Springer. Berlin, pp 191–202
17. Mehmood H, Tripathi NK (2009) Real time self mapping hybrid positioning system. Geospatial Crossroads@ GI_Forum 9:978-3879074815
18. Gallagher T et al (2009) Trials of commercial Wi-Fi positioning systems for indoor and urban canyons. In: IGNSS 2009 Symposium on GPS/GNSS. Citeseer
19. De Maesschalck R, Jouan-Rimbaud D, Massart DL (2000) The mahalanobis distance. Chemometr Intell Lab Syst 50(1):1–18
20. Skyhook. http://www.skyhookwireless.com/
21. Indoor Google Maps. https://maps.google.com/help/maps/indoormaps/

Chapter 5
Multi-core Scheduling Scheme for Wireless Sensor Nodes with NVRAM-Based Hybrid Memory

Seokho Oh and Yeonseung Ryu

Abstract In recent years, multi-core processor technology and next generation non-volatile memory technology have developed dramatically to boost performance while minimizing power consumption. In this paper, we study a hybrid memory-aware multi-core scheduling scheme for wireless sensor nodes which use DRAM/NVRAM hybrid main memory. The proposed HAMC (Hybrid memory-Aware Multi-Core) scheduling scheme considers different access latency of hybrid memory medium and tries to reduce total execution time of tasks. We showed through simulation that proposed scheme outperforms legacy scheduling scheme.

Keywords Multi-core scheduling · Wireless sensor node · NVRAM · Hybrid main memory

5.1 Introduction

In what's called Internet of Things (IoT), sensors and actuators embedded in physical objects are linked through wired and wireless networks. These networks churn out huge volumes of data that flow to central servers for analysis. With this trend, sensor nodes has become complex computer systems, with a CPU, main memory, storage, operating system and a suite of sensors [1–4]. Sensor nodes collect not only sensed data from the environment, but also stream of mass media data like videos and images. In order to perform large data processing, sensor nodes are expected to require high processing capability and much more memory than

S. Oh · Y. Ryu (✉)
Department of Computer Engineering, Myongji University, Yoingin, Korea
e-mail: ysryu@mju.ac.kr

J.J. (Jong Hyuk) Park et al. (eds.), *Ubiquitous Computing Application and Wireless Sensor*, Lecture Notes in Electrical Engineering 331,
DOI 10.1007/978-94-017-9618-7_5

legacy sensor nodes. Furthermore, the power dissipation has become one of the critical design challenges in a sensor network since battery is the main power source in a sensor node [5–7].

In recent years, multi-core processor technology and next generation non-volatile memory technology have developed dramatically to boost performance while minimizing power consumption. Intel has recently put 80 cores in a chip and ClearSpeed has developed a 96-core processor. Also, the recent advance of memory technology has ushered in new non-volatile RAM (NVRAM) designs such as PRAM (Phase change RAM) and STT-MRAM (Spin-Torque Transfer Magnetic RAM) [8–10]. Some studies introduced hybrid main memory technology with DRAM and NVRAM to tackle the energy consumption problem in DRAM-based main memory [11–14].

Though multi-core processors offer a performance increase over single-core processors, it is well known that the interconnect bandwidth between processors and shared main memory has become a major bottleneck. In order to tackle this problem, bandwidth-aware task scheduling has been proposed as an effective strategy [15–19]. It predicts the bandwidth requirement of a task and selects concurrent tasks to avoid bandwidth saturation while keeping full utilization of the available bandwidth. Most bandwidth-aware task scheduling schemes define job segments as segments of code executed during a scheduling quantum. They try to select and schedule job segments that could maximize their total bandwidth requirement as close to the peak memory bandwidth.

However, previous studies of bandwidth-aware scheduling schemes have focused on DRAM-only main memory. With DRAM-only main memory, access latencies are independent of access request type (i.e., read and write). In the hybrid main memory with DRAM and NVRAM, on the contrary, access latency highly depends on the target memory. In general, NVRAM read latency is longer than DRAM access latency and NVRAM write latency is much longer than NVRAM read latency. Hence, DRAM-based bandwidth-aware scheduling schemes cannot be used for next generation wireless sensor nodes that use hybrid main memory.

In this paper, we study a hybrid memory-aware multi-core scheduling scheme for wireless sensor nodes which use DRAM/NVRAM hybrid main memory. Figure 5.1 illustrates the system configuration considered in this paper. The proposed HAMC (Hybrid memory-Aware Multi-Core) scheduling scheme considers different access latency of hybrid memory medium and tries to reduce total execution time of tasks by avoiding memory bandwidth saturation. In doing so, HAMC scheme predicts memory access patterns of each task for next scheduling quantum using observed information during the last quantum. It performs two phases to determine the next task to run. On the first phase, it selects a task at the first position of the global ready queue in order to avoid starvation problem. On the second phase, it finds the remaining tasks that could maximize the total bandwidth requirement. We show, through trace-driven simulation, that the proposed scheme outperforms the legacy round-robin scheduling scheme.

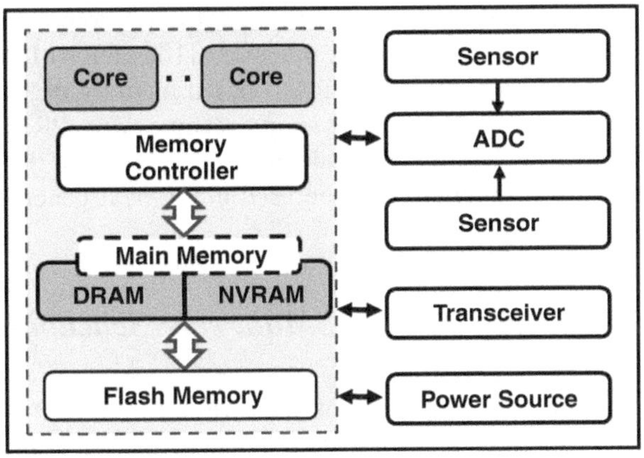

Fig. 5.1 Sensor node architecture in this paper

The rest of this paper is organized as follows. In Sect. 5.2, we describe the characteristics of non-volatile memories such as PRAM and STT-MRAM. Also, we introduce some bandwidth-aware scheduling schemes. Section 5.3, we present a novel hybrid memory-aware multi-core scheduling scheme (HAMC). Section 5.4 presents the experimental results. Finally, Sect. 5.5 concludes the paper.

5.2 Background

5.2.1 Non-volatile Memories

A PRAM cell uses a special material, called phase change material, to represent a bit [8]. PRAM density is expected to be much greater than that of DRAM (about four times). Further, PRAM has negligible leakage energy regardless of the size of the memory. Though PRAM has attractive features, the write access latency of PRAM is not comparable to that of DRAM. Also, PRAM has a worn-out problem caused by limited write endurance. Since the write operations on PRAM significantly affect the performance of system, it should be carefully handled.

STT-MRAM is a next generation memory technology that takes advantage of magnetoresistance for storing data [9, 12, 13]. It uses a Magnetic Tunnel Junction (MTJ), the fundamental building block, as a binary storage. An MTJ comprises a three-layered stack: two ferromagnetic layers and an MgO tunnel barrier in the middle. One of the biggest weaknesses of STT-MRAM is long write latency compared to SRAM or DRAM. Since the fast access time of memories on a chip must be guaranteed and cannot be negotiable, the slow write operations of STT-MRAM limit its popularity, even though it shows competitive read performance. Another serious drawback of STT-MRAM is high power consumption in write operations.

Among the NVRAMs, PRAM and STT-MRAM are becoming promising candidates for main memory because of their high density, comparable read access speed and low power consumption. Hence, some studies have introduced PRAM-based main memory organization [10], DRAM/PRAM hybrid main memory organization [11], STT-MRAM-based memory organization [13]. Also, there have been some buffer management schemes for PRAM based main memory and DRAM-PRAM hybrid main memory [14]. It is highly expected that NVRAM-based main memory will be used in the next generation low-power sensor nodes.

5.2.2 Memory-Aware Multi-core Scheduling Schemes

There have been a lot of bandwidth-aware task scheduling approaches as an effective strategy to reduce the memory bandwidth bottleneck. They attempt to predict the bandwidth requirement of a task, and select synergistic concurrent tasks to avoid bandwidth saturation.

Koukis and Koziris [17] profiled bandwidth usage information of each task and calculate average available bandwidth for new task. Xu et al. [18] showed the impact of memory bandwidth fluctuation on overall performance for tasks on multi-core system. They proposed a new scheme that maintains the total bandwidth requirement at a steady level instead of maximizing the bandwidth utilization.

However, these methods are not applicable with the DRAM/NVRAM hybrid main memory. This is because bandwidth requirement cannot be represented by a unique memory transaction type. With hybrid main memory of DRAM and NVRAM, memory transactions should be classified into three categories based on access latency: DRAM access, NVRAM read, and NVRAM write. Hwang and Park [20] studied a hybrid main memory-aware task scheduling scheme. They collected effective memory bandwidth usage of each task using the memory access monitoring and classified tasks (i.e., latency-sensitive and bandwidth-sensitive) according to how much tasks access memory. Then they selected next tasks according to the task characteristics.

5.3 HAMC Scheduling Algorithm

5.3.1 Overall Architecture

We assume that there are N cores in our system and there is a global ready queue. When a task is created, it is inserted to the tail of the ready queue. Whenever a scheduling quantum is expired, the scheduler fetches N tasks from the ready queue and assigns each task to a core for a time interval of 1 scheduling quantum. When the next scheduling quantum is expired, the running tasks are preempted and inserted to the tail of the ready queue again.

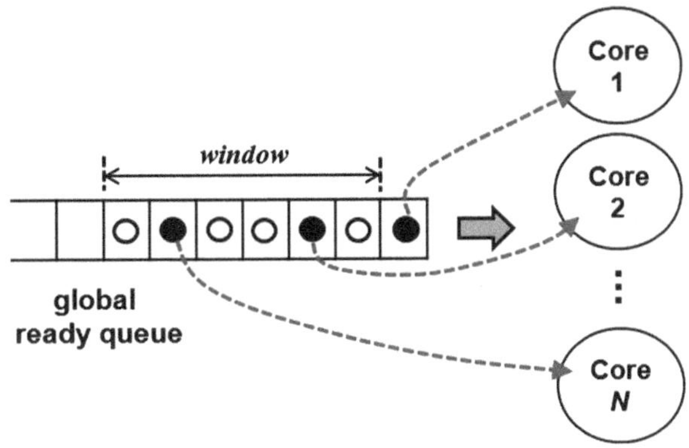

Fig. 5.2 Our scheduling model

When the scheduler determines the next tasks to run, it performs two phases: on the first phase, it selects a task at the first position of the global ready queue in order to avoid starvation problem (i.e., cannot run for a long time). On the second phase, it finds the remaining tasks that could maximize the total bandwidth requirement. That is, the scheduler finds K tasks satisfying the following:

$$\sum_{i=1}^{K} \mathrm{BW}(Task_i) < TotalBW \tag{5.1}$$

Further, HAMC defines a search window as shown in Fig. 5.2 in order to limit search overhead.

5.3.2 Prediction of Bandwidth Requirement

Most existing scheduling framework use the observed bandwidth, or the average of the observed bandwidth of the last few time quanta, as the predicted bandwidth requirement for the next quantum. In this work, we employ a common technique called exponential averaging which predicts a future value on the basis of a time series of past values. In order to predict the bandwidth requirement of a task, we use the following Equation:

$$B_{n+1} = \alpha R_n + (1 - \alpha) B_n \tag{5.2}$$

where

B_n = predicted bandwidth value for the nth instance of this task
R_n = real bandwidth value for the nth instance
α = a constant weighting factor $(0 < \alpha < 1)$

We measure R_n every nth execution of tasks using Eq. (5.3).

$$\begin{aligned} R &= R_D + R_{NV} \\ R_D &= C_D + T_D \\ R_{NV} &= V_R \times T_{NVR} + C_W \times T_{NVW} \end{aligned} \tag{5.3}$$

where

R_D = bandwidth for DRAM
R_{NV} = bandwidth for NVRAM
C_D = number of DRAM access transactions
C_R = number of NVRAM read access transactions
C_W = number of NVRAM write access transactions
T_D = memory access latency of DRAM
T_{NVR} = memory access latency of NVRAM read
T_{NVW} = memory access latency of NVRAM write

5.4 Experiment

In order to evaluate the proposed scheme, we have developed a trace-driven simulator. For the workload, we generated synthetic workload as following:

$$Job_{i,j} = \left(T_{i,j}, C_{D,i,j}, C_{R,i,j}, C_{W,i,j}\right)$$

where

$Job_{i,j}$ = j-th instance of task i
$T_{i,j}$ = execution time of $Job_{i,j}$
$C_{D,i,j}$ = number of DRAM access transactions of $Job_{i,j}$
$C_{R,i,j}$ = number of NVRAM read access transactions of $Job_{i,j}$
$C_{W,i,j}$ = number of NVRAM write access transactions of $Job_{i,j}$

We used total execution time as a performance metric and compared the proposed scheduling algorithm with traditional Round Robin (RR) scheduling algorithm. Table 5.1 shows parameter values used in our simulation.

Figure 5.3 shows normalized execution time of proposed HAMC by that of RR scheme. We can see that the HAMC outperforms RR about 12–15 %.

Table 5.1 Parameters for our experiment

Parameter	Value
The number of cores	2, 4
Scheduling quantum unit	100 ms
Peak memory transaction rate	35 transactions/us
DRAM read/write latency	25 ns
NVRAM read latency	67.5 ns
NVRAM write latency	215 ns
$\propto$ in Eq. (5.2)	0.5

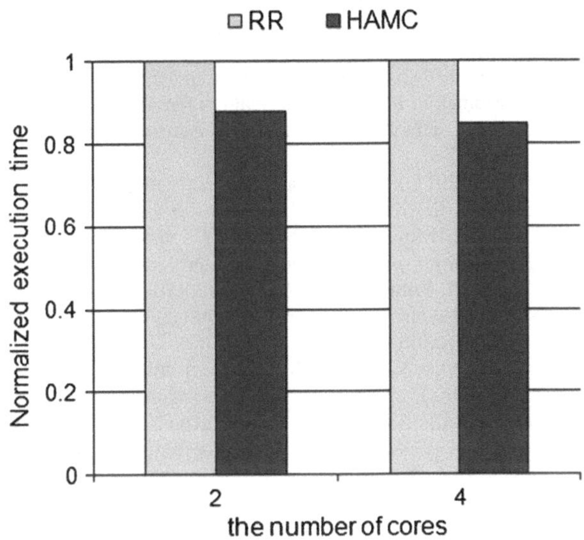

Fig. 5.3 Comparison of execution time

5.5 Conclusion

As wireless sensor nodes have become complex computer systems and require high processing capability and much more memory than legacy sensor nodes, multi-core processors and NVRAM-based hybrid main memory are expected to be used in sensor nodes in near future.

In this paper, we study a multi-core scheduling scheme for sensor nodes which use NVRAM-based hybrid main memory. The proposed scheme considers different access latency of hybrid memory medium and tries to reduce total execution time of tasks by avoiding memory bandwidth saturation. We showed through trace-driven simulation that proposed scheme outperforms legacy round-robin scheme.

Acknowledgments This research was supported by Basic Science Research Program through the National Research Foundation of Korea (NRF) funded by the Ministry of Education, Science and Technology (2010-0021897).

References

1. Lin N, Dong Y, Lu D (2013) Providing virtual memory support for sensor networks with mass data processing. Int J Distrib Sens Netw 2013. doi:10.1155/2013/324641
2. Lachemann A, Marron P, Gauger M, Minder D, Saukh O, Rothermel K (2007) Removing the memory limitations of sensor networks with flash-based virtual memory. SIGOPS Oper Syst Rev 41:131–144
3. Farooq M, Kunz T (2011) Operating systems for wireless sensor networks: a survey. Sensors 11:5900–5930
4. Madden S, Franklin M, Hellerstein J, Hong W (2005) TinyDB: an acquisitional query processing system for sensor networks. ACM Trans Database Syst 30(1):122–173
5. Law Y, Palaniswami M, Hoesel L, Doumen J, Hartel P, Havings P (2009) Energy-efficient link-layer jamming attacks against wireless sensor network MAC protocols. ACM Trans Sens Netw 5(1)
6. Gu Y, Hem T (2010) Bounding communication delay in energy harvesting sensor networks. In: Proceedings of IEEE international conference on distributed computing systems, pp 837–847
7. Barroso L, Holzle U (2007) The case for energy-proportional computing. IEEE Comput 40 (12):33–37
8. Yuan X (2011) Modeling, architecture, and applications for emerging memory technologies. IEEE Des Test Comput 28(1):44–51
9. Charles A, Mojumder N, Fong X, Choday S, Park S, Roy K (2012) Spin-transfer torque MRAMs for low power memories: perspective and prospective. IEEE Sens 12(4):756–766
10. Qureshi M, Srinivasan V, Rivers J (2009) Scalable high performance main memory system using phase-change memory technology. In Proceedings of international symposium on computer architecture
11. Park H, Yoo S, Lee S (2011) Power management of hybrid DRAM/PRAM-based main memory. In: Proceedings of design automation conference
12. Jang H, An B, Kulkarni N, Yum K, Kim E (2012) A hybrid buffer design with STT-MRAM for on-chip interconnects. In Proceedings of ACM/IEEE international symposium on networks-on-chip, pp 193–200
13. Kultursay E, Kandemir M, Sivasubramaniam A, Mutlu O (2013) Evaluating STT-RAM as an energy-efficient main memory alternative. In Proceedings of IEEE international symposium on performance analysis of systems and software
14. Seok H, Park Y, Park K (2012) Efficient page caching algorithm with prediction and migration for a hybrid main memory. Appl Comput Rev 11(4):38–48
15. Antonopoulos CD, Nikolopoulos DS, Papatheodorou TS (2003) Scheduling algorithms with bus bandwidth considerations for SMPs. In: Proceedings of international conference on parallel processing, p 547
16. Merkel A, Bellosa F (2008) Memory-aware scheduling for energy efficiency on multicore processors. In: Proceedings of conference on power aware computing and systems, pp 1–5
17. Koukis E, Koziris N (2006) Memory and network bandwidth aware scheduling of multiprogrammed workloads on clusters of SMPs. In: Proceedings of international conference on parallel and distributed systems, pp 345–354
18. Xu D, Wu C, Yew P-C (2010) On mitigating memory bandwidth contention through bandwidth-aware scheduling. In: Proceedings of international conference on parallel architectures and compilation techniques, pp 237–248
19. Kim H, Niz D, Andersson B, Klein M, Mutlu O, Rajkumar R (2014) Bounding memory interference delay in COTS-based multi-core systems. In: IEEE real-time and embedded technology and applications symposium
20. Hwang W, Park K (2013) HMMSched: hybrid main memory-aware task scheduling on multicore system. In: Proceedings of international conference on future computational technologies and applications

Chapter 6
Security Scheme for LTE Initial Attach

Uijin Jang, Hyungmin Lim and Hyungjoo Kim

Abstract Long Term Evolution (LTE) is a fourth-generation mobile communication technology implemented throughout the world. There is still, however, an exposed vulnerability with regard to the identification parameters such as IMSI and RNTI, as they are transmitted in plain text during the initial attach procedure in the access to the LTE network. The vulnerability has existed since the initial release of the LTE Standards, and is still present in Release 12. To prevent a leak of the identification parameters during the initial attach process and a possible third party attack, the relevant parameters should be encrypted. This paper proposes a security scheme to safely transmit identification parameters in different cases of the initial attach. The proposed security scheme solves the exposed vulnerability by encrypting the parameters in transmission. Using an OPNET simulator, it is shown that the average rate of delay and processing ratio are efficient in comparison to the existing process.

Keywords GUTI · IMSI · Initial attach · LTE · Security · RNTI

6.1 Introduction

The LTE is an abbreviation for Long-Term Evolution, which is a fourth generation mobile communication technology. LTE is designed for high-speed transmission, reduced cost per bit, low transmission delay, and applicability to existing frequency

---

U. Jang
Korea Copyright Commission, Seoul, Republic of Korea
e-mail: neon7624@gmail.com

H. Lim · H. Kim (✉)
Department of Computer Science and Engineering, Soongsil University, Seoul, Republic of Korea
e-mail: Hyungjoo.kim@ssu.ac.kr

H. Lim
e-mail: atskyo@gmail.com

J.J. (Jong Hyuk) Park et al. (eds.), *Ubiquitous Computing Application and Wireless Sensor*, Lecture Notes in Electrical Engineering 331,
DOI 10.1007/978-94-017-9618-7_6

bands. It is currently implemented worldwide. The LTE standard was introduced in 2004. Release 12 is the current version of LTE. South Korea and Japan have been quickly introduced LTE and currently, they are working with Release 9. Presently, LTE is at the market introduction and implementation stage. Ongoing technology upgrades and security vulnerability discoveries are expected as the technology matures. The most serious issue among the known vulnerabilities is the plain text exposure vulnerability of the identification parameter values of the UE (User Equipment) that exists in the initial attach process [1, 2].

In the LTE technical documentation, according to the "Technical Specification Group Services and System Aspects; Rationale and track of security decisions in Long Term Evolved (LTE) RAN/3GPP System Architecture Evolution (SAE) (Release 9)," there is a vulnerability during the initial attach process in the access to the LTE Network, in which the UE identifying parameters are transmitted in plain text. Problems such as tracing and privacy infringement could occur [3].

This paper proposes a plan for safely transmitting identification parameters by classifying the initial attach processes into two tasks, initial attach with IMSI and initial attach with GUTI.

Please see Table 6.1 for definitions and terms used in this paper. The proposed paper consists of six sections. Section 6.2 analyzes the structure of the LTE, initial attach process, security process, and threats. Section 6.3 proposes a security scheme by classifying the initial attach process into multiple cases in order to safely transmit identification parameters. Section 6.4 carries out a security analysis of the proposed scheme and Sect. 6.5 compares and evaluates the performances between the proposed process with a security scheme and the existing process. Section 6.6 concludes the discussion.

Table 6.1 The terms and symbols used in proposed security scheme

UE	User Equipment
eNB	Evolved Node B
MME	Mobility Management Entity
HSS	Home Subscriber Server
IMSI	International Mobile Subscriber Identity
RNTI	Radio Network Temporary Identities
GUTI	Global Unique Temporary Identifier
PLMN ID	Public Land Mobile Network ID (MCC + MNC)
MCC	Mobile Country Code
MNC	Mobile Network Code
RN	Random Number
h()	Hash Function
F	4n bits String by h()
C	Challenge Bits

6.2 LTE

6.2.1 LTE Network Structure

The LTE network consists of LTE entities dealing with wireless access network technology and EPC entities dealing with core network technology [7].

Of the LTE entities, UE accesses to eNB through LTE-Uu wireless interface. eNB, serving as the base station provides the user with wireless interface and provides wireless Remote Resource Management (RRM) features such as radio bearer control, wireless admission control, dynamic wireless resource allocation, load balancing and Inter Cell Interference Control (ICIC) (Fig. 6.1).

EPC entities consist of MME, S-GW, P-G and HSS. MME is an E-UTRAN control plane entity, communicating with HSS for user authentication and user profile download, and through NAS signaling, it provides the user terminal with EPS rambling management (EMM) and EPS Session management (ESM) features. S-GW is the termination point between E-UTRAN and EPC and the anchoring point in the handover with eNB and the handover with 3GPP system. P-GW connects UE to external PDN network and provides packet filtering. In addition, P-GW allocates an IP address to the user terminal, and serves as the mobile anchoring point in the handover between 3GPP and non-3GPP. Lastly, HSS manages the users' personal profiles.

IMS/Internet domain is the domain commonly calling external internet services.

6.2.2 LTE Initial Attach for UE

'Initial Attach for UE' process is a case of the first access to the network by the user subscribing the LTE network using UE [3, 4] (Fig. 6.2).

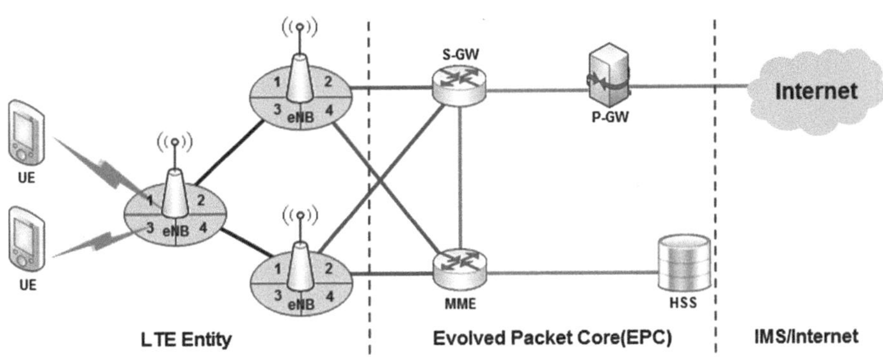

Fig. 6.1 LTE network

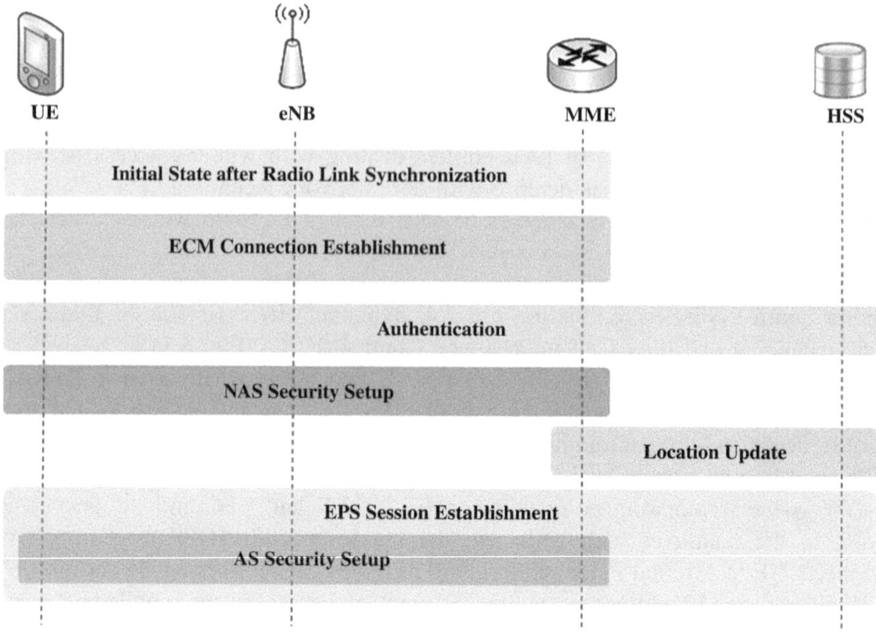

Fig. 6.2 Initial attach process

'Initial State after Radio Link Synchronization' process is one in which UE selects eNB and synchronizes wireless link. 'ECM Connection Establishment' process is a NAS layer, that is, a process of transmitting IMSI to request MME for net access. Through the relevant process, RRC connection and S1 signaling connection are established.

'Authentication' process is a mutual authentication procedure between UE and MME using EPS-AKA and 'NAS Security Setup' process is a key setting process to safely transmit NAS messages between UE and MME.

'Location Update' process is one to receive personal profile information from HSS after registering location, and 'EPS Session Establishment' process is one to allocate network resources so that the users can be provided with the services.

6.2.3 LTE Security

After 'ECM Connection Establishment' process between UE and eNB, UE starts mutual authentication by transmitting IMSI to MME. Centering around the LTE security layer, the LTE network carries out mutual authentication based on EPS-AKA. The LTE Security is divided broadly into three processes: UE-HSS mutual

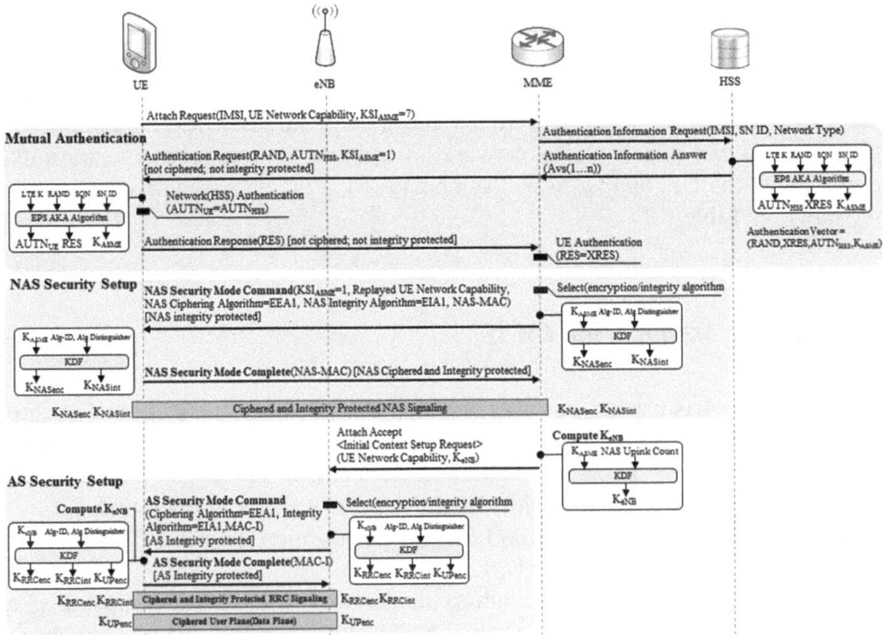

Fig. 6.3 LTE security

authentication process, UE-MME NAS Security Setup process and UE-eNB AS Security Setup process [5–10] (Fig. 6.3).

6.2.4 LTE Threats

IMSI refers to the unique ID requested to each user when the net administrator registers the user to service, and this value refers to the unique number of identifications saved in the USIM in the user device [3, 4, 7, 9].

Yet, when 'Initial Attach for UE' is carried out, in 'ECM Connection Establishment' process, the UE transmits IMSI to MME in plain text. The IMSI transmitted in plain text is transmitted to MME through a number of eNB, which has a vulnerability in which it is leaking to an attacker through malicious eNB. In addition, a user tracking attack using the leaked IMSI, a device tracking attack and a privacy abuse attack may take place.

RNTI, the unique ID to differentiate UE from eNB and GUTI used instead of IMSI after a series of the process, also, are transmitted in a variety of initial attach process in plain text, so the same vulnerability and the threat of attack of IMSI may occur [1, 2, 3, 7].

6.3 Proposed Security Scheme for Initial Attach in LTE

The proposed security scheme was designed to protect unique information about identification such as IMSI and RNTI transmitted in plain text when the UE attempts an initial attach to the network. It consists of four types by the initial connection of the UE, and terms and symbols used in the proposed security scheme are shown in Table 6.1.

6.3.1 Initial Attach with IMSI

The first protocol is carried out after the 'Initial State after Radio Link Synchronization' process in the Initial attach with IMSI Case. It was designed to protect IMSI leaked from the 'ECM Connection Establishment' process in plain text and RNTI leaked from the 'EPS Session Establishment' process in plain text.

After the 'Initial State after Radio Link Synchronization' process, UE and MME start an 'ECM Connection' process.

The UE transmits a generated random number and 'UE Network Capability' to MME for attach request. MME receiving the attach request MME generates a random number and transmits it to the UE, and the UE and the MME carries out a series of arithmetic operation to safely transmit IMSI.

The UE and the MME enter the transmitted and received random numbers and PLMN ID to hash function secretly shared according to MNC and generate an F string with 4n bits. The generated F string is divided into four numerical progressions with n bits each. After this process, the MME generates a random number progression used as challenge bits, the UE generates the second random number and through lr numerical progression and exclusive OR arithmetic operation, it generates RN_{UE_2}'.

The MME generates challenge bits C_i using lr_i, ad_i and c_i. If lr_i is 0, $C_i = c_i \| ad_i$ and if lr_i is 1, $C_i = ad_i \| c_i$. The MME transmits C_i to UE to verify it through the response value and the UE verifies the MME through C_i (Fig. 6.4).

The UE is aware of lr, so it can differentiate C_i transmitted by the MME into $C_i = c_i \| ad_i$ and $C_i = ad_i \| c_i$. The UE generates $R_i = RNUE\_2_i' \| r_i^0$ or $R_i = RNUE\_2_i' \| r_i^1$ if lr_i is 0 and $R_i = r_i^0 \| RNUE\_2_i'$ or $R_i = r_i^1 \| RNUE\_2_i'$ if lr_i is 1. At this time, r_i^0 and r_i^1 transmits r_i^0 if c_i transmitted by the MME is 0 and r_i^1 if c_i is 1. At this time, the MME receive the transmission of RN_{UE_2} of the UE and saves it.

For ad_i transmitted by the MME ($ad_i \neq ad_i$), the UE detects an error and transmits the response value as a random value. The MME, too, halts the attach process if an error is detected through $r_i^0 \neq r_i^0$ and $r_i^1 \neq r_i^1$.

After the challenge-response process, the UE uses unused r_i^0 and r_i^1 concatenated value as a key to encrypt IMSI to transfer it to the MME. The MME generates a key through the same process as that of the UE and then decrypts the transmitted cypher text to get the IMSI.

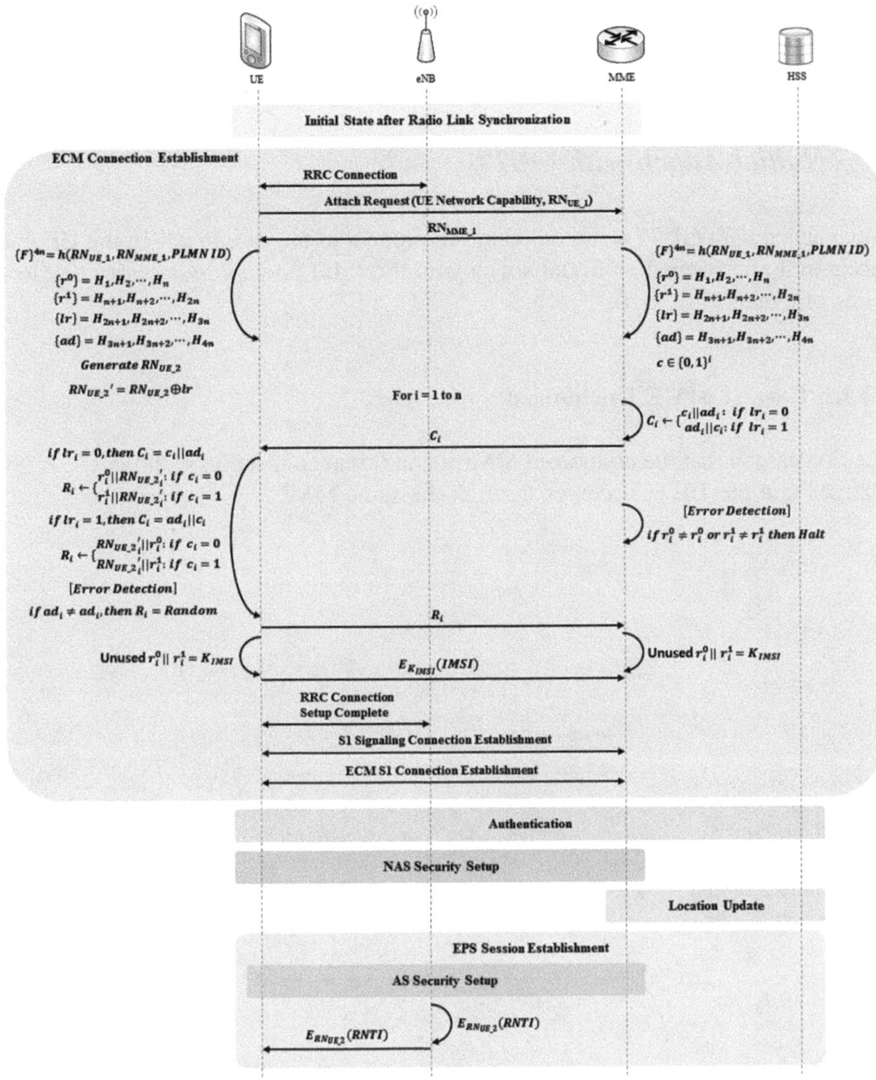

Fig. 6.4 Proposed protocol: initial attach with IMSI

After the IMSI is safely transmitted, UE, eNB, MME and HSS carry out up to 'AS Security Setup' process during 'ECM Connection Establishment', 'Authentication', 'NAS Security Setup', 'Location Update' and 'EPS Session Establishment' process.

After the 'AS Security Setup', the eNB encrypts RNTI to MME using the secret key of the 'AS Security Setup' to allocate the RNTI to the UE. The MME encrypts the transmitted RNTI to RN_{UE_2} saved in the 'ECM Connection Establishment' process to transmit to the eNB, and the eNB allocates the RNTI by transmitting the relevant value to the UE.

6.3.2 Initial Attach with GUTI

Initial attach with GUTI is the initial attach process of the case in which the UE that successfully performed an initial attach with the IMSI process re-accesses due to a series of events (Fig. 6.5).

6.3.2.1 Case 1: MME Unchanged

The first case is that the connected MME in an initial connection with the UE is not changed and the UE re-accesses through the same MME.

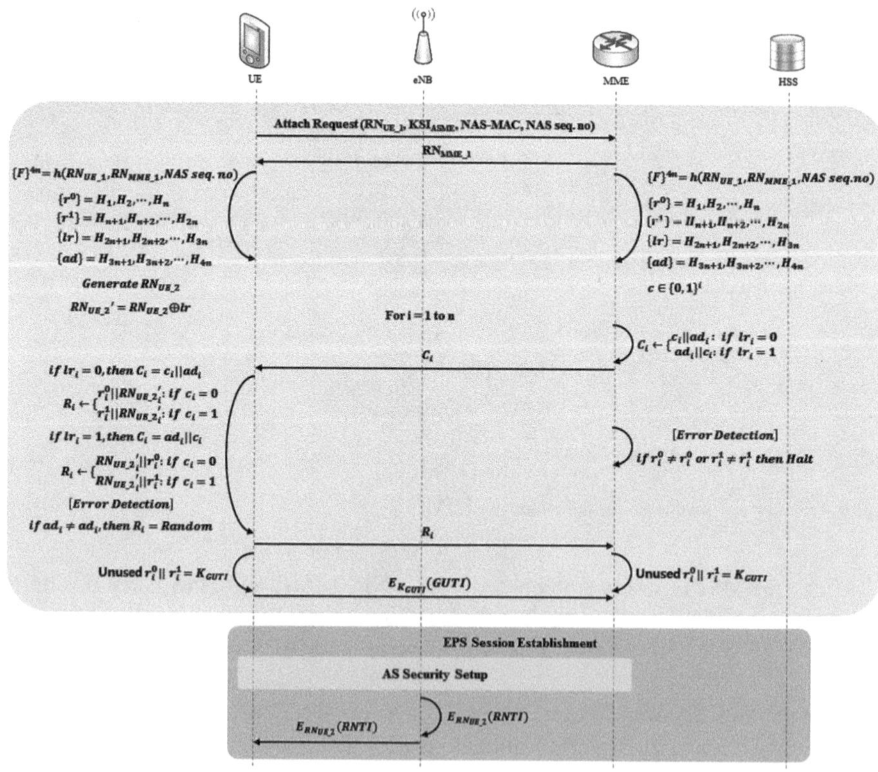

Fig. 6.5 Case 1: MME unchanged

In a re-access, initial attach process carries out authentication using GUTI to protect the IMSI. The process of transmitting the GUTI is the same as "3.1." IMSI transmission process, and the initial attach process is carried out using existing information saved in the MME according to information such as GUTI, NAS-MAC and NAS Seq. No. Since a series of information about the UE has already been saved in the MME, no 'Authentication', 'NAS Security Setup' and 'Location Update' process are carried out, but the 'EPS Session Establishment' process only is carried out.

6.3.2.2 Case 2: MME Changed

The second case is that the MME is changed, but the old MME saves the information about the UE, so it transmits the information about the UE to the new MME.

The process of transmitting GUTI is same as other cases, and the new MME receives the information about the UE from the old MME using GUTI, NAS-MAC and NAS Seq. No.

6.3.2.3 Case 3: MME Changed and IMSI Needed

The third case is that the MME connected during the initial connection has been changed to a new MME and there is no information about the UE in the MME connected during the initial connection.

In a re-access, initial attach process carries out authentication using GUTI to protect the IMSI. The process of transmitting the GUTI is the same as "3.1." IMSI transmission process.

When the MME is changed, the new MME requests the old MME for the information about the UE according to the information such as GUTI, NAS-MAC and NAS Seq. No. At this time, if there is no information about the relevant UE in the old MME, the new MME requests the UE for the IMSI.

In this process, to safely transmit the IMSI, this proposed protocol transmits the GUTI and encrypts IMSI and transmits it using the generated series of value. For the encryption, the UE hashes K_{GUTI} and RN_{UE_2} to generate a key and encrypts the IMSI using the generated key K_{IMSI} to transmit to the MME. After the transmission of the IMSI, 'Authentication', 'NAS Security Setup', 'Location Update' and 'EPS Session Establishment' process are carried out (Figs. 6.6 and 6.7).

6.4 Security Analysis

The initial attach for UE process specified in the LTE Standards Release 10 transmits the parameters to identify the UE in plain text, and the proposed security scheme encrypts and safely transmits the relevant identification parameters (Table 6.2).

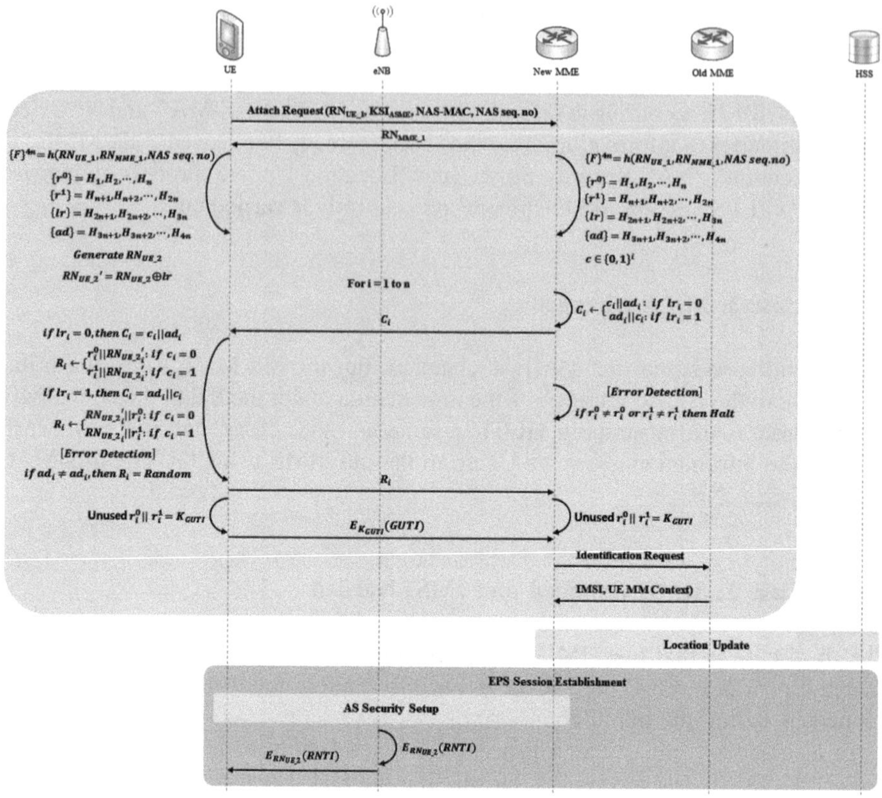

Fig. 6.6 Case 2: MME changed

6.4.1 Key Definition, Encryption and Decryption

The proposed security scheme generates a key value used to encrypt identification parameters through challenge-response process. The key used to encrypt IMSI and GUTI is defined as the value not used in the challenge-response process during the numerical progression generated through the hash function of the secret sharing between the UE and the MME. The relevant key value is defined only in the UE and MME, but not transmitted through the communication process to outside. Therefore, an attacker can find the identification parameters only through the attack on the encryption algorithm like AES-256 to know the encrypted IMSI and GUTI.

The key value encrypting the RNTI is transmitted through the challenge-response process, but the location of the bit continues to change for the transmission. Even if an attacker collects the bit string through an attack like tapping, the probability to find the key value with a total of n bits is (1/4)^n.

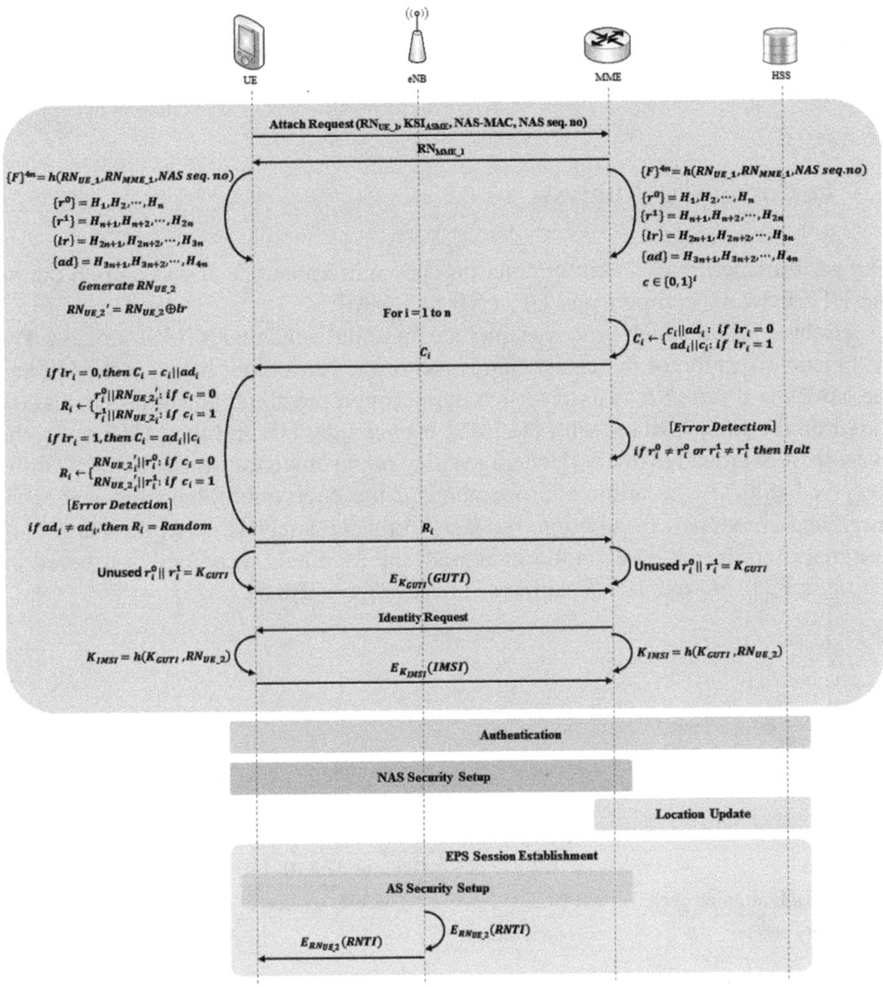

Fig. 6.7 Case 3: MME changed and IMSI needed

Table 6.2 Security comparative analysis

	Original initial attach	Proposed initial attach
IMSI	Plain text transmission	Cipher text transmission
RNTI	Plain text transmission	Cipher text transmission
GTUI	Plain text transmission	Cipher text transmission

6.4.2 *Error Detection and Verification*

In the proposed security scheme, since the challenge-response and encryption are carried out through hash function of the secret sharing between the UE and the

MME and key definition method, the UE and MME can verify if the other is a legitimate entity. In addition, in the challenge-response process, the UE can detect errors through $ad_i \neq ad_i$ and the MME can detect errors through $r_i^0 \neq r_i^0$ and $r_i^1 \neq r_i^1$.

6.5 Performance Analysis

The performance analysis environment is shown in Table 6.3. It was carried out on the LTE network components, UE, eNB and MME.

All three initial attach processes include an initial attach with IMSI process. The arithmetic operation of the encryption algorithm is carried out between the UE and the MME, and the eNB is used for passing through the communication process, so based on the Initial Attach with the IMSI between the UE and the MME node, the average of overhead (Time (s)), average delay (s), average transmission rate (Traffic Received (bits/s)) the arithmetic operation of the encryption algorithm was measured and compared. In addition, for a performance analysis when the number of terminal users in the base station increased, the terminals were analyzed based on the units of 1, 50 and 100 machine(s) (Tables 6.4 and 6.5).

Table 6.3 Performance analysis environment

	Values	Description
Cell	1	–
Number of UE	1, 50, 100	
Test time	9,000 s	Initial data (0–150 s) not reflected
Initial attach attempt cycle	150 s	The UE service uses initial attempt only
Portability model	Deactivation	–

Table 6.4 Performance analysis: original process

Device	Original initial attach		
	Time (s)	Delay (s)	Traffic received (bits/s)
1	–	0.004600	1.214
50	–	0.007210	1.233
100	–	0.010302	1.033

Table 6.5 Performance analysis: proposed process

Device	Proposed initial attach		
	Time (s)	Delay (s)	Traffic received (bits/s)
1	0.0110	0.007615	1.156
50	0.0103	0.010510	1.301
100	0.0109	0.014006	1.100

Table 6.6 Summary of performance analysis

Device	Summary (based on VoIP 100 %)		
	Time (s) (%)	Delay (s) (%)	Traffic received (bits/s) (%)
1	36	39	1
50	34	31	0.9
100	36	26	0.9
Total avg.	35.3	32.0	0.93

Since the existing initial attach does not perform encryption, time (s) was excluded. It was found that the delay according to the encryption process in the proposed algorithm decreased a little (somewhat) and since there was no packet switching other than the initial access process, it was analyzed that it did not have an impact on the trans-mission rate (Table 6.6).

The maximum permissible delay in the LTE access network, based on the QCI defined in 3GPP TS 23.203, is 100 ms (0.1 s) for FTP and web browsing; 50 ms (0.05 s) for video streaming; and 30 ms (0.03 s) for VoIP. Based on VoIP with the lowest delay (0.03 s 100 %), the overhead and delay turned out to be 35.3 and 32.0 % respectively for the arithmetic operation of encryption, so it was found that the initial attach of the proposed algorithm had less overhead. Since the transmission rate dealt with the initial entry process only prior to the security setup, it turned out to have no impact.

6.6 Conclusion

This paper proposed a security scheme to solve the vulnerability of plain text transmission of unique identification value which has continuously been pointed out in the LTE standards and related technical documents. The proposed security scheme was designed to safely transmit IMSI, RNTI and GUTI in a variety of initial attach process when the UE accesses to the LTE network. The proposed security scheme generated a key through challenge-response method to encrypt and transmit the unique ID and support error detection and verification.

As a result of a performance analysis, the security scheme encrypted and safely transmitted vulnerability parameters and turned out to use the performance of an average of 32.0 % based on VoIP 100 % demanding a less rate of delay, so the safety and performance of the proposed encryption algorithm were found to be efficient.

References

1. Bikos AN, Sklavos N (2013) LTE/SAE security issues on 4G wireless networks. IEEE Secur Priv 11(2):55–62
2. Escudero-Andreu G, Raphael CP, Parish DJ (2012) Analysis and design of security for next generation 4G cellular networks. In: The 13th annual post graduate symposium on the convergence of telecommunications, networking and broad-casting (PGNET)
3. Netmanias (2012) EMM Procedure: 1. Initial attach for unknown UE (Part 1)—case of initial attach. NMC Consulting Group Technical Specifications
4. Netmanias (2011) EMM Procedure: 1. Initial attach for unknown UE (Part 2)—call flow of initial attach. NMC Consulting Group Technical Specifications
5. Netmanias (2012) LTE Security I: LTE security concept and authentication. NMC Consulting Group Technical Specifications
6. Netmanias (2012) LTE Security II: NAS and AS security. NMC Consulting Group Technical Specifications
7. Kim S (2013) A design of MILENAGE algorithm-based mutual authentication protocol for the protection of initial identifier in LTE. Master's thesis, Soongsil University
8. GPP TR 33.821 (2010) Technical specification group services and system aspects; rationale and track of security decisions in long term evolved (LTE) RAN/3GPP system architecture evolution (SAE). Release 9
9. GPP TS 32.371 (2012) Telecommunication management; security management concept and requirements. Release 10
10. GPP TS 23.203 (2009) Policy and charging control architecture. Release 10

Chapter 7
ASiPEC: An Application Specific Instruction-Set Processor for High Performance Entropy Coding

Seung-Hyun Choi, Neungsoo Park, Yong Ho Song and Seong-Won Lee

Abstract Entropy coding becomes a performance bottle neck in video codecs because it requires a large amount of bitwise computation. Many video systems accelerate the coding process by implementing it as a hardwired accelerator or by executing it on an extra general-purpose processor to meet performance requirement. However, the variety of video formats causes a high degree of complication to a hardwired accelerator, which increases implementation cost and complexity. When an extra processor is used for the coding process, the property of using variable-length operands in the coding process significantly causes computation inefficiency. This paper presents a novel processor architecture which provides an instruction set suitable for efficient execution of entropy coding process as well as supports for multiple video formats.

Keywords Application specific instruction set processor (ASIP) · Entropy coding · Video codec

7.1 Introduction

Recently, major contents consumed by smart mobile devices are image data, which is characterized by its very large size. For reducing the data size while maintaining the image quality, various compression techniques are used. Those techniques are

S.-H. Choi · S.-W. Lee (✉)
Department of Computer Engineering, Kwangwoon University,
Seoul 139-701, Korea
e-mail: swlee@kw.ac.kr

N. Park
Department of Computer Science and Engineering, Konkuk University,
Seoul 143-701, Korea

Y.H. Song
Department of Electronics and Computer Engineering, Hanyang University,
Seoul 133-791, Korea

J.J. (Jong Hyuk) Park et al. (eds.), *Ubiquitous Computing Application and Wireless Sensor*, Lecture Notes in Electrical Engineering 331,
DOI 10.1007/978-94-017-9618-7_7

combined and form an image compression standard. There are many different image compression standards depending on the purpose of uses. Although each standard adopts slightly different technologies, basically it first removes spatial and temporal redundancy in the image data and then reduces statistical redundancy (aka entropy) in the sequence of image data. Since resent smart mobile devices are so versatile, they support most of the image compression standards at the expense of implementation complexity.

Entropy coding encodes incoming source symbols of fixed size into variable-sized output symbols. This technique achieves the space efficiency by translating frequently used symbols into less number of bits. Variable length coding (VLC) is a well-known entropy coding technique used in MPEG-1, MPEG-2, MPEG-4, and H.264/AVC. Golomb coding and arithmetic coding are another coding techniques which are also used in MPEG-4 and H.264/AVC.

Because of the demand on high variety in many video applications, the entropy coding is often implemented in software using high-throughput processors, or in hardware designed for the specific operation. Some software implementations utilize the high computation capability of general purpose processors or digital signal processors (DSP) [9, 12]. However, in this approach, the implementation is not cost-effective because general purpose processors are optimized for the applications using fixed-length operands in instructions. When the coding module is implemented as a dedicated accelerator [1, 6, 11], it can yields high performance. However, it becomes costly and complicated when it should support many different entropy coding schemes. Recently, application specific instruction set processors (ASIPs) are proposed to execute entropy coding module in software. [5, 2, 7, 4] They provide instruction sets customized for the efficient execution of entropy coding. However, most of them have limited support only for VLC in specific codecs (e.g. CAVLC in H.264/AVC). If they are requested to support various video formats, their design should be augmented to provide additional instructions and hardware components, which complicates the internal organization of processor architecture. In particular, each entropy codec uses its own variable-length code table which consumes a large space in the processor. Another ASIP used for implementation of H.264 decoder [10, 3, 8] utilizes some special bit-patterns found in the entropy decoding tables of H.264 to accelerate its processing speed. However, the use of such special properties hinders the processor from being used for entropy encoding.

In this paper, we propose a novel ASIP architecture for entropy coding, called application specific instruction-set processor for entropy coding (ASiPEC).

7.2 ASiPEC

A. ASiPEC architecture

ASiPEC provides two groups of instructions: (1) a conventional instruction set of general purpose processor and (2) an entropy coding specific instruction set. These two instruction sets share common datapath to avoid the increase of hardware complexity. Figure 7.1a shows the internal architecture of ASiPEC. In this architecture, once fetched, instructions take one of the two execution paths depending on the group they belong to: general purpose instructions are executed using the general purpose register (GPR) and ALU, while entropy coding specific instructions are processed using both GPR and bit stream engine (BSE). The internal organization of BSE is shown in Fig. 7.1b. Since the encoding process which converts well-aligned fixed length symbols to variable length symbols is relatively easy comparing to the decoding process which converts mixed variable length symbols to fixed length symbols, the major concerns of the ASiPEC architecture are focused to variable length decoding.

B. Variable length decoding

The step-by-step operation of variable length decoding process by ASiPEC is as follows. It is assumed that the variable length code table (VLCT) has been stored in the data memory and its base address has been loaded in one of the general purpose register (GPR), respectively, before the execution. In addition, the bitstream to be decoded is also stored in the data memory. Once the decoding process starts, the bitstream is sequentially loaded into the upper barrel shifter (UBS) of the BSE word by word. The UBS has two 16-bit outputs each linked to sequence registers (SR0, SR1). At first, the leading bits of the bitstream are loaded in SR1. Once a variable length symbol is decoded, the next symbol may start at a bit position either in SR1 or in SR0. If the starting bit locates at SR0, the ASiPEC shifts data bits from SR0 to SR1 and then shifts new data from UBS to SR0 as if SR1, SR0 and UBS form a large shift register altogether. To determine the shift amount, the remainder register (REM) keeps track of the number of bits in SR left after the current decoding operation finishes. When decoding a symbol, the ASiPEC generates a table offset by feeding the effective symbol bits in SR to the table offset formatter (TOF). The table offset, once calculated, is added to the base table address in a GPR in order to find a corresponding entry for a decoded symbol as well as its length from VLCT.

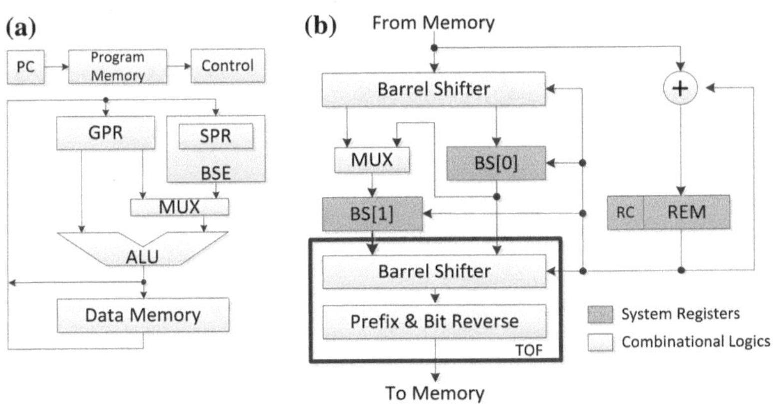

Fig. 7.1 Datapath of ASiPEC. **a** Datapath for the whole ASiPEC. **b** Datapath for BSE

After the table lookup, the ASiPEC moves the decoded symbol to the GPR and then store the length to REM to update a starting bit position of the next symbol in SR. This operation repeats until the entire bitstream is decoded completely.

TOF is designed to calculate a table offset to the VLCT, as shown in Fig. 7.2. The inputs to TOF are SR0 and SR1. If the starting bit of a symbol to decode is not at the most significant bit (MSB) position of the lower barrel shifter (LBS), TOF shifts LBS left by the amount in REM so that the starting bit reaches at the MSB position.

In order to decrease the size of VLCT, we propose a new technique, called zero count and bit reverse. In this technique, the number of leading zeros of a symbol is counted and translated into a 3-bit prefix code, and then, the prefix code is reversed and masked to decrease the effective bits of index. In fact, the length of the current symbol is unknown until the table lookup completes. So, TOF pessimistically assumes the length of the current symbol to be the maximum, which then makes VLCT have an entry for every index. In this case, many of the entries are meaningless because the most symbols are shorter than the maximum length.

(1) Zero count operation: the current symbol in LBS could have leading zeros from the MSB. The symbol length is reduced by replacing leading zeros with a prefix indicating the number of leading zeros. Using the prefix, we can significantly reduce the size VLCT. Figure 7.2 illustrates how to generate a prefix for a given input symbol. However, the zero count and prefix conversion operation requires a lot of overheads in terms of execution cycle and hardware complexity. The overheads result from the followings: (1) the operation that counts leading zeros and (2) the operation to merge a prefix and the remaining effective bits to form a table index. In order to reduce the overheads, the proposed technique makes 4 groups of bits out of a given symbol by separating 4, 4, 4, and 2 bits from the MSB, and then counts the number of leading zeros by checking whether all bits are zero or not via the use of NOR gates, as shown in Fig. 7.2.
(2) Bit reverse operation: The number of leading zeros consequently determines the number of remaining effective bits in the current symbol. So, if table indexes are the same in length, they might contain a part of the next symbol.

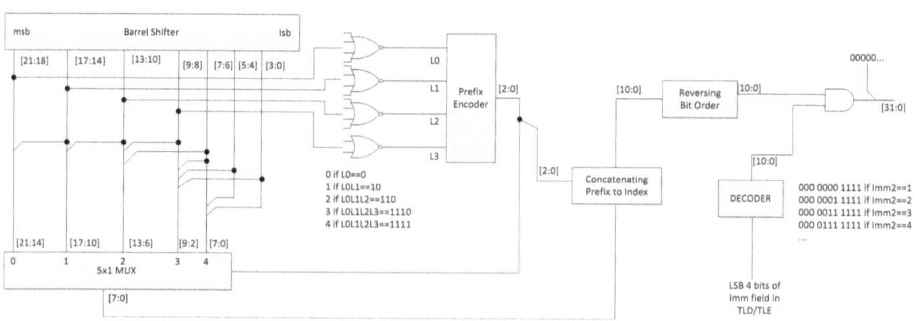

Fig. 7.2 The circuit diagram of table offset formatter (TOF)

One of the solutions to handle this problem is to shift the table index bits toward the LSB of the LBS after the zero count and prefix conversion operation. However, it requires an additional barrel shifter. In the technique proposed in this paper, TOF reverses bits in the index and masks unnecessary bits that do not belong to the current symbol.

The effect of the zero count operation and the bit reverse operation on the size of the table of VLC decoding is analyzed with the exp-golomb code. We assume that the maximum number of leading zeros of the code is eight. We analyze the size of the table on three architectures, which generates table index offset. The undescribed features of these architectures are the same as ASiPEC:

(1) Non-edit: non-edit uses codes, which is left shifted to be started at MSB on barrel shifter, itself to table index offset, so a non-edit includes neither a bit-reverse, bit shift, nor zero count. Note that the architecture that uses a zero count in addition to a Non-edit has an identical effect as a non-edit in terms of the size of table for the variable length code of VLC.
(2) Reversed: Reversed operates bit reverse and bit masking on the codes on the barrel shifter in order. Note that the architecture that uses a bit shift instead of a bit reverse has an identical effect in terms of the size of table for the VLC.
(3) Proposed: Ours is presented as a TOF, which has zero counting and bit reversing. Each row represent several variable length codes for the VLC that has the same number of leading zeros, and each column except the leftmost one represents the number of address offset bits of the table based on the three architectures. Also, the last row represents the maximum number of table address offset bits of the code. The X in the bit pattern represents a bit of the variable length code, and p represent a bit of the zero prefix. From the viewpoint of the number of table entries for the exp-golomb code, Non-edit, Reversed, and Ours requires 2^{32}, 2^{11}, and 2^{10} table entries respectively.

The format and operation of the TLD instruction that executes TOF for VLC decoding is below.

```
TLD dr,sr,(imm1),imm2
; mem[index] <= sr + {L0(BS)[2],bs < 31 - imm2 : 31 >},L0(BS)[0:1] : cycle 1
(TLD generate table index and load table value.)
; dr <= code(mem[index]) : cycle 2
(Lower 24 bits of loaded table value is code. TLD copy the code to dr register)
; rc/rem <= rem + length(mem[index]) : cycle 2
(Upper 8 bits of loaded table value is length. TLD add the length value to REM)
```

The operand imm1 of TLD instruction is maximum number of zero prefix and is optionally used when the number of leading zero is more than 14.

An example of the TOF operations is below. The number of operation is the same as that explained above.

e.g. 1: TLD dr, sr, 4

(1) 00000000000111100110001 (After shifting)
(2) 010 00111100 (After adding zero prefix, left side 010 is prefix)
(3) 0 0011110001 (After bit reversing)
(4) 0 0000110001 (After masking)

e.g. 2: TLD dr, sr, 4, 4

(1) 00000000000111100110001 (After shifting)
(2) 001 00000011 (After adding zero prefix, left side 001 is prefix)
(3) 0 1100000010 (After bit reversing)
(4) 0 0000000010 (After masking)

C. Variable length encoding

The Variable length encoding process of ASiPEC is explained here. The bitstream is located in the data memory as in Fig. 7.1, and the bits to be encoded are moved from the data memory to the GPR that is the SR2 operand in TLE instruction.

The bits in in the GPR and the table base address in another GPR, that is the SR1 operand in TLE instruction, are summed, and then the result is the table index address that indicates the corresponding variable length symbol. The symbol code at the table index address is moved to BS, and length value of encoded code at the same address is added to REM. These operations are done by TLE. The format and operations of TLE instruction is below.

```
TLE sr1, sr2
; mem[index] <= sr1 + sr2 : cycle 1
; bs <= code(mem[index]) : cycle 2
; carry/rem <= rem + length(mem[index]): cycle 2
```

D. Other instructions

ASiPEC has other VLC specific instructions such as LZS, REM, LBS, LBC, and STC. These instructions are described in Table 7.1.

7.3 Experimental Results

Since the proposed processor has a different instruction set architecture from conventional microprocessors, a direct comparison of performance is not feasible. Instead, we compare the processor effectiveness in terms of the number of instructions executed; instruction count. The number of instructions of variable length encoder and decoder are compared with those of two conventional

Table 7.1 Parts of application specific instructions of ASiPEC

Instruction	Operation	Remarks
LZS dr,sr	; dr <= leading0 s(BS)	–
	; carry/rem <= rem + leading0 s(BS)	
REM sr	; carry/rem < 5:0 > <= rem < 5:0 >+sr < 5:0>	–
LBC sr, imm	; if (RC), bs <= mem[sr]	Effectively move the data on bitstream buffer to bitstream register
	; sr <= sr + imm//auto-indexing	
STC sr, imm	; if (RC) mem[sr] <= bs	Effectively move the data on bitstream register to bitstream buffer
	; sr <= sr + imm//auto-indexing	

microprocessor cores, ARM and IA32, because these cores are popular in desktop and mobile applications. The code based on the IA-32 instruction set is compiled using GNU gcc 4.4.0, and the code for the ARM core is compiled using GNU arm-linux-gcc 4.0.0. Both compilations are done with the default compile options.

The two C language codes are the essential kernel codes of the variable length encoder and decoder, respectively, and are used to compare the code lengths of the proposed ASiPEC to the conventional microprocessors such as ARM and IA-32.

Figure 7.3 shows how many instructions can be saved while processing variable length coding with the proposed ASiPEC. ASiPEC reduces the number of instructions for encoding up to 75 and 70 %, compared with ARM and IA-32, respectively. Likewise, it reduces the number for decoding by 78 and 68 %, respectively. The numbers of memory access instructions for both encoding and decoding on ASiPEC are significantly reduced as compared to both ARM and IA-32; and those of branch instructions on ASiPEC are the same or a bit increased when compared to ARM and IA-32, respectively. However, the absolute number of branch instructions is not relevant to the overall performance of the coding.

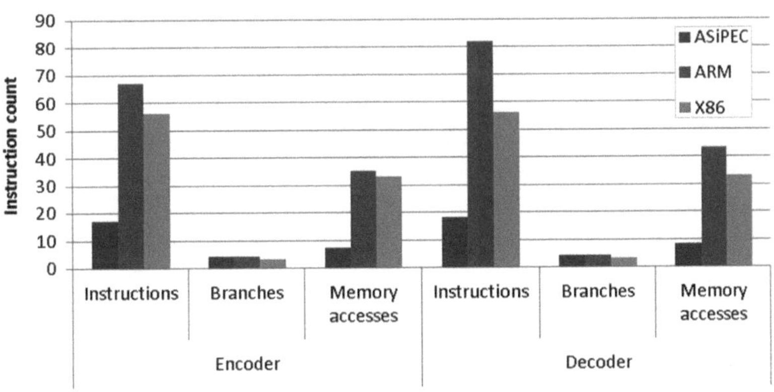

Fig. 7.3 Instruction count comparisons of VLC code for ASiPEC, ARM, and IA-32 instruction set

Table 7.2 Performance of ASiPEC on H.264/AVC CAVLC

Function	Test sequence	VLC code count	Average cycles per VLC code		
			ASiPEC	ARM	IA-32
Encoding	Spider	1,672,725	61.53	976.10	55.85
	Foreman	780,837	68.33	1449.41	76.76
	Violet	229,323	70.98	1936.13	93.80
Decoding	Spider	1,672,725	42.28	1050.08	65.00
	Foreman	780,837	50.68	1390.82	164.71
	Violet	229,323	54.53	1625.44	312.01

Table 7.2 shows the performances of ASiPEC, ARM, and IA-32 for each test sequence. The resolution of the test sequences is set to 177 × 144. In terms of the average cycles per VLC code, ASiPEC outperforms both ARM and Intel processors in most cases up to 29.8 and 5.7 times, respectively.

7.4 Conclusion

This paper presents an ASIP with efficient entropy coding instructions. Therefore, advantage of the proposed ASiPEC is support the existing standards, and the future standard by replacing the standard VLC. The proposed ASiPEC reduces the size of the tables of the VLC decoding by using a prefix and bit reverse. Compared to ARM and IA-32, the instruction count for entropy coding processing is reduced by up to 78 %, and the average cycles per VLC code is improved up to 29.8 times.

The proposed ASiPEC has advantages in terms of scalability and speed compared with other design methods. Because it is entropy codec can be configured very low cost than to build a dedicated hardware for each standard and shows faster processing speeds compared to general purpose processors. Current ASiPEC is tested by entropy coding based on VLC, but ASiPEC structure will be expanded to support arithmetic coding.

Acknowledgments This chapter is based upon work supported by the MSIP (Ministry of Science, ICT & Future Planning), Korea, under the ITRC (Information Technology Research Center) support program supervised by the NIPA (National IT Industry Promotion Agency)" (NIPA-2014-H0301-14-1018).

References

1. Alle M, Biswas J, Nandy SK (2006) High performance VLSI architecture design for H.264 CAVLC decoder. In: International conference on application-specific systems, architectures and processors, pp. 317–322
2. Han JH, Lee MY, Bae Y, Cho H (2005) Application specific processor design for H. 264 decoder with a configurable embedded processor. ETRI J 27(5):491–496

3. Jo HH, Seo JH, Sim DG, Kim DH, Song JH, Kim D, Lee S (2013) Bitstream parsing processor with emulation prevention bytes removal for H.264/AVC decoder. In: IEEE international conference of consumer electronics, pp. 27–28
4. Kim KW, Park SH, Paek YH (2009) Application-specific instruction set processor for H.264 on-chip encoder. ISOCC, pp. 373–376
5. Kim SD, Sunwoo MH (2008) ASIP approach for implementation of H. 264/AVC. J Sign Proces Syst 50(1):53–67
6. Lee BY, Ryoo KK (2010) A design of high-performance pipelined architecture for H.264/AVC CAVLC decoder and low-power implementation. IEEE Trans Consum Electron 56 (4):2781–2789
7. Lee JJ, Park SM, Eum, NW (2008) Design of application specific processor for H.264 inverse transform and quantization. ISOCC, II.57–II.60
8. Seo JH, Jo HH, Sim DG, Kim DH, Song JH (2013) Fast CAVLD of H.264 on bitstream decoding processor. EURASIP J Image Video Process 2013(1):23
9. Sohm O (2002) Variable-length decoding on the TMS320C6000 DSP platform. Application report TMS320C6000 DSP platform, Texas Instruments
10. Song JH, Lee WC, Kim DH, Kim D (2012) Low power video decoding system using a reconfigurable processor. In: IEEE international conference of consumer electronics, pp. 532–533
11. Tsai TH, Fang TL, Pan YN (2011) A Novel design of CAVLC decoder with low power and high throughput considerations. IEEE Trans Circuits Syst Video Technol 21(3):311–319
12. Werda I, Kossentini F, Ayed B, Massmoudi N (2006) Analysis and optimization of UB Video's H.264 baseline encoder implementation on texas instruments' TMS320DM642 DSP. In: International conference on image processing, pp. 3277–3280

Chapter 8
An Enhanced Cooperative Spectrum Sensing Scheme Based on New Rule of Combining Evidences in Cognitive Radio

Muhammad Sajjad Khan and Insoo Koo

Abstract Cognitive radio (CR) is a technology that enables the solution of the spectrum scarcity problem in wireless communication and is based on the concept of opportunistic spectrum access. Spectrum sensing is an essential functionality that enables CRs to detect spectral holes and opportunistically use under-utilized frequency bands without harmfully interfering to licensed users (LU). Spectrum sensing is highly affected by fading and shadowing. To overcome this, cooperative spectrum sensing was introduced. The decision of cooperative users can be combined using centralized or decentralized method. In this paper, we propose a new rule of combining evidences of users when they are highly conflicting, because in this situation the conventional Dempster-Shafer (D-S) combination rule produces illogical results. With our proposed system, we show that the reliability of the system increases as compared to the conventional D-S theory of combination rule. The simulation results show that our proposed system outperform the conventional D-S theory of combination.

Keywords Cooperative spectrum sensing · Energy detection · Dempster-Shafer (D-S) theory · New rule of combination

M.S. Khan · I. Koo (✉)
School of Electrical and Computer Engineering, University of Ulsan,
93 Daehak-ro, Nam-gu, Ulsan 680-749, South Korea
e-mail: iskoo@ulsan.ac.kr

M.S. Khan
e-mail: sajjad.khan@iiu.edu.pk

J.J. (Jong Hyuk) Park et al. (eds.), *Ubiquitous Computing Application and Wireless Sensor*, Lecture Notes in Electrical Engineering 331,
DOI 10.1007/978-94-017-9618-7_8

8.1 Introduction

Frequency spectrum is a limited resource in wireless communication, and is rendered insufficient for the increasing number of wireless services. According to the Federal Communication Commission (FCC) spectrum task report, the usage of the allocated spectrum varies depending on temporal and geographic situations [1].

Cognitive radio (CR) is arising as a tempting solution to the spectral congestion problem by introducing opportunistic usage of frequency bands that are not heavily occupied by the licenses user (LU). CR is characterized by the fact that it can adapt to the environment by changing its transmitting parameters such as modulation, frequency, frame format, etc. [2]. In CR, spectrum sensing is one of the most important functions whose goal is to detect the presence or absence of the licensed users (LU). In CR, un-licensed users constantly sense the spectrum and utilized them opportunistically. Different techniques are being used for spectrum sensing. The popular among them are based on energy detection, matched filtering and cyclo-stationary detection [3, 4].

In cognitive radio (CR), if one CR is used for the sensing of the spectrum, then it is single node sensing. In the case of cooperative sensing, multiple users are sensing the presence or absence of LU. Single node sensing is unreliable and reduces the detection probability due to fading and shadowing. The cooperative spectrum sensing improves the performance of spectrum sensing using multiple CR by aggregating their decision to one decision statistics at the fusion center (FC). In conventional cooperative spectrum sensing each unlicensed user makes a local decision which is sent to the fusion center (FC) for final decision; various fusion rules can be used at FC such as AND, OR, majority rules, etc. Unlike conventional sensing, the detection performance significantly improved with the data fusion scheme based on the Dempster-Shafer (D-S) theory of evidence without any prior knowledge of the LU signal [5]. An outstanding performance can be obtained by cooperative spectrum sensing over a single node sensing method [6]. The reliability of the spectrum sensing is observed by using double threshold energy detection [7]. A weighted evidences fusion rule on the basis of statistical features of observed energy value is considered for the fast and accurate detection of spectrum resources [8]. An ordered sequential cooperative spectrum sensing scheme based on evidence theory is proposed, which provide more feasible and efficient way in using the communication resources [9]. To the best of author knowledge, no one has yet considered the case when the evidences are highly conflicting in cognitive radio.

In this paper, we propose a new rule for combing evidences from multiple users when they are highly conflicting. First, SUs perform the local sensing and assessed its credibility which is equivalent to basic probability assignment (BPA). Then, we combine the BPA evidences of SUs using new rule of combination. Finally, these evidences are send to fusion center (FC), and the global decision is taken regarding the presence or absence of the LU.

The rest of this paper is organized as follows. In Sect. 8.2, we describe the system model of the proposed system. In Sect. 8.3, we explain cooperative spectrum sensing for the proposed rule of combination. Section 8.4, presents the simulation results. The Paper is concluded in Sect. 8.5.

8.2 System Model

In our proposed system model as shown in Fig. 8.1, there is one licensed user (LU), multiple cognitive users (CU), and data fusion center (FC).

Each CU performs local sensing for the presence or absence of the LU locally, and then this result is transmitted sequentially to the fusion center (FC). At the fusion center the sensing data is combined on the basis of their reliability. The whole process of cooperative spectrum sensing is divided in two steps, the CUs performing local spectrum sensing and data fusion at the fusion center (FC).

In the first step, each CU performs local spectrum sensing for the presence or absence of the LU's signal. The local spectrum sensing of each CU can be formulated as binary hypothesis as follows:

$$X = \begin{cases} w(n) & : H_0 \\ h(n)s(n) + w(n) & : H_1 \end{cases} \tag{8.1}$$

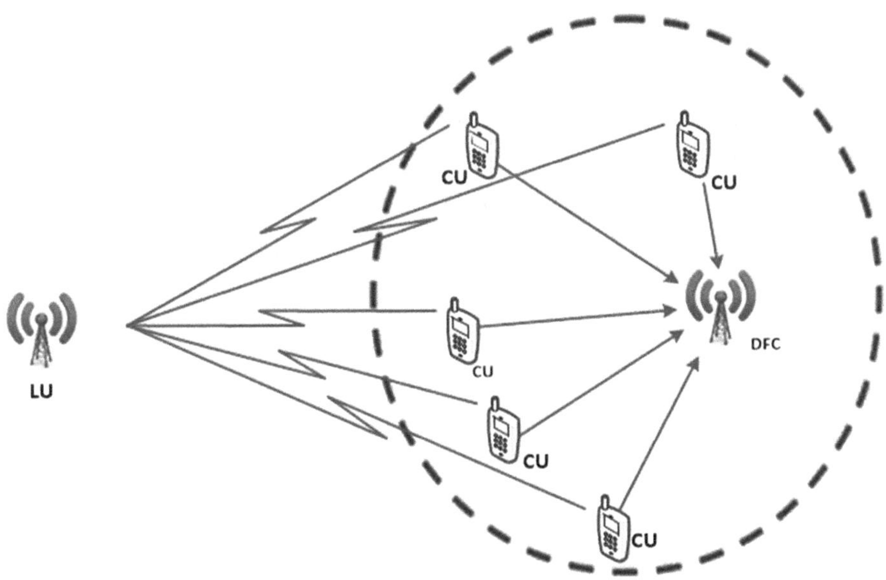

Fig. 8.1 System model of cooperative spectrum sensing

where H_0 and H_1 presents the absence or presence of the LU signals respectively, $s(n)$ represents the LU, hn represent the channel gain, and $w(n)$ is the additive white Gaussian noise (AWGN).

In the proposed system, we consider the energy detection method for the presence or absence of the LU. We use energy detection, because it do not require any prior information about the LU, and easy to implement.

Energy signal received at the CU is measured as:

$$E = \sum_{i=1}^{N} |X_i|^2 \tag{8.2}$$

where X_i is the sample of the received signal, and N = 2TW. T is the detection time and W is the signal bandwidth. When N is large enough, then E can be well approximated as a Gaussian random variable under H_0 and H_1, and follow a chi-square distribution with N degree of freedom [10]. According to the central limit theorem (CLT) the detection problem can be written as:

$$\begin{cases} \mu_0 = N, & \sigma_0^2 = 2N & : H_0 \\ \mu_1 = N(\gamma + 1), & \sigma_1^2 = 2N(2\gamma + 1) & : H_1 \end{cases} \tag{8.3}$$

where μ_0, μ_1 are mean and σ_0^2, σ_1^2 are variances of H_0 and H_1, respectively and γ is the signal to noise ratio (SNR) of licensed user (LU) at CU.

8.3 Cooperative Spectrum Sensing for the Proposed Combination Rule

In this section, we present the distributed sequential fusion for final decision about the presence or absence of LU by the FC using new rule of combination as shown in Fig. 8.2. After the local spectrum sensing, each CU assessed it credibility which is equivalent to the Basic Probability Assignment (BPA) for two hypothesis. The discernment frame can be defined as $\{H_0, H_1, \Omega\}$, where Ω is the ignorance hypothesis, which describe whether hypothesis $\{H_0, H_1\}$ can be true.

BPA function is defined in the form of cumulative function as follows [6].

$$m_i(H_0) = \int_{E_i}^{+\infty} \frac{1}{\sqrt{2\pi\sigma_{0i}}} \exp\left(-\frac{(X - \mu_{0i})^2}{2\sigma_{0i}}\right) dx \tag{8.4}$$

$$m_i(H_1) = \int_{-\infty}^{E_i} \frac{1}{\sqrt{2\pi\sigma_{1i}}} \exp\left(-\frac{(X - \mu_{1i})^2}{2\sigma_{1i}}\right) dx \tag{8.5}$$

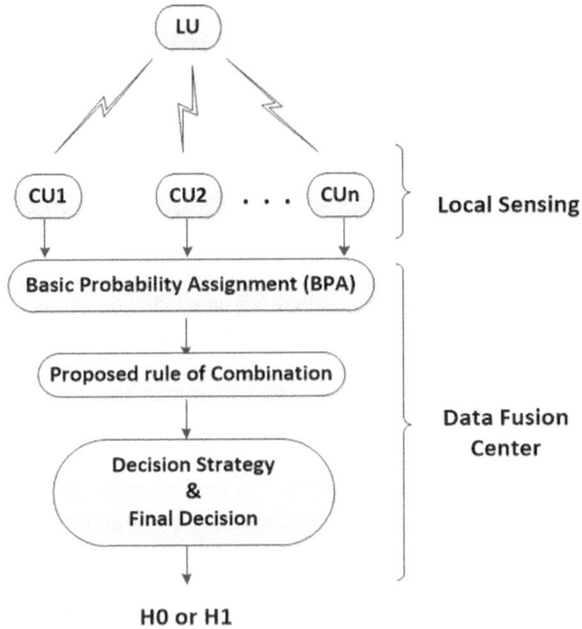

Fig. 8.2 Proposed sequential cooperative spectrum sensing scheme

$$m_i(\Omega) = 1 - m_i(H_0) - m_i(H_1) \tag{8.6}$$

where $m_i(H_0)$, $m_i(H_1)$, and $m_i(\Omega)$ are the BPA of H_0, H_1, and Ω, respectively.

8.3.1 Proposed Combination Rule

The D-S rule of combination produces very unreasonable and unrealistic results, when the evidences are highly conflicting. Let $m_1(A)$ = 0.01, $m_1(B)$ = 0.99, $m_1(C)$ = 0 and $m_2(A)$ = 0.01, $m_2(B)$ = 0, $m_1(C)$ = 0.99. According to Dempster combination rule, the BPA evidences become: $m(A) = 1$, $m(B) = m(C) = 0$, which is highly illogical results.

We propose a new rule of combining evidences coming from multiple sources, which resembles to the algebraic sum of fuzzy sets and the probability rule for union of two events. The combination of the BPA can be obtained using the following equation [11]:

$$m(H_0) = \sum_{A1 \cap A2 \ldots An = H_0} \prod_{i=1, j=1}^{n} \frac{m_i(A_i) + m_j(A_j) - m_i(A_i)m_j(A_j)}{1 + [1 - K]} \tag{8.7}$$

$$m(H_1) = \sum_{A1 \cap A2...An = H_1} \prod_{i=1,j=1}^{n} \frac{m_i(A_i) + m_j(A_j) - m_i(A_i)m_j(A_j)}{1 + [1 - K]} \quad (8.8)$$

where

$$K = \sum_{A1 \cap A2...An = \emptyset} \prod_{i=1,j=1}^{n} m_i(A_i) + m_j(A_j) - m_i(A_i)m_j(A_j) \quad (8.9)$$

where A is focal element of frame of discernment $\{H_0, H_1, \Omega\}$.

The BPA evidences are sequentially combined in order of the data arrival as follows:

$$m_{k,global}(H_j) = m_{k-1,global}(H_j) \oplus m_{k,global}(H_j) \quad (8.10)$$

where $j = 0$, 1, and $m_{k,global}(H_j)$, and $m_{k-1,global}(H_j)$ are the k-th and (k-1)-th global BPA of hypothesis H_j, respectively, and the combination operator $\oplus$ is defined on the basis of new rule of combining evidences, as follows:

$$m_a \oplus m_b(H_j) = m_a(H_j) + m_b(H_j) - m_a(H_j)m_b(H_j) \quad (8.11)$$

where $j = 0$, 1, and a and b denotes the two arbitrary combining sources.

BPA evidence values $M_{k,global}(H_j)$ can be obtained to find out the evidences of a CU about the presence or absence of LU as:

$$M_{k,global}(H_j) = \frac{m_{k,global}(H_j)}{\sum_{H_i \in H} m_{k,global}(H_i)}; H_i \in H = \{H_0, H_1, \Omega\}, \quad (8.12)$$

The global decision will be based on the global decision ratio $r_{k,global}$ of the k reports available at the DFC, which is defined as:

$$r_{k,global} = \log \left[\frac{M_{k,global}(H_1)}{M_{k,global}(H_0)} \right] \quad (8.13)$$

The proposed scheme follows the following strategies for making the final decision.

1. When the number of reports (k) available at the DFC is less than the total number of CUs, the global decision D_{global} can be calculated as:

$$D_{global} = \begin{cases} H_0 & ; if\ r_{k,global} < -\lambda \\ H_1 & ; if\ r_{k,global} > \lambda \\ no\ decision & ; if < -\lambda\ r_{k,global} < \lambda \end{cases} \quad (8.14)$$

where $-\lambda < r_{k,global} < \lambda$, represents that the reports at the DFC is not enough to take decision, and wait for the next data report.

2. In the second case, when the total number of reports (k) available at the DFC is equal to the number of CUs, the global decision D_{global} can be measured as:

$$D_{global} = \begin{cases} H_0 & \mathit{If}\ r_{k,global} < 0 \\ H_1 & \mathit{If}\ r_{k,global} > 0 \end{cases} \tag{8.15}$$

In the next section, we discuss the simulation results.

8.4 Simulation Results and Discussion

In this section, we present the simulation results of our proposed system based on the new rule of combination and compare their performance with a conventional D-S theory of combination in order sequential fusion. The simulation is done by considering the following parameters. We considered one LU (Primary user), 50 and 100 SUs nodes, SNR = −16 dB, while the value of threshold λ is varied from 0 to 18.

Figure 8.3 shows the probability of detection of the proposed method and conventional D-S theory of combination for SNR = −16 dB and number of nodes

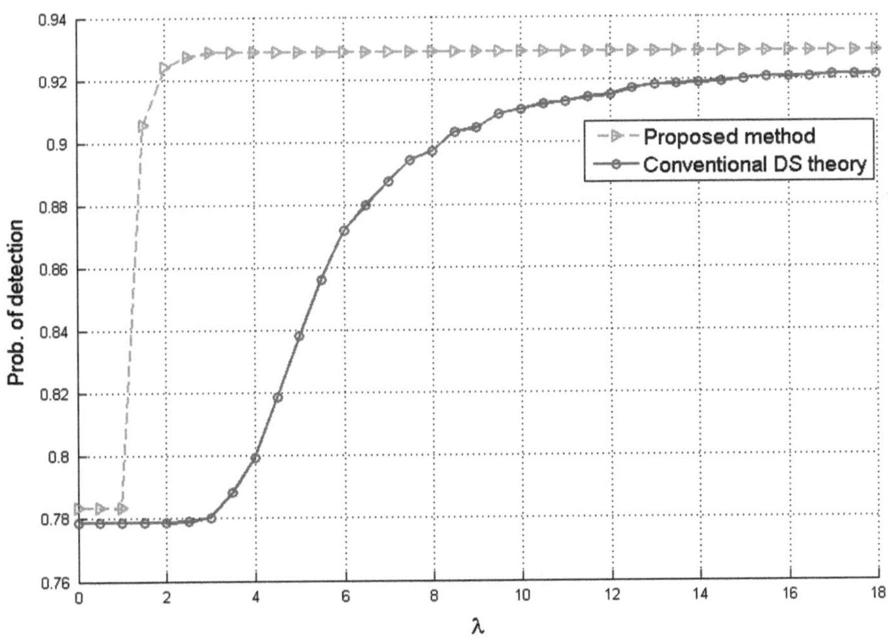

Fig. 8.3 Probability of detection versus threshold λ, when M = 50

M = 50. In Fig. 8.3 it is shown that the probability of detection of the proposed method of combination is better than conventional D-S theory of combination, which is clear indication of the advantage of the proposed method over the conventional D-S theory combination in highly conflicting evidences.

Figure 8.4 shows the performance comparison of the probability of false alarm of the proposed method with the conventional D-S theory of combination for SNR = −16 dB and number of nodes M = 50. The simulation results show that by using the proposed rule of combination the probability of false alarm is less than the conventional D-S theory of combination, which indicate that the proposed method outperforms the conventional D-S rule of combination when the evidences are highly conflicting.

Figures 8.5 and 8.6, presents the probability of detection and probability of false alarm, when SNR = −16 dB and number of nodes M = 100. Both results shows that our proposed rule of combination perform better than the conventional D-S combination rule. Along with this it is shown that with the increase in the number of nodes the probability of detection increases and probability of false alarm decreases, which show that the performance of the system increase with the cooperative spectrum sensing.

Figure 8.7 shows the probability of detection for changing the number of cooperative users. Simulation results show that by increasing the number of users

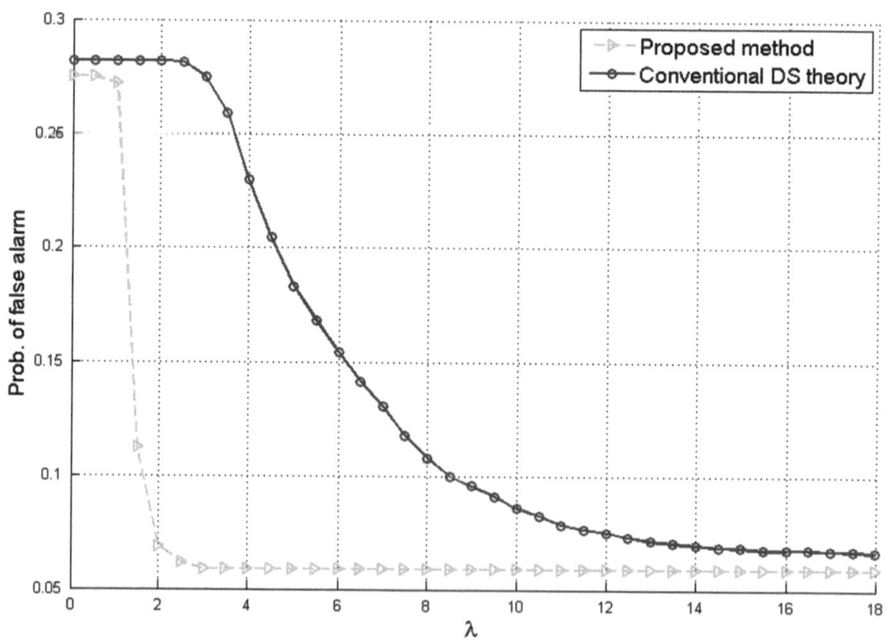

Fig. 8.4 Probability of false alarm versus threshold λ, when M = 50

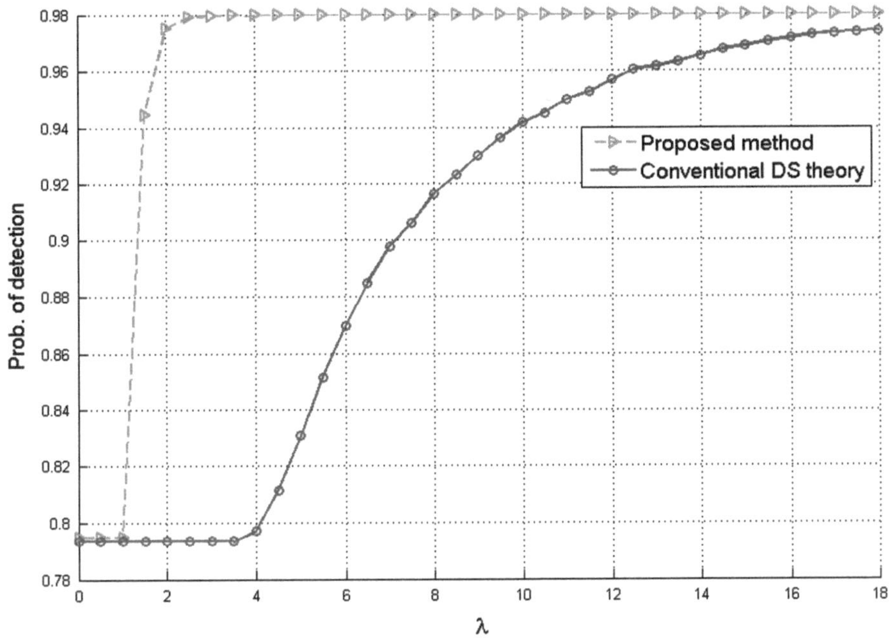

Fig. 8.5 Probability of detection versus threshold, when M = 100

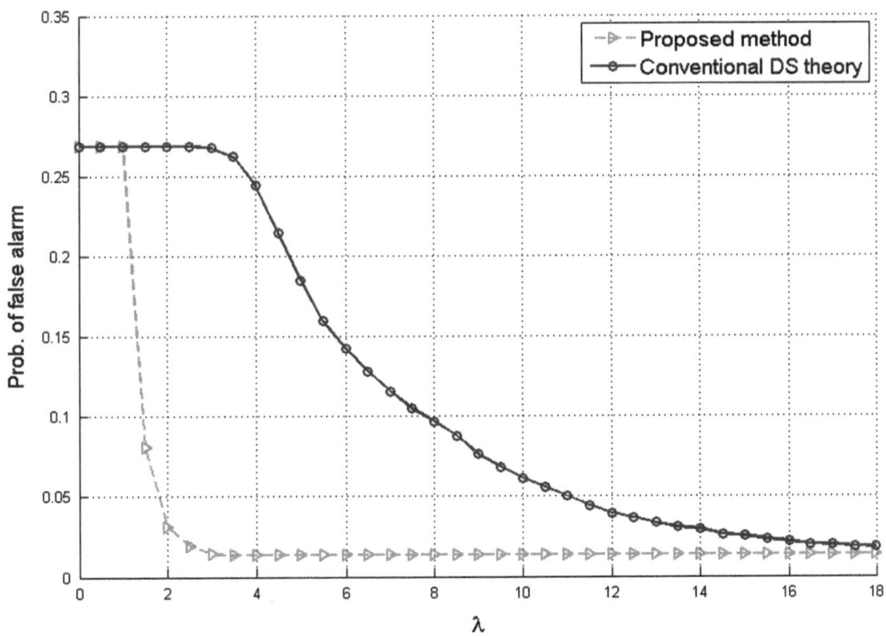

Fig. 8.6 Probability of false alarm versus threshold, when M = 100

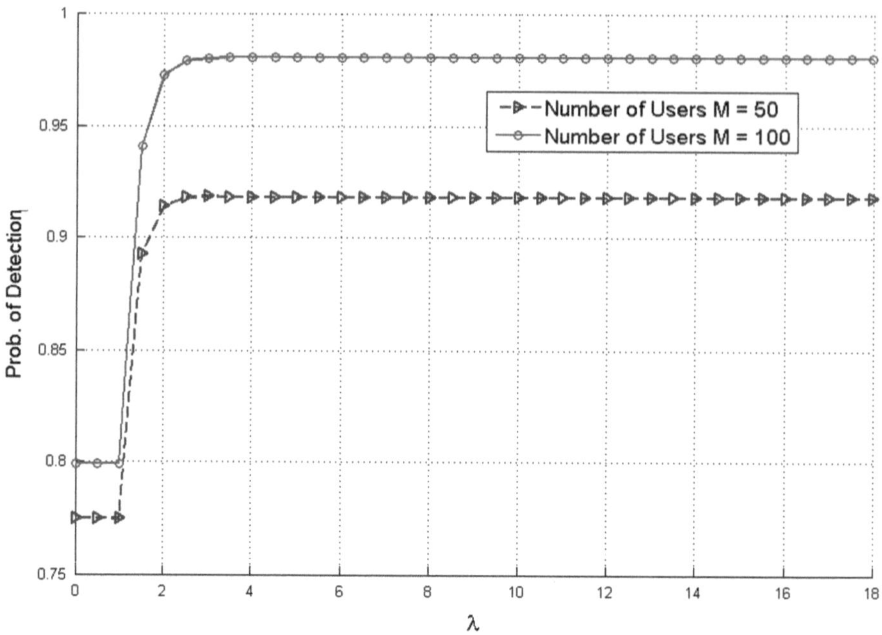

Fig. 8.7 Probability of detection versus threshold λ, when M = 50 and 100

the spectrum sensing increases, but it also increases the processing time and complexity of the system. From the simulation results, it is shown that our proposed rule of combination outperforms the conventional D-S theory of combination.

8.5 Conclusion

In this paper, we have proposed a new rule of combination for cooperative spectrum sensing. In the simulation results, we have shown that when the evidences are highly conflicting then our proposed new rule of combination is more reliable than the conventional D-S theory of rule of combination, and the probability of detection increases and probability of false alarm decreases. Thus our proposed method of combining evidences outperforms the conventional D-S theory of combination in highly conflicting evidences.

Acknowledgement This work was supported by the KRF funded by the MEST (NRF-2012R1A1A2038831).

References

1. FCC (2001) Notice of proposed rule making and order. ET, Docket No. 03-322
2. Gorein A, Thiagarajan B (2007) A signal identification application for cognitive radio. In: SDR forum technical conference
3. Yucek T, Arsalan H (2009) A survey of spectrum sensing algorithm for cognitive radio application. IEEE Commun Surv Tutor 11:116–130
4. Srinu S, Sabat SL (2012) FPGA implementation and performance study of spectrum sensing based on entropy estimation using cyclic feature. Comput Electr Eng 38:1658–1669 (Elsevier)
5. Peng Q, Zeng K, Wang J, Li S (2006) A distributed spectrum sensing based on credibility and evidence theory in cognitive radio context. In: 17th international symposium on personal, indoor and mobile radio communication. IEEE, Helsinki, pp 1–5
6. Nguyen-Thanh N, Koo I (2009) An enhanced cooperative spectrum sensing scheme based on evidence theory and reliability source evaluation in cognitive radio context. IEEE Commun Lett 13:492–494
7. Chen H, Liu J (2010) Cooperative spectrum sensing based on double threshold detection and Dempster-Shafer theory. In: 12th international conference on communication technology (ICCT). IEEE, Nanjing, pp 1212–1215
8. Xing X, Wang W, Wang Z, Liu K (2013) Weighted cooperative spectrum sensing based on D-S evidence theory and double threshold detection. In: 5th international symposium on microwave, antenna, propagation and EMC technologies for wireless communication (MAPE). IEEE, Chengdu, pp 145–149
9. Nguyen-Thanh N, Koo I (2007) An efficient ordered sequential cooperative spectrum scheme based on evidence theory in cognitive radio. IEICE Trans Commun E93-B:3248–3257
10. Urkowitz H (1967) Energy detection of unknown deterministic signals. Proc IEEE 55:523–531
11. Ali T, Dutta P, Boruah H (2012) A new combination rule for conflict problem of Dempster-Shafer evidence theory. Int J Energy Inf Commun 3:35–40

Chapter 9
Trajectory Prediction for Using Real Data and Real Meteorological Data

Yong Kyun Kim, Jong Wook Han and Hyodal Park

Abstract Trajectory prediction is basic work for 4D-trajectory modeling, conflict detection and air traffic flow management. This paper proposes a novel algorithm based distance calculation formula (vincenty formula) for trajectory prediction. We demonstrated through simulations with real flight plan and real meteorological data and experimental results show that our trajectory prediction exhibits good performance in real air traffic management environment.

Keywords Air traffic management · Trajectory prediction · Non-radar control

9.1 Introduction

Air Traffic Management (ATM) system improve safety and efficiency of air traffic by preventing collisions with other aircraft and obstacles and by managing aircraft's navigation status [1].

This research was supported by a grant (code# 07aviation-navigation-03) from Aviation Improvement Program funded by Ministry of Construction and Transportation of Korean government.

Y.K. Kim (✉) · J.W. Han
Cyber Security Research Division, Electronics and Telecommunications Research Institute, SW•Content Research Laboratory, Daejeon, Republic of Korea
e-mail: ykkim1@etri.re.kr

J.W. Han
e-mail: hanjw@etri.re.kr

H. Park
Department of Electronics and Electrical Engineering, Inha University, Incheon, Republic of Korea
e-mail: hdpark@inha.ac.kr

J.J. (Jong Hyuk) Park et al. (eds.), *Ubiquitous Computing Application and Wireless Sensor*, Lecture Notes in Electrical Engineering 331,
DOI 10.1007/978-94-017-9618-7_9

The accuracy of trajectory predictions in En-route airspace impacts ATM conflict predictions and Estimated Times of Arrival (ETA) to control waypoints. For the airspace user, inaccurate trajectory predictions may result in less-than-optimal maneuver advisors in response to a given traffic management problem [2].

ATM of the future allows for the possibility of free flight, in which aircraft choose their own optimal routes, altitudes and velocities. The safe resolution of trajectory conflicts between aircraft is necessary to the success of such a trajectory prediction [3].

In this paper we propose trajectory prediction method to prepare for abnormal operating conditions and collision avoidance. The remainder of this paper is structured as follows. In the next section, trajectory prediction techniques and theoretical background about trajectory prediction algorithm is presented. We present some experimental results of our proposed scheme in Sect. 9.3. And finally conclude with the conclusion in Sect. 9.4.

9.2 Trajectory Prediction Techniques and Theory

9.2.1 Trajectory Prediction Theory [4]

Aircraft's trajectory prediction generate flight path of the aircraft from departure aerodrome to destination aerodrome and predict passage time and altitude accurately at each passage point. In order to aircraft's trajectory prediction, it is essential needed to route information, departure and destination aerodrome information, aircraft's performance data based on aircraft type and runway information. The key elements for trajectory prediction is shown in Fig. 9.1.

(1) Flight Envelope
Flight Envelope considered maximum speed, altitude and minimum speed. In particular, minimum speed considered approach configuration, cruise configuration, initial climb configuration, landing configuration and take-off configuration.

(2) Aerodynamics
Aerodynamics deals with aerodynamic drag and low speed buffeting limit.

(3) Fuel Consumption
Fuel Consumption calculation is related to engine type. We considered three engine types jet, turboprop and piston engine. Fuel consumption calculation is most safety-critical aspect of flight planning.

(4) Aircraft Type
Aircraft type is divided by engine types jet, turboprop and piston. And aircraft type also can divided by four wake turbulence categories jumbo, heavy, medium and light.

(5) Global Aircraft parameters

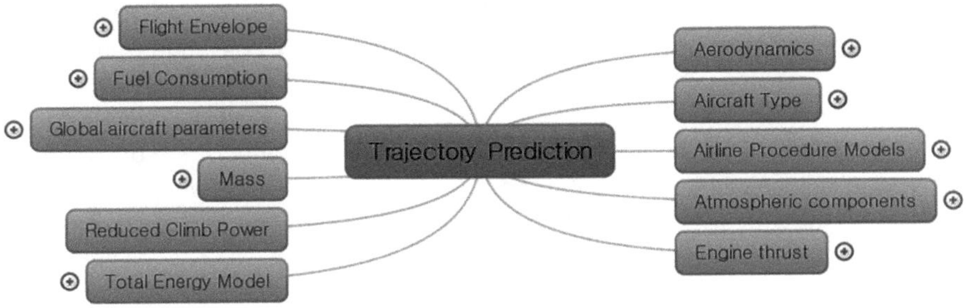

Fig. 9.1 The elements for trajectory prediction

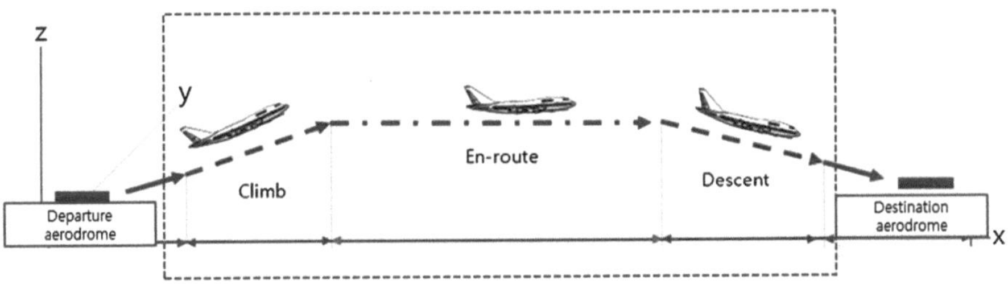

Fig. 9.2 Airline procedure model

Global parameters deals with maximum acceleration, bank angles, expedited descent, thrust factors(take-off thrust coefficient, maximum cruise thrust coefficient), configuration altitude threshold(maximum altitude threshold for take-off, initial climb, approach and landing), minimum speed coefficients (minimum speed coefficient for take-off and all other phases), speed schedules, holding speeds, ground speeds and reduced power coefficient (maximum reduction in power for turboprops/jets/pistons)

(6) Airline procedure model
Typical model of airline procedure model deals with 5 flight phases: take-off, climb, cruise, descent, landing. But we deals with 3 flight phase: climb, cruise and descent for simulating standard or normal aircraft operations using different simulation and modeling tools for various ATM applications.
The flight phase of airline procedure model is shown in Fig. 9.2.

(7) Mass
Mass is divided into minimum mass, maximum mass, reference mass and maximum payload mass.

(8) Atmosphere model
Atmosphere model is related to calculation of aircraft performance and movements. Conversions from Calibrated Air Speed (CAS) to True Air Speed (TAS) and Mach number also require the determination of several atmospheric properties as a function of altitude.

Atmosphere model deals with determination air pressure, air density and speed of sound, CAS/TAS conversion, Mach/TAS conversion and mach/CAS transition altitude.

(9) Reduced Power

The reduced climb power has been introduced to allow the simulation of climbs using less than maximum climb setting. In day-to-day operations, many aircraft use a reduced setting during climb in order to extend engine life and save cost. The correction factors that are used to calculate the reduction in power have been obtained in an empirical way and have been validated with the help of air traffic controllers.

(10) Engine Thrust

Engine thrust deals with coefficients that the calculation of the following thrust levels: maximum climb and take-off, maximum cruise and descent.

(11) Total Energy Model

The total-energy model equates the rate of work done by forces acting on the aircraft to the rate of increase in potential and kinetic energy, that is calculated according to the formula as:

$$(T - D) \cdot V_{TAS} = mg_0 \frac{dh}{dt} + mV_{TAS} \frac{dV_{TAS}}{dt}$$

The symbols are defined below specified.

T thrust acting parallel to the aircraft velocity vector
d aerodynamic drag
m aircraft mass
h geodetic altitude
g_0 gravitational acceleration
V_{TAS} true airspeed
d/dt time derivative

9.2.2 Airspeed Theory

Airspeed is the speed of an aircraft relative to the air. Among the common conventions for qualifying airspeed are: indicated airspeed (IAS), calibrated airspeed (CAS), true airspeed (TAS), and ground speed (GS).

Indicated airspeed (IAS) is the airspeed indicator reading (ASIR) uncorrected for instrument, position, and other errors. From current EASA definitions: Indicated airspeed means the speed of an aircraft as shown on its pitot static airspeed indicator calibrated to reflect standard atmosphere adiabatic compressible flow at sea level uncorrected for airspeed system errors.

Most airspeed indicators show the speed in knots (i.e. nautical miles per hour). Some light aircraft have airspeed indicators showing speed in miles per hour. Calibrated airspeed (CAS) is indicated airspeed corrected for instrument errors, position error and installation errors.

Calibrated airspeed values less than the speed of sound at standard sea level (661.4788 kn) are calculated as follows:

$$V_c = A_0 \sqrt{5\left[\left(\frac{Q_c}{P_0} + 1\right)^{\frac{2}{7}} - 1\right]}$$

where

V_C is the calibrated speed. Q_C is the impact pressure sensed by the pitot tube. P_0 is 29.92126 in. Hg; static air pressure at standard sea level. A_0 is 661.4788 kn; speed of sound at standard sea level

This expression is based on the form of Bernoulli's equation applicable to a perfect, compressible gas. The values P_0 and A_0 are consistent with the International Standard Atmosphere (ISA).

True airspeed (TAS) is the physical speed of the aircraft relative to the air surrounding the aircraft. The true airspeed is a vector quantity. The relationship between the true airspeed (V_t) and the speed with respect to the ground (V_g) is

$$V_t = v_g - V_w$$

where V_w is Wind speed vector.

Aircraft flight instruments, however, don't compute true airspeed as a function of groundspeed and wind speed. They use impact and static pressures as well as a temperature input. Basically, true airspeed is calibrated airspeed that is corrected for pressure altitude and temperature. The result is the true physical speed of the aircraft plus or minus the wind component. True Airspeed is equal to calibrated airspeed at standard sea level conditions.

The simplest way to compute true airspeed is using a function of Mach number

$$V_t = A_0 \cdot M\sqrt{\frac{T}{T_c}}$$

where M is Mach number, T is Temperature (kelvins) and T_0 is Standard sea level temperature (288.15 K)

Second, speed variation by altitude changing means that when aircraft are climb or descent.

The rate of climb (RoC) is the speed at which an aircraft increases its altitude. This is most often expressed in feet per minute and can be abbreviated as ft/min. Else where, it is commonly expressed in meters per second, abbreviated as m/s. The rate of climb in an aircraft is measured with a vertical speed indicator (VSI) or instantaneous vertical speed indicator (IVSI). The rate of decrease in altitude is referred to as the rate of descent or sink rate. A decrease in altitude corresponds with a negative rate of climb.

There are two airspeeds relating to optimum rates of ascent, referred to as Vx and Vy. Vx is the indicated airspeed for best angle of climb. Vy is the indicated airspeed for best rate of climb. Vx is slower than Vy.

Climbing at Vx allows pilots to maximize the altitude gain per unit ground distance. That is, Vx allows pilots to maximize their climb while sacrificing the least amount of ground distance. This occurs at the speed for which the difference between thrust and drag is the greatest (maximum excess thrust). In a jet airplane, this is approximately minimum drag speed, or the bottom of the drag versus speed curve. Climb angle is proportional to excess thrust.

Climbing at Vy allows pilots to maximize the altitude gain per unit time. That is, Vy, allows pilots to maximize their climb while sacrificing the least amount of time. This occurs at the speed for which the difference between engine power and the power required to overcome the aircraft's drag is the greatest (maximum excess power). Climb rate is proportional to excess power.

Vx increases with altitude and Vy decreases with altitude. Vx = Vy at the airplane's absolute ceiling, the altitude above which it cannot climb using just its own lift.

Last, we consider about wind parameters. Wind parameter can divide two components (weather fonts and thermal wind) on a large scale.

Weather fronts are boundaries between two masses of air of different densities, or different temperature and moisture properties, which normally are convergence zones in the wind field and are the principal cause of significant weather. Within surface weather analyses, they are depicted using various colored lines and symbols.

The air masses usually differ in temperature and may also differ in humidity. Wind shear in the horizontal occurs near these boundaries. Cold fronts feature narrow bands of thunderstorms and severe weather, and may be preceded by squall lines and dry lines.

Cold fronts are sharper surface boundaries with more significant horizontal wind shear than warm fronts. When a front becomes stationary, it can degenerate into a line which separates regions of differing wind speed, known as a shear line, though the wind direction across the feature normally remains constant. Directional and speed shear can occur across the axis of stronger tropical waves, as northerly winds precede the wave axis and southeast winds are seen behind the wave axis.

Horizontal wind shear can also occur along local land breeze and sea breeze boundaries.

Thermal wind is a meteorological term not referring to an actual wind, but a difference in the geostrophic wind between two pressure levels p_1 and p_0, with $p_1 < p_0$; in essence, wind shear. It is only present in an atmosphere with horizontal changes in temperature.

In a barotropic atmosphere, where temperature is uniform, the geostrophic wind is independent of height. The name stems from the fact that this wind flows around areas of low (and high) temperature in the same manner as the geostrophic wind flows around areas of low (and high) pressure.

$$f_{\tau t} = K \times \nabla(\phi_2 - \phi_0)$$

where the ϕ_x are geopotential height fields with $\phi_2 > \phi_0$, f is the Coriolis parameter, and $\boldsymbol{k}$ is the upward-pointing unit vector in the vertical direction. The thermal wind equation does not determine the wind in the tropics. Since f is small or zero, such as near the equator, the equation reduces to stating that $\nabla(\phi_2 > \phi_0)$ is small. This equation basically describes the existence of the jet stream, a westerly current of air with maximum wind speeds close to the tropopause which is (even though other factors are also important) the result of the temperature contrast between equator and pole.

9.2.3 Distance Calculation Between Two Points

For calculating distance between waypoints, first of all, we need extract route information from flight plan. Flight plan message format is shown in the Fig. 9.3.

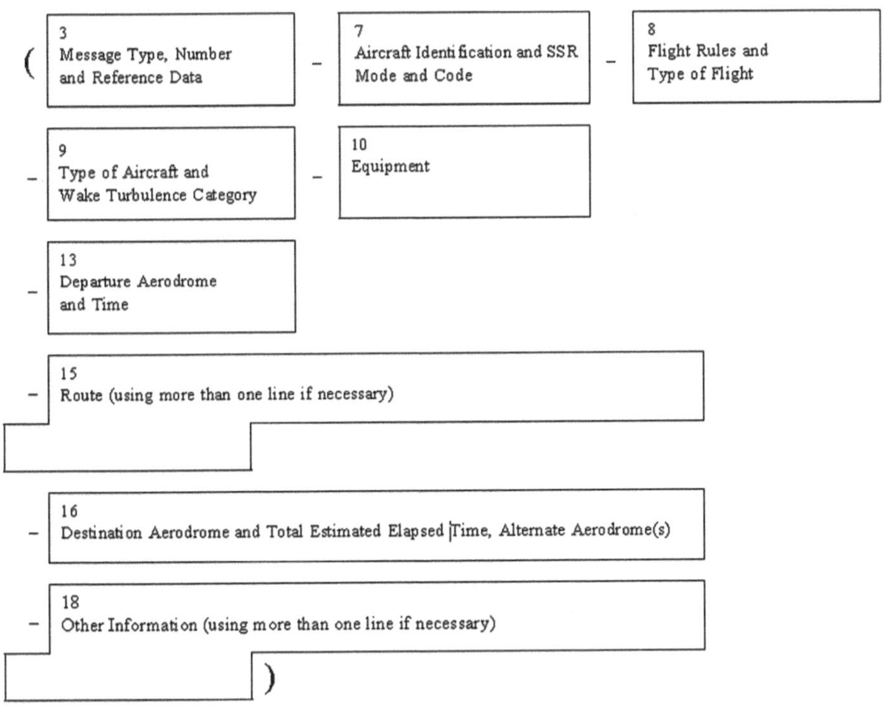

Fig. 9.3 Filled flight plan message format

By enumerating type 15 route information among parsed flight plan, we can find that confirmed aircraft's waypoints and altitude (or speed) variations.

Waypoints consists of latitude and longitude. In accordance, for measuring the distance between waypoints over a flight information region, the curvature of the Earth must be taken into consideration.

As a rough estimate, we could assume the Earth is a sphere.

$$dN = R\theta\phi$$
$$dE = R\cos\theta\phi\lambda$$

where R is the radius of the Earth (average of 6378.1 km) and the differences in latitude and longitude are in radians, the distance is 447.47 km. This method is valid assumption over very small distances, however, over large distances we need to account for the non-uniformity of the Earth. The Earth is not actually a sphere, it is an ellipsoid of revolution, 21 km shorter on the North–South direction than the East–West direction.

The flattening at the poles is caused by the centrifugal force of the spinning Earth. Because of this flattening, the radius if the Earth is not a constant value as we assume for Spherical Earth coordinates.

A more accurate method of measuring the distance between two points on the surface of the Earth is Vincenty's Formula. It is accurate to 0.5 mm over a distance of a few centimeters to nearly 20,000 km.

Given the coordinates of the two points (φ_1, λ_1) and (φ_2, λ_2) the vincenty's inverse method finds the azimuths α_1, α_2 and ellipsoidal distance s.

Calculate reduced latitude $U_1(\arctan[(1-f)\tan\varphi_1]$, $U_2(\arctan[(1-f)\tan\varphi_2]$, and L, and set initial value of $\lambda = L$. Then iteratively evaluate the following equations until λ converges.

$$\sin\sigma = \sqrt{(\cos U_2 \sin\lambda)^2 + (\cos U_1 \sin U_2 - \sin U_1 \cos U_2 \cos\lambda)^2}$$

$$\cos\sigma = \sin U_1 \sin U_2 + \cos U_1 \cos U_2 \cos\lambda$$

$$\sigma = \arctan(\sin\sigma / \cos\sigma)$$

$$\cos(2\sigma_m) = \cos\sigma - (\sin U_1 \sin U_2 / \cos^2\alpha)$$

$$\sin\alpha = [(\cos U_1 \cos U_2 \sin\lambda)/(\sin\sigma)]$$

$$\cos^2\alpha = 1 - \sin^2\alpha$$

$$C = (1/16)\cos^2\alpha[4 + f(4 - 3\cos^2\alpha)]$$

$$\lambda = L + (1 - C)f \sin\alpha\{\sigma + C\sin\sigma[\cos(2\sigma m) + C\cos\sigma(-1 + 2\cos^2(2\sigma m)]\}$$

When λ has converged to the desired of accuracy, evaluate the following:

$$U^2 = COS^2\alpha[(a_2 - a_1)/b_2]$$

$$A = 1 + (u^2/16384)\{4096 + u^2[-768 + u2(320 - 175u^2)]\}$$
$$B = (u^2/1024)\{256 + u^2[-128 + u^2(74 - 47U^2)]\}$$

$$\Delta\sigma = B\sin\sigma\{\cos(2\sigma m) + 0.25B[\cos\sigma(-1 + 2\cos^2(2\sigma m)$$
$$(-3 + 4\sin 2\sigma)(-3 + 4\cos 2(2\sigma m))]\}$$

$$s = bA(\sigma - \Delta\sigma)$$

$$\alpha_1 = \arctan[(\cos U_2 \sin\lambda)/(\cos U_2 \sin U_2 - \sin U_2 \cos U_2 \cos\lambda)$$
$$\alpha_2 = \arctan[(\cos U_2 \sin\lambda)/(-\sin U_2 \cos U_2 + \sin U_2 \cos U_2 \cos\lambda)$$

We can compute azimuths (α_1, α_2) and distance s.

9.2.4 *Aircraft's Position in Compliance with the Wind Parameter*

Consider a fairly simplified model for aircraft location in compliance with the wind parameter uncertainty. Wind parameter uncertainty have an impact on the direction of the aircraft. Typical model of the aircraft position in compliance with wind parameter is as following (Figs. 9.4 and 9.5).

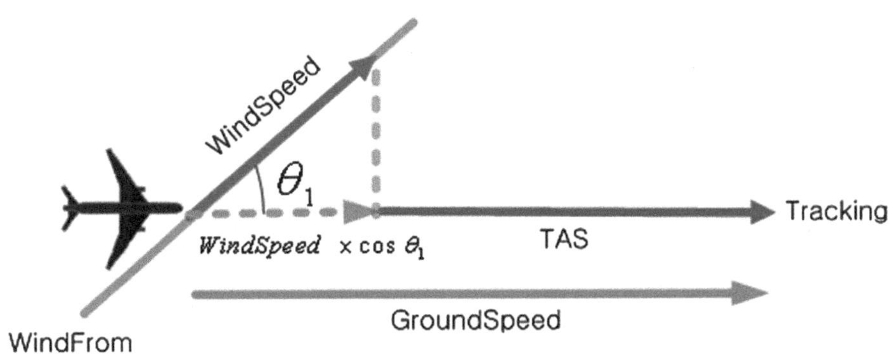

Fig. 9.4 Typical model of the aircraft position in compliance with the wind parameter

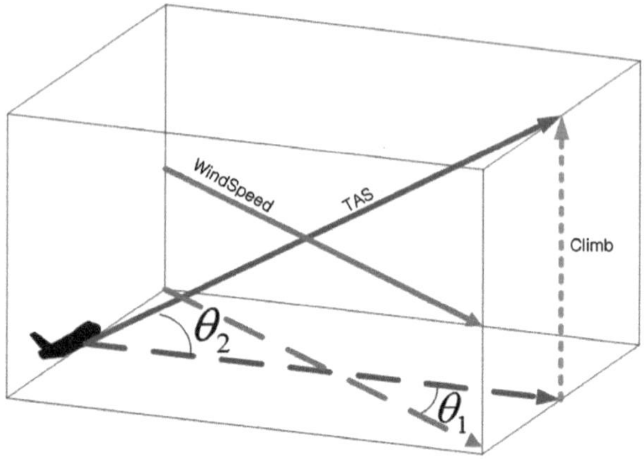

Fig. 9.5 Ground speed calculation using TAS, track angle, climb rate and wind parameter

9.3 Experimental Results of Proposed Scheme

First of all, we measured point-to-point distance performance.

Two points are shown below Table 9.1 and Fig. 9.6.

Table 9.1 Two points information

Waypoint name	Latitude	Longitude
TGU	35.48.35N	128.35.27E
KALOD	35.30.12N	128.46.26E

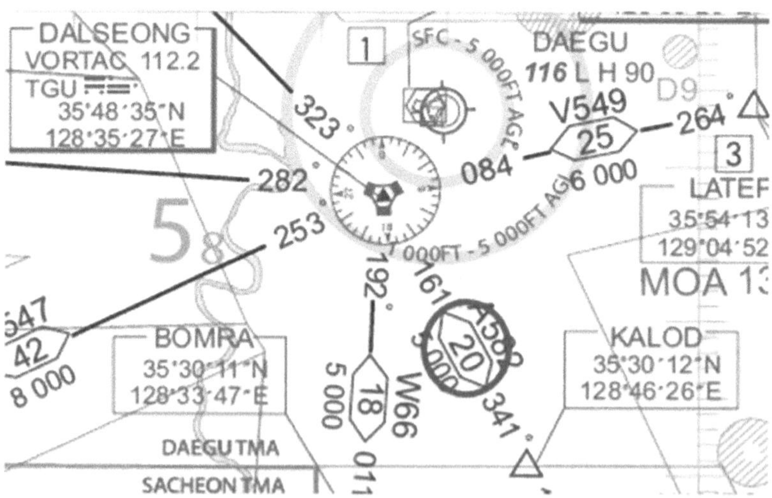

Fig. 9.6 Air traffic service chart (TGU-KALOD: 20 NM)

Using distance calculation algorithm, we calculate point-to-point distance as below Figs. 9.7 and 9.8.

Proposed point-to-point distance calculation algorithm can verify using the arctic regions coordinates. Verification result is shown Fig. 9.9.

Now we measured using real flight plan. Original message is shown in the table below (Table 9.2).

And meteorological data is shown Figs. 9.10, 9.11 and 9.12

Simulation result is shown Figs. 9.13 (Database) and 9.14 (Graphical Result)

Fig. 9.7 Distance simulation result (altitude: 0 ft)

Fig. 9.8 Distance simulation result (altitude: 5,000 ft)

Fig. 9.9 Two-points distance in the arctic regions

Table 9.2 Original flight plan message

Flight plan message
Aircraft ID: KAL9275
Route: DCT SEL G597 KAE B467 TENAS/N0486F350 L512 GTC Y512 OATIS PUTER A590 POXED/N0480F370 A590 HAMND NEELL2
Departure aerodrome: RKSI
Destination aerodrome: PANC
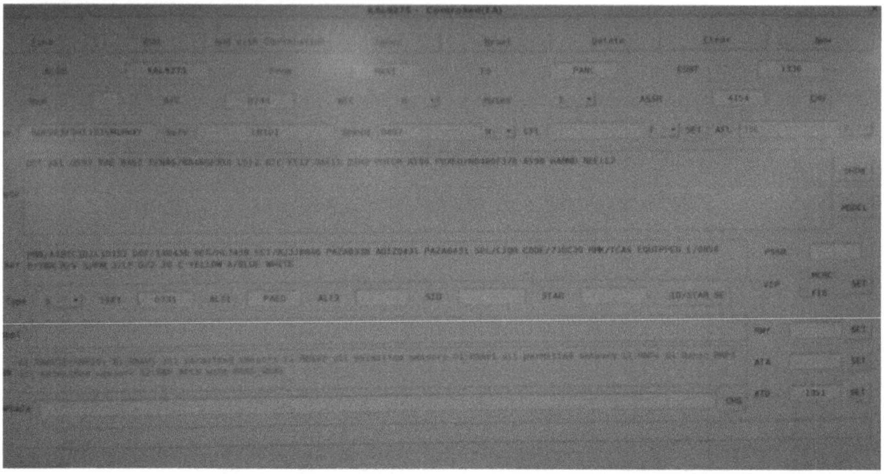

D_MSG_TYPE	D_MSG_SIZE	D_LATITUDE	D_LONGITUDE	D_STRATA	D_NORTHERN_WIND	D_EASTERN_WIND	D_HIGH_TEMP	D_HIGH_PRESSURE
GRD	0025	003	001	000	-009	0005	-005	????
GRD	0025	003	001	002	-013	0036	????	????
GRD	0025	003	001	003	-007	0042	????	????
GRD	0025	003	001	004	-027	0047	????	????
GRD	0025	003	001	005	0000	0062	????	????
GRD	0025	003	001	008	0000	0114	-058	????
GRD	0025	003	002	000	-017	0006	-006	????
GRD	0025	003	002	002	-026	0032	????	????
GRD	0025	003	002	004	0000	0053	????	????
GRD	0025	003	002	005	0000	0091	????	????
GRD	0025	003	002	006	0000	0089	????	????
GRD	0025	003	002	007	0000	0095	-052	????
GRD	0025	003	002	009	0000	0103	????	????
GRD	0025	003	003	000	-017	0006	-006	????
GRD	0025	003	003	001	-016	0014	-009	????
GRD	0025	003	003	002	-013	0036	????	????
GRD	0025	003	003	003	-007	0042	????	????
GRD	0025	003	003	004	-027	0047	????	????
GRD	0025	003	003	005	0000	0062	????	????
GRD	0025	003	003	006	-014	0080	????	????
GRD	0025	003	003	007	0000	0095	-052	????
GRD	0025	003	003	008	0000	0101	-060	????
GRD	0025	003	003	009	0000	0101	????	????
GRD	0025	003	004	000	-009	0005	-005	????
GRD	0025	003	004	001	-016	0014	-009	????
GRD	0025	003	004	002	-013	0036	????	????
GRD	0025	003	004	003	-021	0037	????	????
GRD	0025	003	004	004	0000	0053	????	????
GRD	0025	003	004	005	0000	0062	????	????
GRD	0025	003	004	006	0000	0089	????	????
GRD	0025	003	004	007	0000	0095	-052	????
GRD	0025	003	004	008	0000	0114	-058	????
GRD	0025	003	004	009	0000	0095	????	????
GRD	0025	004	001	000	-009	0005	-005	????

Fig. 9.10 upper airspace meteorological database

D_PLACE	D_TIME	D_TRANSITIONLEVEL	D_TEMPERATURE	D_QNH	D_QNH_STATE
RKSI	301400	140	12	Q1012	
RKPK	300600	140	17	Q1009	
RKJB	301300	140	(null)	Q1013	
RKJJ	301400	140	10	Q1013	
RKTU	301400	140	(null)	Q1011	
RKTN	301400	140	13	Q1011	
RKNY	301300	140	09	Q1011	
RKSO	301400	140	13	Q1011	
RKSW	301400	140	12	Q1012	
RKTP	300300	140	21	Q1010	
RKSS	301400	140	(null)	Q1012	
RKPC	301400	140	(null)	Q1014	
RKSM	301400	140	13	Q1011	
RKTY	300600	140	20	Q1008	
RKTH	301300	140	14	Q1013	
RKPS	301300	140	15	Q1013	
RKNW	301200	140	14	Q1009	
RKTI	300600	140	21	Q1008	
RKNN	301400	140	11	Q1011	
RKPU	300600	140	15	Q1009	
RKJY	301400	140	15	Q1012	
RKJK	280200	140	07	Q1024	

Fig. 9.11 Aerodrome meteorological database

D_PLACE	D_TIME	D_TRANSITIONLEVEL	D_TEMPERATURE	D_QNH	D_QNH_STATE
RKSI	301400	140	12	Q1012	
RKPK	300600	140	17	Q1009	
RKJB	301300	140	(null)	Q1013	
RKJJ	301400	140	10	Q1013	
RKTU	301400	140	(null)	Q1011	
RKTN	301400	140	13	Q1011	
RKNY	301300	140	09	Q1011	
RKSO	301400	140	13	Q1011	
RKSW	301400	140	12	Q1012	
RKTP	300300	140	21	Q1010	
RKSS	301400	140	(null)	Q1012	
RKPC	301400	140	(null)	Q1014	
RKSM	301400	140	13	Q1011	
RKTY	300600	140	20	Q1008	
RKTH	301300	140	14	Q1013	
RKPS	301300	140	15	Q1013	
RKNW	301200	140	14	Q1009	
RKTI	300600	140	21	Q1008	
RKNN	301400	140	11	Q1011	
RKPU	300600	140	15	Q1009	
RKJY	301400	140	15	Q1012	
RKJK	280200	140	07	Q1024	

Fig. 9.12 Air pressure database

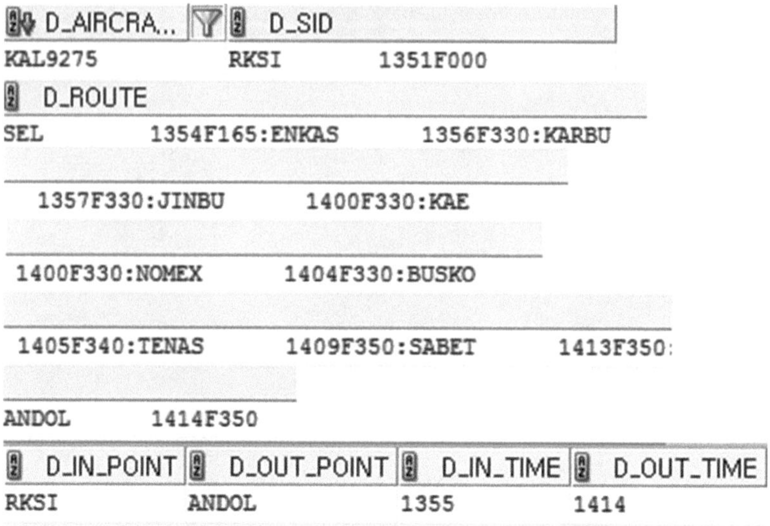

Fig. 9.13 Trajectory prediction result

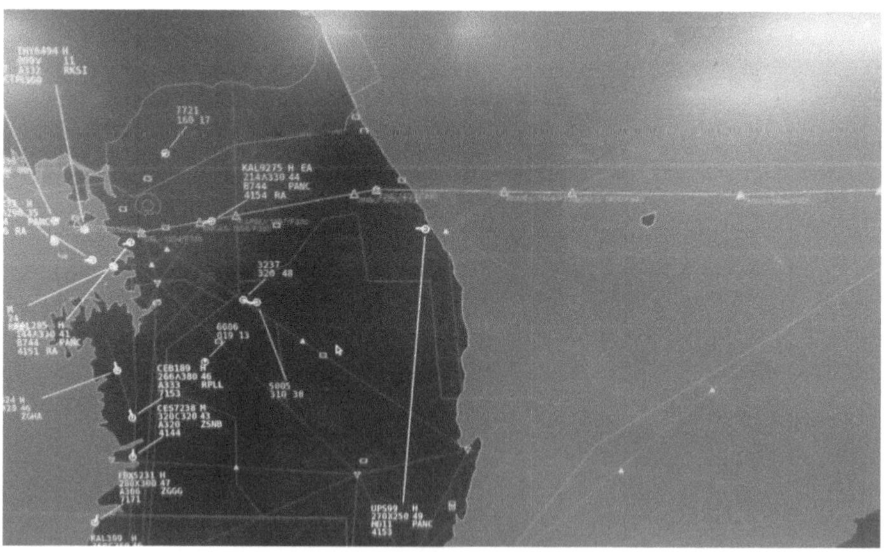

Fig. 9.14 Trajectory prediction results

9.4 Conclusion

In this paper propose trajectory prediction method using wind data and real flight plan data.

The applicability of the proposed algorithm is such as rocket positioning control, aeronautical traffic flow management system.

From now on, it is further suggested that the proposed algorithm may be extended to the non-radar object tracking, which may further improve air traffic advisory system.

Acknowledgments Foundation item: This research was supported by a grant (code# 07aviation-navigation-03) from Aviation Improvement Program funded by Ministry of Construction and Transportation of Korean government.

References

1. FAA (2012) Air traffic control Chapter 2. General control. FAA order JO7110.65U, Feb 9 2012
2. ICAO (2008) Threat and error management (TEM) in air traffic control. ICAO Cir 314 QN/178 2008
3. Tomlin C, Pappas GJ, Sastry S (1998) Conflict resolution for air traffic managements: a study in multiagent hybrid systems. IEEE Trans Autom Control 43(4):509–521
4. EUROCONTROL (2012) User manual for the base of aircraft data (BADA) revision 3.10. Report No. 12/04/10-45, Apr 2012

Chapter 10
The ADS-B Protection Method for Next-Generation Air Traffic Management System

Seoung-Hyeon Lee, Jong-Wook Han and Deok-Gyu Lee

Abstract ADS-B data is an important data to effectively utilize limited airspace as it contains main flight information including the location, altitude, speed, and others of plane but it is very vulnerable to hacking attack such as Ground Station Flood Denial, Ground Station Target Ghost Injection, etc. Therefore, the purpose of this study lies in proposing ground-to-ground security framework using SPKI certificate to protect ADS-B data received from the plane. In security framework proposed in this study, each ADS-B sensor connected to ATC is authenticated using SPKI certificate and creates encrypted ADS-B data using the authenticated data. Also, the verification on validity of transmission delayed ADS-B is conducted using timer information.

Keywords ATC · ADS-B · SPKI · Authentication

10.1 Introduction

With a dramatic increase in demand for air traffic, supplementary procedures have been added to air traffic control work and ATC (Air Traffic Control system) to secure safe and effective flight of more planes in limited airspace and it has been developed to become a system which can support air traffic control work [1]. ADS-B (Automatic Dependent Surveillance-Broadcast) is a core of next generation ATC and it is a system which broadcasts real-time main flight information including the accurate location, altitude, speed, and others of plane using GNSS (Global Navigation

---

S.-H. Lee (✉) · J.-W. Han
Industrial Control System (ICS) Security Research Section, ETRI, Daejeon, Korea
e-mail: duribun@etri.re.kr

J.-W. Han
e-mail: jwhan@etri.re.kr

D.-G. Lee
Department of Information Security, Seowon University, Cheongju-si, Korea
e-mail: deokgyulee@seowon.ac.kr

J.J. (Jong Hyuk) Park et al. (eds.), *Ubiquitous Computing Application and Wireless Sensor*, Lecture Notes in Electrical Engineering 331,
DOI 10.1007/978-94-017-9618-7_10

Satellite System). Plane or ground air traffic control center uses ADS-B data to assists safe flight in limited airspace and effective use of air space through the minimization of distance between planes and prevention of plane crash [2].

Although ADS-B data contains important flight information of planes, it is vulnerable to external attack including Aircraft Reconnaissance, Ground Station Flood Denial, Aircraft Target Ghost Inject, etc. [3]. For instance, it was reported in DEFCON Hacking Conference held in 2012 at Las Vegas that it would have severe effect on air traffic control work when 50 ghost planes are created and displayed on controller working position (CWP) and similar cases to cheat system using fabricated ADS-B signal were demonstrated and reported [4].

Therefore, ground-to-ground security framework which can protect ADS-B data through real-time authentication of ADS-B sensor using light SPKI certificate and use of authenticated information as such is proposed in this study. The composition of this study is as following. Chapter 2 introduces ADS-B and describes security vulnerability of ADS-B and SPKI certificate. Chapter 3 proposes a security framework adequate for ground-to-ground environment. Chapter 4 points out the significance of this study and task for follow-up studies.

10.2 Related Work

10.2.1 ADS-B

10.2.1.1 Overview

ADS-B is a broadcast surveillance system with air-to-ground (aircraft-to-ATS) and air-to-air (aircraft-to-aircraft) applications. ADS-B avionics broadcast identification, position, altitude, velocity and other data automatically about every half second. The system 'depends' on other aircraft systems, such as a barometric encoder and GNSS equipment for the data [5] (Fig. 10.1).

Advantages of introducing ADS-B to air traffic control are as below.

- Airservices' policy is to give priority to ADS-B equipped aircraft (when doing so gives an operational advantage to air traffic control
- Position reports by voice no longer required for identified ADS-B aircraft
- Ability to approve continuous rather than stepped climbs and descents to and from cruising level
- Greater flexibility in allocating appropriate flight levels at the request of pilots. (That is, to climb to optimum flight level, as aircraft weight decreases with fuel burn)
- Airspace which previously had no radar, and only procedural separation services, can now have ATC surveillance service
- Greater ability for ATC to grand clearances to fly requested routes or levels
- Aircraft are easier to locate for search and rescue

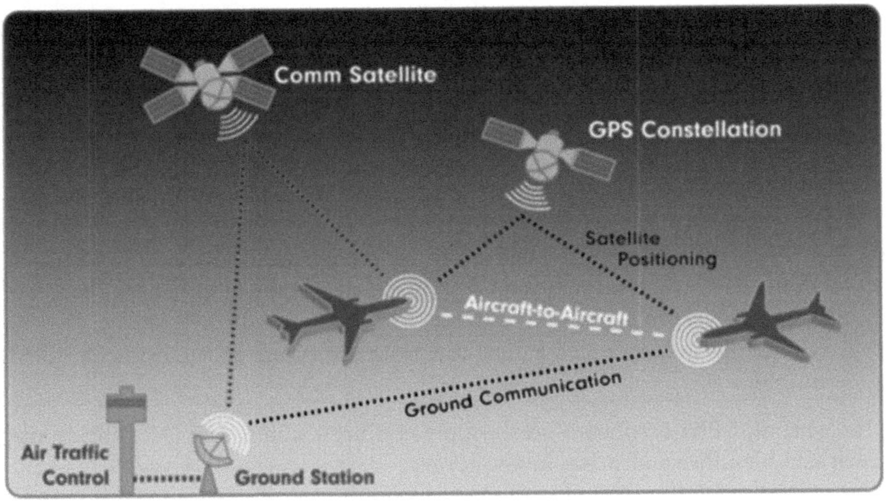

Fig. 10.1 ADS-B diagram

10.2.1.2 Security Threats

During the ADS-B introduction, development and deployment, both academic and industrial communities tried to come up with threat and vulnerability models in order to better understand the security impacts and possible mitigation techniques and solutions. Below is a summary of broad categories of identified and described threats throughout the literature [6].

- jamming, denial of service
- eavesdropping
- spoofing, impersonation
- message injection/replay
- message manipulation

10.2.2 SPKI

The SPKI (Simple Public Key Infrastructure) certificate is the standard proposed for the application of the PKI (Public Key Infrastructure) certificate. The SPKI certificate binds the authority of a user with the public key and provides the access control. The SPKI certificate calls the other word to the "authority certificate". As to the SPKI certificate, a server publishes. A client possesses the SPKI certificate which a server publishes. When a client need to use the resources provided by a server, a client is permitted about the use which is restrictive according to the access authority policy which submits the SPKI certificate to a server and is set up. The

SPKI certificate has a feature as follows compared to the X.509 user certificate [7–10].

- The SPKI certificate indicates an issuer and subject by using the hash-value of the public key or the public key. Therefore, the anonymity about a user can be ensured.
- The SPKI certificate has without the modification of the server Database it easily delegate the other user.
- The SPKI certificate independently can operate about the specific service.
- As to the SPKI certificate, a publication and management are easy. Therefore, maintenance cost is expensive.
- A restrict and multiple delegation can be easily supplied by using the SPKI certificate.

In general, SPKI certificate uses 4-tuple certificate containing a certificate possessor and identifier and it has below format.

<Private Key(I), Pubic Key(S),
Subject(ADS-B Sensor Name),
Validity(20/May/2014)>Signature(I)

10.3 Proposed ADS-B Security Framework

10.3.1 ADS-B Sensor Authentication

Components of ADS-B sensor authentication mechanism proposed in this study include model to perform authentication as Fig. 10.2 and CA (Certificate Authority) to issue SPKI certificate.

Authentication module is a core module to authenticate ADS-B sensor and it is installed to ADS-B sensor and ATC. As Fig. 10.3, the authentication module is composed of a unit in charge of parsing and creation of signed XML, a unit for verification of key and certificate status, and a unit for creation of certificate request message creation. Data flow and data of each module are controlled at execution environment. In order to apply the method proposed in this study, it was assumed that SPKI certificate has been distributed to each ADS-B at least once offline.

10.3.1.1 Signed XML Document Creation and Verification for Authentication

Figure 10.4 illustrates a module to create electronically signed XML document using ADS-B sensor data and certificate. ADS-B sensor data creates a value for verification through hash and creates signature value through encryption to private key of sensor to request for authentication in XML signature process. Series of such process is as below.

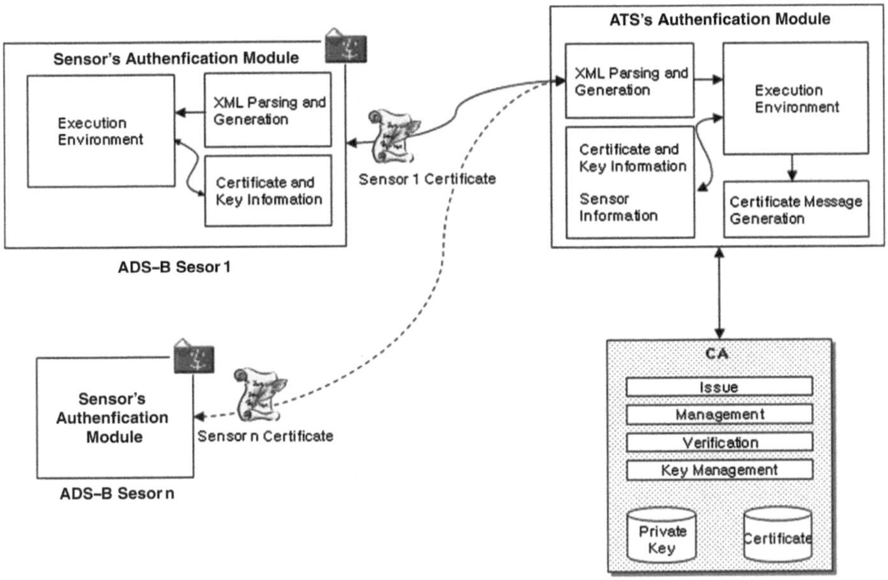

Fig. 10.2 Configuration of security module for proposed ADS-B sensor authentication

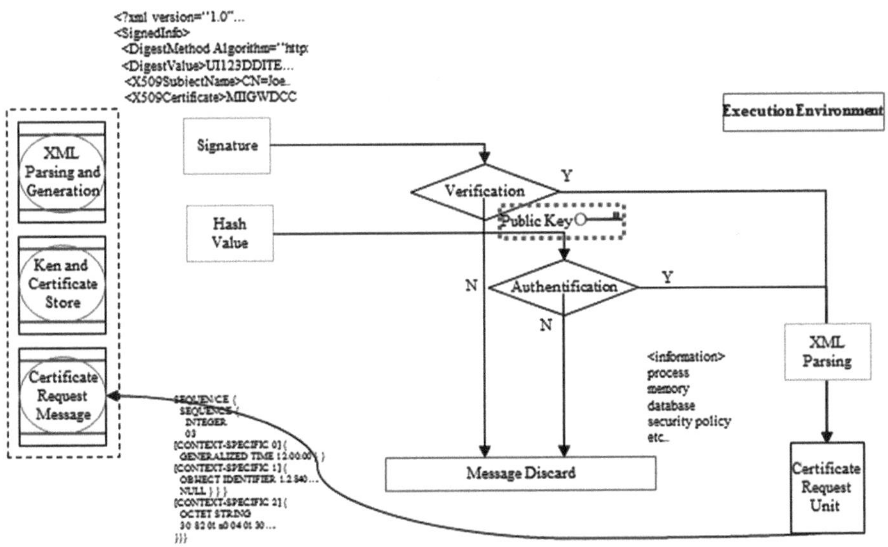

Fig. 10.3 Structure and operation of authentication module

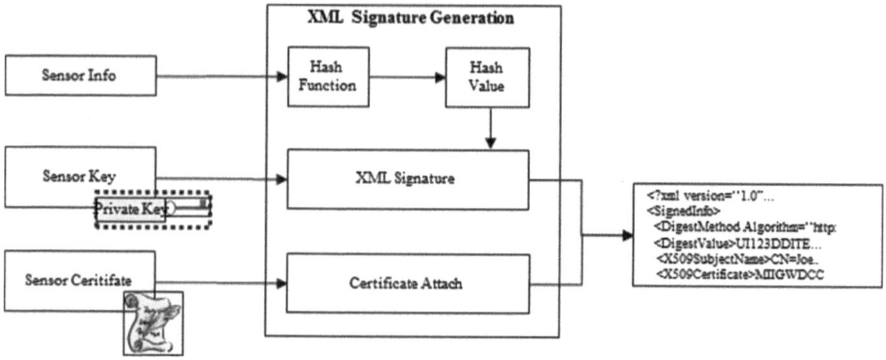

Fig. 10.4 Creation of electronically signed XML document

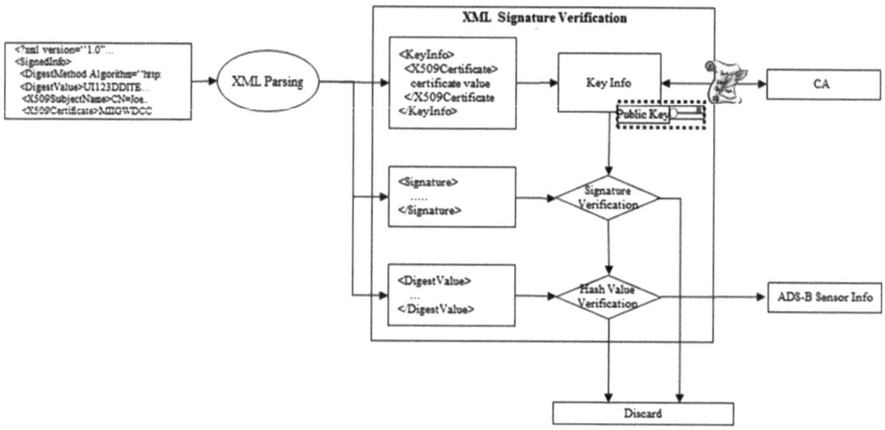

Fig. 10.5 Verification of electronically signed XML document

1. Create document by collecting ADS-B sensor data.
2. Sign with private key initially distributed to ADS-B sensor and add digest value.
3. Public key data for signature verification creates <KeyInfo> which includes SPKI certificate distributed to sensor.
4. Create electronically signed XML document including the value of above process.

Figure 10.5 is a module to examine the validity of electronically signed XML document through verification and extract ADS-B sensor data in request for authentication. Series of such process is as below.

1. Separate electronically signed XML document to each attribute tag using a parser.
2. Examine the validity of certificate contained in <KeyInfo> tag through inquiry to CA and acquire public key value.

3. Verify the signature by decoding the signature value contained in electronically signed XML document.
4. Verify the integrity by comparing digest value contained in electronically signed XML document with the hash value created through signature verification.
5. Acquire ADS-B sensor data in request for authentication.

10.3.1.2 Creation and Transmission of SPKI Certificate

Certificate request message creation for ADS-B sensor is composed including communication and encryption module to request/acquire SPKI certificate for ADS-B to CA and transmit created SPKI certificate to each ADS-B sensor. Description on each component illustrated in Fig. 10.6 is as below.

- ASN.1 Parsing Unit: A data structure creation unit to create data for certification creation in certificate request message format which is international standard regulation
- Encryption Unit: A unit to create encrypted data for security service toward created certificate request data
- Transmission Unit: A transmission/reception unit of encrypted data
- Certificate and Private Key: A unit to acquire created SPKI certificate and private key for ADS-B from CA and transmit them to ADS-B sensor

10.3.2 Security Framework for ADS-B Data Protection

Figure 10.7 illustrates ADS-B authentication and encrypted data processing procedure of proposed security framework. First, the client creates signed XML document with location and identification information of ADS-B sensor using initially

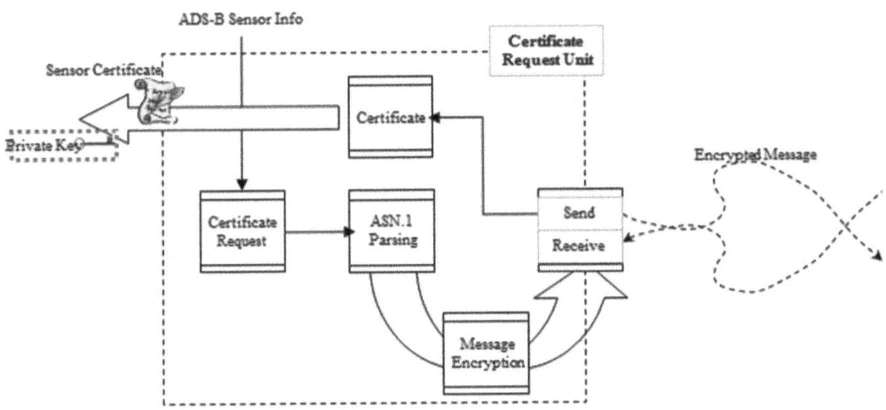

Fig. 10.6 Composition and operation of SPKI certificate requesting unit

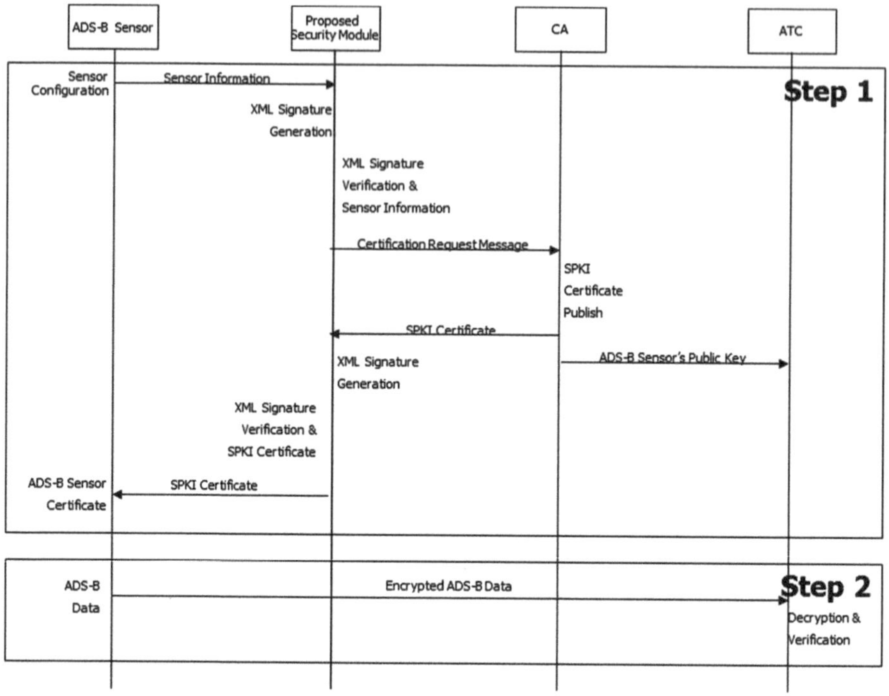

Fig. 10.7 Communication process at proposed security framework

distributed private key of SPKI certificate and transmits it to the server. Then, the server verifies the XML document and requests a certificate for contained ADS-B sensor to CA and CA issues the certificate for ADS-B sensor and transmits SPKI certificate and private key to server through encrypted channel. Also, it transmits public key for ADS-B to ATC. The server creates signed XML document using the private key of server for acquired SPKI certificate and transmits it to the client. Then, the client verifies received XML document and saves SPKI certificate and private key for ADS-B sensor. Data received through ADS-B sensor is transmitted upon completion of encryption using private key of SPKI certificate distributed to each sensor and ATC decodes the received information using public key of ADS-B sensor and use it for air traffic control.

The client manages SPKI certificate until the expiration date of registered SPKI certificate and repeats Step 1 upon the termination of expiration date. The expiration date of SPKI proposed in this study is set very short due to real-time use of ADS-B data and mass creation of ADS-B data.

10.4 Conclusion

Next generation ATC is developing in a direction toward safe flight of more planes in limited airspace. ADS-B is a core system of next generation ATC which accurately provides main flight information such as the location, altitude, speed, and others of plane using GNSS data. However, it is vulnerable to security threats including the alteration in GPS data, ADS-B data falsification, etc. and it was actually proven that ADS-B data can be used to interfere with air traffic control work.

Most effective method to correspond to security threat toward ADS-B is issuing X.509 certificate to all planes and providing certificate based security service but it is difficult in reality. Therefore, a method to protect ADS-B data by authenticating each ADS-B sensor connected to ATC with the use of SPKI certificate and using the authenticated data was proposed in this study. In case of applying the method proposed in this study, effective correspondence to security attack such as ghost plane creation using ADS-B data spoofing, plane location change using falsified ADS-B data, and others that can occur at ground-to-ground would be available.

As research direction of follow-up studies, advantages of proposed method in problem resolution and application results shall be verified after implementing the proposed security framework and applying it to currently run ATC.

References

1. Lee SH et al (2013) The route adherence monitoring method for performance based navigation. In: 2013 summer conference of Korea information and communications society, June 2013
2. Australian Government Civil Aviation Safety Authority (2012) ADS-B, Civil Aviation Safety Authority, Nov 2012
3. McCallie D, Butts J, Mills R (2011) Security analysis of the ADS-B implementation in the next generation air transportation system. Int J Crit Infrastruct Protect 4:78–87
4. RTCA Inc (1994) VHF air-ground communications system improvements alternative study and selection of proposals for future action. In: RTCA/DO-255
5. Australian Government CASA (Civil Aviation Safety Authority), ADS-B(Automatic Dependent Surveillance-Broadcast, CASA, Nov 2012
6. Costin A, Francillon A (2012) Ghost in the Air (traffic): on insecurity of ADS-B protocol and practical attacks on ADS-B devices. In: Black Hat, USA
7. Ellison C et al (1999) SPKI certificate theory. Request for comments, RBC 2693, Sept 1999
8. Hwa Shin J et al (2003) An access control using SPKI certificate in peer-to-peer environment. J Inf Process Syst 10-C(6):793–798
9. Lee YK et al (2002) SPKI/SDSI HTTP secure server to support role-based access control & confidential communication. J Korea Institut Inf Secur Cryptol 12(6):256–261
10. Lee SH et al (2008) The research about the grid delegation service using the SPKI certificate. Inf J 2(6):215–230

Chapter 11
Interactive Drawing Based on Hand Gesture

Tae-Eun Kim

Abstract In this paper, a system that draws drawings on a real-time camera input image is proposed. First, a background subtraction operation is performed on the camera input image. A skin area detection algorithm is applied to the image that has undergone a background subtraction operation. After applying a labeling algorithm to the detected skin region image, a pen is augmented on the position of the hand. In addition, based on the obtained coordinates, a hand gesture-dependent drawing is drawn simultaneously on the camera image, in real-time.

Keywords Hand gesture · Interactive drawing · Augmented reality

11.1 Introduction

Attributable to the advances in computers, a variety of HCI (Human Computer Interface) systems are being developed. This paper proposes a system that draws hand-drawn drawings on a real-time camera image via HCI. There are various drawing methods and there are many systems that provide assistance for creating drawings using the computer; however, such systems require devices such as a mouse or a tablet.

The proposed system facilitates easy drawing using only a computer and a camera. Furthermore, it provides augmentation in real-time to the drawing that the user is drawing right on the camera input image. In previous systems, the background is fixed when a drawing is drawn. The system proposed in this paper draws drawings right on the image inputted from the camera; therefore, it offers the user a greater sense that he/she is drawing in the same space that the user is occupying.

T.-E. Kim (✉)
Department of Multimedia, Namseoul University, Cheonan-City, Choongnam 330-707, South Korea
e-mail: tekim5@empas.com; tekim@nsu.ac.kr

J.J. (Jong Hyuk) Park et al. (eds.), *Ubiquitous Computing Application and Wireless Sensor*, Lecture Notes in Electrical Engineering 331,
DOI 10.1007/978-94-017-9618-7_11

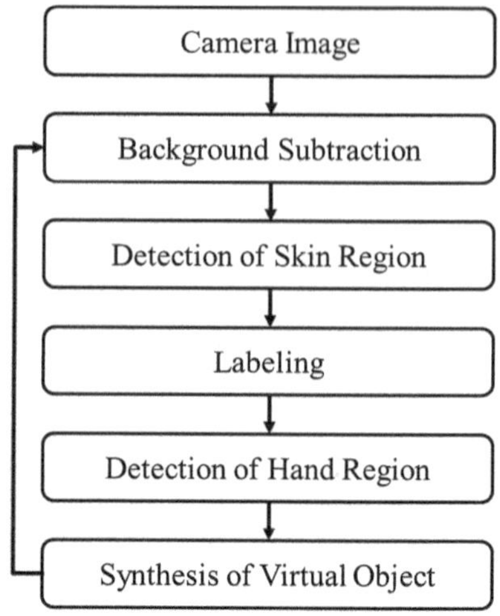

Fig. 11.1 Configuration (flowchart) of proposed system

In order to enhance the sense of realism when drawing on a real-time screen in the proposed system, a pen is augmented on top of the hand. In previous studies, markers were used to augment virtual objects at specific locations [1]. However, in recent years, studies in which markers are not used in the augmented reality have been pursued. By using a specific color wristband so as to avoid using a specific marker, Choi [2] conducted a study on hand tracking and augmentation. In this present paper, a pen is augmented without the use of a marker. Figure 11.1 shows the proposed system's schematic diagram. In this paper, a method in which the human hand is searched for and drawings are made on the screen in accordance the hand movement is proposed. Configuration of the system proposed in this paper is comprised of, in order: background subtraction, skin region detection, labeling, pen augmentation and figure synthesis.

11.2 Main Discussion

11.2.1 Background Subtraction

As images are inputted from the camera, first, background subtraction is performed. For background subtraction, a pre-treatment process transforms the camera input image and the reference image into a grey image. To accomplish background subtraction, first, the first frame of the camera input image is captured and background subtraction is applied, as this image becomes the reference image. There are two stages in the background subtraction operation. First, there is a step where the camera input image is subtracted from the reference image. In addition, there is a step in which the reference image is subtracted from the camera input image [3]. The subtracted image from the two process steps is used to obtain the final

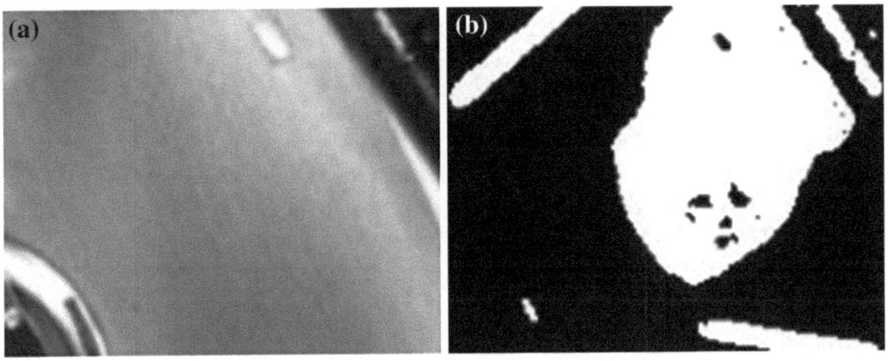

Fig. 11.2 Input image and difference image

difference image via OR logic operation. Equation (11.1) represents the subtraction operation.

$$
\begin{aligned}
B_1(x,y) &= \begin{cases} 0, & if\,(C(x,y) - R(x,y) < threshold \\ 1, & otherwise \end{cases} \\
B_2(x,y) &= \begin{cases} 0, & if\,(R(x,y) - C(x,y) < threshold \\ 1, & otherwise \end{cases} \\
B_3(x,y) &= B_1(x,y) \cup B_2(x,y)
\end{aligned} \tag{11.1}
$$

B1 and B2 represent the subtraction image of the two steps. The images of C and R represent the camera input image and reference image, respectively. B3 represents the operations of B1 and B2.

Figure 11.2a is a reference image obtained from the camera input image to obtain the difference image, and Fig. 11.2b shows the difference image.

11.2.2 Detection of Skin Region

To conduct detection of the skin region, a pretreatment process of background subtraction is carried out. In order to extract the hand from the camera input image that has undergone background subtraction, detection of the skin region is carried out. In performing extraction of the skin region, in the B3 image of Eq. (1), for parts whose pixel values that correspond to each pixel position are non-zero RGB color images inputted from the camera are shown. Detection of the skin region is carried out from the image inputted in such a way. For performing detection of the skin region, Eq. (2) is used to convert the RGB color image into YCbCr color image [4].

$$Y = 0.299R + 0.587G + 0.114B$$
$$C_b = B - Y \tag{11.2}$$
$$C_r = R - Y$$

In the image converted into YCbCr color image with the values of Cb and Cr as reference, detection of the skin region is carried out. Equation (3) is the reference region of Cb and Cr [5].

$$77 \leq C_b \leq 127$$
$$133 \leq C_r \leq 173 \tag{11.3}$$

Figure 11.3 shows the extracted skin region using Eq. (3).

11.2.3 Labeling

Looking at the skin region extraction image in Fig. 11.3b, it can be seen that not just the person's skin but also other objects are detected as well.

This is due to the fact that a skin color detection algorithm was used in the skin region detection color image. As such, in order to eliminate other objects that have colors similar to the skin color, a labeling and post-treatment processes are performed.

In this paper, for real-time processing One-Pass run length labeling algorithm was used as a labeling algorithm. In order to carry out One-pass run length labeling, a conversion process for the final image that has undergone skin region detection is needed. In the image for which labeling was performed, the determined regions are sorted based on the size of each region and the largest size region is determined as the hand. One-Pass run length labeling algorithm has faster speed of operation than the existing Grass-fire labeling method. Whereas Grass-fire labeling method detects

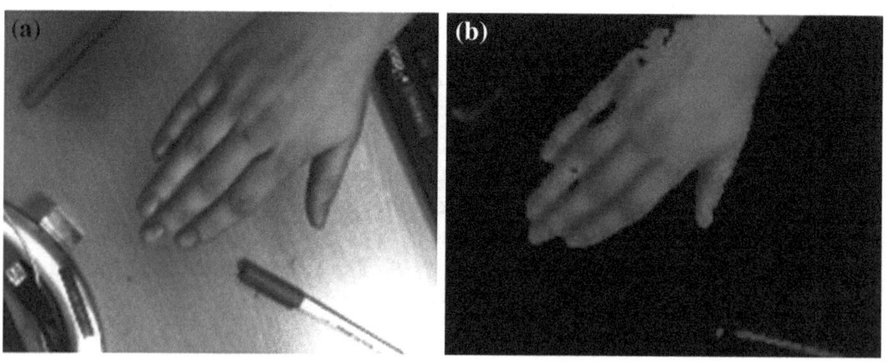

Fig. 11.3 Real-time camera image and skin region extraction image

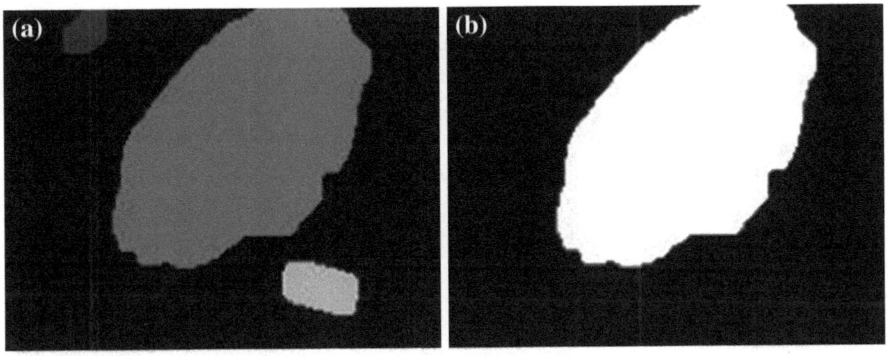

Fig. 11.4 Labeling image and image after size comparison

from the starting pixel of a region to the finishing pixel in one time, One-Pass run length labeling algorithm uses a method of detecting one line at a time in an image and, after one line is detected, while detecting the next line it assigns a sequential number to the region in each line that is deemed as an object. For regions determined in such a way, by comparing with the determined regions of upper lines, when there are contiguous regions this algorithm makes a final determination that the regions are the same region [6]. Figure 11.4a is the image after labeling the skin region detection image, and Fig. 11.4b is the image that underwent only hand region extraction through size comparison after labeling.

11.2.4 Synthesis of Hand Tracking and Virtual Objects

With the detected hand image through size comparison after labeling as the reference, for the synthesis of drawings the center of gravity of the hand (c) and the endpoint of the hand (a) are first obtained. When the center of gravity of the hand (c) and the endpoint of the hand (a) are obtained, a point between the two points is detected by using Eq. (4).

$$\begin{aligned} Mp \cdot x &= (CP \cdot x - FP \cdot x)/2 + FP \cdot x \\ Mp \cdot y &= (CP \cdot y - FP \cdot y)/2 + FP \cdot y \end{aligned} \tag{11.4}$$

The point between the center of gravity (CP) and the endpoint of the hand (FP) is used as the reference coordinate for representing the drawing through the hand on the real-time camera input image and in addition it becomes the reference coordinate for pen augmentation. The reason for augmenting the pen with the point between the center of gravity (CP) and the endpoint of the hand (FP) and representing the drawing is due to the position of the pen tip when a person is holding the pen. Figure 11.5 shows an image representing all the points.

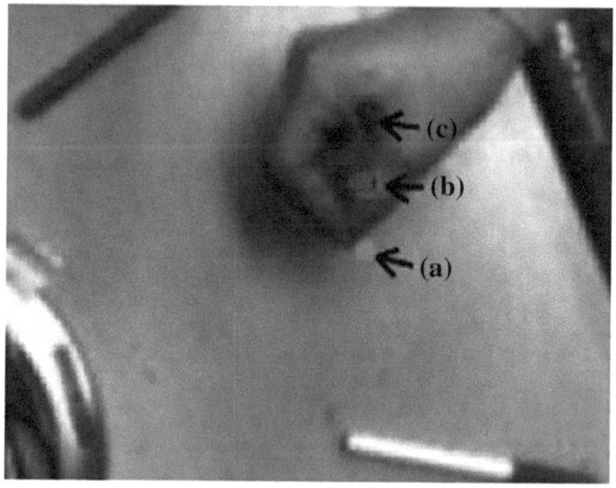

Fig. 11.5 Representation image of the points: (*a*) Endpoint of the hand, (*b*) Point between endpoint of the hand and center of gravity, (*c*) Center of gravity

The superscript numeral used to refer to a footnote appears in the text either directly after the word to be discussed or—in relation to a phrase or a sentence—following the punctuation mark (comma, semicolon, or period). Footnotes should appear at the bottom of the normal text area, with a line of about 5 cm set immediately above them.

11.3 Experiment Results

One 6 mm lens 1394b camera was used in this study. The first frame acquired from the real-time camera image is set as the reference image. The reference image is used in background subtraction in two steps; which entails, first, subtraction of the camera input image from the reference image, and a method of subtracting the reference image from the camera input image in a different step, and by using OR logical operation, the background subtraction image is obtained. Detection of the skin region is carried out on the parts whose pixel values of the acquired background subtraction image are non-zero. In order to solve the issue of detecting as regions the colors that are similar to the skin color during detection of the skin region, after labeling the hand region is extracted through comparison of region sizes. From the extracted hand region the center of gravity and endpoint of the hand are detected, and afterwards the point between the two points is detected. With the point detected in this way as the reference the pen is augmented and a drawing that follows the hand movement is drawn on the real-time camera input image.

Figure 11.6 shows images of implementation of the system. In Fig. 11.6, it can be seen that a penis drawn over the pen, and the pen is augmented with the center of gravity and the endpoint of the hand as the reference. The highlighted portions (black circles) in Fig. 11.6a are the problem areas caused by a flickering phenomenon of the reference coordinate for making a drawing. of the area indicated

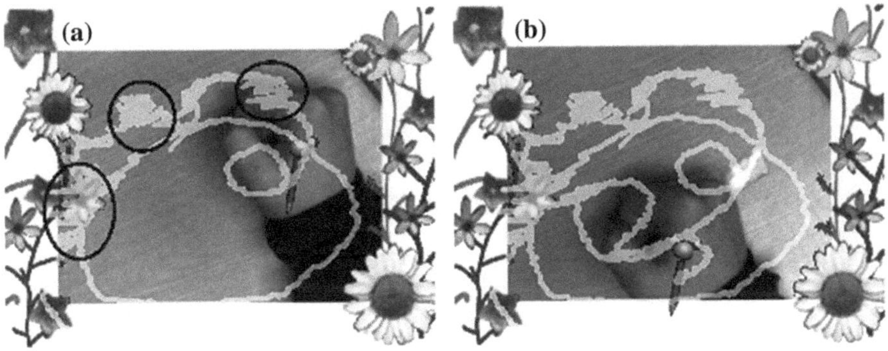

Fig. 11.6 Augmentation of pen and proposed sketch system

reference coordinates for drawing the picture blur due to duplicate the problem was being drawn . Figure 11.6b shows a human face drawn on the desktop through the system.

11.4 Conclusion and Future Research Directions

In this paper, the hand was estimated from the camera input image and, by tracking the hand position, three coordinates of the hand were extracted [6].With the extracted coordinate values as the reference, a drawing, which follows pen augmentation and hand gestures, was drawn. Although a slight flickering phenomenon was seen, the results of overall, stable augmented reality were obtained through an experiment. The slight flickering phenomenon seen in the existing system will be compensated for in future studies. In addition, a research that represents drawings drawn on the screen in a more realistic 3D will be pursued.

Acknowledgments Funding of this paper was provided by Namseoul University.

References

1. http://www.hitl. washington.edu/artoolkit/
2. Choi JY, Park HH, Park JI (2008) Interface for augmented reality using efficient hand gesture recognition. In: FCV 2008, pp. 439–444
3. Piccardi M (2004) Background subtraction techniques: a review. Int Conf Syst Man Cybern 4:3099–3104
4. Vezhnevets V, Sazonov V, Andreeva A (2003) A survey on pixel-based skin color detection techniques. In: Proceeding. Graphicon-2003, Moscow, Russia pp. 85–92
5. Gasparini F, Schettini R (2006) Skin segmentation using multiple thresholding. In: Proceeding. of SPIE Internet Imaging VII, 6061:60610F.1–60610F.8
6. Anton H (1987) Elementary liner algebra. Wiley, New York

Chapter 12
Nullifying Malicious Users for Cooperative Spectrum Sensing in Cognitive Radio Networks Using Outlier Detection Methods

Prakash Prasain and Dong-You Choi

Abstract A number of cooperative spectrum sensing techniques have been purposed in cognitive radio networks. However, collaboration between multiple cognitive radio (CR) users also raises a number of security issues. It has been shown that the cooperative gain can be severely affected by malfunctioning or malicious CR users in cooperative sensing. One of them is spectrum sensing data falsification (SSDF) attack, where malicious users transmit false information instead of real detection results and thereby affecting the final decision. In this paper, we study the detection and suppressing the malicious users using different outlier detection methods based on Grubb's test, Boxplot method and Dixon's test. We have compared their performance through simulation and receiver operating characteristics (ROC) curve shows that Boxplot method outperforms both Grubb's and Dixon's test for the case where multiple malicious users are present.

Keywords Cognitive radio · Energy detection · Cooperative sensing · Security

12.1 Introduction

Currently, frequency spectrum is statically allocated to licensed users, i.e., primary users (PUs) only, in a traditional wireless communication system. Since licensed users may not always occupy the allocated radio spectrum, this static spectrum allocation results in spectrum underutilization. This was confirmed in a report published in 2002 from the FCC (Federal Communications Commission) where it

P. Prasain · D.-Y. Choi (✉)
Department of Information and Communication Engineering, Chosun University, Gwangju 501-759, Republic of Korea
e-mail: dychoi@chosun.ac.kr

P. Prasain
e-mail: prakashprasain@gmail.com

J.J. (Jong Hyuk) Park et al. (eds.), *Ubiquitous Computing Application and Wireless Sensor*, Lecture Notes in Electrical Engineering 331,
DOI 10.1007/978-94-017-9618-7_12

was shown that even in a crowded area; more than half of the radio spectrum is not occupied at any given time [1]. Thus, new spectrum allocation policies were introduced to allow unlicensed users, i.e., secondary users (SUs) to access radio spectrum when it is not occupied by PUs. However, when PU comes back into operation, the SU should vacate the spectrum instantly to avoid interference with the primary one.

Spectrum sensing in cognitive radio (CR) has the great role in order to utilize the idle spectrum opportunistically, since it is responsible for making available dynamic spectrum access efficiently. In this literature, cooperation among multiple cognitive radio users has been proposed for the betterment of detection reliability [2]. It involves many SUs and they can share their sensing information for making combined decision more accurate than individual decisions. They send their local sensing results to fusion center (FC) through a control channel. Then, the FC combines the received local sensing information and determines the presence of PU. Even though collaboration among them improves the spectrum sensing performance, some falsely reported local sensing results by malicious users degrade the performance rigorously [3]. This is what; we call spectrum sensing data falsification (SSDF) attack. Hence, existing cooperative spectrum sensing algorithm has to be modified so that it can identify those malicious users and nullify their harmful effects in final decision making to ensure the reliability of the sensing decisions.

In [4], authors have compared several outlier detection methods for low SNR scenario. In [5], the weighted SPRT with a reputation-based mechanism is proposed as the robust cooperative sensing scheme to address the data falsification problem. In [6], simple outlier detection is proposed for the pre-filtering of the extreme values in sensing data. The trust factor that measures the CR user's reliability is then evaluated as the weights in calculating the mean value of received sensing data. In this paper, we have studied different outlier detection techniques based on Grubb's test, Boxplot method and Dixon's test [7] to detect the presence of malicious users and compared their performances.

The rest of the paper is organized as follows. In Sect. 12.2, we discuss the system model for cooperative spectrum sensing using energy detection technique. Similarly, we state three outlier detection techniques based on Grubb's test, Boxplot method and Dixon's test in Sect. 12.3. The performance of each outlier detection method is analyzed through simulation results in Sect. 12.4. Finally, we draw some conclusions of this study in Chap. 5.

12.2 System Model

We consider a network composed of N SUs and a fusion center. We assume that each SU consists of an energy detector and performs sensing independently. Then, the local sensing data are sent to the FC which can fuse all available information to decide the absence or presence of PU. The essence of spectrum sensing for PU detection is a binary hypothesis-testing problem:

H0: primary user is absent;
H1: primary user is present;

The sensing method is to decide between the two hypotheses,

$$y_i(n) = \begin{cases} u_i(n), & H0 \\ h_i(n) \cdot s(n) + u_i(n), & H1 \end{cases} \tag{12.1}$$

where, $y_i(n)$ is the received signal at the *i*th CR, $s(n)$ is the signal from PU, each sample is assumed to be an independent identically distributed (i.i.d.) random process with zero mean and variance $\mathrm{E}\left[|s(n)|^2\right] = \sigma_s^2$. Similarly, $u_i(n)$ is the additive white Gaussian noise (AWGN) with zero mean and variance $\mathrm{E}\left[|u_i(n)|^2\right] = \sigma_u^2$, and $h_i(n)$ denotes the channel gain of the sensing channel between PU and the *i*th CR. It has the same variance $\mathrm{E}\left[|h_i(n)|^2\right] = \sigma_h^2$. The area of coverage of the cognitive radio system is assumed to be small enough so that the variations in path loss can be neglected. The average received SNR at each SU is given as $\gamma = \sigma_s^2\sigma_h^2/\sigma_n^2$.

The energy detector output Y_i at the *i*th SU is given by

$$Y_i = \left(\frac{1}{\mathrm{M}}\right)\sum_{\mathrm{n}=1}^{\mathrm{M}} |y_i(n)|^2 \text{ for i} = 1, 2, \ldots, \mathrm{N}. \tag{12.2}$$

where, M is the number of signal samples that are collected at each SU during the sensing period, which is the product of the sensing time τ and the sampling frequency f_s. We assume perfect channel conditions for the control channels between SUs and FC.

We consider that each SU sends their received energy values through an error free control channels to the FC. Then, FC runs algorithms for detecting malicious users. For this, we follow average combination scheme due to simplicity. The mean received energy in dB by all SUs is calculated and FC compares it with a fixed threshold. Then the decision *D* made by FC is given by

$$D = \begin{cases} H0, & \mathit{if}\ Y < \lambda_{FC} \\ H1, & \mathit{if}\ Y \geq \lambda_{FC} \end{cases} \tag{12.3}$$

where, λ_{FC} is threshold at FC and *Y* is the mean of received energies which is given by

$$Y = \left(\frac{1}{N}\right)\sum_{i=1}^{N} Y_i \tag{12.4}$$

12.3 Outlier Detection Techniques

There are several well studied methods to determine outliers. We select only Grubb's test, Boxplot method and Dixon's test [7].

12.3.1 Grubb's Test

Grubb's test is one of the most commonly used for the detection of a single outlier in univariate data. This test for outliers compares the deviation of the suspect value from the sample mean with the standard deviation of the sample. The suspect value is the value that is furthest away from the mean. In order to use Grubb's test for an outlier, the statistic G is calculated:

$$G = \frac{|Suspect\ value - \bar{x}|}{s} \tag{12.5}$$

where, $\bar{x}$ and s are mean and standard deviation respectively. They are calculated with the suspicious value included. If the calculated value of G exceeds the critical value, the suspicious value is taken as an outlier, and it is rejected. A table for critical values at specified significance level for different sample size has been provided in [7]. This test is used to detect single outlier. To detect more than one outlier, we have applied this test iteratively so that it can test one value at a time until and unless the sample data set of received energies is free from the extreme values produced by malicious users.

12.3.2 Boxplot Method

In this method, different energy values obtained from different SUs are arranged in ascending order from smallest to largest $Y_1 \leq Y_2 \leq \ldots Y_N$. Then, lower and upper bounds are calculated as follows:

$$Q_{lower} = Q_1 - 1.5Q_{intqtr} \tag{12.6}$$

$$Q_{upper} = Q_3 - 1.5Q_{intqtr} \tag{12.7}$$

where, Q_{lower} and Q_{upper} are lower and upper threshold respectively. Q_1 is first quartile, Q_3 is third quartile and Q_{intqtr} is interquartile range. The values of obtained energies below Q_{lower} and above Q_{upper} are considered as outliers.

12.3.3 Dixon's Test

It is based on the ratios of differences between the observations, and the calculation of the ratio depends on the number of observations. As in the previous two techniques, it avoids the calculation of mean and standard deviation. This test is also for detecting a single outlier. In this method, outlier factors for each SU are calculated based on their local sensing results to detect the presence of malicious users. The received energy values are arranged in ascending order $Y_1 \leq Y_2 \leq \ldots Y_N$, and outlier factor f_i for ith SU is calculated as:

For $3 \leq N \leq 7$,

$$f_i = \begin{cases} \frac{Y_2 - Y_1}{Y_N - Y_1}, & \text{if smallest value is suspected} \\ \frac{Y_N - Y_{N-1}}{Y_N - Y_1}, & \text{if largest value is suspected} \end{cases} \tag{12.8}$$

For $8 \leq N \leq 10$,

$$f_i = \begin{cases} \frac{Y_2 - Y_1}{Y_{N-1} - Y_1}, & \text{if smallest value is suspected} \\ \frac{Y_N - Y_{N-1}}{Y_N - Y_2}, & \text{if largest value is suspected} \end{cases} \tag{12.9}$$

For $11 \leq N \leq 13$,

$$f_i = \begin{cases} \frac{Y_3 - Y_1}{Y_{N-1} - Y_1}, & \text{if smallest value is suspected} \\ \frac{Y_N - Y_{N-2}}{Y_N - Y_2}, & \text{if largest value is suspected} \end{cases} \tag{12.10}$$

For $14 \leq N \leq 25$,

$$f_i = \begin{cases} \frac{Y_3 - Y_1}{Y_{N-2} - Y_1}, & \text{if smallest value is suspected} \\ \frac{Y_N - Y_{N-2}}{Y_N - Y_3}, & \text{if largest value is suspected} \end{cases} \tag{12.11}$$

where, N is the number of statistical data i.e., number of SUs in our case. The calculated outlier factor f_i is compared with a critical value Q, which depends on N and the significance level. The table for the critical values for different values of N for three significance levels can be found in [8]. If the outlier factor is less than the critical value Q, this energy value is assumed to be normal, otherwise if it exceeds the critical value Q, it is assumed to be high energy reported by corresponding SU. Outlier factor for the smallest suspect value and largest value are calculated individually.

12.4 Simulation Results

We carried out simulations to test and compare all the outlier detection methods discussed above. In the simulations, we assume Additive White Gaussian Noise (AWGN) channel and primary user signal is BPSK modulated. In addition, we have taken $N = 20$ cooperating SUs. The distance between any two SUs is small in comparison with the distance from any SU to PU, and the received PU signal at each SU experiences almost identical path loss. Moreover, we assume that the SUs use the same threshold λ, so that $\lambda_1 = \lambda_2 = \ldots \lambda_N = \lambda$. It means that probability of false aram $P_{f,i}$ is independent of i and we can denote it as P_f. In case of AWGN channel, the probability of detection $P_{d,i}$ is independent of i, we denote this as P_d.

A SU might be malicious due to device malfunctioning or due to selfish reasons. We consider the different kind of malicious users. One is "Always Yes" user, and another is "Always No" user. An "Always Yes" node gives a value above the threshold which means it declares that a PU is present all the time. Similarly, an "Always No" node gives a value below the threshold which means PU is absent all the time. An "Always Yes" user increases the probability of false alarm P_f and an "Always No" user decreases the probability of detection P_d. Additionally there might be other malicious users that provide extreme false value once in a while and produce correct values in rest of the time. In simulation, we have assumed that it reports energy 5 dB lower than the normal SU for those providing "Always No" decision. Similarly, "Always Yes" user reports energy 5 dB higher than the normal SU. The significance level is taken 0.05.

In Fig. 12.1, we consider a cooperative spectrum sensing scenario in which 10 % of SUs are 'Always No' malicious users. We provide the ROC curve of cooperative spectrum sensing, which shows the degradation in performance when we add 'Always No' malicious users. We can see that 'Always No' type malicious users have decreased the probability of detection, i.e., increased the probability of miss detection. It also shows the performance of cooperative spectrum sensing after nullifying the malicious effects by applying Grubb's test, Boxplot method and Dixon's test. Initially, we assume one 'Always No' malicious user in which case; all the three tests show almost same performance, i.e., they successfully removed the effect of one malicious user. Later, when we introduced multiple 'Always No' users, we found some differences in their performances. We can see that probability of detection after applying Boxplot method is closest among three to that of without any malicious user. We have applied Grubb's test iteratively two detect the multiple malicious users since this test is supposed to detect only one malicious user at a time. On the other hand, the performance of Dixon's test is better than the Grubb's test and worse than Boxplot method in case of multiple malicious users introduced. Dixon's test cannot be easily implemented for detecting multiple malicious users. If the first three users observed almost same energy, the numerator of Eq. (12.11) for the case of lowest suspected, becomes so small and the outlier factor will be smaller than the critical value. This results in not detecting the malicious users present. This

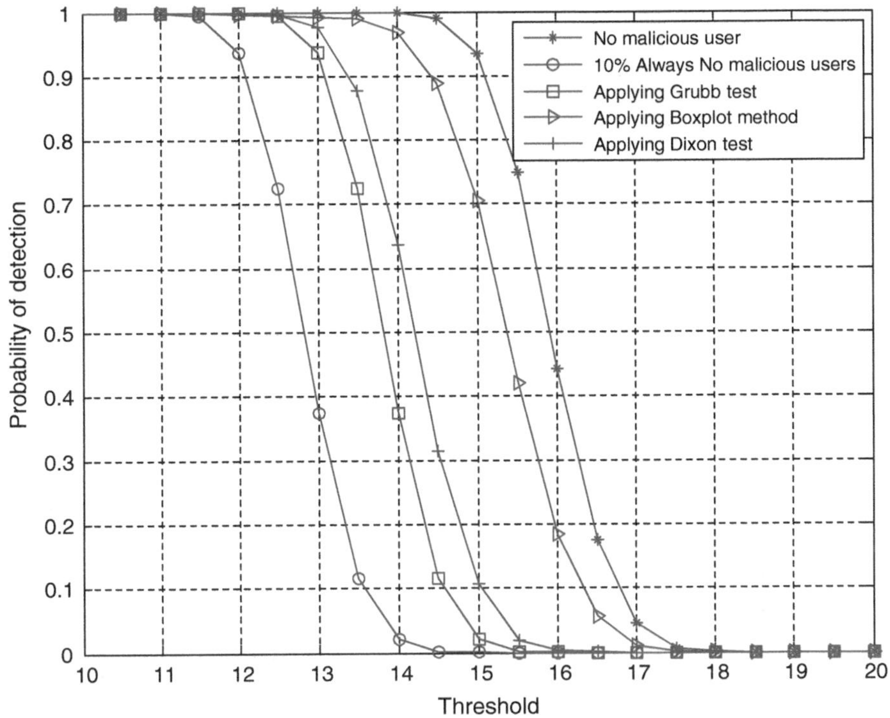

Fig. 12.1 Probability of detection versus threshold for average normal SNR = 5 dB for always no malicious user

is because of method can detect multiple malicious users. This will degrade the performance of cooperative spectrum sensing.

Similarly, we plot the ROC curve as shown in Fig. 12.2. It shows the performance of cooperative spectrum sensing after adding 10 % of 'Always Yes' malicious users. 'Always Yes' type malicious users degrade the performance by increasing the probability of false alarm of the system. Similar to the previous case of adding 'Always No' malicious users, all the three outlier detection methods did not success to nullify the effects of malicious users completely. However, out of these three outlier detection methods, the Boxplot method performs better than that of Grubb's test and Dixon's test, and it succeeded to bring the probability of detection and probability of false alarm of the system closer to that of cooperative spectrum sensing system without any malicious user.

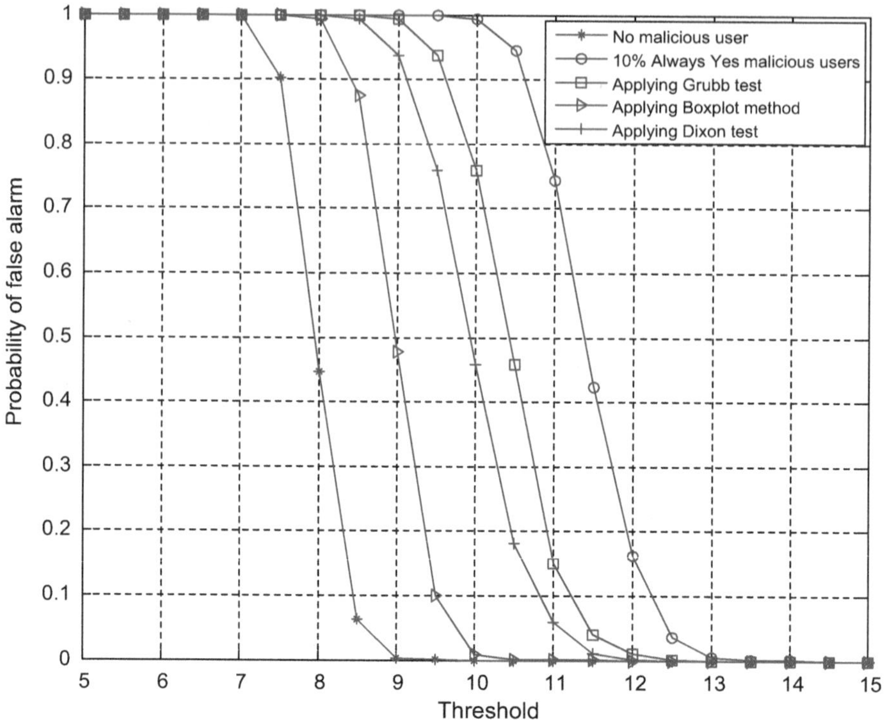

Fig. 12.2 Probability of detection versus threshold for average normal SNR = 5 dB for always no malicious user

12.5 Conclusion and Future Work

In this paper, we first studied the energy detection technique for spectrum sensing in cognitive radio networks. Later, it is applied in cooperative spectrum sensing and concentrated on the detection and nullifying the effects of falsely reported sensing data by malicious users in final decision making. We first studied techniques that detect the outliers in a statistical data and compared their performance applying in cooperative spectrum sensing. Even though, the first two techniques, i.e., Grubb's and Dixon's are supposed to detect one malicious user at a time, we iterated it to remove all possible malicious users. Through Monte Carlo simulations, we analyzed their performances and observed that Boxplot method technique performed better than the other two in the presence of multiple malicious users. However, none of them could able to nullify the negative effect of falsely reported sensing data completely. Hence, it is essential to develop more reliable algorithms to identify and suppress the harmful effect of multiple outliers efficiently to ensure the reliability of cooperative spectrum sensing decision.

Acknowledgments This research was supported by the MSIP (Ministry of Science, ICT and Future Planning), Korea, under the C-ITRC (Convergence Information Technology Research Center) support program (NIPA-2014-H0401-14-1009) supervised by the NIPA (National IT Industry Promotion Agency).

References

1. FCC (2002) Spectrum policy task force report ET docket, 02–155
2. Visotsky E, Kuffner S, Peterson R (2005) On collaborative detection of tv transmissions in support of dynamic spectrum sharing. In: First IEEE symposium on new front in dynamic spectrum access networks, pp 338–345
3. Mishra S, Sahai A, Brodersen R (2006) Cooperative sensing among cognitive radios. In: IEEE ICC, pp 1658–1663
4. Le HV (2010) Outlier detection methods of low SNR nodes for cooperative spectrum sensing. In: ISWCS, pp 966–970
5. Chen R, Park JM, Bian K (2008) Robust distributed spectrum sensing in cognitive radio networks. In: IEEE conference on computer communications
6. Kaligineedi P, Khabbazian M, Bhargava VK (2008) Secure cooperative sensing techniques for cognitive radio systems. In: IEEE international conference on communications, pp 3406–3410
7. Barnett V, Lewis T (1994) Outliers in statistical data, 3rd edn. Wiley, Chichester
8. Grubbs F (1969) Procedures for detecting outlying observations in samples. Technometrics 11 (1):1–21

Chapter 13
A Study on Electronic-Money Technology Using Near Field Communication

Min-Soo Jung

Abstract Recently, NFC (Near Field Communication) and its related technologies are being loaded into smartphones. Consequently, payment and data exchange are easily made through NFC. However, smartphone can be permitted an attack such as a user impersonation by exposure to information in the communication process. Therefore, the communication between mobile devices in NFC environments must be made safely with leaving any information at the other party. Moreover, it should be light-weight for wireless communication. In this paper, we propose an authentication method and a scheme of key agreement that can reduce the weight by using a hash function and XOR operation algorithms for mobile payment systems. We confirm that the proposal method leaves no information at the other party.

Keywords NFC · Key agreement · Mobile payment

13.1 Introduction

Nowadays, due to the prevalence of smartphones, many people can share information and can process finance payments anytime and anywhere [1–3]; Even though these information sharing can be performed the communication with a server in the network, it also can be shared the information through the communication between devices. Especially in recent years, due to prevalence of NFC equipped smartphones, a variety of information has become to obtain more easily [4, 5]. NFC-SEC as NFC-related security standards have been published 2010 [6, 7]. The standards present a way to perform key-agreement process by using an elliptic curve algorithm [8]. This way is also a public key-based encryption algorithm. NFC supports functions like as 'Tag to Mobile', 'Mobile to Mobile'. Due to these functions, there can be compatible with existing RFID, and also can be traded or transferred freely large amounts of data and content [9, 10]. NFC provides a new environment that it

M.-S. Jung (✉)
Department of Computer Engineering, Kyungnam University, Changwon, Korea
e-mail: msjung@kyungnam.ac.kr

J.J. (Jong Hyuk) Park et al. (eds.), *Ubiquitous Computing Application and Wireless Sensor*, Lecture Notes in Electrical Engineering 331,
DOI 10.1007/978-94-017-9618-7_13

can be produced and traded the data of finance payments by the user actively [11]. Especially, the electronic wallet will be highly activated such as 'mobile wallet'. Accordingly, it must provide a safe communication in these environments (Kim et al. 2013; [4]). However, the important data can be easily exposed to a malicious user [9]. In earlier papers, to solve above problem, they presented method to perform the authentication by using a public key based encryption. And they also proposed way to perform the key agreement by using a secret key-based encryption scheme. However, their ways are not acceptable in wireless environment, because these schemes require complex and time-consuming processing. And also, the other side of the user information can be remained on the reader or other mobile devices in the process of performing the communication. Thus, it is possible that user impersonation attack using this information. In this paper, in order to solve these problems, our method performs a secure user authentication and key agreement using XOR operations and hash function algorithms. And especially, it is robust against user impersonation attack, because it performs the authentication with leaving no information on the other's mobile or reader. We are described characteristics of NFC, structural environment and finally, research the about existing studies in Sect. 13.2. Section 13.3 gives description of our proposed method. And the safety and efficiency analysis are given in Sect. 13.4. And conclusion is given in Sect. 13.5.

13.2 Related Works

NFC performs the communication in the 13.56 MHz frequency band for compatibility with the RFID [9]. NFC communications are shown in Fig. 13.1. The NFC-equipped mobile devices can perform the communication with trusted third party or the bank, the AuC, market server and web server. This is also same with reader.

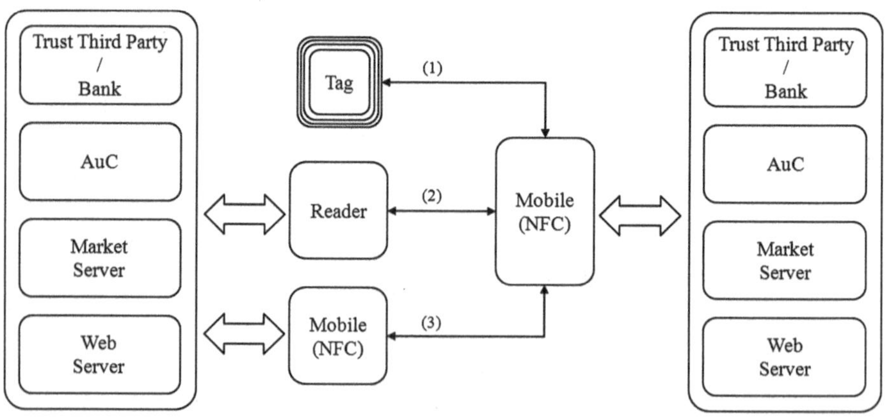

Fig. 13.1 NFC environments

Especially, the core feature of NFC is to be able to communicate on device to device, in other word communicate between mobile devices [5]. The existing Tag has drawbacks that it can not be operated high-performance encryption algorithms. Hence, it is possible to resolve above the security vulnerabilities. Therefore NFC has the advantage that it is possible to perform the communication without exposing the important secrets of the personal information and financial transaction data. However, the completed standards do not exist yet. There are some authentications by using a shared secret key in the wireless band have been proposed, however, their works have a disadvantage that user secret information is exposed at the other side of the node.

There are many studies for support secure payments including such things as *SET* [6], *iKP* (i-Key-Protocol [i-1,2,3], [1]), *Kungpisdan* [2], NFC-SEC [12], Hasoo [13] and *Sekhar* [7] method.

These methods are performed the authentication and encrypted communication by using a public key-based way to ensure the safety. *SET* is a typical public key-based financial transaction [6]. As it is the communication method consisting of a pair of 'Request' and 'Response', it proceeds to the transaction by obtaining a certification. This method is mainly used on payment with a credit card but, the transaction process is complex. And also, *iKP* method is an another public key-based method [1]. It performs to the communication of the encryption and decryption between the digital signature and the participants by using a public key pair. This way also has complex operations as a public key-based method on the transaction process. The ways of SET and *iKP* is useful in wired payment environments. However, some delays can occur because complex payment methods. *Kungpisdan* is a method that reduces usage count of computation algorithm [2]. It has improved the efficiency by using a symmetric key and hash-function. However, the number of using of the symmetric key in customer and merchant is high relatively. Therefore, ways to relieve this problem are needed. In the case of a way of *Sekhar*, it is a symmetric key-based financial payment system [7]. And the number of using of symmetric key operation algorithm is reduced 1 step than *Kungpisdan* scheme. The number of the Keyed-hash functions is reduced 1 step, however, these methods in the NFC environment are too heavy for mobile communication.

13.3 Our Proposed Method

For NFC environments, we propose a high-speed processing of proper authentication and key agreement. Also we present a way to perform secure communications with leaving any information at the other Mobile or Reader. Our proposed method can be summarized as in Fig. 13.2. First, all mobile devices must go through the registration process once. And authentication is performed by using the issued secret information in the registration phase. A detailed description is as follows. Some terminologies in this paper are shown in Table 13.1.

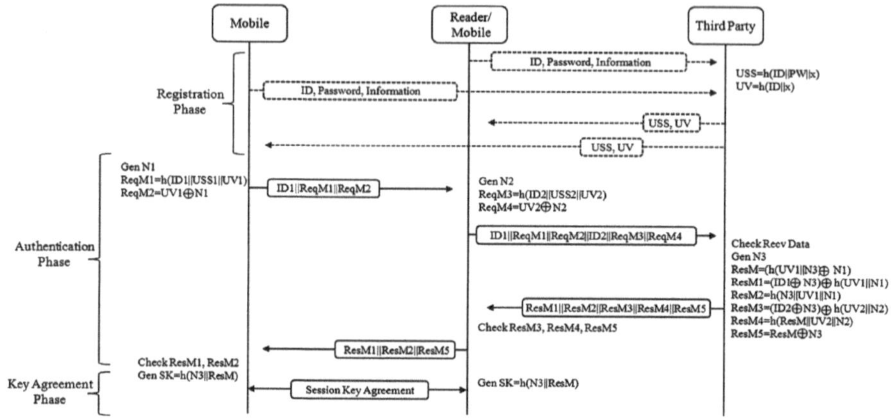

Fig. 13.2 Our proposed scheme

Table 13.1 Terminologies in the paper

Sign	Description
TP	Third party
H()	One way hash function
N	Nonce
||	Connection
USS	User shared secret number
UV	User value
ReqM	Request message
ResM	Response message
SK	Session key

Our method consists of 4 phases of Registration, Authentication, Key Agreement, and Payment.

(1) Registration Phase

All mobile devices send an its own *ID* and a password to the trusted TP (Third Party) or the AuC. After this, TP stores the received information on the secure DB, and generates following operations. This generated *USS* and *UV* is sent to the user through the secure channel and then, is stored securely.

$$USS = h(ID||PW||x)$$
$$UV = h(ID||x)$$

(2) Authentication Phase

In the authentication phase, initiating user generates *N1* as nonce value, and also generates the following operation.

$$ReqM1 = h(ID1||USS1||N1)$$
$$ReqM2 = UV \oplus N1$$

This generated information is sent to the other side of a reader or a mobile device with own *ID1*, and the receivers performs to operate the following.

$$ReqM3 = h(ID2||USS2||N2)$$
$$ReqM4 = UV2 \oplus N2$$

After this, send the information that is generated received information with own *ID2* to the TP. The TP verifies *ReqM1*, *ReqM2*, *ReqM3* and *ReqM4*. And then generates nonce *N3*, if it is right. And it performs various operations as the following.

$$\begin{aligned} ReqM &= h(UV1||N3) \oplus N1 \\ ReqM1 &= (ID \oplus N3) \oplus (UV1||N1) \\ ReqM2 &= h(N3||UV1||N1) \\ ReqM3 &= (ID2 \oplus N3) \oplus h(UV2||N2) \\ ReqM4 &= h(N3||UV2||N2) \\ ReqM5 &= ResM \oplus N3 \end{aligned}$$

After this, TP sends *ResM1*||*ResM2*||*ResM3*||*ResM4*||*ResM5* to the receiving equipment. And this verifies *ResM3*, *ResM4* and *ResM5*.

(3) Key Agreement Phase

The device that verifies the *ResM3*, *ResM4* and *ResM5* sends *ResM1*, *ResM2* and *ResM5* to other device that it is access to own device. The receiver of these data generates a session key as follows, after verifying relevant information.

$$SK = h(N3||ResM)$$

After this, two devices can communicate securely through the common session key.

(4) Payment Phase

This process is a symmetric key-based way. Based on the previously agreed SK, performs the encrypted communication between each mobile, reader/mobile and bank or TP as you can see in Fig. 13.2.

In this process, some information can be used flexibly. And mainly, each different type of information may be formed such as Invoice, Invoice information, Price and User information. Replay or user impersonation attack is impossible because that information should be performed based on *SK* after agreement the key previously. The process of this operation is as follows.

$$E_{SK}(Invoice||Invoice\ information||Price||User\ information\ \ldots.)$$

13.4 Analysis

We analyze our proposed method with two perspectives. There are included an analysis of the safety and efficiency.

(1) Safety Analysis

   (a) Replay Attack
   In our proposed scheme, a key idea that prevents replay attacks is the usage the nonce. As the value of the nonce varies in every session, participate in the operation of all phase with the exception of the registration process. In other words, the exposable information from the communication phase are values of the *ID1*, *ID2*, *ReqM1*, *ReqM2*, *ReqM3*, *ReqM4*, *ResM1*, *ResM2*, *ResM3*, *ResM4* and *ResM5*. And above statements data can be protected by using the nonce.

   (b) The verification about for leaving no user information in the other device
   To authenticate with leaving no their information on the device of the other party, our proposed scheme is performed the authentication through the TP. However, it is essential element to ensure the safety. In this paper, to solve these problems, remaining information on the other side are just *ID*, *ReqM1* and *ReqM2*. And the essential *USS* and *UV* do not be opened, if anyone gets received *ResM1*, *ResM2* and *ResM5* through the TP. And also he can not open *N1* as the generated value of nonce by the user. Therefore, all values, except *ID* are meaningless. Thus, anyone can not know the information to disguise.

   (c) User Impersonation Attack
   User impersonation attack is performed in conjunction with a replay attack. And also, attack that it is disguise for users often occurs by using the received information of other user from the other party. Firstly, in order to disguise the user, if you acquired the information in the previous communication process or wireless communication band, it can be acquired *ID1*, *ReqM1* and *Re2M2*. In order to disguise the other user, the user impersonation attack that sends relevant information to other user *U3* is possible. In other words, there is a case that user *U2* access to *U3* as through other user *U1*. In this case, a user *U2* sends received information

to *U3*. And then *U3* sends the relevant information to the TP. And the TP verifies the right value, sends generating data to the U3, after generating a *ResM1* ~ *ResM5*. After *U3* receive relevant information, sends *ResM1*, *ResM2* and *ResM5* to *U2*. To generate a session key, *U2* that received this information should be able to perform *h(N3||ResM)*. At this point, because the *ResM* is *h(UV1||N3)N1*, even if you mix a *ResM1* and *ResM2*, you can not create a *ResM*. Hence, attacker can not create a session key because it can not know changing values of *UV1* and a *N1* in every session. And even if you want to use the previous session key in the previous session, you can not achieve a match of a session key, because *N3* is changed in every session. In conclusion, the user impersonation attack is impossible.

(2) Efficient Analysis
The main crypto algorithm in our proposed authentication and key agreement method are hash function and XOR operation. Thus basically, it is faster than the crypto algorithm of based on public key and secret key relatively. Like this, we improved the efficiency using a crypto algorithm that can perform a high-speed operation. And the overall number of uses is as follows.

$$15T_{hash} + 16T_{xor}$$

At this point, T_{hash} is the time to perform the hash function once. T_{xor} is time to perform the XOR operation once. And this point, T_{sec} is run time of the crypto algorithms based symmetric key. It requires nearly 6 T_{sec}, because it performs the operation using a cryptography system for financial payment. Financial payment is based a symmetric key after getting the agreement of the session key. Each node is involved in encryption process time. And it is also involved in decryption process time. Therefore, all run-time for payment processing is '$6T_{sec}$ + $15T_{hash}$ + $16T_{xor}$'. The time required in earlier study process is as follows.

$$NFC - SEC : 2T_{KDF} + 4T_{hash}$$
$$SET : 6T_{puk} + 10T_{sig} + 3T_{sec} + 5T_{hash}$$
$$iKP : 2T_{puk} + 11T_{sig} + 7T_{hash} + 1T_{khash}$$
$$Kungpisdan : 11T_{sec} + 2T_{hash} + 5T_{khash} + 4T_{kgen}$$
$$Sekhar : 11T_{sec} + 4T_{hash} + 4T_{khash} + 4T_{kgen}$$
$$Hasoo : 4T_{KDF} + 4T_{hash} + 2T_{numberGen}$$

At this time, T_{puk} refer to the operating time of the public key based crypto algorithms. And T_{sig} is the time required that generates and verifies the signature. And T_{khash} means operating time of Keyed-hash functions. T_{kgen} is required time for generating a key. T_{KDF} is the abbreviation of "Key Derivation Function". And finally, $T_{numberGen}$ is a new operation algorithm like multiplying. We greatly reduced a public key based and a symmetric key based crypto algorithms than earlier study. Therefore, it is judged that we improved the safety and efficiency.

However, in earlier studies, the process of pre-registration is simple than our method or may be unnecessary.

13.5 Conclusion

A variety of data can be easily obtained using a smartphone. Thus today, due to the introduction of NFC is performed more easily the communication between users. However, there is a need to overcome the user impersonation attack and inefficiency of authentication phase due to the user data that is left on the reader or device of the other side. This paper solves these problems by using only an XOR operation and hash function. In the future, it can be used efficiently in secure transactions and data exchange between users.

Acknowledgments This work was supported by Kyungnam University Foundation Grant, 2012.

References

1. Bellare M, Garay JA, Hauser R, Herzberg A, Krawczyk H, Steiner M, Tsudik G, Herreweghen EV, Waidner M (2000) Design, implementation and deployment of the iKp secure electronic payment system. IEEE J Sel Areas Commun 18(4):611–627
2. Kungpisdan S, Srinivasan B, Dung Le P (2004) A secure account-based mobile payment protocol. In: ITCC
3. Juniper Research (2012) NFC mobile payments and retail marketing business models and forecasts 2012–2017
4. Michahelles F, Thiesse F, Schmidt A, Williams JR (2007) Pervasive RFID and near field communication technology. IEEE Pervasive Comput 6(3):94–96
5. Kuspriyanto EH, Basjaruddin N, Purboyo T, Purwantoro S, Ubaya H (2011) Efficient tag-to-tag near filed communication(NFC) protocol for secure mobile payment. In: ICICI-BME
6. Secure Electronic Transaction Specification, version 1.0 (1997). http://www.cl.cam.ac.uklresearch/security/resources/set/
7. Sekhar VC, Sarvabhatla M (2012) Secure lightweight mobile payment protocol using symmetric key techniques. In: ICCCI
8. ISO/IEC 13157-2010 (2010) Information technology telecommunications and information exchange between systems—NFC Security—Part 2: NFC-SEC cryptography standard using ECDH and AES. ISO/IEC
9. Vincent J, Limi V, Aude P, Gaber C, Pasquet M (2012) A mobile payment evaluation based on a digital identity representation. In: CTS Conf
10. Xi L, Ping HH (2007) Efficient protocol of secure mobile payment. J Commun Comput 4 (5):366
11. Bodhani A (2013) New ways to pay [communications near field]. Eng Technol 8(7):32–35
12. ISO/IEC 13157-1 (2010) Information technology telecommunications and information exchange between systems—NFC Security—Part 1: NFC-SEC NFCIP-1 security service and protocol. ISO/IEC

13. Eun H, Lee H, Son J, Kim S, Oh H (2012) Conditional privacy preserving security protocol for NFC applications. In: IEEE international conference on consumer electronics (ICCE), pp 380–381
14. Eun H, Lee HJ, Oh HK (2013) Conditional privacy preserving security protocol for NFC applications. IEEE Trans Consum Electron 59(1):153–160

Chapter 14
A Novel Android Memory Management Policy Focused on Periodic Habits of a User

Jang Hyun Kim, Junghwan Sung, Sang Yun Hwang and Hyo-Joong Suh

Abstract Responsiveness and energy saving are key issues on the Smartphones due to its limited battery capacity. For gaining of execution delay and energy saving, the Android OS manage the main memory by LRU-based algorithm to maintain some terminated applications which expected to be re-execution soon. However, this LRU-based history does not correspond with the user's behavior because most applications may be invoked by the user's periodic habits. In this paper, we propose a novel memory management policy to improve such mismatch. The proposed policy based on user behavior outperforms about 10–30 % compare to the LRU-based policy.

Keywords LRU · User periodic habits · Android, memory management · Responsiveness

14.1 Introduction

Recently, powerful embedded processor and high-bandwidth communication become basic features of the Smartphones. However, these requirements consume lots of energy while the power source is based on the secondary batteries. Thus, various methods are revised to suppress the energy consumption. Voltage and frequency scaling of the processor is one of the common methods to saving the

J.H. Kim · J. Sung · S.Y. Hwang · H.-J. Suh (✉)
Department of Computer Science and Engineering, The Catholic University, Seoul, Korea
e-mail: hjsuh@catholic.ac.kr

J.H. Kim
e-mail: newskyzzzz@gmail.com

J. Sung
e-mail: superstar9999999@gmail.com

S.Y. Hwang
e-mail: syhwang@catholic.ac.kr

J.J. (Jong Hyuk) Park et al. (eds.), *Ubiquitous Computing Application and Wireless Sensor*, Lecture Notes in Electrical Engineering 331,
DOI 10.1007/978-94-017-9618-7_14

energy of the Smartphone. But the limiting of processor power incurs some slack of user responsiveness, while the responsiveness is important factor of user interactivity. Thus, there are various countermeasures which maintain an acceptable responsiveness with sustainable energy consumption.

Voltage and clock scaling [1, 2] are one of the basic power balancing methods which implemented on most of the embedded processors. The Android adopts this method from *governor* in OS level [3]. However these clock scaling policy inhibit the user responsiveness due to the additional tracking delay for measurement of required computing power.

Main memory is another key factor that deeply related to the user responsiveness and re-useable memory. The Android maintains terminated applications in the main memory for possible re-execution of the terminated applications. The Android uses various techniques to control the main memory. Among them, *low memory killer* (LMK) is a key method which controls the main memory [4]. The LMK added in the Linux kernel driver modules for more efficient processes and memory management, and thus it selects and removes applications by LRU in the memory when new application executed by the user. In terms of user responsiveness, if an application was kicked out in the main memory by the LMK and if it is re-executed by the user request, it requires not only additional latency for memory loading but also wastes of additional energy.

In this paper, we addressed to solve this issue. We propose a novel memory management policy to replace the LRU scheme in the Android system.

14.2 Android System Memory Management

The Android system has Out of Memory (OOM) killer in kernel level for memory management. When out of memory state, the OOM killer terminates some processes to free of space in memory with its priority order [5]. However, OOM killer operates in a kernel level without consideration of user interactivity and reusability of processes. Thus user interactive process may be terminated by the OOM based on the LRU. In this case, user may feel some inconvenience by interaction of the system.

In order to solve this problem, the Android system has LMK memory manager for supplement of the OOM. When out of memory state, the LMK attempt to remove processes/memory with priorities and the process state. First attempt is remove a process relatively have a LRU-based low priority and largest memory consumer. Figure 14.1 shows the processing sequence of the LMK and OOM.

As the OOM killer and LMK are performs memory management based on the LRU algorithm, some periodic applications (e.g. alarm, news reader) does not controlled efficiently, while these applications can be predicted when it needed. Furthermore, the Android is preloading frequent applications to reducing the initial latency of its execution. Thus it may push the periodic applications to kick out from memory.

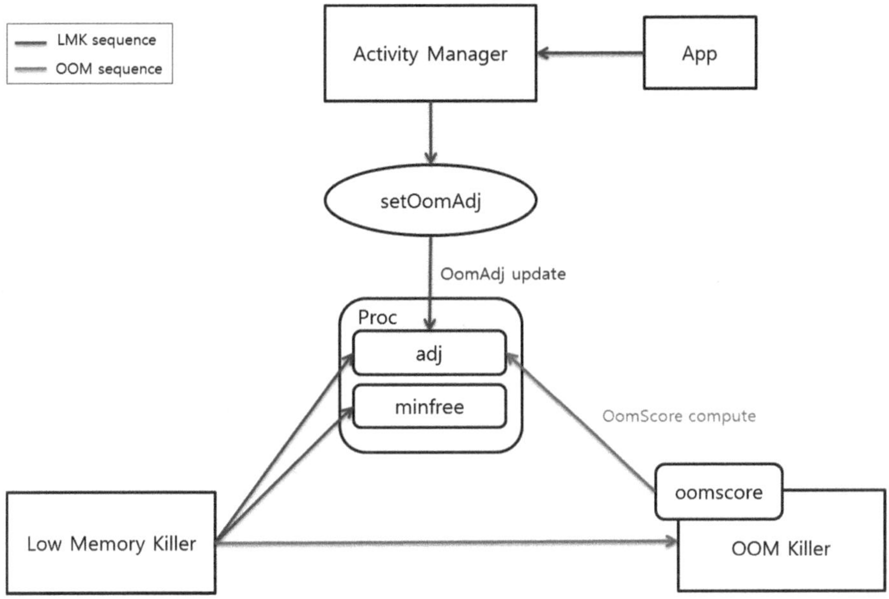

Fig. 14.1 LMK and OOM processing sequence

14.3 Memory Management Policy Based on Periodic Habits of a User

From the viewpoint of human habits, a user executes an application depending on temporal situation, meanwhile the Android manages various applications by the LRU. A Human is under the Time condition (e.g. morning, night, weekday and weekend) and he executes an application depend on this circumstances. It informs the human behaviors have a close relation with time and social environment. For instance, a man can goes to work on every weekday, he can exercise in the park on weekend morning, and he can go to college for language learning on every Friday night. In this case, behavior of the man can be extracted by the time factor. Thus we will be able to build an effective memory management policy if we keep the Track of relations between time and executed applications. For example, when a user takes a ride on subway in every weekday morning and he tends to read some news in the subway, it implies a high probability of news reader application for every weekday morning.

The proposed memory management policy has two components: *habit data extraction unit* and *priority control unit*. The LRU in the LMK and OOM is replaced simply by the priority control unit. Habit data extraction unit is a small application which implanted in the Android framework. Its function is extracting the correlation between time and applications. Priority control unit controls the priority of applications by modifying the *adj* value. Figure 14.2 shows the two components and priority control.

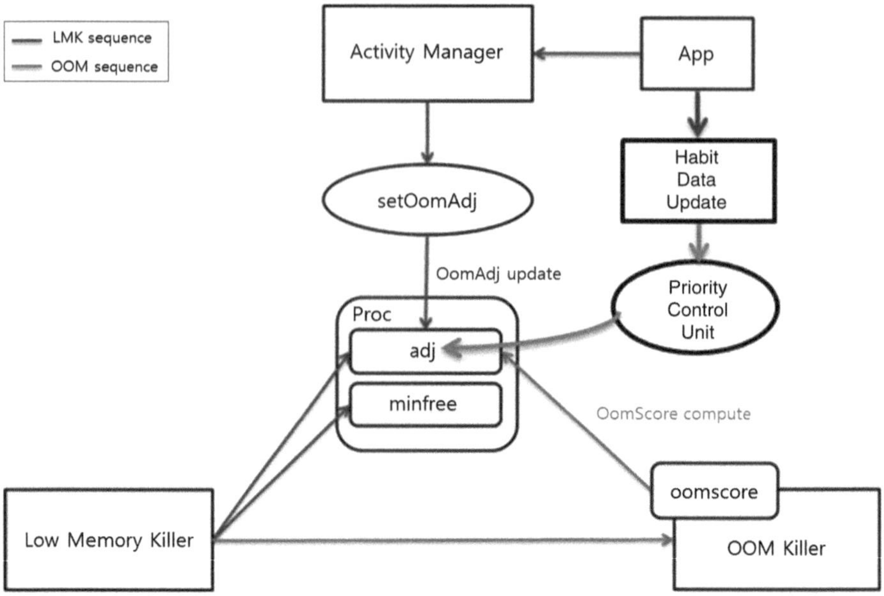

Fig. 14.2 LMK and OOM with priority control

14.4 Experiment Results

Table 14.1 shows collected data and its characteristics by its implied meaning. We collected data elements by relative and absolute time information. Some applications have executed in adjacent time, thus this distance of time is handled as relative. Other relative terms are native meaning of time, i.e. holiday, workday, the office-going hour, and quitting time, etc. The absolute time information means execution date, execution time, and execution day of the week of applications.

Data collection methods are implemented in android framework level with time factor. Data collection process should be naturally carried out as far as possible while user is not aware data collection. In experiment, we developed app by the Android API and collected data with app in android framework level as shown in Fig. 14.3.

Table 14.1 Collected information elements

Time characteristics	
Absolute elements	Relative elements
Execution date	Application execution time interval between applications
Execution time	Holiday/workday
Execution day	Office-going hour/quitting time

```
c =Calendar.getInstance();

int curYear = c.get(Calendar.YEAR);
int curMonth = c.get(Calendar.MONTH)+1;//0~11
int curDay = c.get(Calendar.DAY_OF_MONTH);
String Date = Integer.toString(curYear)+Integer.toString(curMonth)+Integer.toString(curDay);

int Hour = c.get(Calendar.HOUR);
int Minute = c.get(Calendar.MINUTE);
int Sec = c.get(Calendar.SECOND);
String Time = Integer.toString(Hour)+Integer.toString(Minute)+Integer.toString(Sec);

int PmAm = c.get(Calendar.AM_PM); //am=0, pm=1
int curDayWeek = c.get(Calendar.DAY_OF_WEEK);
int week = DayWeekCheck(curDayWeek);
```
①

```
ScanResult result = mScanResult.get(0);
if(week==0)
    mdadapter.firstinsert(3,result.SSID, Time, PmAm, Date, week, 0, 0, 0, toptask);
if(week==1)
    mdadapter.firstinsert(7,result.SSID, Time, PmAm, Date, week, 0, 0, 0, toptask);
```
②

Fig. 14.3 Implemented codes for data collection

Table 14.2 Characteristics of 'Naver Webtoon' application (2 weeks)

App name	Naver Webtoon
Weekday/weekend	Weekday
Preceding app	Naver home
AM characteristics	Office-going hour
PM characteristics	Quitting time
Behavioral meaning	User execute app for spend time when you go to school or home
Longest term after preceding app	0:01:09
Total count	107
Hit count	84
Hit rate	79 %

Figure 14.3 shows part of the codes in the android framework-level. ① is the data type of time data, and ② is the database archiving part in the Smartphone.

Table 14.2 is one of the application characteristics which titled 'Naver Webtoon' after 2 weeks data collection, and we can confirm some periodic characteristics of it.

Collected data shows that the user repeatedly executes 'Naver Webtoon' during office-going hour and quitting time. Also, table appears as a 79 % portion of execution can be predicted by our policy. Therefore, it means our memory management policy could preserve this application by upgrading of priority to avoid kick-out from memory.

By control according to the collected information in repeatedly to the multiple applications, our proposed policy shows improved responsiveness by preserving these applications in the memory as shown in Fig. 14.4. It shows the hit rate of applications by the LRU sequence and our proposed policy during 1 month.

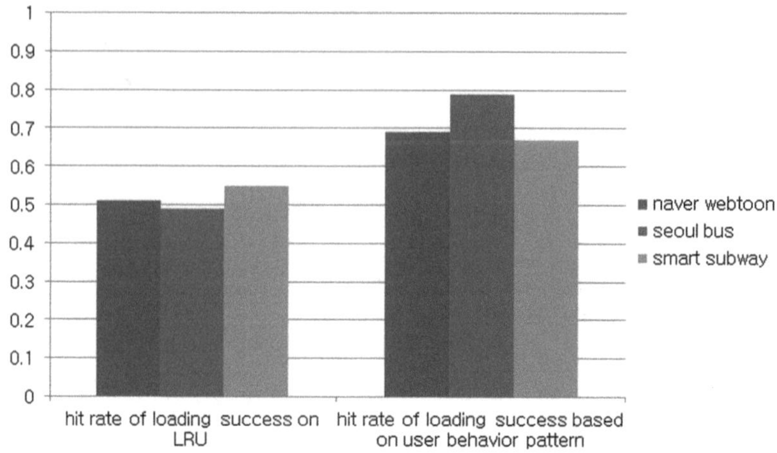

Fig. 14.4 Application hit rate during 1 month

Experimental results show that the hit rate of proposed policy is higher than the LRU due to many execution of the application can be predicted by our policy. The differences are around 10–30 % which portion is preserved by gaining priority promotion. Furthermore, it implies our policy gains additional responsiveness as well as power saving.

14.5 Conclusion

Recently, the Smartphones consume more energy due to powerful processors and high-bandwidth communications. Thus, the Smartphones suppress its peak performance for saving energy such as clock scaling, while this scheme has demerit in terms of the user responsiveness.

User responsiveness is one of the key factor of the Smartphones, thus the Android maintain some defunct applications in the memory to exhibit quick execution of some applications. This memory management policy is LRU-based, but there is some inefficiency due to users executes application by the user's habit that mismatch to the LRU.

In this paper, we propose a novel memory management policy which based on the tracking of user's periodical habits, and replace the Android component to evaluate the proposed policy.

By the experiment results, our proposed policy outperforms about 10–30 % compare to the LRU-based policy, and it gains not only quicker response but also energy saving.

Acknowledgments This research was supported by Basic Science Research Program through the National Research Foundation of Korea (NRF) funded by the Ministry of Education (2013R1A1A2057967).

References

1. Grunwald D, Morrey III CB, Levis P, Neufeld M, Farkas KI (2004) Policies for dynamic clock scheduling. In: Proceedings of the 4th conference symposium on operating system design and implementation, p 4
2. Pouwelse J, Langendoen K, Sips H (2001) Dynamic voltage scaling on a low-power microprocessor. In: Proceedings of the international conference on mobile computing and networking, pp 251–259
3. Mercati P, Bartolini A, Paterna F, Benini L, Rosing T (2014) A Linux-governor based dynamic reliability manager for android mobile devices. In: Design, automation and test in Europe conference and exhibition, pp 1–6
4. Haase C, Guy R (2012) Android UI toolkit engineers for butter or worse smoothing out performance in Android UIs. Google Developers, pp 1–135
5. Kook J, Hong S, Lee W, Jae E, Kim JY (2011) Optimization of out of memory killer for embedded Linux environments. In: ACM symposium on applied computing, pp 633–634

Chapter 15
Care Record Summary Validation Tool with Social Network Service

Jae Woo Sin, Joon Hyun Song, Do Yun Lee and Il Kon Kim

Abstract Health Information Exchange (HIE) is regarded as an important requirement which can enhance the quality of healthcare services in many countries. In Korea, Care Record Summary (CRS) has been developed as a clinical document guideline for national HIE service since 2013, which includes medical information like laboratory, medication, diagnosis, and plan of care etc. And, in 2014, CRS almost reached the final standardized step of Korean Industrial Standard (KS) for exchanging health information among healthcare providers. Before CRS documents being widely populated, it has to be guaranteed that only validated CRS documents are exchanged. As a result, we develop CRS Validation Tool (VT) which can contribute to the improvement of the level of conformance to CRS KS. Furthermore, in order to help the healthcare professionals like physicians, nurses, and healthcare software developers to obtain knowledge about CRS KS as well as VT, we combine CRS VT with Social Network Service (SNS).

Keywords Care record summary (CRS) · Validation · Health information exchange (HIE) · Socialization

15.1 Introduction

High-income countries such as United States, Australian, and Canada have adopted HIE as a way of innovating the quality of healthcare services in their countries. Those countries established an organization which takes exclusive charge of HIE-related projects with full responsibility. The followings are the list of cooperation: Office of the National Coordinator for Health Information Technology (ONC) in United States, National Electronic Health Transition Authority (NEHTA) in Australian, and Health Infoway in Canada [1].

J.W. Sin · J.H. Song · D.Y. Lee · I.K. Kim (✉)
College of IT Engineering, Daegu Campus, Kyungpook National University, Daegu, Korea
e-mail: com3352001@naver.com; ikkim@knu.ac.kr

J.J. (Jong Hyuk) Park et al. (eds.), *Ubiquitous Computing Application and Wireless Sensor*, Lecture Notes in Electrical Engineering 331,
DOI 10.1007/978-94-017-9618-7_15

In 2006, the Korean Ministry of Health and Welfare also started Electronic Health Record (EHR) project in order to develop the core common components such as Clinical Contents Model (CCM), Clinical Document Repository (CDR), Detailed Clinical Model (DCM) in terms of health informatics. HIE was one part which belongs to subprojects. After developing HIE framework based on core modules, in 2008, they launched HIE service to the Seoul National University Bundang Hospital (SNUBH). The name of HIE service has been usually called as EHR service in SNUBH interestingly. As time goes on, EHR service of SNUBH gains popularity from the healthcare professionals as well as the general public.

As a result, the Ministry of Health and Welfare decided to deploy the national HIE project again, named as 'Establishing Medical-IT convergence industry upbringing infra', of which goal is spreading HIE service into nationwide hospitals regardless of the location and the size. In the project, they plan to not only develop several necessary components, but also make a government regulation and security policy for HIE. Furthermore, they define clinical elements such as medication, laboratory, and observation which can be used among hospitals when exchanging healthcare information [2].

Those clinical items for exchanging are selected through a process of many discussions among experts, specifically physicians and then are represented by the usage of Health Level Seven (HL7) Clinical Document Architecture (CDA). CRS is an outcome which includes those clinical items as well as follows the structure of HL7 CDA. And one of important results of the HIE project is CRS, which is currently stepping forward into Korean Industrial Standard (KS). Many organizations, in Korea, started to recognize CRS as as a basic structure for exchanging clinical information semantically.

When HL7 CDA documents are exchanged among healthcare providers, they have to be electronically validated against CDA Implementation Guideline (IG), which includes a set of constraints [3]. The CDA IG is a document which describes a set of rules in natural languages with the aim of letting the public clearly know the purpose and contents of CDA document. Furthermore, before processing CDA documents electronically, it must be checked out that it follows every constraint properly. Consequently, we develop CRS Validation Tool (VT) in order for software developers and clinicians to make sure that CRS document conforms to the CRS IG.

In spite of distribution of CRS VT with web user interface, when users visit CRS VT, they do not know exactly how to perform CRS validation. In order to not only help them acquire knowledge about CRS validation process and CRS itself, but also make them communicate with each other, we expand CRS VT to a socialized VT with Social Network Service (SNS). We expect people ranging from beginners to experts easily to understand CRS contents and share their in-depth knowledge.

15.2 Materials and Methods

The overall design of CRS validation system consists of two parts: backend server and foreend client. Basically, server system provides validating function which tests the level of conformance for CRS document. Additionally, it supports a couple of useful services such as CRS Schematron Transformer and CRS Validation History Repository (shown in Fig. 15.1). Client system provides three functions which mainly focus on socialization among users such as developers, clinicians and hospital staffs. In this paper, server system will be described first and client system will be introduced later.

1. CRS Schematron transformer: This component was implemented to resolve limitation of validation process which uses schematron file. When CRS validation process begins, schematron file checks the presence or absence of patterns in XML trees and this file expresses constraints in hierarchical way. Although schematron file provides constraints during validation process, in this case schematron file cannot be used for validation because the structure of validation result is not optimized for rendering XML file which performs validation and saves validation result. For this reason, there need some file formation that can render XML documents. So we change schematron file to eXtensible Stylesheet Language (XSL) file for rendering and validating CRS document. To change file formation to XSL, there need some transformation engine in system [4]. So this system use eXtensible Stylesheet Language Transformation (XSLT) engine. This transformer use XSLT engine that ISO recommends.
2. CRS Document validator: CRS validator validates CRS documents by using XSL file which is created by CRS Schematron transformer. To validate and save

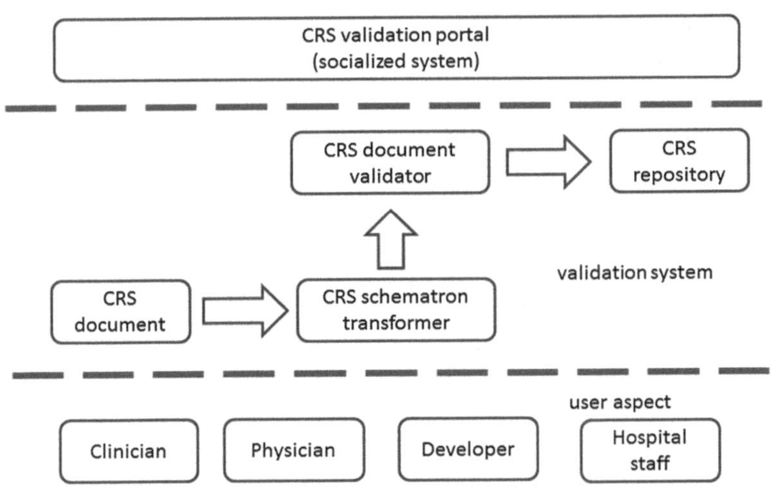

Fig. 15.1 The system architecture of socialized CRS validation tool

CRS document, rendering process will be performed first. In this work, XSL file will be used to render CRS document which consists of xml format [5]. After rendering process validator saves rendered result in xml format. The validation result contains error messages and warning messages of CRS document. Validator divides error messages and warning messages by electronic method. If there is no error or warning, empty xml file will be created.
3. CRS Validation History Repository: It stores CRS validation results in database. When validation process is finished by transformer and validator, various data which related with validation will be stored in this repository. For example, repository stores CDA type that is validated by this tool, CDA file location where validated CDA file is stored in server system, Schematron file location which used to validate file, validation time and so on. Remaining these records in server will be useful for checking validation information. With this data, we can check whether validation is performed by using this system. Furthermore, this system can check when CRS document validation has done. The files saved in server repository and it is ordered in folder categorized with validation date.

We explained server system by looking components which consists of overall system. But server system only provides basic functions for validation and it doesn't provide socialized service. To provide this service there need some points to consider when developing service. Users who participate to service can be important point for this purpose. So to operate service well, understanding users is very important because user is focal point of service including objective, range and degree of service [6].

So we analyze who use this validation service. First, the target of this validation tool is developer who works in hospital or IT-medical conversion domain. They need to manage medical data on HIE. For this reason, to spread CRS document exchange and validation, participation of developers are essentially needed.

Clinicians and Physicians are also important participants. They are performing main role in medical domain. In addition, they have final right to decide whether apply CRS document exchange and validation.

In totalized view, users participate to validation tool and server provides core function. Client gives simple interface for users and activates social network about validation tool. Harmony with these roles is key point for entire validation tool operation. In other words, server side of validation system is component for basic operation and client provides user interface and social network by reflecting user demand.

For this reason, client side of validation system is also important issue as much as server system. This client system contributes on CRS validation social network and helps server to give accessible service to users. So we will introduce several services which are provided by client.

To socialize CRS validation tool we constructed client side by making validation portal that gives socialized place to medical related people. This portal gives various methods to communicate between users and CRS validation tool developers. The portal will provide the following services (shown in Fig. 15.2).

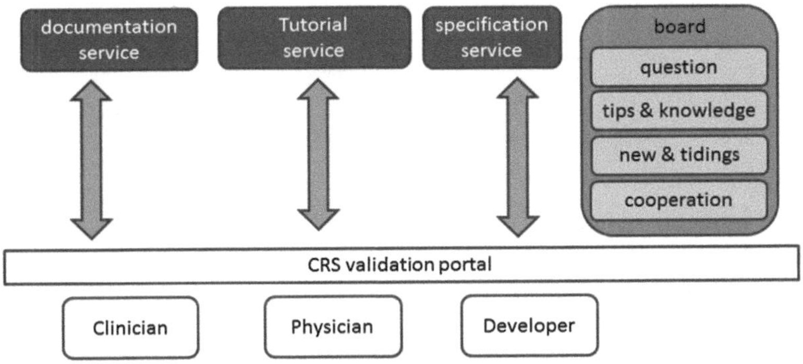

Fig. 15.2 The socialized service structure of the CRS validation tool

Documentation service: Sometimes CRS validation tool is required refactoring, debugging or managing for the reason of difference between development environment, network error among the server and client and so on. Without socialized activities between developers, the cost of using CRS validation tool will increase when developers resolve problems of validation tool. But if development process is recorded on document, activities of socialization between medical systems will be activated. It will help developers to resolve problems with lower cost and more efficiency. In the other words, socialized developer network will resolve most of CRS validation tool issue in the view of implement side.

Tutorial service: Most of clinicians and physicians don't have common sense about CRS document validation and it is the reason why they feel difficult about using CRS validation tool. To solve this problem we choose socialization between beginners and experts of CRS validation tool. The first step to achieve this purpose is making tutorial about usage of CRS validation tool. By giving tutorial document to non-IT staff, it will increase their attendance to socialized group. Second step is visualizing this tutorial for the purpose of providing specified process of validation step by step.

Specification service: CRS implementation guideline is scheduled to become KS standard on June 2014. In spite of standardization of CRS, most developers are not familiar with what this guideline is. In other words, when developing their own applications which are tightly connected to CRS VT, they will spend a huge amount of time and effort. As a result, this part, named as a Specification service, provides a link for developers or IT staffs to easily access to KS standard specification. Furthermore, this service will promote the developers not only to participate in social activities, but also to get knowledge about CRS.

Portal gives information and environment for socialization about CRS validation tool. But these services don't provide enough environments to communicate and socialize between users on internet circumstances. So to fulfill the lack of socialized CRS validation tool we add user board to discuss and to perform socialized activities between users.

Table 15.1 Social media comparison to measure compatibility for social network

	Google Groups	Twitter	Facebook
Notability	Good	Normal	Normal
Clarity	Good	Poor	Poor
Accessibility	Normal	Good	Good

In this research, ‘Google Groups’, ‘Twitter’, ‘Facebook’ were considered as candidate for user board. We discussed about what internet social media is most appropriate with the purpose of socializing CRS validation tool. The discussion result is figured out in Table 15.1.

The analyzing was performed by following criteria. First, to socialize CRS validation tool on internet, board should support notability when some contents are updated or other participants responded [7]. Without this characteristic, communication between users in social network will take long time due to delay of exchanging opinion. For example, you ask a question about validation process and need emergent respond. But if you can’t notify when respond comment is provided, you have to observe board all the time or you will miss responded comment in time.

Second, board should give clarity that can distinguish between different subjects [8]. In this case clarity means how contents are divided inside board. If there is no clarity on board, searching specific contents will become more difficult.

Finally, board can be accessible to various people as much as possible. If there is few person who access board, social network of board will have limited socialization effect [9]. In this case, CRS validation social network will have small scale group and limited domain participant. This will reduce the intellectual power of social network.

In this research, ‘Google Groups’ has been used for user board. ‘Google Groups’ do not have competitiveness at accessibility because ‘Twitter’ and ‘Facebook’ have high awareness proportion for their success in business view. However, the strength of ‘Google Groups’ is providing clarity between different subjects because ‘Facebook’ and ‘Twitter’ is optimized to updating personal experience so their interface can’t give clear division for the view of group working.

In addition ‘Google Groups’ have high notability than other two social media. ‘Google Groups’ provides direct links and folders when contents are updated. Also this media sends email that contains progressive status about discussion. But ‘Twitter’ and ‘Facebook’ only gives notice when new contents are uploaded. In other words, they don’t totalize discussion progress. The people who use CRS validation tool works in professional domain, so ordering discussion result in folder or recording entire discussion to e-mail is very useful for them.

CRS validation board is constructed by ‘Google Groups’ for providing electronic place to discuss and learn CRS validation tool by socialized way. Socialized method includes exchange of ideas, asking a question, replying answer to question and so on. In the clinician view, CRS validation tool is unfamiliar because understanding usage of CRS validation tool needs some knowledge about xml document, concept of standard exchange document and so on. If these topics can be

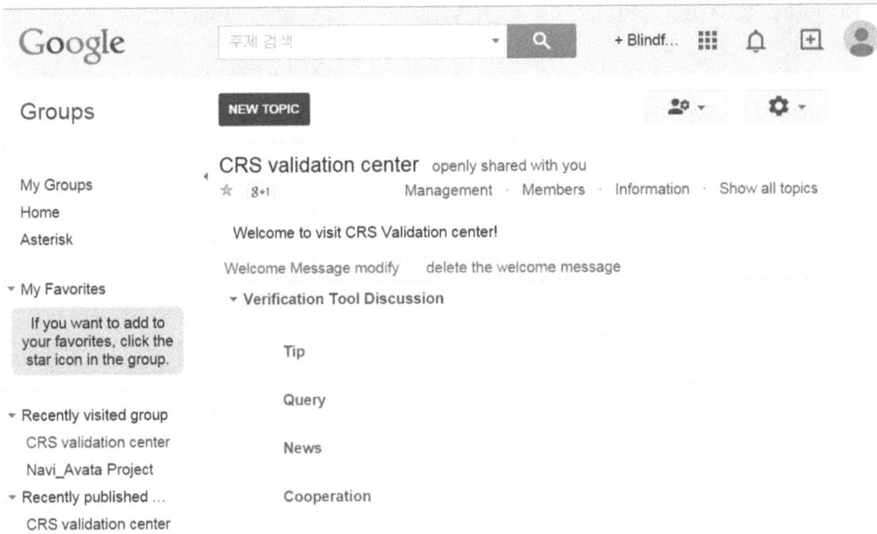

Fig. 15.3 The CRS validation board for socialization

discussed and created in some socialized way, it will contribute so much on HIE based on KS standard [10].

So to attain these purpose we constructed CRS validation board (shown in Fig. 15.3) in four parts: question board, tips and knowledge, news and tidings and cooperation board. Although some other topics can be added to board, new boards are not insulted because too many boards can give confusion to users in early stage. In other words, clarity is important issue for this professional society so simple interface is suitable to support socialization of CRS validation tool [11].

First, question board provides place to ask about CRS validation. In this board, clinicians, nurses, physicians that are not familiar with CRS validation tool can learn about validation by performing communication with developer, other HIE expert or other clinicians that became familiar with using CRS validation tool. We managed question board to accelerate socialization between beginner and expert, to harden foundation of HIE environment based on KS standard.

Next, tips and knowledge board contributes producing knowledge about CRS validation and HIE document based on KS standard. This board was made to give socialized place for CRS document experts and validation tool developers. By giving socialized place to experts, they produce intellectual network for HIE based on KS standard. This network notices CRS validation process to clinicians and IT developers. It makes CRS validation tool more accessible to medical related people.

News and Tidings board notices new notable events about KS standard, validation tool and so on. This board doesn't contribute direct effect about socialization. But this board generates background of network between world standard document validation society and Korean standard document validation society. By transporting fresh issue to CRS validation society, this board also contributes extending knowledge pool.

Cooperation board provides opportunity to collaborate between clinicians, physicians, IT developers and so on. This board helps to make social network integrating various domains related on HIE. The integrated network contributes activating communication among various domains.

15.3 Results

This system has been implemented by using Java language, Maven dependencies, several open source packages and open services. The client portal that provides socialized service is implemented by JSF V2.1 open source package, Richface api V4.2.2 and xhtml V1.0 to interact between users and server. The board and survey service is constructed with google open cloud services. The server side is using Saxon V9.1.0.8 XSL transformation engine and MySQL V5.5 data repository to validate CRS documents and to save validation results.

This system has been deployed by Seoul National University Bundang Hospital (SNUBH), providing socialization place by portal which contains 4 user boards, survey and other social functions. Additionally, we have refactored 9 section templates of CRS schematron based on KS standard. Using our CRS validation tool, it will be more accessible to perform HIE in Korea.

To analysis our CRS validation tool whether this tool can provide socialized services in Korea, this research compare validation tool with other validation services such as lantana and NIST validator [12]. The comparison has been performed by the view of socialization. The result of comparison is shown in Table 15.2.

We choose four criteria to compare validator how efficient their social network. First, accessibility is important factor for socialized validation tool that activates social network more dynamic by connection of more people [13]. For example, Facebook is famous service which is exposed by various media and device. Because of this recognition effect, social network of this service widely spread by increase of access, interact with other media and so on.

User interaction contributes developing collective intelligence of social network [14]. In other words, social network can produce creative ideas and opinions which based on various different viewpoints. So user interaction is also essential issue for evaluating socialization degree of validators.

Table 15.2 Comparison between social networks of validation tool

	CRS validator	NIST	Lantana
Accessibility	Normal	Good	Good
User interaction	Good	Normal	Normal
Language compatibility	Good	Poor	Poor
Effective tutorial	Good	Normal	Poor

Language compatibility is considerable issue when evaluating social network of services [15]. The obstacle of language disturbs socialization between people. It makes misunderstanding and dissonance when communicate each other due to difference between languages: grammar, cultural metaphor and unfamiliar words. So ensuring language compatibility is important factor.

Finally, effective tutorial is important issue for establishing socialized service. Without this most of users will feel difficulty because they don't have enough knowledge and experience about CRS document exchange and validate. For this reason, the tutorial that easily accepted to common medical staff is required for active participation from developers, clinicians and physicians.

NIST validator and Lantana validator has better accessibility than IHIS validator. The NIST validator has been provided service since 11/21/2007 and Lantana validator also serviced several years. So these validators have recognition to people with reliability based on Long-term standard document validation service experience. This recognition helps to access validation service by search engine and communication with acquaintances. IHIS validator doesn't have recognition like other validation tool so this tool has less accessibility compared with other two validators.

But, CRS validator has strength to other three criteria when it used for Korea HIE: user interaction, language compatibility and effective tutorial. First, this validation tool has social gate for users. Portal contains board to communicate between users for various themes so this socialized square makes users more interact with each other. NIST and lantana validation tool also have portal to validate however they don't afford user board for communication and socialization between people.

Language compatibility is also strength of CRS validator than other two validators. CRS validator is consisted by user interface which uses Korean language. Although validation result is represented by English, localized interface helps users adopting validation process and participating CRS validation network. Lantana and NIST validator don't provide interface which is represented by Korean so CRS validator have better compatibility about socialization in Korea.

CRS validator has effective tutorials about validation process. First, the explanation about detailed validation course is provided by google docs. This tutorial explains one by one step in Korean, so Korean users can learn easily adopt CRS validation service. Second, tutorial which represents validation process by video is provided. This tutorial gives expressive information transport by visualization. NIST and Lantana validation tool only provide text guide and this guides not provide total process of validation. This is why CRS validator has possibility to activate social network of CRS validation.

By analyzing validators with these four factors, we find that CRS validator is the most reasonable choice in Korea HIE circumstances. In other words, NIST and Lantana can provide better social network in global stage, CRS validation tool can provide optimized method for CRS document validation in the view of localization.

15.4 Conclusion

In this paper, we developed CRS validation tool (VT) based on KS standard and constructed social network for VT. The focus of this research was validating CRS document and making social network for CRS document validation. The overall system is constructed well but there need some more works to do.

First, this validation tool was made use only in SNUBH and not in other hospital. So we need to spread this system toward other hospitals. Second, user opinion and suggestion should be collected by using social network that is provided with this validation tool. Then by analyzing these results, validation tool needs to be more efficient and localized for Korea HIE environment.

For this reason, we will continue our research on final half of this year. Research will focus extending CRS validation tool to field application and it will aid standard HIE more easily.

Acknowledgments This work was supported by the the IT R&D program of MSIP/KEIT [10047273, Development of EHR Platform and Application Systems based on International Standards for Multi-Countries Hospital Information Systems] and the IT R&D program of MSIP/KEIT [10041145, Self-Organized Software platform(SOSp) for Welfare Devices] and the BK21 Plus project (SW Human Resource Development Program for Supporting Smart Life) funded by the Ministry of Education, School of Computer Science and Engineering, Kyungpook National University, Korea (21A20131600005).

References

1. Paterson G, Shaw N, Grant AM, Delisle E, Leonard K, Corley SM, Kraetschmer N (2011) Cross-Canada EMR case studies: analysis of physicians' perspectives on benefits and barriers. Electron J Health Inf 6(4):e34
2. Vest JR, Gamm LD (2010) Health information exchange: persistent challenges and new strategies. J Am Med Inform Assoc 17(3):288–294
3. Kussaibi H, Macary F, Kennedy M, Booker D, Brodsky V, Schrader T, Daniel C (2009) HL7 CDA implementation guide for structured anatomic pathology reports methodology and tools. Stud Health Technol Inform 160(1):289–293
4. Chen CM, Chu FS, Chen PS (2012) Compiler support for effective XSL transformation. Concurr Comput Pract E 24(14):1572–1593
5. Kerji VK (2011) Decorator pattern with XML in web application. In: IEEE 2011 3rd international conference on electronics computer technology (ICECT), vol 5, pp 304–308
6. Schefels C (2013) Computing user importance in web communities by mining similarity graphs. Int J Adv Internet Technol 6(1, 2):79–89
7. Persson RS (2011) Ability climates in Europe as socially represented notability. High Abil Stud 22(1):79–101
8. Tang C, Liu Y, Oh H, Weitz B (2014) Socialization tactics of new retail employees: a pathway to organizational commitment. J Retail 90(1):62–73
9. Fang R, Duffy MK, Shaw JD (2011) The organizational socialization process: review and development of a social capital model. J Manag 37(1):127–152
10. Li SM, Yuan LT (2013) On the socialized service of university libraries in small and medium-sized cities. J Hubei Radio Telev Univ 7:91

11. Chua AY, Balkunje RS, Chang K (2012) Rendezvous-a social web-based application for knowledge sharing and entertainment. In: Proceedings of the international multi conference of engineers and computer scientists, p 1
12. Heymans S, McKennirey M, Phillips J (2011) Semantic validation of the use of SNOMED CT in HL7 clinical documents. J Biomed Seman 2(1):2
13. Stutzman F (2006) An evaluation of identify-sharing behavior in social network communities. J Int Digital Media Arts Assoc 3(1):10–18
14. Viswanath B, Mislove A, Cha M, Gummadi K P (2009) On the evolution of user interaction in facebook. In: Proceedings of the 2nd ACM workshop on online social networks, pp 37–42
15. Lee RM, Grotevant HD, Hellerstedt WL, Gunnar MR (2006) Cultural socialization in families with internationally adopted children. J Fam Psychol 20(4):571

Chapter 16
Performance Study of an Adaptive Trickle Scheme for Wireless Sensor Networks

Yen-Wen Lin and Po-Hsiang Wang

Abstract In this paper, an adaptive Trickle scheme is proposed for RPL-based networks. When the network conditions are changed and/or the minimal residual energy on the sensor node gets less than a pre-determined threshold, the Trickle-related parameters are dynamically adjusted to save more battery power on the nodes in the WSN (Wireless Sensor Network). The simulation results show that the proposed scheme improves the routing performance and effectively prolongs the network lifetime of the WSN.

Keywords Iot · WSN · RPL · Trickle · Adaptive

16.1 Introduction

IoT (Internet of Things) applications [1] recently attract great interests. WSN [2] is one of the most important technologies to empower the IoT vision. Though plenty efforts have been proposed for WSN, quite a few challenges [3] newly emerge from the context of IoT. Among others, the sensor nodes are usually battery-powered. Therefore, the energy efficiency practically decides the lifetime of the WSN and the availability of IoT services. It is vital to effectively prolong the network lifetime of a WSN.

IETF proposes the RPL (IPv6 Routing Protocol for Low Power and Lossy Networks) [4] for routing packets in LLN (Low Power and Lossy Networks). In RPL [5], the sensor nodes are arranged as a DAG (Directly Acyclic Graph) rooted at the DAG root which is typically the data sink node of the DAG. The root periodically pushes DIO (Destination Oriented DAG Information Object) messages into the networks. Each node computes the cost of the potential paths after

Y.-W. Lin (✉) · P.-H. Wang
Department of Computer Science, National Taichung University of Education, Taichung, Taiwan, ROC
e-mail: ywlin@mail.ntcu.edu.tw

J.J. (Jong Hyuk) Park et al. (eds.), *Ubiquitous Computing Application and Wireless Sensor*, Lecture Notes in Electrical Engineering 331,
DOI 10.1007/978-94-017-9618-7_16

receiving the DIO messages sent from different ancestor nodes. The node selects its parent node and then sends DIO messages to its downward neighbors. Specifically, a node selects its parent node according to the routing metrics. The preference of these parent nodes is evaluated with kinds of OFs (Objective Functions).

The Trickle algorithm [6] determines the rate of sending the DIO messages in RPL. The rate is dynamically adjusted according to the conditions of the network. Generally, the rate will be increased when the network gets unstable. And, less control messages are sent when the network becomes stable. Though the basic operations of the Trickle algorithm are described in [6], several issues [6] need to be solved.

In this paper, to inspect the effects of different settings of these Trickle-related parameters, associated simulations are implemented. To stretch the lifetime of RPL-based WSN, an adaptive Trickle scheme is proposed in this paper. Trickle-related parameters are dynamically set when the network conditions are changed and/or the minimal remaining energy on the sensor node becomes lower than a pre-determined threshold. Consequently, the battery power is saved and the network lifetime of the WSN is extended accordingly. As will be proven in the simulation results, the proposed scheme improves the routing performance and extends the network lifetime of the WSN.

The rest of the paper is organized as follows. Section 16.2 overviews the system. The simulations are described in Sect. 16.3. Section 16.4 briefly concludes this paper.

16.2 System Overview

16.2.1 Basic Operations

In the system, the sensor nodes are organized as a DAG. Only one root is set as the sink node in this DODAG (Destination-Oriented DAG). RPL is used for the packets routing. Traditionally, the root periodically sends the DIO messages to maintain the DAG. And, the rate of sending the DIO messages is determined by the Trickle algorithm [6]. Specifically, related parameters, including I_{min} and k, are dynamically adjusted.

In this paper, a series of simulations is carried out to investigate the effects of these parameters. Inspired by the simulation results (described later), an adaptive Trickle scheme is proposed. Basically, to save the overhead caused by sending unnecessary DIO messages, the Trickle-related parameters are dynamically set according to the network conditions and/or the remaining energy on the nodes. Initially, the value of I_{min} is set small to achieve quick network convergence. Then, when the minimum node residual energy gets lower than a pre-determined threshold and/or the network gets converged, the value of I_{min} and k is properly altered to lessen the sending rate of DIO messages and save more battery power.

For now, in our design, at the beginning, the value of I_{min} is smaller (i.e. 2^8) to obtain quick network convergence. Later, to save energy, the value of I_{min} is set larger (i.e. 2^{14}) after the network becoming converged. Besides, when the minimum node residual energy gets lower than the pre-defined threshold, I_{min} is set larger (i.e. 2^{14}) and k is set smaller to save more battery energy by reducing the DIO sending rate. For now, three energy thresholds including 30, 50, and 70 % of the initial battery energy on the node, are considered.

16.2.2 Convergence Time

The convergence time [5] here is defined as the time needed for all the sensor nodes joining the DAG. In this paper, the convergence time is collected via the Cooja simulators [7]. The DAG turns into inconsistent for various reasons [6] including loop, or routing cost changes. The transmission interval will be reset to I_{min}. Thence, the DIO messages are sent more often to achieve quick network convergence. In our design, I_{min} will be set larger to lessen the transmission rate of the DIO messages after the network getting converged. As will be displayed in the simulations, the proposed adaptive Trickle scheme is able to quickly converge with less power consumption.

16.2.3 Energy Model

The energy model used in [8] is adapted in this paper to estimate the power consumption of a sensor node. Specifically, for now, the power consumption [8, 9] caused by the external sensing modules is not considered. The power consumption of a sensor node (namely E_{pc}) is computed in expression (16.1). Where, E_{cpu}, E_{lpm}, E_{tx}, and E_{rx} is the power consumption of the processor in execution mode, the processor in LPM (Low Power Mode), the communication module in transmitting mode (TX), and the communication module in receiving mode (RX) respectively.

$$E_{pc} = E_{cpu} + E_{lpm} + E_{tx} + E_{rx} \tag{16.1}$$

16.2.4 Node Residual Energy Estimation

The residual energy [8, 9] on a sensor (namely $E_{residual}$ in mAh) is computed in expression (16.2). Where, $E_{initial}$ is the initial capacity of battery energy.

$$E_{residual} = E_{initial} - E_{pc} \tag{16.2}$$

Routing-MC-Type(8bits)	Res Flags (5bits)	P (1bits)	C (1bits)	O (1bits)	R (1bits)	A (3bits)	Precedence (4bits)	Object Body Length(8bits)

Fig. 16.1 DAG metric container (modified from [10])

As depicted in Fig. 16.1, OF-related data is carried by the 32-bits *Option* field of the DIO message [10]. Three energy-related fields are described next. Firstly, the *Routing-MC-Type* field (Routing Metric Constrain Type, 8 bits) brings the data collected from the routing table of the nodes on the route, including the ETX value, the hop count, the rank, and the remaining energy. It can offer 256 OFs (i.e. 0–255). And, "2" is taken by the energy OF. Secondly, the *A* field (3 bits) represents the selection of routing metrics aggregation. The *A* field is set as 0, 1, 2, and 3 for additive, maximum, minimum, and multiplicative aggregation respectively. For now, the *A* field is set as 0 in this paper. Thirdly, the *Object Body field* (8 bits) records the remaining energy. In our design, it can be 0–100 for representing the percentage of the initial energy.

16.3 Simulations

In the following simulations, the effects of various settings of Trickle-related parameters are studied. Also, the performance of the proposed scheme is compared with that of the method proposed in the study [11].

16.3.1 Test Bed

The simulation is implemented with Cooja [7] on the operating system Contiki 2.6 [12]. The test bed used in the study [11] is referenced for setting the simulations in this paper. Two grids with different scales (i.e. 5*4 and 10*9) used in the simulations are deployed in related sensing areas (i.e. 375*300 m^2 [11] and 750*675 m^2). The transmission range and interference range are 120 m [11] and 140 m [11] respectively. In the simulations, distance between two neighbor nodes is 75 m [11]. The transmission success ratio is 100 % [13] and the reception success ratio is 40 % [13]. Sensor node uses Tmote sky [14]. The Contiki MAC and RPL are used for MAC layer and routing. Throughput is 20 packet/min. The energy consumption [14] of LPM, CPU, listen mode, and transmit mode are 0.1635, 5.4, 60.0, and 53.1 mW respectively.

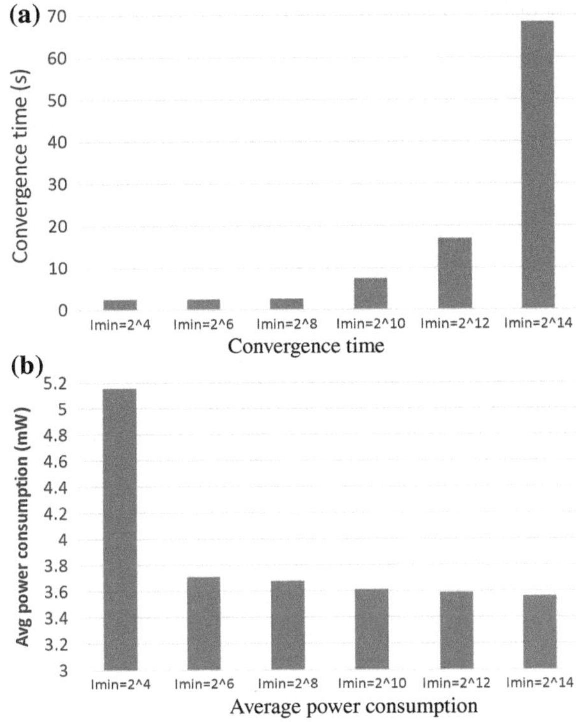

Fig. 16.2 The effects of I_{min}. **a** Convergence time. **b** Average power consumption

16.3.2 The Effects of I_{min}

As described in [6], more DIO messages are sent when the I_{min} is set smaller. As shown in Fig. 16.2a, the convergence time increases when the I_{min} increases. Also, in Fig. 16.2b, average power consumption decreases when the I_{min} increases.

16.3.3 The Effects of k

More DIO messages are sent when the k is set larger [6]. In Fig. 16.3a, given the $I_{min} = 2^{12}$, the convergence time decreases when the k increases. In Fig. 16.3b, given the $I_{min} = 2^3$, average power consumption increases when the k increases.

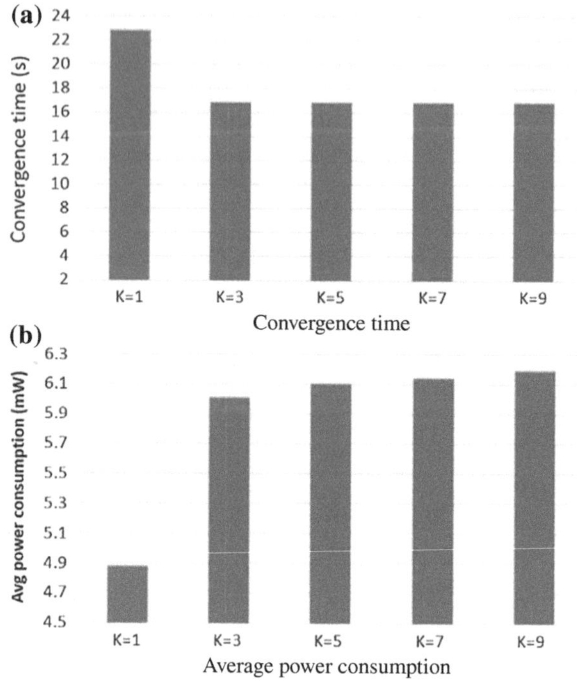

Fig. 16.3 The effects of k. **a** Convergence time. **b** Average power consumption

16.3.4 The Effects of Adaptability

16.3.4.1 Network Lifetime

The network lifetime of a WSN in this paper is defined as the time before the first sensor node in the WSN using up its battery power [15]. With the proposed adaptive Trickle scheme, when the network becomes converged, the I_{min} is set larger to save more battery power on the nodes. As displayed in Fig. 16.4, initially, the I_{min} is set as 2^{12}, the network lifetime of the proposed adaptive Trickle scheme (for all three thresholds) is longer than that of the Energy scheme [11]. Besides, the effects of various energy thresholds (70, 50, and 30 %) are compared. That is, when the residual energy gets lower than these thresholds, the I_{min} is set larger (i.e. changed from 2^{12} to 2^{14}). In Fig. 16.4, the network lifetime is longer when the I_{min} is set larger earlier (i.e. 70 % remaining energy).

For both 70 % (in Fig. 16.5a) and 50 % (in Fig. 16.5b) residual energy threshold, the proposed adaptive Trickle scheme (for various k settings) presents longer network lifetime than that the Energy scheme [11]. And, smaller k offers longer network lifetime. In Fig. 16.5c, for 30 % residual energy threshold, the effects of k settings are not obvious.

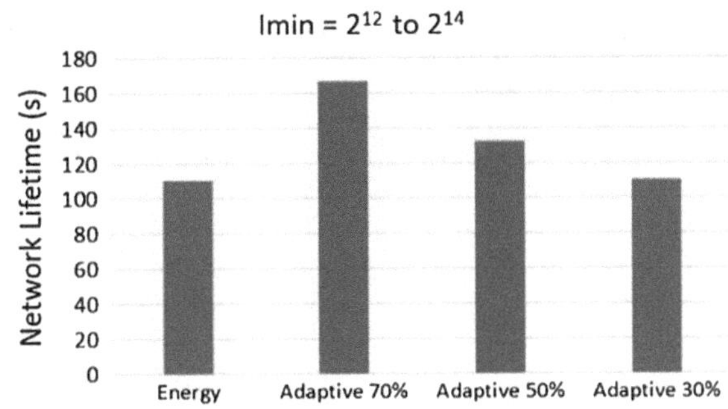

Fig. 16.4 Adaptive I_{min} and network lifetime

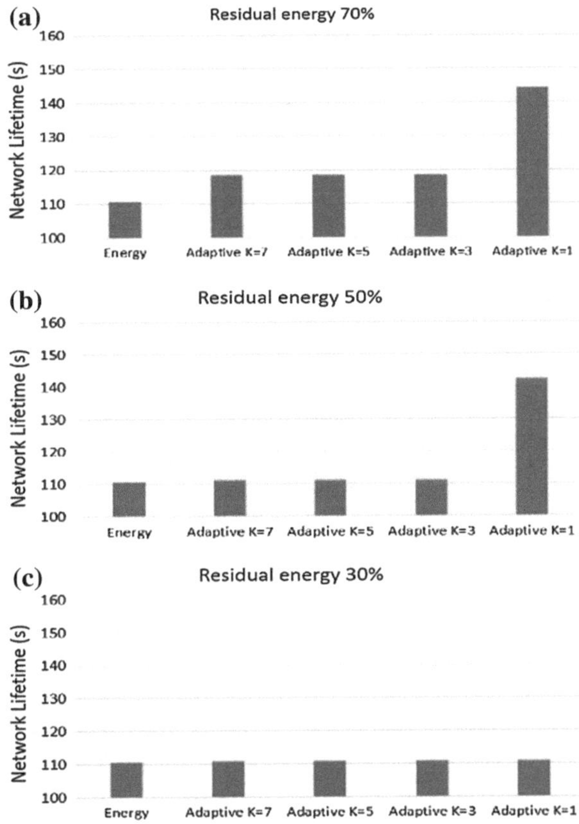

Fig. 16.5 Adaptive k and network lifetime. **a** 70 % Residual energy. **b** 50 % Residual energy. **c** 30 % Residual energy

16.3.4.2 Scalability

For a small WSN, 20 (i.e. 5*4) sensor nodes are deployed in a 375*300 m^2 area in this experiment. The proposed scheme initially sets the I_{min} as 2^8 for quick network convergence. In Fig. 16.6a, the convergence time of the proposed scheme is much

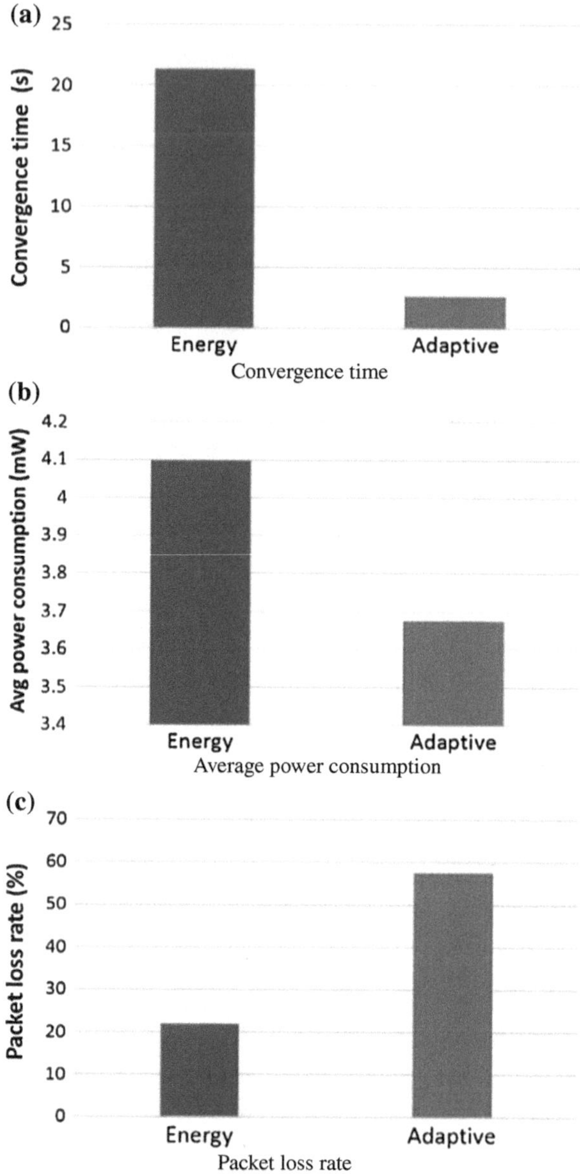

Fig. 16.6 Performance comparison in a small WSN. **a** Convergence time. **b** Average power consumption. **c** Packet loss rate

shorter than that of the Energy scheme [11]. And, in Fig. 16.6b, the average power consumption of the proposed scheme is less than that of the Energy scheme [11]. However, in Fig. 16.6c, the packet loss rate of the proposed scheme is higher than that of the Energy scheme [11]. Because that quick convergence is likely to find the sub-optimal routes; that causes more packet loss.

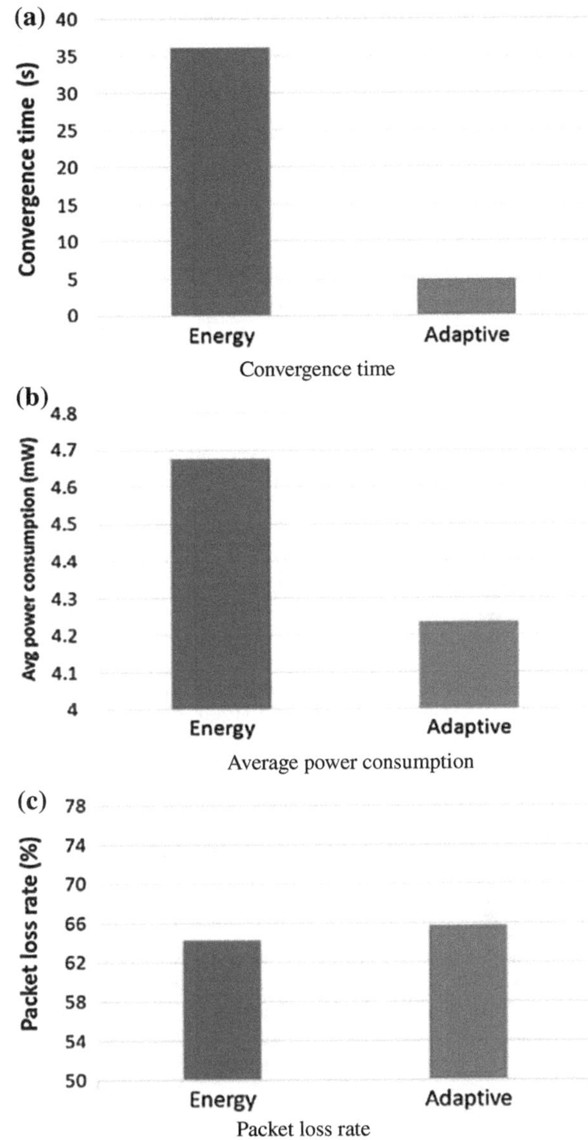

Fig. 16.7 Performance comparison in a large WSN. **a** Convergence time. **b** Average power consumption. **c** Packet loss rate

Additionally, for a large WSN, 90 (i.e. 10*9) sensor nodes are deployed in a 750*675 m^2 area in this experiment. The proposed scheme initially sets the I_{min} as 2^8 for quick convergence. In Fig. 16.7a, the convergence time of the proposed scheme is much shorter than that of the Energy scheme [11]. And, in Fig. 16.7b, the average power consumption of the proposed scheme is less than that of the Energy scheme [11]. Specially, as shown in Fig. 16.7c, the packet loss rate of the proposed scheme is a little higher than that of the Energy scheme [11]. Particularly, as compared with the results of Fig. 16.6c, the proposed scheme performs well in a large scale WSN.

16.4 Conclusions

An adaptive Trickle scheme is proposed in this paper; which is able to dynamically set the Trickle-related parameters according to the network conditions and/or the remaining energy on the sensor nodes. Simulation results display that the proposed scheme can improve the routing performance and elongate the network lifetime of the WSN.

Acknowledgments This work was supported in part of by the R.O.C. National Science Council under grant number NSC 101-2221-E-142-006, NSC 102-2221-E-142-007, and MOST 103-2221-E-142-008.

References

1. Palattella M et al (2013) Standardized protocol stack for the internet of (important) things. IEEE Commun Surv Tutor 15(3):1389–1406
2. Atzori L, Iera A, Morabito G (2010) The internet of things: a survey. Int J Comput Telecommun Netw 54(15):2787–2805
3. Perera C, Zaslavsky A, Christen P, Georgakopoulos D (2013) Context aware computing for the internet of things: a survey. IEEE Commun Surv Tutor 99:1–41
4. Winter T et al (2012) RPL: IPv6 routing protocol for low-power and lossy networks. IETF, RFC 6550
5. Gaddour O, Koubâa A (2012) RPL in a nutshell: a survey. Elsevier Comput Netw 56 (14):3163–3178
6. Levis P, Clausen T, Hui J, Gnawali O, Ko J (2011) The trickle algorithm. IETF, RFC 6206
7. Osterlind F, Dunkels A, Eriksson J, Finne N, Voigt T (2006) Cross-level sensor network simulation with Cooja. In: Proceedings of 31st IEEE conference on local computer networks, pp 641–648
8. Dunkels A et al (2007) Software-based on-line energy estimation for sensor nodes. In: Proceedings of the 4th workshop on embedded networked sensors
9. Wang PH, Lin YW (2014) Performance evaluation of sink selection strategies for extending network lifetime in wireless sensor networks. Accepted by international conference on information management
10. Vasseur J, Kim M, Pister K, Dejean N, Barthel D (2012) Routing metrics used for path calculation in low power and lossy networks. IETF, RFC 6551
11. Kamgueu PO, Nataf E, Djotio T, Festor O (2013) Energy-based routing metric for RPL. Research report (hal-00779519, version 1, 22 Jan 2013), pp 1–17. http://hal.inria.fr/hal-00779519
12. Contiki-2.6. http://www.contiki-os.org/

13. Takizawa S, Komuro N, Sakata S (2013) Energy efficient routing control for 6LoWPAN WSN with power-supplied and battery-powered nodes. Multi J Sci Technol, J Sel Areas Telecommun 2:10–16
14. Tmote Sky Datasheet. http://www.eecs.harvard.edu/~konrad/projects/shimmer/references/tmote-sky-datasheet.pdf
15. Dietrich I, Dressler F (2009) On the lifetime of wireless sensor networks. ACM Trans Sens Netw 5(1):1–38

Chapter 17
A New Distributed Grid Scheme Utilizing Node-based Preprocessing Technique for Supporting k-NN Queries in Location-based Services

Hyunjo Lee, Min Yoon and Jae-Woo Chang

Abstract Because moving objects usually move on spatial networks in location-based service applications, their locations are updated frequently, leading to the degradation of retrieval performance. To manage the frequent updates of moving objects' locations in an efficient way, we propose a new distributed grid scheme which utilizes node-based pre-computation technique to minimize the update cost of the moving objects' locations. Because our grid scheme manages spatial network data separately from the POIs (Point of Interests) and moving objects, it can minimize the update cost of the POIs and moving objects. Using our grid scheme, we propose a new k-nearest neighbor (k-NN) query processing algorithm which minimizes the number of accesses to adjacent cells during POIs retrieval in a parallel way. Finally, we show from our performance analysis that our k-NN query processing algorithm is better on retrieval performance than that of the existing S-GRID.

Keywords Distributed grid scheme · Query processing algorithm · Road network · Moving objects

17.1 Introduction

With the advancements on GPS and mobile device technologies, it is required to provide location-based services (LBSs) to moving objects which move into spatial networks, like road networks [1]. That is, geo-information and geo-processing

H. Lee · M. Yoon · J.-W. Chang (✉)
Department of Computer Engineering, Chonbuk National University, Chonju Chonbuk 56-756, South Korea
e-mail: jwchang@chonbuk.ac.kr

H. Lee
e-mail: o2near@chonbuk.ac.kr

M. Yoon
e-mail: myoon@chonbuk.ac.kr

J.J. (Jong Hyuk) Park et al. (eds.), *Ubiquitous Computing Application and Wireless Sensor*, Lecture Notes in Electrical Engineering 331,
DOI 10.1007/978-94-017-9618-7_17

services are delivered to users with mobile phones, according to their current locations and their points of interests. Such services as automatic vehicle location delivering, tourist services, transport management, and traffic control are all based on mobile objects and the management of their continuous change of location data [2]. Several types of location-dependent queries are significant in LBS, such as range queries [3], k-nearest neighbor (k-NN) queries [3–6], reverse nearest neighbor queries [7], and continuous queries [8]. Among them, the most basic and important queries are k-NN ones. The k-NN query processing algorithms for moving objects in spatial networks make use of real network distance instead of Euclidean distance. However, they suffer from the overhead of distance calculation between POIs (Points of Interests) and moving objects on road networks. To solve this problem, the existing k-NN query processing algorithms use pre-computation techniques for improving performance [9–12]. First, VN3 [9], PINE [10], and islands [11] were proposed to pre-compute the distance between POIs and nodes (or border points) in road networks. However, when POIs need to be updated, they are inefficient because distances between new POIs and nodes should be re-computed. To solve it, S-GRID [12] divides a spatial network into two-dimensional grid cells and pre-compute distances between nodes which are hardly updated. However, S-GRID cannot handle a large number of moving objects which is common in real application scenario. As the number of moving objects increases, a lot of insertions and updates of location data are required due to continuous changes in the positions of moving objects. Because of this, a single server with limited resources shows low performance for handling a large number of moving objects. However, to the best of our knowledge, there exists no work to consider a distributed processing technique using multiple servers for spatial networks. Therefore, we, in this paper, propose a distributed grid scheme which manages the location information of a large number of moving objects in spatial networks. Based on our grid scheme, we propose new k-NN query processing algorithm which minimize the number of accesses to adjacent cells during POIs retrieval in a parallel way. The rest of the paper is organized as follows. In Sect. 17.2, we present related works. In Sect. 17.3, we describe the details of our distributed grid scheme. Section 17.4 presents a new k-NN query processing algorithm based on our grid scheme. In Sect. 17.5, we provide the performance analysis of our k-NN query processing algorithm. Finally, we conclude this paper with future work in Sect. 17.6.

17.2 Related Work

In this section, we describe some related works on k-NN query processing in spatial networks. The existing k-NN query processing algorithms in spatial networks can be divided into two types: online computation technique by incrementally searching the network until k nearest POIs are found, and pre-computation technique by looking up k nearest POIs collected in pre-computed data structure. First, the early works [3–6] and one recent work [7] belong to the online computation technique.

They model the spatial network as a graph structure and store the connectivity of the network. To find the k-nearest POIs, they use a network expansion technique that expands the network from a given query point incrementally by visiting adjacent nodes and examines POIs in the order they are encountered during the network expansion. It terminates when the range of expansion exceeds the network distance of the kth nearest POI. This technique works well for dense POIs, but it requires excessive accesses to the network data when the POIs are spare. Secondly, to resolve the problem of online computation technique, pre-computation techniques have been proposed [9–12]. We can clarify the existing pre-computation techniques into two types: POI-based one and node-based one. The POI-based pre-computation techniques utilize the pre-computed distance between POI and node (or border point). First, Kolahdouzan and Shahabi [9] proposed VN3 (Voronoi-based Network Nearest Neighbor) to process k-NN query in spatial networks. VN3 first generates a Network Voronoi Diagram [13] based on a given set of POIs and pre-computes the network distances within each Voronoi polygon. The network expansion within each Voronoi polygon can be replaced by the pre-computed distances. Secondly, Safar [10] proposed a novel approach, termed PINE (Progressive Incremental Network Expansion), to address KNN queries in SNDB (Spatial Network Databases). PINE partitions a large network into Voronoi cells like VN3 and performs across-the-network computation for only the border points of the neighboring regions. To find the K nearest neighbors, it finds the first nearest POI by simply locating the Voronoi cell that contains a query point q. Then, starting from the query point q, it uses INE algorithm to find POIs. Finally, Huang et al. [11] proposed the Islands approach. Starting from each data point, the Islands approach first pre-computes "islands". When nodes being within a given radius rmin to the POI are part of the POI's island, the distance for such POI is recorded. A k-NN query processing algorithm performs network expansions from the query point by using the pre-computed "islands" encountered during the expansions. The POI-based pre-computation techniques have in common that the pre-computations are POI dependent because distances to POI are pre-computed and the network is subdivided based on the positions of the POIs. Although they reduce the cost of expensive network expansions by pre-computing network distances between nodes and POIs, they have a disadvantage that the frequent updates of the POI causes performance degradation.

To resolve the problem of the VN3, PINE and Island approaches, the node-based pre-computation technique is described. For this, Huang et al. [12] proposed S-GRID (Scalable Grid) which represents a spatial network into two-dimensional grids and pre-computes the network distances between nodes and POIs within each grid cell. To process k-NN query, they adopt the INE algorithm [3] which consists of inner expansion and outer expansion. The inner expansion starts a network expansion from the cell where a given query point is located and continues processing until the shortest paths to all data points inside the cell have been discovered or the cell holds no data points. Whenever the inner expansion visits a border point, the outer expansion is performed from that point. The outer expansion finds all POIs in the cells sharing the border point. This process continues until k nearest

POIs are found. In S-GRID, the updates of the pre-computation data are local and POI independent. However, S-GRID have a critical problem that it is not efficient in handling a large number of moving objects, which are common in real application scenario, because it focuses on a single server environment. That is, when the number of moving objects is great, a lot of insertions and updates of location data are required due to continuous changes in the positions of moving objects. Thus, a single server with limited resources shows bad performance for handing a large number of moving objects.

17.3 Distributed Grid Scheme

To support a large number of moving objects, we propose a distributed grid scheme, by extending S-GRID. Our new grid scheme employs a two-dimensional grid structure for a spatial network and performs pre-computations on the network data, such as nodes and edges, inside each grid cell. In our distributed grid scheme, we assign a server to each cell for managing the network data, POIs and moving objects. Each server stores the pre-computed network data and manages two cell-level indices, one for POIs and the other for moving objects. To assign a unique ID (identifier) to each cell, we define CellID as follows.

Definition 17.1 Let assume a spatial network is partitioned into n*n two-dimensional grid structure. A unique ID of a cell being located in i-th row and j-th column, CellIDi,j, is defined by $CellID_{i,j} = (i-1)^{*}n + j$.

Figure 17.1 shows an overall structure of our distributed grid scheme. Each cell consists of seven data structures: a cell table (Cell Table), border point table (BP table), POI R-tree, MO index, Vertex-Edge component, Vertex-Border component, and Cell-Border component. In the limited resource environment, a small number of servers can be allocated to the large number of grid cells where each server creates the same number of threads as the number of cells to deal with.

We describe each component of our DS-GRID. First, to make network communication with other servers for accessing both network data and POI information in a grid cell, each server maintains a cell table whose The entry of the cell table is <CellIndex, IP:port, #_POI> where CellIndex is ID of a grid cell, IP:port is a pair of IP address and the port number of the corresponding server, and #_POI is the total number of POIs within the cell boundary. Secondly, our DS-GRID maintains BP table to store ID of adjacent cells which share the border point of a cell. The entry of BP table is <BPID, AdjCell> where BPID is ID of a border point in a cell and AdjCell is ID of adjacent cell sharing the border point. Thirdly, our DS-GRID uses an R-tree to maintain POIs in the cell. While S-GRID has an overhead of a series of split and merge operations due to the update of POIs in a cell, DS-GRID shows good retrieval performance because the update of one cell does not affect the updates of POIs in other cells. Fourthly, our MO index consists of edge table and

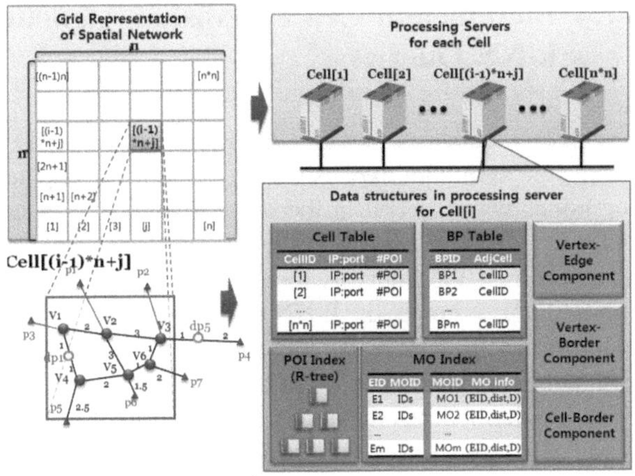

Fig. 17.1 Overall structure of our distributed grid scheme

MO table in order to maintain moving objects' location information. The entry of edge table is <EdgeID, MOID list> where EdgeID is ID of an edge in a cell and MOID list is the list of IDs of moving objects in the related edge. On the other hand, the entry of MO table is <MOID, EdgeID, dist, D> where MOID is ID of moving object, EdgeID is ID of an edge containing the moving objects, 'dist' is the distance from the starting node of the edge to the moving object, and D is the movement direction of moving object. Finally, our DS-GRID manages an adjacent node list by using the vertex-edge component, a distance from a node to a border point by using the vertex-border component, and a distance from a border point to another border point by using the cell-border component. Unlike S-GRID, our DS-GRID deals with the vertex-edge, vertex-border and cell-border components in a cell-wise manner.

In our DS-GRID, the updates of POIs and nodes in a cell lead to their updates in the other cells of the network because the information of POIs, moving objects and network data is stored separately in a cell. When two nodes of an edge lie in different cells due to the edge's update, the edge is divided into two parts by inserting a border point and each part is stored separately in respective cells. When a new border point is inserted or an existing border point is repositioned due to the updates of nodes and edges, the update of border point should be reflected to an adjacent cell sharing the border point. That is, the cell in which the border point is updated sends an update message to its adjacent cells so that they can update their information. If the location of a POI in a cell is changed, our DS-GRID updates its location by using R-tree operations. However, only when the total number of POIs within a cell is changed, a server assigned to the cell sends a new total number of POIs to the rest of the grid cells so that their cell tables can be updated. Our DS-GRID uses R-tree operations to handle the update in a cell.

17.4 Incremental Cell Expansion Algorithm for Processing k-NN Queries

The k-NN query processing algorithm of S-GRID consists of two steps: inner expansion and outer expansion. First, the inner expansion retrieves POIs by visiting the nodes within a cell in the order of distance from a query point. Secondly, the outer expansion retrieves POIs from adjacent cells sharing border points with the cell in which the query is located. To apply the expansion approach to our distributed grid scheme, the k-NN query processing algorithm should visit an adjacent cell repeatedly if there are more than one shared border points between the query cell and the adjacent cell. Figure 17.2 shows an example of k-NN query processing by using the outer expansion. Let us assume that server 1 and server 2 manage cell 1 and cell 2, respectively. Because a query point Q is located in cell 1, the algorithm first visits nodes within the cell 1 by using inner expansion. When the next node to visit is border point p2, the algorithm sends the query to server 2 because cell 2 shares the border point. The server 2 computes the distances between the query point and POIs, i.e., dp1 and dp2, by using the border point p1 provided by the server 1. Then, the server 2 returns the result to the server 1. Next, the server 1 inserts the POIs retrieved from adjacent cells and continues to perform inner expansion. When the next node to visit is a border point p1, the algorithm should send the query to the server 2 again. The server 2 has to repeat the same process as in the case of the border point p2. That is, the process is repeated for the number of shared border points between cell 1 and cell 2. Therefore, the retrieval performance for k-NN query processing is decreased as the number of shared border points increases. To solve this problem, it is necessary to send a query once by maintaining all the shared border points.

For this, we propose a new k-NN query processing algorithm, namely Incremental Cell Expansion (ICE) algorithm. Although ICE algorithm also uses both inner expansion and outer expansion, the process of expanding a cell is quite different from S-GRID. First, our ICE algorithm finds all the border points and creates a list of cells containing the border points by doing the inner expansion. Next, a coordinate (server) sends the query to all the cells in the cell list. Secondly, servers receiving the query retrieve both POIs and other border points by doing

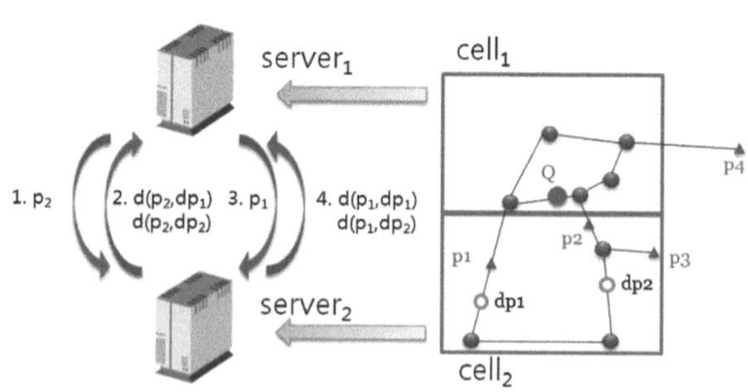

Fig. 17.2 k-NN query processing by using outer expansion

outer expansion. Then, they send the retrieved POIs and border points to the coordinate from which query is originated. Thirdly, the algorithm checks whether or not there is a border point being nearer than the k-th POI. If true, the process is repeated until no border point is nearer than the k-th POI. To find k nearest neighbors, our ICE algorithm performs both inner expansion and outer expansion by using two priority queues, Qv and Qdp. Qv stores both relevant nodes and the distance between a query point and the nodes while Qdp stores both retrieved POIs and their distances from a query point. Thus, our ICE algorithm can improve retrieval performance by minimizing unnecessary visiting of adjacent cells.

Figure 17.3 shows an example of our ICE algorithm. Let us assume that our algorithm finds 3-NN (nearest POIs) with a given query being located in the cell 5 where a spatial network is partitioned into a 3*3 two-dimensional grid structure. First, our ICE algorithm expands the network starting from a given query point in the cell 5 and finds the border points bp4, bp6, bp7, bp9 and the POI P5. It inserts the retrieved border points and the POI into Qv and Qdp respectively, and it sets the value dmax as ∞ because the number of retrieved POIs is less than k. Secondly, our ICE algorithm creates a list of cells which share the retrieved border points. A generated cell list is {cell 2, cell 4, cell 6, cell 8} because the border points bp9, bp6, bp7, bp4 are shared by a cell 2, a cell 4, a cell 6, and a cell 8, respectively. Thirdly, our algorithm sends the query to those cells in the cell list so as to perform its outer expansion. By performing the expansion, it finds the POI P6 in the cell 2 and computes the distance from bp9 to the other border points, i.e., bp10 and bp11, in the cell. Similarly, it finds the POIs P3, P4 in the cell 4 and P2 in the cell 8, and it computes the distances from the query point to bp3 in the cell 4, to bp5 and bp8 in the cell 6, to bp1 and bp2 in the cell 8, respectively. The retrieved POIs and the border points are sent to the query cell and are inserted into Qdp and Qv, respectively. As a result, the items of the Qdp are {(P5,4), (P2,7), (P3,9)} and dmax is set to the distance of the third nearest POI, i.e., 9. Because the Qv has the border point bp1 whose distance from the query is less than the dmax, our ICE algorithm creates another cell list for performing the second round of outer expansion. Fourthly, our

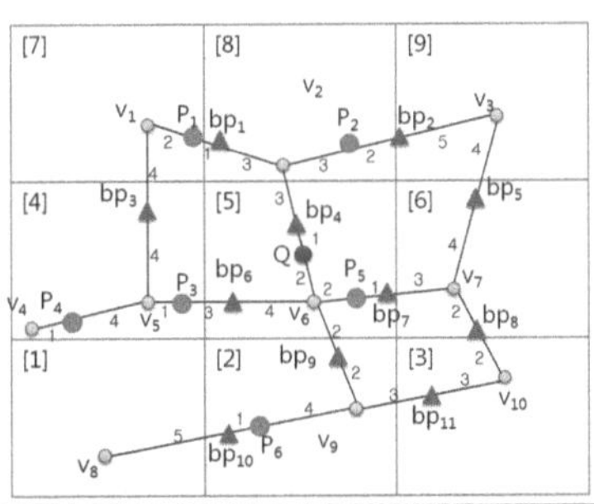

Fig. 17.3 Example of our ICE algorithm

Step		Qv	Qdp	d_{max}	CellList
1	Create CellList	$(bp_4,1)$ $(bp_9,4)$ $(bp_7,5)$ $(bp_6,6)$	$(P_5,4)$	∞	Cell[2] Cell[4] Cell[6] Cell[8]
2	1st Outer Expansion	$(bp_1,7)$ ~~$(bp_2,9)$ $(bp_{11},9)$ $(bp_8,10)$ $(bp_{10},11)$ $(bp_5,12)$ $(bp_3,14)$~~	$(P_5,4)$ $(P_2,7)$ $(P_3,9)$	9	Cell[7]
3	2nd Outer Expansion	-	$(P_5,4)$ $(P_2,7)$ $(P_1,8)$	8	-

algorithm finds the POI P1 in the cell 7 and the border point bp3, and it sends them to the query cell. As a result, P3 is replaced by P1 and dmax is set to 8. Finally, our algorithm returns the POIs of Qdp as 3-NN because no border point in the Qv has less distance from the query than dmax.

17.5 Performance Analysis

We present performance analysis of k-NN query processing algorithm for our grid scheme. We implement our grid scheme by using visual studio 2003 under HP ML 150 G3 server with Intel Xeon 3.0 GHz dual CPU, 2 GB memory, HP 250 GB SATA 7,200 rpm HDD, and BCM 5703 Gigabyte Ethernet. In our experiments, we used multiple processes in a single server and each process manages a single cell. To provide an environment appropriate to a distributed grid scheme, we let each process use a different port number to communicate with other processes by using TCP/IP protocol. For spatial network data, we use San Francisco Bay map consisting of 220,000 edges and 170,000 nodes, and generate four sets of POIs (i.e., 2200, 4400, 11000, 22000) by using Brinkhoff algorithm [14]. These POIs are indexed by using R-trees. Moreover, we randomly select 100 nodes from San Francisco Bay map as query points. To measure the retrieval performance of k-NN queries, we average response times for all the 100 query points. Because the existing works VN3 [8], PINE [9], island [10] are very inefficient for the update of POIs due to their POI-based pre-computation techniques, they are not appropriate for dealing with a large number of mobile objects in spatial networks. Thus we compare our ICE algorithm with S-GRID algorithm in terms of POI retrieval time. For this, we conducted three types of experiments by varying (i) the number of grid cells, (ii) the value of k, (iii) the density of POI. To show the efficiency of our algorithm according to the update of POIs, we also measured retrieval time after the update of POIs.

Figure 17.4 shows the performance of k-NN query processing according to the number of grid cells when k = 20 and POI density = 0.01. We observe that as the number of grid cells increases, the retrieval time of all the two methods is increased

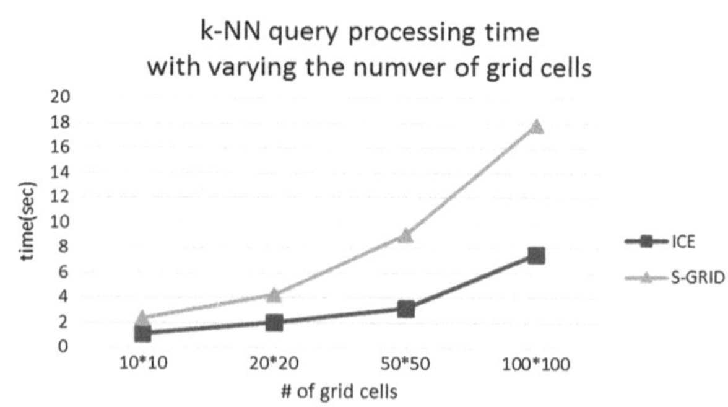

Fig. 17.4 k-NN query processing time retrieval time with different number of grid cells

because the number of POIs per cell can be reduced. As a result, the number of cells to be visited is increased to find the same number of POIs. The performance of our algorithm is better than that of S-GRID when the number of grid cells is more than 10*10. This is because our ICE algorithm performs outer expansion in a parallel way.

Figure 17.5 shows the retrieval time of k-NN query with the varying value of k when the density of POI is 0.01 and the number of grid cells equals 20*20. It is shown that as the value of k increases, the retrieval times of the two algorithms are increased because when the number of disk I/Os of the cell-border and vertex-border components are increased to visit adjacent cells. When k = 20, the retrieval times of the S-GRID, our ICE algorithm is about 4.5 and 1.9 s, respectively. We can say the performance result that our ICE algorithm is better because it can reduce a query propagation step by sending a query to a list of cells at a time.

Figure 17.6 shows the retrieval time of k-NN query with the varying density of POIs where the number of grid cells equals 20*20 and k = 20. It is shown that as the density of POI increases, the retrieval time of all the three methods is decreased because the number of POIs within a cell is increased. As a result, we can reduce the cost of inner expansion within a cell and the number of adjacent cells to be visited. When the density of POIs equals to 0.05, the retrieval times of S-GRID and our algorithm are about 2.5 and 0.4 s to retrieve k nearest POIs, respectively.

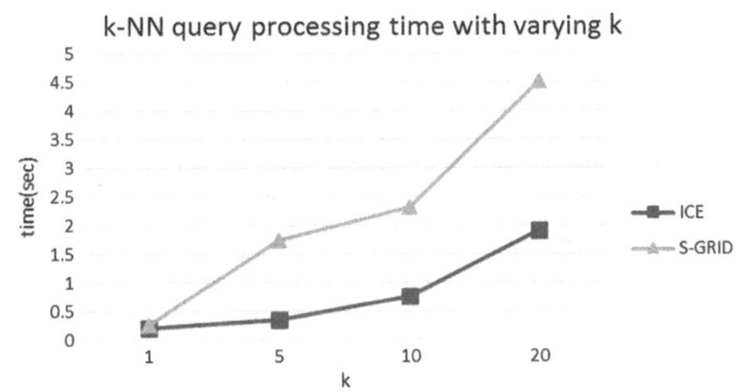

Fig. 17.5 k-NN query processing time with varying the value of k

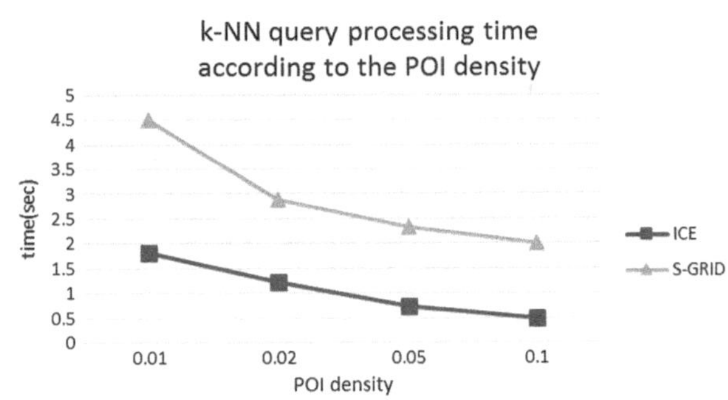

Fig. 17.6 k-NN query processing time according to the density of POIs

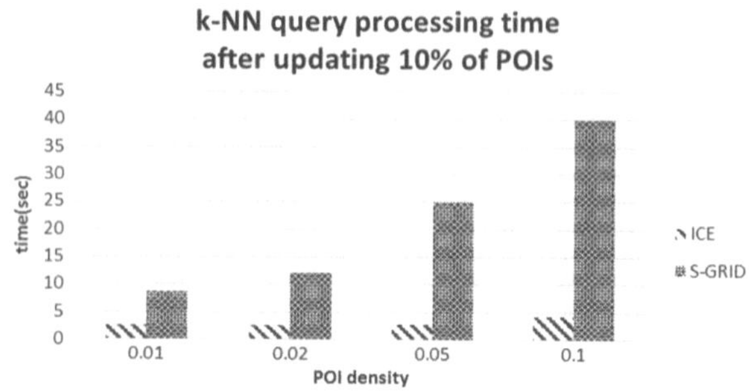

Fig. 17.7 k-NN query processing time after updating 10 % of POIs

Figure 17.7 shows the retrieval time of k-NN query after updating POIs. For this experiment, we measure the search time of k-NN query when the 10 % of POIs is updated. In the case of S-GRID, the retrieval time is exponentially increased as the density of POIs increases. This is because S-GRID uses one R-tree to index all the POIs of the network and so the update of POIs in a cell affects the whole system. Whereas, because our grid scheme uses a separate R-tree per each grid cell to index POIs within it, the update of POIs in a cell does not affect all the grid cells globally. As a result, even though the number of updated POIs increases, the retrieval performance of our grid scheme is not dramatically increased.

17.6 Conclusion and Future Work

In this paper, we proposed a grid scheme to manage the location information of a large number of moving objects in spatial networks. Our grid scheme makes use of a node-based pre-computation technique so that it can minimize the update cost of the moving objects' locations. Our grid scheme splits a spatial network into two-dimensional grid cells so that it can update network data locally. Based on our grid scheme, we proposed a new k-NN query processing algorithm. Our algorithm improves the retrieval performance of K-NN queries because it decreases the number of adjacent cells visited by transmitting a query to all the shared border points. Our experimental results show that our algorithm is better on retrieval performance than that of S-GRID. As a future work, we need to extend our grid scheme to handle a spatial network with dense and sparse regions in an efficient manner by using non-uniform grid cells.

Acknowledgments This work was supported by the Ministry of Education (MOE) and National Research Foundation of Korea (NRF) through the Human Resource Training Project for Regional Innovation(2014065816).

References

1. Speičys L, Jensen CS (2008) Enabling location-based services-multi-graph representation of transportation networks. Proc GeoInformatica 12(2):219–253
2. Ilarri S, Mena E, Illarramendi A (2010) Location-dependent query processing: where we are and where we are heading. ACM Comput Surv 42(3):1–73 (Article 12)
3. Papadias D, Zhang J, Mamoulis N, Tao Y (2003) Query processing in spatial network databases. In: Proceedings of the VLDB 29: pp 802–813
4. Shahabi C, Kolahdouzan MR, Sharifzadeh M (2003) A road network embedding technique for K-nearest neighbor search in moving object databases. Proc GeoInformatica 7(3):255–273
5. Cao H, Wang S, Li L (2003) Location dependent query in a mobile environment. Proc Inf Sci 154(1–2):71–83
6. Jensen CS, Pedersen TB, Speicys L, Timko I (2003) Data modeling for mobile services in the real world. In: Proceedings of the SSTD, vol 2750, pp. 1–9
7. Benetis R, Jensen C S, Karčiauskas G, Šaltenis S (2006) Nearest and reverse nearest neighbor queries for moving objects. Proc VLDB 15: 229–250
8. Huang YK, Chen C-C, Lee C (2009) Continuous K-nearest neighbor query for moving objects with uncertain velocity. Proc GeoInformatica 13(1):1–25
9. Kolahdouzan MR, Shahabi C (2004) Voronoi-based nearest neighbor search for spatial network databases. In Proceedings of the VLDB 30: pp 840–851
10. Safar M (2005) K nearest neighbor search in navigation systems. Mobile Inf Syst 1(3):207–224
11. Huang X, Jensen CS, Saltenis S (2005) The islands approach to nearest neighbor querying in spatial networks. In Proceedings of the SSTD, LNCS 3633: 73–90
12. Huang X, Jensen CS, Lu H, Saltenis S (2007) S-GRID: a versatile approach to efficient query processing in spatial networks. In Proceedings of the SSTD, LNCS 4605: 93–111
13. Okabe A, Boots B, Sugihara K, Chiu SN (2000) Spatial tessellations, concepts and applications of voronoi diagrams, 2nd edn. Wiley, Chichester
14. Brinkhoff T (2002) A framework for generating network-based moving objects. Proc GeoInformatica 6:153–180

Chapter 18
A Centroid-GPS Model to Improving Positioning Accuracy for a Sensitive Location-Based System

Md. Rashedul Islam and Jong-Myon Kim

Abstract This paper proposes a centroid global positioning system (GPS) model to improve the positioning accuracy of low-cost GPS receivers of a sensitive location-based system. The proposed model estimates the precise movement position by a centroid sum of the individual improved positions of three GPS receivers. Each GPS receiver's position is improved by using a direction and velocity averaging technique based on combining the vehicle movement direction, velocity averaging, and distance between the waypoints of each GPS receiver using coordinate data (latitude, longitude, time, and velocity). Finally, the precise position is estimated by calculating a triangular centroid sum with distance threshold of the improved positions of three GPS receivers. In order to evaluate the performance of the proposed approach, we used three GARMIN GPS 19x HVS receivers attached to a car and plotted the processed data in Google map. The proposed approach resulted in an improved accuracy of about 2–12 m compared to the original GPS receivers. In addition, we compared the proposed approach to two other state-of-the-art methods. The experimental results show that the proposed approach outperforms the conventional methods in terms of positioning accuracy.

Keywords GPS accuracy · Long-term averaging · Direction averaging · Location-based system

18.1 Introduction

Location-based services (LBSs) have been an innovative information technology to identify the location or geographical position of users [1]. Several location-based systems have been introduced such as GPS, geographical information system (GIS),

---

Md.R. Islam · J.-M. Kim (✉)
Department of Electrical, Electronics, and Computer Engineering,
University of Ulsan, Ulsan, South Korea
e-mail: jongmyon.kim@gmail.com

Md.R. Islam
e-mail: rashed.cse@gmail.com

J.J. (Jong Hyuk) Park et al. (eds.), *Ubiquitous Computing Application and Wireless Sensor*, Lecture Notes in Electrical Engineering 331,
DOI 10.1007/978-94-017-9618-7_18

Wi-Fi fingerprinting, wireless sensor network (WSN), wireless local area network (WLAN), Bluetooth, and sensors for identifying indoor and outdoor location [2–4]. Among these techniques, GPS has been widely used for a wide range of location-based services due to its cost effectiveness and energy consumption [5–8].

GPS was developed in early 1960 [9], and it has been used to measure any desired position on the earth. Currently, GPS is a popular general-purpose positioning system [10, 11] that consists of three major segments: a space segment, a control segment, and a user segment [12]. The space segment is composed of the orbiting GPS satellites over 20,000 km from the earth. The control segment monitors the operation and position of GPS from a ground station. The user segment calculates the position of GPS through the signal from satellites.

Although GPS is the most popular positioning system in the field of LBS, positioning accuracy improvement of GPS is a challenging issue. Several techniques have been developed to enhance the accuracy of GPS positioning. The conventional researches are categorized into three groups [5]. Firstly, expensive devices and technologies including the wide area augmentation system (WAAS), differential GPS (DGPS), and assisted GPS (AGPS) have been developed to enhance positioning accuracy by from 3 to 15 m. However, these technologies require an expensive infrastructure. The second group uses an additional peripheral module for a GPS receiver to improve the positioning accuracy. In the third group, several researchers have developed software methods to improve the accuracy of GPS positioning [9, 11]. Refan et al. proposed auto regressive moving average (ARMA) interpolation methods to improve the accuracy of a low-cost GPS positioning and showed satisfactory performance [9]. Islam et al. [11] proposed an effective direction averaging method to improve the positioning accuracy of a low-cost standard GPS by estimating direction angle, velocity, and distance between two waypoints.

To improve the accuracy of a low-cost GPS using a sensitive location-based system, we propose a new centroid-GPS model. The proposed model precisely estimates the position as a centroid sum of the improved positions of three GPS receivers based on the vehicle movement direction, velocity averaging, and distance between waypoints. The experimental results demonstrate that the proposed model results in an improvement of about 2–12 m in several different experiments.

The rest of this paper is organized as follows. Section 18.2 describes the proposed model with related terminologies, and Sect. 18.3 presents the experimental results. Finally, Sect. 18.4 concludes the paper.

18.2 Proposed Approach

This paper focuses on improving the positioning accuracy of a location-based system using low-cost standard GPS receivers. This proposed approach consists of the following main functional blocks to improve the positional accuracy of GPS: (i) improvement of the individual GPS positioning accuracy using an effective direction averaging technique based on combining the movement directions and averaging speed and distances of the waypoints of past and current states of each individual GPS receiver [11] and (ii) estimating the vehicle's position using a triangular centroid sum of the improved GPS positions of the three GPS receivers. To enhance the performance of the proposed model, this study also utilized a precise reference point and invalid data check.

Three GPS receivers were placed at three corners of a vehicle to form a triangle with the centroid of the triangle located at the approximate center of the vehicle. Figure 18.1 represents the physical layout of this model, where D_1, D_2, and D_3 are the actual (initial) distances of the 1st–2nd, 2nd–3rd, and 3rd–1st GPS receiver, respectively.

Figure 18.2 represents a basic flow diagram of the proposed approach. In the proposed approach, the direction and velocity averaging module of each receiver

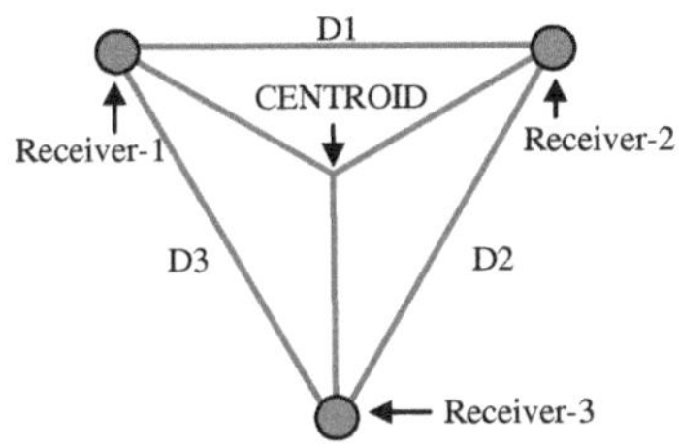

Fig. 18.1 Physical layout of the GPS receiver position in the proposed model

Fig. 18.2 Basic flow diagram of the proposed method

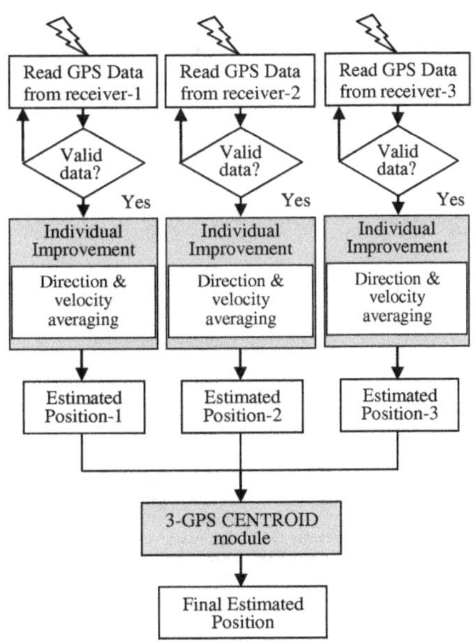

receives valid data, processes the data, and estimates the improved position for each individual receiver. Finally, the proposed approach calculates the centroid sums of the improved positions of the three GPS receivers. In following sections, we describe the details of the different functional blocks.

18.2.1 Accurate Calculation of the Reference Point

To estimate a precise position using the proposed method during navigation, an accurate reference point is needed. This paper utilizes a long-term averaging technique [11] to effectively calculate a reference point by using the Eq. (18.1):

$$AVG_x = \frac{1}{N}\sum_{i=1}^{N} x_i \quad \text{and} \quad AVG_y = \frac{1}{N}\sum_{i=1}^{N} y_i, \tag{18.1}$$

where x is latitude, y is longitude and N is the number of timestamps.

18.2.2 Invalid Data Check

Table 18.1 shows an example of the NMEA (National Marine Electronics Association) sentence information of GPS [13]. A GPS receiver sometimes provides an invalid sentence that contains null latitude, longitude, and altitude data or invalid fixed quality or a deficient number of required satellites, which makes invalid waypoints. This step filters those error data by checking the valid data flag, the number of connected satellites, latitude, and longitude values.

Table 18.1 Example of NMEA (national marine electronics association) sentence information of GPS [13]

$GPGGA,123519,4807.038,N,01131.000,E,1,08,0.9,545.4,M,46.9,M,,*47

Where:

GGA	Global Positioning System Fixed Data
123519	Fix taken at 12:35:19 UTC
4807.038,N	Latitude 48 deg 07.038' N
01131.000,E	Longitude 11 deg 31.000' E
1	Fix quality: 0 = invalid, 1 = GPS fix (SPS), 2 = DGPS fix, 3 = PPS fix ...
08	Number of satellites being tracked
0.9	Horizontal dilution of position
545.4,M	Altitude, Meters, above mean sea level
46.9,M	Height of geoid (mean sea level) above WGS84 ellipsoid
(empty field)	time in seconds since last DGPS update
(empty field)	DGPS station ID number
*47	the checksum data, always begins with *

18.2.3 Position Improvement of Each Individual GPS

Before calculating a centroid sum of three GPS receivers, the proposed approach improves the positioning accuracy of each individual GPS by using the direction and velocity averaging technique [11]. The direction and velocity averaging technique enhances the positioning data by estimating direction angle, velocity, and distance between two waypoints, as shown in Fig. 18.3.

New coordinate values are estimated by Eqs. (18.2) and (18.3) [11]:

$$X'_{n+1} = X'_n + D \times \cos\left(\tan^{-1}\frac{y_{n+1} - y_{n-1}}{x_{n+1} - x_{n-1}}\right) \quad (18.2)$$

$$Y'_{n+1} = Y'_n + D \times \sin\left(\tan^{-1}\frac{y_{n+1} - y_{n-1}}{x_{n+1} - x_{n-1}}\right) \quad (18.3)$$

where the distance is calculated by $D = \sqrt{(x_{n+1} - x_n)^2 + (y_{n+1} - y_n)^2} \times \frac{V_n}{V_{n-1}}$. X' and Y' are the new enhanced coordinate values.

18.2.4 Distance Threshold and Centroid Calculation of Three GPS Receivers

The final step in the proposed approach utilizes the previously estimated positions of the three GPS receivers and then estimates a more precious position by calculating the triangular centroid of the data. Figure 18.4 shows a detailed flow diagram of the centroid calculation of the 3 GPS receivers.

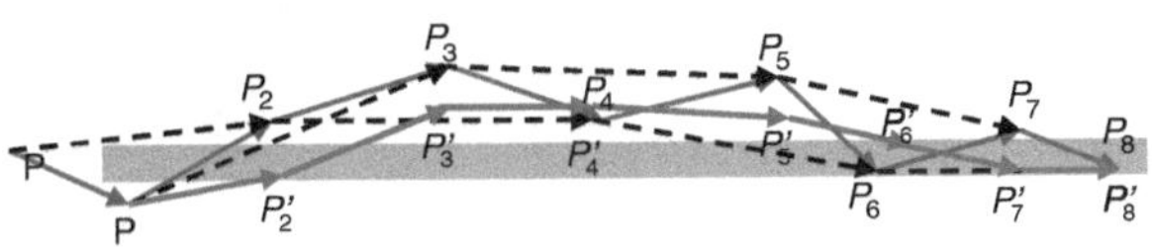

Fig. 18.3 Direction and velocity averaging process [11]

Fig. 18.4 Flow diagram of the centroid calculation of the 3 GPS receivers

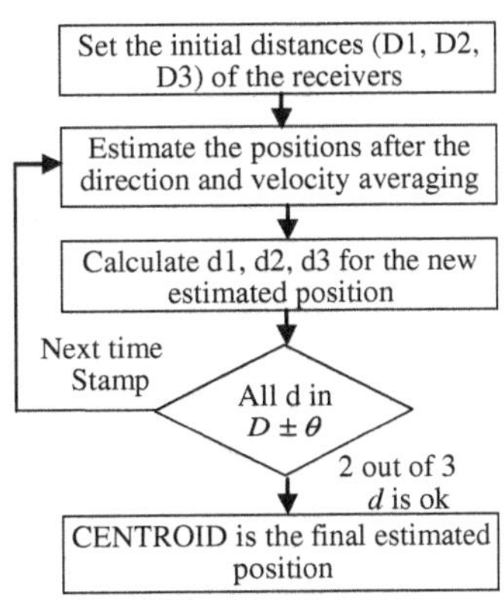

During the initial setup, the three GPS receivers are fixed at the known distances, D_1, D_2, and D_3. This process receives three estimated position values and recalculates the distances (d_1, d_2, and d_3) between the three new estimated points. The new distances (d_1, d_2, and d_3) are compared with the initially fixed distances with an error tolerance threshold ($+\theta/-\theta$). If at least 2 out of the 3 distances meet the condition, this step calculates the centroid of the three estimated points. The calculated position value is the final estimated position using the proposed model.

18.3 Experimental Results

In the experiment, we used three GARMIN GPS 19x HVS receivers for data collection. We attached the three receivers on a car, as shown in Fig. 18.5. The 1st and 2nd GPS receivers were attached at the left and right sides on the front of the car and the 3rd one was placed in the middle of the back of the vehicle, making a triangle configuration. Inside the car, we set up the developed simulation software on a laptop as a data processing terminal and plotted the data in Google maps.

A serial-to-USB convertor was used for the connection between the GPS receivers and the data processing terminal (laptop). Table 18.2 describes the specifications of the receivers.

We collected the GPS data while driving a car and performed several simulations at different locations. In the experiment, we ignored invalid data such as null values of latitude, longitude, time, and velocity. Figure 18.6 shows the results of the proposed approach.

Fig. 18.5 Experimental setup for an autonomous vehicle design

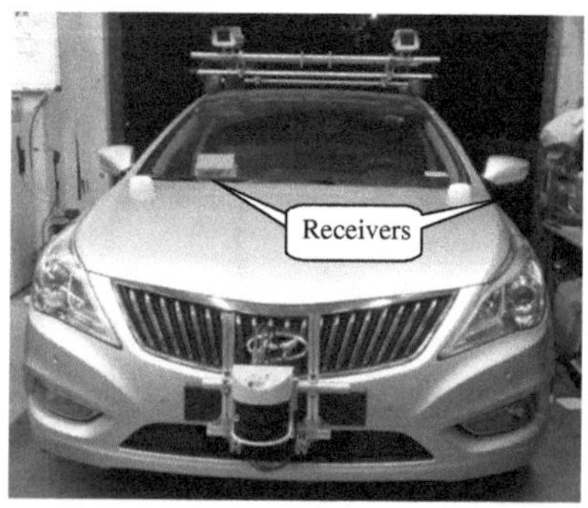

Table 18.2 The specifications of the receivers

Type	Updating rate	Accuracy	Provided data
Standard GPS receiver	1, 5, 10 records per second (we used 1 for steady data)	Less than 15 meters with 95 % typical	Pseudo range, integrated carrier phase, Doppler shift, satellite ephemeris, and processed data

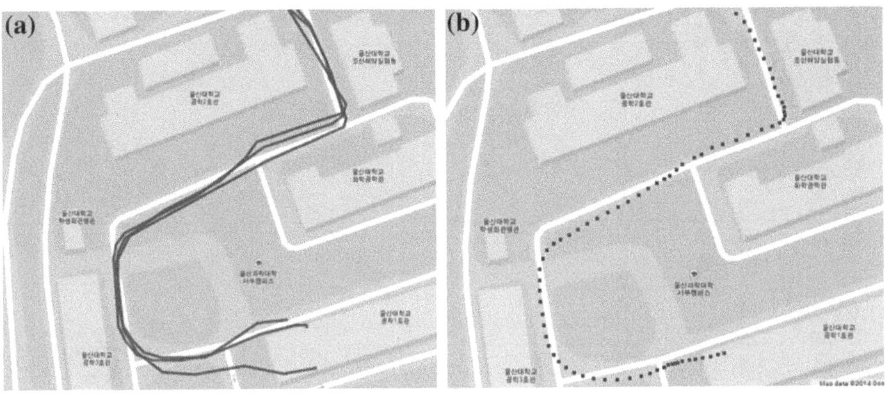

Fig. 18.6 Experimental results: **a** real data of the three GPS receivers and **b** processed data obtained using the proposed approach

Figure 18.7 and Table 18.3 demonstrate the improvement of the latitude and longitude values with the new estimated waypoints over the original GPS receiver's data, where the red dashed line represents the improvement of GPS receiver 1, the blue stared line is for receiver 2, and the black solid line corresponds to receiver 3.

The proposed approach resulted in an accuracy improvement of about 2–12 m during driving. In addition, we compared the results obtained from the proposed approach with two conventional methods, ARMA interpolation [9] and direction averaging [11]. Table 18.4 shows the average improvement (in meters) obtained using the proposed approach and other state-of-art models. The ARMA interpolation has two coefficient parameters which impact the accuracy of the new estimated points. On the other hand, the direction averaging method estimates the new position based on the last 2 steps of one GPS receiver. In this case, it cannot improve the positional data due to significant errors. Overall, the proposed method outperforms the other methods in terms of positioning accuracy.

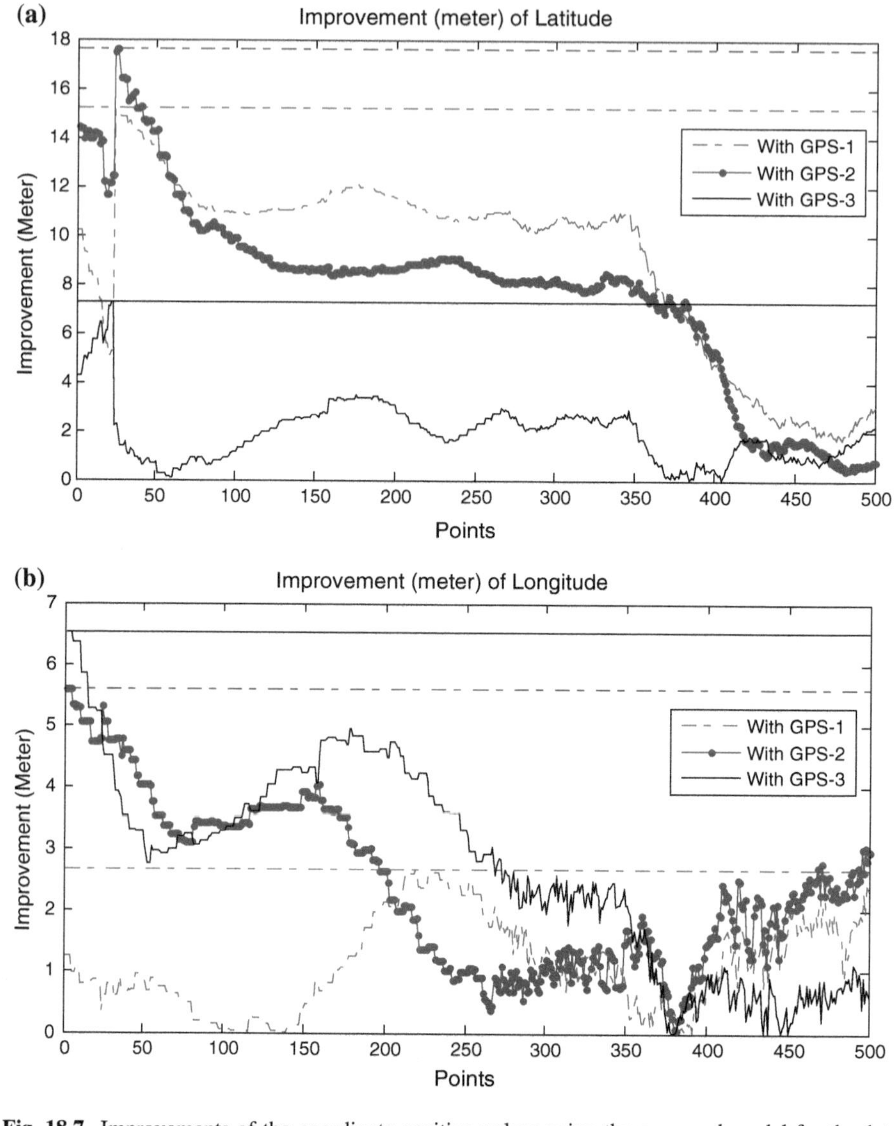

Fig. 18.7 Improvements of the coordinate position values using the proposed model for the 1st, 2nd, and 3rd GPS receivers: **a** latitude and **b** longitude

Table 18.3 Improvement of the maximum and average coordinate values (in meters) utilizing the three receiver's data

	With GPS-1		With GPS-2		With GPS-3	
	MAX	AVG	MAX	AVG	MAX	AVG
Latitude	15.203	11.731	17.589	8.672	7.271	2.562
Longitude	2.665	1.825	5.607	2.855	6.536	3.152

Table 18.4 Improvement of the average coordinate values (in meters) of the proposed model and other state-of-art methods

Experiment number	Proposed approach		ARMA interpolation		Direction averaging	
	Latitude	Longitude	Latitude	Longitude	Latitude	Longitude
1	8.439	9.312	5.167	6.212	7.535	6.813
2	6.977	5.747	3.236	3.571	4.512	3.563
3	11.731	1.825	7.718	4.544	8.955	4.452
4	9.856	8.594	2.497	4.981	6.253	6.125
5	6.945	4.221	1.416	4.204	4.265	5.565

18.4 Conclusion

This paper proposed a centroid-GPS approach to improve the positioning accuracy using three low-cost GPS receivers. The proposed approach estimates positions by employing the direction angle, speed, and distance of three GPS receivers with the help of an accurate reference point by using long-term averaging and an invalid data check. In the experiment, we used three GARMIN GPS 19x HVS receivers to evaluate the positioning accuracy of the proposed approach and two other conventional methods. The experimental results showed that the proposed approach improves the coordinate position more efficiently and outperforms other conventional methods in terms of positioning accuracy by significantly reducing data fluctuation.

Acknowledgments This work was supported by 2014 Funds of LG Yonam Foundation.

References

1. Mahmood FM, Salam ZABA (2013) A conceptual framework for personalized location-based Services (LBS) tourism mobile application leveraging semantic web to enhance tourism experience. In: IEEE 3rd international advance computing conference (IACC 2013), Ghaziabad, India, pp 287–291
2. Luo Y, Hoeber O, Chen Y (2013) Enhancing Wi-Fi fingerprinting for indoor positioning using human-centric collaborative feedback. Hum-Centric Comp Info Sci 3(2):1–23
3. Lin P, Li Q, Fan Q, Gao X, Hu S (2014) A real-time location-based services system using WiFi fingerprinting algorithm for safety risk assessment of workers in tunnels. Math Prob Eng 2014:1–10
4. Brković M, Simić M (2014) Multidimensional optimization of signal space distance parameters in WLAN positioning. Sci World J 2014:1–6
5. Huang J, Tsai C (2008) Improve GPS positioning accuracy with context awareness. In: First IEEE international conference on Ubi-Media computing, Lanzhou, China, pp 94–99
6. Hong S, Chang J (2013) A new k-NN query processing algorithm based on multicasting-based cell expansion in location-based services. J Convergence 4(4):1–6

7. Bisio I, Lavagetto F, Marchese M, Sciarrone A (2013) GPS/HPS-and Wi-Fi fingerprint-based location recognition for check-in applications over smartphones in cloud-based LBSs. IEEE Trans Multimedia 15(4):858–869
8. Zarazaga FJ, Álvarez PJ, Guillo J, López R, Valiño J, Muro-Medrano PR (2000) Use Cases of vehicle location systems based on distributed real-time GPS data. In: TeleGeo'2000: second international workshop on telegeoprocessing, Zaragoza, Spain, pp 53–61
9. Refan MH, Palangi H (2012) Positioning error reduction of a low-cost GPS receiver for kinematical applications. Am J Sci Eng 37:2221–2230
10. Garcia J, Zhou C (2010) Improving GPS precision and processing time using parallel and reduced-length wiener filters. J Telecom 2(2):91–98
11. Islam R, Kim J (2014) An effective approach to improving low-cost GPS positioning accuracy in real-time navigation. Sci World J 2014:1–8
12. Pike J (2009) GPS III operational control segment (OCX). http://Globalsecurity.org. Accessed 8 Dec 2009
13. GPS—NMEA sentence information. http://aprs.gids.nl/nmea/. Accessed 20 June 2014

Chapter 19
Intelligent Evaluation Models Based on Different Routing Protocols in Wireless Sensor Networks

Ning Cao, Russell Higgs and Gregory M.P. O'Hare

Abstract This paper aims to introduce some key parameters for the tracking application in wireless sensor networks. This paper has compared three different routing protocols which have been implemented in J-Sim simulation platform, and proposed several evaluation models. Based on these evaluation models, simulations are not necessary for the users to deploy sensors.

Keywords J-Sim · Routing protocols · Evaluation model

19.1 Introduction

A wireless sensor network consists of a number of sensors deployed either randomly or in a pre-determined state in a given two or three dimensional space, thus forming a network of sensors. Such sensors are designed to measure one or more physical quantities in the space, such as temperature or location. The sensors need to transmit this collected data to the end-user, who often will be remote from the space being measured, which could constitute a dangerous environment. Since the sensors concerned are wireless they are typically powered by a battery with a finite lifetime and power output, it may be impossible or indeed impracticable to recharge or replace such batteries. Thus in a real wireless sensor network a number of parameters naturally need to be considered such as energy consumption, network lifetime and network throughput. The network may be assigned a routing protocol,

---

N. Cao (✉) · R. Higgs · G.M.P. O'Hare
CLARITY: The Centre for Sensor Web Technologies, University College Dublin, Belfield, Dublin 4, Ireland
e-mail: ning.cao@ucd.ie

R. Higgs
e-mail: russell.higgs@ucd.ie

G.M.P. O'Hare
e-mail: gregory.ohare@ucd.ie

J.J. (Jong Hyuk) Park et al. (eds.), *Ubiquitous Computing Application and Wireless Sensor*, Lecture Notes in Electrical Engineering 331,
DOI 10.1007/978-94-017-9618-7_19

so that rather than the sensors transmitting directly to the end-user (single-hop protocol), they instead transmit their data via a number of other sensors with the data eventually arriving at the end-user (multi-hop protocol). Not using a single-hop protocol inevitably means that parameters such as network delay and latency will have to be considered, a crude measure of both would be the average number of hops needed for the data to reach the customer. This work seeks to establish relationships between these parameters and indeed others so as to show that varying a given parameter does or does not significantly affect another.

Wireless sensor networks are typically used to monitor the environment. One significant example of this is the location-tracking problem, whose goal is to trace the objects in the area in which the sensors are deployed. Clearly data sensed by sensors that are far from the area of activity is of little use, on the other hand, if the sensors are densely deployed, then the data sensed by some of these nodes close to the area of activity may be redundant. One of the widely used approaches to solve this problem is to place each sensor into either an active or sleep state and adjust these two modes as required.

Strategies by which to control the sensor state is one of the main research areas in wireless sensor networks, the aim being to conserve sensor energy and consequently extend network lifetime. In a tracking application, tracking accuracy and energy consumption are normally the key evaluation parameters. Thus obtaining a trade-off between energy consumption and tracking accuracy is a significant problem in the theory of wireless sensor networks.

Evaluation models based on different routing protocols in wireless sensor networks will be proposed in this paper. These evaluation models would allow the deployment of the sensor nodes to be accomplished without simulations to satisfy the demands of the customer.

The rest of this paper is structured as follows: Sect. 19.2 introduces the background of this research. Section 19.3 describes J-Sim simulation tool. Section 19.4 focuses on routing protocols. Section 19.5 defines the evaluation parameters. Section 19.6 focuses on the experimental set-up information and simulation results. Section 19.7 proposes several evaluation models and analyzes these models. Section 19.8 concludes this paper.

19.2 Background

Many researchers [1–4] have performed—simulations for target tracking applications in WSNs given their significant role in the efficiency of military and civil applications such as environment monitoring, target surveillance. These researchers have analyzed the relationship among some evaluation parameters, but most results in their published work are to minimize energy consumption or analyze the relationship between two parameters. No mathematical model with three or more parameters has been proposed, so far.

The following sub-sections will detail some of the published work on the evaluation parameters for target tracking.

19.2.1 Radius and Number of Sensors

In [5], Maity and Gupta considered a randomly distributed wireless sensor network covering a large area. They wished to find an estimate for the number of nodes required with the minimum critical communication distance to ensure network connectivity and stability. Using results from graph theory certain mathematical formula-based algorithms already existed for a relatively small number of sensors; however, the authors proposed a new formula based on mathematical simulation, which minimized the inter-node critical radius prediction for a large number of nodes. They chose MATLAB as their simulation tool and constructed a regression equation between the radius and number of nodes. The theoretical model provided better results if the number of nodes is less than 250, but their regression equation provided smaller answers for the radius to preserve connectivity for larger numbers of nodes.

19.2.2 Coverage and Lifetime

One of the key concepts considered in [6] is the coverage problem. As pointed out in [7], the coverage concept is a measure of the quality of service (QoS) of the sensing function and is subject to a wide range of interpretations due to a large variety of sensors and applications. The aim in this work is to make each location in the physical space within the sensing range have at least one sensor node. The goal for the authors to address the target coverage problem is to prolong the network lifetime of an energy constrained wireless sensor network. The sensor nodes are deployed randomly around the target and if the target enters into the sensor's sensing range, then the sensor will get the target location information and send it to a central processing node.

In order to prolong the lifetime of the network, the authors divide the sensor nodes into several sets, in which all the targets can be covered by the sensors in each set. To conserve energy and extend the lifetime of the sensor network, these sensor sets are activated successively, such that at any time instant only one set is active (the active state contains transmit, receive and idle). The nodes in the other sets will be in a low-energy sleep state. Sensor nodes can adjust their state between active and sleep, which will extend network lifetime compared with the case when all sensors are active. At the same time the number of active sensors in the application area will decrease, which will result in reducing contention at the MAC layer.

19.2.3 Energy, Density, Latency, Accuracy Trade-offs

In the literature [8] the researchers concluded that different parameters could be 'tweaked' in order to achieve better accuracy or longevity. The relationships between these parameters are not limited to tracking applications. The researchers examined the localization accuracy of a wireless sensor network with different data latency, target speed and application deadline. In order to get good accuracy, the routing protocol needed to be selected. In turn with a suitable choice of routing protocol in place, the tracking application parameters were adjusted to obtain better accuracy. In the simulations, the researchers showed the relationship between accuracy, radius, target speed and lifetime of the application.

This section has included a review of work on target-tracking applications. In some of the papers, the researchers have performed some simulations and obtained some useful results with related evaluation parameters. However, none of these have constructed a mathematical model for those related parameters nor have the trade-off approaches been discovered. The mathematical models will put forward a management tool between the evaluation parameters. Thus, simulation tools will not be necessary for the deployment of sensors resulting in a saving of both money and time.

19.3 J-Sim

J-Sim [9] (formerly known as JavaSim) is an open-source, component-based compositional network simulation environment. The system is based on the IEEE 802.11 [10, 11] implementation provided with J-Sim. IEEE 802.11 is the first wireless LAN (WLAN) standard proposed in 1997. J-Sim is implemented on top of the autonomous component architecture (ACA), components from the basic elements in this architecture and through these J-Sim implements the data transmission process. J-Sim provides a script interface that allows its integration with Tcl and has been developed entirely in Java. Java is a general purpose object-oriented computing language that is specifically designed to have as few implementation dependencies as possible.

J-Sim was selected as the simulation tool of choice for the following reasons:

The authors of J-Sim have performed detailed performance comparisons in simulating several typical WSN scenarios in J-Sim and NS-2. The simulation results indicate J-Sim and NS-2 incur comparable execution time, but the memory allocated to carry out simulation in J-Sim is at least two orders of magnitude lower than that in NS-2. As a result, while NS-2 often suffers from out-of-memory exceptions and was unable to carry out large-scale WSN simulations, the proposed WSN framework in J-Sim exhibits good scalability.

J-Sim models are easily reusable, so users can combine the components in the framework freely. J-Sim also provides a GUI, which makes it easy to operate the simulation.

J-Sim is a Java-based platform. The Java-based sensors could be integrated with Java-based simulation tools in the future.

Base on J-Sim simulation platform, several routing protocols have been implemented. Section 19.4 will analyze the routing protocols.

19.4 Routing Protocols

Flat protocols, hierarchical protocols and location-based protocols are the three main types of routing protocols in wireless sensor networks. Single-hop [12], LEACH [13] and Nearest Closer are typical and basic routing protocols for these three types respectively. So in this paper, Single-hop, LEACH and Nearest Closer protocols have been integrated into the simulation tool called J-Sim.

19.4.1 Single-hop Protocol

This is a simple protocol in which sensors transmit directly in one hop to the sink node. Latency will thus be minimized, but data collision may occur at the sink node and sensors situated a long way from the sink node may expend a large amount of energy in transmitting to the sink node.

19.4.2 LEACH Protocol

LEACH is a TDMA-based MAC protocol which is integrated with clustering and a simple routing protocol in wireless sensor networks.

The LEACH protocol differentiates itself in that it uses randomized rotation of the cluster-heads and it reduces the amount of data that needs to be transmitted to the sink node. Consequently, the use of clusters decreases the number of intermediate nodes. Furthermore, using rotating cluster-heads and adaptive clusters, the energy requirements of the system are in general distributed among all the sensors.

LEACH is an instance of a hierarchical protocol, in which most sensor nodes transmit sensed data to cluster head, and the cluster head thereafter aggregates and compresses the data and forwards this data to the sink node. Each node uses a stochastic algorithm at each round to determine whether it will become a cluster head in that round. Consequently, if the remaining energy for each node can be measured, it will make great contributions to this research area.

19.4.3 Nearest Closer Protocol

The NC protocol relies on the strategy of greedy forwarding, which tries to bring a transmitted packet closer to the sink node in each step or hop using only local information. Thus each sensor forwards the message to its neighbour that is most suitable from a local point of view.

19.5 Evaluation Parameters

In this section, the evaluation parameters of Reliability, Lifetime, Coverage and Density will be defined for use in the following sections. This paper will focus on the parameters of Energy, Density, Coverage and Reliability, based on the results of experiments measuring these parameters, an intelligent evaluation model will eventually be constructed.

19.5.1 Reliability

In this work, experiments are conducted using the concept of reliability, defined by:

Reliability = the number of packets received by the sink node/the number of packets sent to the sink node.

19.5.2 Lifetime

The definition for network lifetime we have taken in our experiments is the time when the last packet is received by the sink node.

19.5.3 Coverage

In a wireless sensor network, all the nodes are deployed with a predefined communication radius. This limits the area that the sensors can detect phenomena in, so that this paper will also evaluate the effect that this limitation on coverage has on lifetime and reliability by varying the radius.

19.5.4 *Density*

The number of sensors deployed in a fixed area will be taken as the Density parameter in this paper. Obviously as the number of sensors goes up, so does the average number of sensors per square metre i.e. the density of the sensors.

19.6 Experimental Set-up and Simulation Results

The simulated area for the experiments in the following sections is a 10 m × 10 m^2 with randomly deployed nodes. The sink node for this application is located in the middle of this square. One of the primary reasons for selecting this setup is to allow the results to be generalized to large areas by concatenation of networks similar to this. For example, a 70 m × 70 m^2 region could be configured using 49 instances of the setup used here in a 10 × 10 grid formation. All the points (in the figures) in the following section are the average value from at least 5 separate experiments.

Figures 19.1, 19.2 and 19.3 show the main results for the Single-hop, LEACH and NC protocols respectively.

19.6.1 *Single-hop Protocol*

19.6.1.1 LEACH Protocol

Nearest Closer Protocol

In Fig. 19.1, Lifetime reached its lowest value when the number of sensors equalled 40, whereas the highest value of Lifetime occurred when there were 20 sensors. All in all, Lifetime is between 14,500 and 16,500. The reason why the Lifetime is around 15,600 is that any sensor in this network could transmit data to the sink node directly, that is to say almost all the sensors do not have enough energy to transmit packets to the sink node when the first sensor fails to transmit packets to the sink node. Thus, the Lifetime doesn't change too much as the number of sensor nodes increases.

In Fig. 19.1, the relationship between the number of sensors and Reliability is very clear. Reliability for this application decreased as the number of sensors increased. Although the curve is essentially a straight line, the Reliability is above 99 % in all cases considered.

In Fig. 19.2, the Lifetime for this experiment increases from 1541 to 2097, 2503, 2779, 3077, 3307, 3427, 3475, 3601, 3707, 3903 and finally to 4043. Figure 19.2 also shows the relationship between Reliability and number of sensors. The Reliability for this experiment decreased from 22.86 to 16.35, 12.44, 9.31, 7.46, 6.45,

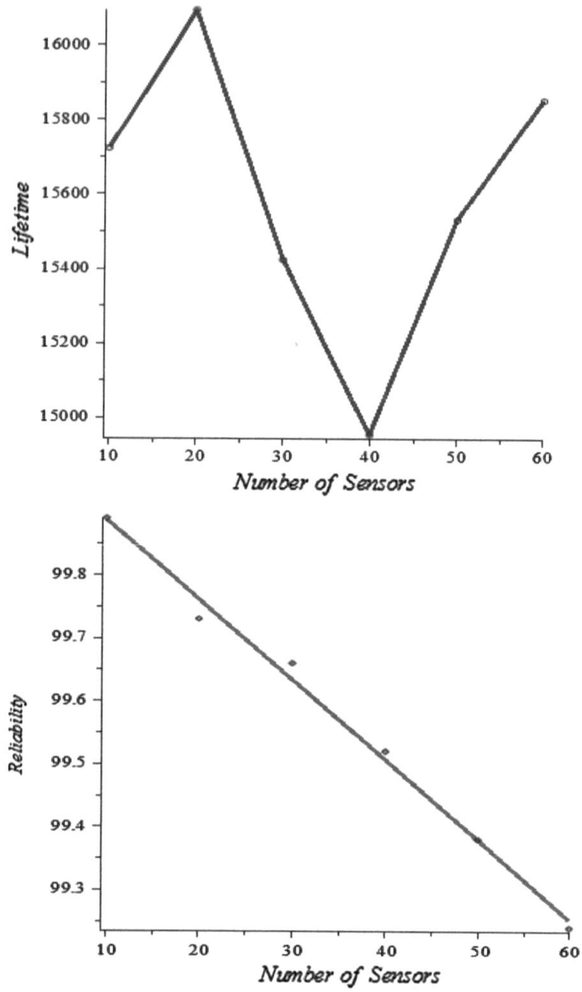

Fig. 19.1 Density, lifetime and reliability relationship for single-hop

5.13, 4.67, 4.38, 3.23, 2.40 % and finally to 1.89 %. Thus the relationship between the number of sensors and Reliability is one of negative correlation. This may be explained as follows: As density increases, the cluster heads will consume much more energy to communicate with the sensor nodes. Then the remaining energy for each cluster head will decrease rapidly with the increasing density. In addition, in order to transmit data to the sink node will also consume a lot of energy as well, and then more and more elected cluster heads cannot transmit data to the sink node. Thus it is reasonable to expect Reliability to decrease with Density.

When the number of sensors equalled 60, Reliability was found to be 1.89 %. Compared to the previous data, Reliability has reached a very low level, but at this

Fig. 19.2 Density, lifetime and reliability relationship for LEACH

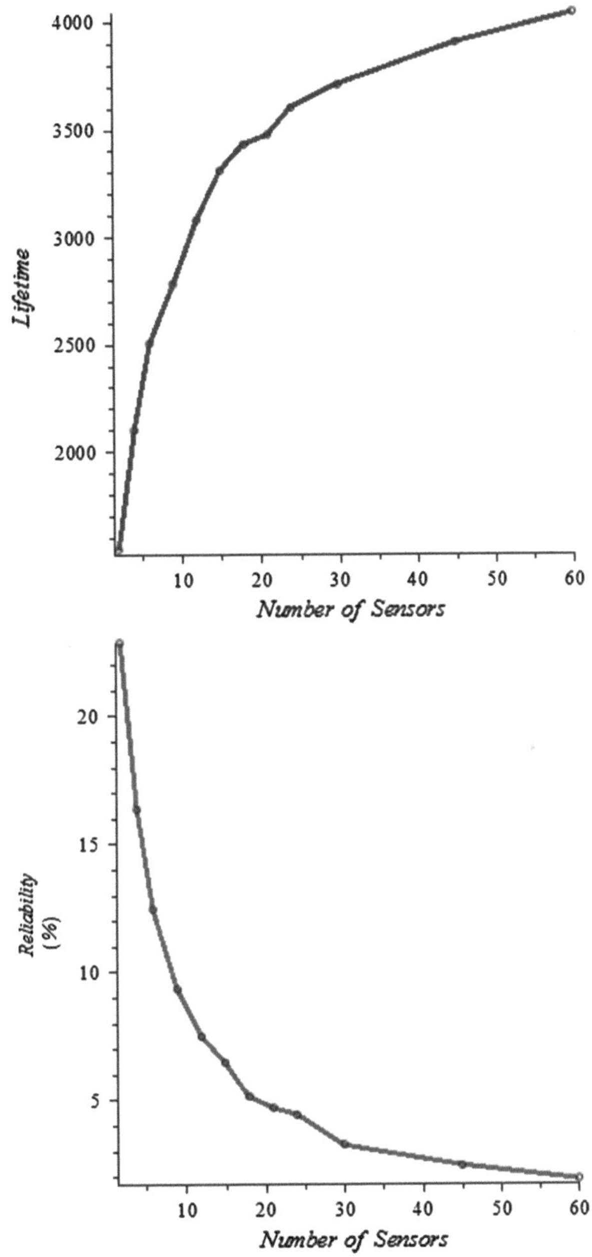

point, the network Lifetime reached the highest value 4,043. Thus it is possible for users to choose an optimum Density value for this application depending on the Reliability required.

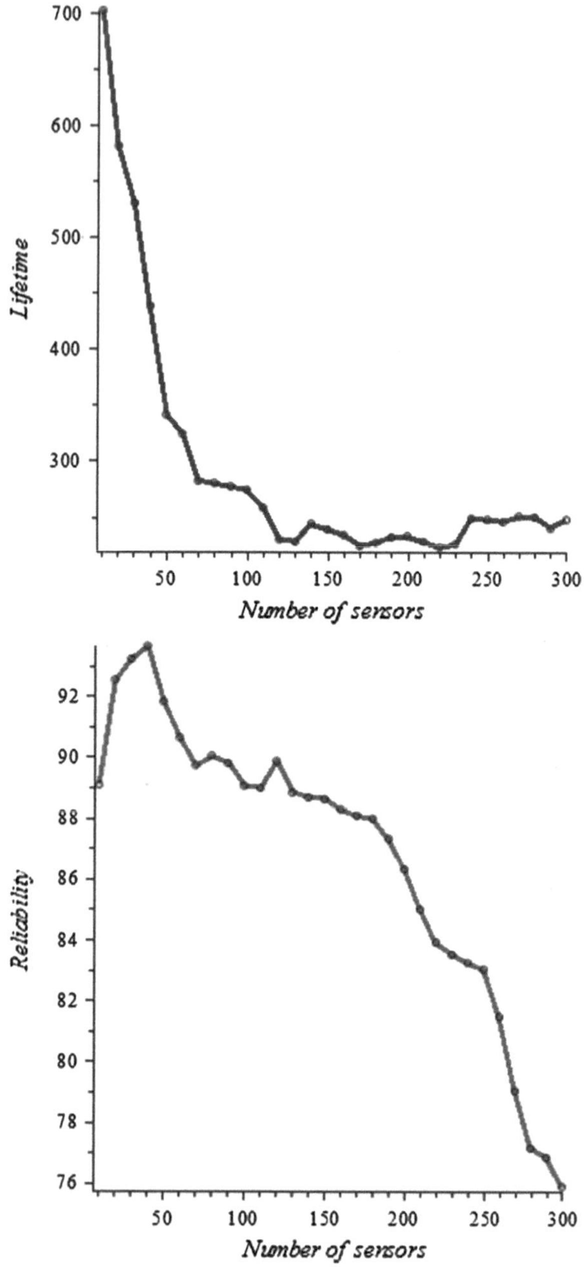

Fig. 19.3 Density, lifetime and reliability relationship for NC

In Fig. 19.3, Lifetime reached its lowest value when the number of sensors equalled 220, whereas the highest value of Lifetime occurred when there were 10 sensors.

Note that when the number of sensors equals 40, Reliability was 93.64 %, its highest level, but with this number Lifetime is fairly short. This illustrates that it is possible for users to choose an optimum Density value for this application depending on the Reliability and Lifetime required.

In Fig. 19.3, Reliability for this application increased as the number of sensors increased to 40, when it reached its highest value. It then essentially decreased as the number of sensors increased from 40 to 300, when it reached its lowest value. This may be explained by observing that as the density of nodes increases more sensors will join the data transmission process and consequently communication among the sensors will become more and more complex. So dropped data due to data collision and latency cannot be ignored. Consequently it is reasonable to expect Reliability to decrease with Density.

19.7 Evaluation Models and Analysis

This work has obtained several simulation results for Single-hop, LEACH and NC protocols and the simulation results are based on the same simulation setup. Based on above results, this work has proposed several important mathematical equations.

For the Single-hop protocol:

$$\mathrm{Lifetime} = 15600,$$

and

$$\mathrm{Reliability} = 100.014 - 0.0126857142857145\mathrm{n}.$$

where n denotes the number of sensors.

For the LEACH protocol:

$$\mathrm{Lifetime} = 1135.237983\mathrm{n}/[1 + 0.24225\mathrm{n}],$$

and

$$\mathrm{Reliability} = 0.396385551/[2 + 0.326372516^{*}(\mathrm{n} - 2)],$$

where n denotes the number of sensors.

$$\mathrm{Reliability}/\mathrm{Lifetime}^{*}\mathrm{Number\ of\ sensors} = \mathrm{Constant}$$

For the NC protocol:

$$\mathrm{Lifetime} = 49181416/\left(66.53\mathrm{n}^2 + 686750\mathrm{n} + 1250578\right)$$

for $10 \le n \le 130$, and

$$\text{Reliability} = 0.126^{*}\text{Lifetime(n)} + 0.0022^{*}\text{n}^{*}\text{Lifetime(n)}$$

for $10 \leq n \leq 120$.

where n denotes the number of sensors.

Based on the above equations, people can predict the Lifetime and Reliability directly. The following equation can then be utilized to set weight values for the Lifetime and Reliability:

$$\text{Selection} = \text{a}^{*}\text{Lifetime} + (1 - \text{a})^{*}\text{Reliability}$$

for $0 \leq a \leq 1$.

If the users only care about the Lifetime and not the Reliability, then the coefficient a in the above equation should be set as 1. Conversely, if the users' only concern is the Reliability and not the Lifetime, then the coefficient a in the above equation should be set as 0. In more generality users can assign their own desired weight values for the Lifetime and Reliability in the above equation depending on how much they wish to emphasize each of these two parameters. Thus, the users can obtain a Selection value for each routing protocol. After comparing these Selection values, the users can select the best routing protocol for their sensor network. Consequently, the deployment of sensor nodes could be achieved without further simulations.

19.8 Conclusions

Flat protocols, hierarchical protocols and location-based protocols are the three main types of routing protocols in wireless sensor networks. Single-hop, LEACH and Nearest Closer are typical and basic routing protocols for these three types respectively. So in this paper, Single-hop, LEACH and Nearest Closer protocols have been integrated into the simulation tool called J-Sim.

Energy consumption (measured by Lifetime in this paper), density, coverage, and accuracy (measured by Reliability in this paper) are key parameters in wireless sensor networks.

Several simulation results have been proposed in this paper. Based on these results, this paper has developed some mathematical models to organize a high efficiency wireless sensor network without prior use of simulations. The most important formula which has been obtained is the following one in LEACH:

$$\text{Reliability/Lifetime}^{*}\text{Number of sensors} = \text{Constant}$$

This equation says that the rate of successful packet reception per unit time is independent of the number of sensors. The constant in the equation will depend on the parameters in the simulation and it is a reasonable model if the number of cluster heads is small. The number of sensors in the cluster is one way of measuring packets received, but again ignores data lost from sensors before it reaches the cluster head. Such lost data will increase with the number of sensors, so that the constant on the right hand side of the equation may have to be found for a small, medium or large number of sensors.

These evaluation models could allow the deployment of the sensor nodes to be accomplished without simulations to satisfy the demands of the customer.

Acknowledgments This Research is supported by Science Foundation Ireland under grant 07/CE/1147.

References

1. Lee W, Xu Y (2003) On localized prediction for power efficient object tracking in sensor networks. In: Proceedings of the 23rd international conference on distributed computers systems workshops, pp 434–439
2. Niyogi K, Mehrotra S, Venkatasubramanian N, Yu X (2004) Adaptive target tracking in sensor networks. In: Proceedings of the communication networks and distributed systems modelling and simulation conference, pp 253–258
3. Huang CF, Lee H, Kuo SP, Tseng YC (2004) Location tracking in a wireless sensor network by mobile agents and its data fusion strategies. Comput J 47(4):448–460
4. Zhao F, Shin J, Reich J (2002) Information-driven dynamic sensor collaboration for tracking applications. Signal Process Mag 19(2):61–72
5. Maity C, Gupta A (2010) Critical communication radius prediction with random distributed nodes in wireless sensor network. In: Proceedings of the annual seminar of C-DAC Noida technologies, pp. 31–38
6. Cardei M, Thai MT, Li Y, Wu W (2005) Energy-efficient target coverage in wireless sensor networks. In: Proceedings of the 24th annual joint conference of the IEEE computer and communications societies, vol 3, pp 1976–1984
7. Meguerdichian S, Koushanfar F, Potkonjak M, Srivastava M (2001) Coverage problems in wireless ad-hoc sensor networks. In: Proceedings of the 20th annual joint conference of the IEEE computer and communications societies, vol 3, pp 1380–1387
8. Tynan R, O'Grady MJ, O'Hare GMP, Muldoon C (2009) Benchmarking latency effects on mobility tracking in WSNs. In: Proceedings of the second international workshop on applications of Ad-hoc and sensor networks, pp 768–774
9. Sobeih A, Chen W, Hou J C, Kung L, Li N, Lim H, Tyan H, Zhang H (2005) J-Sim: a simulation environment for wireless sensor networks. In: Proceedings of the 38th annual symposium on simulation, pp 175–187
10. Bianchi G (2000) Performance analysis of the IEEE 802.11 distributed coordination function. IEEE J Sel Areas Commun 18(3):535–547
11. Garg S, Kappes M (2003) An experimental study of throughput for UDP and VoIP traffic in IEEE 802.11b networks. Wireless Commun Networking 3:1748–1753
12. Cao N, Higgs R, O'Hare GMP (2014) Intelligent evaluation models based on the single-hop protocol in wireless sensor networks. In: Proceedings of the first international workshop on engineering energy efficient WSNs in conjunction with AINA 2014
13. Cao N, Higgs R, O'Hare GMP (2012) An intelligent evaluation model based on the LEACH protocol in wireless sensor networks. In: International conference on cyber-enabled distributed computing and knowledge discovery, pp 381–388

Chapter 20
Implementation of Personalized Wellness Service

Marie Kim, Namje Park and Hyo-Chan Bang

Abstract In this paper, we propose a secure framework for mobile based RFID services using personal policy-based access control for personalized ultra-high frequency (UHF) tags employing the Electronic Product Code (EPC). The framework provides a means for safe use of mobile phone-based RFID services by providing security to personalized wellness service. This new technology aims to provide absolute condentiality with only basic tags.

Keywords Wellless · Personalized · RFID · Profile

20.1 Introduction

Radio frequency identification (RFID) technology is widely used in supply chain management and inventory control, and is recognized as a strong potential vehicle for ubiquitous computing. However, continued development and global adoption has also raised fears of the potential for exploiting such tags for privacy infringement in 'Big Brother' type scenarios. Thus, information security and privacy protection are as important as standardization of the technology behind the tag, reader, middleware, etc. Conventional authentication algorithms and protocols are not applicable in the greatly resource-limited RFID paradigm, and so new technology should

M. Kim · H.-C. Bang
Electronics and Telecommunications Research Institute (ETRI), 218 Gajeong-Ro, Yuseong-Gu, Daejeon 305-700, Korea
e-mail: mariekim@etri.re.kr

H.-C. Bang
e-mail: bangs@etri.re.kr

N. Park (✉)
Department of Computer Education, Teachers College, Jeju National University, 61 Iljudong-Ro, Jeju-Si, Jeju 690-781, Korea
e-mail: namjepark@jejunu.ac.kr

J.J. (Jong Hyuk) Park et al. (eds.), *Ubiquitous Computing Application and Wireless Sensor*, Lecture Notes in Electrical Engineering 331,
DOI 10.1007/978-94-017-9618-7_20

be developed in terms of general interconnection among elements and their characteristics of RFID to such technology that meets the RFID circumstances.

The typical architecture of an RFID system, as defined by EPCglobal [1], comprises tags embedded or attached to objects, tag readers that read tag information, and a backend Information Services (IS) server that provides the required information. The tag reader can be designed to be portable or handheld, which allows for several possible applications. While RFID is most commonly used in business-to-business (B2B) commerce for managing supply channels, distribution, and logistics, there is also growing interest in the integration of tag readers with mobile phones, allowing individuals to collect and use tag data, such as in business-to-customer (B2C) marketing. And although current implementations have been limited to movie promotions and museums, where information security is not a major concern, continued development will see more frequent adoption in such fields as retail, medical care, and electrical drafts, where security and privacy are indispensable.

Researchers have developed many techniques to address the various security flaws in RFID systems [6, 8, 9], the simplest being to "kill" tags before they are in the hands of the user [13]. However, because these low-cost tags have numerous applications, the user may want the tag to remain active. As one solution, Rieback et al. proposed the *RFID Guardian* system [12], which relies on a strong proxy device—a mobile phone or PDA—to mediate access by an external reader to tags for auditing of scans and tags, key management, access control, and user authentication. The *RFID Enhancer Proxy* (REP) proposed by Juels et al. requires the use of a similar higher-computing-power proxy device [6] but provides better tag acquisition and ownership transfer. Kim et al.'s *Mobile Agent for RFID Privacy Protection* (MARP) is a further development that aims to provide high-level privacy [8] by employing a public key center that manages keys for the readers, tags, server, and proxy. Kim et al. also recently proposed a scheme suitable for mobile-phone-based reader systems [3], but the method only provides reader-based authentication, which is not sufficient.

Here we propose a secure framework for mobile-phone based RFID services using personal privacy-policy-based access control for personalized ultra-high frequency (UHF) tags employing the Electronic Product Code (EPC). The framework, called M-RPS, has dynamic capabilities that extend upon extent trust-building service mechanisms for RFID systems. This new technology aims to provide absolute confidentiality with only basic tags.

20.2 Strategic Security Framework Architecture

20.2.1 Multilateral Approaches for Improved Privacy

This technology is aimed at RFID application services like authentication of tag, reader, and owner, privacy protection, and non-traceable payment system where stricter security is needed.

- Approach of Platform Level: This technology for information portal service security in offering various mobile RFID applications consists of application portal gateway, information service server, terminal security application, payment server, and privacy protection server and provides a combined environment to build a mobile RFID security application service easily [2, 3].
- Approach of Protocol Level: It assists write and kill passwords provided by EPC (Electronic Product Code) Class1 Gen2 for mobile RFID tag/reader and uses a recording technology preventing tag tracking. Information technology solves security vulnerability in mobile RFID terminals that accept WIPI as middleware in the mobile RFID reader/application part and provides E2E (End-to-End) security solutions from the RFID reader to its applications through WIPI based mobile RFID terminal security/code treatment modules.
- Approach of Privacy Level: This technology is intended to solve the infringement of privacy, or random acquisition of personal information by those with RFID readers from those with RFID attached objects in the mobile RFID circumstance except when taking place in companies or retail shops that try to collect personal information. The main assumptions are privacy in the mobile RFID circumstance when a person holds a tag attached object and both information on his/her personal identity (reference number, name, etc.) and the tag's information of the commodity are connected. Owners have the option to allow access to any personal information on the object's tag by authorized persons like a pharmacist or doctor but limit or completely restrict access to unauthorized persons [4–6].

20.3 Implementation of Personalized Wellness Service

20.3.1 Privacy Policy for Patient Care at a Hospital

We implemented the proposed system for tracking patient care at a hospital. Context-relevant information is important in a ubiquitous computing environment for providing medical care. Different user policies are necessary for patient tags and product tags in EPC global's enterprise application. This ubiquitous sharing system for medical information poses a serious threat to the privacy of personal medical information such location, health, and clinical history. Standards such as Health Level Seven (HL7) do not allow customization and do not include rigorous privacy mechanisms. Therefore, we propose a mechanism that manages privacy policy in a user-centric manner for ubiquitous medical care. It is flexible, secure, and can be integrated with a cryptographic algorithm for mitigating the aforementioned problems.

20.3.2 Design M-RPS Based Customized Service

In a hospital, tags can be used for asset management for location finding. Patient tags are effective in preventing medical accidents, but must be properly designed and constructed to avoid massive collateral damage to user privacy. Hence, we define three-step privacy-aware service architecture for our mobile RFID-based medical application service [14]. The first step is setting the default level of access control over patient information in the default policy. The second step is user-controllable profile-based privacy protection, and the third step is auditable privacy management. Furthermore, we introduce a new RFID-based service model and mobile phone application.

The mobile RFID reader requests for information related to a tag attached to a patient from the backend IS via the middleware system. The mechanism allows individuals to control who can access their personal information. For privacy management, we apply the proposed profile-based privacy management architecture by the addition of a privacy bit to the tag, which is a simple and cost effective mechanism. The privacy bit is the only reference for the privacy service. The medical RFID information server check the privacy guaranteed service or not from the privacy policy. To illustrate how the privacy policy works on the IS, let us consider its use in the application and content information system of the service provider. The privacy level is stored in M-RPS. The RFID code format for the application is defined in the mobile RFID application data format as standard. The default privacy level follows the privacy applied standard of each application service; and if there is no standard, the privacy level is determined based on the results of a privacy impact assessment. The privacy level consists of a 10-tuple of information, where 'L = L1, L2, …, L10' as the default privacy policy. It also protected by a secure tag area and privacy server. We also define privacy weights for medical information, as shown in Table 20.1.

Classify the personal medical information by patient's policy and make personal's profile. The patient can control his privacy level. Encrypted information can be transferred between the hospital and an emergency transportation service in XML format with security (WS Security) and also can be subject to the standard access control technology for Web services (XACML) (Fig. 20.1).

Table 20.1 Examples of a default privacy weight level

Privacy related people	Privacy weight	Privacy information
Doctor	L4–L9	Medical history treatment information
Nurse	L4–L9	Medical history treatment information
General doctor	L3–L7	Medical treatment
General nurse	L3–L7	Medical treatment
Family	L2–L6	Medical tracking information
Emergency agency	L2–L6	Medical tracking information
Others	L1	All cut off

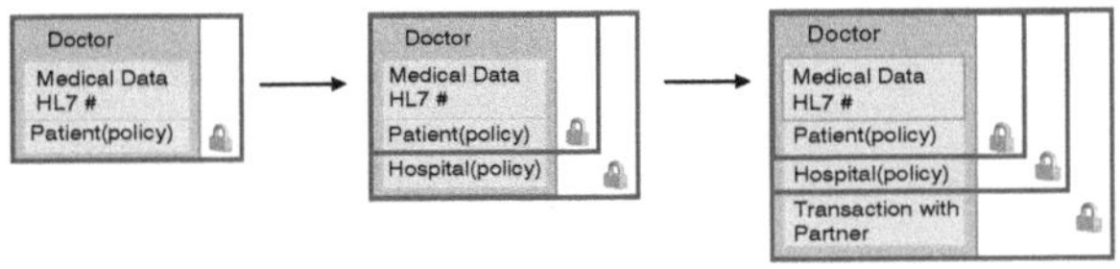

Fig. 20.1 Electronic signature and authentication

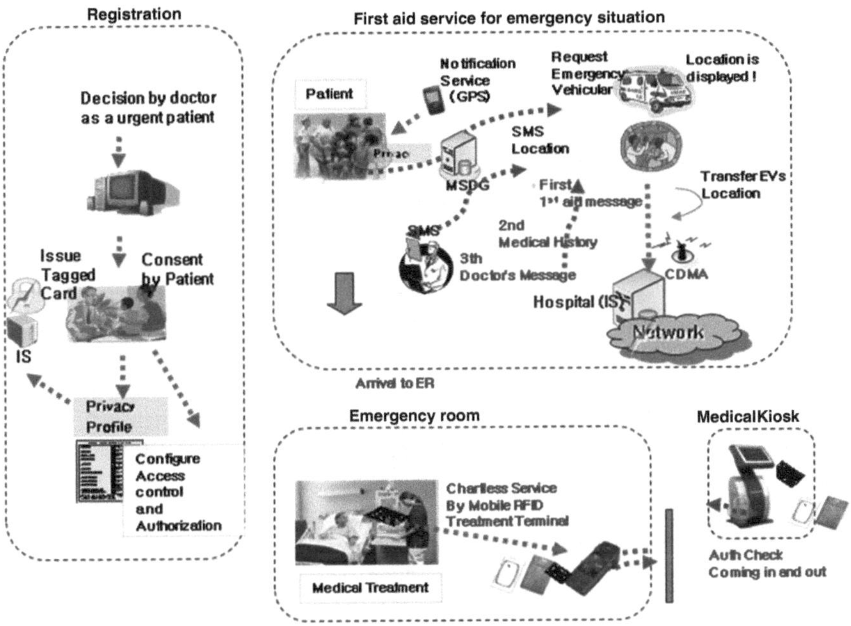

Fig. 20.2 Medical examination with proposed system

In the proposed hospital data management system, RFID-tagged medical card are given to patients on registration. Patients with sensitive conditions, for example, heart disease or cerebral hemorrhage, can use the medical card to rapidly provide medical history that can used for fast application of first aid. Further, biosensors can be incorporated to provide real-time data to the doctor for each specific patient. The RFID patient tags also can be used to verify patient identity to ensure the correct treatment is administered. Thus, the system allows chartless service (Fig. 20.2).

20.3.3 Implementation

The hospital generated an initial set of control data, which included the patient code, medical ID, and related information. The default privacy level was used and

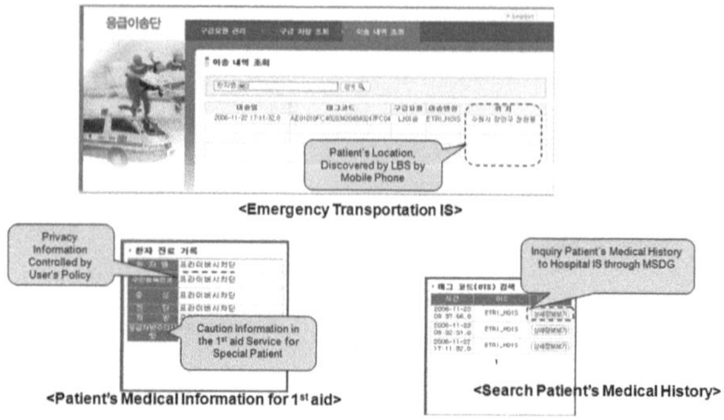

Fig. 20.3 Proposed customized ubiquitous hospital model

the patient was not allowed to control security policy. In order to provide authentication and privacy interface to patient as a agent in medical discovery gateway and hospital's information server system. Essentially, each bit of sensitive data was initially classified by the default privacy weight, which was then modified by the end user's detailed policy. The user-controllable privacy policy in this system evaluation is considered a basic part of RFID privacy management. The compatibility and scalability may be limited, which will hamper system migration, but the mechanism is suitable for policy based privacy control. The proposed privacy management mechanism was implemented in an actual medical emergency room, including a networked medical information RFID kiosk, RFID networked emergency rescue system, and medical examination service, as shown in Fig. 20.3. There is some approach applying the RFID to medicine and hospital. From above, proposed privacy scheme has advantages in custom centric approach aspect for constructing a privacy aware ubiquitous medical system [7, 10, 11, 15].

20.4 Conclusion

RFID technology will evolve to become ubiquitous, allowing automatic detection and delivery of information on the surrounding environment, and interconnecting them through the network. This will require RFID implementation of security measures as the technology is vulnerable to privacy infringement via counterfeiting, falsification, camouflage, tapping, and tracking. Therefore, it is necessary to enact laws and regulations that meet the expectations of consumer protection organizations that are sensitive to individual privacy, and develop and apply secure technologies that can follow such laws and regulations.

Mobile RFID readers are being actively researched and developed throughout the world, and more efforts are underway for the development of related service technologies. Though legal and institutional systems endeavor to protect privacy and encourage data protection, the science and engineering world must also provide suitable technologies. Seemingly, there are and will be no perfect security/privacy protection methods. The technologies proposed in this paper, however, would contribute to the development of secure and reliable RFID systems.

Acknowledgments This paper is extended and improved from accepted paper of UCSWSN 2014, ICHIT2011 conferences. This work was supported by Electronics and Telecommunications Research Institute (ETRI) grant funded by the Korea government. And, this research was supported by Basic Science Research Program through the National Research Foundation of Korea (NRF) funded by the Ministry of Education (2013R1A1A4A01013587).

References

1. Kang T, Lee H, Park D, Bang H, Park N (2013) Creation mechanism for access group based on user privacy policy-based protection. In: MUSIC, pp 125–130
2. Park N (2011) Implementation of terminal middleware platform for mobile RFID computing. IJAHUC 8:205–219
3. Park N (2011) Customized healthcare infrastructure using privacy weight level based on smart device. ICHIT 2:467–474
4. Park N, Lee K, Yoo S, Lee J, Kim Y, Kim H (2011) Secure RFID personal data management using privacy reference profile. In: FGIT, pp 268–276
5. Park N, Kwak J, Kim S, Won D, Kim H (2006) WIPI mobile platform with secure service for mobile RFID network environment. In: Shen HT, Li J, Li M, Ni J, Wang W (eds) APWeb Workshops 2006. LNCS, vol 3842. Springer, Heidelberg, pp 741–748
6. Park N (2010) The implementation of open embedded s/w platform for secure mobile RFID reader. J Korea Inf Commun Soc 35(5):785–793
7. Park N (2011) Secure data access control scheme using type-based re-encryption in cloud environment. In: Katarzyniak R, Chiu T, Hong C et al (eds), vol 381. Springer, Heidelberg, pp 319–327
8. Park N (2010) Security scheme for managing a large quantity of individual information in RFID environment. In: Zhu R, Zhang Y, Liu B et al (eds), vol 106. Springer, Heidelberg, pp 72–79
9. Park N (2012) Mobile RFID/NFC linkage based on UHF/HF dual band's integration in U-sensor network era. In: Park JH, Kim J, Zou D et al (eds), vol 180. Springer, Netherlands, pp 265–271
10. Park N (2014) Design and implementation of mobile VTS middleware for efficient IVEF service. J Korea Inf Commun Soc 39C(6):466–475
11. Park N (2011) Secure UHF/HF dual-band RFID: strategic framework approaches and application solutions. In: Computational collective intelligence. Technologies and applications, lecture notes in computer science, vol 6922. Springer, Heidelberg, pp 488–496
12. Park N (2011) Customized healthcare infrastructure using privacy weight level based on smart device. In: Communications in computer and information science, vol 206. Springer, Heidelberg, pp 467–474
13. Park N (2008) Reliable system framework leveraging globally mobile RFID in ubiquitous era. Ph. D. Thesis. Sungkyunkwan University, South Korea
14. Park N, Kim M (2014) Implementation of load management application system using smart grid privacy policy in energy management service environment. Cluster Computing, vol 17. Springer, New York, pp 653–664
15. Park N (2012) Cell phone based mobile RFID: models, mechanisms and its security. Int J Radio Freq Identif Technol Appl 4(1):67–101

Chapter 21
SOM Clustering Method Using User's Features to Classify Profitable Customer for Recommender Service in u-Commerce

Young Sung Cho, Song Chul Moon and Keun Ho Ryu

Abstract This paper proposes an SOM clustering method using user's features to classify profitable customer for recommender service in e-Commerce. In this paper, it is necessary for us to classify profitable customer with RFM (Recency, Frequency, and Monetary) score, to use the purchase data to join the customers using SOM with input vectors of different features, RFM factors in order to do the recommending services in u-commerce, to reduce customers' search effort for finding items, and to improve the rate of accuracy. To verify improved performance of proposing system, we make experiments with dataset collected in a cosmetic internet shopping mall.

Keywords RFM · Collaborative filtering · SOM (self-organizing map)

21.1 Introduction

Due to the advent of ubiquitous networking environment, it is becoming a part of our common life style that the demands for enjoying the wireless internet using intelligent portable device such as smart phone, are increasing anytime or anyplace without any restriction of time and place. Data mining is useful in finding knowledge from huge amounts of data. Deboeck and Kohonen describe how SOM (Self-Organizing Map) can be used for effective clustering and segmentation of

Y.S. Cho · K.H. Ryu
Database and Bioinformatics Laboratory, Computer Science in College of Electrical and Computer Engineering, Chungbuk National University, Cheongju, Korea
e-mail: youngscho@empal.com

K.H. Ryu
e-mail: khryu@dblab.chungbuk.ac.kr

S.C. Moon (✉)
Department of Computer Science, Namseoul University, Cheonan, Korea
e-mail: moon@nsu.ac.kr

J.J. (Jong Hyuk) Park et al. (eds.), *Ubiquitous Computing Application and Wireless Sensor*, Lecture Notes in Electrical Engineering 331,
DOI 10.1007/978-94-017-9618-7_21

financial data [1]. Clustering algorithm is a kind of customer segmentation methods commonly used in data mining. In this paper, SOM network is applied to segment the purchase data to join user data and finally forms clusters of the purchase data to join user data with different features, RFM factors in order to do the recommending services in u-commerce. The recommendation system helps customers to find easily items and helps the e-commerce companies to set easily their target customer by automated recommending process. Therefore, customers and companies can take some benefit from recommendation system. The possession of intelligent recommendation system is becoming the company's business strategy. A recommendation system using RFM segmentation analysis technique to meet the needs of customers, it has been actually processed the research [2–5]. We can make the solution for an efficient purchase pattern clustering based on SOM. Finally, we can improve the performance of personal ontology recommender system through SOM learning method based on the purchase data to show customer's buying patterns. The next chapter briefly reviews the literature related to studies. The Chap. 3 is described a new method for personalized recommendation system in detail, such as system architecture with sub modules, the procedure of processing the recommendation, the algorithm for proposing system. The Chap. 4 describes the evaluation of this system in order to prove the criteria of logicality and efficiency through the implementation and the experiment. In Chap. 5, finally it is described the conclusion of paper and further research direction.

21.2 Relative Works

21.2.1 RFM

RFM is generally used in database marketing and direct marketing and has received particular attention in retail. RFM consists of three initial characters. R means recency-"How recently a customer has purchased?". F means frequency-"How often she purchases?". M means monetary-"How much does she spend?". The general way to use RFM model in customer behavior analysis is to sort the customer data by each dimension of RFM variables and then divide the data into five equal quintiles. For recency, the customer database is sorted by purchase dates by descending order. So, the top segment is given a value of 5 and the others are discerningly assigned of 4, 3, 2, and 1. For frequency and monetary, sorting customer visiting frequency data and the customer data related to the amount of the money spent in descending order, respectively. These three variables belong to behavioral variables and can be acted as the segmenting variables by observing customers' attitudes toward the product, brand, benefit, or even loyalty from the database. We can suggest that using average purchase amount instead of total accumulated purchase amount is better in order to reduce co-linearity of frequency and monetary. Finally, all customers are presented by 555, 554, 553, …, 111, which thus creates 125 (5 × 5 × 5) RFM cells. Moreover, the best customer segment is 555, while the worst customer segment is 111. Based on the assigned RFM behavior scores, customers can be classified into segments and their profitability can be further analyzed. The RFM score can be a basis factor how to determine purchasing behavior on the internet shopping mall, is helpful to buy the item which they really want by the personalized recommendation [5, 6].

21.2.2 Neural Network

The SOM introduced by Kohonen, is an unsupervised learning algorithm for clustering [8]. Also SOM is called as a neural networks model based on competitive learning. SOM can convert a high dimensional input space into a simpler low dimensional discrete map. It has two layers which are input and feature layers. We can cluster all elements by feature map with two dimensions. Firstly SOM performs clustering with input vector X and weight matrix W. The data point X_i is treated one at a time. Also the closest W_j to X_i is found by Euclidean distance, and then M_j is updated as the following [9].

$$W_k = W_j + \alpha(X_i - W_j). \tag{21.1}$$

where W_j and W_k are current and new weights. So W_k moves to X_i This learning is repeated until given conditions such as change rate of weights and the number of repeat. In this paper, we can use the SOM learning algorithm [9], where M_j and M_k are current and new weights. So M_k moves to X_t. This learning is repeated until given conditions such as change rate of weights and the number of repeat. In this paper, we can use the SOM learning algorithm [9] as the following (Table 21.1).

Table 21.1 SOM learning algorithm

```
Input   Set of N dimension vector, X // input node
Output  Subset of input data (M subsets)
begin
  Randomly initialize Mi = Mi1, Mi2 , ...Min for each node ;
   for (t=0; unless a stopping condition is reached; Increase t)
     for (for all input data)
       for (i=0 to M)
      Compute  Di = || Xt − Mi (t) || ;
       endfor
      Find the winner j=i such that
             Di(t) is minimum for over all I ;
      Update the winner j (and its neighbors) ;
      endfor
endfor
end
```

21.3 Our Proposal for Recommender System in U-Commerce

21.3.1 Clustering Method Using SOM to Classify Profitable Customer

This proposal SOM clustering method in this paper is better than k-means clustering the data directly, is depicted. First, a large result sets of prototyping for clustering user data (much larger than the expected number of output count, purchase pattern in clusters) is formed using the SOM or some vector quantization algorithm. We can apply a SOM clustering to purchase data to join user data in order to classify profitable customer with RFM score having RFM factors for recommender service in u-commerce. Finally the prototyping application is made and the prototyping result is classified to make clusters in order to classify profitable customer with RFM score. The system can use the code of classification (54 bits), demographic variables such as age, gender, an occupation, skin type, region and customer's RFM factors as input vectors for pre-processing so as to be possible to recommend the items with efficiency. The system can make clusters with neighborhood customer-group using a new clustering method, which is classified by the code of classification and customer's RFM score in customer information. The system can take the preprocessing task which is able to use the whole purchase data by each rank of the RFM score and then makes cluster of purchase data sorted by item category, joined cluster of user data called by customer DB, neighborhood user group [5]. As a matter of course, the system can use the whole purchase data (sale). After that, the system using SOM algorithm, can recommend the items by each rank of the RFM score. The SOM learning algorithm for clustering of user's information to join user's score is depicted as the following Table 21.2.

21.3.2 The Procedural Algorithm for Recommendation

The system can search cluster selected by using the code of classification and customer's RFM score in users' information. It can scan the preference of brand item in cluster, suggest the brand item with the highest score in item category selected by the highest probability for preference of item category as the average of brand item. This system can create the list of recommendation with TOP-N of brand item with the highest score to recommend the item with purchasability efficiently. This system can recommend the items with efficiency, are used to generate the recommendable item according to the basic of loyalty of RFM factors through clustering method using SOM algorithm. It can recommend the associated item to TOP-N of recommending list. This system takes the cross comparison with purchase data in order to avoid the duplicated recommendation which it has ever taken.

Table 21.2 SOM learning algorithm for clustering of user's information to join user's score

```
Step 1  Initialize parameters of SOM model
// Representative pattern of bits
for demographic variable(54bits), RFM(15bits) factors)
Step 2  Set input value vector
Step 3  Calculate Output value
```

$$C_j = \sum_{i=0}^{n} w_{ij} x_i \quad (2)$$

```
Step 4  Select winner node
```

$$O_j^* = Max(O_j) \quad (3)$$

```
Step 5  Readjust connection weights
```

$$w_{ij}(t+1) = w_{ij}(t) + \alpha\,(x_i(t) - w_{ij}(t)) \quad (4)$$

```
Step 6  Completion of learning
// IF Reach the learning  cycles
then Make the result of SOM
otherwise GO to Step 2
Step 7  Calculate output value
Step 8  Calculate winner node
Step 9  Result of pattern
```

21.4 The Environment of Implementation and Experiment and Evaluation

21.4.1 Experimental Data for Evaluation

We make the implementation for prototyping of e-shopping mall which handles the cosmetics professionally and do experiments. It is the environment of implementation and experiments in Apache2.2.14, j2sdk 1.7.0_11 as Java environment, JSP/PHP 5.2.12 as server-side script, JQuery*mobile, XML/XHTML4.0/HTML5.0/CSS3/JAVASCRIPT as client-side script, C# .net framework 2.0, jakarta-tomcat, apache 5.0.28 as web server under Windows O.S.

21.4.2 Experimental Data for Evaluation

We used 319 users who have had the experience to buy items in e-shopping mall, 580 cosmetic items used in current industry, 1,600 results of purchase data recommended in order to evaluate the proposing system [4]. In order to do that, we make the implementation for prototyping of the internet shopping mall which handles the cosmetics professionally and do the experiment. We have finished the system implementation about prototyping recommendation system. We'd try to carry out the experiments in the same condition with dataset collected in a cosmetic internet shopping mall. It could be evaluated in MAE and output count by RFM score level for the recommendation system in clusters. It could be proved by the experiment through the experiment with learning data set for 12 months, testing data set for 3 months in a cosmetic cyber shopping mall [4]. The 1st system of SOM clustering method using user's features based on purchase data to join user data in order to classify profitable customer with RFM score having RFM factors, is proposing system called by "SOM", the 2nd system is other previous system (k-means) using k-means clustering algorithm based on the whole data.

21.4.3 SOM Results for Application of Neural Network

SOM network is applied to classify purchase data to join the user data, finally forms clusters of purchase pattern groups of user data with different features, demographic variables and RFM factors as input vectors. In order to segment purchase data join to user data into appropriate number of clusters, SOM is applied to determine the number of clusters. Figures 21.1 and 21.2, nine clusters are recommended among 1,600 purchase data, 319 customers when recency, frequency, monetary are the three input variables and then divide the data into five equal quintiles beside off demographic variables. From the SOM result, we can find 5 level based on RFM score of customer so as to recommend the items in real-time environment [3]. The following Fig. 21.3, show the result with statistics of output counts based on

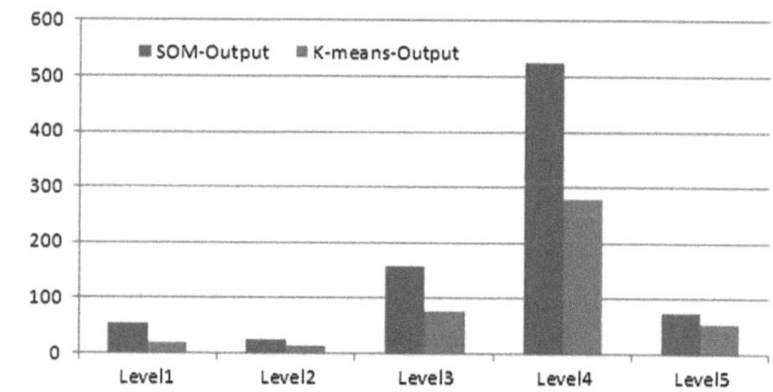

Fig. 21.1 Descriptive statistics of output counts by the level of RFM score with comparing SOM and k-means

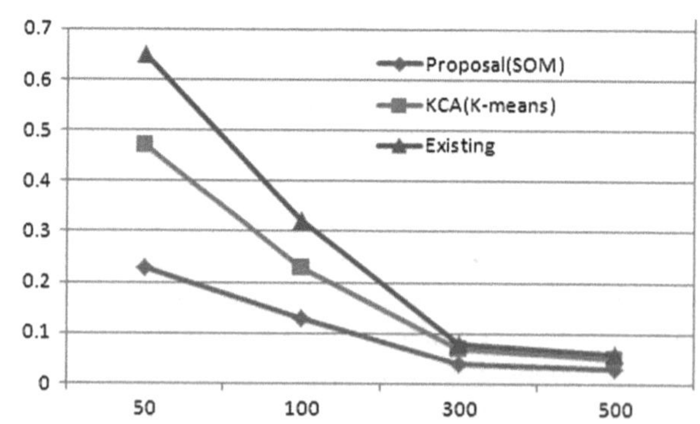

Fig. 21.2 The result of MAE by comparing proposal system (SOM) with previous system (k-means)

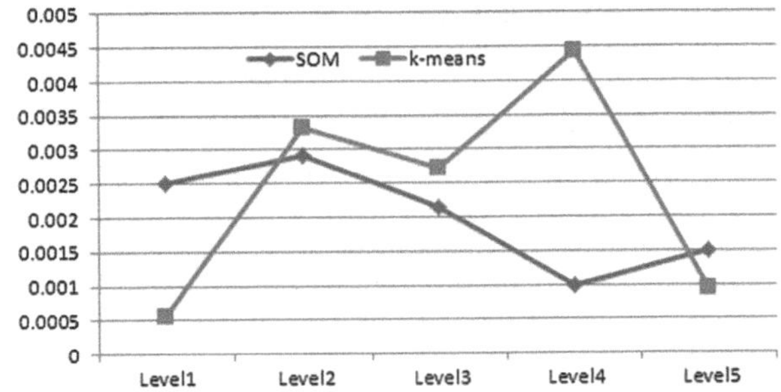

Fig. 21.3 The result for the graph of MAE by RFM score level

purchase data for the segmentation as comparing SOM and k-means. It is depicted in the result, that level 1 is the RFM score of customer is more 90 points, level 2 is the range of RFM score (score ≥80 and score <90), level 3 is the range of RFM score (score ≥60 and score <80), level 4 is the range of RFM score (score ≥40 and score <60), and level 5 is the range of RFM score (score ≥20 and score <40). The purchase data is not at the range of RFM score (score <20). It shows the improvement on the number of output purchase pattern count in the result of evaluation levels for the proposal system (SOM) comparing with previous system (k-means). The proposal is higher on the number of output purchase pattern count than the previous system from level 1 to level 5. As a result of that, the performance of the proposal system is improved better on the number of output purchase pattern count than previous system from level 1 to level 5.

21.4.4 Experiment and Evaluation

The proposing system's overall performance evaluation was performed by dividing the two directions. The first measurement is output counts of purchase pattern in the (Table 21.2) The second evaluation is mean absolute error (MAE). The mean absolute error between the predicted ratings and the actual ratings of users within the test set. The mean absolute error is computed the following expression (21.5) over all data sets generated on purchased data.

$$\mathrm{MAE} = \frac{\sum_{i=1}^{N} |\epsilon_i|}{N} \tag{21.5}$$

N represents the total number of predictions, ε represents the error of the forecast and actual phase i represents each prediction.

Above Table 21.3 shows the result of evaluation metrics (MAE) for recommendation system. It shows the improvement in the result of evaluation rates for proposal system comparing with previous system (k-means). At the Table 21.4, the proposal system is better than the previous system from level 2 to level 4 in the part

Table 21.3 The result of MAE by comparing proposal system with existing system

	P_count	Proposal (SOM)	Previous (KCA)	Existing
MAE	50	0.23	0.47	0.65
	100	0.13	0.23	0.32
	300	0.04	0.07	0.08
	500	0.03	0.05	0.06

Table 21.4 The result of MAE by RFM score level

RFM score level	SOM output count	K_means output count	SOM MAE	k-means MAE
1	52	18	0.0025	0.00056
2	24	12	0.00292	0.00333
3	158	77	0.00215	0.00273
4	524	280	0.00099	0.00443
5	74	54	0.00149	0.00093

Fig. 21.4 The site of recommendation of cosmetics

of large purchase count. As a result of that, the performance of the proposal system is improved better than previous system from level 2 to level 4 although it is not good on level 1 and level 5 in the part of small purchase count. The following Fig. 21.4 is shown in the site of recommendation of cosmetics on a smart phone. This system can be used immediately in e-Commerce under ubiquitous computing environment which is required by real time accessibility and agility because of finishing particular tasks such as clustering and calculating the probability of preference for pre-processing to reduce the processing time in real-time environment.

21.5 Conclusion

Recently u-commerce as a application field under ubiquitous computing environment required by real time accessibility and agility, is in the limelight [4]. We proposed an SOM clustering method using user's features to classify profitable customer for recommender service in e-Commerce.in order to improve the accuracy of recommendation with an immediate effect We have described that the performance of the proposal system with SOM clustering method using user's features is improved better than previous system (k-means) from level 2 to level 4 in the part of large purchase count although it is not good on level 1 and level 5 in the part of small purchase count. It could make appropriate recommendation for each user's level possible based on neural network in real-time environment. We could simulate the SOM Clustering Method using user's features to classify profitable customer, generate recommending items to be possible to measure the purchasability for the future [3]. Thus, we could make the level of RFM score for the measurement of accuracy and efficiency, and validate the system by our results, then we can recommend items by each user's level according to the loyalty of RFM factors. To verify improved better performance of proposing system, we carried out the experiments in the same dataset collected in a cosmetic internet shopping mall. It is meaningful to present an SOM Clustering Method using user's features to classify profitable customer for recommender service in e-Commerce The following research will be looking for ways of a personalized recommendation using fuzzy clustering method to increase the efficiency and scalability.

Acknowledgments This is work was supported by funding of Namseoul University.

References

1. Deboeck G, Kohonen T (1998) Visual explorations in finance with self-organizing maps. Springer, London
2. Cho YS, Moon SC, Ryu KH (2012) Mining association rules using RFM scoring method for personalized u-commerce recommendation system in emerging data. In: International conferences, SecTech, CA, CES3 2012, held in conjunction with GST 2012, communications in computer and information science, vol 341, pp 190–198
3. Cho YS, Moon SC, Ryu KH (2014) Clustering analysis by customer feature based on SOM for predicting purchase pattern recommendation system. J Korea Soc Comput Inf 19(2)
4. Cho YS, Moon SC, Noh SC (2013) Weighted mining association rules based quantity item with RFM score for personalized u-commerce recommendation system. In: The 8th international conference GCP2013, LNCS, vol 7861, pp 367–375
5. Cho YS, Moon SC, Jeong SP, Oh IB, Ryu KH (2012) Clustering method using item preference based on RFM for recommendation system in u-commerce. Ubiquit Inf Technol Appl LNEE 214:353–362
6. Wei J-T, Lin S-Y, Wu H-H (2010) The review of the application of RFM model. Afr J Bus Manage 4(19):4199–4206
7. Smith KA, Gupta JND (2001) Neural network in business techniques and applications. Idea Group Publishing, Hershey
8. Kohonen T (2000) Self-organizing maps. Springer, Berlin
9. Hastie T, Tibshirani R, Friedman J (2001) The elements of statistical learning—data mining, inference, and prediction. Springer, Berlin

Chapter 22
A Performance Improvement Scheme for Discovery Service

Peng Liu, Ning Kong, Ye Tian, Xiaodong Lee and Baoping Yan

Abstract Radio Frequency Identification (RFID) tags have been widely used for identifying objects on a global scale. Discovery service (DS) acting as a broker between objects and their information sources needs to handle a huge amount of query requests from the user. So the performance of the node of DS must be powerful enough to handle lots of query requests sent to it. However, no performance improvement schemes for the node of DS have been proposed until now. Therefore this paper proposes a novel performance improvement scheme for the node of DS based on anycast and Distributed Hash Table (DHT) network. By analysis, the performance improvement scheme proposed in this paper can not only significantly improve the performance of the node of DS but also has the advantages of shortening its response time, avoiding the adverse influence of ISP-unfriendly policies and being transparent to the user, which satisfies the actual requirements pretty well.

Keywords Radio frequency identification · Discovery service · Anycast · Distributed hash table

P. Liu · N. Kong (✉) · Y. Tian · X. Lee · B. Yan
Computer Network Information Center, Chinese Academy of Sciences, Haidian, Beijing 100190, China
e-mail: nkong@cnnic.cn

P. Liu
University of Chinese Academy of Sciences, Shijingshan District, Beijing 100049, China
e-mail: liupeng@cnnic.cn

Y. Tian
e-mail: tianye@cnnic.cn

X. Lee
e-mail: xl@cnnic.cn

B. Yan
e-mail: ybp@cnic.cn

J.J. (Jong Hyuk) Park et al. (eds.), *Ubiquitous Computing Application and Wireless Sensor*, Lecture Notes in Electrical Engineering 331,
DOI 10.1007/978-94-017-9618-7_22

22.1 Introduction

Radio Frequency Identification (RFID) [1] tags have been widely used in many industries such as manufacturing, logistics and retailing for identifying objects on a global scale in recent years. Discovery Service (DS) is a pivotal system that is designed to serve the following lookup function: Given the RFID identifier of an object, it returns a list of Internet addresses of servers across the supply chain which offer detailed information about this object [2]. Without DS acting as a broker between objects and their information sources, the user can't get detailed information about the object across the supply chain.

Because RFID tags have been widely used for identifying objects on a global scale, DS acting as a broker between objects and their information sources needs to handle a huge amount of query requests from the user. Take top-level domain ".cn" of Domain Name System (DNS) for reference, there are about 11 million domain names of ".cn" and more than 2 billion query requests for ".cn" should be handled every day [3]. To handle the query requests for ".cn" and supply reliable service for the user, CNNIC built 35 data centers around the global based on anycast [4] technique. According to the report of IDTechEx, more than 2.3 billion RFID tags were sold only in year 2010 which were used to identify the objects across supply chains [5]. Compared with DNS, DS needs to handle more query requests.

The existing researches on DS mainly focus on the architecture and access control policy of DS [2, 6–13]. According to the difference of DS architecture, the existing DS solutions can be divided into two categories: DS based on Tree topology and DS based on Distributed Hash Table (DHT). To handle lots of query requests, the performance of the node of DS must be powerful enough, especially the upper level nodes of DS based on Tree topology and the hot nodes of DS based on DHT. If the performance of a node (a server) of DS is not enough to handle the users' query requests sent to it, that will result in service unavailable of this node or even DS system crash. Although this is a severe problem, no performance improvement schemes for the node of DS have been proposed until now.

In addition, the response time of DS is very important for user experience. If the user accesses a node of DS across one or multiple Internet Service Provider (ISP) networks, that may lead to long response time which seriously influences user experience. Moreover, because there is competition among ISPs, to avoid the adverse influence of ISP-unfriendly policies such as bandwidth-limited between different ISP networks is an important requirement.

Therefore this paper focuses on those problems and proposes a novel performance improvement scheme based on anycast and DHT network to solve them. By analysis, the performance improvement scheme proposed in this paper can not only significantly improve the performance of a node of DS but also shorten its response time, avoid the adverse influence of ISP-unfriendly policies and be transparent to users, which satisfies the actual requirements pretty well.

The remainder of this paper is structured as follows: Related work is given in Sect. 22.2. The novel performance improvement scheme based on anycast and DHT is proposed in Sect. 22.3. Then this scheme is analyzed in Sect. 22.4. Finally, the contribution of this paper is summarized in Sect. 22.5.

22.2 Related Work

DS acting as a broker between objects and their information sources needs to handle a huge amount of query requests. The existing researches on DS mainly focus on the architecture and access control policy of DS [2, 6–13]. The DS solutions which have been proposed until now can be divided into two categories: DS based on Tree topology such as BRIDGE DS [6], as shown in Fig. 22.1 and DS based on DHT such as DS proposed by Nina et al. [10], as shown in Fig. 22.2. To handle lots of query requests, the performance of the node of DS must be powerful enough, especially the upper level nodes of DS based on Tree topology and the hot nodes of DS based on DHT. However, no performance improvement schemes for the node

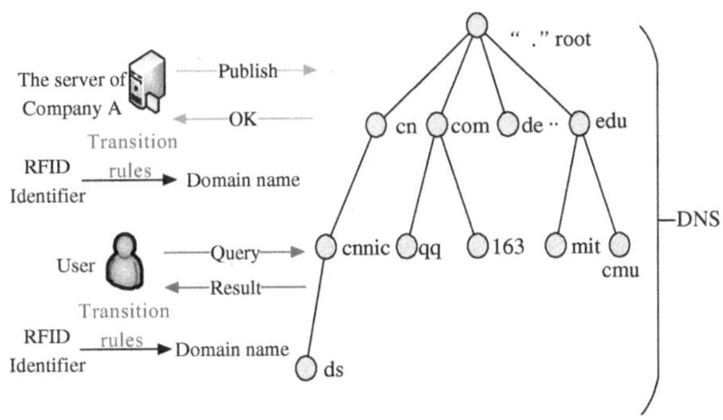

Fig. 22.1 The architecture of BRIDGE DS

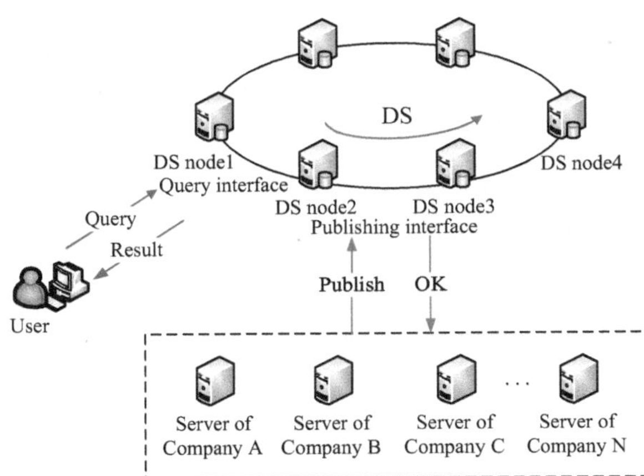

Fig. 22.2 The architecture of DS proposed by Nina et al.

of DS have been proposed until now. In addition, how to shorten the response time of DS and avoid the adverse influence of ISP-unfriendly policies are also the problems to be solved.

Anycast is a network addressing and routing methodology in which datagrams from a single sender are routed to the topologically nearest node in a group of potential receivers, though it may be sent to several nodes, all identified by the same destination address [4]. Anycast is widely used in many applications such as DNS and Content Delivery Network (CDN), which has the advantages of improving system performance, shortening the response time, avoiding the adverse influence of ISP-unfriendly policies and being transparent to users. Most DHTs resolve lookups in O(log N) hops through the overlay network, where N is the number of servers in the DHT network [14]. In addition, DHT networks support dynamic join and departure of servers, offering excellent performance scalability. Considering the advantages of anycast and DHT network, in this paper, a novel performance improvement scheme for DS based on anycast and DHT network is proposed to solve the above problems.

22.3 The Performance Improvement Scheme for DS

22.3.1 Overview of the Performance Improvement Scheme for DS

To solve the problems mentioned above, a novel performance improvement scheme for the node of DS based on anycast and DHT network is proposed in this section, as shown in Fig. 22.3. The node of DS of which the performance needs to be improved is considered as the origin server, which is responsible for data update and data distribution. If the origin server receives a registration request including the RFID identifier of an object and the address of the server which stores detailed information of this object from an authenticated and authorized user such as the server of a company, it will update its data according to this request. The format of each record stored in the origin server is $\langle$RFID identifier; $\mathrm{URI}_1, \mathrm{URI}_2, \ldots, \mathrm{URI}_n\rangle$, which reflects the addresses of the servers about the RFID identifier across supply chain. The origin server distributes updated data to its replica servers which distribute in different ISP networks in real time. To reduce Internet traffic and shorten the time of data transfer, updated data is compressed by the origin server before sent to its replica servers.

The replica servers of an origin server distribute in different ISP networks. In addition, the replica servers of an origin server distributing in the same ISP network constitute a DHT network called replica system, as shown in Fig. 22.3. In this paper, anycast is implemented by using BGP (Border Gateway Protocol) [15] to simultaneously announce a same destination Internet Protocol (IP) address from different ISP networks. The origin server and all of its replica servers are identified

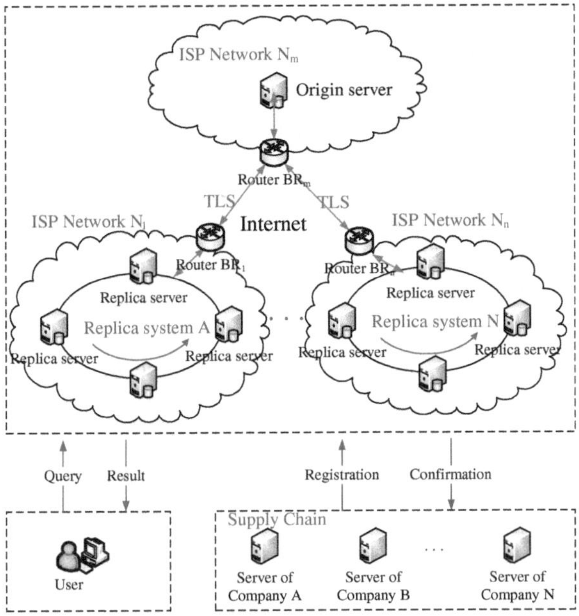

Fig. 22.3 Overview of the performance improvement scheme for DS

by a same anycast IP address. By this way, the requests addressed to an anycast IP address from an ISP network will be routed to the topologically nearest replica system identified by this anycast address. Most DHT networks resolve lookups in O (log N) hops through the overlay network, where N is the number of servers in the DHT network [14]. Therefore a query request can be handled efficiently by replica system. Since replica system is deployed as a DHT network, it can support dynamic join and departure of servers, which offers excellent performance scalability. The identifier of each replica server in the same replica system is the hash value of its anycast IP address and a random number provided by its owner. For each record $\langle$RFID identifier; $URI_1, URI_2, \ldots, URI_n\rangle$ sent from the origin server, replica system first hashes the RFID identifier included in this record as key and then stores this record in the replica server of this replica system responsible for this key. The replica system is responsible for handling the users' query requests, but for the users' registration requests, it forwards them to its origin server. In order to communicate between the origin server and its replica servers distributing in different ISP networks, each of them also has a different unicast IP address besides the anycast IP address of them. In addition, the communication between them is encrypted using Transport Layer Security (TLS) to guarantee security of the data traveling over the Internet.

22.3.2 Data Update and Distribution Process

The data update and distribution process will be illustrated by an example. After the server of company M (denoted as $server_M$) stores detailed information about an object, it will send a registration request including the RFID identifier of this object (denoted as $RFID_{object}$) and its address (denoted as URI_M) to the corresponding origin server (denoted as S) of DS. The origin server S will update its data according to this registration request and distribute updated data to its replica servers in real time. The data update and distribution process is described as follows, as illustrated in Fig. 22.4.

(1) $Server_M$ stores detailed information about an object and then sends a registration request including the object identifier $RFID_{object}$ and its address URI_M to the corresponding origin server S of DS whose anycast IP address is denoted as $IP_{anycast}$.
(2) The registration request of $server_M$ is forwarded to the border router BR_1 of the ISP network N_1.
(3) The border router BR_1 redirects this registration request to replica system A according to its routing table.
(4) Replica system A receives the registration request and then forwards it to the origin server S.
(5) The origin server S adds a piece of record $\langle RFID_{object};\ URI_M \rangle$ to its database according to the registration request of $server_M$ and returns a confirmation to $server_M$.
(6) The origin server S distributes updated data to each replica system distributing in different ISP networks in real time.

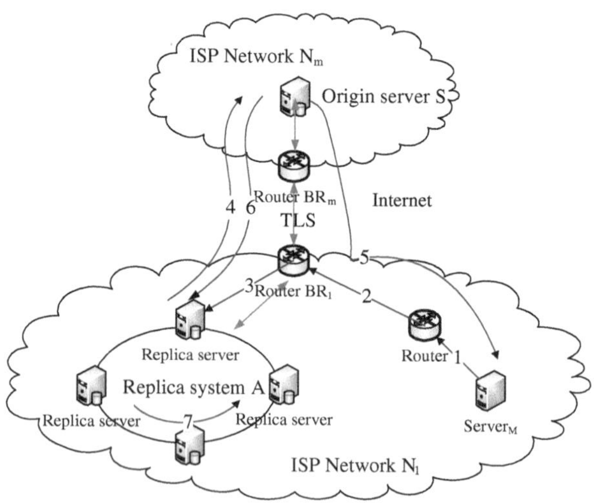

Fig. 22.4 Data update and distribution process

(7) Each replica system receives the updated data including the record $\langle RFID_{object};\ URI_M \rangle$, then it calculates the hash value of $RFID_{object}$ as key and stores the record $\langle RFID_{object};\ URI_M \rangle$ in the replica server responsible for key.

22.3.3 The User's Lookup Process

If a user such as a trace and track application needs to obtain an object's detailed information across the supply chain, it will query the corresponding node of DS to get the addresses of servers about this object first. The assumptions are the same as mentioned above. Then the lookup process of a user (denoted as U) to get the addresses of the servers related to an object across the supply chain is described as follows, as illustrated in Fig. 22.5.

(1) The user U sends a query request including the RFID identifier of an object (denoted as $RFID_{object}$) to the origin server S.
(2) The query request of user U is forwarded to the border router BR_1 of the ISP network N_1.
(3) The border router BR_1 redirects this query request to replica system A according to its routing table.

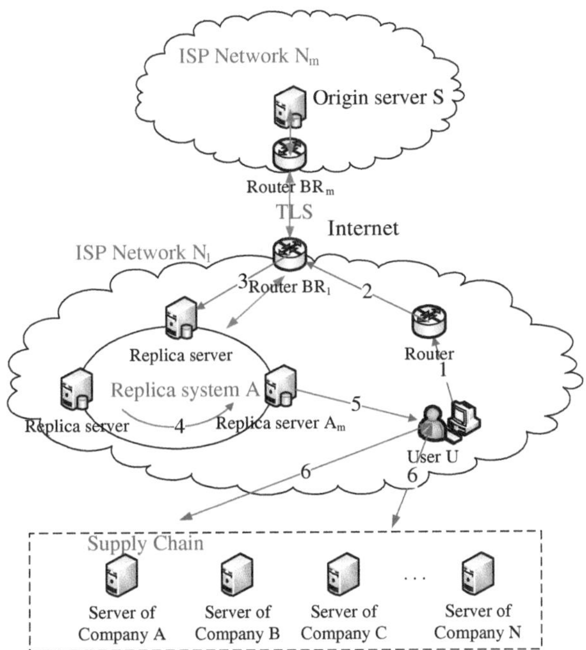

Fig. 22.5 The user's lookup process

(4) Replica system A receives the query request of user U and calculates the hash value of $RFID_{object}$ as key. Finally this query request is forwarded to the replica server (denoted as A_m) of replica system A which is responsible for the key.
(5) Replica server A_m returns the addresses of servers $\langle URI_1, URI_2, \ldots, URI_n \rangle$ about $RFID_{object}$ included in the query request to the user U.
(6) The user U obtains detailed information about this object by accessing the relevant servers across the supply chain.

22.4 Analysis

To solve the problems mentioned above, this paper proposes a novel performance improvement scheme for DS based on anycast and DHT network. By analysis, it is concluded that the performance improvement scheme for DS presented in this paper has the following advantages, which satisfies the actual requirements pretty well.

22.4.1 Improving the Performance of the Node of DS

RFID tags have been widely used for identifying objects on a global scale, DS acting as a broker between objects and their information sources needs to handle a huge amount of query requests from the user. To handle lots of query requests, the performance of the node of DS must be powerful enough, especially the upper level nodes of DS based on Tree topology and the hot nodes of DS based on DHT. A single server may not satisfy this requirement. In our performance improvement scheme, an origin server distributes data to its replica servers distributing in different ISP networks which are responsible for handling query requests. Moreover, the replica servers of an origin server distributing in the same ISP network are organized into a DHT network, which supports dynamic join and departure of replica servers, offering excellent performance scalability. Therefore, by this way, the performance of an origin server can be significantly improved.

22.4.2 Shortening the Response Time

To obtain detailed information about an object, the user needs to access DS to get the addresses of servers across supply chain first. The response time of DS is very important for user experience. If a user accesses a node of DS across one or multiple ISP networks, the response time may be long which seriously influences user experience. By our scheme, there are multiple replica servers of an origin server distributing in different ISP networks and the query request of a user addressed to the anycast IP address of this origin server will be routed to the topologically nearest replica system of it, which significantly shortens its response time.

22.4.3 Avoiding the Adverse Influence of ISP-Unfriendly Policies

Because there is competition among ISPs, to avoid the adverse influence of ISP-unfriendly policies such as bandwidth-limited between different ISP networks is an important requirement. In our scheme, there are multiple replica servers of an origin server distributing in different ISP networks and the query request of a user sent to this origin server will be routed to the topologically nearest replica system of the origin server. Therefore, the adverse influence of ISP-unfriendly policies can be avoid by our scheme.

22.4.4 Being Transparent to Users

In our scheme, anycast is implemented by using BGP to simultaneously announce a same destination IP address from different ISP networks. The origin server distributes data to its replica servers distributing in different ISP networks in real time, which are identified by a same anycast IP address. By this way, the query request of a user addressed to an anycast IP address will be routed to the topologically nearest replica system identified by this address while it is transparent to the user.

22.5 Conclusion

There are no performance improvement schemes for the node of DS have been proposed until now. So this paper proposes a novel performance improvement scheme for the node of DS based on anycast and DHT network. By analysis, the performance improvement scheme proposed in this paper can not only significantly improve the performance of the node of DS but also has the advantages of shortening the response time, avoiding the adverse influence of ISP-unfriendly policies and being transparent to the user, which satisfies the actual requirements pretty well. Next step, the performance improvement scheme for the node of DS proposed in this paper will be applied in our projects.

Acknowledgments Foundation item: The National Development and Reform Commission of China Project: 2012 The IOT (Internet of Things) Technology Development and Industrialization—The IOT Identifier Management Platform for Public Service. The Chinese Academy of Sciences—Guangdong Province Joint Project: The IOT (Internet of Things) Identifier Management Platform for Public Service (No. 2011BY100363). Around Five Top Priorities of "One-Three-Five" Strategic Planning, CNIC (No. CNIC_PY-1403). Authors are grateful to the National Development and Reform Commission of China, Chinese Academy of Sciences, Guangdong province and CNIC for financial support to carry out this work.

References

1. Wikipedia (2014) Radio frequency identification. http://en.wikipedia.org/wiki/Radio-frequency_identification
2. Evdokimov S, Fabian B, Kunz S, Schoenemann N (2010) Comparison of discovery service architectures for the internet of things. In: 2010 IEEE international conference on sensor networks, ubiquitous, and trustworthy computing (SUTC), pp 237–244
3. CNNIC (2014) The statistical report of china internet network development situation
4. Partridge C, Mendez T, Milliken W (1993) Host anycasting service. RFC 1546
5. IDTechEx (2011) RFID forecasts, players and opportunities 2011–2021
6. BRIDGE (2007) High level design for discovery services
7. Zhao W, Li XP, Liu DX, Zhang SK, Wang LF (2010) A distributed RFID discovery service for supply chain. Acta Electronica Sinica 38:99–106
8. Rezafard A (2008) Extensible supply-chain discovery service problem statement draft-rezafard-esds-problem-statement-03. IETF Internet-Draft
9. Thompson F (2008) Extensible supply-chain discovery service schema draft-thompson-esds-schema-04. IETF Internet-Draft
10. Nina S, Kai F, Detlef S. P2P Architecture for Ubiquitous Supply Chain Systems. 17th European Conference on Information Systems (ECIS'09); 2009.
11. Fabian B, Ermakova T, Muller C (2012) SHARDIS: a privacy-enhanced discovery service for RFID-based product information. IEEE Trans Ind Inf 8:707–718
12. Shi J, Li YJ, Deng RH (2012) A secure and efficient discovery service system in EPCglobal network. Comput Secur 31(8):870–885
13. Shi J, Li YJ, He W, Sim D (2012) SecTTS: a secure track and trace system for RFID-enabled supply chains. Comput Ind 63:574–585
14. Balakrishnan H, Kaashoek MF, Karger D, Morris R, Stoica I (2003) Looking up data in P2P systems. Commun ACM 46(2):43–48
15. Rekhter Y, Li T, Hares S (2006) A border gateway protocol 4 (BGP-4). RFC 4271

Chapter 23
A Cross Cloud Authorization Mechanism Using NFC and RBAC Technology

Jun-Fu Chan, Ta-Chih Yang and Horng Twu Liaw

Abstract Because of the popularity of the internet, a variety of communications devices that make people live more convenient. The people generate information faster than before. The need of server rises gave birth to cloud computing. In recent years, the mobile device manufacturers launched aboard Near Field Communication, NFC chip device, and get large number of development the NFC application, such as Electronic Wallet, Electronic Ticket, Authentication function and so on. In North America has google wallet service, in England has mobile payment service, in Taiwan, we have prepare to work up NFC authentication center. So that our research plan to combine NFC device with RBAC authorization let users uses cloud service through internet any time. In this research we consider the widespread of mobile device and cross cloud application, we plan to use NFC technology combine RBAC mechanism to simulation the cross cloud authorization management.

Keywords NFC · Authorization · RBAC · Cloud computing

23.1 Introduction

In the rapid development of technology and internet, there are a lot of communication device between you and me, let us more convenient, but also produce more needs of information.

Cloud computing is a new technology and also a new concept. The cloud is combine Internet and Client, users can get powerful computing services to finish

J.-F. Chan · T.-C. Yang (✉)
Department of Information Management, National Taiwan University of Science and Technology, Taipei, Taiwan
e-mail: tcyang@livemail.tw

H.T. Liaw
Department of Information Management, Shih Shin University, Taipei, Taiwan

J.J. (Jong Hyuk) Park et al. (eds.), *Ubiquitous Computing Application and Wireless Sensor*, Lecture Notes in Electrical Engineering 331,
DOI 10.1007/978-94-017-9618-7_23

works. The cloud computing can be divided according to function and purpose such as Public Cloud, Private Cloud, Community Cloud and Hybrid Cloud [1–4].

With the cloud computing technology matures, the use of cross cloud service has become increasingly widespread. Enterprise become more difficult to manage authorization. Therefore, in order to enable enterprises and managers can become easy to manage permissions without the information security, we propose a RBAC authorization mechanism combine NFC mobile devices using in cross cloud authorization management mechanism.

In this research we consider in cross cloud service environment, propose a NFC device combine with RBAC authorization mechanisms, and achieve the following objectives:

i. Help user proceed a secure login and authorization procedure: User can login the cross cloud authentication system through NFC device, and using the NFC device encrypt the login and authorization information, to ensure the login and authorize are safe.
ii. To easier authorization management system: Due to large number of users using in cross cloud, manager may be cause the setting error about the separation of duty. In this research, we propose a NFC device combine with RBAC mechanism help manager can be easier to manage the authorization system.
iii. Using in cross cloud environment: Due to the complex of cross environment, therefore this research propose a suit for cross authorization environment process and architecture.

23.2 Related Work

Near Field Communication, NFC is a short distance and high frequency wireless communication standard in band 13.56 MHz, which allow electronic device conduct untouched peer to peer communication, the transmission rate are 106, 212 and 424 KBps. The NFC has become ISO/IEC IS 18092 international standard, EMCA 340 and ETSI TS 102 190 standard.

The NFC is evolution from Radio Frequency Identification, RFID, which allow device obtain information from tag or other device within 10 cm. The NFC device unlike Bluetooth technology organize the PICONET has to pass the authentication code, NFC has a feature that allow devices connection quickly without mapping another devices. For the mobile devices, using NFC is more convenient cause the short distance communication and low power consumption, NFC can have higher confidentiality to archive the information transmission security. Such as in protect the credit card transaction from malicious attackers.

The Role-Based Access Control has widely use in recent years. It's different from Mandatory Access Control and Discretionary Access Control, RBAC is assign permission to role. In 1992, Ferraiolo and Kuhn proposed a role based access control, the authorization mode has a feature, and the whole access control are assign by role. The role is a permission set, users are get permission from the role or inherit role from another user. In 1996, Ravi Sandhu et al. based on 1992 RBAC theory, publish a new model of RBAC mechanism. And National Institute of Standards and Technology, NIST has redefined the RBAC named NIST RBAC as a standard.

In the main secure standard, an important component named separation of duty, it could be define as below:

i. Static Separation of Duty, SSD: It mainly define exclusive role member, they are conflict each other, to avoid benefit fraud.
ii. Dynamic Separation of Duty, DSD: It mainly let exclusive role in some limit time allow the weak exclusive role assign to same user in same time. In this case, need more complete define.

23.3 Proposed Mechanism

23.3.1 System Introduce

In this research, we assume the system environment construct on plural private cloud and third party authentication cloud environment. Modifying NIST RBAC mechanism to proposed a new RBAC mechanism for cross cloud authorization standard. Users using NFC mobile device to make a secure login process and get the RBAC token. According to previous researcher's mechanism the RBAC mechanism has modification for a single distribute environment. But there are no existing cross distribute environment, so in this research we improve the weakness and strength in previous studies, in cross cloud environment providing a more safety and complete RBAC authorization mechanism.

Due to this research construct on the cross cloud environment, so far, every researches are inflexible in cross cloud authorization, so it will happen role mapping problem. In this research we focus on design authorization mechanism, so in authentication we use the sample password authentication mechanism.

23.3.2 System Architecture

This research proposed a token based RBAC authorization mechanism in plural private cloud and one third party authentication cloud. To ensure the hybrid cloud safety, this research suppose the environment organize by two private cloud combine with one thirty party authentication cloud as below in Fig. 23.1.

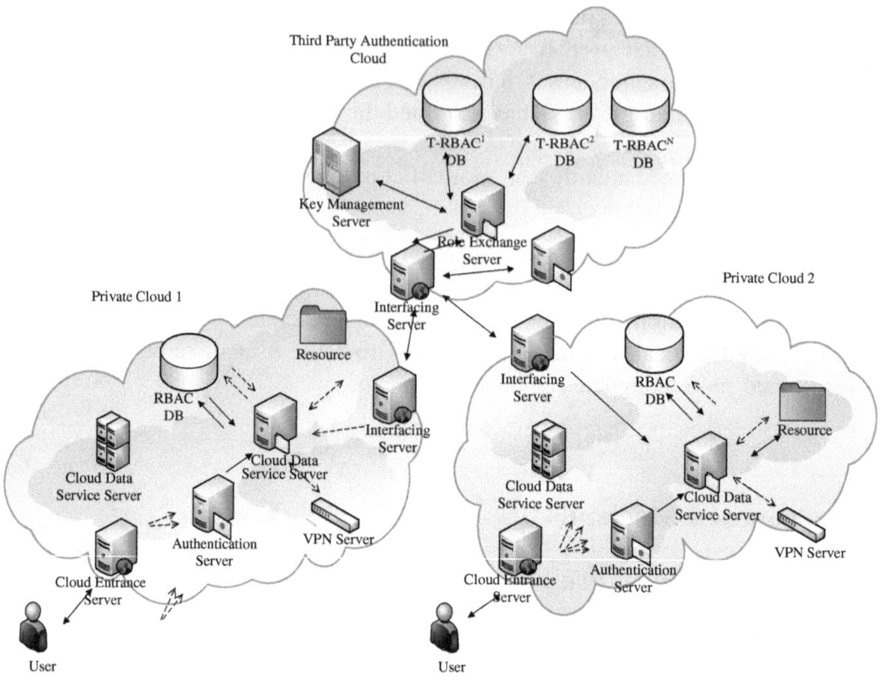

Fig. 23.1 The system architecture

The component are descript as below:

i. User: A legal private cloud user.
ii. Cloud login server: For each cloud service login server.
iii. Authentication server: The identity authentication server.
iv. NFC device: The NFC device user owned.
v. Token: The hashed user ID and password xor with role P and department code D. It's a unique number, for user to use cloud resource.
vi. Virtualize management server: Be responsible for the virtualize authentication server, to lower burden of authentication.
vii. Interfacing server: Be responsible for interfacing private cloud and third party authentication cloud.
viii. RBAC database: Stored role information database. There are token, role, user ID, permission in the database column.
ix. Third party RBAC database: Stored the thirty party role, users can login from the private or public cloud which certificate by the third party authentication server and get the role from third party RBAC database, users can login to other private cloud or public cloud by the role in third party RBAC database.
x. Third party registration server: Registration by private cloud or public cloud.
xi. Key management server: It stored all public cloud or private cloud symmetric key.

xii. VPN server: After other cloud user has get the role, user can direct access the resource through VPN.
xiii. Cloud data service server: This server is to manage the resource, every access action are check by this server whether if user's role can legal using the request target resource.
xiv. TP-RBACn database: It stored all private and public cloud information. Users must get the role from the database and use this role to login another cloud.
xv. Role exchange server: It located on thirty party authentication cloud, each private or public cloud need to access another cloud's resource, user must certificate token by the server to exchange role and login to anther cloud.
xvi. Resource: It's private or public cloud's storage, calculate capability.

This research use token to mapping role for exchange cross cloud role. The token are organize as below:

Each user based on the number of roles they have in RBAC database in a private cloud, while the same number token amount each pay table is only a token role. In this mechanism, each role are granted a large prime number, and department belongs pay code, so each character will have different large prime numbers, the manager responsible for the management RBAC responsibilities will be to distinguish between static (Static Separation of Duty, SSoD) and Dynamic Separation of Duty (Dynamic Separation of Duty, DSoD) problem exists, therefore, this study RBAC data bank increased static power. Conflict tables and dynamic power conflicts table, static power conflicts put each table memory conflicts arise out roles large prime numbers Table; managers in granting the user role, the system will automatically power conflicts within a static table to check given to managers whether the user's role in conflict. Dynamic power of large prime numbers sucked conflict table will be mutually conflicting roles. Exclusion or computing, storage in dynamic power conflict table. Because as long as the user has multiple roles, therefore, when the user starts roles, will get a token; If users start other roles, in this case the system will self-moving the user's token be exclusive or operation, because the user account and password belonging to a single department and after hashing string values are unique, after the operation, leaving only two token of large prime numbers, then will put this value and dynamic power conflicts do comparison table, so if this time a weak mutually exclusive, the system will warn the user not to let get roles.

Authorization for cross-cloud when the user's token will be sent to a third-party certification servers, third-party certification server. After the service is authenticated and confirmed the user's token, will enable the corresponding T-RBAC database, and then use the latter sector pay D Gaussian operations, obtain integer department users pay table, then the user angle color downgrade its role as the lowest level of the department, and referring users to establish a VPN server belongs to secure access channel, access to limited resources, this way, to complete the cross-domain authorization, and is due to obtain permission to take a minimum. Permission inheritance way, and because the role is mapped to a token, each cross-cloud authorized use only one token, and so. There is no distinction between static

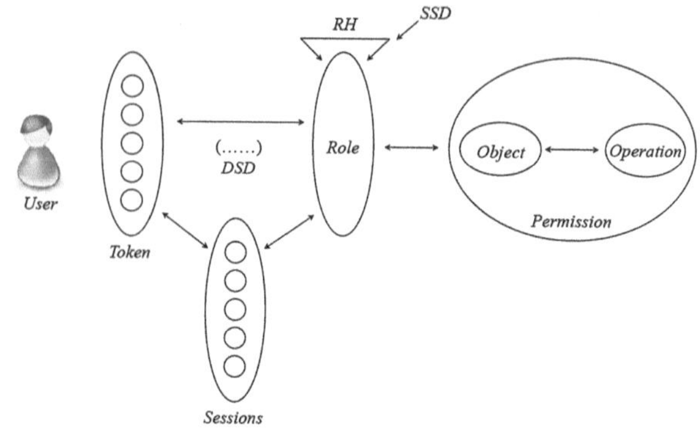

Fig. 23.2 Token-based RBAC architecture

powers problem, and the minimum permissions inheritance way there will not be a problem distinguishing dynamic powers, study of cross-domain license can also minimize the risk of information security, so this study can take into account security and cross-domain authorization convenience, so that users can smoothly access different cloud resources. T-RBAC mechanism of this study is shown in Fig. 23.2, each user has a complex token, each token corresponds to single role, so there may be more than one group in the same token enabled session within Therefore, this study on the use of inspection Charles large prime number test whether there is a way to distinguish between dynamic powers asked occur distinction is static in terms of making managers. That is when the system is managed through the power of the conflict table to verify conflict.

23.4 Security Analysis

This study will focus on cloud services environment and authorization mechanisms of the more common vulnerabilities and methods of attack, to explore whether this mechanism can effectively compensate for vulnerabilities or attack effectively preventing its practices and relevant mechanisms to compare the advantages and disadvantages.

Currently on the cloud service environment is more common methods of attack and security omission, namely: replay attacks, forgery attacks, middle attacks, eavesdropping attacks, malicious attacks reader, and offline guessing attacks, user data privacy protection and so on.

Replay attacks: the attacker cloud services for end packet network eavesdropping and gathering, will re-send the packet to the authentication device for authentication. In this study, the mechanism in order to prevent replay attacks, are using the time stamp in the external connection, once the authentication server does not belong to the time stamp found within a reasonable time frame, it will reject the user's private cloud or a request, so the attacker access to cloud resources cannot be correct.

Counterfeiting attacks: an attacker masquerading as a legitimate attack or a legitimate user equipment, although fake cloud server there are technical difficulties, but to forge a legitimate user is not difficult. In the present study, has to send a message in advance before using the server public key is transmitted encrypted; forged in the user part, this study also uses two-factor authentication, must have a good NFC devices and pre-registration users know the password to be able to authenticate correctly and get access to cloud resources roles.

Middle attack: Message attacker unprotected network environment between the transmitter and the receiver will pass tampering or analyze the content for this attack technique, the mechanism of the user or server messages sent by adding a time stamp T, combined with asymmetric encryption and symmetric encryption, so the attacker cannot know the status of the next key, the message cannot crack and tampering, even cracked, according to the discrete number of degrees of difficulty cannot be within a reasonable time stamp subsisting range, crack, so the message cannot be tampered with after validation authentication server.

23.5 Conclusion

Due to the rapid development and diversification of the Internet now, and driven by software technology, the use of cloud services across Increasingly widespread, cloud systems increasingly large number of users in the case of rising in the authorized management relative. Difficulties manager for role-based authorization management complexity. Therefore, this study proposes mechanisms RBAC (Role-Based Access Control) to study cloud services across the authority. Since most of the current mobile devices have the near-field wireless communication.

Telecommunications (Near Field Communication, NFC) chip for mobile devices, and begin the development of a large number of applications, such as electricity Wallet, electronic ticket coupons, identity authentication, etc. applications. Therefore, this study considered mobile devices equipped with near-field wireless communication. Popularity, and the development of cross-cloud service environment, the application of NFC technology with RBAC authorization mechanism used across. Cloud services license management mechanism, the following simple description of the column and the contribution of this study:

i. The study was conducted through a secure NFC mobile device user login and obtain authorization assisted, so that users can logon authentication system across the cloud via NFC mobile devices, and use the computing power of NFC mobile devices with encryption for authentication and authorization information, be sure to use Login and obtain authorization actors are safe.
ii. In this study, managers can simplify the management of the authorization system, due to the number of users across many environmental cloud, so prone to negligence managers manage authorization management responsibilities led to distinguish setting errors, so the study of conflict through dynamic and static

power to distinguish responsibilities table strengthen the authorization mechanism to help administrators manage RBAC authorization system.

iii. Applied across the cloud environment: As the cloud across a very complex environment, this study proposes a cross-cloud environment suitable authorization process and architecture, and in the future be able to simulate the mechanism to ensure the viability of this research.

References

1. National Institute of Standards and Technology (NIST) (2011) The NIST definition of cloud computing
2. Ferraiolo DF, Barkley JF, Kuhn DR (1999) A role-based access control model and reference implementation within a corporate intranet. ACM Trans Inf Syst Secur 2(1):34–64
3. NFC Forum. http://www.nfc-forum.org/home
4. Liu S, Huang H (2009) Role-based access control for distributed cooperation environment: International conference on computational intelligence and security, pp 455–460

Chapter 24
A Study on High Security Authentication Mechanism for 3G/WLAN Networks

Wei-Chen Wu and Horng-Twu Liaw

Abstract Because of the more and more services wireless communication technology can offer nowadays, the quality of wireless communication became an important key. In this research, 3G/UMTS and WLAN, which are two major wireless communication techniques will be mentioned mainly. The former offers a wide-range, high-movability, complete and safe record of accounting; the latter offers a narrow-range, low-movability, high-speed-transmission access on the Internet. The complementary between these two techniques can not only enhance the quality of wireless communication but offer more services for customers to choose, and customers can use wireless application services regardless of any environmental limit. This research will focus on the problem of fast-handover when 3G/UMTS and WLAN is interworking, such as authentication and authorization. About the two formers, we will use W-SKE to accomplish authentication procedure, and achieve safer Mutual Full Authentication and Fast-Authentication.

Keywords 3G/UMTS · WLAN · Interworking · Authentication

24.1 Introduction

The mobile communication technologies have become more and more popular in recent years, and cell phone service is an important example among kinds of mobile communication. People can contact with each other by cell phone anytime and anywhere even if they are traveling by train or walking on the street. Third-generation

W.-C. Wu (✉)
Computer Center, Hsin Sheng College of Medical Care and Management,
Taipei, Taoyuan County, Taiwan, R.O.C
e-mail: wwu@hsc.edu.tw

H.-T. Liaw
Department of Information Management, Shih Hsin University,
Taipei, Taiwan, R.O.C
e-mail: htliaw@cc.shu.edu.tw

J.J. (Jong Hyuk) Park et al. (eds.), *Ubiquitous Computing Application and Wireless Sensor*, Lecture Notes in Electrical Engineering 331,
DOI 10.1007/978-94-017-9618-7_24

mobile system (3G) like Universal Mobile Telecommunication System (UMTS) will improve data capacity and make data rates up to 2 Mb/s. In addition, 3G systems are expected to provide new applications such as videophone or video streaming. However, the users cannot be actually offered the full theoretical capacity because the deployment cost is high. On the other hand, wireless local area networks (WLAN) are adopted widely in the small areas called hotspots because of their cost-effectiveness and ease of deployment and high data rates (802.11 g up to 54 Mb/s) in an unlicensed frequency band. Nevertheless, WLAN cannot provide the wide coverage cost-efficiently while 3G exactly can. In order to provide large varieties of services at high data rate in the hotspots and campus-wide areas, 3G service providers regard WLAN as a technology to complement their 3G system. Thus, the idea of integrating 3G and WLAN networks to unify the advantages of the two systems as well as to minimize the disadvantages arise as a great market opportunity. They can't replace each other. When WLAN and 3G/UMTS coexist, the handoff mechanism should be created and provided. Many researches [1–6] have proposed about it actually, but it is insufficient to the security requirements. Therefore we will focus on the security of the communication sessions of 3G/UMTS and 802.11 WLAN when a handoff mechanism between them is trigged.

For Integrated 3G/UMTS and WLAN networks, there are several authentication protocols has been proposed [6–9], but some mechanism are not secure or efficient. Hence, EAP-SIM [10] and EAP-AKA [11] have been proposed some authentication mechanism for 3G/UMTS and WLAN interworking. Both EAP-AKA and EAP-SIM provide user with anonymity through pseudonyms or temporary identities called Temporary Mobile Subscriber Identities (TMSI). However, the mobile subscriber called Mobile Node (MN) of real identity is exposed to the air when authenticating MN at the first time. This might cause the real identity of the user to be exposed and traced at some time periods. Moreover, EAP-AKA and EAP-SIM do not minimize the number of exchanges between the foreign domain and home domain. Such problems incur long latency and some packet loss when mobile nodes roam into foreign environment. Salgarelli proposed W-SKE to reduce the number of message exchanged and to minimize the latency. The existing mechanisms are not so suitable for 3G/UMTS and WLAN interworking.

In this paper, we propose a secure vertical handoff policy between 3G/UMTS and 802.11 WLAN networks. To achieve this goal, our scheme is proposed to create a secure communication channel from UMTS to WLAN. Also, a security vertical handoff scheme from WLAN to 3G/UMTS is presented. On the other hand, we propose a robust authentication protocol which can perform efficient localized re-authentication procedure and provide non-repudiation service. Our scheme refers to Keyed-Hash Message Authentication Code (HMAC) [13], Hash-chaining techniques [12], Challenge/Response and Symmetric key Encryption which mention how to withstand the replay attack, guessing attack, impersonation attack and WEP (Wired Equivalent Privacy) weakness attack.

The summary of these articles will be presented in the following sections. The proposed our mechanism will be presented in Sect. 24.2. Moreover, the security analysis and performance of the proposed scheme will be mentioned in Sect. 24.3 and Sect. 24.4. Finally, we will make the conclusions and come up with some future research directions in Sect. 24.5.

24.2 Proposed Mechanism

In this section we propose a new authentication mechanism based on challenge/response, HMAC and one-way hashed chain. Our protocols greatly improve the security and the communication performance.

24.2.1 Notation

$IMSI$	International Mobile Subscriber Identity
PID_A	Pseudonym Identity of A
ID_A	Real Identity of A
TID_A	Temporary Identity of A
$ASID$	Unique identity of Authentication System
k	A Secret Key pre-shared between the H-AAA and the MN
f	A Secret Key pre-shared between the H-AAA and the F-AAA
Ks	A Secret Key produced between the H-AAA and the MN at authentication time
K_{AB}	A Session key between the A and the B
$RAND$	A random seed/value
MAC_{AB}	Message Authentication Codes Function between the A and the B
$E_k()$	A symmetric function with key k
$PRF_k()$	A Pseudo Random Function with key k
$f_k()$	Produce K_S function with key k
$Hash()$	One way hash function
AHC_A	Authentication Hash-Chain value of A

24.2.2 Network Architecture

The network architecture is considered for 3G/UMTS and WLAN interworking in this study, MN denotes mobile node, H-AAA denotes home AAA server of a mobile user MN, and F-AAA denotes foreign AAA server of the WLAN that a MN wants to visit. The F-AAA and the H-AAA belong to separate providers called AAA Brokers; those should be the association between the H-AAA and the

F-AAA. The AAA Brokers sets up reliable security associations and routes AAA messages to the H-AAA.

Our authentication model is based on Salgerelli's work. The authentication model directly corresponds to network architecture in previous section. In order to authenticate and/or protect data in transit between X and Y, a security association AX, Y should be set up and can be defined as the combination of the nodes' identity information (e.g. IMSI, NAI), some form of cryptographic key (e.g. public keys, pre-shared symmetric key), and information on cryptographic algorithms to be used. Each AS maintains a preconfigured security association AAS, F-AAA with its F-AAA server, other AX, Y same meanings. In the 3G/WLAN interworking, F-AAA and H-AAA may belong to separate service providers, and then an association has to be set up via an AAA broker or pair-wise relationship should be setup part of roaming agreement.

24.2.3 Protocols

The characteristic of our mechanism is that it doesn't need the security channel, so every node passes itself legal information of authentication to the H-AAA verity. There are four proposed protocols in our proposal: the full authentication protocol, the WLAN AS fast re-authentication protocol, the 3GPP F-AAA network fast re-authentication protocol, and the 3GPP H-AAA network fast-authentication.

24.2.3.1 Full Authentication Protocol

Here we depict the detailed successful authentication and key exchange process of the full authentication protocol which involves a Mobile Node (MN), Authentication System (AS), Foreign AAA Server (F-AAA), and Home AAA Server (H-AAA). The steps of the authentication message exchange are as follows:

1. The MN sends an EAPOL start to AS after the WLAN association process.
2. The AS response an EAPOL-EAP request/identity to the MN.
3. The MN generate a random seed $RAND_M$, and computes $MAC_{HM} = HMAC_k (RAND_M, IMSI)$.
4. The MN send the EAP Response/Identity message to AS, which involves ID_H, PID_M, $RAND_M$, and MAC_{HM}.
5. The AS sends the EAP Response/Identity message to F-AAA, the *ASID* append to the message.
6. F-AAA computes $MAC_{HF} = HMAC_f (RAND_F, ID_F)$, in order to make the MN easy to verity H-AAA legally.
7. The F-AAA sends the EAP Response/Identity message to H-AAA.
8. The H-AAA first checks whether the MN access profile is available. If not, the H-AAA was rejected by the MN.

- The H-AAA uses the pre-shared key and the received $RAND_M$, $RAND_F$, ID_F, $IMSI$ to verity MN and F-AAA legally.
- If verity failed, the will be rejected. Otherwise, the H-AAA generates random seed $RAND_H$ to compute $k_s = f_k(RAND_M \oplus RAND_H)$.
- The H-AAA generates a new temporary identity of MN, with TID_M have replace PID_M for next time of full authentication protocol.
- The H-AAA computes the first authentication hash-chain value $xAHC$ of the H-AAA and the F-AAA. In order to make the MN easy to re-authentication by the H-AAA and the F-AAA, and keep track of the MN spent based on $xAHC$. Show as follows:

$$xAHC_H^1 = HMAC_{Ks}(ID_H)$$
$$xAHC_F^1 = HMAC_{Ks}(ID_F)$$

- After generating the authentication hash chaining, the H-AAA computes $xAUTH$ purpose to avoid falsity message.
- The H-AAA computes the authentication hash-chain value $xAHC_H^i$ and $xAHC_F^j$, in order to make the MN easy to re-authentication by the H-AAA and the F-AAA (the i and j indicates the hash time; and it can be adjusted on demand). Show as follows:

$$xAHC_H^i = Hash^i(xAHC_H^1)$$
$$xAHC_F^j = Hash^j(xAHC_F^1)$$

- The H-AAA computes the session key between the MN and the F-AAA, shown as follows:

$$K_{MN-F} = PRF_{Ks}\left(xAHC_H^i\right)$$

- The H-AAA computes $xMAC_{FH}$ for the purpose of the F-AAA avoid falsity message from the malice attacker.
- The H-AAA keeps TID_M and Ks, which have replaced PID_M and k for next time of full authentication protocol.

9. The H-AAA sends the EAP success message to the F-AAA.
10. The F-AAA preserve $xAHC_F^j$, K_{MN-F} and TID_M after receiving the EAP success message. Among $xAHC_F^j$ is the hash-chain value when MN and AS process re-authentication protocol, K_{MN-F} is a session key between the MN and the F-AAA; TID_M will not be using PID_M at the time of full-authentication next-time, in order to be anonymous.

- The F-AAA proves whether $xMAC_{FH} = ?MAC_{FH}$ is equal from the H-AAA. If the F-AAA is not with the secret key pre-shared f, it will fail to verity.
- The F-AAA computes $xMAC_{FA}$ for the purpose that make MN avoid falsity message from the malice attacker.

- The F-AAA computes the session key between the MN and the AS as shown follows:

$$K_{MN-AS} = PRF_{K_{MN,F}}(xAHC_F^j)$$

11. After the F-AAA forwards successful authentication message to the AS.
12. The AS preserve the $K_{MN\text{-}AS}$, TID_M for the ease of transmission between the MN and the AS.
13. The AS forwards the EAP success message to the MN.
14. The MN obtains the $RAND_H$ from the H-AAA and the $RAND_M$ producted when MN requests for authentication. Then computes to the secret key K_S produce between the H-AAA and the MN of authentication time, shown as follows: $K_S = f_k(RAND_M \oplus RAND_H)$
    - The MN generates a new temporary identity TID_M = $Hash(RAND_M \oplus RAND_H, IMSI)$.
    - The MN computes the first authentication hash-chain value *xAHC* o as per Eq. (8.4)
    - The MN computes the authentication hash-chain value $xAHC_H^i$ and $xAHC_F^j$ as per Eq. (8.6)
    - The MN computes the session key $K_{MN\text{-}F}$ as per Eq. (8.7)
    - The MN computes the session key $K_{MN\text{-}AS}$ as per Eq. (10.3)
    - The MN verity *xAUTH* in order to avoid falsity message from the malice attacker, show as follows as per Eq.(8.5)
    - The MN computes $xMAC_{FA}$ for the purpose that make the MN avoid falsity message from the malice attacker, show as follows as per Eq. (9.2)
    - The MN keeps TID_M and *Ks* which have replace PID_M and *k* for next time of full authentication protocol.

15. In this step, the MN and the H-AAA successfully authenticate each other.

24.2.3.2 Fast Re-authentication Protocol of the F-AAA

Here we depict the detailed successful re-authentication of the F-AAA. The MN and the F-AAA share a session-key $K_{MS,F}$ which made the re-authenticate key. In the step of the *n*-th re-authentication, *j* is limited for the number of F-AAA re-authentication times, and *j–n* number of re-authentication times once left. When the MN accesses the F-AAA which belongs to the 3GPP visit network, the authentication mechanism is also based on the hash chaining technique. The MN presents its identity TID_M,and the MN computes AHC_F^{j-n}, then sends the result to the F-AAA. After this F-AAA verifies the $Hash\,(xAHC_F^{j-n+1}) = ?AHC_F^{j-n}$; If passing, it means the F-AAA has authenticated the MN. The AHC_F^{j-n} is stored for the next authentication and for the non-repudiation evidence. Afterwards, the F-AAA responses to a challenge $xMAC_{FA} = \boldsymbol{HMAC}_{K_{MN,F}}\ (xAHC_F^{j-n}, ASID^*)$, and computes new session key $K_{MN-AS} = \boldsymbol{PRF}_{K_{MN-F}}\ (xAHC_F^{j-n})$. On the other hand, the $xMAC_{FA}$ and $K_{MN\text{-}AS}$ are sent to the WLAP AS; the AS keeps the $K_{MN\text{-}AS}$ which is used as dynamic WEP key, and forwarded the $xMAC_{FA}$ to the MN. The MN first verifies the $xMAC_{FA}$. If passing, it means the MN has authenticated the F-AAA server. Next, the MN derives the $K_{MN-AS} = \boldsymbol{PRF}_{K_{MN-F}}\ (xAHC_F^{j-n})$. Eventually, the mutual authentication has been successfully completed and the WEP key has been confidentially delivered.

24.2.3.3 Fast Re-authentication Protocol of the H-AAA

This is a roaming reference model. When the MN accesses the H-AAA which belongs to the 3GPP visit network, the authentication mechanism is also based on the hash chaining technique. The Fast Re-authentication protocol of the H-AAA is just the same as the Fast Re-authentication protocol of the F-AAA. The only difference is that the Authentication Hash-Chain Value is added to AHC_H^{i-m}, and the access control is charged by the H-AAA server. The detailed authentication message exchange of the H-AAA fast re-authentication protocol is demonstrated by using the AHC_F^{j-n} and AHC_H^{i-m} sent to the H-AAA, the re-authentication method is based on the hash chaining technique result to the mutual authentication and key agreement can be achieved.

24.2.3.4 Fast Re-authentication Protocol of the AS

This case is under the non-roaming reference model, so the authentication traffic is routed through the New AS and Old AS. The MN computes *Ticket* in order to help Old AS prove whether New AS is a legal node. The MN produces and offer the random value *RAND* to the New AS computes the New Session Key $K_{MN\text{-}AS**}$ so that it can take precautions of the backward to security attack.

Because both sides have agreements of roaming, the other side shares the private session key which can decrypt message of encrypt. The New AS forwards the request message to the Old AS, then the Old AS verifies *Ticket*; if unsucceding, it will reject to serve. Otherwise, represent authentication of the New AS is legal node and produces new random key K_{RAND}. The Old AS responses to a challenge $Ticket_2$ and computes new session key $K_{MN\text{-}AS*}$, making use of private session key $K_{AS\text{-}OLD}$ to encrypt $K_{MN\text{-}AS*}$ and old session key $K_{MN\text{-}AS}$ to encrypt K_{RAND}. Then it produces new session $K_{MN\text{-}AS**}$ so that the New AS makes XOR operation with $K_{MN\text{-}AS*}$ and *RAND*. With that, The New AS forwarded $Ticket_2$ and encrypt K_{RAND} to the MN. The MN decrypt message obtains K_{RAND} at first, and verifies $Ticket_2$ as Eq. (7.2). If passing, it means the MN has authenticated the New AS. Next, the MN produces the new session $K_{MN\text{-}AS**}$. Finally, the mutual authentication has been successfully completed and the WEP key has been confidentially delivered.

24.3 Security Analysis

In this section, we will show our mechanism can preclude several attacks, according to Byzantine insiders, which indicates the network elements belong to independent service provider that are not trusted fully because it have a direct or indirect security association between each other.

Prevent Guessing Attack: In full authentication protocol, the Secret Key Pre-shared k between the H-AAA and the MN, are for authentication purpose. Therefore, it is possible for an attacker to reveal the Secret key Pre-shared k from the known information. However, the k is impossible to derive it during a reasonable time which is at least 128 bits. Utilizing one time password of AHC^{i-m} and AHC^{j-n} to upgrade session key in fast authentication, it is invalid to obtain session key K_{WEP}.

Prevent Replay Attack: In full authentication protocol, it is the situation where an attacker intercepts $\{ID_H, PID_M, RAND_M, MAC_{HM}\}$ sent by the MN in step 4 and uses it to masquerade as the MN to send the authentication request next time. Though $RAND_M$ is generated by the MN, the malice attacker don't knowing the Secret Key Pre-shared k between the H-AAA and the MN, and it can't respond the correct MAC_{HM} and $AUTH$ to the H-AAA and the MN both. On the other hand, the authentication hash chaining value AHC of fast re-authentication will be used only once, to replay the AHC will not pass the authentication.

Prevent Impersonation Attack: The malice attacker attempts to impersonate the MN to access the WLAN. In full authentication protocol, $MAC_{HM} = HMAC_k(RAND_M, IMSI)$ is encrypted with a pre-shared secret k; hence without secret key k, it can't impersonate the MN. In fast re-authentication, the attacker cannot compute $xAHC_H^i = Hash^i(xAHC_H^1)$ or $xAHC_F^j = Hash^j(xAHC_F^1)$ to impersonate the MN, because the Pre-shared Secret Key k is only known by the MN-self and the AHC^1 has been securely sent to the H-AAA by the MN in full authentication. In this case, the attacker can't compute backward the authentication hash chaining value AHC.

Prevent WEP weakness attack: The WEP key congenital disadvantage in the gold key IV value is not enough and easy to analyze and explain for the Brute-Force attack in length, since the WEP key is also renewed in each full authentication of fast re-authentication protocol. Therefore, our mechanism can overcome the weakness of the original WEP.

Prevent Forward Secrecy and Backward Secrecy to possible attacks: One session key/secret key will not lead to the compromise of the past session key/secret key and the corresponding transmission because one key follows the form of randomness, the one-way property of hashing chains and the session key pre-shared between each other. Thus, each session key/secret key is random and independently, and is fairly controlled by the MN and AAA Server or AS. It can prevent Forward/Backward secret attack then accomplishes the resistance to the known-key attack, the impersonate attack, and the replay attack.

Legal evidence for use-bill: In the billing process, the AS and F-AAA have to submit all latest hash chain values sent by the MN after each full authentication to

H-AAA. *xAHC* will record the usage of MN so that WLAN ISP and the 3G ISP will charge H-AAA for fees according to *xAHC*. Because of having one-way characteristic of hash chaining function, the ISPs is unable to compute to the $xAHC^{i-n-1}$ value so that it is also unable to cheat H-AAA with incorrect data of the usage of the evidence, then reach both sides' mutually beneficial fairness.

Mutual Authentication: The $xAHC_H$ and $xAHC_F$ hidden in *xAUTH* is resulted from computing the AAA Server Identity. The MN will fail to authenticate if there is no legal AAA Server. According to the principle of Transitive, when the authentication between MN and H-AAA, H-AAA and F-AAA, F-AAA and AS all succeed, the one between MN and AS will success, too. The MN will prove legal node of the AS, if the mobile node computes to MAC_{FA} equals with $xMAC_{FA}$ from the H-AAA.

Secret Key Establishment: The first secret key K_s is produced after the H-AAA and the MN accomplish full authentication. By $RAND_M$, $RAND_H$ and k compute K_s which needn't pass K_s to the MN, the MN will computes K_s by itself. The main purpose for this is to improve its security, which will replace the secret key pre-share k in the full authentication protocols next time.

Non-repudiation: Our mechanism will complete secret key k with registering in advance. To put *RAND* and *ID* in the Message Authentication Codes Function can produce *MAC* value, then will can use *MAC* to verify the legitimacy of both sides with by its characteristic of Challenge and Response

Message Integrity: Guarantee mainly the content in the course of transmission has not been falsified. Our mechanism check out the equality between *MAC* and *xMAC*; so does between *AUTH* and *xAUTH*, in order to confirm the integrality of the message.

Protect Transmit Session Key: The Session Key $K_{A,B}$ is transmitted to other communication apparatus under the protection of the symmetric function, preventing $K_{A,B}$ from being stolen in the course of transmission.

Perfect User Anonymity: H-AAA and MN will figure out TID_M, and TID_M will replace PID_M for next time of full authentication protocol. Only the issuer (MN or H-AAA) is able to produce the temporary identifier TID_M. Our scheme, different with EAP-SIM and EAP-AKA, is not transmitted for each time when the temporary identifier is not available; it adopts the dynamic way to produce TID_M. Therefore, perfect anonymity is achieved.

24.4 Performance Analysis

In this section, we will evaluate the efficiency of our mechanism in terms of authentication latency in more details. Let $T_{F\text{-}AAA,H\text{-}AAA}$ denote the one trip latency between H-AAA and F-AAA, $T_{F\text{-}AAA,AS}$ denote the one between F-AAA and AS, and $T_{MS,AS}$ denote the one between MS and AS. According to the number of authentication time, we can see that $T_{F\text{-}AAA,H\text{-}AAA} > T_{F\text{-}AAA,AS} > T_{MS,AS}$. EAP-AKA, W-SKE and IDKE. The authentication latency of our full authentication is $2T_{F\text{-}AAA,}$

${}_{H\text{-}AAA} + 2T_{F\text{-}AAA,AS} + 4T_{MS,AS}$, EAP-AKA is $4T_{F\text{-}AAA,H\text{-}AAA} + 8T_{F\text{-}AAA,AS} + 10T_{MS,AS}$; W-SKE is $2T_{F\text{-}AAA,H\text{-}AAA} + 4T_{F\text{-}AAA,AS} + 6T_{MS,AS}$; but IDKE doesn't point out this method. Moreover, in terms of fast re-authentication, our scheme is $2T_{F\text{-}AAA,AS} + 4T_{MS,AS}$; EAP-AKA is $6T_{F\text{-}AAA,AS} + 8T_{MS,AS}$; IDKE is $2T_{F\text{-}AAA,AS} + 4T_{MS,AS}$; while W-SKE doesn't mention it. If the time of handover for authentication or re-authentication protocol is overlong, it will cause to lose the packages. Our mechanism can shorten the authentication time delay.

24.5 Conclusions and Future Works

In our mechanism, we discuss about the security and authentication protocol for WLAN and 3G/UMTS interworking. Three state-of-the–art authentication protocols for integrated 3G/WLAN networks: EAP-AKA, W-SKE and IDKE, have been examined, and shown the security weaknesses of W-SKE, the in-efficiency of EAP-AKA, and integrate localized re-authentication of IDKE. Based on Keyed-Hash Message Authentication Code (HMAC), Hash-chaining techniques, Challenge-Response and Symmetric key Encryption, we have figured out a new authenticated key exchange protocol. We propose a robust authentication protocol which can perform efficient localized re-authentication procedure, provide non-repudiation service, solve the problems of losing packages, shorten the authentication time delay and greatly improve the security.

In our future work, we hope we will be able to solve Denial of Service (DoS) attacks to improve the security. On the other hand, we also expect to do a more in-depth research focused on the handover mechanism, roaming management, packet forwarding and transmission in the future days.

References

1. GPP TR 22.934 (2003) Feasibility study on 3GPP system to wireless local area network (WLAN) interworking (release 6)
2. GPP TS 22.234 (2004) 3GPP system to wireless local area network (WLAN) interworking, system description (release 6)
3. GPP TS 33.234 (2006) 3G securtiy; wireless local area network (WLAN) interworking security (release 7)
4. Buddhikot M, Chandranmenon G, Han S, Lee YW, Miller S, Salgarelli L (2003) Integration of 802.11 and third-generation wireless data networks. In: Proceedings of the IEEE INFOCOM'03
5. Zhu J, Ma J (2004) A new authentication scheme with anonymity for wireless environments. Member, IEEE
6. Salgrelli L, Buddhikot M, Garay J, Patel S, Miller S (2003) Efficient authentication and key distribution in wireless IP networks. Bell Laboratories, Lucent Technologies, IEEE Wireless Communication, Dec 2003

7. Kambourakis G, Rouskas A, Kormentzas G, Gritzalis S (2004) Advanced SSL/TLS-Based authentication for secure WLAN-3G interworking. In: Communications, IEE proceedings, vol 151
8. Prasithsangaree P, Krishnamuthy P (2004) A new authentication mechanism for loosely coupled 3G-WLAN integrated networks. In: IEEE vehicular technology conference
9. Tsen Y-M, Yang C-C, Su J-H (2004) An efficient protocol for integrating WLAN and cellular networks. In: The 6th international conference on advanced communication technology
10. IETF Draft (2003) IETF internet draft EAP-SIM authentication. http://www.ieft.org/internet-draft-haverinen-appext-eap-sim-10.txt
11. IETF Draft (2006) Extensible authentication protocol method for 3rd generation authentication an key agreement (EAP-AKA). RFC 4187
12. Lamport L (1981) Password authentication with insecure communication. Commun ACM 24 (11):770–772
13. Krawczyk H, Bellare M, Canetti R (1997) Keyed-hashing for message authentication. Request for comments RFC2104, IETF

Chapter 25
A Simple Authentication Scheme and Access Control Protocol for VANETs

Wei-Chen Wu and Yi-Ming Chen

Abstract In 2011, Yeh et al. proposed a PAACP: A portable privacy-preserving authentication and access control protocol in vehicular ad hoc networks. However, PAACP in the authorization phase is breakable and cannot maintain privacy in VANETs. In this paper, we present a cryptanalysis of an attachable blind signature and demonstrate that the PAACP's Authorized Credential (AC) is not secure and private even if the AC is secretly stored in a tamper-proof device. An eavesdropper can construct an AC from an intercepted blind document. Any eavesdropper can determine who has which access privileges to access which service. For this reason, this paper copes with these challenges and proposes an efficient scheme. We conclude that a simple authentication scheme and access control protocol for VANETs not only resolves the problems that have appeared, but also is more secure and efficient.

Keywords VANET · Cryptanalysis · Privacy · Authentication · Access control

25.1 Introduction

In the recent years, several researches on VANETs have been conducted by academics and various industries. Recently, some of these works addressed the security issues. As an instance of MANET, VANETs might suffer any malicious user's behaviors, such as bogus information and replay attacks on the disseminated

---

W.-C. Wu
Computer Center, Hsin Sheng College of Medical Care and Management,
Taoyuan, Taiwan, ROC
e-mail: wwu@hsc.edu.tw

Y.-M. Chen (✉)
Department of Information Management, National Central University,
Taipei, Taiwan, ROC
e-mail: cym@cc.ncu.edu.tw

J.J. (Jong Hyuk) Park et al. (eds.), *Ubiquitous Computing Application and Wireless Sensor*, Lecture Notes in Electrical Engineering 331,
DOI 10.1007/978-94-017-9618-7_25

messages. Among various security threats, privacy preservation in VANETs is one of the new challenges of protecting users' private information. For instance, Chen and Wei [2, 10] proposed a safe, distance-based location privacy scheme called SafeAnon. By simulating vehicular mobility in a cropped Manhattan map, they evaluated the performance of the SafeAnon scheme under various conditions to show that it could simultaneously achieve location privacy as well as traffic safety. However, as Chen and Wei focused on the issues of the vehicles' location privacy, little emphasis was put on the initial authentication phase of communications among vehicles.

In 2005 [7], Raya et al. first proposed a solution that mentioned both the security and privacy issues of safety-related applications. Wang and others reviewed Raya and Hubaux's communication scheme in 2008 [9] and argued that Raya and Hubaux paid a great deal of attention to safety-related applications, such as emergency warnings, lane changing assistance, intersection coordination, traffic-sign violation warnings, and road- condition warnings [9], but non-safety-related applications were neglected. In Raya and Hubaux's communication scheme, Safety messages do not contain any sensitive information. However, VANETs also provide non-safety applications that offer maps [4, 12], advertisements, and entertainment information [13].

Similar to safety applications, non-safety applications in VANETs have to take both security and privacy issues into consideration. In addition, designing a practical non-safety application for VANETs should take the following requirements into consideration [15, 11]:

Mutual Authentication: Providing mutual authentication between the two communicating parties, such as a vehicle-to-roadside communications device.
Context privacy: Allowing mobile vehicles to anonymously interact with roadside devices to access services.
Lower computation cost: A system must have light overhead in terms of computational costs and high efficiency.
Session key agreement: Generating dynamic session keys to secure the communications between nodes in VANETs.
Differentiated service access control: Providing several services with different levels of access privileges for different users' requirements.
Confidentiality and integrity: Providing data confidentiality and integrity in applications of communications.
Preventing eavesdropping: An intruder cannot be allowed to discover valuable information from communications between members in VANETs.
Scalability: Coping with the large scale and dynamic environment presented by VANETs.

In 2008, Li et al. proposed a secure and efficient communication scheme named SECSPP [15] that employs authenticated key establishment for non-safety applications in VANETs. SECSPP is the first security scheme with explicit authentication procedures for non-safety applications. However, the speed of a vehicle can be extremely high in SECSPP. It is possible that the response sent from the Service Provider (SP) has not yet arrived, but the requesting vehicle has passed the RSUs' transmission range. Moreover, all requests made by non-safety applications must first be verified by the proper SP, which will become a bottleneck of SECSPP. The scalability issue rises in a popular SP if a large number of requests are made.

In 2011, Yeh et al. proposed a PAACP: A portable privacy-preserving authentication and access control protocol for vehicular ad hoc networks [11]. However, in the authorization phase, a PAACP is breakable and cannot maintain privacy in VANETs. Recently, Wu et al. presented a cryptanalysis of an attachable blind signature and demonstrate that the PAACP's Authorized Credential (AC) is not secure and private even if the AC in secretly stored in a tamper-proof device [10]. This is because an eavesdropper is able to construct an AC from an intercepted blind document. Consequently, PAACP in the authorization phase is breakable and cannot maintain privacy in VANETs. Any outsiders can determine who has which access privileges to access which service. In addition, this paper efficiently copes with these challenges and proposes an efficient scheme. We conclude that a simple authentication scheme and access control protocol for VANETs will not only resolve the problems that have appeared, but also is secure and efficient.

The remainder of this paper is organized as follows. Section 25.2 reviews the cryptanalysis of a PAACP. Section 25.3 introduces an efficient scheme. In Sect. 25.4, we compare the performance of our schemes with PAACP and SECSPP and analyze various aspects of the security of our scheme. Finally, we conclude this paper and indicate some directions for future research in Sect. 25.5.

25.2 Cryptanalysis of a PAACP

In 2011, Yeh et al. proposed a novel portable privacy-preserving authentication and access control protocol for vehicular ad hoc networks [11]. To eliminate the communication with service providers, they proposed a novel portable access control method to store a portable service right list (SRL) into each vehicle, instead of keeping the SRLs with the service providers. In order to assure the validity and privacy of an SRL and prevent privilege elevation attacks, an attachable blind signature is used by PPACP. Recently, Wu et al. proposed a cryptanalysis of an attachable blind signature and demonstrate that the PAACP's Authorized Credential (AC) is not secure and private even if the AC is secretly stored in a tamper-proof device [10]. Their analysis showed that in PAACP, an eavesdropper can construct the AC from an intercepted blind document. As a result, PAACP in the authorization phase is breakable, and as any outsider can determine who has which access privileges to access which service, the privacy of users in PAACP's scheme is jeopardized. Wu et al. presented cryptanalysis 1 shows that m' cannot keep privacy and cryptanalysis 2 shows that an intruder can use public key PK_{S_t} of the S_t to compute authorized credential AC_{S_t}. The notation used throughout the remainder of this paper is shown in Table 25.1.

Table 25.1 Notation used in the remainder of the paper

Notation	Description
V_i	The *i*-th vehicle
VID_i	*i*-th vehicular node's real identification
S_t	The *t*-th service provider
SID_t	*t*-th service provider's real identification
$SVID_k$	k-th service's identification
AR_k	The access privilege of $SVID_k$
AC_i	Authorized credential for vehicle V_i
$AC_i^{S_t}$, $AC_i^{V_i}$	Authorized credential made by S_t and V_i, respectively
AC_i^*	Portable authorized credential for vehicle V_i
$SRL_i^{S_t}$, $SRL_i^{V_i}$	Service right list made by S_t and V_i, respectively
$D_k()$	A corresponding symmetric cryptosystem that uses the secret key *k* for decryption
$E_k()$	A secure symmetric cryptosystem that uses the secret key *k* for encryption
N_i	Fresh nonce, random generated by VID_i
N_s	Fresh nonce, random generated by service provider
$h()$	A collision-free and public one-way hash function
$\|\|$	A string concatenation
$X \rightarrow Y{:}Z$	A sender *X* sends a message *Z* to receiver *Y*

25.2.1 Cryptanalysis 1

To acquire a message m/, an intruder can eavesdrop on the two blind documents $BD_1{}'$ BD_2 in the (*User* $\rightarrow$ *Signer*) channel and also eavesdrop on $BD_1{}'$, $BD_2{}'$ in the (*Signer* $\rightarrow$ *User*) channel. After stealing BD_1, BD_2, $BD_1{}'$ and $BD_2{}'$, the intruder can use public key *e* of the signer to compute the following equation:

$$(BD_1' \cdot BD_2')/(BD_1 \cdot BD_2) = m'$$

25.2.2 Cryptanalysis 2

Similarly, to acquire authorized credential $AC_i^{V_i}$ and $AC_i^{S_t}$, an intruder can eavesdrop on the two blind documents $BD1_i$, $BD2_i$ in the (*Vehicle* $\rightarrow$ *Service Provider*) channel and also eavesdrop on $BD1_i'$, $BD2_i'$ in the (*Service Provider* $\rightarrow$ *Vehicle*) channel. After stealing $BD1_i$, $BD2_i$, $BD1_i'$ and $BD2_i'$, the intruder can use public key PK_{S_t} of the Service Provider to compute the following equation:

$$(BD1'_i \cdot BD2'_i)^{PK_{S_t}} / (BD1_i \cdot BD2_i) = AC_i^{S_t}$$

Finally, according to $((AC_i^*)^{PK_{S_t}})^{1/2} = AC_i^{V_i} = AC_i^{S_t}, AC_i^{S_t}$ is equal to $AC_i^{V_i}$, where AC_i^* consists of both $AC_i^{V_i}$ and $AC_i^{S_t}$. Yeh et al. claimed that an attachable blind signature can keep privacy, no one could comprehend the access privileges in $AC_i^{V_i}$and no one can realize who is accessing those services [11]. On the basis of our cryptanalysis, $AC_i^{S_t} = \{SID_t || T_{expired} || SRL_i^{S_t}\}$ and $AC_i^{V_i} = \{SID_t || T_{expired} || SRL_i^{V_i}\}$ could be comprehended by outsiders who could then decode the service right lists $SRL_i^{S_t}$ and $SRL_i^{V_i}$ respectively. In a previous description, the service right list is as the following equation:

$$SRL_i^{V_i} = \{SVID_1 || AR_1 || SVID_2 || AR_2 || \cdots || SVID_k || AR_k ||\}$$

where $SVID_k$ denotes the index of the kth service, and AR_k represents the granted access privileges of $SVID_k$. Hence, anyone can determine who has which access privileges to access which service even if AC_i^* is secretly stored in a tamper-proof device.

25.3 A Simple Scheme

In this section, we propose a simple scheme and offer an efficient authentication and access control protocol for VANETs. The security of this scheme depends on a secure one-way hash function, not the use of an attachable blind signature. This scheme consists of three phases: the registration phase, the authentication phase, and the access phase. We demonstrate our scheme as follows:

25.3.1 The Registration Phase

A vehicle V_i creates a service right list $SRL_i^{V_i}$and an authorized credential $AC_i^{V_i}$, just as Yeh et al. proposed. Let x be a secret key maintained by the service provider S_t, and let $h()$ be a secure one-way hash function with a fixed-length output. The registration phase is performed over a secure channel.

- $V_i \rightarrow S_t : VID_i, AC_i^{V_i}$

  A V_i who submits his/her identity VID_i and his/her $AC_i^{V_i}$ to the S_t for registration.

- $S_t \rightarrow V_i : h(), e_i$

The S_t also creates $SRL_i^{S_t}$ and $AC_i^{S_t}$ as Yeh et al. proposed. The S_t then computes V_i's secret information $y_i = h(VID_i, x)$ and $e_i = y_i \oplus AC_i^{S_t} \oplus AC_i^{V_i}$ and writes $h()$ and e_i into the smart card of on-board units (OBUs) and issues the card to V_i.

- $S_t \rightarrow R_j : y_i, AC_i^{S_t}$

The S_t also performs a multicast to send messages y_i and AC_i^{St} to their road side units (RSUs) R_j.

25.3.2 The Authentication Phase

After V_i sends an authentication request message to the S_t, the S_t and V_i will execute a mutual authentication between the vehicle and the service provider. First, let $E_k()/D_k()$ be a symmetric encryption/decryption function with secret k, respectively.

- $V_i \rightarrow S_t : VID_i,\ C,\ N_i$

When V_i wishes to access services provided by S_t, V_i generates a nonce N_i, where N_i is a random and fresh number. Then V_i computes $C = h(e_i \oplus AC_i^{V_i}, N_i)$ and sends an authentication request message (VID_i, C, N_i) to the S_t.

- $S_t \rightarrow R_j : M$

After receiving the authentication request message (VID_i, C, N_i), the S_t and V_i execute the following steps to facilitate a mutual authentication between the vehicle and the service provider. The S_t performs the following operations:

- Verifies that VID_i is a valid vehicle identity. If not, the authentication request is rejected.
- Computes $y_i' = h(VID_i, x)$ and verifies whether $y_i = y_i'$. If the verification fails, the request is rejected.
- Check whether received $C = h(y_i' \oplus AC_i^{St\prime}\ N_i C = h(y_i' \oplus AC_i^{St\prime}\ N_i)$. If not, the request is rejected; otherwise, the request proceeds to a next step.
- Generates a nonce N_s, where N_s is a random and fresh number.
- Encrypts the message $M = E_{yi \oplus AC_i S^t}\{N_s, N_i, AC_i^{S_t}\}$ and sends it back.
- After V_i receives the message M, V_i will decrypt the message $D_{ei \oplus} ACiVi\{M\}$ to derive $(N_s', N_i', AC_i^{S_t}\prime)$ and verify whether $N_i' = N_i$. If the answer is yes, the mutual authentication is done. The portable authorized credential is $AC_i = AC_i^{Vi} \oplus AC_i^{St}$, and we propose $AC_i^{V_t}$is not equal to $AC_i^{S_t}$. Either S_t may reduce access privileges for some reason (for example, not paying before the deadline or breaking a contract) or V_i may disable access privileges himself/herself for some reason (for example, privacy issue or lower communication costs). So AC_i is $AC_i^{V_t}$ performs an exclusive operation with $AC_i^{S_t}$ that is reasonable and makes sense.

25.3.3 The Access Phase

This phase is based on the key exchange protocol proposed by Diffie et al. [3]. It is used to encrypt an individual conversation with a session key. The lifespan of a session key is the period of a particular communication session. A new session phase involves two public parameters, q and α, where q is a large prime number and α is a primitive element mod q. After V_i sends a service request to its neighboring R_j, R_j will verify the authorized credential AC_i by itself without further communication with S_t. According to the access privileges stored in the authorized credential AC_i^{St}, R_j could decide whether V_i's request is accepted or not. Furthermore, R_j could detect whether V_i is launching an Elevation of Privilege (EoP) attack.

- $V_i \rightarrow R_j : W_i$

  V_i computes $W_i = \alpha^{r_{vi}}\ mod\ q$ and sends W_i to R_j, where r_{vi} is a random number.

- $R_j \rightarrow V_i$: Si

  Similarly, R_j computes $S_i = \alpha^{r_{Rj}}\ mod\ q$ and sends S_i to V_i, where r_{Rj} is a random number. V_i computes $K_V = (S_i)^{r_{vi}}\ mod\ q$ and R_j computes $K_R = (W_i)^{r_{Rj}}\ mod\ q$. Then, both of them check whether $K_V = K_R$. If yes, a new session will be created. That is because:

$$\begin{aligned} \text{Session key} &= (S_i)^{r_{vi}}\ mod\ q \\ &= (\alpha^{r_{Rj}}\ mod\ q)\ mod\ q \\ &= (\alpha^{r_{Rj}\, r_{vi}})mod\ q \\ &= (\alpha^{r_{vi}}\ mod\ q)^{r_{Rj}}\ mod\ q \\ &= (W_i)^{r_{vi}}\ mod\ q \end{aligned}$$

- $V_i \rightarrow R_j$: *(Service request message)*

  If V_i wants to access service, it encrypts $E_{KV}(SVID_1||AC_i)$ with K_V as the service request message and sends it to R_j. After R_j receives the message, R_j will decrypt the message:

$$D_{KR}(E_{KV}(SVID_1||AC_i))$$

with K_R to gain $(SVID_1||AC_i)$ and then derive AC_i, and $SVID_1$ because of $K_V = K_R$. When R_j derives AC_i, R_j verifies it and is then convinced that V_i is a legal user.

- $V_i \rightarrow R_j$ *(Service request message)nth*

  When V_i continues to access the *n*th service, it encrypts the *n*th service request message $E_{KV+n}(SVID_n||AC_i)$ with K_V + n and sends it to R_j. After R_j receives the *n*th service request message, R_j will decrypt the message

$$D_{KR+n}(E_{KV+n}(SVID_n||AC_i))$$

with K_R + n to gain ($SVID_n||AC_i$) and then derive AC_i, and $SVID_n$. R_j examines whether SID_t as well as $SVID_n$ are included in AC_i^{St}, and checks the validity of the authorized credential by $T_{expired}$. If the verification succeeds, AC_i is legitimate and V_i is authorized; otherwise, R_j terminates this session.

25.4 Analysis of the New Scheme

In this section, we roughly compare the security properties and performance of the related mechanisms discussed. The security properties comparisons between PAACP, SECSPP and our scheme in the authentication phase and access phase are shown in Table 25.2. The performance comparisons are shown in Table 25.3.

25.4.1 Comparison

Table 25.2 lists important security properties in VANETs based on Yeh et al.'s proposals. As mentioned, with PAACP, an attachable blind signature is breakable and cannot maintain privacy, and the PAACP's AC is not secure even if the AC is secretly stored in a tamper-proof device. An eavesdropper is able to construct the AC from an intercepted blind document. Any outsiders in VANETs can know who has which access privileges to access which service. Consequently, PAACP cannot still satisfy context privacy properly.

Table 25.2 Comparison of security features

Requirements	Our scheme	PAACP	SECSPP
Mutual authentication	Yes	Yes	Yes
Context privacy	Yes	No[a]	Yes
Session key agreement	Yes	Yes	Partially Yes[b]
Differentiated service access control	Yes	Yes	No
Confidentiality and integrity	Yes	Yes	N/A
Preventing eavesdropping	Yes	No[a]	Yes
Scalability	Fully distributed	Fully distributed	Bottleneck at service
Computation cost	Lower cost	High cost	High cost

[a] In PAACP, Authorized Credential (AC) is not secure and private

[b] In SECSPP, the session key *TSK* is determined by *V* and *S*, not *V* and *R*

25.4.2 Performance

Since the computation load of the PKI cryptosystem is the heavy burden of all communicating nodes in the PPACP and SECSPP, we propose an efficient version without PKI cryptosystems. Furthermore, the speed of encryption/decryption with symmetric encryption schemes is faster than with asymmetric ones, namely PKI cryptosystems. For instance, it is known that DES is 100 times faster than RSA in software and 1,000 times faster in hardware [6]. An exponential operation is approximately equal to 60 symmetric en-/decryption [6, 14]. Consequently, we treat the computation load of a PKI operation as that of 100 symmetric operations. As listed in Table 25.3, the PPACP needs nearly 702 symmetric operations and SECSPP needs 740 symmetric operations in the related work, while it requires about 124 symmetric operations in our scheme. Moreover, it takes 0.0005 s to complete a one-way hash operation and 0.0087 s to finish a symmetric en-/decryption. We hence ignore the computation load of the one-way hash function since it is quite lighter than that of a symmetric en-/decryption [1]. As a result, computation loads can be reduced to 1.0788 s in our scheme.

The following is based on the computation method in PAACP. Assume that n vehicles in the VANET request the services of the same services provider at the same time and the locations where these service requests are invoked are uniformly distributed within m RSUs. The transmission delay $T_{trans-delay}$ is the time in seconds to deliver a message from a vehicle, that is forwarded to the service provider by an RSU. The waiting time $T_{waiting}$ consists of the round-trip transmission delay and the time spent on verification by the service provider. In SECSPP, the average waiting time $T_{waiting}$ for a requesting vehicle can be estimated as

$$T_{waiting} = 2 \cdot T_{trans-delay} + ((n+1) \cdot T_{Access\, verification})/2$$

Table 25.3 Comparison of efficiency

Requirements	Our scheme	PAACP	SECSPP
Authorization	$2T_{sym} + 2T_{hash} + 5T_{xor}$	$4T_{asym} + T_{hash}$	$2T_{asym} + 2T_{exp} + 3T_{hash} + 4T_{xor}$
Access service	$2T_{sym} + 2T_{hash} + 3T_{xor}$	$3T_{asym} + 2T_{sym} + T_{hash}$	$3T_{asym} + 2T_{exp} + 6T_{hash} + 5T_{xor}$
Total T_{sym}	$\approx 124T_{sym}$	$\approx 702T_{sym}$	$\approx 740T_{sym}$
Rounds	4	3	5
Authorization costs (s)	≈0.0174	≈3.48	≈2.784
Access costs (s)	≈1.0614	≈2.6274	≈3.654
Total costs (s)	≈1.0788	≈6.1074	≈6.438

T_{hash}: Computation cost of one-way function
T_{xor}: Computation cost of Exclusive-OR operation
T_{sym}: Computation cost of symmetric encryption
T_{asym}: Computation cost of asymmetric operation
T_{exp}: Computation cost of modular exponentiation

In PAACP and our scheme, the average waiting time $T_{waiting}$ for a requesting vehicle can be estimated as:

$$T_{waiting} = ((n/m + 1) \cdot T_{Access\,verification})/2, \quad \text{if } n > \text{m}$$
$$T_{waiting} = T_{Access\,verification}, \quad \text{otherwise}$$

In a uniform distribution of locations, the average number of requests pending in each RSU will be n/m. Therefore, the average time spent for request verification in a RSU is $((n/m + 1) \cdot T_{Access\ verification})/2$. Figure 25.1 shows that when m is equal to 10, the average waiting time $T_{waiting}$ for a service request from vehicle n increases 1–50. Figures 25.2, 25.3 and 25.4 show that the average waiting time $T_{waiting}$ for a service request from vehicle n increases 1–100 when m is equal to 10, 30 and 50 respectively. As Fig. 25.2 shows, when 100 vehicles are requesting the desired services, the average waiting time $T_{waiting}$ to finish the authentication in PAACP is 14.32 s. In our scheme, the average waiting times $T_{waiting}$ is about 5.73 s. Similarly, as shown in Fig. 25.3, our scheme takes about 2.28 s, compared to about 5.65 s for

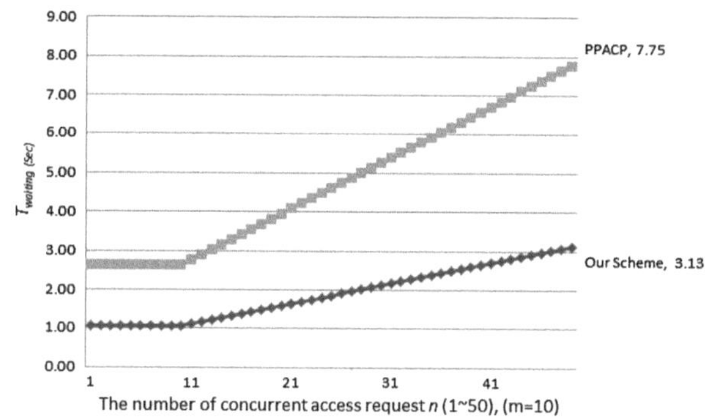

Fig. 25.1 Average waiting time when m is equal to 10

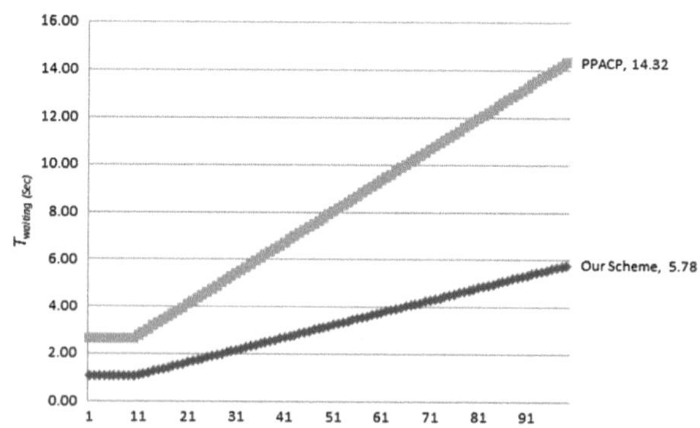

Fig. 25.2 Average waiting time when m is equal to 10

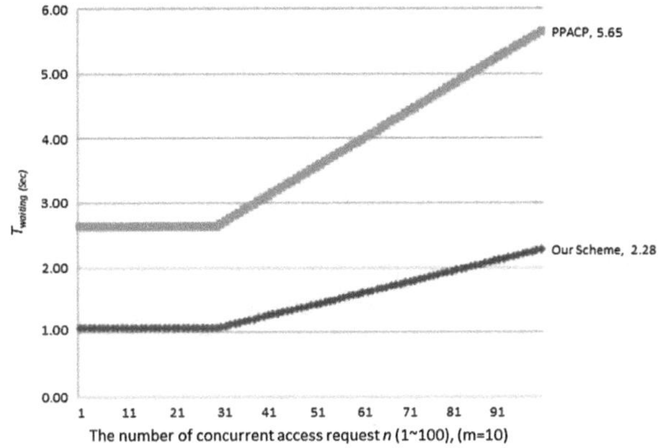

Fig. 25.3 Average waiting time when m is equal to 30

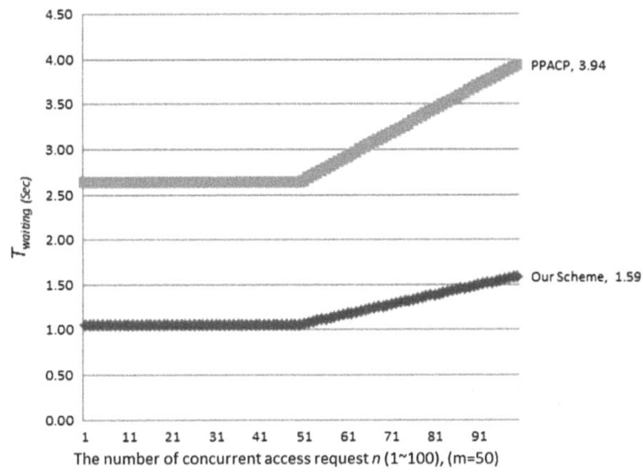

Fig. 25.4 The average waiting time when m is equal to 50

PAACP. Finally, our scheme takes about 1.59 s, compared to PAACP's average of about 3.94 s, as shown in Fig. 25.4. In summary, the average waiting time $T_{waiting}$ decreases when RSU increases.

25.4.3 Security Analysis

The other security features of our new scheme are also discussed below:

25.4.3.1 Forward Secrecy

This security means that before a V_i wants to access the $(n + 1)$th service, he/she cannot decrypt the service request message that existed prior to his/her session key $K_V + n$. Our scheme can attain forward secrecy because if a V_i requests next $(Service\ request\ message)_{(n+1)\text{th}}$, then a new $K_V + (n + 1)$ will be generated by the $(n + 1)$th service.

25.4.3.2 Backward Secrecy

After a V_i accesses the nth service, he/she cannot decrypt the service request message that existed the newest his/her session key $K_V + (n + 1)$. Our scheme can attain backward secrecy because after a V_i requests next $(Service\ request\ message)_{(n+1)\text{th}}$, the session key $K_V + (n + 1)$ will be generated, and the $K_V + (n)$ will be invalid.

25.4.3.3 Authentication

A V_i must submit his or her authentication request message (VID_i, C, N_i) to the service provider S_t and then the S_t acknowledges the V_i. After receiving the authentication request message, the S_t encrypts the message $M = E_{yi \oplus AC_i S_t}\{N_s, N_i, AC_i St\}$ to facilitate a mutual authentication between the vehicle and the service provider.

25.4.3.4 Authorization

In the registration phase, service provider creates service right list is as the following equation:

$$SRL_i^{Vi} = \{SVID_1||AR_1||SVID_2||AR_2||\cdots\cdot||SVID_k||AR_k||\}$$

where $SVID_k$ denotes the index of the kth service, and AR_k represents the granted access privileges of $SVID_k$. Hence, anyone can determine who has which access privileges to access which service. Only valid V_i can encrypts $E_{KV}(SVID_1||AC_i)$ with K_V. After R_j receives $E_{KV}(SVID_1||AC_i)$, R_j will decrypt the message: $D_{KR}(E_{KV}(SVID_1||AC_i))$ with K_R to gain $(SVID_1||AC_i)$ and then derive AC_i, and $SVID_1$ because of $K_V = K_R$.

25.4.3.5 Replay Attack

In the registration phase, a V_i submits his/her registration information over a secure channel so there are no any replay attack issues. In the authorization phase, an old message was eavesdropped by an attacker. He/she may try to replay the old message (VID_i, C, N_i). It may fail because it is not always the same and the Nonce N_i is fresh that not be used before.

25.5 Conclusion

In this paper, we review a cryptanalysis of an attachable blind signature and demonstrate that the PAACP's AC is not secure and private even if the AC secretly stored in a tamper-proof device. An eavesdropper can construct the AC from an intercepted blind document. Consequently, during the authorization phase, PAACP is breakable and cannot maintain privacy in VANETs. Consequently, any outsiders can determine who has which access privileges to access which service.

Furthermore, this paper efficiently copes with these challenges and proposes an efficient scheme. We conclude that a simple authentication scheme and access control protocol for VANETs not only resolves the documented problems, but also is secure and efficient. Compared with PAACP and SECSPP, our scheme achieves more functionality and satisfies security features required by VANETs.

References

1. Chen H, Hsueh S (2003) Light-weight authentication and billing in mobile communications. In: Proceedings of the IEEE 37th annual 2003 international Carnahan conference on security technology, pp 245–252
2. Chen Y, Wei Y (2012) Safeanon: a safe location privacy scheme for vehicular networks. Telecommun Syst 50:1–16
3. Diffie W, Hellman M (1976) New directions in cryptography. IEEE Trans Inf Theor 22:644–654
4. Isaac J, Camara J, Zeadally S, Marquez J (2008) A secure vehicle-to-roadside communication payment protocol in vehicular ad hoc networks. Comput Commun 31:2478–2484
5. Lee J, Chang C (2007) Secure communications for cluster-based ad hoc networks using node identities. J Netw Comput Appl 30:1377–1396
6. Li C, Hwang M, Chu Y (2008) A secure and efficient communication scheme with authenticated key establishment and privacy preserving for vehicular ad hoc networks. Comput Commun 31:2803–2814
7. Raya M, Hubaux J (2005) The security of vehicular ad hoc networks. In: Proceedings of ACM workshop on security of ad hoc and sensor networks
8. Schneier B (1996) Applied cryptography: protocols, algorithms, and source code in C, 2nd edn. Wiley, New York
9. Wang N, Huang Y, Chen W (2008) A novel secure communication scheme in vehicular ad hoc networks. Comput Commun 31:2827–2837
10. Wei YC, Chen YM (2010) Safe distance based location privacy in vehicular networks. In: IEEE 71st vehicular technology conference (VTC 2010-Spring), pp 1–5
11. Wischhof L, Ebner A, Rohling H (2005) Information dissemination in self-organizing intervehicle networks. IEEE Trans Intell Transp Syst 6:90–101
12. Wu W, Chen Y (2012) Cryptanalysis of a PAACP: a portable privacy-preserving authentication and access control protocol in vehicular ad hoc networks. Appl Math Inf Sci 6:463S–469S
13. Yeh L, Chen Y, Huang J (2011) PAACP: a portable privacy-preserving authentication and access control protocol in vehicular ad hoc networks. Comput Commun 34:447–456

14. Yousefi S, Mousavi M, Fathy M (2006) Vehicular ad hoc networks (VANETs): challenges and perspectives. In: Proceedings of the 6th international conference on ITS telecommunications, pp 761–766
15. Zhang C, Lin X, Lu R, Ho P, Shen X (2008) An efficient message authentication scheme for vehicular communications. IEEE Trans Veh Technol 57:3357–3368

Chapter 26
Decision Tree Approach to Predict Lung Cancer the Data Mining Technology

Jui-Hung Kao, Hui-I Chen, Feipei Lai, Li-Min Hsu and Horng-Twu Liaw

Abstract The purposes of this research is using the Data Mining techniques, and explore the potential information from the National Health Insurance Research Database, analysis lung cancer patients which has the highest mortality in Taiwan, as a reference of the analysis of Bureau of National Health Insurance files and the clinical index of hospitals. By using statistical software, "SPSS Clementine 12.0", this research merged "The detail file of hospital medical expenses inventory" and "The basic data file of medical institution" between 06' and 09' as the study data, then hospitalized lung cancer patients screened for the study sample, and using Two-Step clustering technique to produce the results that are effective. The hospitalized patients that with lung cancer and death patients, males have a higher percentage than females, lung cancer people are usually accompanied with comorbidities, especially for pneumonia; most lung cancer patients go to the center of medical, and most patients make booking as out-patient. The treatment of lung cancer patients should notice the factor that lead to pneumonia, and this research shows that most patients make booking as out-patient, therefore, taking drug treatment for patients can regarded as the main dependence for curing.

Keywords Data mining · Health insurance database · Lung cancer · Two-Step · Decision tree

J.-H. Kao (✉) · F. Lai
Graduate Institute of Biomedical Electronics and Bioinformatics, National Taiwan University, Taipei, Taiwan
e-mail: kao.jui.hung@gmail.com

H.-I. Chen · H.-T. Liaw
Department of Information Management, Shih Hsin University, Taipei, Taiwan

L.-M. Hsu
National Taiwan University Hospital, Taipei, Taiwan

J.J. (Jong Hyuk) Park et al. (eds.), *Ubiquitous Computing Application and Wireless Sensor*, Lecture Notes in Electrical Engineering 331,
DOI 10.1007/978-94-017-9618-7_26

26.1 Introduction

Thanks to the evolution of medical and biological technology as well as the introduction of new drugs and equipment, an aged population structure has developed which is has resulted in a higher level of mortality associated with catastrophic illnesses, rare disorders and cancer. In Taiwan, cancer was blamed for 41,046 deaths in 2010, accounting for 28.4 % of deaths due to all causes that year, leading to a standardized death rate of 131.6 per 100,000 individuals. As lung cancer is associated with the highest prevalence and mortality [1], the latter has become one of the most important issues in epidemiology and the development of national health insurance. The realization of computerized NHI information makes timely collection of medical records possible and this has enabled the building of a more comprehensive and robust NHI database. As lung cancer is the most deadly cancer in Taiwan, we plan to perform explorations and analyses based on the two-step (K-means clustering and Wards method) automation model guided by data mining techniques and a clustering design with high intra-cluster similarity but low inter-cluster comparisons, to define cluster characteristics and determine a clinical index while only few had such experience.

26.2 Literature Review

26.2.1 Lung Cancer

Lung cancer may be divided into two types, small cell lung cancer (SCLC) and non small cell lung cancer (NSCLC). SCLC, a.k.a. oat cell cancer, proliferates rather rapidly and readily spreads to other organs. NSCLC is more common, proliferates slower and includes three subtypes: adenoma, squamous cell carcinoma (SCC, epithelial cancer) and large cell cancer (LCC). Typically, SCC and SCLC are presented as centralized tumors, while peripheral nodules or tumors are usually seen in adenoma and LCC. For patients with advanced or metastatic lung cancer, systemic chemotherapy is still the mainstay treatment. The survival of NSCLC patients without medical intervention is usually no more than 4–5 months, and the 1-year survival rate is 10 %. Therapeutic agents covered by NHI but requiring pre-review include Alimta, Tarceva and Iressa. However, the drug used may vary because the prescription policies adapted by each hospital and doctor differ [2, 3].

26.2.2 Introduction of Data Mining

Data mining is a procedure of searching information from countless sources, much like mining precious ores from mountains [4]. This method was originally applied

Table 26.1 Definition of data mining

Expert	Definition
Lin [5]	Data mining is to discover more important information required for making decision from historical archives
Chen [6]	Data mining is the analysis during a series of knowledge discovery and "data mining" has replaced "knowledge discovery" itself
Hsieh [9]	Data mining is searching for information hidden in data, including trends, characteristics and relationships i.e. KDD
Kumari and Godara [10]	Advanced data mining techniques can be used to discover hidden pattern in data
Yoo et al. [11]	Data mining can help researchers gain both novel and deep insights and can facilitate unprecedented understanding of large biomedical datasets

in military settings and censuses. In recent years data mining has become a popular information collection method and been widely utilized in all kinds of professions [5]. Thus, we define data mining to be: "creating meaningfulness from data by searching for useful information and its relationships among a vast and unprocessed sea of data".

This method is known in academics and industry as know discovery in databases (KDD), knowledge extraction, data pattern analysis, data archaeology and data dredging [6] (Table 26.1).

26.2.3 Two-Step Clustering

Two-step clustering is the combination of stratified and unstratified clustering. The first step includes the determination of a coefficient of concentration and tree diagram with Wards method as well as clustering. The second step involves the distribution of observed values by clustering with K-means method. The validity of clustering variables is explored after securing clustering principles with Wards method used in the first step (to determine the number of clusters) followed by unstratified clustering with K-mean method [7].

26.2.4 Decision Tree

For a decision tree, the most significant difference from the standard statistical models is the characterization based on logical grouping. It is good for the processing of non-value data (for example, texts, categories or discrete data) and performs faster calculations as it promotes model accuracy with boosting and takes up fewer system resources and memory. C5.0 has four advantages. First of all, it is stable when there are missing values and problems related to data entering.

Also, estimations can be performed without too much training and it is easier to comprehend. Lastly, the enhancement technology of the C5.0 model improves accuracy of categorization [8].

26.3 Materials and Methods

26.3.1 Data Pre-processing

"Detailed files on hospital medical expenses inventory" and "basic data files of medical institutions" between 2006 and 2009 were used as study data. To reduce the effect of incomplete data on the result of data mining, this process was preceded by data purification, merging and conversion. Data-value pairing is ensured by checking the presence of null or blank values and "99" was assigned for these values. Since the data comes from two files, merging is required to establish the columns essential for this study. New columns were added and included "MED_SUM" (total hospitalization fee), "AGE" (age of patient), "OP" ("1" refers to operated and "0" as unoperated), "TRAC" ("1" refers to "Y" and "0" as "N"), "SEX" ("1" refers to "M" and "2" as "F"), ESB_AMT (number of days hospitalized). See Table 26.2 for the information of ESB_AMT column (Table 26.3).

The conversion of texts into values ensures the applicability of data in the model and creates new variables. A total of 45 variables (Table 26.4) and 22,553 samples was obtained at the conclusion of data pre-processing (Figs. 26.1 and 26.2).

26.3.2 Results and Analysis

The statistical software SPSS Clementine 12.0 was used to select patients with a history of hospitalization from "Detail files on hospital medical expenses inventory" from 2006 to 2009. 249,232 of 251,996 patients selected had no lung cancer history and 2,764 (1.10 %) had a history of lung cancer. For the lung cancer population, 63.1 % were male and 36.9 % female (as shown in Fig. 26.3). The mortality of the

Table 26.2 Table of coding for ESB_AMT column

Column variable	Code	Column variable	Code
EB_APL30_AMT	11	SB_APPL30_AMT	11
EB_PART30_AMT	12	SB_PART30_AMT	12
EB_APPL60_AMT	21	SB_APPL90_AMT	23
EB_PART60_AMT	22	SB_PART90_AMT	24
EB_APPL61_AMT	31	SB_APPL180_AMT	41
EB_PART61_AMT	32	SB_PART180_AMT	42
SB_APPL181_AMT	52	SB_PART181_AMT	51

Table 26.3 The 10 most significant comorbidities of lung cancer patients from 2006 to 2009

Ranking	ICD-code	English name of comorbidities
1	486	Pneumonia, organism unspecified
2	1985	Secondary malignant neoplasm of bone and bone marrow
3	4019	Unspecified essential hypertension
4	51881	Acute respiratory failure
5	1983	Secondary malignant neoplasm of brain and spinal cord
6	1970	Secondary malignant neoplasm of lung
7	1972	Secondary malignant neoplasm of pleura
8	25000	Diabetes mellitus without mention of complications, Type II or unspecified type, not stated as uncontrolled
9	389	Hearing loss
10	5119	Unspecified pleural effusion

patients with lung cancer was 8 % (222 individuals). 75 % of deaths were male, 25 % were female, as shown in Fig. 26.4. The factors leading to more males dying of lung cancer than females are considered to be smoking or prolonged exposure to pollution.

"Detailed files on hospital medical expenses inventory" from 2006 to 2009 shows that lung cancer patients tended to have comorbidities. An examination of the 10 most significant comorbidities of lung cancer showed pneumonia to be the most common and to have a direct relationship with lung cancer. Also on the list are secondary malignant neoplasm of bone and bone marrow as well as essential hypertension, but these conditions are usually overlooked as it is more difficult to associate them with lung cancer. It is worth noting that carcinoma is the most important comorbidity because metastases may develop in bone and bone marrow, brain and spinal cord, lungs and pleura. Type II diabetes is also included and may be a risk factor of lung cancer development. A total of 1,221 diagnostic codes were discovered. The risks of the 10 most significant comorbidities are similar except for that of pneumonia, which is much higher than the others, see Fig. 26.5. Moreover, the proportion of patients with more than one comorbidity is high, which means that comorbidities are usually presented in different combinations and that simple combinations are relatively rare.

Three clusters were arranged, see Fig. 26.1. The significance of clustering factors indicates that variables other than the number of days hospitalized, other diagnostic code IV and psychiatric treatment fees are important factors based on clustering sensitivity. Clustering characterization values were used to designate cluster 1 a high grade cluster (including medical centers and educational institutes graded as "best" or "good" via new hospital accreditation), cluster 2 an average cluster (including all grades) and cluster 3 a non-high grade cluster (cluster 1 excluded). Pruning was performed on the decision tree with a pruning significance of 90 and a minimal subbranch of 80 (see Fig. 26.2). The pruned decision tree diagram was obtained and the accuracy of model prediction is 98.74 %. See Fig. 26.6.

Table 26.4 Codes of variables

Variables				
AGE_CLUSTER	AMIN_CLUSTER	HD_CLUSTER	ICD9CM_CODE_1	AREA_NO_H
Age	Examination fee	Hemodialysis fee	Other diagnostic code I	Area code
CASE_TYPE	ANE_CLUSTER	INJT_CLUSTER	ICD9CM_CODE_2	HOSP_CONT_TYPE
Case type	Subcluster: anesthesia fee	Injection fee	Other diagnostic code II	Contract type
ESB_AMT	BLOD_CLUSTER	MEAL_CLUSTER	ICD9CM_CODE_3	HOSP_GRAD_ID
Number of days hospitalized	Blood and plasma fee	Tube feeding cost	Other diagnostic code III	Institute grading
EXT_CODE_1	CHARG_CLUSTER	METR_CLUSTER	ICD9CM_CODE_4	HOSP_TYPE_ID
External factor I	Service fee	Special material fee	Other diagnostic code IV	Type
EXT_CODE_2	DAIG_CLUSTER	NRTP_CLUSTER	ICD_OP_CODE	TRAC
External factor II	Diagnostic cost	Psychiatric treatment fee	Major operation (management)	Traffic accident status
FUNC_TYPE	DRUG_CLUSTER	PHSC_CLUSTER	ICD_OP_CODE_1	TRAN_CODE
Specialty visited	Drug cost	Rehabilitation fee	Major operation (management) I	Transition code
GAVE_KIND	DSVC_CLUSTER	RADO_CLUSTER	ICD_OP_CODE_2	APPL_CLUSTER
Insurance covering status	Pharmaceutical service cost	X-Ray fee	Major operation (management) II	Applied amount
OP	SGRY_CLUSTER	ROOM_CLUSTER	ICD_OP_CODE_3	PART_CLUSTER
Operation status	Operation cost	Ward fee	Major operation (management) III	Copayment
SEX	THRP_CLUSTER	ICD9CM_CODE	ICD_OP_CODE_4	MED_CLUSTER
Sex	Treatment cost	Major diagnostic code	Major operation (management) IV	Total hospitalization cost

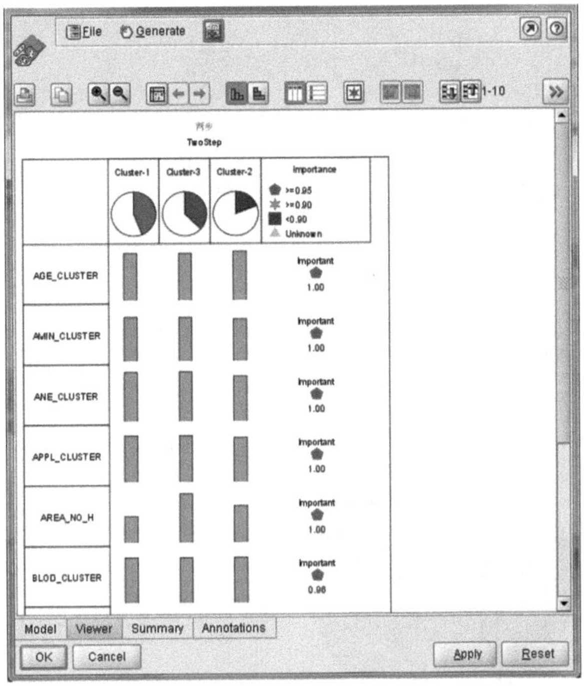

Fig. 26.1 Two-step monitor output

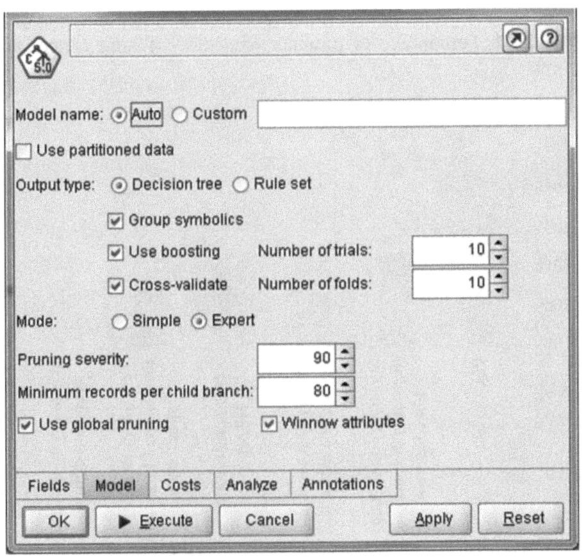

Fig. 26.2 Pruning of decision tree model

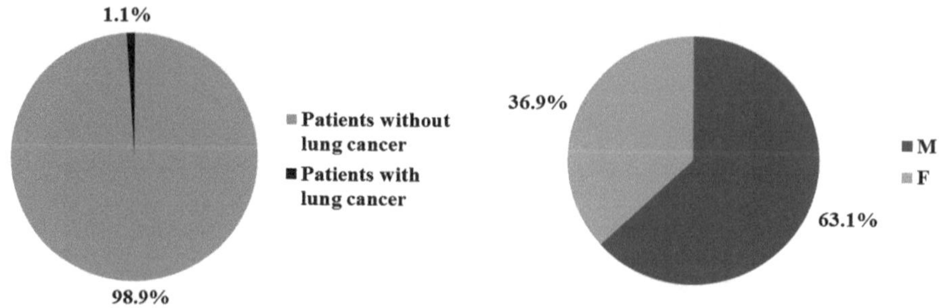

Fig. 26.3 Percentage of hospitalized lung cancer patients and male-female ratio from 2006 to 2009

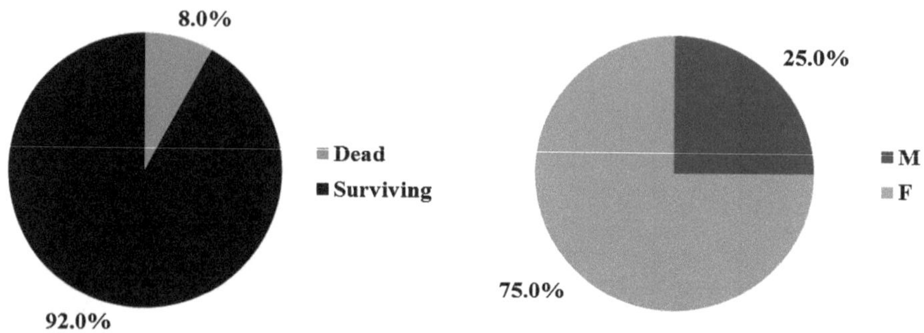

Fig. 26.4 Percentage of patients who died of lung cancer and male-female ratio from 2006 to 2009

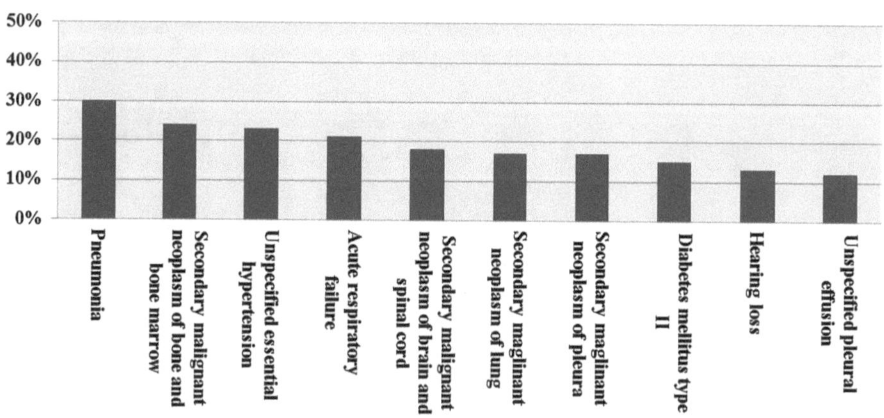

Fig. 26.5 Percentage of the 10 most ten significant comorbidities for each hospitalized lung cancer patient during from 2006 to 2009

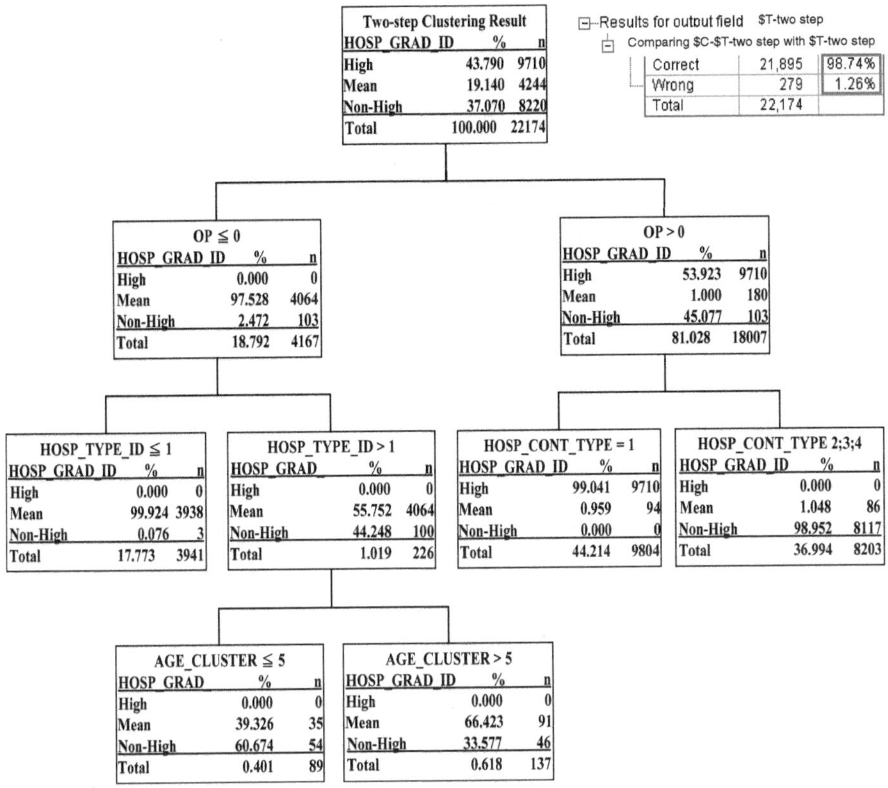

Fig. 26.6 Pruned decision tree diagram

26.4 Conclusion and Recommendations

The result of analysis shows that the proportion of male lung cancer patients who are hospitalized and died was higher than that of female. Patients with a single diagnosis of lung cancer are relatively rare and comorbidities included malignant neoplasm of bone and bone marrow, brain and spinal cord, lung and pleura, which may be metastatic diseases from lung cancer. However, the large number of patients with diabetes may be overlapping with lung cancer in Taiwan, but Type II diabetes mellitus is associated with essential hypertension.

The median age of patients is 72, with a range of 27–114. Most patients get medical help from hospitals that are graded as medical centers and most hospitalized lung cancer patients continue their treatment under ambulatory settings.

The decision tree indicates that when considering unoperated patients visiting hospitals, specialty hospitals and chronic hospitals, patients older than 75 tend to select average grade medical institutes, while those younger than 75 prefer non-high grade ones. This reflects the fact that older patients are more likely to visit better grade hospitals but this may be a result of a family decision.

In conclusion, although most lung cancer patients are elderly individuals those who are younger than 30 should not be ignored. Considering the high proportion of comorbidities, it is recommended that medical institutions emphasize comorbidity prevention as well as health promotion for elderly individuals (i.e. those aged 70 or older) and patients with diabetes mellitus when developing therapeutic guidelines. It is also necessary to clarify to the public that males are more at risk from lung cancer and that exposure and habituation to environment and behavior that are detrimental to the respiratory system should be avoided. As most patients receive ambulatory care, recent medication history can be a good guidance to treatment of lung cancer patients.

References

1. National Health Insurance Administration Ministry of Health and Welfare: National Health Insurance Statistical Trends (2009). http://www.nhi.gov.tw/webdata/webdata.aspx?menu=17&menu_id=661&WD_ID=689&webdata_id=3352
2. Chuang T-N (2006) Establishment of the patient guide using data mining techniques. J Taiwan Assoc Med Inform 15(1):17–44
3. National Health Insurance Administration Ministry of Health and Welfare: Pharmaceutical Benefit Provision (2010). http://www.cancernews.com.tw/index.php?REQUEST_ID=bW9kPXdtMiZwYWdlPWRldGFpbCZOZXdzSUQ9MjM=&pn=0
4. Chen Y-F (2003) Data mining technique researching on evidence-based medicine: case study of appendectomy, hernia, diabetes, gastric hemorrhage, Department of Health Services
5. Lin Y-J (2008) Applying data mining in health management information system for chronic disease, Department of Information Management Providence University
6. Chen M-Y (2009) Analysis on emergency care triage of a regional hospital in Taiwan by data mining technique, Department of Industrial Engineering and Management Chin-Yi University of Technology
7. Shia B-C (2009) Overview of data mining an example in clementine 12.0, China Certification of Disability Management Specialists
8. Liao S-H (2009) Data mining for business intelligence, Yeh Yeh Book Gollery
9. Hsieh C-J (2010) Understanding the medical cost structure through data mining techniques: a case study of regional hospital, Department of Information and Computer Education Kaohsiung Normal University
10. Kumari M, Godara S (2011) Comparative study of data mining classification methods in cardiovascular disease prediction 1
11. Yoo I, Alafaireet P, Marinov M, Pena-Hernandez K, Gopidi R, Chang J-F et al (2012) Data mining in healthcare and biomedicine: a survey of the literature. J Med Syst 36:2431–2448

Chapter 27
Study of Customer Value and Supplier Dependence with the RFM Model

Jui-Hung Kao, Feipei Lai, Horng-Twu Liaw and Pei-hua Hsieh

Abstract In this study, data mining is applied to the use of an Enterprise Resource Planning (ERP) database to explore the core value of business operations by revising the RFM model into an RFGP model through "customer value analysis". This model is used as a tool by enterprises to make proper adjustment to business strategies in a timely manner. By using an RFM model to "analyze dependence on suppliers", enterprises can properly plan to reduce internal procurement costs and suppliers can properly manage their risks. With this two-prong focus on "profit maximization" and "cost reduction", enterprises can maximize their profits. The findings of this study indicate that high-value customers are stable customers and accounted for 31 % of total customers. The statistics also show that customers of LED backlight modules constituted 18 % of the total and were the primary source of profit for the enterprises. However, there remains much uncertainty in the LED lighting market and the rate of return was low. With limited operational resources, enterprises should explore technologies related to backlight products, purchase automation equipment, train professionals in related disciplines and seek to enhance their overall competitive power. Suppliers of key materials are suppliers of high dependence, accounting for 11 % of the total and including suppliers of heavy reliance and exclusive suppliers. Enterprises are strongly advised to establish new sources of supplies to ensure a proper procurement balance and reduce materials supply related risks. In addition, enterprises should take the top 10 key materials suppliers as targets for price reductions in order to cut costs and maximize profits. Enterprises are also strongly advised to invest in the suppliers of key materials. The recycled use of enterprise resources could feedback to the enterprises themselves and helps to maximize enterprise profits.

J.-H. Kao (✉) · F. Lai
Graduate Institute of Biomedical Electronics and Bioinformatics,
National Taiwan University, Taipei, Taiwan
e-mail: kao.jui.hung@gmail.com

H.-T. Liaw · P. Hsieh
Department of Information Management, Shih Hsin University, Taipei, Taiwan

J.J. (Jong Hyuk) Park et al. (eds.), *Ubiquitous Computing Application and Wireless Sensor*, Lecture Notes in Electrical Engineering 331,
DOI 10.1007/978-94-017-9618-7_27

27.1 Introduction

Profit making is the ultimate goal of business operations. The sourcing of profits for the enterprises can be realized in two ways. "Source optimizing" and "Cost Saving". In this study, an ERP database will be used to "analyze customer value" with the RFGP model to explore "source optimizing." This better enables enterprises to accurately control high value customers, understand customer needs, develop new products and proactively to keep abreast of the changes in market. In so doing, the enterprises can adjust their business strategies and investment plans accordingly, and effectively allocate their limited marketing resources. In terms of "cost saving" the "analysis of dependence on suppliers" examines suppliers of high dependence and the materials they supply. The findings will reference designing a cost reduction plan for internal procurement and management of supplier risk. With the dual strategy of "source optimizing" and "cost saving", enterprises are in a better position to maximize their profits.

ERP aims to enhance corporate operational efficiency in response to rapid changes in global competition. However, most of the system functions focus on the creation of data files, management and the processing of account transactions. Despite the huge volume of data being accumulated, little has been further data mined or used as a point of reference for enterprises in corporate decision-making.

27.2 Literature Review

27.2.1 Customer Value

We can view customer value from two perspectives. One is the value perceptible to the customers. From this perspective, we can explore the value customers place on products and services provided by enterprises. Customers expect top value at low cost. Another is the enterprise perspective, which looks at the value contribution of customers and is calculated by subtracting related cost from the expected return on customers. The difference is profit. Returns in excess of costs are called Customer Lifetime Value (CLV). As such, customers of value can bring in profits for a company that far exceed the cost of the customers to the company. The 80/20 rule of Vilfredo Pareto states that "80 % of an enterprise's profits come from 20 % of its customers, while 80 % of marketing expenses are spent on 20 % of sales profit". If a firm could track down the 20 % of customers of value, that would sustain the fundamental operations of the enterprise. This group of customers is vital to the enterprise and it is that customer value this study will examine [1–3].

We could review and analyze customer value on the basis of historical transaction data. The RFM model has been used extensively for such purposes. Using customer time of purchase, purchase frequency and purchase amount, we can develop a customer value matrix to be used as a foundation for the market segmentation of customer value. The customer value matrix is a two dimensional

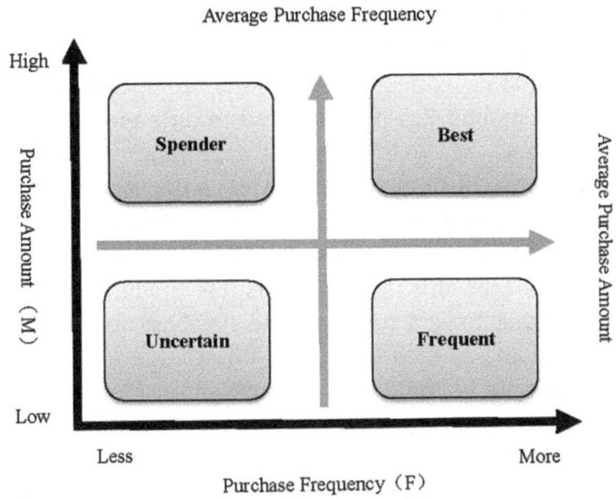

Fig. 27.1 Two dimensional customer value matrix

matrix and the mean of the two indicators is the demarcation line that forms four quadrants, including Best Customers, Spender, Frequent Customers and Uncertain Customers, as shown in Fig. 27.1 and explained below [1, 4]:

I. Best Customers
Customers falling in this quadrant are those who buy more frequently and at an amount higher than the mean value. They are the enterprises' core customers and shall be kept effectively.

II. Spenders
This group of customers buys at an amount higher than the mean value but at a low frequency. Enterprises should introduce promotional plans to intensify the frequency these customers buy.

III. Frequent Customers
Customers falling in this quadrant buy small amounts very frequently. Enterprises should apply cross-sales method to increase the amount of purchases these customers make.

IV. Uncertain Customers
This group purchases low amounts infrequently. Enterprises should identify suitable groups of customers, develop new customers, or explore the customers with specific types of products for the effective allocation of marketing resources.

27.2.2 Data Mining

Data mining is one step of the Knowledge Discovery in Database (KDD) process and is extensively adopted by enterprises for the effective extraction of information and knowledge from massive databases. This technique has been listed by the

Massachusetts Institute of Technology as one of the top 10 innovative technologies that change the future of mankind. Enterprises tend to adopt data mining when they make vital decisions to bolster their competitive power [5, 6].

Data mining has been defined in different ways by scholars. Ting [7] translates the term as the "extraction and exploration of data", which highlights the process and essence of the technique. This is one step in the analysis. Frawley defines data mining as the extraction of data from databases, which is not obvious but implies the possibility of a process of data use that has never been used before. Hall [8] suggests that this technique is the integration of data visualization, machine learning, statistical methods and database forms, retrieving knowledge from massive volume of data to present knowledge in matrix form or other models. Grupe and Owrang [9] hold that data mining is the relationship between existing data and the anatomy of new facts, the discovery of which is unknown to the experts. Fayyad [10] suggests that this is a multi-step process and that the objective is to help people determine understandable data models that should be able to induce some useful behavior. Cabena [11] holds that this is the process of sorting out previously unknown and effective information from a mass database, and that the data being sorted are vital references for the decision-making process. Berry and Linoff [12] define data mining as the exploration and analysis of data from mass databases automatically or semi-automatically, to discover meaning models or rules.

27.2.3 RFM Model

The RFM model is the most commonly used segmentation tool for the analysis of customer value. With this tool, data can be easily accessed and calculated. With the rapid development of computer technology and the advancement of database systems, this tool is extensively used in customer relations management and marketing strategy. Hughes [13] suggested that, "good customers of the enterprises are those who spend all the time, have the highest frequency of spending and spend the highest amount of money". As such, he developed the RFM model made up of these indicators: time of recent purchase (R), frequency of purchase (F) and amount of purchase (M). These are used to measure the importance of customer value. R, F, and M are explained below in Table 27.1:

The RFM model of analysis is extensively used by enterprises. This model was developed by Hughes using the 5 equal intervals method. Hughes held that the importance of the three RFM indicators is consistent and the weighting is the same. He divided the whole database from the day of last purchase to the day of analysis into 5 equal intervals. The most recent interval, the top 20 %, which is also the largest assigned number, is "5" or 0–20 %. Likewise, number "4", is 20–40 %; number "3" is 40–60 %; number "2", is 60–80 %; and number "1" is 80–100 %. The greater the value of the number, the higher the customer response rate and the higher the probability a purchase will be made. Purchase frequency and purchase amount data is also coded in accordance with the 5 equal intervals method.

Table 27.1 RFM analysis sheet

Indicator	Customer value assessment
Time of purchase	This refers to the day of the last purchase and the distance from the day of analysis. The closer the day of the last purchase to the analysis day, the higher the value of the customer to the enterprise
Frequency of purchase	The total number of purchases made by the customer within a stipulated period. The higher the frequency of purchases made the higher the value the customer to the enterprise
Purchase amount	The total amount the customer spent within the stipulated period. The higher the amount of customer spending, the stronger the need of the customer for products, and the higher his/her contribution value to the enterprise. As such, the customer has a higher value to the enterprise

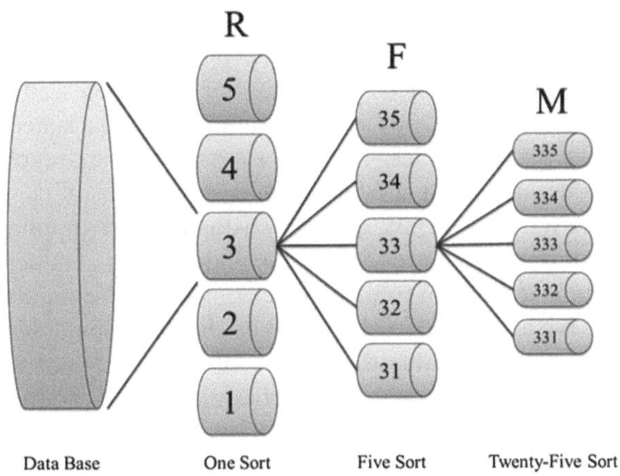

Fig. 27.2 5 equal intervals of the RFM model

After the code, each customer is assigned a set of three-digit numbers, from 555, 554, 553, 552, 551, 545, 544…112, 111. Accordingly, the customers are grouped into 125 combinations. The grouping of customers in this manner helps to sort out the value of customers and hence serves as a foundation for proposing different marketing resources [1, 13], as shown in Fig. 27.2.

27.3 Methods

27.3.1 RFGP Model and the RFM

The RFGP model was proposed to assess customer contribution value. The customer sales data contained in the ERP database was used in the study with time of purchase, frequency of purchase and gross profit being used to group customers. In this manner, the positioning of customers is more precise, enabling enterprises to sort out top priority customers for marketing. In addition, the RFM model was applied to determine dependence on suppliers. The time of purchase, frequency of purchase, and purchase amount of enterprises were used as indicators to group customers in order to find out the type of materials being supplied by the suppliers and hence prioritize primary concerns in the purchase cost reduction plan.

Basic customer data and data on sales history is retrieved for sorting and convergence. The RFGP model, a data mining technique, is used for customer value analysis to determine customers of high value and products for these customers. A conclusion and recommendations are provided based on the findings as a reference point for the enterprises mapping out marketing strategies and customer relations management.

The data from the ERP database used in this study: basic customer data, shipment details, basic information on suppliers, list of incoming materials. In the preliminary processing of data, statements and reports in Taiwan and Mainland China were retrieved from the ERP database. The database stores massive amounts of data on daily sales but this has not been processed and is not suitable for analysis. Microsoft Excel 2007 was used to merge all the statements in the required fields and unify the currencies. Data on sales returns and exchanges, gifts, samples and on purchases, which unit price is zero, was insufficient and therefore deleted on a selective basis. Only data on customers and suppliers that could be used in data mining was retained. Statistical analysis was conducted on this data to sort out the shipment status of individual customers and the number of purchases from suppliers and the amount involved based on the data inputted in data mining for RFM related analysis. Currently, the data sources are: 3,410 shipment data, 90 customers, 27 products on sale. In purchasing, 13,521 counts of data, 191 suppliers and 30 materials were designated.

Most of the products for the enterprises are customized in high variety. The prices of these items varied significantly due to variations in product functions and dimensions. When introducing RFM, the "purchase amount" was revised as "Gross Profit, GP" to avoid significant variations in unit price that could interfere with the result of data analysis. The scoring standards for the other two indicators were last day of shipment and shipment frequency. A new model was developed on the basis of customer value that we called "RFGP" in this study. In this model, the R, F, GP indicators are divided into 1–5 points, from 555, 554, 553, 552, 551, 545, 544… 112, and 111 and grouped into 125 combinations in customer grouping. The scoring standard is described below:

Last shipments: customer shipment dates were arranged in chronological order, and differentiated by distance from the current day. The customized products analyzed in this study required preliminary development costs before customers could be successfully attracted. These involved fees for molding tool development. If the tooling is successful and the enterprise has not been negligent in ways that impact the interest of customers, they seldom switch suppliers and incur new development costs. The products involved were mostly 3C or electronic products, and the cycle time for each shipment was usually 1–3 months. Shipments that take 3–9 months were uncommon and beyond 9 months unnecessary as the products are finished or customers have switched to other suppliers. As such, the scoring standard of R is: (1) for “5” marks: less than 1 month; (2) for “4” marks: 1–3 months; (3) for “3” marks: 3–9 months; (4) for “2” marks: 9–18 months; and (5) for “1” mark: more than 18 months.

Shipment frequency: statistical analysis on the total frequency of shipments was undertaken and scores given. In this study, historical sales data covering 29 months was used. Customers were categorized as shipment once monthly to shipment once every 9 months. The F scoring standard is: (1) for “5” marks: more than 30 times; (2) for “4” marks: 30–15 times; (3) for “3” marks: 15–5 times; (4) for “2” marks: 5–2 times; (5) for “1” mark: 1 time.

Gross profit: statistical analysis of total gross profit was undertaken for the data period and scores were given based on the results. The GP scoring standard was determined by the gross margin of the subject of study. For strategic marketing and return, the enterprise may have negative gross profits. As such the GP scoring standard was: (1) for “5” marks: more than 500,000; (2) for “4” marks: 500,000–300,000; (3) for “3” marks: 300,000–50,000; (4) for “2” marks: less than 50,000; and (5) for “1” mark; 0. The scoring standard under the “RFGP model” is shown in Table 27.2 below:

With the RFGP model applied to the analysis 90 customers were divided into 47 groups. However, only 4 groups of customers, 555, 322, 222, and 212, contained more than 5 accounts, which indicated that the characteristics of transactions and shipment requirements of these customers were very different, which is shown in Table 27.3:

The proportion of customer distribution in the RFGP indicators is approximately 20 %. The determination of characteristics and scoring standard of subjects in this study for verification was almost identical to that seen in the 5 equal intervals method of Hughes, as shown in Table 27.4:

27.3.2 Customer Value Analysis

The shipment characteristics of targeted customers in this study will be analyzed. Customers with active shipments in the last 9 months, a shipment frequency of more than 5 times and gross profits in excess of NT$50,000 are considered customers of high value, which are the R, F, GP indicators: a score of 1–2 marks constitutes low value, a score of 3–5 is high value. Likewise, the three-dimensional customer value matrix is divided into 8 quadrants for customer type and customer value. Figure 27.3: details the definitions and R, F, GP quadrants in high and low value:

R Indicator: this represents time of shipment from the day of analysis. Customers under this indicator are of higher value. The subjects of this study are customized products and mostly products under special projects or for special types of machines. As such, customers with no shipments for more than 9 months represent

Table 27.2 RFGM model scoring standard

Score	5	4	3	2	1
Indicator					
Last purchase (Recnecy, R)	Less than 1 month (≤30 days)	1–3 months (>30 ≤90 days)	3–9 months (>90 ≤270 days)	9–18 months (>270 ≤540 days)	More than 18 months (>540 days)
Shipment frequency, F	More than 30 times (≥30)	30–15 times (≥15 <30)	15–5 times (≥5 <15)	5–2 times (≥2 <5)	1 time (≥1 <2)
Gross profit, GP	More than NT $500,000 (≥500,000)	NT$500,000–300,000 (≥30 <500,000)	NT$300,000–50,000 (≥5 <300,000)	Less than NT$50,000 (≥1 <50,000)	NT$0 (≥0 <1)

Table 27.3 RFGP model group results

RFGP grouping	No. of customers	RFGP grouping	No. of customers	RFGP grouping	No. of customers
555	11	442	1	241	1
553	2	433	1	235	1
552	2	432	3	232	2
551	1	422	4	231	1
545	1	413	1	223	1
544	1	412	2	222	5
543	1	353	1	221	1
542	2	343	1	212	5
535	2	342	1	211	1
532	1	335	1	133	2
531	1	333	3	123	1
522	1	332	3	122	2
455	1	323	2	121	1
454	1	322	6	113	1
451	1	321	1	112	3
445	1	242	1	TOTAL	90

Table 27.4 RFGP—distribution of various customer indicators

Score	5	4	3	2	1
Indicator					
Last purchase Recnecy, R	Less than 1 month	1–3 months	3–9 months	9–18 months	More than 18 months
No. of customers	26	16	19	19	10
Proportion (%)	29	18	21	21	11
Shipment frequency, F	More than 30 times	30–15 times	15–5 times	5–2 times	1 time
No. of customers	20	11	21	25	13
Proportion (%)	22	12	23	28	14
Gross profit, GP	More than NT $500,000	NT $500,000–300,000	NT $300,000–50,000	Less than NT$50,000	NT$0
No. of customers	18	2	17	44	9
Proportion (%)	20	2	19	49	10

the end of the product life cycle (End of Life, EOL), or customers may have switched orders to competitors. Accordingly, a score of 3–5 is defined as high value while a score of 1–2 is defined as low value.

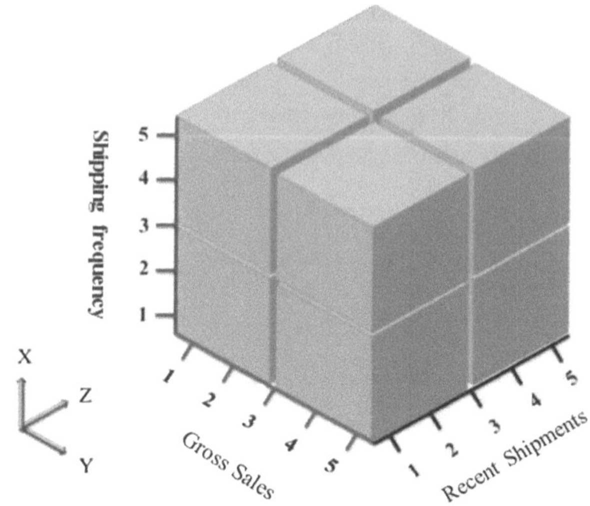

Fig. 27.3 Three-dimensional customer value matrix quadrants

F Indicator: this represents the shipment frequency. In this study, customers with high shipment frequency frequently demand products or greater shipment volume the enterprise cannot supply in one go and has to deliver in batches. Historical data covering the period of 29 months indicated that customer of high value are those with a frequency of 5 times shipments or more every 9 months. Under the F indicator a score of 3–5 constituted high value and 1–2 low value. Yet, the enterprise has to bear the cost of each shipment irrespective of the size of the shipment, including customs declarations, truck fees and processing fees. As such, whether or not frequent shipments are beneficial to the enterprise must be combined with the GP indicator for precise analysis.

GP Indicator: This indicates gross profits, which also constitutes the state of the enterprise's profitability. The purpose of business operations is profit. As such, the GP indicator helps to analyze the contribution of each customer. Enterprises may adopt a low pricing strategy to enlarge their market share. The result may be negative gross profit and a negative contribution. As such, NT$50,000 in gross profit was set as the standard limit. Value in excess of this amount is high value. Likewise, a score of 3–5 on the GP indicator is a high value and 1–2 a low indicator. The three-dimensional customer value matrix is divided into 8 quadrants for different types of customers and values, which is shown in Table 27.5. The characteristics of different types of customers are shown below:

Stable customers: These are in the 1st quadrant, with R, F and GP indicators and falling in the high value zone. This indicates they have high frequency recent shipments with high gross profits, making them stable customers and good partners for business growth. They are one of the preferred customer groups.

Potential customers: These are in the 2nd quadrant, with R, GP indicators falling in the high value zone, but their F indicator falls into the low value zone. This indicates they have recent shipments with high gross profits, and that the frequency

Table 27.5 Customer type in RFGP quadrants

Item	Customer type	R indicator	F indicator	GP indicator	Customer value
Quadrant					
Quadrant I	Stable type	5–3	5–3	5–3	High value
Quadrant II	Potential type	5–3	1–2	5–3	High value
Quadrant III	Strategic type	5–3	5–3	1–2	Low value
Quadrant IV	Stable customer in losing	1–2	5–3	5–3	High value
Quadrant V	Potential customers in losing/project customers	1–2	1–2	5–3	High value
Quadrant VI	Thin margin customers	1–2	5–3	1–2	Low value
Quadrant VII	New customers	5–3	1–2	1–2	Low value
Quadrant VIII	Lost customers	1–2	1–2	1–2	Low value

of shipments can help to maintain high gross profits. The indicators also show that these customers have shipments of high gross profit products. They are potential type customers and also one of the preferred customer groups.

Strategic customers: these are in the 3rd quadrant, with R, F indicators falling in the high value zone, but their GP indicator falls into the low value zone. This indicates they have high frequency recent shipments, but that gross profits are very thin and possibly even negative. This group of customers tends to be price-oriented or big firms. There is keen competition on the supply side, as there are a large number of competing suppliers. Shipment volume is high which compels supplying enterprises to adopt a low pricing strategy to sell in high volume. This group of customers is suitable for enterprises only when there is a lack of stable orders or an economic downturn in order to maintain normal operations. Enterprises should cut shipments or give up this group of customers when economic recovery takes hold.

Stable customers in losing: these are in the 4th quadrant, with F, GP indicators falling in the high value zone but the R indicator falling in the low value zone. This indicates that shipment frequency and gross profit are high, but this group of customers has not placed any orders in the last three quarters and may have lost their confidence in the enterprise and switched to a competitor. Alternatively, they elected to terminate the business cooperation with the enterprise at EOL of products and place no new orders to the enterprise. Under such circumstance, the enterprise should proactively contact this group of customers and determine the reason for their decision and make winning them back a top priority.

Potential customers in losing/project customers: these are in the 5th quadrant, with a GP indicator that falls in the high value zone and R, F indicators falling in the low value zone. This indicates these customers did not place any orders in the last three quarters and shipment frequency is low but gross profits high. This also indicates that the customers may make shipments once or a few times for specific project, or, they are potential customers and may have already switched orders to competing suppliers. For this group of customers, enterprises should adopt the same strategy as for potentially lost stable customers by making a positive effort to contact them, determine the cause of their decision and make a attempt to win them back.

Thin margin customers: these are in the 6th quadrant, with the F indicator falling in the high value zone and the R, GP indicators in the low value zone. This indicates the customers did not place any orders in the last three quarters and that gross profit is very low but shipment frequency is very high. This group of customers contributes very thin profit margins. The enterprise must use sizable development resources with frequent shipment for weak profits. For the more efficient use of resources the enterprise should give up this group of customers.

New customers: these are in the 7th quadrant, with the R indicator falling in the high value zone and F, GP indicators in the low value zone. This indicates that these customers have had a recent shipment, but shipment frequency and gross profit are low. They could be new customers, or former customers that the enterprise has just won back for business. The contribution of this group of customers to the enterprise is not high, but the enterprise can upgrade F and GP indicators by making a further effort to collaborate with these customers.

Lost customers: these are in the 8th quadrant, with R, F and GP indicators falling in the low value zone. This indicates that these customers have not had a recent shipment and that gross profit is low. This group of customers has not made contribution to the enterprises for a long time, which essentially means the enterprise has lost them. However, the cost of developing new customers is much higher than keeping old customers and as such, the enterprise should determine the reason these customers were lost, understand their attributes, sort out the few high contribution customers, win them back and upgrade their value (Fig. 27.4).

Fig. 27.4 Proportion of customers in the 8 quadrants

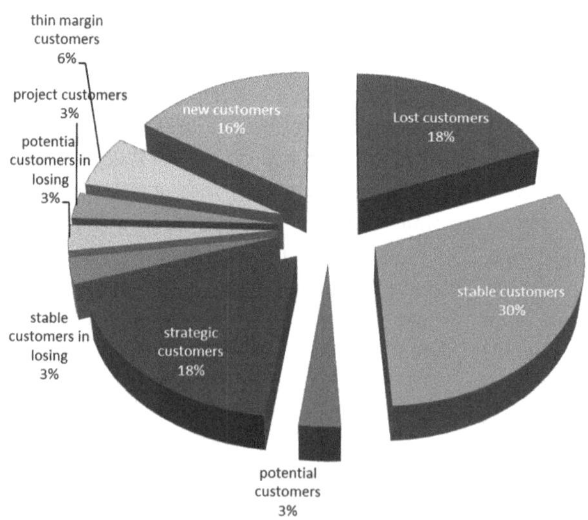

27.3.2.1 Result

The analysis of customer value is conducted by using an RFGP model and three-dimensional matrix. For the 8 groups of customers in the 8 quadrants, the analysis indicated that the primary source of profit for the enterprise is the group of high customer value stable customers. Of the top 10 customers in this group, there were indications of losing two over the last 6 months. The enterprise should conduct customer satisfaction surveys on targeted customers to prevent such losses. In addition, the enterprise should make a proactive effort to development high value customers as potential customers, as the average profit of these customers is higher than that of stable customers but at a much lower proportion. As such, the sales team should target these customers when developing new accounts. In addition, more effort should be made to win back lost stable customers and lost potential customers/project customers. It is recommended that enhanced service quality is necessary for these 4 major categories of high value customers in order to establish positive customer relations.

This study uses an RFM model and three-dimensional matrix of dependence on suppliers. The findings indicate that suppliers of high dependence are also suppliers of key materials and accounted for 11 % of total suppliers. There are 3 suppliers that the enterprise relies on for more than 60 % of total supply and there are 4 exclusive suppliers. The higher the dependence of the enterprises on the supplier, the higher the risk of purchase the enterprise is bound to assume. In the event of supply problems, the enterprise's production scheduling and delivery will be impacted. It is recommended that the enterprise should develop 1–2 new suppliers to balance the allocation of purchase and reduce the risk of materials supply.

There are 10 key materials on which the enterprise is heavily dependent. The top 10 items by purchase value are ranked from A to J. The total purchase value amounts to NT$100 million. It is recommended that these 10 items should be the subject of reduced purchase as a top priority. Cost reduction is indeed profit for the enterprise and a reduction of 1 % can contribute more than NT$600,000 in terms of profit. With this contribution, the enterprise could still enjoy growth even under conditions of economic adversity.

Furthermore, the enterprise should consider investing its suppliers of key materials and planning direct investment. As the enterprise has stable demand for the aforementioned 10 items, investing in suppliers of these materials could be profitable when these suppliers make profits. As a shareholder in these suppliers, the enterprise could enjoy preferential discounts in purchases and better service quality. It could also recycle and reuse its resources with eventual feedback to itself, and hence optimize profitability.

References

1. Huang H-H (2012) The application of data mining to customer value analysis in the food industry. Thesis, Southern Taiwan University of science and Technology, Department of Business Administration
2. Cheng C-H, Chen Y-S (2009) Classifying the segmentation of customer value via RFM model and RS theory. Expert Syst Appl 36(3):4176–4184
3. Lo C-C (2011) Research on the application of data mining of bookstore customer relationship management. Thesis, National Chin-Yi University of Technology, Department of Distribution Management
4. Huang J, Zhou C, Han W (2013) Assessing competitive advantage based on customer satisfaction and customer value, pp 12–17
5. Atmani B, Beldjilali B (2012) Knowledge discovery in database: induction graph and cellular automaton. Comput Inf 26(2):171–197
6. Kaas Q, Yu R, Jin A-H, Dutertre S, Craik DJ (2012) ConoServer: updated content, knowledge, and discovery tools in the conopeptide database. Nucleic Acids Res 40(D1):D325–D330
7. Ting I-H (2005) Data mining
8. Frawley WJ, Piatetsky-Shapiro G, Matheus CJ (1992) Knowledge discovery in databases: an overview. AI Mag 13(3):57
9. Grupe FH, Mehdi Owrang M (1995) DATA BASE MINING discovering new knowledge and competitive advantage. Inf Syst Manag 12(4):26–31
10. Fayyad UM, Piatetsky-Shapiro G, and Smyth P "Knowledge discovery and data mining: towards a unifying framework." pp. 82-88
11. Cabena P (1998) Discovering data mining: from concept to implementation. Prentice Hall PTR
12. Linoff GS, Berry MJ (2011) Data mining techniques: for marketing, sales, and customer relationship management. Wiley, Chichester
13. Hughes AM (2005) Strategic database marketing. McGraw-Hill Publishing Co., New York

Chapter 28
Feature Selection for Support Vector Machines Base on Modified Artificial Fish Swarm Algorithm

Kuan-Cheng Lin, Sih-Yang Chen and Jason C. Hung

Abstract Feature selection is a search process to find the optimal feature subset to describe the characteristics of dataset exactly. Artificial Fish Swarm Algorithm is a novel meta-heuristic search algorithm, which can solve the problem of optimization by simulate the behaviors of fish swarm. This study proposes a modified version of Artificial Fish Swarm Algorithm to select the optimal feature subset to improve the classification accuracy for Support Vector Machines. The empirical results showed that the performance of the proposed method was superior to that of basic version of Artificial Fish Swarm Algorithm.

Keywords Artificial fish swarm algorithm · Feature selection · Support vector machine · Swarm intelligence

28.1 Introduction

Rapid advances in information and communications technology have created the huge amount of data in diverse domain, such as industrial, medical, financial, marketing and demographic. Data mining is the concept and technology that help

K.-C. Lin (✉) · S.-Y. Chen
Department of Management Information Systems,
National Chung Hsing University, Taichung 40227,
Taiwan, R.O.C
e-mail: kclin@nchu.edu.tw

S.-Y. Chen
e-mail: g101029017@mail.nchu.edu.tw

J.C. Hung
Department of Information Management,
Overseas Chinese University, Taichung 40721,
Taiwan, R.O.C
e-mail: jhung@ocu.edu.tw

J.J. (Jong Hyuk) Park et al. (eds.), *Ubiquitous Computing Application and Wireless Sensor*, Lecture Notes in Electrical Engineering 331,
DOI 10.1007/978-94-017-9618-7_28

people process and analysis the data in different area. Feature selection is a key technology in the area of data mining, especially in the analysis of the "big data" of diverse domains. The feature selection algorithms shift out the unnecessary features that noise, redundant and irrelevant. Usually, feature selection algorithms are divided into two categories: filter and wrapper methods [1]. Filter methods evaluate the optimality of the feature subset by using the predetermined criteria that independent of the classifier. Wrapper methods directly use the adopted classification algorithm to evaluate and find the optimal feature subset. Filter methods have the advantage of computational simplicity since they are independent of the classifier. However, wrapper methods can provide better performance than filter methods. For example, higher prediction accuracy of a classifier can be achieved.

There are many common classifiers such are Support Vector Machine (SVM) [2] and Artificial Neural Network [3]. SVM is a supervised learning algorithm proposed by Vapnik and Cortes [4] and Vapnik [5]. It can be used to solve the problem of classification or regression. The concept of SVM is finding the optimal hyperplane to classify the data. Since SVM can achieve high classification accuracy with a small training set, it becomes more attractive. And in this study, the SVM is adopted as the classification algorithm.

Therefore, the feature selection algorithm is a search process to find the optimal feature subset to describe the characteristics of dataset exactly. Hence, the problem of feature subset selection is viewed as the problem of combinatorial optimization. When the number of features increases, the number of possible combinations exponentially grows. To overcome the problem of curse of dimensionality, a meta-heuristic algorithm is used to search the problem space to obtain the optimal solution for diverse domain, such as function optimization [6], intrusion detection [7] and Schedule Management [8]. There are many well-known meta-heuristic algorithms, such as Genetic Algorithm (GA) [9], Particle Swarm Optimization (PSO) [10] and Artificial Fish Swarm Algorithm (AFSA) [11]. Results in research [12] indicate that the AFSAs outperform the PSOs in the problem of function optimization. It shows the potential of the AFSAs in different field of optimization. However, they still have a few defects such as the solution might fall into local optimal or be lack of multiplicity.

Hence, this study proposes a modified version of Artificial Fish Swarm Algorithm (MAFSA). And, we use the MAFSA as the search algorithm to find the optimal feature subset. Classification capabilities of SVM are used to measure the performance of the proposed scheme and the basic version of AFSA. Experimental results showed that the performance of the proposed method was superior to that of basic version of Artificial Fish Swarm Algorithm.

28.2 Feature Selection Using the AFSA

The AFSA was first proposed by Li et al. [11]. It simulates the intelligence and behaviors of fish swarm by three main steps including Follow, Swarm and Prey. AFSA use those three steps repeatedly to find the best solution. To solve the

F_1	… F_i …	F_n

Fig. 28.1 Representation of feature set

problem of feature subset selection, each fish represents their own feature subset by binary coding; 0 represents feature was not be selected and 1 represent feature was be selected. Figure 28.1 shows the feature subset of each fish.

The main process of the AFSA for feature selection is outlined as follows:

(1) Initialization.
Encode the feature subset and define the fitness function. Randomly initialize the value of position and the parameters of distance and vision for N fishes. The distance is calculated by formula (28.1) and the vision is the average distance of each fish.

$$Distance(F_i, F_j) = \sum_{k=1}^{k} |F_i(k) - F_j(k)| \tag{28.1}$$

(2) Evaluate fitness.
Define the variation of classification accuracy for SVMs as the fitness function. And, evaluate fitness for each N fish by using the computing of fitness function.

(3) Movement of fish swarm
Execute the step of Follow, Swarm and Prey movement for each fish to find the optimal feature subset.
Follow: Each fish searches the other fish in its own neighbor to find the fish which has the best fitness. If the neighbor fish has better fitness and the crowded degree of this fish isn't greater than the maximum crowded degree, then the original fish replace its own feature subset by the neighbor fish's features subset. If the replacement is indeed executed, the step of Follow is successful and the algorithm proceeds to the next fish. If there are no such fish which has better fitness in the neighbor, the step of Follow is fail and this fish proceeds to the step of Swarm. The neighbor was calculated by formula (28.2) and the crowded degree was calculated by formula (28.3).

$$\text{Neighbor}(\text{F}_i) = \{\text{F}_k | 0 < distance(\text{F}_i, \text{F}_k) \le vision\} \tag{28.2}$$

$$\text{Crowded Degree}(F_i) = \frac{Neighbors\ of\ F_i}{Total\ number\ of\ Fishes} \tag{28.3}$$

- **Swarm**: The fish compares the fish in its own center with fitness value. If the center fish has the better fitness and the crowded degree of center isn't greater than the maximum crowded degree, then the original fish replace its feature subset by center's features subset. If the replacement is indeed executed, the step of Swarm is successful and the algorithm proceeds to the next fish. Otherwise, the step of Swarm is fail and this fish proceeds to the step of Prey. The center was calculated by formula (28.4)

$$F_{center}(i) = \begin{cases} 0, \sum_{k=1}^{k} F_k(i) < \frac{k}{2} \\ 1, \sum_{k=1}^{k} F_k(i) \geq \frac{k}{2} \end{cases} \quad (28.4)$$

- **Prey**: The fish randomly change its own features subset, and the random number of changed feature will not greater than its own vision. When a feature was selected as the changed feature, the value was changed. For example, the value of feature was 0 will be changed to 1 and vice versa. If the fitness of changed fish is greater than the fitness of original fish, then the original fish replace its own feature subset by the changed fish's features subset. If the replacement is indeed executed, the step of Prey is successful and the algorithm proceeds to the next fish. Otherwise, the algorithm keeps trying the step of Prey until the try number over the predefined maximum.

(4) Algorithm stops.
If the terminal conditions are satisfied, then the algorithm stops and outputs the best feature subset. Otherwise, returns to (28.2) to keep proceeding to the next iteration until the terminal condition are satisfied. Figure 28.2 shows the process for feature selection of SVM by using the AFSA schemes.

28.3 Feature Selection Using MAFSA

The parameter vision has played an important role in AFSA. If the vision was set too small, the search spaces of each fish were limited. And, it might lead the AFSAs fall into local optimal. On the other hand, if the vision was set too wide, the search spaces of each fish were too extensive. Hence, it might result weak local search capability of the AFSAs. Therefore, we need a mechanism to dynamically control the parameter vision of each fish. If the fitness value of the fish was better than the other fishes, the parameter vision of this fish should be decreased to enhance the ability of local search. In contrast, if the fitness value of this fish was lower than the others, the parameter vision should be increased to enhance the global search ability. To achieve the above mechanism, we adopt the endocrine-based formula, referencing to [13], to control the parameter vision. The endocrine-based formula is outlined as (28.5) and (28.6).

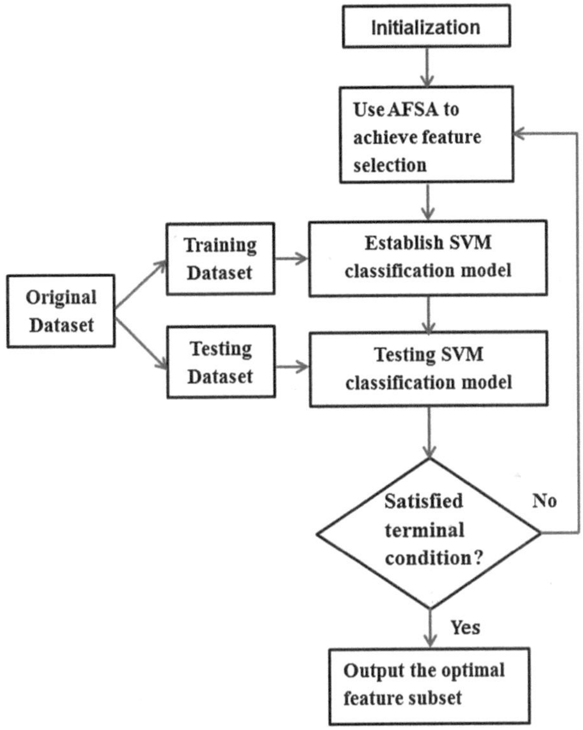

Fig. 28.2 Feature selection of SVM using AFSA

$$EA = f_1\left(\frac{f_{max} - f_i}{f_{max} - f_{avg}}\right) \cdot \left[\frac{\pi}{2} + f_2(f_i - \frac{f_{i-1} + f_{i+1}}{2})\right] \tag{28.5}$$

$$Vision(f) = Vision(f) \cdot ES(f) \tag{28.6}$$

In formula (28.6), f_{max} represents the best fitness of fish swarm, f_{avg} represents the average fitness of fish swarm and f_i represents the fitness of fish f_i. $f_1(x) = atan(x), f_2(x) = atan(-x)$ Parameter f1 and f2 are used to adjust the range of update mechanism. The MAFSAs can dynamically adjust the parameter vision by the above mechanism. Hence, the value of parameter vision for the MAFSA is different from that of the original AFSA. In AFSA, the vision for all the fishes in swarm is unique. However, the MAFSA uses dynamical vision depending on the fitness of each fish to make suitable search scope for every fish. Figure 28.3 shows an example for the dynamical vision.

Original vision demonstrates the vision size for the original AFSA and the fish *i* and *j* has different fitness value but the same vision size. Modified vision demonstrates the vision size for the MAFSA and vision size is adjusted by using endocrine-based formula. The fitness of fish *i* has better fitness, 80; the vision of fish *i* decreases to enhance local search. On the other hand, fish *j* has lower fitness value, 65; the vision of fish *j* increases to enhance global search.

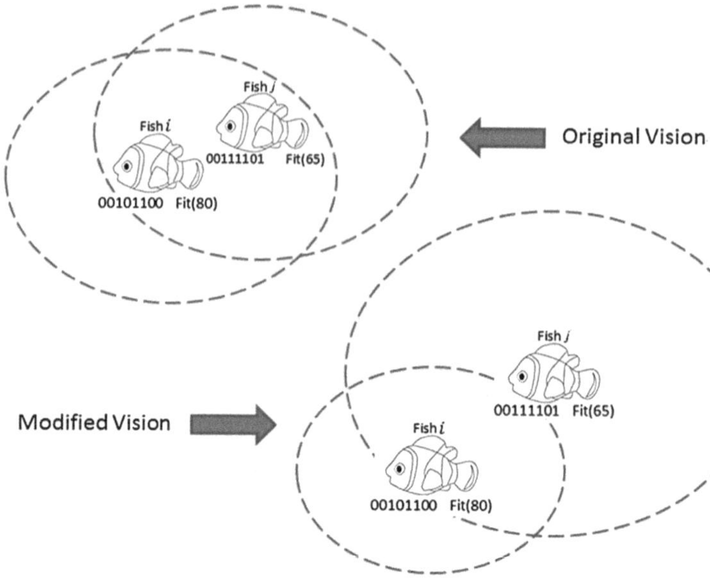

Fig. 28.3 Original vision and modified vision

Similar to the description of the AFSA in Sect. 28.2, we use the MAFSA as the search algorithm to find the optimal feature subset. Classification capabilities of SVM are used to measure the performance of the proposed scheme. For estimating the accuracy of the SVM classifiers, this study conducted the fivefold cross validation test.

28.4 Experimental Result

To verify the effectiveness of the proposed MAFSA schemes, the UCI dataset are adopted to measure the classification accuracy of AFSA and MAFSA. The UCI dataset are used to be analysis for matching learning, we select three different dataset for the experiment as show in Table 28.1.

The software used to run the AFSA and MAFSA includes the Microsoft Window 7 operate system and the LIBSVM [14] library for SVM classifier. And, the hardware are Intel Core(TM)2 Quad Q8400 2.66 GHz coprocessor and 2 GB memory.

The algorithm parameters we used in both algorithms are presented as follows. Number of fish are 30, the maximum try number is 30, the maximum crowded degree is 0.5. The algorithm uses five-fold cross-validation and the terminal condition of each fold was that the optimal feature subset was not update in 3,600 s.

Table 28.1 The UCI dataset

Dataset	No. of classes	No. of features	No. of instances	Area
German	2	24	1,000	Financial
Pima	2	8	768	Life
Sonar	2	60	208	Physical
Vehicle	4	18	846	Image

Table 28.2 The experimental results of AFSA and MAFSA

Datasets	AFSA-SVM			MAFSA-SVM		
	No. of selected features	Average accuracy rate (%)	Executed time (s)	No. of selected features	Average accuracy rate (%)	Executed time (s)
German	13.8	80.4	31,755	13.2	82.1	36,416
Pima	5	81.1	28,015	4.8	81.8	31,522
Sonar	31.2	98.1	23,375	29.8	99.5	24,146
Vehicle	11.8	87.7	22,503	11.4	89.9	31,503

Table 28.2 compares the average classification accuracy, average number of selected features of the optimal feature subset and total time of AFSA and MAFSA. Experimental results indicate that the classification accuracy of the proposed MAFSA is higher than basic version of AFSA. And for the four dataset, the MAFSA also has less number of selected features of the feature subset than AFSA. However, the MAFSA spend more time than AFSA.

28.5 Conclusion

This study proposes a modified version of Artificial Fish Swarm Algorithm to select the optimal feature subset to improve the classification accuracy for Support Vector Machines. The MAFSA enhance the searching ability by integrating an endocrine-based dynamic vision mechanism. The experimental results indicate that MAFSA is not only preform higher classification accuracy than AFSA, but also obtain lesser number of optimal features subset. Although MAFSA spend more time in the algorithm process, but the classification accuracy usually is the first priority. This MAFSA-SVM method performs great result in feature selection and expects to be used in many different applications.

References

1. Liu H, Motoda H (1998) Feature selection for knowledge discovery and data mining. Kluwer international series in engineering and computer science. Kluwer Academic Publishers, Boston
2. Furey TS, Cristianini N, Duffy N, Bednarski DW, Schummer M, Haussler D (2000) Support vector machine classification and validation of cancer tissue samples using microarray expression data. Bioinformatics 16:906–914
3. Zhang GP (2000) Neural networks for classification: a survey. IEEE Trans Syst Man Cybern Part C-Appl Rev 30(4):451–462
4. Vapnik VN, Cortes C (1995) Support-vector networks. Mach Learn 20(3):273–297
5. Vapnik VN (1995) The nature of statistical learning theory. Springer, New York
6. Houck CR, Joines JA, Kay MG (1995) A genetic algorithm for function optimization: a Matlab implementation NCSU-IE TR 1995
7. Liu T, Qi Al, Hou YB, Chang XT (2009) Feature optimization based on artificial fish-swarm algorithm in intrusion detections. Int Conf Netw Secur 57:542–545
8. Saeed F (2009) Efficient job scheduling in grid computing with modified artificial fish swarm algorithm. Int J Comput Theory Eng 1(1):13–18
9. Cheng LH, Chieh JW (2006) A GA-based feature selection and parameters optimization for support vector machines. Expert Syst Appl 31:231–240
10. Kennedy J, Eberhart RC (1995) Particle swarm optimization. In: IEEE international conference on neural networks
11. Li XL, Shao ZJ, Qian JX (2002) An optimizing method based on autonomous animats: fish-swarm algorithm. Syst Eng-theory Pract 22(11):32–38
12. Chen H, Wang S, Li J, Li Y (2007) A hybrid of artificial fish swarm algorithm and particle swarm optimization for feedforward neural network training. IEEE advanced intelligence system research
13. Hsu SH, Lin KC (2012) Feature selection and parameter optimization based on improved EPSO for support vector machines. National Chung Hsing University, Taichung
14. Chang CC, Lin CJ (2001) LIBSVM: a library for support vector machines. http://www.csie.ntu.edu.tw/~cjlin/libsvm/

Chapter 29
A New Remote Desktop Approach with Mobile Devices: Design and Implementation

Teng-Yao Huang, Hai-Hui Wang, Chun-Lung Peng and Hsin-Mao Huang

Abstract As the mobile device industry grows rapidly, smart phones and tablet computers have become the indispensable devices in our daily lives. Even more, one could have owned many devices at the same time. Although, the traditional personal computers or laptops has more powerful computing performance than mobile devices, but, compared with convenience and portability, the mobile device would be a better choice. Eventually, this leads to the increase of the variety of devices in our living; however, each of these devices has its own functions, which is not always replaceable from other devices, and how will these devices interact has become one of the most important discussion in the field of computer science. In this paper, we will focus on the integration of personal computer, which has better computing capacity, and mobile device, so it allows the user to control the mobile device through their personal computer via Wi-Fi or 3G. This allows the user to work with their personal computer and mobile device at the same time. The integration of both devices will improve the usage of the resources provided by these devices, such as phone call communications, GPS and G-sensor, so that we will be able to use our PC sufficiently, and contribute to computer technology in the future.

Keywords Mobile device · Remote control · Remote desktop

T.-Y. Huang (✉) · H.-H. Wang · C.-L. Peng · H.-M. Huang
Department of Computer Science and Information Engineering,
Taipei Campus, Aletheia University, Tamsui, Taipei, Taiwan
e-mail: am001871@au.edu.tw; cswang@mail.au.edu.tw

H.-H. Wang
e-mail: am001909@au.edu.tw

C.-L. Peng
e-mail: am001959@au.edu.tw

H.-M. Huang
e-mail: xmhuang@au.edu.tw

J.J. (Jong Hyuk) Park et al. (eds.), *Ubiquitous Computing Application and Wireless Sensor*, Lecture Notes in Electrical Engineering 331,
DOI 10.1007/978-94-017-9618-7_29

29.1 Introduction

29.1.1 Research Backgrounds

Since the increase in the usage of mobile devices, the functions of these devices are no longer confined in phone calls and SMS. The developers enhance the device by adding more tools that will satisfy customers' needs. These hardware such as GPS, Bluetooth, E-compass and Ambient light sensors, are exactly what lakes in our personal computers.

However, the computing capacity of mobile devices is limited, and we still focus our work on personal computer at most of the time. This paper will integrate the features of both devices, and hope to solve problems corresponds to convenience. In scenarios like, typing an SMS with a small keyboard on the touchscreen or taking a call while your phone is not around.

Our system allows you to remote control your mobile device from your personal computer through wireless network. Therefore, are able to make phone calls, SMS, and all other applications in the mobile device from personal computer.

29.1.2 Research Problem

Due to that the main operating system for most mobile devices are based on Android. So, we will focus on Android version 4.0.3, which is the general version, as a research platform.

The main functions in our research are divided into audio, visual interface and input event. Wherein, the audio streaming and screenshot are restricted in the Android system.

Therefore, our research encounters two major problems:

(1) The audio cannot be recorded without consent of the caller based on legal norms from certain countries or regions. So, in order to avoid the illegal behavior, Google does not allow applications to access audio streams.
(2) Additionally, obtaining the screenshot of the mobile devices is restricted by operating system, to avoid unworthy applications getting device information, such as account numbers, passwords, etc. Thus, the screen features of our system will be limited.

In order to complete the functions used in our system, some adjustments in the android operating system are needed.

29.1.3 Research Purpose

Our research aims to combine the advantages of mobile devices and personal computers. We could enjoy the larger screen and speakers of the desktop computers or type with an actual keyboard instead of the virtual keyboard, which is not so smooth, on the mobile device. Thus, let the mobile device to retrieve the functions which are removed due to convenience.

The research platform for our experiment consists of a device based on Android operating system version 4.0.3, and a personal desktop computer or notebook equipped with Windows 7 operating system. The screen display of the mobile device is transferred to the screen of the personal computer through network connection. The users will be able to view the screen display of the mobile device on the personal computer instantly, as well as using the mouse and keyboard to simulate the input touch event on a mobile device. Even if using the keyboard to directly type the text into the device's input box. In the case during receiving or making a phone call, the audio streaming will continue to send to the personal computer and broadcast by speaker, at the same time, the voice streaming captured through the microphone of the personal computer will be send to the mobile device.

Figure 29.1 shows the system flow chart, our system is separated to PC control end and mobile device end. After running the program both ends connects via TCP. The application on the mobile device will constantly send screenshot as well as the output audio streaming of the device. Then, these data are passed to the PC control

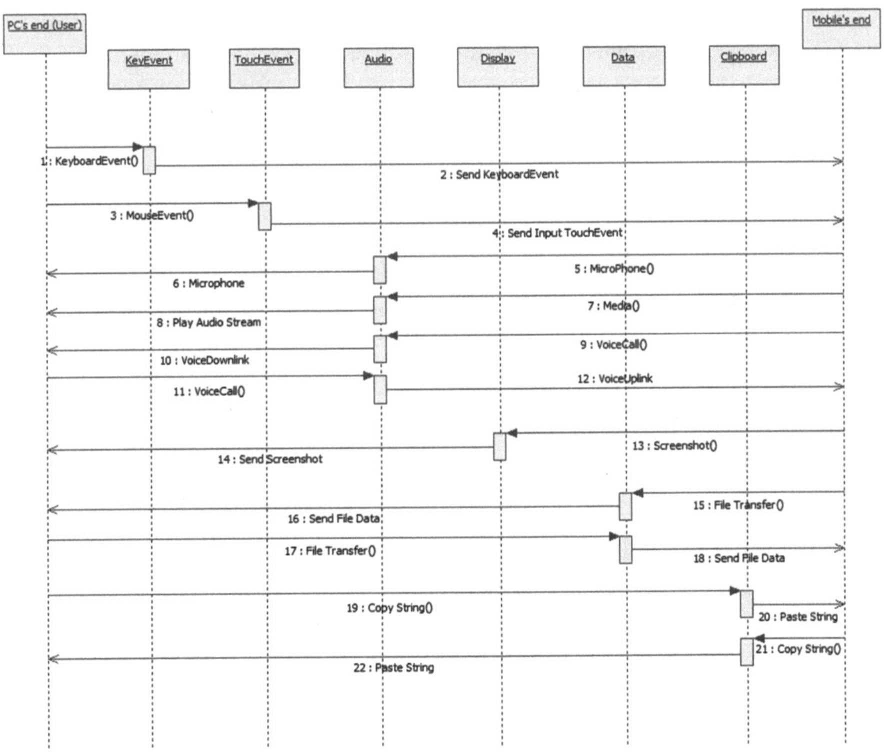

Fig. 29.1 Sequence diagram

end to do the corresponding outputs. The user can control the mobile device's keyboard and touch event through the graphical user interface on the PC control end.

29.1.4 Framework

In Sect. 29.2, we will briefly discuss on researches that are related to our work. In the following chapter, we will look through the Android kernel, permission of Superuser on the device, screen shot transfer, synchronization of audio streaming, as well as touch event and keyboard event simulation.

Finally, in Sects. 29.4 and 29.5, we will present the experimental results and our conclusions.

29.2 Related Work

Connection architecture and data transmission of Remote Desktop is released by VNC (Virtual Network Computing) protocol during the earliest 2002 by American Telephone and Telegraph Company (AT & T). The architecture is mainly divided into the controlled end (client) and manipulation end (server) as shown in Fig. 29.2.

Moreover, researches that send the data of the mobile device using remote control back for cloud computing are implemented in [1–3].

Researches which take into account that the user's PC might shutdown due to unexpected reasons, Priyanka developed a system that allows the user to boot, shutdown or restart the personal computer through network connections [4]. Other than that, there are two researches which uses RFB (remote frame buffer) transport protocol, to display the graphical user interface [5, 6].

Grant's research allows multiple users to input control simultaneously, by using cloud computing concepts, the multi-control input events is then integrated [7]. Research which focus on resolution, frame rate, compression ratio and color depth, particularly on the calculation of the transmission rate, while transferring the image [8]. In [9], it uses Wi-Fi Adapter as an example to show how to develop kernel files in an Android system.

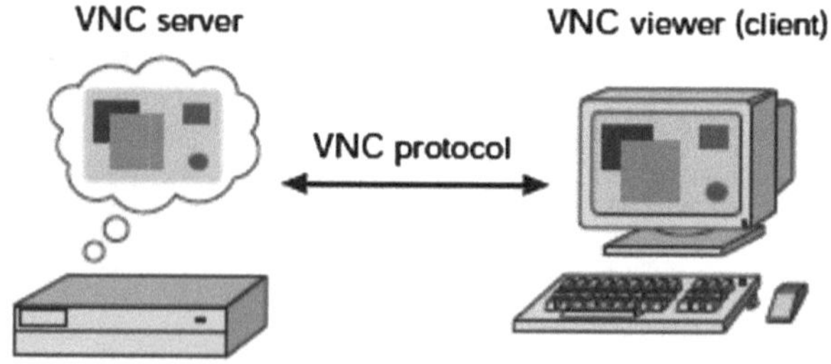

Fig. 29.2 Structure of VNC [11]

Table 29.1 Functions of related development

Functions		AirDroid	TeamViewer: QuickSupport	Droid VNC server	Our system
Touch event			V	V	V
Phone data		V	V	V	V
Display		–	V	V	V
Media	Audio	V	–	–	V
	Video	–	V	–	V
Camera		V	V	–	V
Clipboard		–	V	V	V
Phone call	Uplink	–	–	–	V
	Downlink	–	–	–	V

In the past, there are many studies based on remote control between PC's desktop and mobile devices [10–12]. These systems can connect through the Internet to obtain real-time desktop of the remote computer, and controls made on the mobile device's end is then passed to the desktop computer to do the corresponding actions.

In 2012, Leo Sicard proposed a study that transmits the screen display of mobile device to a secondary screen instantly [13]. This study completes a similar real-time video transmission function of the remote desktop, but it does not handle the user's control and event transmission, the system is only used for viewing the screen of mobile devices on the secondary display.

In addition, the opposite control that control mobile devices' desktop from PC also have proposed, such as TeamViewer:QuickSupport [14], AirDroid [15], Droid VNC Server [16]. In [15], AirDroid cannot display real-time screen of mobile devices on PC. In [16], Droid VNC Server is unable to provide the functions of real-time image of camera and video streaming. In [14], Team Viewer:QuickSupport cannot process the voice. Therefore, in this paper, we proposed a new approach to implement a remote desktop system that completely controls the mobile device from PC. In our system, it not only provides the above functions, but also allows to dial/receive phone call on the desktop of PC. Table 29.1 show the comparison of functions between our system and [14–16].

29.3 Methods

29.3.1 Permission of Superuser on Device

In experimental environment, we use HTC One V smartphone with Android 4.0.3 operating system. In order to achieve functions such as screen capture, input touch event and input keyboard etc., we must obtain the highest user privileges (root) of the device, root su executable file under the system/bin folder of your mobile device, and install the application superuser to obtain the highest authority and control all applications on the device.

Thus, the application in our research will be able to access the complete data within the operating system through the highest authority obtained, to fulfill the Remote Desktop purpose of this paper. But, not all mobile device users has or willing to get the highest authority on their mobile devices (Fig. 29.3).

Figure 29.4 shows the diagram of the system structure scheduled to completion. It allows the user to control the mobile device completely via personal computers. This includes access of data of the mobile device, input events and obtaining information of peripheral devices.

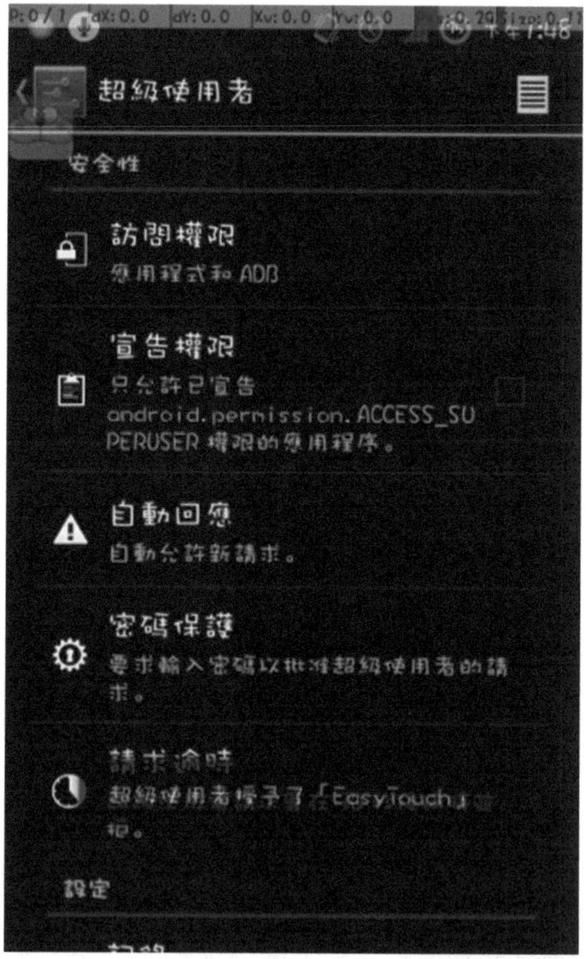

Fig. 29.3 Superuser obtained on the mobile device

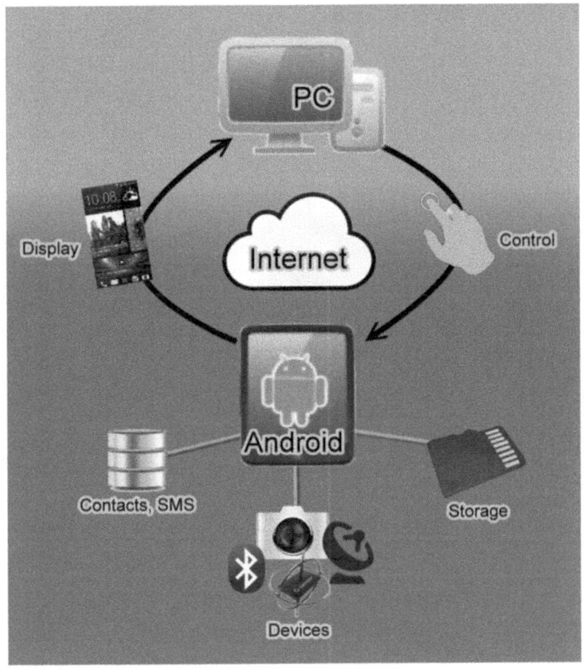

Fig. 29.4 System structure

29.3.2 Screenshot Transfer

After analyzing the structure of the Android operating system, our system displays from the libui's EglWindows in the Libraries, then, transferred from display composition in the Surface flinger of the Framework, and then communicated between Framebuffer and hardware drivers.

Surface flinger communicates with hardware-drivers through the process of saving files, and at the same time, data transfer is made. Surface flinger will save the new screenshot in the system's streaming file (/dev/graphics/fb0), each time the new screen image is calculated. However, the file format of the content is stored in order to drive the hardware, and it's directly written on the equipment used. So, to be able to make use of this file in this system, this file will be converted into visible view for the users to see. The method first converts the pixel array (raw data) into RGB mode as shown in Table 29.2, and then uses Android's Bitmap to convert to png.

29.3.3 Synchronization of Audio Streaming

Generally, audio structure can roughly be seen as three-parts, user, audio driver and hardware devices. The Android system divides user layer into Applications, Framework, Audio library and Audio HAL, shown in Fig. 29.5. Due to the audio

Table 29.2 RAW to RGB

```
for (int m = 0; m < colors.length; m++) {
  int r = (piex[m * 4] & 0xFF);
  int g = (piex[m * 4 + 1] & 0xFF);
  int b = (piex[m * 4 + 2] & 0xFF);
  int a = (piex[m * 4 + 3] & 0xFF);
  colors[m] = (a << 24) + (r << 16) + (g << 8) + b;
}
```

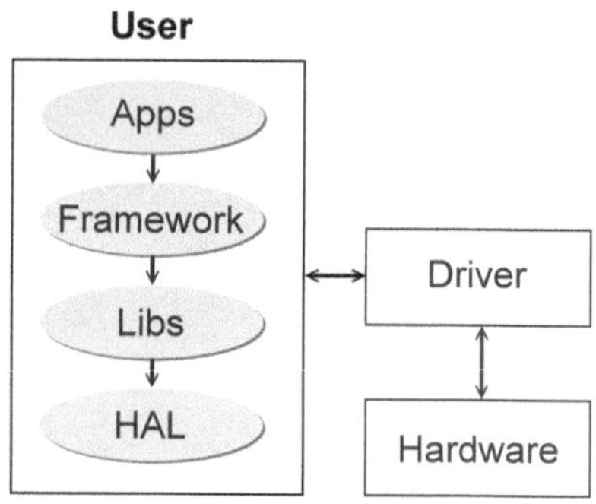

Fig. 29.5 Android audio architecture

processing used in the system cannot be done with Android API (Application Programming Interface), some adjustments in the Audio library and Audio HAL layer is needed.

Thus, in dealing with the audio, the Android kernel processes the audio input and output most on AudioTrack/AudioRecorder, AudioFinger and AudioPolicyService.

Wherein, AudioFinger played a role in continuously process the requests of AudioTrack/AudioRecorder from upper programs, or manage the audio equipment through Audio HAL, so, it is able to get the audio stream needed for our system if AudioFinger is modified, and then by writing JNI and Application framework, which will return the data to the Application layer.

29.3.4 Touch Event and Keyboard Event Simulation

Processing Touch Event and Keyboard Event in Android:

1. Hardware driver reads input data from hardware
2. Write in streaming file (/dev/input/eventX)
3. Operating system detects the event files nonstop and processes it.

Therefore, to handle the input events, the remote input of the user is written in the streaming file.

Table 29.3 Event argument

Event	Type	Code	Value
Key down	0x01	Key code	1
Key up	0x01	Key code	0
Touch set	0x03	–	x, y co-ordinates
Touch down	0x01	330	1
Touch up	0x01	330	0

The only difference between Touch Event and Keyboard Event is that they have different data code, so both can be used in the same way. The code format can be found corresponds to the file (kernel/include/linux/input.h) in the Linux kernel. There are three parameters needed to be send, sequentially as: Event type, Event code and Event value. Table 29.3 shows the parameters our system will use.

29.4 Experiments and Results

29.4.1 Devices

The hardware used for our research is as the follows:

1. ASUS Desktop PC, with INTEL®Core™ i5 CPU 2.80 GHz, 4 GB memory and 500 GB hard disk drive.
2. HTC One V Smart phone, with Qualcomm MSM8255 CPU 1 GHz, 3.7 inches (480*800) display, 512 MB RAM, 4 GB ROM, Android 4.0.3 Operating System.

29.4.2 Remote Control Mobile Device

29.4.2.1 Initialization and Touch Simulation

Start up the program on the PC which then enters the on-line state, it will detect the PC's IP address. Afterwards, start the application on the mobile device, enters the IP address to connect, as shown on Figs. 29.6 and 29.7.

When initialize connections on both ends complete, the mobile device will synchronize two threads, one for continuously capture the screenshot from the Android operating system, it is then converted to the view back on the PC's end, the other thread will continue to capture the output of audio streaming on the device, and send to PC's end for uninterrupted playback.

Fig. 29.6 Connection on PC's end. **a** Initialization on PC's end. **b** Waiting for connections

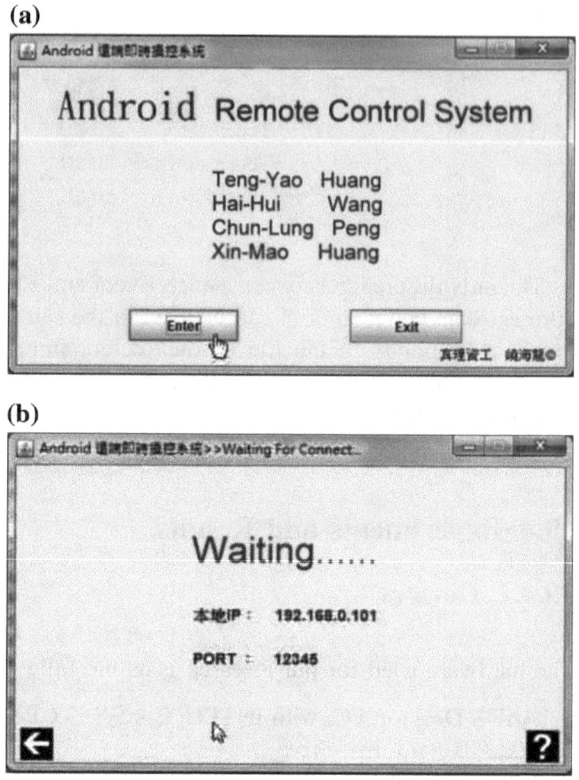

Figure 29.8 shows the real-time screen of the mobile device that showed on the PC's end after connection initializes. Figure 29.9 shows the actual device after connection. Users can directly control this screen by clicking or dragging with the mouse. The system will immediately pass input events to the mobile device's end, in order to complete the simulation of touch capabilities, and reach a remote control with a more humane graphic interface.

Figure 29.10 shows the real-time view stage when the action dragging is being done on the mouse from the PC's end. It is known that the smoothness of the screenshot on the PC's end will differ due to each operating system's update frequency (Fig. 29.11).

29.4.2.2 Remote Phone Call

In the outgoing calls section, click the call button on the phone screen, after entering the general phone dialer screen, now, you could already use the mouse on the screen to the view history, or the numeric keypad to enter the number using the mouse. Since it has been controlled by a PC, of course, you can use the keyboard on

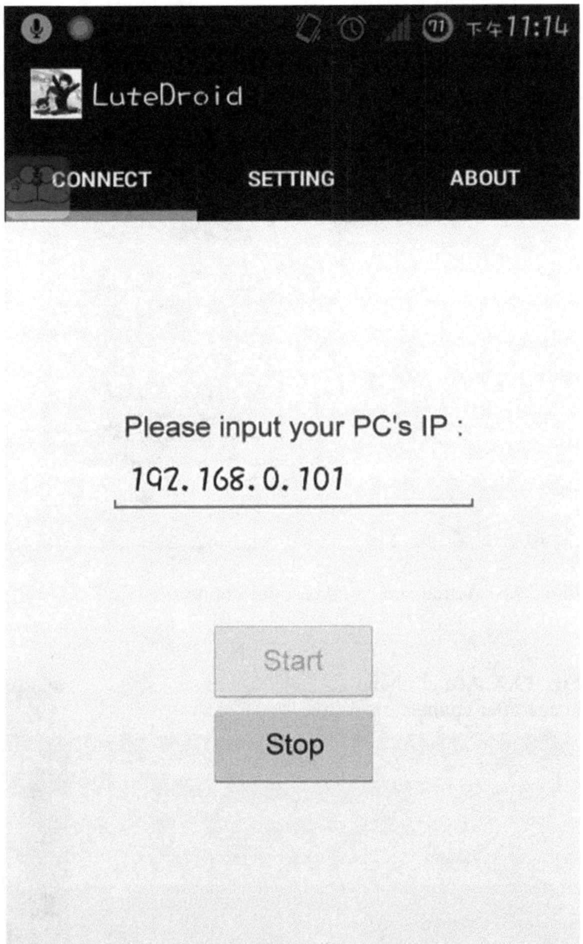

Fig. 29.7 Connection on device's end

the PC to type in phone numbers, or even copy a phone number from the web, it is allowed to copy a piece of numbers from the PC's end and paste it on the mobile device's end.

Moreover, the user can complete the phone call using the audio output of the computer, by connecting the PC to the mobile device via A2DP in the communication range of Bluetooth.

29.4.2.3 GPS via Remote Control

Through real-time simulation of touch screen operations and display, you can completely control the mobile device through computer, including running applications directly on the device.

Fig. 29.8 Actual screen taken after connection

Fig. 29.9 Actual phone image after connection

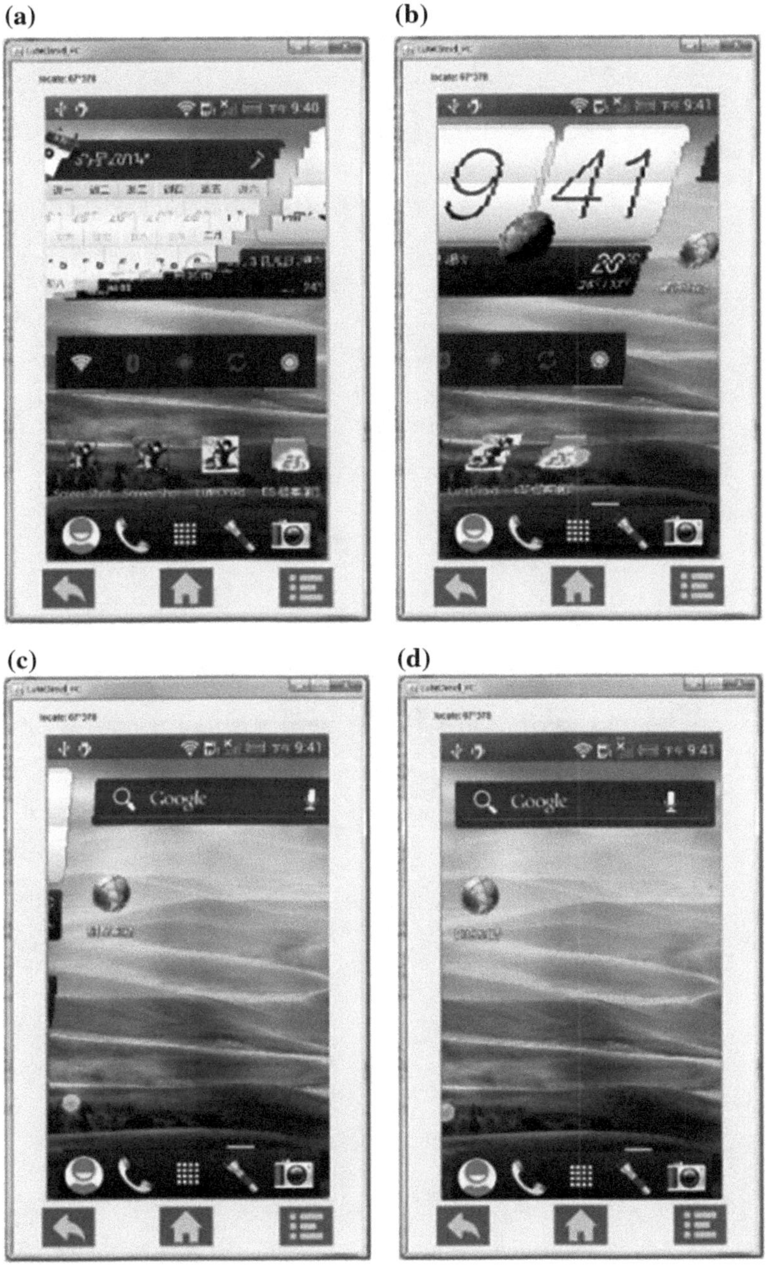

Fig. 29.10 Drag view on PC's end

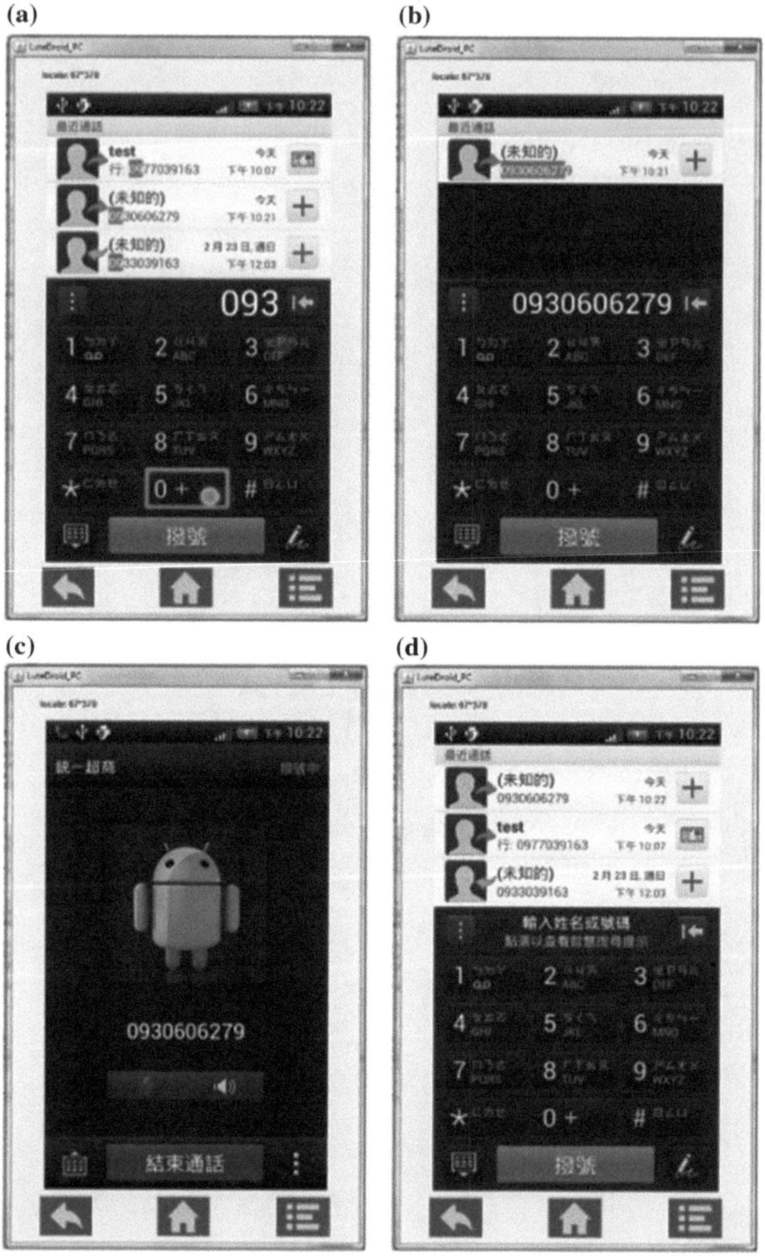

Fig. 29.11 Phone call via remote control

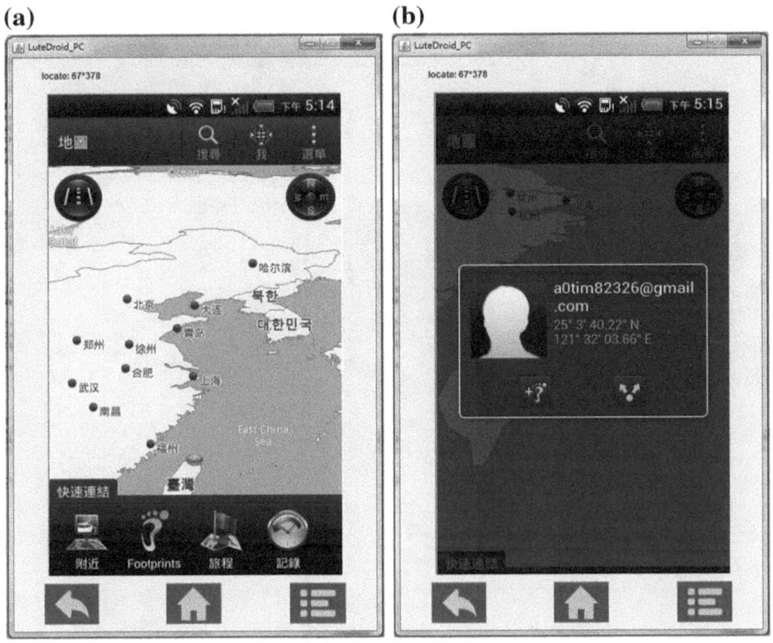

Fig. 29.12 GPS application on PC's end. a Enter map app. b Get GPS information

In the experiments, we use the map applications for example, allows users to obtain accurate GPS location of mobile devices in order to determine the location of the users and mobile device. Figure 29.12 shows using the GPS program from the mobile device's applications on personal computers.

29.4.2.4 File Transfer

Figure 29.13 shows the bidirectional file transfer user interface, click the file transfer function keys, both PC and mobile device will access to system files structure tree, and displayed on the window's sides on the screen.

In our experiments, we transfer file from PC to mobile device for example, the file structure tree of the PC is shown on the left side of the window, the user can select the file wished to be transferred, and select the destination folder using the file structure tree of the device on the right side. Finally, click the button with the right arrow, then, the system begins to transfer the file; it will show a notification once it is completed.

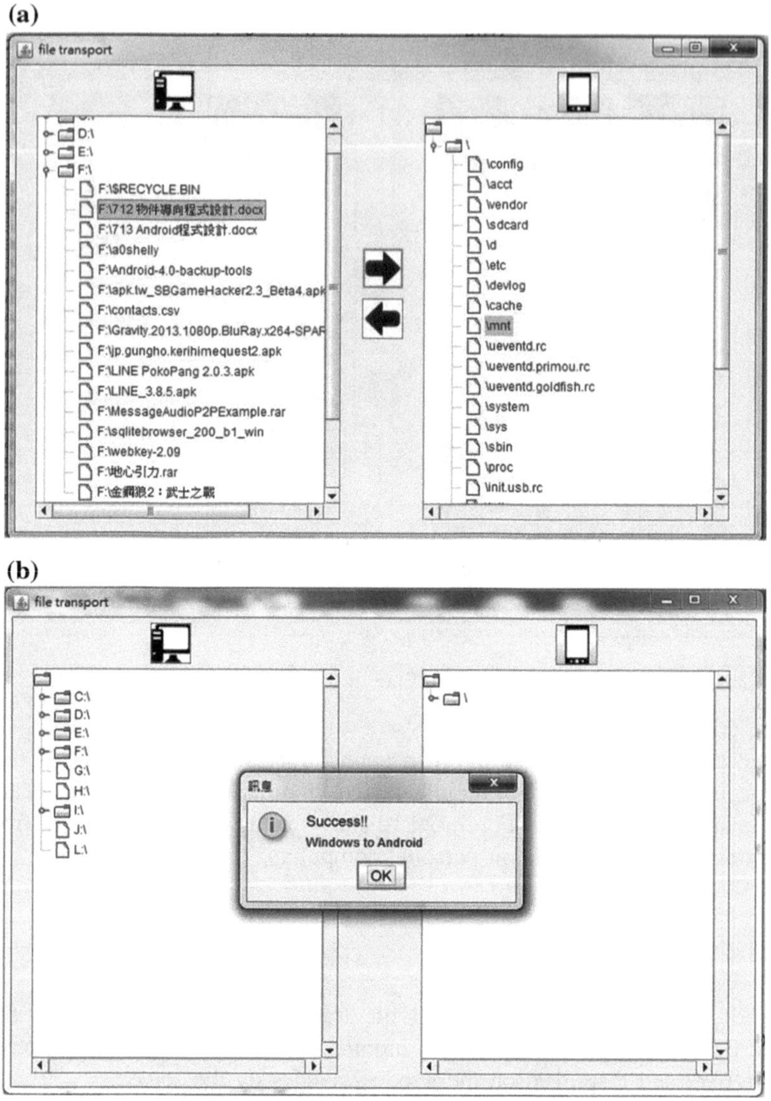

Fig. 29.13 File transfer from PC to device. a Choose file and destination. b Notification

29.5 Conclusion and Future Work

In this paper, we proposed a new approach to implement a remote desktop system that controls the mobile device from PC. This system mainly focuses on developing for Android operating system, which successfully connects mobile device to personal computer through a stable internet communication that completes the integration between both devices. In addition, we also proposed the solution to solve three significant problems which control mobile devices completely: real-time screenshots from mobile device, input event control and audio synchronization. Therefore, we successfully and effectively integrate many devices that surround in personal, and adequately use various kinds of functions that they provide.

Currently, the real-time remote control system for mobile devices is complete, but it is needed to improve the transmission of data through the Internet. Such as to match with the increasing resolution of the mobile devices, if no additional processing are made, and the screenshot is directly transferred to the personal computer, then this may led to a large data transfer which takes longer for the image to show that will cause the program to delay. In the future, will focus on the data processing before transmission; find the most suitable compression method for the current Internet speed, let the users to have the best operating experience.

References

1. Divya VL (2011) Mobile application platform on cloud server. In: 2011 international conference on advancements in information technology, vol 20. IACSIT Press, Singapore
2. Baig Md. S, Rajasekar M, Balaji P (2012) Virtual network computing based remote desktop access. Int J Comput Sci Telecommun 3(5)
3. Dharsan A. Mahesh., A. Manju (2012) Desktop viewer solution for mobile computing using virtual network computing. Int J Commun Netw Syst 01(02)
4. Kampasi PV, Kulkarni YC (2013) Creating an intelligent environment in mobile technology. Int J Eng Sci Innov Technol 2(6)
5. Gopikrishnan B, Mani A (2012) Desktop solution for mobile environment using mobile cloud computing. Int J Adv Res Comput Commun Eng 1(1)
6. Deepak M, Visalakshi P Remote desktop access using remote frame buffer in mobile cloud environment. Int J Eng Res 30–34
7. Li K, Wallace G Virtually shared displays and user input devices. Princeton University, Princeton
8. Ciminiera L, Lamberti F, Paravati G, Sanna A, A novel approach to support quality of experience in remote visualization on mobile devices. Politecnico di Torino, Dipartimento di Automatica e Informatica, Italy
9. Li C (2013) The development of Android software and kernel files by using example of Wi-Fi adapter. Int J Comput Sci Issues 10(2):3
10. Aggarwal H, Kadhane M, Kadam S, Inamdar A (2013) COMPDROID—remote desktop access through Android mobile phone. Int J Sci Modern Eng 2(1). ISSN: 2319–6386
11. Balkhande BW, Gavhane S, Phanse R, Sadafule M Remote desktop on mobile. Int J Innov Eng Technol
12. Baravkar N, Kumari S, Shaikh M, Thadani H , Kale S (2013) Monitoring PCs using Android. Int J Sci Eng Res 4(4)
13. Sicard L, Bardram JE (2012) TIDE—using device composition on tabletop computers to extend the smartphone experience. IT University of Copenhagen, Copenhagen
14. TeamViewer-QuickSupport. http://www.teamviewer.com/zhtw/products/mobile-device-support.aspx#android
15. AirDroid. http://web.airdroid.com/
16. Droid VNC Server. http://opensourceexcedio.wordpress.com/2010/10/28/droid-vnc-server/

Chapter 30
Implementation of Low Power LED Display Controller Using Adiabatic Operation

Kyung-Ryang Lee, Sung-Dae Yeo, Seung-il Cho and Seong-Kweon Kim

Abstract Conventional fluorescent light source, as well as incandescent light source is gradually being replaced to LED for reducing power consumption in image display area for multimedia application. LED light source requires the controller with a low-power operation. In this paper, a low-power technique using adiabatic operation is applied for the implementation of LED controller with a stable constant-current, a low-power and low-heat function. From the simulation result, the power consumption of the proposed LED controller using adiabatic operation was reduced to about 87 [%] in comparison with conventional operation with a constant V_{DD}. The proposed circuit is expected to be an alternative LED controller which is sensitive to external conditions such as heat.

Keywords LED · Adiabatic · Constant current · Low power

K.-R. Lee · S.-D. Yeo · S.-K. Kim (✉)
Department of Graduate of NID, Seoul National University of Science and Technology, Seoul, Korea
e-mail: kim12632@seoultech.ac.kr

K.-R. Lee
e-mail: corenc@seoultech.ac.kr

S.-D. Yeo
e-mail: ysd1009@seoultech.ac.kr

S. Cho
Innovation Center for Organic Electronics, Yamagata University, Yonezawa, Japan
e-mail: cho_si@yz.yamagata-u.ac.jp

J.J. (Jong Hyuk) Park et al. (eds.), *Ubiquitous Computing Application and Wireless Sensor*, Lecture Notes in Electrical Engineering 331,
DOI 10.1007/978-94-017-9618-7_30

30.1 Introduction

There have been a number of methodological attempts for energy conservation in image display area for multimedia application. In lighting and image display area, light emitting diode (LED) system has been studied. Also conventional fluorescent light source, as well as incandescent light source is gradually being replaced to LED. LED light source is often used with controller for low-power operation. Increasing LED channel, LED controller has been required to be operated with a low-power design. Especially, low power design of display controller becomes essential block in a study of image display [1].

In this paper, it is introduced that the power consumption of LED lighting controller can be reduced through low-power adiabatic methodology. So far, a various attempts have been tried in the limited condition of a constant power supply voltage without any control [2]. But, there was no dramatic reduction of power consumption. General energy consumption of Complementary Metal Oxide Semiconductor (CMOS) circuit is caused by the channel ON resistance in switching time of transistor.

Typical LED driving system generates a large driving current with analog or pulse width modulation (PWM) dimming to represent the brightness of LED device [3]. Because large current flows through channel resistance R_{ON} of the MOS transistor, large power consumption is occurred. The power consumption of ON resistance degraded the performance of LED device. LED device with a temperature-sensitive characteristics can make the designed output characteristics worse. In many cases, this effect caused the destruction of LED driving system.

Low power circuit design method using adiabatic operation is to transmit adiabatic signal of ON resistance [4]. When this method for designing a low power LED application is applied, the power consumption of a large current driving LED system can be reduced. And instability resulted from overheating of LED system can be overcomed. Therefore, in this paper, adiabatic method for low-power operation is utilized for implementing a LED controller.

30.2 Adiabatic Operation

Figure 30.1 shows voltage variation of source (V_{DD}), capacitance (V_C) and resistance (V_{RON}) respectively, in conventional MOS pull-up network. A transistor is assumed a constant current source that can move the energy of CV^2 on the period of time. V_R represents the variation of the transistor channel voltage supplied from the drain. V_C represents load voltage variation from the charged voltage. Rapid and large voltage variation at resistance cause a circuit a large power dissipation due to the thermal loss [5].

Figure 30.2 shows the voltage variation of source (V_{TRI}), capacitance (V_C') and resistance (V_{RON}') respectively, in adiabatic MOS pull-up network. When voltage

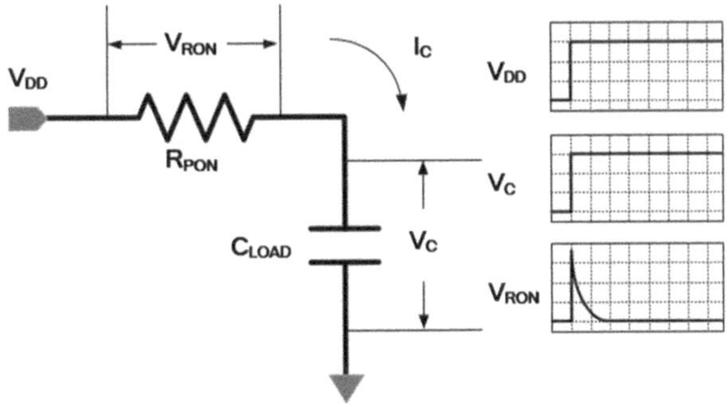

Fig. 30.1 MOS pull-up network V_{RON} energy consumption

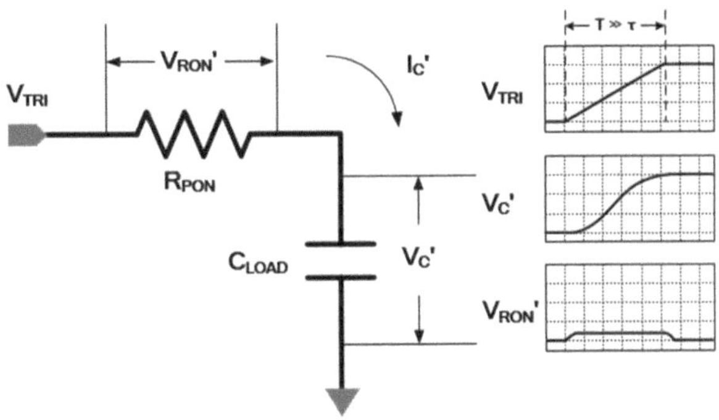

Fig. 30.2 Adiabatic pull-up network V'_{RON} energy consumption

source with long time constant of τ, such as ramped voltage V_I, was applied to transistor through drain, there was a small voltage variation that shows a little power consumption in the transistor. Theoretically, if time constant is approaches infinity, the energy consumption of resistance converges to zero [6]. V_R is the change of voltage on the channel resistance R_{ON} from step voltage source V_{DD} at pull-up network in Fig. 30.1. V_R can be expressed as:

$$V_{DD}(t) = R_{PON}I_C(t) + V_C(t)$$

where, I_C is the displacement current across the load capacitance voltage V_C. I_C can be obtained from:

$$I_C(t) = C\frac{dV_C(t)}{dt}$$

Load capacitance voltage, V_C and pull-up path current, $I_C(t)$ is calculated as:

$$V_C(t) = V_{DD}(1 - e^{-\frac{t}{CR_P}}),$$
$$I_C(t) = \frac{V_{DD}}{R_P}e^{-\frac{t}{CR_P}}$$

Power dissipation at load capacitance can be calculated as $V_C(t)*I_C(t)$. MOS R_{ON} power dissipation can be expressed as:

$$P_{LOADCAP} = V_C(t) \cdot I_C(t) = \frac{V_{DD}^2}{R_P} e^{-\frac{t}{CR_P}} (1 - e^{-\frac{t}{CR_P}})$$

$$P_{R_{ON}} = \frac{V_{DD}^2}{R_P} e^{-\frac{2t}{CR_P}}$$

Therefore, total energy dissipation of pull-up path can be calculated as:

$$w_C(t) + w_{R_{ON}}(t) = \frac{1}{2} CV_{DD}^2 + \frac{1}{2} CV_{DD}^2 = CV_{DD}^2$$

where, total energy dissipation is equals to supply energy.

V_R' is the change of voltage on the channel resistance R_{ON}' from the ramped voltage source V_{TRI}, with the delay time of τ, at pull-up network in Fig. 30.2. And I_C' is displacement current from load capacitance voltage V_C'. V_C' and I_C' are calculated as:

$$V_C(t)' = \frac{V_{TRI}}{\tau} [t - CR_P'(1 - e^{-\frac{t}{CR_P'}})$$
$$- [(t - \tau) - CR_P'(1 - e^{-\frac{t}{CR_P'}})]u(t - \tau)]$$
$$I_C(t)' = \frac{CV_{TRI}}{\tau} [(1 - e^{-\frac{\tau}{CR_P'}}) - (1 - e^{-\frac{\tau}{CR_P'}})(t - \tau)]$$

where, τ is the ramped time constant of V_{TRI} voltage.

Therefore, total energy consumption of adiabatic pull-up path can be expressed by using [7] as:

$$w_C(t) + w_{R_{ON}'}(t) = \frac{CV_{TRI}^2}{2} + \frac{R_P' C^2 V_{TRI}^2}{\tau} [1 - \frac{CR_P'}{\tau} (1 - e^{-\frac{\tau}{CR_P'}})] = CV_{TRI}^2$$

where, total energy dissipation is equals to supply energy.

When ramped voltage source V_{TRI} with delay time of τ, is supply to adiabatic circuit design, energy consumption, $CV^2/2$ in conventional MOS R_{ON} resistance can be reduced to the amount of CV_{TRI}^2.

30.3 Current LED Controller Using Adiabatic Operation

Figure 30.3 shows the proposed design of adiabatic digital current LED controller. Current channel columns of the MOSFET are constituted according to power of the two for LED drive current with exponential function. Linearity of the output current

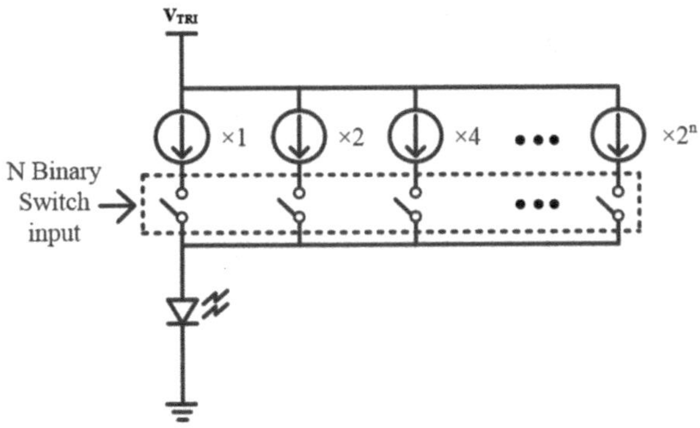

Fig. 30.3 Current LED control design circuit using adiabatic operation

can be achieved by the ON/OFF switch. The amount of the LED dimming can be determined in each binary switch current. Output current is determined with the LED specification. The current rating of the proposed design is determined with 25 mA by typical intermediate current of LED.

Figure 30.4 shows the basic circuit of the proposed LED controller with adiabatic operation. The basic circuit was designed with a high output impedance of two-stage P_{CH}. Because LED has a feature of function that the current is determined with a voltage change exponentially. Utilized MOS transistor is saturated when the switch is high. Utilized load as the LED is an equivalent circuit which is constituted with 100 Ω resistance in the condition of typical voltage of 2.5 V with the current of 25 mA, and zener diode that is used for safe operation of the LED is represented with 1 nF capacitance.

Figure 30.5 shows the I/V characteristic with 0.35 um CMOS process technology of the Dong-Bu company. Figure 30.6 shows the characteristics of V_{DS}/I_{DS} = R_{ON}. Time constant is 2^{-15} s in the condition of a constant V_{DS} = 5 V, and the on resistance of the used MOS transistor is 2 mΩ. And load capacitance is 1 pF. Ramped V_{DS} is characterized with a cycle with 0.5^{-6} s and the time constant can be

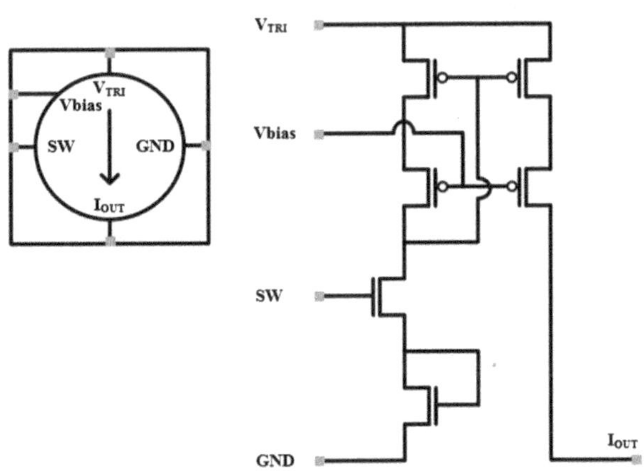

Fig. 30.4 Basic circuit of LED controller with adiabatic operation

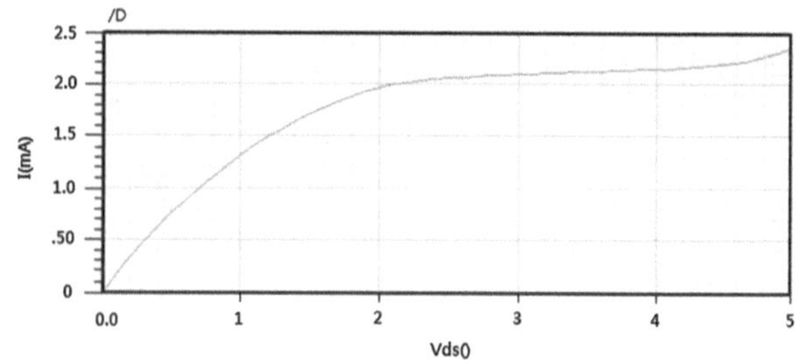

Fig. 30.5 I/V characteristic of the used MOS transistor

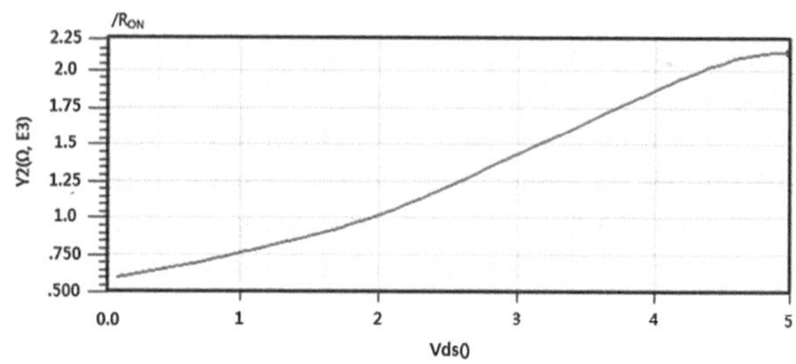

Fig. 30.6 R_{ON} characteristic of the used MOS transistor

expressed as 0.25^{-6} s as a maximum value. If V_{DS} is not a peak value, V_{DS} is always smaller than 5 V, therefore, the time constant of the ramped voltage of V_{DS} is calculated as 0.25^{-6} s*63 % = 0.15^{-6} s.

Figure 30.7 shows the power dissipation in comparison with constant V_{DD} and ramped V_{DD} operation of the designed circuit. Energy dissipation of constant V_{DD} operation in the period time of 0–100 us is 16 uW. Energy dissipation of ramped V_{DD} operation is 3 uW. The basic circuit with adiabatic operation in comparison

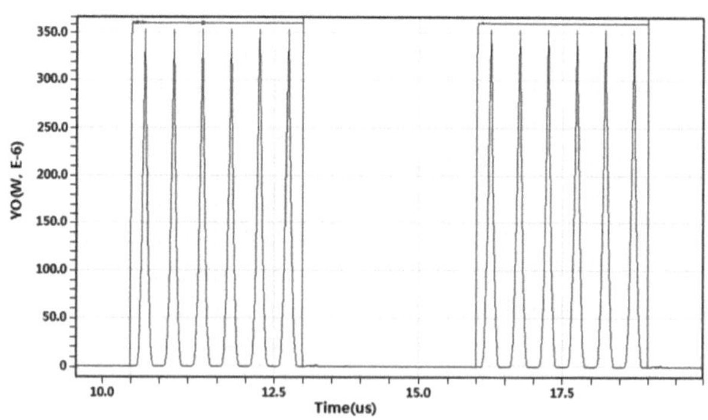

Fig. 30.7 Power consumption comparison between constant V_{DD} operation and ramped V_{DD}

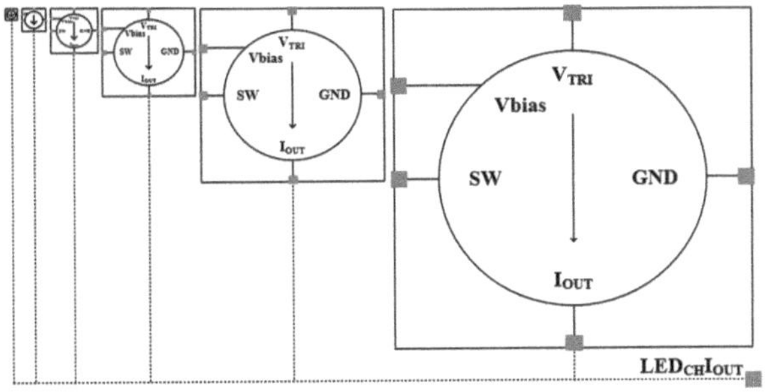

Fig. 30.8 6-bits current LED controller structure using adiabatic operation

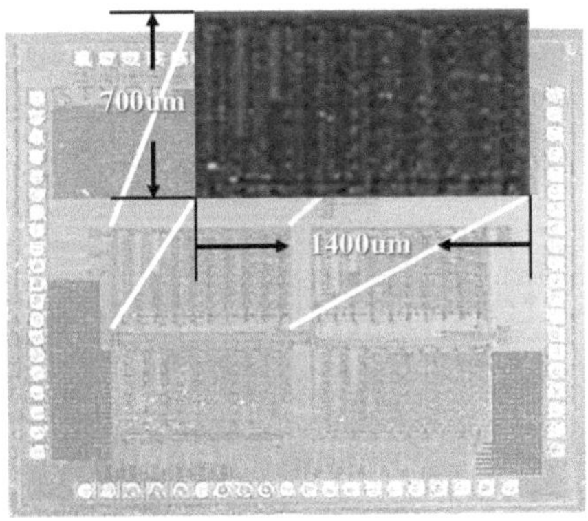

Fig. 30.9 Photograph of LED controller

with conventional operation shows the energy reduction of 82 %. Figure 30.8 shows the LED controller structure that has 6-channel switches to the 63 basic circuits. Output current, $LED_{CH}I_{OUT}$ shows the linear characteristic. Figure 30.9 shows the photograph of the designed LED controller. The LED 4-channels with adiabatic operation was implemented with 0.35 um CMOS process technology of the Dong-Bu company and the size was 1,400*700 um.

Figure 30.10 shows the power consumption of proposed adiabatic design with constant V_{DD}. The energy dissipation of conventional operation is observed with 1.13 μJ in the period time of 150 us from the simulation result.

Figure 30.11 shows the power consumption of the proposed adiabatic design with V_{TRI}. The energy dissipation of adiabatic operation is evaluated with 0.21 μJ in the period time of 150 us. From the simulation result, power consumption of the proposed LED controller using adiabatic operation was reduced to about 87 % in comparison with conventional operation with constant V_{DD}.

Specific heat of Silicon material as the main components of LED system is 0.713 −0.8 J/g °C [8]. When LED channel is used in same condition of proposed design. Temperature of the LED controller can be limited to 0.26/1.41 u °C, 18 %.

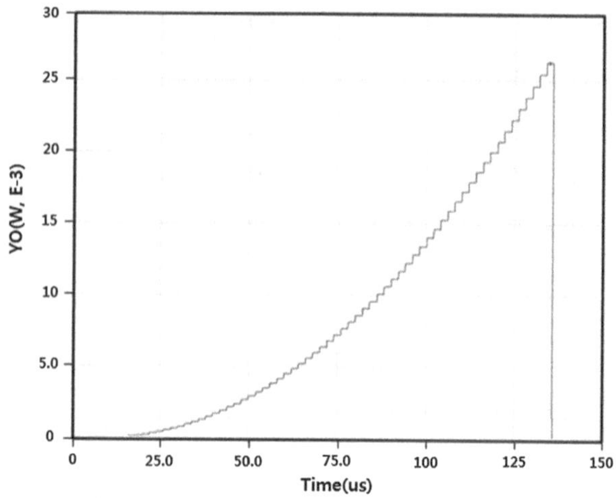

Fig. 30.10 Power consumption of conventional 6-bits LED controller

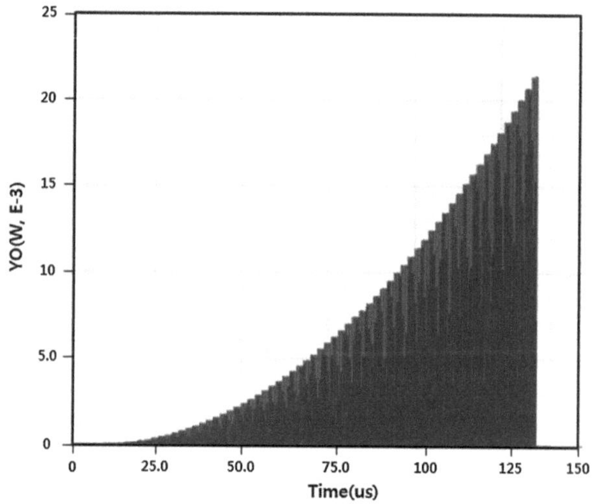

Fig. 30.11 Power consumption of adiabatic 6-bits LED controller

30.4 Conclusion

LED channel increasing, LED controller block has been required to be operated with low-power design. In this paper, a low-power technique using adiabatic operation was applied to the LED controller for a stable constant-current, a low-power and low-heat function. From the simulation result, the power consumption of the proposed LED controller can be reduced to about 87 % in comparison with conventional operation. The proposed circuit design method is expected to be a solution for low- power LED controller which is sensitive to external conditions such as heat.

Acknowledgments Foundation item: This research was partially funded by the MSIP (Ministry of Science, ICT & Future Planning), Korea in the ICT R&D Program 2014.

References

1. Tanimoto M, Panahpour M (2011) Free-viewpoint TV. IEEE Signal Process 1053(11):67–76
2. Panda PR, Shrivastava A (2010) Power efficient system design. Springer, New York
3. Jayakumar N, Paul S (2010) Minimizing and exploiting leakage in VLSI design, vol 405, issue 1. Springer, New York, pp 220–227
4. Athas WC, Svensson LJ (1994) Low-power digital systems based on adiabatic-switching principles. IEEE Trans Very Large Scale Integr (VLSI) Syst 2(4):398–407
5. Dickinson AG, Denker JS (1995) Adiabatic dynamic logic. IEEE J Solid-States Circuits 30 (3):311–315
6. Moon Y, Jeong DK (1996) An efficient charge recovery logic circuit. IEEE J Solid-States Circuits 31(4):514–522
7. Anuar N, Takahashi Y (2009) Adiabatic logic versus CMOS for low power applications. In: Proceedings of the 24th international technical conference on circuits/systems, pp 302–305
8. Sze SM (2007) Physics of semiconductor devices, 3rd edn. Wiley, New York, p 790

Chapter 31
Hand Gesture Recognition Using 8-Directional Vector Chains in Quantization Space

Seongjo Lee, Sohyun Sim, Kyhyun Um, Young-Sik Jeong and Kyungeun Cho

Abstract This paper proposes a hand gesture recognition technique that allows users to enjoy uninterrupted interaction with a variety of multimedia applications. Hand gestures are recognized using joint information acquired from a Kinect sensor, and the recognized gestures are applied to multimedia content. To this end, hand gestures are quantized in the grid space, expressed using an 8-directional vector chain, and finally recognized on the basis of a hidden Markov model. To assess the proposed approach, we define the hand gestures used in the "Smart Interior" multimedia application, and collect a dataset of gestures using the Kinect. Our experiments demonstrate a high recognition ratio of between 90 and 100 %. Furthermore, the experiments identify the possibility of applying this approach to a variety of multimedia content by verifying its superior operation in actual applications.

Keywords Hand gesture recognition · Kinect sensor · Hidden Markov model · Multimedia content

31.1 Introduction

The recent evolution of computing functions has given rise to the era of ubiquitous computing, whereby computers and networks can be used in any place and at any time in a simple and convenient manner. Accordingly, ubiquitous computing has been extensively studied.

In such a context, interest has focused on a simpler, more intuitive interaction between humans and computers. The use of a mouse or keyboard, which has been common up to now, requires practice and has a spatial limit. To overcome the

S. Lee · S. Sim · K. Um · Y.-S. Jeong · K. Cho (✉)
Department of Multimedia Engineering, Dongguk University-Seoul,
30 Pildong-ro 1-gil, Seoul, Jung-gu 100-715, Republic of Korea
e-mail: cke@dongguk.edu

J.J. (Jong Hyuk) Park et al. (eds.), *Ubiquitous Computing Application and Wireless Sensor*, Lecture Notes in Electrical Engineering 331,
DOI 10.1007/978-94-017-9618-7_31

limitations of existing methods, natural user interfaces, which employ natural human gestures, have been actively examined.

Gestures are the natural actions performed by humans, and can be made by anyone. Moreover, users can interact with computers via intuitive actions, without the need to learn how to use existing device-based approaches. In particular, gesture recognition has been intensively investigated since the launch of Microsoft's Kinect sensor in 2010. The Kinect provides detailed joint information about the human operators [1–6]. Because gestures can replace a variety of existing input devices, it is highly likely that the approach proposed in this paper could be applied in the field of green IT.

This paper examines the recognition of gestures using human joint information from a Kinect sensor. In particular, we focus on the recognition of gestures for controlling multimedia content. The system proposed in this paper creates a vector chain by expressing the movement direction of joints using an 8-directional vector, and then recognizes the gestures by applying a left-right hidden markov model (HMM). To evaluate the gesture recognition function proposed in this paper, we present results using the "Smart Interior" application.

The remainder of this paper is organized as follows. In Sect. 31.2, we present an overview of some related research. Section 31.3 describes the 8-directional vector chain proposed in this paper for expressing hand movements, and explains the HMM-based hand gesture recognition algorithm. In Sect. 31.4, we apply our hand gesture recognition technique and assess its performance. Finally, Sect. 31.5 discusses the experimental results and some ideas for further.

31.2 Related Work

Gesture recognition has been extensively investigated for several years. In particular, several studies have used gestures as inputs in multimedia, games, and medical applications.

An algorithm is proposed to control multimedia content by recognizing human actions without input devices [1]. Hands were identified from images based on the YCbCr color and depth information from a Kinect, and the tracks of the hands were symbolized into an 8-directional chain code. The hand gestures were then identified using an HMM algorithm.

The authors configured a specific vector with the coordinate values of arm joints given by the Kinect, and converted the feature vector into the angle between the joints [3]. Hand gestures were recognized on the basis of an HMM in a real environment by reducing and separating the dimension using a K-means clustering method.

The authors proposed a hand gesture recognition technique using depth information and visual images [7]. The hand region was identified from a depth image from the Kinect sensor, and this enabled the analysis of hand gestures. Next, the hand region of the depth information and the visual image were mapped. Hand gestures were recognized by defining the number of open fingers. An experiment to control medical images on a monitor was conducted to assess the performance of this approach.

The authors adopted a process of mathematical morphology for gesture recognition [8]. Their work followed the center trajectory of a hand gesture sequence containing important information on the forms of hand gestures. Next, an 8-directional chain code edge vector was acquired from the center trajectory, and the hand gesture sequence was recognized.

The above research proposed gesture recognition systems for multimedia, game, and medical applications. However, such methods cannot be applied to a wide variety of applications, because the gestures they consider are very simple. Accordingly, this paper proposes a general-purpose gesture recognition system that can be applied to all kinds of applications.

31.3 Hand Gesture Recognition

The proposed method converts a user's gesture into an 8-directional vector chain using hand joint information from a Kinect sensor. In addition, we present a gesture model to classify the types of gestures using the converted vector chain.

31.3.1 Identification of 8-Directional Vector Chain

A vector is created using the difference in relative locations between a hand in a previous frame and that in the present frame [1]. However, consecutive frames can mistakenly reflect even momentary hand shaking or negligible hand gestures. To overcome such problems, we create a grid space and quantize the hand joints. The 8-directional relative vector can then be applied to this space.

The algorithm for extracting the 8-directional vector chain comprises the three steps shown in Fig. 31.1. The first projects 3D coordinates as planar movement on

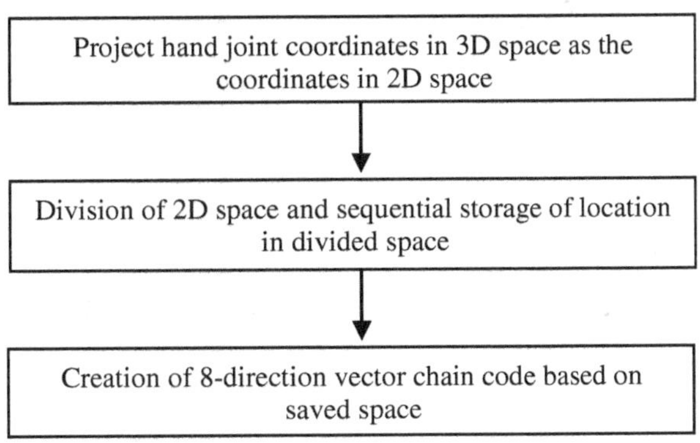

Fig. 31.1 8-direction vector chain extraction flow chart

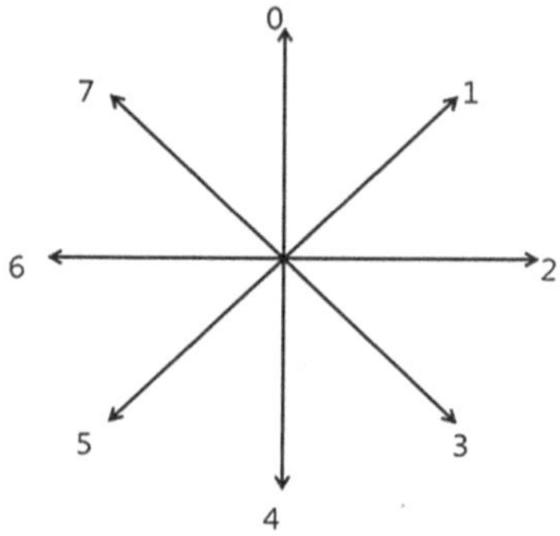

Fig. 31.2 8-direction vector chain encoding code

the 2D space, thus expressing hand gestures as an 8-directional vector, as presented in Fig. 31.2.

The second step divides these 2D hand joint coordinates into several grids. In this paper, we consider grids of size 10 × 6. The location in each grid at which hand coordinates meet is saved in sequential order whenever a grid is changed.

The third step estimates the relative vector to the space sequentially saved in the previous location, and converts this into a vector chain code. As the grid is rectangular, vectors 1, 3, 5, and 7 are expressed as a combination of vectors 0, 2, 4, and 6.

When a hand moves in direction 1 in Fig. 31.2, the grid spaces that the hand joint coordinates pass are saved in the sequence 5, 2, 3 or 5, 6, 3, as presented in Fig. 31.3. At this point, the relative vector to the previous grid is either a combination of 0.2 or 2.0. To solve such problems, the algorithm recognizes this as the vector for direction 1.

31.3.2 Hand Gesture Recognition Using HMM

This paper considers the six one-handed gestures presented in Fig. 31.4.

HMM is an appropriate model for processing sequential data. This paper adopts a left-right HMM. The number of states is from 2 to 8, in accordance with the pattern of gestures shown in Fig. 31.4. The HMM was learned using the Baum–Welch

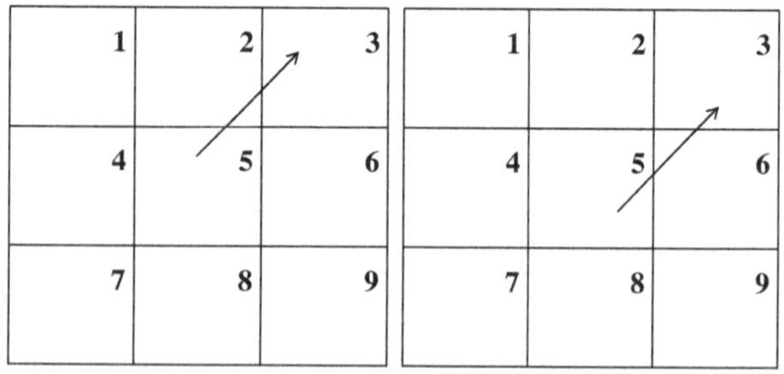

Fig. 31.3 Hand movement in grid space

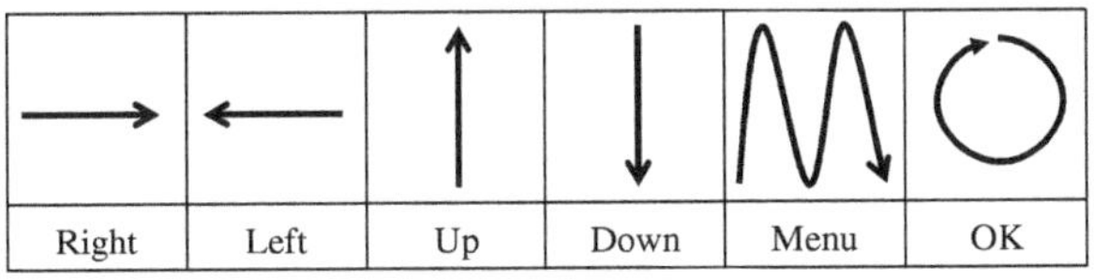

Fig. 31.4 Kinds of gestures to recognize

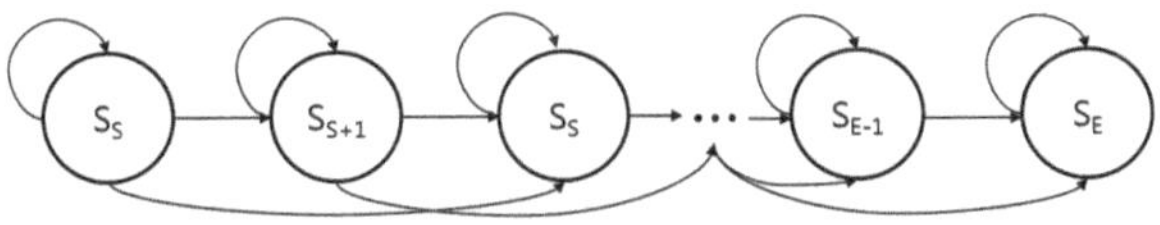

Fig. 31.5 State transition diagram in *left-right* HMM

algorithm. The number of training data is 40 per each gesture (fig. 31.5). Figure 31.5 presents LR-HMM used in this paper.

It is necessary to distinguish those hand gestures that are used to express a gesture from meaningless hand motion. To this end, we classify the start and end points of a gesture by sending signals such as 'grip' and 'release.'

In gesture models, the relevant likelihoods are computed using the vector chain created by a user's hand gesture. The model with the maximum likelihood is then recognized as the gesture. This can be represented as:

$$\mathrm{q} = \arg\max_{i} P(o|\Theta_{g^i})P(g^i) \tag{31.1}$$

where o is the vector chain for a hand gesture and Θ_{g^i} is the learned gesture model. Hand gestures are rejected when the gesture model in Eq. (31.1) has a likelihood below some threshold, that is:

$$P(0|\Theta_{g^q})P(g^q) < Threshold_q \tag{31.2}$$

31.4 Experiments

31.4.1 Experimental Environment

To assess the recognition of hand gestures, we used 50 images per gesture recorded by the Kinect. The Kinect was set at a height of 1 m. The distance from the Kinect sensor to the user was 1.5–2 m.

Table 31.1 Gesture performance assessment table

	N	N_{Rej}	N_E	N_{Hit}	R (%)	R_{ext} (%)
Right	50	1	2	47	94	95.92
Left	50	1	2	47	94	95.92
Up	50	3	2	45	90	95.74
Down	50	0	2	48	96	96
OK	50	0	0	50	100	100
Menu	50	4	1	45	90	97.83

31.4.2 Performance Assessment

Table 31.1 presents the performance results given by the approach proposed above. N is the number of gestures, N_{Rej} is the number of rejected gestures, N_E is the number of wrongly recognized gestures, N_{Hit} is the number of correctly recognized gestures, and R is the recognition ratio:

$$R = \frac{N_{Hit}}{N} \times 100\ \%$$

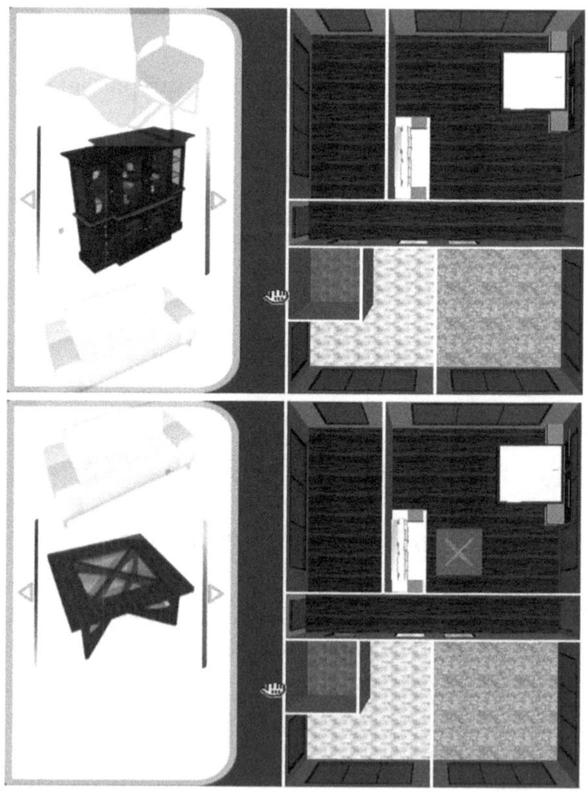

Fig. 31.6 Gesture recognition in smart interior

Table 31.2 Tasks corresponding to each gesture

Gesture	Tasks
Menu	Open and close list of furniture
OK	Select a furniture and load it into a room
Right	Rotate the selected furniture to the right
Left	Rotate the selected furniture to the left
Up	Scale the selected furniture up
Down	Scale the selected furniture down

R_{ext} is the accuracy with which the (non-rejected) gestures were identified:

$$R_{\mathrm{ext}} = \frac{N_{\mathrm{Hit}}}{N - N_{\mathrm{Rej}}} \times 100\ \%$$

Figure 31.6 illustrates the application of our gesture recognition technique to the "Smart Interior" application. This application is used to arrange furniture in a virtual space before changing the interior of the real space. In this application, we used these gestures as presented in Table 31.2.

31.5 Conclusion

This paper presented an improved hand gesture recognition method. Our approach extracts the vector chain using movement in a grid space, rather than according to the time-lapse using joint coordinates from the Kinect sensor. Furthermore, we created a gesture recognition model by applying an HMM.

The experimental results demonstrate that the proposed technique has a recognition ratio of over 90 % for all gestures. We verified that the hand gesture recognition worked very well as the interface of an actual program environment. Furthermore, the results in this paper will enable the production of a gesture recognition library.

Acknowledgments This research was supported by the MSIP (Ministry of Science, ICT and Future Planning), Korea, under the ITRC (Information Technology Research Center) support program (NIPA-2014-H0301-14-1021) supervised by the NIPA (National IT Industry Promotion Agency).

References

1. Kim YS, Park SY, Ok SY, Lee SH, Lee EJ (2012) Human gesture recognition technology based on user experience for multimedia contents control. J Korea Multimedia Soc 15(10):1196–1204
2. Cho SY, Byun HR, Lee HK, Cha JH (2012) Arm gesture recognition for shooting games based on Kinect sensor. J KIISE Softw Appl 39(10):796–805

3. Heo SK, Shin YS, Kim HS, Kim IC (2013) Design of an arm gesture recognition system using feature transformation and Hidden Markov models. KIPS Trans Softw Data Eng 2(10):723–730
4. Sohn MK, Lee SH, Kim DJ, Kim B, Kim H (2013) 3D hand gesture recognition from one example. In: IEEE, 2013 IEEE international conference on consumer electronics (ICCE), pp 171–172
5. Biswas KK, Basu SK (2011) Gesture recognition using Microsoft Kinect®. In: IEEE, 2011 5th international conference on automation, robotics and applications (ICARA), pp 100–103
6. Wang Y, Yang C, Wu X, Xu S, Li H (2012) Kinect based dynamic hand gesture recognition algorithm research. In: IEEE, 2012 4th international conference on intelligent human-machine systems and cybernetics (IHMSC), vol 1, pp 274–279
7. Park KS, Lee DH, Park YT (2013) Hand gesture recognition using depth information and visual image. J KIIT 11(7):57–65
8. Lee KH, Choi JH (2004) Hand gesture sequence recognition using morphological chain code edge vector. J Korea Soc Comput Inf 9(4):85–91

Chapter 32
Forensic Artifacts in Network Surveillance Systems

Kyung-Soo Lim, Jeong-Nye Kim and Deok-Gyu Lee

Abstract The network video surveillance system is the latest trend in surveillance technology and has been rapidly surpassing the older analogue cameras used in most surveillance systems nowadays. The network surveillance system is mostly similar to contemporary computer systems such as personal computer, server system, and embedded system. In means, acquiring or recovering video evidence in network surveillance are quite similar to traditional evidence collection techniques in digital forensics and these operations are needed to be considered proving the integrity of evidence. However, a lack of research on evidential video management in network surveillance system is able to damage reliability and admissibility in a court of law. In this paper, we defines forensic artifacts for performing video evidence retrieval in network surveillance systems. The purpose of this work is to provide reliable video evidence collection from each network surveillance systems with admissibility and integrity of evidence.

Keywords Network video surveillance · Video evidence · Digital evidence management

K.-S. Lim · J.-N. Kim
Electronics and Telecommunications Research Institute, 128 Gajeong-ro, Yuseong-gu, Daejeon 305-700, Korea
e-mail: lukelim@etri.re.kr

J.-N. Kim
e-mail: jnkim@etri.re.kr

D.-G. Lee (✉)
Seowon University, 377-3 Musimseoro, Heungdeok-gu, Cheongju, Chungbuk 361-742, Korea
e-mail: deokgyulee@gmail.com

J.J. (Jong Hyuk) Park et al. (eds.), *Ubiquitous Computing Application and Wireless Sensor*, Lecture Notes in Electrical Engineering 331,
DOI 10.1007/978-94-017-9618-7_32

32.1 Introduction

Video surveillance system has a wide variety of application domain both in public and private environment, such as homeland security, traffic control, accident prediction and detection. In spite of these various applications, there are two main purposes to the video surveillance system. The first one is to detect a crime occurred through real-time monitoring or event alarm at the remote area. The second is to help catch a suspect when a crime has been committed from CCTV video records. A typical intelligent video surveillance system has provided the various video analysis functions such as background modeling, object detection, object tracking, object classification, event detection, and event or object retrieval etc. [1-6]. As the intelligent surveillance system requires a wide area monitoring, in addition, video analysis functions for the non-overlapping view multi-camera have been considered together lately [7]. Especially, the object tracking and video retrieval from a number of recorded in multiple cameras are an important issue [8, 9].

Meanwhile, network-based surveillance systems are the latest trend in surveillance technology and are rapidly surpassing the older analogue cameras used in most surveillance systems. The adoption of IP network surveillance is perhaps the most significant trend in video surveillance today. In means, acquiring or recovering a CCTV video is the same as evidence collection techniques in digital forensics. And these operations are certainly needed to be considered contemporary digital forensic techniques [10, 11]. However, a lack of research on evidential video management in network surveillance can bring damaged reliability and admissibility in a court of law. The other one is constraints of storage capacity in ordinary archiving method for video records. The amount of data stored and the retention period of the video or pictures are subject to compression ratios, images stored per second, image size and duration of image retention before being overwritten. Recordings are usually kept for a preset amount of duration time and then automatically archived, overwritten or deleted. If the stored video file in the event a crime committed is deleted or overwritten for these reasons, an investigator will not be able to collect crucial evidence [12].

In this paper, we defines forensic artifacts for performing video evidence retrieval in network surveillance systems. The forensic artifacts are important to determine forensic soundness [13]. These could be saved and managed by video forensic server in our proposed system. When an investigator presents video evidence, these information will be contained with video files in digital evidence container. Most of network video surveillance system operates on PC or in embedded systems. It depends on the each purpose of system for video transmission or storage of surveillance video. This paper describes forensic artifacts to collect surveillance video and forensic data in different surveillance devices including network camera, network video recorder, and intelligent video analytic system. It considers forensic artifacts in which generated these devices from transmission, storage, retrieval and presentation process with admissibility of evidence by ensuring the integrity and reliability. It will contribute to produce reliable retrieval of video evidence in the courts and help to reduce unnecessary disputes in related industries [12].

32.2 Network Camera

The forensic artifacts of network camera consist of system, network, authentication, and video settings. Especially, video configurations are important to compare with others, because of these are embroiled in a legal battle, for example, video resolution can be lowered by the defendant for reducing identification rates of an object intentionally. Moreover, mostly VMS (Video Management Software or System) are not able to change video configuration because of security issues or camera-dependent features. Thus, video settings must be collected for admissibility and integrity of video evidence (Table 32.1).

Table 32.1 Forensic metadata in network camera

Category	Information
System	• Device name • Device model • Vendor name
	• Date and time
	• System ID • Camera name • Capabilities
Network	• IP address • Gateway • DNS • MAC address • Hostname • Ports (HTTP, HTTPS, RTSP, RTP)
	• Up/down bandwidth • Zero configuration
Video	• Encoding type • Video resolution • Frame rate • Video rate • Max upload rate
	• Video adjusts (contrast, brightness, saturation, hue)
Authentication	• Logged on account and password
	• 802.1X configurations

32.3 Network Video Recorder

The network video recorder includes NVR (Network Video Recorder) and VMS. Most of network video recorder based on PC or embedded system which are commonly use general operating system [10, 14]. The forensic artifacts of network video recorder consist of system, network, authentication, storage, logs and video settings. Especially, storage configurations are important compared with others because of these affords the means of verifying integrity of video files. Thus, following these settings must be collected for admissibility and integrity of video evidence (Table 32.2).

Table 32.2 Forensic metadata in network video recorder

Category	Information
System	• Device name • Device model • Vendor name
	• Date and time
	• System ID • Camera name • Capabilities
Network	• IP address • Gateway • DNS • MAC address • Hostname • Ports (HTTP, HTTPS, RTSP, RTP)
	• Up/down bandwidth • Zero configuration
Video	• Encoding type • Video resolution • Frame rate • Video rate • Max upload rate
Authentication	• Logged on account and password
	• 802.1X configurations
Storage	• Disk information • Partition information
	Video file name
	• Video file format (file format, codec)
	• Cryptographic hash value of video file
	• Filesystem metadata (MFT entry (NTFs), MAC time, size, path)
	• Recording time
Logs	• System log • Authentication log • Connection log
	• Event log
	• Video analysis log

Table 32.3 Forensic metadata in network video analytics

Category	Information
Event	• Event types (intrusion, abandoned baggage, trespassing, stolen object, illegal parking, etc.)
	• Event arisen date and time
Object	• Object type (human, vehicle, goods)
	• Object feature (face, size, color, behavior)
	• Appearance time • Disappearance time
	• Object routing path

32.4 Network Video Analytic System

The network analytic system provides event related information and object metadata information [6, 9, 15]. Likewise network video recorder, it based on PC which are commonly use general operating system. The forensic artifacts of network video analytic consist of event and object information (Table 32.3).

32.5 Defining Forensic Artifacts in Network Video Surveillance System

The forensic artifacts is defined to be FA, which is pair of type of network surveillance system (NV_{type}) and forensic metadata (FD) according to an each type of network surveillance system.

$$\text{FA} = \left\{NV_{type}|FD\right\}$$

The NV type means the type of network video system (or device), which comprise of network video camera (NVC), network video recorder (NVR), and Network video Analytic (NVA). These are defined to be following.

$$NV_{type} = \{D_{NVC}|D_{NVR}|D_{NVA}\}$$

The forensic metadata are can be defined according to these types of surveillance system. The specific forensic metadata should be collected are described in previous chapter. The FD consist of the forensic settings (FS) and forensic logs (FL).

$$\text{FA} = \{FS|FL\}$$

$$\text{FS} = \sum_{i=1}^{n} FS_i$$

$$\text{FL} = \sum_{i=1}^{n} FL_i$$

where, each FS_i aggregates following materials, and FL_i is the associate updated/ changed logs. The FS includes setting values including system (SYS), network (NTW), video (VID) and authentication (AUTH).

$$\mathrm{FS}_i = \{SYS|NTW|VID|AUTH\}$$

The FL_i are recorded by each category and these logs are recorded by each system logs in most of cases. And these FL are written when it changes compared to previous settings.

$$\mathrm{FL}_i = \left\{ \begin{array}{r} update_{NTW}\,\mathrm{O} \\ update_{sys}\,\mathrm{O}\; update_{VID}\,\mathrm{O} \\ update_{AUTH}\,\mathrm{O} \end{array} \right\}$$

The forensic artifact of each device type is defined as following equation. The following equation means settings and logs are saved into the XML documents we shown before, according to each network surveillance system.

$$\mathrm{FA}_{type} = \left[NV_{type}, \sum_{i=1}^{n} (FS_i + FL_i) \right]$$

Thus, it present and aggregate the forensic artifacts and it consist of forensic artifacts from each network surveillance systems. These forensic artifacts can be described into XML documents (Figs. 32.1 and 32.2).

$$\mathrm{FA} = \bigcup (FA_{NVC}, FA_{NVR}, FA_{NVA})$$


```
 <?xml version="1.0" encoding="UTF-8" ?>
- <ForensicArtifacts des="forensic data for network video sruveillance">
    <CaseInfo beganTime="2012-11-17 14:01:10" completedTime="2012-11-17 14:03:12" investigator="Kyungsoo Lim"
      caseNum="121117-TEST-CASE" comName="VMS_ClientPC" />
  - <ForensicArtifact type="camera" ID="1">
    - <System vender="Axis" product="Q1755" ID="20111029234" ver="1.5">
        <Name systemName="">CAM001</Name>
        <DateTime>2012-10-11, 23:55:8</DateTime>
        <URIs>http://112.220.232.210:8000</URIs>
      </System>
    - <Network IPaddr="129.254.180.111" port="8000">
        <Gateway>129.254.180.1</Gateway>
        <MacAddr>00-07-18-17-00-92</MacAddr>
        <Hostname />
        <Port rtsp="554" RTP="6970">8000</Port>
        <DNS />
      </Network>
    - <VIDEO encoding="H.264" mode="default">
        <Resolution>1280x720</Resolution>
        <FrameRate>30</FrameRate>
        <VideoRate>1000</VideoRate>
      + <Adjust>
      </VIDEO>
    - <Authentication>
        <Accout type="user" password="user1234">user</Accout>
        <Accout type="admin" password="admin1234">admin</Accout>
      </Authentication>
    </ForensicArtifact>
  + <ForensicArtifact type="recorder" ID="2">
 </ForensicArtifacts>
```


Fig. 32.1 An XML document for forensic artifacts in network camera

```xml
<?xml version="1.0" encoding="UTF-8" ?>
- <ForensicArtifacts des="forensic data for network video sruveillance">
    <CaseInfo beganTime="2012-11-17 14:01:10" completedTime="2012-11-17 14:03:12" investigator="Kyungsoo Lim"
        caseNum="121117-TEST-CASE" comName="VMS_ClientPC" />
  + <ForensicArtifact type="camera" ID="1">
  - <ForensicArtifact type="recorder" ID="2">
    - <System vender="Win4net" product="trium-i" ID="" ver="2.12">
        <Name systemName="ServerPC01">VMS Server1</Name>
        <DateTime>2012-10-11, 23:57:11</DateTime>
        <URIs />
      </System>
    + <Network IPaddr="129.254.180.110" port="">
    + <VIDEO encoding="H.264" mode="default">
    - <Authentication>
        <Accout type="user" password="user1234">user</Accout>
        <Accout type="admin" password="admin1234">admin</Accout>
      </Authentication>
    - <Storage>
      - <DiskInfo type="physical_disk">
          <Disk DriveID="PhysicalDrive0" DiskSize="1280347084800(Bytes)=1000(Gb)" Cylinders="125566"
              TrackPerCylinders="255" SectorPerTrack="63" BytesPerSector="512" ID="0" />
        </DiskInfo>
      - <PartitionInfo type="partition">
          <Partition DriveID="1" PartitionID="C" PartitionType="NTFS" Bootable="YES" Length="16056240"
              PartitionNumber="" PhysicalNumber="" StartingOffset="80" />
        </PartitionInfo>
      </Storage>
    </ForensicArtifact>
</ForensicArtifacts>
```


Fig. 32.2 An XML document for forensic artifacts in network video recorder

32.6 Conclusion

The forensic artifacts are important to determine forensic soundness. When an investigator presents video evidence, these information will be important to prove integrity of video evidence. Collecting these forensic artifacts of each network surveillance system could be developed evidence management server, we will develop for future works, for forensic soundness. Moreover, when video evidence is generated, these artifacts and the video evidence could be kept in a portable digital evidence container [14, 16]. These are generated from transmission, storage, retrieval and presentation with admissibility of evidence by ensuring the integrity and reliability. It will contribute to produce reliable retrieval of video evidence in the courts and help to reduce unnecessary disputes in related industries. These research further expand and redefine XML documents using digital forensics XML (DFXML) or digital evidence markup language (DEML).

Acknowledgments Foundation item: This work was supported by the IT R&D program (10043959, Development of EAL 4 level military fusion security solution for protecting against unauthorized accesses and ensuring a trusted execution environment in mobile devices) of KEIT/MOTIE/MSIP (Ministry of Science, ICT and Future Planning), Korea.

References

1. Chung RHY, Chin FYL, Wong KYK, Chow KP, Luo T, Fung HSK (2005) Efficient block-based motion segmentation method using motion vector consistency. In: Proceedings on IAPR conference on machine vision applications, pp 550–553
2. Kalman RE (1960) A new approach to linear filtering and prediction problems. Trans ASME J Basic Eng 82:35–45
3. Hwang T, Cho S, Park J, Choi K (2006) Object tracking for a video sequence from a moving vehicle: a multi-modal approach. ETRI J 28(3):367–370
4. Perrott AJ, Lindsay AT, Parkes AP (2002) Real-time multimedia tagging and content-based retrieval for CCTV surveillance systems. In: Proceeding on SPIE, vol 4862
5. Brown LM (2008) Color retrieval for video surveillance. In: IEEE international conference on AVSS 2008, pp 283–290
6. Tian Y, Hampapur A, Brow L, Feris R, Lu M, Senior A (2009) Event detection query and retrieval for video surveillance. In: Artificial intelligence for maximizing content based image retrieval
7. Montcalm T, Boufama B (2010) Object inter-camera tracking with non-overlapping views: a new dynamic approach. In: Proceedings of the 2010 Canadian conference on computer and robot vision, pp 354–361
8. Park S, Lim K-S, Han JW (2012) Videos analytic retrieval system for CCTV surveillance. In: LNEE future information technology, application, and service, FutureTech 2012, vol 2
9. Yuk JS-C, Wong K-Y, Chung RH-Y, Chow KP, Chin FY-L, Tsang KS-HT (2007) Object-based surveillance video retrieval system with real-time indexing methodology. In: International conference on image analysis and recognition (ICIAR), pp 626–637
10. Lim KS, Choi YS, Kim JS, Lee CH, Lee SJ (2009) CFES: comprehensive framework for forensic analysis of embedded systems. J Int Technol 10(5):523–536
11. Lim KS, Savold A, Lee CH, Lee SJ (2012) On-the-spot digital investigation by means of LDFS: live data forensic system. Math Comput Model 55:223–240
12. Lim K-S, Park S, Han JW (2012) EVM: a new methodology for evidential video management in digital CCTV systems. In: LNEE future information technology, application, and service, FutureTech 2012, vol 2
13. Lee SH, Savoldi A, Lima KS, Park JH, Lee SJ (2010) A proposal for automating investigations in live forensics. Comput Stand Interfaces 32:246–255
14. Lim KS, Lee CH, Park JH, Lee SJ (2012) Test-driven forensic analysis of satellite automotive navigation systems. J Intell Manuf 25:689–690
15. Lipton AJ, Clark JI, Brewer P, Venetianer PL, Chosak AJ (2004) Object video forensics: activity-based video indexing and retrieval for physical security applications. In: IEEE IDSE '04, Feb 2004
16. Lim KS, Lee CH (2013) A framework for unified digital evidence management in security convergence. Electr Commer Res 13(3):371–398. http://link.springer.com/article/10.1007/s10660-013-9119-y

Chapter 33
Routing Protocol for Hierarchical Clustering Wireless Sensor Networks

Alghanmi Ali Omar, ChunGun Yu and ChongGun Kim

Abstract Energy efficient and lifetime aware wireless sensor network design is still a challenging issue in sensor network research community. Cluster based wireless sensor network is a proven approach of power aware routing. A new the challenge of cluster based approach is handling dynamic clusters, selection of cluster heads, balancing of energy consumptions of member nodes of a cluster. In this paper, we propose an energy efficient routing method for dynamic hierarchical clustering architecture WSNs with additional intra-cluster management additional subordinate cluster head (SCH) are selected in a cluster for working as data relay nodes. In our proposal, SCH nodes are selected using multilevel cluster algorithms for energy efficiency and load balancing. The proposed routing approach combines the facilities of proactive and reactive sensing, routing and data transmission and thus suitable for both time-sensitive and time-insensitive applications.

Keywords Hierarchical clustering · Routing protocol · Wireless sensor network · Energy-efficient · Lifetime-aware · Dominating set

33.1 Introduction

Wireless sensor network (WSN) is an emerging communication technology for environmental monitoring and target tracking. A wireless sensor network is one of the most significant research areas throughout the last two decades. As the

A.A. Omar · C. Yu · C. Kim (✉)
Department of Computer Engineering, Yeungnam University,
Gyeongsang, South Korea
e-mail: cgkim@yu.ac.kr

A.A. Omar
e-mail: alghanmia88@gmail.com

C. Yu
e-mail: tonko96@gmail.com

J.J. (Jong Hyuk) Park et al. (eds.), *Ubiquitous Computing Application and Wireless Sensor*, Lecture Notes in Electrical Engineering 331,
DOI 10.1007/978-94-017-9618-7_33

application of wireless sensors increases day by day, form battlefield to healthcare and from underwater to space shuttle the research also shifted in various dimensions.

A wireless sensor networks consists a number of sensor nodes, those are wirelessly communicated to each other and cooperatively pass data towards the base station to accomplish their dispensed responsibilities. The sensor nodes of WSN is small in size, and it consists of tiny dimension of battery for power supply, small memory chip for data storage and routing table storage, and radio interface to send and receive signals. As the sensor nodes have limited battery power, and in many circumstances it is not feasible or possible to recharge the batteries of sensor, like in underwater sensors, battle field sensors, natural disaster prevention and monitoring sensors and implantable bio-sensors. So, energy efficient communication methods are indispensable for WSNs.

Gaining the energy efficient design of sensor networks is still a research issue to enhance the lifetime of sensor networks. The nodes of any wireless sensor networks, generally work as a unit of a system to complete certain obligations. Downing of any sensor nodes from the network creates data deficiency, and as a result the whole sensor network produces erroneous results, incorrect and imperfect vision of environment and network becomes paralyzed. Therefore, the purpose of the research is to propose cluster based power efficient routing protocol, which ensures balanced consumption of energy among the sensor nodes of the WSNs.

In this paper, we propose an additional cluster management representatives namely subordinate cluster head (SCH) for effective energy management of a cluster. We select the SCH nodes by considering residual energy and degree of nodes of any sensor node. Unlike relay node of a cluster, the SCH creates the opportunity of reactive and proactive routing of any clusters while maintaining data relays.

33.2 Previous Studies

Energy efficiency is not only an issue of wireless sensor networks, it is also a challenging issues in all forms of networks to meet the green communication requirements. Efficient routing ensures the efficient and energy-aware communication in wireless sensor networks. Routing protocols of wireless sensor networks have been studied in Ref. [1] with introducing some challenges and future directions. Partial differential equation based geographical routing is proposed by the authors of Ref. [2]. Their model is dependent on a central node, which collects the position information, residual energy information and then determines the routing path based on their proposed algorithm. The proposal is based on centralized control unit, which is not suitable for the WSNs, where there is no central node. A cluster based and threshold sensitive routing protocol is presented in Ref. [3], where they consider power availability, nodes position, and reachability factors to determine the routing path by using cluster head. Though tis proposal achieved energy efficiency but the proposal doesn't concentrate on networking life time.

A hybrid routing protocol for WSNs is presented in Ref. [4], which allows a comprehensive information retrieval of environmental analysis and facilitate user to query of past, present and future data. This is also an application specific and cluster based routing protocol, which is focused on efficient path finding by maintaining energy-efficiency but not concerning about network life time. The greedy perimeter stateless routing approach for wireless networks is proposed by the authors' of Ref. [5], where it considers the position of source and destination to send data packets, they also presented better results than shortest-path and ad hoc routing protocols in respect of routing protocol overhead, packet delivery rate and path length. They didn't consider energy efficiency and energy balancing issues to model their routing protocols. The security gaps, and possible attacks of wireless sensor networks routing are studied in Ref. [6], the authors also presented the countermeasures and challenges of designing routing protocol with ensuring security of the data packets travelling through huge nodes of WSNs.

Power efficient topologies for sensor networks are presented in Ref. [7], where the authors proposed directional source aware routing protocol (DSAP) and deploy it in different 2D and 3D static network topologies to study power efficient network topology. Though the presented DSAP minimizes the energy consumption of the nodes of considered networks, but DSAP could not ensure energy balancing among the nodes of WSNs. The Low-Energy Adaptive Clustering Hierarchy (LEACH) is proposed in Ref. [8], a clustering-based protocol that utilizes randomized rotation of local cluster base stations (cluster-heads) to evenly distribute the energy load among the sensors in the network. It is proactive routing protocol, where the nodes in the network periodically switch for transmitting data by following predefined schedule. Thus it is less energy efficient then TEEN and APTEEN protocol.

The recent trends of wireless sensor networks are presented in [9]. In this paper, the node around the sink node of wireless sensor network depletes their batteries quicker than the other nodes of the network. Thus the mobility of the sink node is a solution for network lifetime extension of the sensor nodes. A Q-learning based intelligent routing protocol is proposed in [10], where on the autonomic exploitation and exploration of routes are focused in wireless body sensor networks ensuring energy-efficiency, minimum hop count and lower packet dropping rate. Another intelligent algorithm for wireless sensor network is proposed by the [11], where they proposed clustering mechanism based on artificial bee colony. The foraging behavior of honey bee swarms are utilized here in this proposal for cluster and routing management. The intelligent and autonomic algorithms are efficient with the cost of their computational complexity, that's why the necessity of simple and efficient algorithms is demands in need for future application and usage of WSN.

33.3 Proposed Methods

The system model and routing procedure of our proposed power efficient and energy balanced routing procedure for wireless sensor network is discussed in this section. Wireless sensors are deployed in various patterns based on application requirements. In this research, we consider that the deployment of sensor nodes is followed two-dimension (2D) hierarchical network topology and general cluster heads (GCH) are selected randomly by the model is shown at Fig. 33.1 base stations (BS) as proposed in Ref. [8]. Here, in this paper, the neighbor nodes are defined on the basis of 1-hop communication model.

33.3.1 Description of the System Model

The cluster of sensor nodes is formed on the basis of nodes within the radio range including multi-hop (suppose 2-hop) communication range from the cluster head. To handle the multi-hop nodes of a cluster, the cluster head which we call here general cluster head (GCH) appoints some subordinate nodes namely subordinate cluster head (SCH) to collect data from other general sensor nodes (SN). The SCH node always remains active to receive data from general sensor nodes (SNs) and send those data immediately to general cluster head (GCH). To save energy the general sensor nodes follow TDMA schedule to be active from sleep mode and sends collected data to SCH within its time slot, if and only if the sensed observation is significantly changed with respect to the hard and soft threshold. GCH sends the collected data towards the base station (BS) directly or through first level GCH.

Fig. 33.1 xxxx

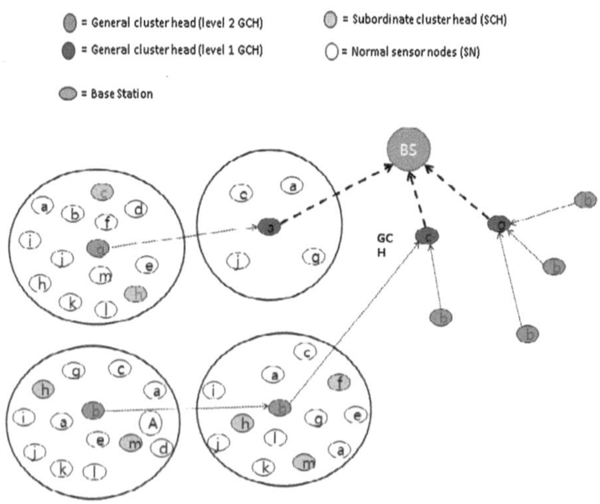

33.3.2 Proposed Routing Procedure

The dynamic cluster is formed firstly based on the general cluster head (GCH) considering the GCH as the center node of the cluster. Soon after the first phase, the GCH select some of its member nodes as the subordinate node (SCH) using dominating set approach presented in Algorithm 3.1, where dominating node is determined by considering the residual energy and degree of the node i.e. the number of eligible neighbors of the subordinate nodes. Here eligible neighbors are the set of neighbor nodes which can act as the relay or forwarder of the subject node.

Furthermore, GCH not only assigns a list of follower nodes of the SCH but also broadcast the ID of SCH with the list of follower nodes and time schedule of all of the follower nodes to send data to SCHs. GCH also broadcast the hard and soft threshold limit to all of its member nodes. Then hearing the time schedule, SCH node ID and threshold values; general sensor nodes (SNs) goes on sleep mode and actives according to the schedule. But SCH nodes remain in active mode in all the time.

Algorithm 33.1 Selection Dominating_Set_SCH_ ()

1. Initially mark GCH node as gray, mark eligible 1 hop neighbor nodes of GCH node as green and all other nodes of a cluster marks as white.
2. Determine eligibility score of each of the white nodes using following formula.

$$EligibleScore(N) = \boldsymbol{\beta} * (\mathbf{1}/N_{\mathbf{e}}) * EnergyLevel + (\mathbf{1}-\boldsymbol{\beta}) * (\mathbf{1}/N_{\mathbf{d}}) * DegreeOfNode + l * (\mathbf{1}/N_{\mathbf{s}}) * Distance$$

   where, β = priority constant; N_e = normalization factor of energy level; and N_d = normalization factor of degree of node; N_s = normalization factor of distance from the GCH node; here the '*DegreeOfNode*' = Number of 1 hop successor neighbors of that node.
3. Select a node among the white adjacent nodes (also neighbors of green nodes) which has maximum eligibility score. Color it reddish.
4. Mark all of its eligible neighbor nodes as green.
5. Repeat step 2 to 5 until all nodes become green or reddish
6. GCH defines all the reddish nodes as the SCH node. (The set of all reddish nodes is also called dominating set shows in Fig. 33.2.)

After receiving data from different sensor nodes, the SCH sends data to GCH along with the self-observed data. So, the nodes, which follow the schedule to send data from multi-hop communication range, functions as like as the proactive nodes.

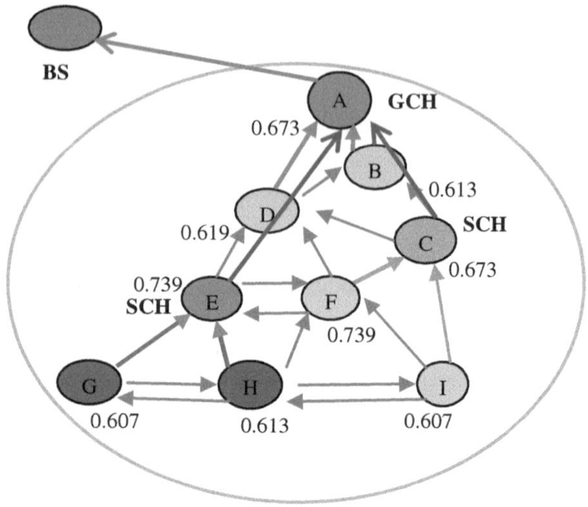

Fig. 33.2 Dominating set based SCH selection of a cluster

Algorithm 33.2 Hierarchical_Clustered_Routing ()

1. Base Station (BS) selects general cluster heads (GCH) randomly to form clusters.
2. GCH Broadcast membership advertisement to the sensor network.
3. Sensor nodes SN transmit membership request to GCH node
4. GCH select SCH nodes using Dominating_Set_SCH_Selection() algorithm, and broad cast the ID of SCH node to its member SNs. It also broadcast hard threshold and soft threshold for the current round of data transmission
5. SN nodes senses environment and transmit data to SCH by following TDMA protocol and following threshold policy
6. SCH nodes senses environment continuously and transmit data to GCH
7. GCH compress data and send to base station (BS)
8. New round begins from step 1 again after base stations redistribution of new set of GCH nodes

The SCH nodes also act as the proactive node because they also send data to the GCH continuously, and remain active all the time. Receiving data from the environment, each of the sensor nodes compare the data to its hard threshold value, if the observation exceeds the hard threshold it sends the inspected observation to SCHs. Then the sensor node begin to monitor the environment constantly and inspect its observations, if it finds the significant change in its inspection with respect to the soft threshold, then it sends the observation results again to the SCHs. This is the reactive nature of the proposed routing procedure. If the sensed data is below the hard threshold value then sensor node will not send anything to SCH in its time slot. But, if the number nothing to send slot time exceeds the time threshold/slot threshold value then the sensor node only sends the beacon to inform that the node is still alive but nothing to send at this moment.

Finally, regardless of the observed values the GCH sends the observed data towards the BS directly or through 1st level GCH nodes. The cluster formation and routing procedure is presented in Algorithm 3.1. The combined proactive and reactive nature of sensor nodes can save huge energy of tiny sensor nodes. The SCH nodes ensure continues monitoring of the environment whereas the dominating set based efficient SCH selection balances the energy consumption of sensor nodes of the cluster, managing dynamicity of clusters e.g. topology change issues.

33.4 Simulation Results

To evaluate the performance of the proposed multilevel cluster based energy efficient routing protocol (MERP), the MATLAB R2010a simulation tools are used to analyze the energy efficiency and lifetime awareness. For simulation purpose, we simulate our proposed algorithm firstly without considering the distance value i.e. considers only the energy level and degree of node to determine the eligibility score of node to be an SCH. Secondly; we consider the distance value with energy level and degree of node to determine eligibility score of a node to be an SCH. We compare our proposed algorithm with the benchmarked LEACH routing protocol. The simulation scenario is presented in Table 33.1.

The simulation scenario is presented in Fig. 33.3, where the green cross mark represents the base station (BS), the blue circles are represented as GCH for a round and the plus signed filled circles are represented as the SCH node and the other red circles are represented as the normal sensor nodes (SNs).

The performance of proposed MERP is studied using two essential performance metric of wireless sensor networks i.e. energy efficiency and network lifetime [12]. Figures 33.4 and 33.5 show the performance of the MERP algorithm.

Table 33.1 Simulation scenario for performance study of MERP algorithm

Simulation parameters	Symbols	Values
Topology	2D	Random number of neighbors within 1 hop
Number of nodes	N	100
Simulation area	WxH	100×100 m^2
Packet size	b	512 bits
Total number of rounds (or time slots)	R	4,500
Transmitter circuity energy	ETxcircuit	50 nJ/bit
Transmitter amplification energy	ETx_ampl[fier	100 pJ/bit/m2
Receiver circuitry energy	ERxtircuit	50 nJ/bit
Initial energy of each node	E_0	0.5 J

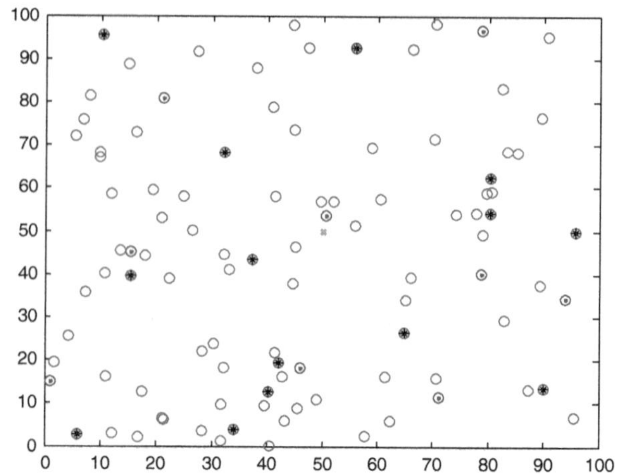

Fig. 33.3 Simulation scenario, considering 100 node random networks

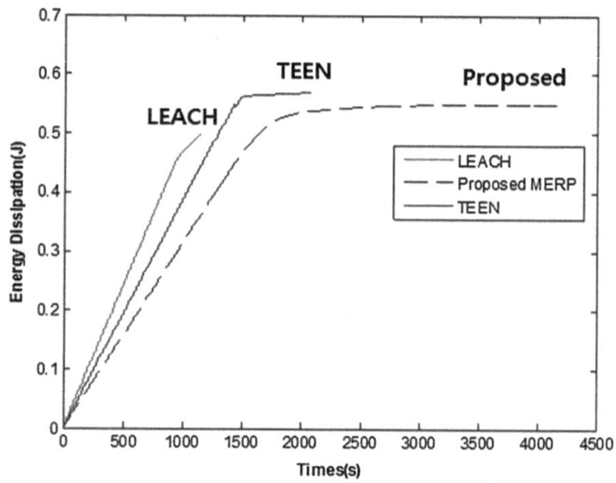

Fig. 33.4 Comparison of average energy dissipation

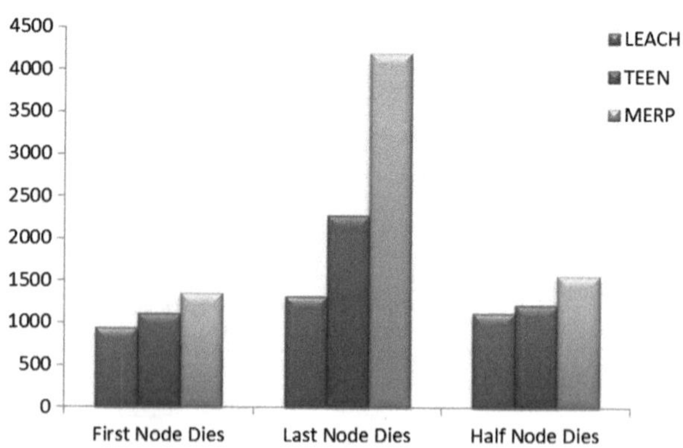

Fig. 33.5 Simulation result regarding lifetime awareness of the sample network with the base station located at (50, 50)

Figure 33.4 shows the average energy dissipation of LEACH, TEEN and the proposed MERP protocols. The proposed protocol always dissipates less energy than the LEACH and TEEN protocol. The logic behind the less energy dissipation is the combination of reactive and proactive sensor nodes, usages of soft, hard and time threshold, letting some nodes in hibernation mode when some other neighboring sensor nodes remain active. Some other factor e.g. the integration of subordinate cluster head and efficient dominating set based SCH selection also plays an important role to reduce the energy dissipation of sensor networks. Whereas the proactive nature of LEACH protocol pushes the sensor nodes to consumes more energy. The TEEN protocol is reactive with soft and hard threshold and consumes less energy than LEACH, but still all nodes should remain active in all the time thus it also consumes more energy than MERP.

In Fig. 33.5, we compare the network lifetime of our proposed MERP algorithm with the popular LEACH and TEEN algorithm. The figure shows that our network remains alive up to 4,189th rounds whereas the LEACH and TEEN remains active up to 1,308th and 2,272nd rounds respectively. In case of LEACH and TEEN first node dies at 935th and 1,107th round respectively, on the other hand, in MERP first node dies at 1,355th round. Though half of the total node dies within 1,112th, 1,215th and 1,567th round following LEACH, TEEN and MERP algorithm respectively, but MERP remains active longer time than LEACH and TEEN.

The performance of proposed MERP is studied using another essential performance metric of wireless sensor networks i.e. energy balancing. We also compare the energy balancing capability of our algorithm with conventional LEACH and TEEN routing protocol. Figure 33.6 shows the 500th round residual energies using LEACH algorithm where 100 nodes were deployed in 100 × 100 m of area. The total energy consumption is 43.172 J and residual energy is balancing between 571 and 824 mJ. In comparison to TEEN and MERP it consumes higher energy. The energy differences among the nodes of the network is (824−571) = 253 mJ. So, energy dissipations are almost balanced in LEACH protocol.

Figure 33.7 shows the 500th round residual energies using TEEN algorithm where 100 nodes were deployed in 100 × 100 m of area. The total energy consumption is 35.91 J and residual energy is balancing between 416 and 926 mJ.

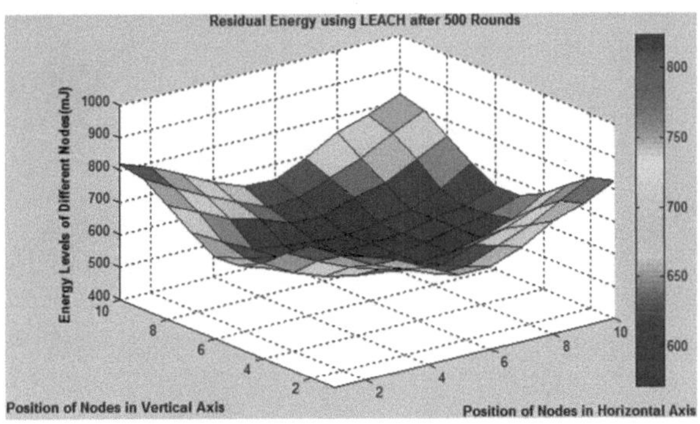

Fig. 33.6 Residual energy of different nodes using LEACH at 500th round

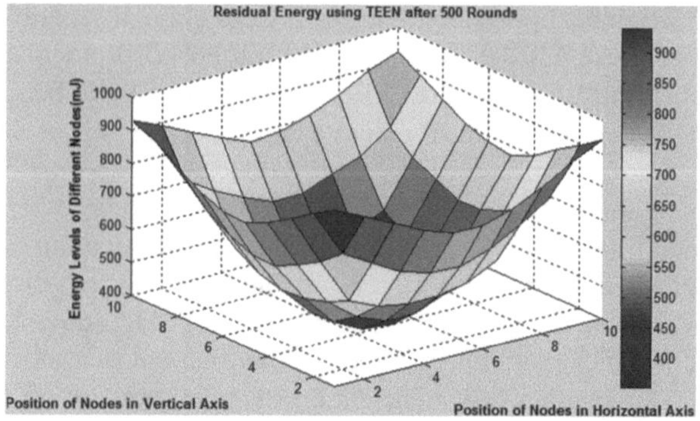

Fig. 33.7 Residual Energy of different nodes using TEEN at 500th round

In comparison to LEACH it consumes lower energy but in comparison to proposed MERP it consumes higher energy. The energy differences among the nodes of the network is (926−416) = 510 mJ. So, energy dissipations are imbalanced in TEEN routing protocol.

Figure 33.8 shows the 500th round residual energies using MERP algorithm where 100 nodes were deployed in 100 × 100 m of area. The total energy consumption is 29.358 J and residual energy is balancing between 504 and 984 mJ. In comparison to both LEACH and TEEN, the proposed MERP consumes much lesser energy. The energy differences among the nodes of the network is (984−504) = 480 mJ. So, energy dissipations are moderately balanced in proposed MERP.

According to the above simulation results, we can see that MERP consumes less energy while balancing the energy dissipation of the each nodes of the network, which plays the key role of the extended lifetime of the proposed MERP routing protocol.

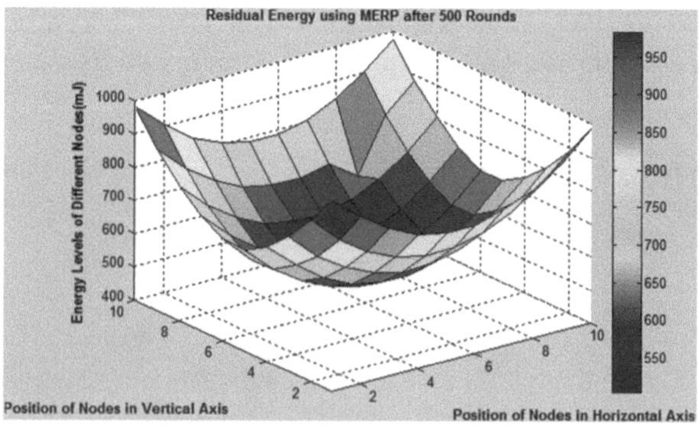

Fig. 33.8 Residual energy of different nodes using MERP at 500th round

33.5 Conclusions

The dominating set based hierarchical clusters of WSN is expected to outperform in energy efficiency and network longevity. Our proposed method hierarchical cluster based routing protocol with the combined proactive and reactive functional mode, we expect less power consumption of the WSNs using the proposed routing protocol then the existing legendary routing protocol for WSNs. Furthermore, the proposal has effective method to select the subordinates of the cluster head for dynamic cluster management. The SCH selection policy ensures the fairness in workload distribution of any clusters that also certifies balanced energy consumption of the nodes of the cluster as well as sensor networks. Thus, we expect longer lifetime of the network than using the legacy routing protocol.

Acknowledgments This work has been funded by the BK21 + program of the National Research Foundation (NRF) of Korea.

References

1. Al-Karaki Jamal N, Kamal Ahmed E (2004) Routing techniques in wireless sensor networks: a survey. Wireless Commun IEEE 11(6):6–28
2. Kalantari M, Mark S (2004) Energy efficient routing in wireless sensor networks. In: Proceedings of conference on information sciences and systems
3. Manjeshwar A, Agarwal DP (2001) TEEN: a routing protocol for enhanced efficiency in wireless sensor networks. In: IPDPS, vol. 1
4. Manjeshwar A, and Agarwal DP (2002) APTEEN: A hybrid protocol for efficient routing and comprehensive information retrieval in wireless sensor networks. In: IPDPS, vol. 2
5. Karp B, Kung H-T (2000) GPSR: Greedy perimeter stateless routing for wireless networks. In: Proceedings of the 6th annual international conference on Mobile computing and networking. ACM
6. Karlof C, Wagner D (2003) Secure routing in wireless sensor networks: attacks and counter measures. Ad Hoc Netw 1(2):293–315
7. Salhieh A et al (2001) Power efficient topologies for wireless sensor networks. In: Parallel processing, international conference on, IEEE
8. Heinzelman RW, Chandrakasan A, Balakrishnan H (2000) Energy-efficient communication protocol for wireless microsensor networks (LEACH) system sciences. In: Proceedings of the 33rd annual Hawaii international conference on, IEEE
9. Tunca C, Isik S, Donmez M, Ersoy C (2014) Distributed mobile sink routing for wireless sensor networks: a survey. Commun Surv Tutorials: 1–21
10. Rabiul Md Golam Alam, Seung Il Moon, Haw R, Hong CS (2013) A Q-learn based routing protoc body area netw 40(5):250–257
11. Karaboga D, Selcuk O, Celal O (2012) Cluster based wireless sensor network routing using artificial bee colony algorithm. Wireless Netw 18(7):847–860
12. Pantazis NA, Nikolidakis SA, Vergados DD (2013) Energy-efficient routing protocols in wireless sensor networks: a survey. Commun Surv & Tutorials, IEEE 15(2):551–591

Chapter 34
The Design of a New Virtualization-Based Server Cluster System Targeting for Ubiquitous IT Systems

Jungmin Lim, Sihoo Song, Soojin Lee, Seokjoo Doo and Hyunsoo Yoon

Abstract To meet the demand for seamless real-time services with limited resources of ubiquitous computing environments, this paper suggests a new virtualization-based server cluster system introducing three schemes: *simplified rotation process* to reduce system overhead, *exposure time adjustment* to prevent the degradation of system performance, and *spare server insertion* to cope with heavy incoming packets. It is verified for ubiquitous computing systems which should satisfy strict operational constraints with limited resources that the proposed schemes enable the systems to achieve higher levels of service quality as well as to stand up to distributed denial of service (DDoS) attacks.

Keywords Species · Beta-diversity · Diversity · Forest

34.1 Introduction

In ubiquitous computing environments, where computing is made to appear everywhere and anywhere, real-time services with sustainability under any circumstances is considered as the most important factor. As the number of ubiquitous devices increases and computing environments broadens, however server cluster systems which should support various services promptly might be suffered from a heavy volume of requests incoming from large scale of devices. For this reason, cost-effective and resilient server systems are inevitable for maintaining real-time services under recent computing environments.

Even though there are many studies to implement those required server systems, a virtualization-based intrusion tolerance system is regarded as a good way of standing against outside attacks or internal faults [1]. First of all, a virtualization-based server

J. Lim (✉) · S. Song · S. Lee · S. Doo · H. Yoon
Department of Computer Science, KAIST, Daejeon, South Korea
e-mail: jmlim@nslab.kaist.ac.kr

J.J. (Jong Hyuk) Park et al. (eds.), *Ubiquitous Computing Application and Wireless Sensor*, Lecture Notes in Electrical Engineering 331,
DOI 10.1007/978-94-017-9618-7_34

system provides advantages for various virtual schemes including hardwares such as AMD-V [2] or ARM series [3], and virtual machine software such as Xen, VMWare, or VirtualBox. Since they support to implement various virtualized server systems with different operating systems, applications, and hardware resources, a virtualization-based intrusion tolerance system is for certain very cost-effective and flexible to implement appropriate server systems needed.

Moreover, a virtualization-based system supports sustainable services even under bad situations. For instance, Self-Cleansing Intrusion Tolerance (SCIT), one of the typical virtualization-based intrusion tolerance systems, gives the system the capabilities of dealing with a variety of unfavorable conditions such as faults by internal errors, system compromise by outside hackers and security failures by distributed denial of service (DDoS) attacks. The effectiveness of SCIT in security and performance has been verified in many parts such as Web service systems [4], DNS systems [5], and firewall systems [6].

In addition to SCIT, several adaptive schemes such as adaptive cluster transformation (ACT) [7] and a historical data-based ITS [8] also showed good performance, especially with proving the possibility of complementing SCIT's weakness of lacking any ability to react to the circumstances while supplying services.

However, more advanced efforts are needed for supporting real-time services and saving server resources under ubiquitous computing environments. In this sense, this paper suggests a new concept of a server cluster system based on virtualization, which simplifies the process of VM rotation, adjusts the exposure time of each VM, and inserts several spare servers with pristine images. All the effectiveness on performance and security is verified with a CSIM 20 simulator.

The rest of this paper consists of as follows: Sect. 34.2 explains related works including virtualization-based ITSs and adpative ITSs. The detailed analyses about these two previous studies are also included. Section 34.3 presents three main proposed schemes of *simplified rotation process*, *exposure time adjustment,* and *spare server insertion for crisis*. Section 34.4 shows the experimental results with the comparison of performance. Finally Sect. 34.5 makes the conclusion remarks.

34.2 Related Works

34.2.1 Virtualization-Based SCIT

A virtualization-based server cluster system is typically represented by Self-Cleansing Intrusion Tolerance (SCIT) [9]. The main principle of SCIT is to remove any opportunity for attackers to detect information about the system's vulnerabilities by doing fast state rotation of each VM: active ⇒ grace ⇒ cleansing ⇒ wait. In SCIT, all servers supply services to online clients for a short period of time, about 2–3 min, and go back to a pristine state by reloading clean image stored in advance. Several VMs constitute a server cluster system by the control of a central controller. The main focus of this scheme is to rejuvenate each VM in order to maintain a pristine state and support seamless request-response services.

34.2.2 *Adaptive Intrusion Tolerance Systems*

In order to allow a system to adjust the performance, an adaptive intrusion tolerance system, named as an advanced adaptive cluster transformation (AACT), is suggested and its efficiency is verified by simulation in [7, 10]. The main considerable parameter of this scheme is response delay, which means the time between the request and the response packets. If response delay exceeds the designated threshold value, the current cluster size would be expanded for maintaining the performance.

On the other hand, if the system maintains the reasonable performance for a certain period of time, the cluster size would be reduced for saving server resources. Furthermore, when massive incoming packets are detected, the system resists against it through the whole cluster substitution which replaces all the VMs with the pristine state of VMs. Figure 34.1 shows that AACT stands against a DDoS attack with the cluster substitution while both fixed size clusters having 5 and 7 VMs suffer from long response delay.

In addition to above the schemes, Ref. [8] suggested to use a historical data-based intrusion tolerance system. After a central controller secures historical data about incoming packet rate from enough learning processes, the cluster transformation can be accelerated.

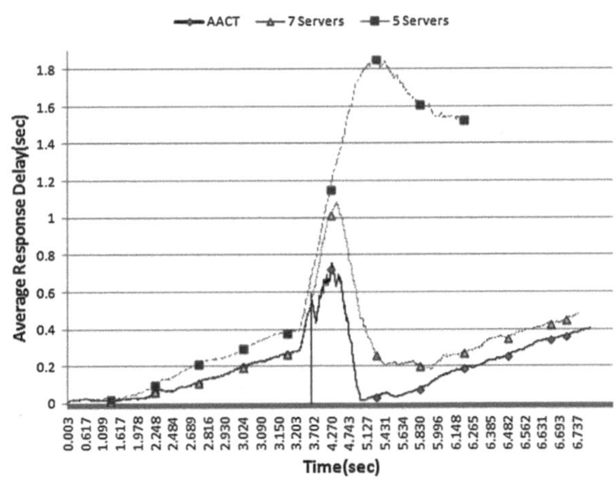

Fig. 34.1 DDoS resistance through cluster substitution

34.2.3 Analysis on Previous Studies

Even though SCIT has been verified as a typical virtualization-based server cluster system, it does not consider other important factors such as the fluctuation of incoming packet rate or outside threats like DDoS attacks. Because the cluster size cannot be varied after setting the exposure time initially, it is hard to expect that SCIT guarantees the required performance under the serious fluctuating ubiquitous computing environments.

The adaptive intrusion tolerance system suggested in [7, 8] retains the capability of maintaining performance while resisting against DDoS attacks to a SCIT server cluster system. However, both two systems are restricted in that hardware server resources should be fully enough like a Cloud server group. This means that they are not suitable for server cluster systems having limited hardware resources.

As considering collectively, in order to acquire an optimal server cluster system, several ingredients should be considered as follows:

- In addition to SCIT, adaptive exposure time should be considered for conserving server resources and maintaining a certain level of performance.
- Limited server resources should be considered. In ubiquitous systems, there is no guarantee that sufficient server clusters for real-time services can be supported within the range of what it needs.
- Suggested server cluster systems should resist against extreme environments such as drastic incoming request packets.

34.3 Proposed Schemes

In this chapter, three main schemes are presented: simplified rotation process, exposure time adjustment, and spare server insertion for crisis. Figure 34.2 shows the overall architecture of our approach. Each HW server dominates several VMs

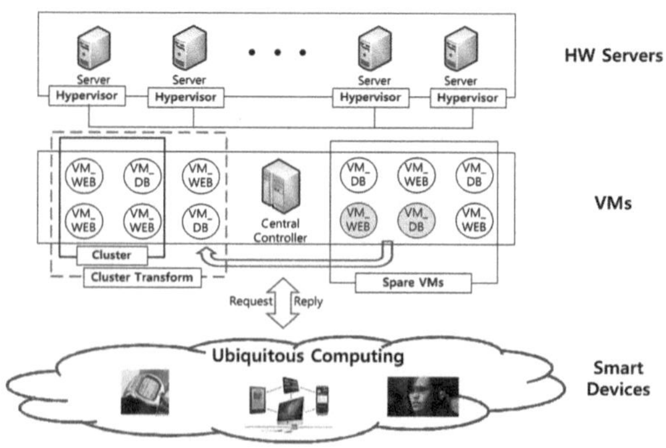

Fig. 34.2 Overall architecture

Table 34.1 Main variables and explanations

Variable	Explanation
Tot_srv	The number of total VMs
CL_srv	The number of VMs included in current cluster
SP_srv	The number of VMs waiting for active state
Cris_srv	The number of spare VMs for crisis
Exp_du	Exposure time of VM
Cl_du	Cleansing time of VM
rm(i)	Remained request volume of ith VM
σ	Multiplying factor for changing exposure time, designated by the operator
rm_{upper}	Upper bound of request volume
rm_{lower}	Lower bound of request volume
rm_{max}	Maximum number of requests in queue
rm_{ext}	Threshold value for spare server
CL_{upper}	Upper bound of cluster size
CL_{lower}	Lower bound of cluster size

controlled by a hypervisor installed on that server. All the VMs supply various services to a variety of devices belonging to ubiquitous environments. A cluster is a server group which supplies online services to smart devices. Each VM rotates its own state periodically as in the case of SCIT, and all the operations of VMs are controlled by the central controller.

Table 34.1 explains the variables used throughout this paper.

34.3.1 Simplified Rotation Process

In order to maximally utilize the restricted resources, the rotation pattern of VMs in SCIT can be simplified from 4 states to 2 states. If the VM finalizing cleansing state does not need to wait for active state, then the state of the VM is changed into active state promptly. It means that there is no spare-state of servers in any case. In addition, if the grace period, the time for processing remained requests without accepting any more requests, is as short as negligible [11], then the grace period can be ignored as shown in Table 34.2.

Periodic cleansing in SCIT often generates dispensable overhead influencing to the utilization of server resources. We named this ‘cleansing overhead’. That is, cleansing overhead happens while each VM resides in cleansing state, as much as

$$\text{Cleansing Overhead} = \frac{\text{Cl\_du}}{\text{Cl\_du} + \text{Exp\_du}} \times \text{Whole resource}$$

Table 34.2 Time period for each state

State in SCIT	Residence time
WAIT	0
ACTIVE	Exp_du
GRACE	0 (negligible)
CLEANSING	Cl_du

It means that if exposure time is reduced to improve security, cleansing overhead increases. This loss is generated due to the reduction of the cluster size caused by the increase of cleansing time.

To validate the effect of simplified rotation process proposed, let us assume that each VM in SCIT system starts Web services at random time, and conduct statistical analysis. At a specific time t, the cluster size would follow binomial distribution like

$$\Pr(X = x) = f(x; N, p),$$

where

$$p = \frac{\text{Cl\_du}}{\text{Cl\_du} + \text{Exp\_du}}$$

In case cleansing time Cl_du is 60 s, the probability density function of cluster size varies depending on the amount of exposure time, as shown in Fig. 34.3. As exposure time increases, the probability that cluster size is larger is high. Because simplified rotation process makes VMs in the cluster maintains online states for a longer time, larger cluster supporting high availability can be obtained from simplified rotation process.

34.3.2 Exposure Time Adjustment and Spare Server Insertion

Section 34.3.1 tells us that exposure time should be longer to make cluster size larger for high availability of a system. However, security of a system can be improved by shortening the exposure time. Between both cases, it is verified again that trade-off between availability and security exists [12].

In ubiquitous environments, it should be noted that the reduction of a cluster size due to short exposure time might cause the temporal decrease of availability and performance.

As an evaluation factor for assessing the availability of the current cluster, the maximum unprocessed requests, rm(i), which is mostly remained in a queue of ith VM among the current cluster, is used as

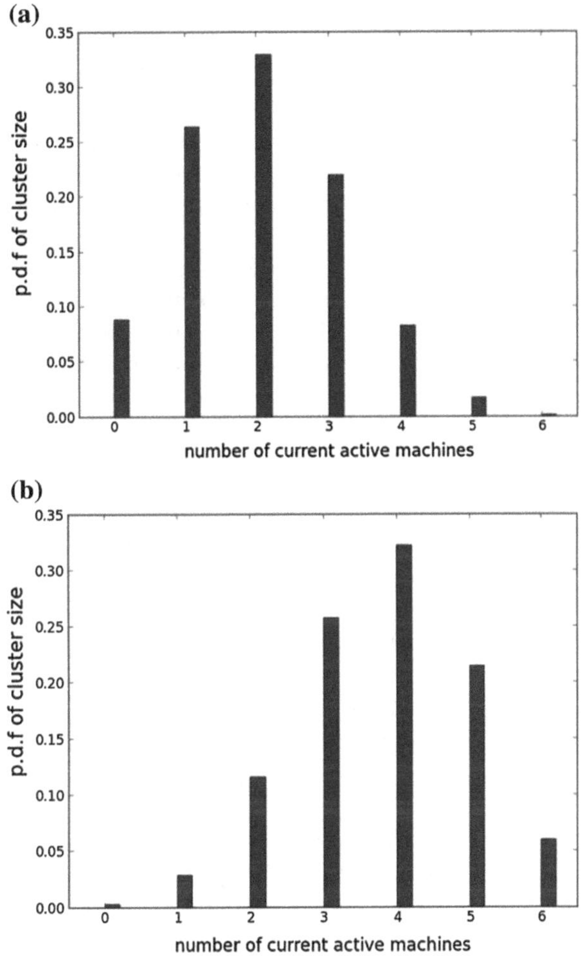

Fig. 34.3 The probability density function comparison depending on exposure time for $\mathrm{Cl_{du}} = 60\mathrm{s}$. **a** Probability density function for $\mathrm{Exp_{du}} = 30$. **b** Probability density function for $\mathrm{Exp_{du}} = 60$

$$\mathrm{rm} = \max(\mathrm{rm}(1), \mathrm{rm}(2), \cdots, \mathrm{rm}(\mathrm{i}) \cdots \mathrm{rm}(\mathrm{CL\_srv}))$$

The adjustment of exposure time is done by using

$$\text{if rm } \geq rm_{upper}, \mathrm{Exp_{du}} = \mathrm{Exp_{du}} \times \sigma$$
$$\text{if rm } \leq rm_{lower}, \mathrm{Exp_{du}} = \frac{\mathrm{Exp_{du}}}{\sigma}$$

If rm exceeds the designated upper bound of request volume, exposure time is increased by a factor of σ. order to expand the current cluster. On the contrary, exposure time is decreased by the inverse of σ. rm is smaller than the lower bound of request volume.

The lower bound of a cluster size, CL_{lower}, is set to guarantee the minimum number of roles like Web and DNS. The upper bound, CL_{upper}, is set to Tot_{srv} − Cris_srv. The detailed explanation about Cris_srv is followed in the next section.

When massive packets such as a DDoS or a buffer-overflow attack income, the current cluster might fail to provide required services because of limited resources. In order to overcome this situation, spare servers in live-spare state can be used. These servers are separated from normal state rotation and inserts into the cluster when the current situation is extremely bad. This process occurs when more than 80 % of available queue is full with requests packets. For example, if the maximum number of packets available in the current queue is 50, spare servers are inserted into the cluster when 40 packets stack up to the current queue. As soon as the situation is over, these servers go back to live-spare state after finishing cleansing. Even if this process causes some burdens to the system due to applying more spare servers, it gives the system the capability to resist against bad situations.

34.4 Simulation Results

34.4.1 Experimental Environments

For the comparison of SCIT which has designated rotation pattern of each VM and maintains the fixed size of clusters, the CSIM 20 simulator [13] is used. Table 34.3 shows the experimental conditions including initial values of main variables.

As shown in Table 34.3, the total number of servers is set to 6. Maximum request volume available in the queue is 50 so that spare servers are activated if request volume exceeds 40. Every packet is set to forward to each VM in a cluster as a round-robin method. Interval time between request and processing time follows

Table 34.3 Experimental conditions

Used values
$CL_{srv} + SP_{srv} = 6$
$rm_{upper} = 10, rm_{lower} = 2$
$rm_{max} = 50$
$rm_{ext} = 40$
$\sigma = 1.1$
$30 \leq Exp\_du \leq 300$
$CL_{upper} = 5, CL_{lower} = 2$

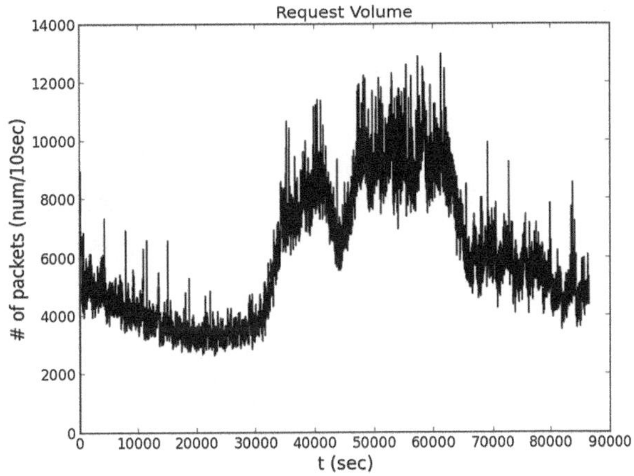

Fig. 34.4 Requests volume with time

the exponential distribution. Processing time for one packet is set to 10 ms. With the designated multiplying factor σ of 1.1, exposure time adjustment occurs when the current number of packets is beyond the upper and lower bounds. The maximum and lower values of exposure time are 30 s and 6 min, respectively. The cluster size supplying services can be changed from 2 to 5.

The request packet graph in Fig. 34.4 is obtained from a real environment in the DNS server of Korea Advanced Institute of Science and Technology (KAIST)'s intra-network during 2013-7-26-00:00 $\sim$ 24:00 for 86,400 s. These data can be used as an appropriate input because the traffic information in the DNS is possible to be implemented with a SCIT system.

34.4.2 Effectiveness of Exposure Time Adjustment

With the request volume in Fig. 34.4 as an input to SCIT, exposure time, dropped packets, and response delay can be obtained. Figures 34.5 and 34.6 show the simulated results for the cluster size of 3 and 4, respectively.

As shown in Fig. 34.5, exposure time is maintained as a small average value of 65 s because this system uses a fixed size of the cluster. However, due to the bounded computing resources such as the queue size, big loss of packets (239,936) occurred between 40,000 and 60,000 s where packet rate is increased drastically. Also, response delay is rising up in the same period, so it can be concluded that clients feel slow responses to their service requests.

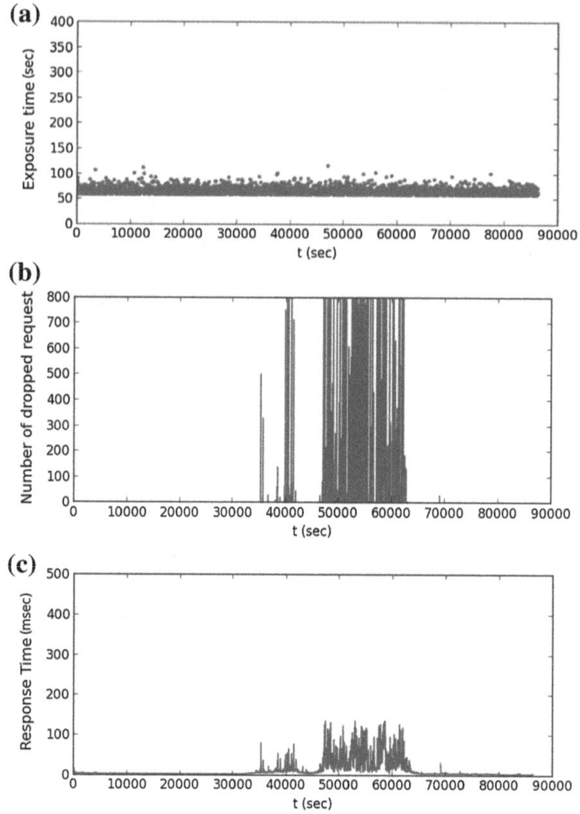

Fig. 34.5 Performance with the fixed cluster size of 3
a Exposure time variation
b Dropped packets with time
c Response delay with time

Conversely, if the cluster size is increased from 3 to 4, completely different results can be obtained as shown in Fig. 34.6. Even if dropped packets and loss of response delay cannot be found, exposure time goes almost double the time in Fig. 34.5.

For the proposed scheme, exposure time adjustment is applied to the system, and Fig. 34.7 shows the simulation results. It is interesting to see that exposure time is varied according to the variation of incoming packet rate described in Fig. 34.4. It means that the proposed exposure time adjustment scheme is applied throughout the whole period. It also showed that dropped packets are not big as 7,369 and stable response delay is maintained for the entire time.

Table 34.4 summarizes the average values of exposure time and the number of dropped packets. With the same resources having 6 VM servers, exposure time adjustment is verified as an appropriate solution for both maintaining performance and security in real-time environments.

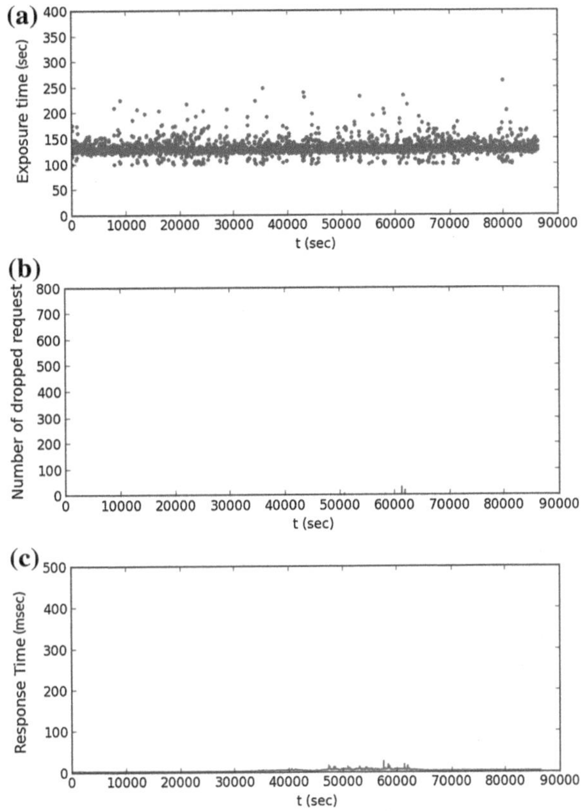

Fig. 34.6 Performance with the fixed cluster size of 4. **a** Exposure time variation. **b** Dropped packets with time. **c** Response delay with time

34.4.3 *Effectiveness of Spare Server Insertion*

To verify the effectiveness of the spare server insertion scheme, incoming packets in Fig. 34.7 are generated artificially. Remarkable point is at t of 2,000 s when the number of incoming packets is rising up drastically. One of 6 VM servers is prepared as a spare server maintaining pristine image in live-spare state (Fig. 34.8).

The Artificially-generated packets were applied to the exposure time adjustment scheme with and without spare server insertion. The results are illustrated in Figs. 34.9 and 34.10. Significant reduction of packet drop is shown in the case of having a spare server, especially at the time of around 2,000 s. Table 34.5 shows the exact reduced number of dropped packets is 6,137. This value reaches about 80 %

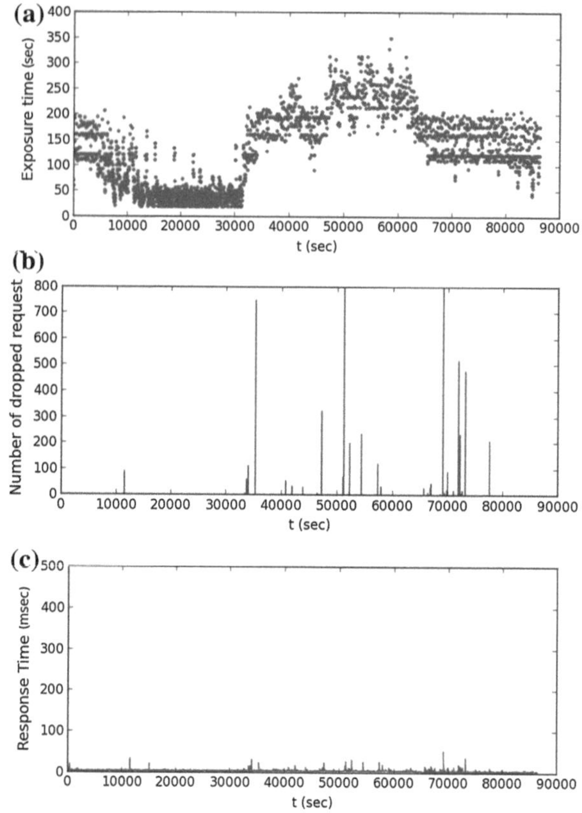

Fig. 34.7 Performance of exposure time adjustment. **a** Exposure time variation. **b** Dropped packets with time. **c** Response delay with time

Table 34.4 Performance comparison between SCIT and exposure time adjustment

Scheme		Exp_du (s)	Dropped packets
SCIT	Cluter size = 3	65	239,936
	Cluter size = 4	131	72
Exposure time adjust		110	7,369

of total dropped packets, so this proves that securing spare servers is very efficient under the seriously bad situation. However, it should be noted that exposure time is increased around 8 % as in Table 34.5 because of preparing a spare server.

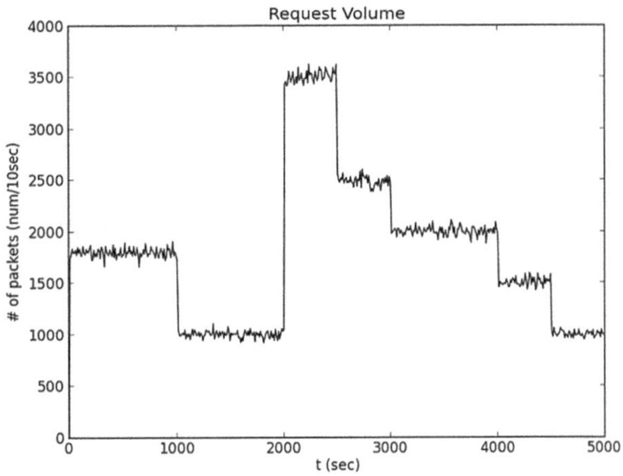

Fig. 34.8 Artificially-generated packet rate

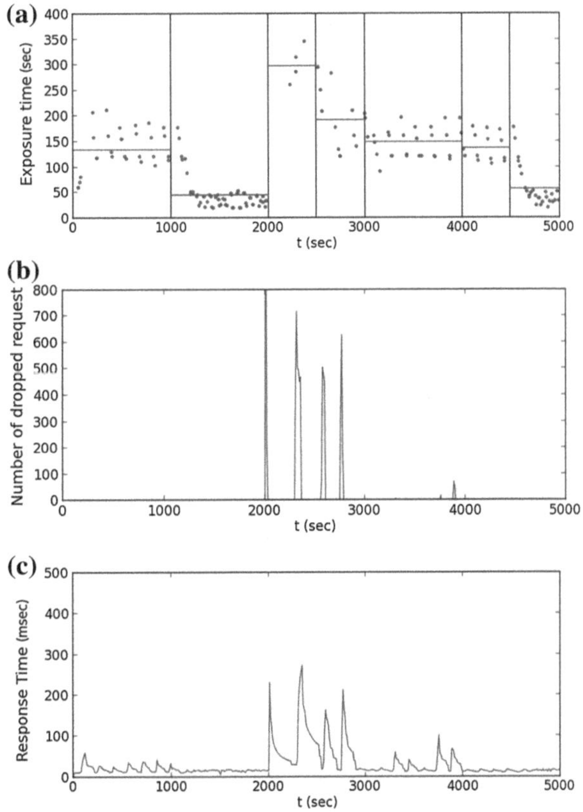

Fig. 34.9 Performance only with exposure time adjustment. **a** Exposure time variation. **b** Dropped packets with time. **c** Response delay with time

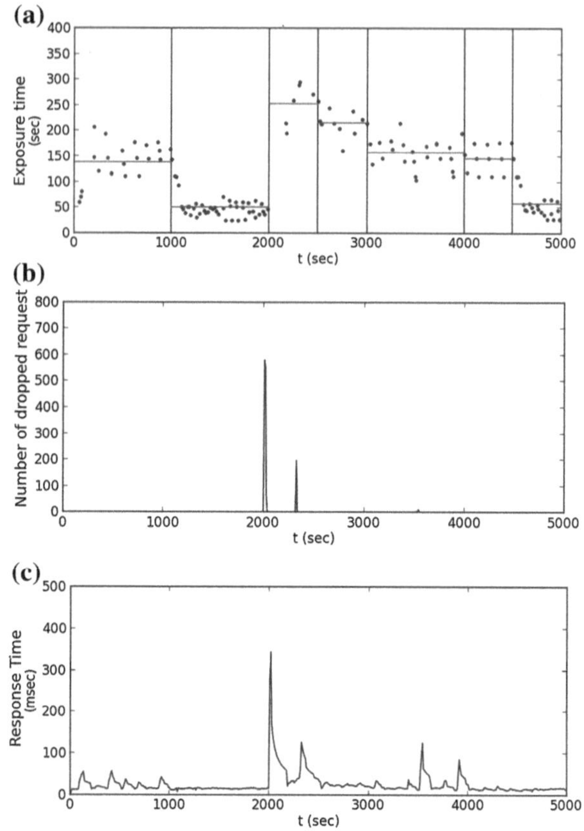

Fig. 34.10 Performance with exposure time adjustment and spare servers for crisis. **a** Exposure time variation. **b** Dropped packets with time. **c** Response delay with time

Table 34.5 Result comparison with exposure time adjustment and spare servers

Scheme	Exp_du (s)	Dropped packets
Exposure time adjust (only)	104	7,564
Exposure time adjust with spare server insertion	112	1,427

34.5 Conclusion

When limited server resources are regarded as an important factor, the proposed three schemes are surely effective to maintain crucial services and resist against extremely bad situations. System overhead due to frequent cleansing process is largely reduced through simplified rotation process. Also, exposure time adjustment is verified that it can prevent packet drop and performance degradation due to the change of environments. Under the bad situation where request packets are drastically incoming, spare servers with pristine image insertion into a cluster also helps to maintain services. All these schemes used in a virtualization-based server cluster system should be applicable to implement ubiquitous systems which have restricted system resources while requiring various real-time services.

Acknowledgments This work was supported by the IT R&D program of MKE/KEIT[10041244, SmartTV 2.0 Software Platform].This work was supported by the National Research Foundation of Korea(NRF) grant funded by the KoreaGovernment(MEST) (No.2014R1A2A2A01006957).

References

1. Smith M, Schridde C, Freisleben B (2008) Securing stateful grid servers through virtual server rotation. In: Proceedings of the 17th international symposium on high performance distributed computing (HPDC '08), pp 11–22
2. Advanced Micro Devices (2008) AMD-V™ nested paging
3. Varanasi P, Heiser G (2011) Hardware-supported virtualization on ARM. In: Proceedings of the second Asia-Pacific workshop on system article no. 11
4. Saidane A, Nicomette V, Deswarte Y (2009) The design of a generic intrusion-tolerant architecture for web servers. IEEE Trans Dependable Secure Comput 6(1):45–48
5. Huang Y, Arsenault D, Sood A (2006) Incorruptible system self-cleansing intrusion tolerance and its application to dns security. J Netw 1(5):21–30
6. Yih H, Arun S (2002) Self-cleansing systems for intrusion containment. In: Proceedings of workshop on self-healing, adaptive, and self-managed systems (SHAMAN)
7. Lim J, Kim Y, Koo D, Lee S, Doo S, Yoon H (2013) A novel adaptive cluster transformation (act)-based intrusion tolerant architecture for hybrid information technology. J Super Comput 66(2):918–935
8. Kim Y, Lim J, Doo S, Yoon H (2012) The design of adaptive intrusion tolerant system (ITS) based on historical data. In: International conference for internet technology and secured transactions, pp 662–667
9. Huang Y, Arsenault D, Sood A (2006) Closing cluster attack windows through server redundancy and rotations. In: Proceedings of the sixth international symposium on cluster computation and the grid workshops (ccgridw '06)
10. Lim J, Doo S, Yoon H (2013) The design of a robust intrusion tolerance system through advanced adaptive cluster transformation and vulnerability-based vm selection. Mil Commun Conf (MILCOM 2013), pp 1422–1428
11. Bangalore AK, Sood AK (2009) Securing web servers using self cleansing intrusion tolerance (scit). In: The Second international conference on dependability (DEPEND)
12. Nguyen Q, Sood A (2010) Realizing S-Reliability for services via recovery-driven intrusion tolerance mechanism. International conference on dependable systems and networks workshops (DSN-W)
13. Schwetman H (2001) CSIM19: a powerful tool for building system models. In: Proceedings of the 2001 winter simulation conference, pp 250–255

Chapter 35
A Conceptualisation of a Strategy to Shiftwork Scheduling Optimisation in an Emergency Department

Muhaimin Noor Azhar, Aida Bustam, Rashidi Ahmad, Rishya Manikam and Shamimi A. Halim

Abstract Shiftwork schedule is required to provide uninterrupted service for the hospital emergency department. There are many limitations with the manual shiftwork scheduling applied at the University of Malaya Medical Centre Emergency Department (UMMC-ED). We propose an alternative shiftwork scheduling strategy to solve these limitations which consists of three modes; Initialization, Sufficient and Insufficient mode. These modes detect the imbalance between workforce and workload which enable the appropriate distribution according to the demands of individual staff. We developed a tool, Trauma Hospital Online Roster (THOR) that implements the above strategy.

Keywords Optimisation strategy · Staff scheduling · Shiftwork · Emergency department · Non-cyclical scheduling

M.N. Azhar · A. Bustam · R. Ahmad · R. Manikam
Trauma and Emergency Department, University Malaya Medical Centre, Kuala Lumpur, Malaysia
e-mail: muhaimin@um.edu.my

A. Bustam
e-mail: aidabustam@um.edu.my

R. Ahmad
e-mail: rashidi@um.edu.my

R. Manikam
e-mail: rishya@um.edu.my

S.A. Halim (✉)
Faculty of Computer and Mathematical Sciences, Universiti Teknologi MARA, Shah Alam, Malaysia
e-mail: shamimi@tmsk.uitm.edu.my

J.J. (Jong Hyuk) Park et al. (eds.), *Ubiquitous Computing Application and Wireless Sensor*, Lecture Notes in Electrical Engineering 331,
DOI 10.1007/978-94-017-9618-7_35

35.1 Introduction

Hospital Emergency Department (ED) is a primary care unit that covers a broad spectrum of illnesses and injuries which require an immediate treatment and the urgency of handing this situation gives a challenge to the physician. For that reason, the appropriate shiftwork scheduling that provides equal distribution of work and meeting staff satisfaction is essential in ED since it influences the workforce physiologically, psychologically and socially [1]. Besides optimises the workforce, it has to be transparently constructed and flexible.

In this paper, we conceptualise a strategy to optimise the shiftwork scheduling in order to produce a fair, ergonomic, flexible to the staff, reducing the risk of human errors, increases work quality and productivity while meeting patient demand. We present our department, University of Malaya Medical Centre Emergency Department (UMMC-ED) in Malaysia as a basis of a shiftwork scheduling study by extracting their scheduling constraints. Based on this strategy, we developed and implemented a tool namely Trauma Hospital Online Roster (THOR) in UMMC-ED.

This paper is organized as follows. In Sect. 35.2, we briefly discussed the established scheduling strategies. Section 35.3 explains the constraints of UMMC-ED in shiftwork scheduling and proposes our strategy in Sect. 35.4. Section 35.5 briefly described the platform of THOR implementation and we discuss about the future work in Sect. 35.6.

35.2 Related Works

Generally there are two methods of staff scheduling: Cyclical and Non-Cyclical [2]. Cyclical scheduling method is where the same schedule is repeated. This method is easier to construct and particularly applies to tactical scheduling. Non-cyclical scheduling method on the other hand, is adaptable to the variability of staff and workload demands. It enables specific staff skill characteristics to be identified to meet staffing targets based on workload demands, scheduling policies, and staff preferences [3]. However, non-cyclical scheduling is complex. A new schedule needs to be constructed regularly to meet the dynamicity of scheduling constraints.

Scheduling constraints are divided into hard and soft constraints [4]. Hard constraints are those restrictions that cannot be violated such as legal working hours, local employment policies and annual leaves. In contrast, soft constraints are demands desired by individual staff such as day-off requests and shift time preferences. Beside the above, the ergonomic constraints concern the circadian rhythm of shift workers. Staff perception of fairness in shift distribution is also an important constraint to be considered [5]. Several approaches have been studied to solve staff scheduling problems such as mathematical programming [6], metaheuristic approach [3], self-selected scheduling [7], particle swarm optimization [8] and genetic algorithm [9].

35.3 UMMC-ED Scheduling Constraints

The workforce of UMMC-ED is provided by Medical Officers (MO) from different backgrounds and experience levels. The majority of the workforces are post-graduate students enrolled in the Emergency Medicine (EM) programme. Other sources of workforce are resident medical officers who are not in the training program and locum doctors.

The UMMC-ED is divided into several zones according to triage levels. The triage levels in order of descending severity are: resuscitation (Triage 1), acute medical care (Triage 2), consultation (Triage 3) and pediatrics. Therefore, the MO must be located to the appropriate triage levels according to their experience.

The shiftwork scheduling in UMMC-ED consider many constraints and variables that affect the process of shiftwork scheduling. The following sub-sections will briefly explain each of the constraints.

35.3.1 The Malaysian Government General Order on Labour Law

The Malaysian Government has imposed a minimum of 41 h and a maximum of 48 h per week.

35.3.2 Numbers of Workforce

The number of training MOs, non-training MOs and locums varies from month to month. This is because different numbers of training MOs are assigned to the ED by the Head of Department each month as part of their training. Locum work, on the other hand, is voluntary and all the MOs are allowed to take leave.

35.3.3 Workforce Allocation

The UMMC-ED's MO in the training program consists of year 1 to year 4 of EM residency. It is expected that the higher EM residency years to have more experience and maturity compared to earlier residency years. Therefore, the senior students need to occupy the higher triage zones and the registrar shifts. In addition, the workforce also consists of MOs that are not in the training program and locums. They have limited ED experience and should only occupy lower triage zones.

Table 35.1 The shift pattern in UMMC-ED

Shift	Time	Hours
Morning	08:00–15:00	7
Evening	15:00–22:00	7
Night	22:00–08:00(+1)	10
Registrar	08:00–08:00(+1)	24

35.3.4 Shift Pattern and Mandatory Rest Periods

The continuity of service is essential. Therefore, the shiftwork is divided into day, evening and night shifts. Table 35.1 shows the shift pattern in UMMC-ED. '(+1)' represents a shift that extends to the next day. The following day after night and registrar shifts, the MO will be assigned a rest period of at least 14–24 h respectively.

35.3.5 Leave Allocation

Each MO is allocated with 28 days of annual leave and varying lengths of study leave as allowed by the Head of Department. Each MO is entitled to have an off day during public holidays or to be granted in lieu compensation. Considering these constraints, the staffing availability in UMMC-ED is dynamic in nature. In spite of this, the shift workload has to be divided fairly among all the MOs working in the ED of the particular month.

35.4 Shiftwork Scheduling Strategy

Considering the constraints in UMMC-ED, we strategized the shiftwork scheduling into three modes: Initialization, Sufficient and Insufficient Mode (presented in a dotted box). Figure 35.1 shows the overall view of our proposed shiftwork scheduling strategy for UMMC-ED.

35.4.1 Initialisation Mode

In the initialization mode, all the available work resources contributing to workforce is calculated and then compared to work demand. Workforce (F) and Demand (L) are calculated units of 'hours'.

The sum of the workforce is calculated based on the Malaysian General Order as the maximum work hours allowed for each of the MOs. It is reduced by the number of public holidays in that month and by leave requests made by the MOs.

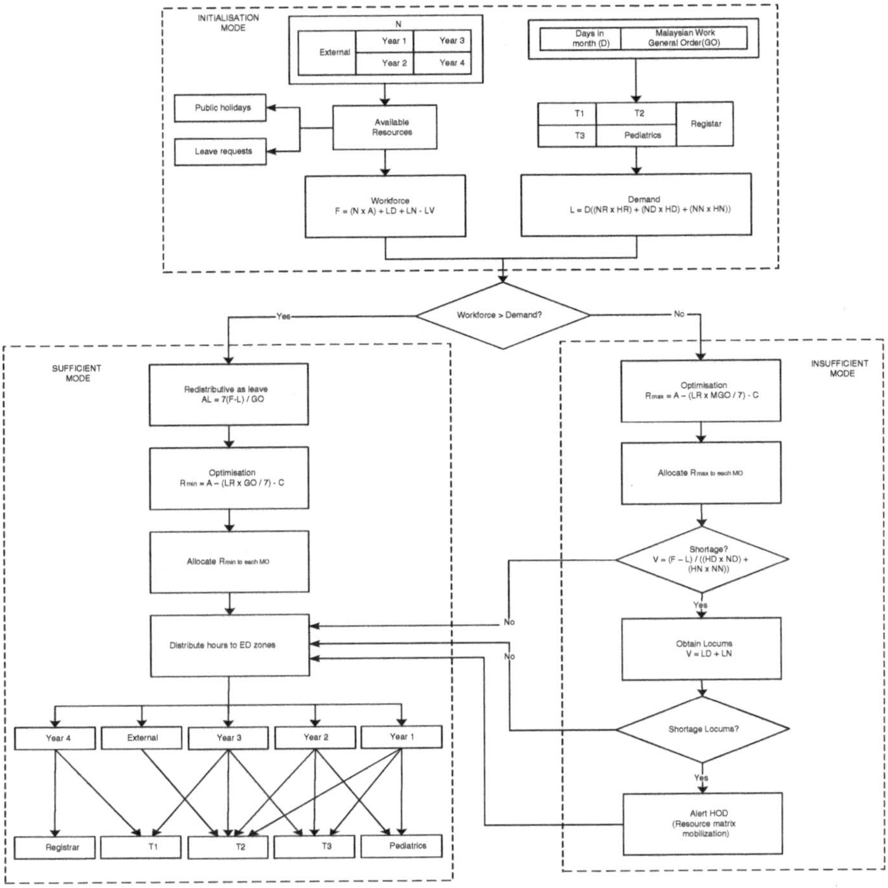

Fig. 35.1 THOR scheduling strategy

Workforce is expressed as follows:

$$F = (N * A) + LD + LN - LV \tag{35.1}$$

where N is the number of MOs available for work for the month that the schedule is made, A is maximum working hours allocated to each MO, LD is the number of locum hours employed to occupy day shifts, LN is the locum hours employed to occupy night shifts and LV is the sum of hours taken by all the MOs as leave.

The allocated maximum working hours (A) for each MO is expressed as follows:

$$A = (GO * (D - PH))/7 \tag{35.2}$$

where GO is the minimum working hours governed by the Malaysian Work General Order (MWGO), D is the number of days in the month of the schedule and PH is the number of public holidays in that month.

Demand is expressed as follows:

$$L = D * ((NR * HR) + (ND * HD) + (NN * HN)) \quad (35.3)$$

where NR, ND and NN is the number of registrar, day and night shifts in 24 h respectively; and HR, HD and HN is the length measured in hours of the registrar, day and night shifts respectively. Workforce is then compared to Demand. If Workforce exceeds or equals to Demand, the Sufficient mode is executed. In contrast, if Demand exceeds Workforce then the Insufficient mode is executed.

35.4.2 Sufficient Mode

In this mode, the minimum working hours governed by the MWGO is implemented (41 h per week). When Workforce exceeds Demand, the extra hours are considered available for MOs to take leave (AL). The extra hours is converted into days of leave using the following formula:

$$AL = 7 * (F - L)/GO \quad (35.4)$$

Fine tuning is performed by calculating the minimum working hours for each MO (R_{min}) by the following formula which considers their leave request:

$$R_{min} = A - (LR * (GO/7)) - C \quad (35.5)$$

where LR is the MO's leave request (in days) and C is the extra hours carried over from the previous month. R_{min} is divided into hours in each triage zones. The zones are occupied based on seniority of the MO. Senior MOs are posted into Registrar posts and higher triage zones, whereas junior MOs occupy lower triage zones and the Pediatrics zone. The distribution process is manually coordinated by the schedule administrator in order to maintain the integrity of the schedule.

35.4.3 Insufficient Mode

The Insufficient Mode is executed when Workforce is inadequate to fulfill Demand. In this mode, the MOs shift to the maximum hours allowed by the MWGO and Shortage is calculated. The calculation for the maximum working hours for an individual MO (R_{max}) is as follows:

$$R_{max} = A - (LR * (MGO/7)) - C \quad (35.6)$$

where MGO is the maximum working hours allowed by the MWGO (48 h per week). Shortage (V) is calculated using the following formula:

$$V = (F - L)/((NR * HR) + (HD * ND) + (HN * NN)) \quad (35.7)$$

where V is the number of vacant shifts resulting from insufficient staff. The schedule administrator is alerted to obtain more locums proportional to Shortage:

$$V = LD + LN \quad (35.8)$$

If Workforce is still insufficient and the quota for locums is exceeded, the Head of Department is alerted in order to recruit more MOs by mobilizing external resources. If Workforce becomes sufficient with the increase in locums and number of MOs, the Sufficient Mode is followed.

35.5 The THOR Implementation

THOR is a tool that utilizing a Google Spreadsheets and is implemented in the UMMC-ED to solve our shiftwork scheduling problem. As it is free and online, it allows a real-time collaboration between staff. Meanwhile, a Googlemail is the official mail back-end used in UMMC. Every staff is provided with a Google account complete with the full suite of Google Drive Apps, Google Calendar, and Contacts. Google Spreadsheets allows calculation of the mathematical formulae outlined in our shiftwork scheduling strategy. It also allows for string searches, array formula calculations and statistical functions.

35.6 Discussion

Many strategies have been proposed to solve the conundrum posed by scheduling issues and different institutions will have different constraints and solutions. Being a teaching institution, UMMC-ED has unique constraints and thus we have to devise our own solutions. We believe our scheduling strategy is a possible solution to our unique problem. We postulate that this strategy can be applied to other fields requiring 24-h work coverage such as factories, the airline industry, police and fire services.

For future work, we would like to explore an automation process of the manual distribution process of the MOs into the various zones based on their calculated work hours. We envision a java-based THOR application and conceive the use of Artificial Intelligence techniques for scheduling optimization such as Genetic Algorithm.

References

1. Knauth P (1996) Designing better shift systems. Appl Ergon 27:39–44
2. Trilling R, Guinet A, Le Magny D (2006) Nurse scheduling using integer linear programming and constraint programming. Inf Control Probl Manufact 12:671–676
3. Carter MW, Lapierre SD (2001) Scheduling emergency room physicians. Health Care Manag Sci 4:347–360
4. Al-Najjar S, Ali S (2011) Staffing and scheduling emergency rooms in two public hospitals: a case study. Int J Bus Adm 2:137–148
5. Zwemer FL Jr, Schneider S (2004) The demands of 24/7 coverage: using faculty perceptions to measure fairness of the schedule. Acad Emerg Med 11:111–114
6. Beaulieu H, Ferland JA, Gendron B, Michelon P (2000) A mathematical programming approach for scheduling physicians in the emergency room. Health Care Manag Sci 3:193–200
7. Binder DS, Crandall CS, Hauswald M (2003) Use of a self-selected scheduling method in a large academic emergency medicine group. Ann Emerg Med 41:653–658
8. Lo CC, Lin TH (2011) A particle swarm optimization approach for physician scheduling in a hospital emergency department.In: Natural Compution (ICNC), seventh international conference on 4, pp 1929–1933
9. Abo-Hamad W, Arisha A (2013) Towards operations excellence: optimising staff scheduling for new emergency department. In: Proceedings of the 20th international annual EurOMA conference—operations management at the heart of the recovery, Dublin, Ireland

Chapter 36
The Evaluation of Pen Gestures in a Digital Painting Environment

Chih-Hsiang Ko

Abstract Academic researchers argued that the use of pen-gesture commands in a pen-based environment could contribute to the improvement of task efficiency. However, previous research on pen-gesture commands tended to be focused on PDAs. A graphics tablet is also a pen-based input device. Therefore, it is possible to get better task efficiency while applying pen-gesture commands on a graphics tablet for a digital painting environment. The effects of pen-based input devices and command operations on subjects' task efficiency and the System Usability Scale (SUS) were explored in this paper. The results indicated that pen-gesture commands provided better task efficiency in file editing, layer control and integrated applications. The result from the SUS showed that pen gestures with a rubber grip pen got better usability.

Keywords Pen gestures · Pen-based system · Graphics tablet

36.1 Introduction

There are a lot of covert interactive characteristics in a pen-based environment that are worthwhile for further exploration. Pen-gesture commands are among those interactions with high development potential. Frankish et al. [1] indicated that gesture commands were easier to remember than keystroke commands. Some existing pen and paper skills could be successfully incorporated into a pen computer interface. User interactions were both faster and more fluent than could be achieved by other means. There might be significant advantages in using pen input for non-standard text, such as mathematical formulae or musical notation. Long, Landay, and Rowe [2] concluded that users appreciated the benefits that currently available

C.-H. Ko (✉)
Department of Industrial and Commercial Design,
National Taiwan University of Science and Technology, Taipei, Taiwan
e-mail: linko@mail.ntust.edu.tw

J.J. (Jong Hyuk) Park et al. (eds.), *Ubiquitous Computing Application and Wireless Sensor*, Lecture Notes in Electrical Engineering 331,
DOI 10.1007/978-94-017-9618-7_36

gestures afforded and wanted applications to support more gestures. However, gestures should be more recognizable and easier to remember.

A liquid crystal tablet is a computer peripheral operated in a pen-based environment. Simple adaptation of software interface from a mouse and keyboard environment might not be appropriate and more operating possibilities are necessary to be explored. Gesture-based interfaces promise to increase the efficiency of user input, however, most conclusions were made from using PDAs and the situation on a liquid crystal tablet was largely unexplored. Under such a context, the author tried to examine the interactive behavior of pen-gesture commands in a pen-based digital painting environment. The main purposes were as follows.

1. To explore the influence of input devices on the use of pen-gesture commands in a pen-based digital painting environment.
2. To evaluate the task efficiency of pen-gesture commands in a pen-based digital painting environment.

36.2 Background

36.2.1 Pen-Based Computing

Using pen as an input device is regarded as natural, because almost everyone could use pen and paper to communicate. Frankish et al. [1] argued that the supposed "naturalness" of the pen interface was almost entirely based on the pen and paper metaphor, and the assumption that users' experience and intuitive knowledge of this medium would be an advantage. The visible difference from other technologies is in the use of a pen or pencil as the primary means of interaction between a user and a machine [3]. As one form of human communication, using pen to communicate with computer is becoming commonplace. Pen-based computing has a rich history of innovative hardware and software solutions as human computer interface [4].

36.2.2 Pen-Gesture Commands

A gesture-based interface is one in which the user specifies commands by simple drawings, typically made with a stylus. A single intuitive gesture can simultaneously specify objects, an operation, and additional parameters, making gestures more powerful than traditional direct-manipulation interfaces [5]. The benefits of the gestural interface include the fewer number of steps required to carry out an operation, the greater ease of remembering gestural commands, and the ability to focus on a single surface for input and output [6]. Kurtenbach et al. [7] presented the design principles of revelation, guidance and rehearsal which promoted the integration of the interactive mechanism and gestures. The notion was that the interactive mechanism was intended for the novice while the gestures were intended for experts. The integration of the two was intended to support the learning transition from novice to expert.

36.3 Methods

An experiment was designed to evaluate the influence of using pen-gesture commands for software command operations, different pen-based input devices on subjects' task efficiency, and the System Usability Scale (SUS). The result could provide further clarification of the method and scope for applying pen-gesture commands to a digital painting environment.

There were two independent variables in the experiment. The first was the software command operations and was classified into four levels by operating methods: graphical user interface (GUI), shortcut keys (SK), pen gestures with a rubber grip pen (PGr) and pen gestures with a rubber grip pen and a keyboard (PGk). The second was the types of pen-based input devices and was classified into two levels: a liquid crystal graphics tablet Cintiq 21 UX (C) and an ordinary graphics tablet Intuos 3 (I). Therefore, there were eight experimental groups, as shown in Table 36.1. A total of 64 design students were selected to participate in the experiment, eight in each group. The number of male and female was even in each group to eliminate the gender bias.

The graphics software used in the experiment was Adobe Photoshop CS6, which was used by most designers in a pre-experimental survey. Five experienced designers selected 27 commands with preset shortcuts from a total of 40 frequently used commands. Together with additional four necessary file commands, these 31 commands were combined with pen-gesture commands and were activated through real buttons to distinguish the data type of pen-gesture commands.

Five tasks were designed for subjects to execute the selected 27 commands and four necessary file commands. The tasks were as follows:

1. Task 1–File Editing: To explore the influence of input devices on the use of pen-gesture commands in a pen-based digital painting environment.
2. Task 2–Layer Control: All commands were related to the layer management commands. The result indicated appropriate software command operations for layer management in a pen-based environment.

Table 36.1 Experiment groups

Variables	GUI[a]	SK[b]	PGr[c]	PGk[d]
C (Cintiq 21UX)	C-GUI	C-SK	C-PGr	C-PGk
I (Intuos 3)	I-GUI	I-SK	I-PGr	I-PGk

[a] Graphical user interface
[b] Shortcut keys
[c] Pen gestures with a rubber grip pen
[d] Pen gestures with a rubber grip pen and a keyboard

3. Task 3–Graphics Toolbar: All commands were located on the graphics toolbar. The result reflected suitable software command operations for the graphics toolbar in a pen-based environment.
4. Task 4–Brushes Adjustment: All commands were related to brushes adjustment. The result indicated appropriate software command operations for brushes adjustment in a pen-based environment.
5. Task 5–Integrated Applications: The combination or previous four tasks to simulate the actual workflow. The result reflected the effect of commands on overall task efficiency.

36.4 Results

36.4.1 Task 1–File Editing

The result of the two-way ANOVA of task 1 indicated that the main effect of command operations was F = 252.07, P value = 0.000 < 0.05, there was a statistically significant difference in four types of command operations. The results of Fisher's least significant difference (LSD) of command operations of task 1 indicated that there was a statistically significant difference in comparing graphical user interface (GUI, M = 17.81, SD = 1.93) to shortcut keys (SK, M = 7.84, SD = 1.24), pen gestures with a rubber grip pen and a keyboard (PGk, M = 7.73, SD = 0.73), and pen gestures with a rubber grip pen (PGr, M = 7.35, SD = 0.69). Graphical user interface (GUI) had the worst task efficiency among the four command operations.

36.4.2 Task 2–Layer Control

The result of the two-way ANOVA of task 2 indicated that the main effect of command operations was F = 50.35, P value = 0.000 < 0.05, there was a statistically significant difference in four types of command operations. The results of Fisher's least significant difference (LSD) of command operations of task 2 indicated that there was a statistically significant difference in comparing graphical user interface (GUI, M = 13.84, SD = 1.29) to shortcut keys (SK, M = 10.28, SD = 1.84), pen gestures with a rubber grip pen and a keyboard (PGk, M = 9.35, SD = 0.93), and pen gestures with a rubber grip pen (PGr, M = 8.90, SD = 0.82). Graphical user interface (GUI) had the worst task efficiency among the four command operations. There was a statistically significant difference in comparing shortcut keys (SK, M = 10.28, SD = 1.84) to pen gestures with a rubber grip pen and a keyboard (PGk, M = 9.35, SD = 0.93), and pen gestures with a rubber grip pen (PGr, M = 8.90, SD = 0.82). The task efficiency of shortcut keys (SK) was worse than pen gestures with a rubber grip pen and a keyboard (PGk), and pen gestures with a rubber grip pen (PGr).

36.4.3 Task 3–Graphics Toolbar

The result of the two-way ANOVA of task 3 indicated that the main effect of command operations was F = 34.79, P value = 0.000 < 0.05, there was a statistically significant difference in four types of command operations. The results of Fisher's least significant difference (LSD) of command operations of task 3 indicated that there was a statistically significant difference in comparing graphical user interface (GUI, M = 8.73, SD = 0.95) to shortcut keys (SK, M = 6.03, SD = 0.67), pen gestures with a rubber grip pen and a keyboard (PGk, M = 7.92, SD = 0.63), and pen gestures with a rubber grip pen (PGr, M = 7.57, SD = 0.74). Graphical user interface (GUI) had the worst task efficiency among the four command operations. There was a statistically significant difference in comparing shortcut keys (SK, M = 6.03, SD = 0.67) to pen gestures with a rubber grip pen and a keyboard (PGk, M = 7.92, SD = 0.63), and pen gestures with a rubber grip pen (PGr, M = 7.57, SD = 0.74). The task efficiency of shortcut keys (SK) was better than pen gestures with a rubber grip pen and a keyboard (PGk), and pen gestures with a rubber grip pen (PGr).

36.4.4 Task 4–Brushes Adjustment

The result of the two-way ANOVA of task 4 indicated that the main effect of command operations was F = 190.83, P value = 0.000 < 0.05, there was a statistically significant difference in four types of command operations. The results of Fisher's least significant difference (LSD) of command operations of task 4 indicated that there was a statistically significant difference in comparing graphical user interface (GUI, M = 26.12, SD = 2.57) to shortcut keys (SK, M = 14.24, SD = 1.31), pen gestures with a rubber grip pen and a keyboard (PGk, M = 16.32, SD = 0.66), and pen gestures with a rubber grip pen (PGr, M = 15.44, SD = 0.84). Graphical user interface (GUI) had the worst task efficiency among the four command operations. There was a statistically significant difference in comparing shortcut keys (SK, M = 14.24, SD = 1.31) to pen gestures with a rubber grip pen and a keyboard (PGk, M = 16.32, SD = 0.66), and pen gestures with a rubber grip pen (PGr, M = 15.44, SD = 0.84). The task efficiency of shortcut keys (SK) was better than pen gestures with a rubber grip pen and a keyboard (PGk), and pen gestures with a rubber grip pen (PGr).

36.4.5 Task 5–Integrated Applications

The result of the two-way ANOVA of task 5 indicated that the main effect of command operations was F = 173.11, P value = 0.000 < 0.05, there was a statistically significant difference in four types of command operations. The results of

Fisher's least significant difference (LSD) of command operations of task 5 indicated that there was a statistically significant difference in comparing graphical user interface (GUI, M = 22.55, SD = 2.50) to shortcut keys (SK, M = 12.83, SD = 1.34), pen gestures with a rubber grip pen and a keyboard (PGk, M = 12.75, SD = 0.98), and pen gestures with a rubber grip pen (PGr, M = 11.92, SD = 0.99). Graphical user interface (GUI) had the worst task efficiency among the four command operations.

36.4.6 System Usability Scale (SUS)

The result of the two-way ANOVA of SUS indicated the main effect of the pen-based input devices was F = 0.07, P value = 0.795 > 0.05, there was no statistically significant difference in two types of pen-based input devices. The main effect of command operations was F = 2.09, P value = 0.112 > 0.05, there was no statistically significant difference in four types of command operations. The interaction between pen-based input devices and command operations was F = 0.48, P value = 0.695 > 0.05, there was no interaction effect.

The above result showed that the subjects shared the same subjective evaluation of the eight experimental groups. However, the average score of pen gestures with a rubber grip pen (PGr) in two pen-based input devices reached 60, while the other three command operations were rated lower than the level of 60. The result indicated that pen gestures with a rubber grip pen (PGr) had the best system usability among four command operations.

36.5 Conclusions

According to the results of the five task performances, the task efficiency of the graphical user interface (GUI) was the worst among the four command operations. The task efficiency of the shortcut keys is worse than pen-gesture commands in task 2. However, the task efficiency of the shortcut keys is better than pen-gesture commands in task 3 and task 4. There was no statistically significant difference between pen gestures with a rubber grip pen and a keyboard (PGk) and pen gestures with a rubber grip pen (PGr) in all five tasks. Furthermore, two types of pen-based input devices showed no statistically significant difference in the task efficiency of command operations. The result of the system usability indicated that there was no statistically significant difference in two types of pen-based input devices and there was no statistically significant difference in four types of command operations. However, the average score of pen gestures with a rubber grip pen (PGr) in two pen-based input devices reached 60, which indicated that pen gestures with a rubber grip pen (PGr) had the best system usability among four command operations.

Acknowledgments This work was sponsored by the National Science Council, Taiwan, under the Grant No. NSC102-2410-H-011-027.

References

1. Frankish C, Morgan P, Noyes J (1994) Pen computing: some human factors issues. In: IEE colloquium on handwriting and pen-based input. The institution of electrical engineers, London 5/1–5/3
2. Long AC Jr, Landay JA, Rowe LA (1997) PDA and gesture use in practice: insights for designers of pen-based user interfaces. Technical report UCB/CSD-97-976. University of California, Berkeley
3. Meyer A (1995) Pen computing: a technology overview and a vision. ACM SIGCHI Bull 27 (3):46–90
4. Wu ZC, Zhang LP, Shen F (2007) Pen-based user interface based on handwriting force information. In: Jacko JA (ed) Human-computer interaction: interaction platforms and techniques, Part II. Springer, Berlin/Heidelberg, pp 496–503
5. Rubine D (1991) Integrating gesture recognition and direct manipulation. In: Summer 1991 USENLY technical conference, USENIX Association, Ann arbor, 95–100
6. Wolf CG (1992) A comparative study of gestural, keyboard, and mouse interfaces. Behav Inform Technol 11(1):13–23
7. Kurtenbach G, Moran T, Buxton W (1994) Contextual animation of gestural commands. Comput Graph Forum 13(5):305–314

Chapter 37
Disease Pattern Analysis Using Electronic Discharge Summary of EMR in HL7 for Computerized Treatment Plan of Cancer Patients in Korea

Young Sung Cho, Song Chul Moon, Kwang Sun Ryu and Keun Ho Ryu

Abstract This paper proposes an efficient recommending method for computerized treatment plan of cancer patients in Korea using electronic discharge summary of EMR in HL7. In this paper, it is necessary for us to classify disease patterns in the medical historical record to join the electronic discharge summaries data in EMR to analyze the disease pattern with input vectors of different features, disease code, to build the medical treatment plan of cancer patients in Korea using recommending service in medical data sets, to reduce inpatients' search effort to get the information of curing procedure, the diagnosis for recovering their health and to improve the rate of accuracy of recommending service. To verify improved performance, we make experiments with dataset collected in medical center.

Keywords EMR · K-means · An artificial neural network

Y.S. Cho · K.S. Ryu · K.H. Ryu (✉)
Database and Bioinformatics Laboratory in Department of Computer Science, Chungbuk National University, Cheongju, Korea
e-mail: khryu@dblab.chungbuk.ac.kr

Y.S. Cho
e-mail: youngscho@empal.com

K.S. Ryu
e-mail: ksryu@dblab.chungbuk.ac.kr

S.C. Moon
Department of Computer Science, Namseoul University, Cheonan, Korea
e-mail: moon@nsu.ac.kr

J.J. (Jong Hyuk) Park et al. (eds.), *Ubiquitous Computing Application and Wireless Sensor*, Lecture Notes in Electrical Engineering 331,
DOI 10.1007/978-94-017-9618-7_37

37.1 Introduction

As new information about the biology of disease emerges, treatments will be developed and modified to increase effectiveness, precision, survivability, and quality of life. Data mining is useful in finding knowledge from huge amounts of medical data. However, human medical data are difficult of all biological data to mine and analyze. And also, the medical record, health record, and medical chart are used an artificial neural network what interchangeably to describe the systematic documentation of a single patient's medical history and care across time within one particular health care provider's jurisdiction [1]. The medical treatment recommending service is required in medical center for medical diagnosis of disease, medical treatment service and the plan of clinical treatment. It is necessary for them to obtain the service of helping information so as to recover their health care. The medical treatment recommending service is the purpose of the function of help desk in medical center. It can suggest how to cure the disease or how to make the plan to recover their health for patient's medical treatment based on successful medical treatment history records. There is increasing interest in an artificial neural network predictive methods for medical diagnosis and medical treatment recommending service. Generally, there are three methods, which have attracted particular attention, logistic regression, classification trees, and artificial neural networks. Classification is an important problem extensively studied in several research areas, such as statistical pattern recognition, machine learning and data mining [2–4]. In this paper, we can propose a new method of patterns analysis using an artificial neural network in medical data sets for medical treatment recommending service. Clustering algorithm is a kind of methods of patterns analysis using an artificial neural network in medical data sets, commonly used in data mining. In this paper, artificial neural network is applied to segment the medical historical data to join patient's data and finally forms clusters of the medical historical data to join patient's information with different features, disease code, input vectors in order to do the medical treatment recommending services in medical data sets. This proposing method helps patient to find easily how to get the medical treatment recommending service and helps the medical center to set easily their target patient by automated process of medical treatment recommending service. Therefore, patients and medical centers can take an artificial neural network benefit from the service. The possession of intelligent medical treatment recommending service is becoming the hospital's health care strategy. A medical treatment recommending service using an artificial neural network, technique for clustering analysis of disease pattern to meet the needs of patient so as to recover their health condition, it has been actually processed the research [6, 7]. We can make a solution for a predictive disease pattern clustering based on artificial neural network. Finally, we can improve the performance of personal ontology medical treatment recommending service through an artificial neural network learning method based on the medical data sets to show patient's disease patterns. The next chapter briefly reviews the literature related to studies. The Chap. 3 is described a new method for personalized medical treatment recommending service in detail, such as system architecture with sub modules, the procedure of processing the medical treatment recommending service, the algorithm for proposing system. The Chap. 4 describes the evaluation of this system in order to prove the criteria of logicality and efficiency through the implementation and the experiment. In Chap. 5, finally it is described the conclusion of paper and further research direction.

37.2 Relative Works

37.2.1 Electronic Medical Record (EMR)

The information in the medical record allows health care providers to determine the patient's medical history and provide informed care. The medical record serves as the central repository for planning patient care and documenting communication among patient and health care provider and professionals contributing to the patient's care. An increasing purpose of the medical record is to ensure documentation of compliance with institutional, professional or governmental regulation. Patients' medical information can be shared by a number of people both within the health care industry and beyond.

The schema of EMR was based on the HL7 reference information model, and HL7 interface engine verse 2.4 was used as the transmission protocol. It founded on the HL7 clinical document architecture and reference information model, to apply international standards to Korean contexts.

It includes an electronic health record (EHR) and a clinical data repository (CDR), and also make possible medical information-sharing among various healthcare institutions. Recently most hospitals have adopted artificial neural network form of electric medical record system that computerizes existing medical records which have been written on a paper without any loss of process structure, scope and content of information [5]. The electric medical record system plays a key role in providing information for connection between all of systems managed in a hospital as well as gathering information for clinical research or strategic business. Personal health records combine many of the above features with portability, thus allowing a patient to share medical records across providers and health care systems [6]. The uniqueness of medical data may be organized as follows, such as heterogeneity of medical data, ethical, legal, social issues, statistical philosophy, and special status of medicine [8].

37.2.2 K-Means Clustering

Clustering can be defined as the process of grouping physical or abstract objects into classes of similar objects. Clustering involves classifying or segmenting the data into groups based on the natural structure of the data. Clustering techniques [10, 11] fall into a group of undirected data mining tools. Clustering algorithm is a kind of customer segmentation methods commonly used in data mining. In this paper, we can do clustering the customers' data using K-means clustering algorithm to segment customers and finally forms groups of customers with different features. Through analyzing different groups of customers, we try to do the recommending service for the target customers of internet shopping mall efficiently. The principle of clustering is maximizing the similarity inside an object group and minimizing the similarity between the object groups. K-means is the most well-known and commonly, used partition methods are the simplest clustering algorithm. In the K-means algorithm, cluster similarity is measured in regard to the mean value of the objects in a cluster, which can be viewed as the cluster's center of gravity. This algorithm uses as input a predefined number of clusters that is the *k* from its name. Mean stands for an average, an average location of all the members of a particular cluster. The euclidean norm is often chosen as a natural distance which customer a between k measure in the K-means algorithm. The a_i means the preference of attribute i for customer a.

$$d_{\alpha \dot{x}} = \sqrt{\sum_i (a_i - k_i)^2}. \tag{37.1}$$

In this paper, we can use the K-means algorithm [9, 12] for evaluation of proposing system using recommending service with weight based on the survival rate for medical treatment plan of cancer patients in Korea.

37.2.3 An Artificial Neural Network

The Self Organizing Map introduced by Kohonen, is an unsupervised learning algorithm for clustering [13]. Also Self Organizing Map is called as an artificial neural networks model based on competitive learning. An artificial neural network can convert a high dimensional input space into a simpler low dimensional discrete map. It has two layers which are input and feature layers. We can cluster all elements by feature map with two dimensions. Firstly an artificial neural network performs clustering with input vector X and weight matrix W. The data point X_i is treated one at a time. Also the closest W_j to X_i is found by Euclidean distance, and then W_j is updated as the following [9, 13].

$$W_k = W_j + \alpha(X_i - W_j) \tag{37.2}$$

where W_j and W_k are current and new weights. So W_k moves to X_i. This learning is repeated until given conditions such as change rate of weights and the number of repeat. In this paper, we can use the an artificial neural network learning algorithm [13, 14] where M_j and M_k are current and new weights. So M_k moves to X_i. This learning is repeated until given conditions such as change rate of weights and the number of repeat. In this paper, we can use the an artificial neural network learning algorithm [13].

37.3 Our Proposal of Recommending Service for Medical Treatment Plan Using Electronic Discharge Summary of EMR in HL7

37.3.1 Clustering Method Using an Artificial Neural Network for Predictive Pattern Analysis Based on Medical Data Sets to Join the Electronic Discharge Summaries Data of EMR in HL7

This clustering using clustering of disease code with weight based on the survival rate in the an artificial neural network for predictive pattern analysis in this paper had better than clustering the data directly, using K-means is depicted so as to analyze the disease pattern with input vectors of different features, disease code, to build the medical treatment plan of cancer patients in Korea using recommending service in medical data sets. First, a large set of prototyping for clustering patient data (much larger than the expected number of output count, disease pattern in clusters) is formed using clustering of disease code with weight based on the survival rate in the an artificial neural network. We can apply to make clustering of disease code with weight based on the survival rate in an artificial neural network to medical data sets to join the electronic discharge summaries data in order to classify disease pattern with survival rate for medical treatment plan of cancer patients in Korea. Finally the prototyping application is made and the prototyping result is classified to make clusters in order to classify disease pattern with survival rate score. The system can use the code of classification, demographic variables such as age, gender, an occupation, blood type, region and patient's data factors as input vectors (including symptoms, signals, clinical history, chief complaint, history of the present illness, systems review, allergies, medications, past medical history, surgical history, family history, social history, psychiatric history, and so on.) for pre-processing so as to be possible to provide how to recommend for the medical treatment plan of cancer patients in Korea efficiently. The system can make clusters with neighborhood patient-group using a new clustering of disease code with weight based on the survival rate, which is classified by the code of classification and patient's disease code with survival rate using electronic discharge summary of EMR in HL7. The system can take the preprocessing task which is able to use the whole medical data sets by preferred curing clinical rate of the disease code with survival rate and then makes cluster of medical data sets sorted by category of disease code, joined cluster of patient data called by patient DB, neighborhood patient group. As a matter of course, the system can use the whole medical data sets (medi_rd: called by medical_record). After that, the system using an artificial neural network algorithm, can provide how to get the medical treatment recommending service by preferred curing clinical rate of the disease code with survival rate.

37.3.2 Clustering of Disease Code with Weight Based on the Survival Rate in an Artificial Neural Network

See Table 37.1.

Table 37.1 The learning algorithm of an artificial neural network

Step 1 : Initialize parameters of an artificial neural network K model

// Representative pattern of bits

for demographic variable(age, gender, an occupation, blood type, region), patient's data factors as input vectors (including symptoms, signals, clinical history, chief complaint, history of the present illness, systems review, allergies, medications, past medical history, surgical history, family history, social history, psychiatric history, and so on.)

Step 2 : Set input value vector

Step 3 : Calculate Output value

$$O_j = \sum_{i=0}^{n} w_{ij} x_i \quad (2)$$

Step 4 : Select winner node

$$O_j^* = Max(O_j) \quad (3)$$

Step 5 : Readjust connection weights

$$w_{ij}(t+1) = w_{ij}(t) + \alpha\,(x_i(t) - w_{ij}(t)) \quad (4)$$

Step 6 : Completion of learning
// IF Reach the learning cycles
then Make the result of an artificial neural network
otherwise GO to Step 2
Step 7 : Calculate output value
Step 8 : Calculate winner node
Step 9 : Result of pattern

37.3.3 The Procedural Algorithm of Recommending Service for Computerized Treatment Plan of Cancer Patients in Korea

The recommending service for computerized treatment plan of cancer patients in Korea is the purpose of the function of help desk in medical center. It can suggest how to cure the disease or how to make the plan to recover their health for patient's medical treatment based on successful medical treatment history records. The system can search cluster selected by using the code of classification and patient's disease code with survival rate in patients' information. It can scan the preference of curing disease, which is preferred curing clinical rate of the disease code with survival rate, in cluster, suggest the preferred curing clinical treatment of disease code with the highest survival rate in category of disease code selected by the highest probability of preference of category of disease code as the average of preferred curing clinical rate of disease code. This system can provide recommending service for medical treatment plan by preferred curing clinical rate of the disease code with survival rate predictively. This system can generate recommending service for medical treatment plan efficiently through clustering method using disease code with weight based on the survival rate in an artificial neural network algorithm based on successful medical treatment history records. It can provide the associated recommending service for medical treatment plan to the best recommending service with medical treatment list using electronic discharge summary of EMR in HL7 for medical treatment plan of cancer patients in Korea.

37.4 Experimental Result

We make the implementation for prototyping of medical center which handles the medical service professionally and do experiments. It is the environment of implementation and experiments in Apache2.2.14, j2sdk 1.7.0_11 as Java environment, JSP/PHP 5.2.12 as server-side script, JQuery*mobile,XML/XHTML4.0/HTML5.0/CSS3/JAVASCRIPT as client-side script, C#.net framework 2.0, jakarta-tomcat (5.0.28) as web server under MS Windows.

37.4.1 Experimental Data for Evaluation

We used the data of 250 patients who had the experience to have had the clinic for cancer treatment in medical center, the data of 7 medical doctors, the data of 29 categories of disease codes about cancer category used in medical center, and results of 497 medical records for clinic in order to evaluate the proposal system. It could be evaluated in precision, recall, F-measure in clusters by preferred curing clinical rate of the disease code with survival rate based on successful medical history records for the medical treatment recommending service. It could be proved by the experiment through the experiment with learning data set for 9 months, testing data set for 3 months in a medical center. We try to carry out the experiments in the same condition of the previous system using K-means with dataset collected in medical center.

37.4.2 Experiment and Evaluation

We can make the task of clustering of disease code category based on disease code for medical treatment recommending service. The proposing system's overall performance evaluation was performed by dividing the two directions. The first evaluation is mean absolute error (MAE). The mean absolute error between the predicted ratings and the actual ratings of users within the test set. The mean absolute error is computed the following expression-3 over all data sets generated on medical record.

N represents the total number of predictions, ε represents the error of the forecast and actual phase i represents each prediction (Table 37.2 and Fig. 37.1).

In order to use the clustering algorithm, a set of discharge summary data has been selected and transformed into an equivalent set of simple sentence in temporal order by applying converting techniques. Our experiment shows an artificial neural

Table 37.2 The result for table of MAE by comparing proposal system with previous system

	P-count	Proposal	K-means
MAE	50	0.47	0.65
	100	0.23	0.32
	300	0.07	0.08

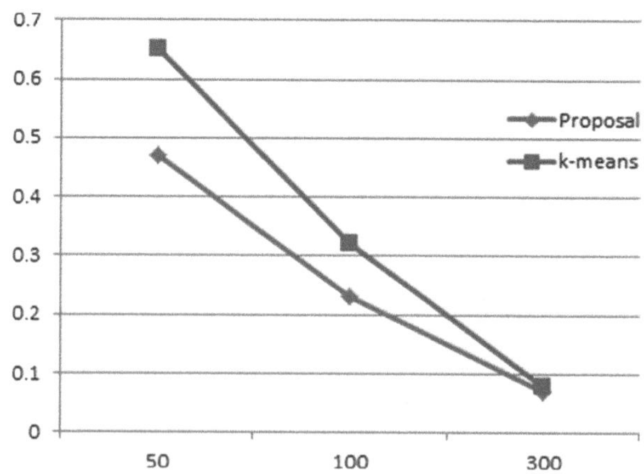

Fig. 37.1 The result for the graph of MAE by comparing proposal system with previous system

Fig. 37.2 The result of preferred curing clinical rate by precision

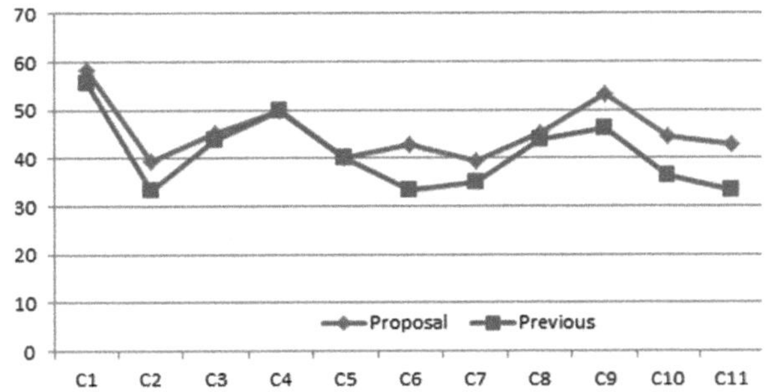

Fig. 37.3 The result of preferred curing clinical rate by recall

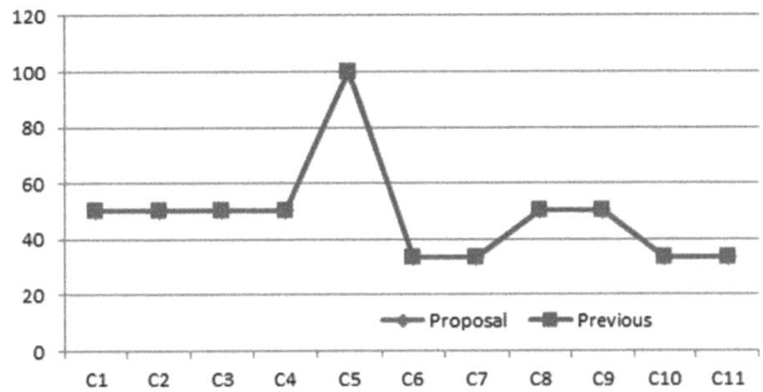

Fig. 37.4 The result of recommending rate by F-measure

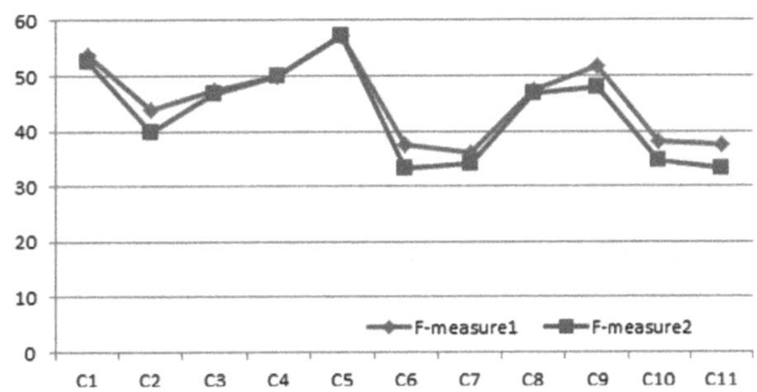

network meaningful predictive patterns of treatment procedure among cancer patients could be found when it can take effective measure as the threshold of probability is greater than 24.9 %. The next evaluation is precision, recall and F-measure for proposing system in clusters. The performance was performed to prove the validity of recommending service and the system's overall performance evaluation. The metrics of evaluation for medical treatment recommending service in our system was used in the field of information retrieval commonly [15]. The proposing system's overall performance evaluation was performed by evaluation metrics (precision, recall and F-measure) (Figs. 37.2, 37.3 and 37.4).

Above Table 37.3 presents the result of evaluation metrics (precision, recall and F-measure) of recommending service for medical treatment plan of cancer patients

Table 37.3 The result for table of precision, recall, F-measure for medical treatment recommending service with preferred clinical rate by each cluster

Cluster	Proposal (SOM)			Previous (k-means)		
	Precision1	Recall1	F-measure1	Precision2	Recall2	F-measure2
C1	58.33	50.00	53.84	55.56	50.00	52.63
C2	39.29	50.00	44.00	33.33	50.00	40.00
C3	45.00	50.00	47.37	43.75	50.00	46.67
C4	50.00	50.00	50.00	50.00	50.00	50.00
C5	40.00	100	57.14	40.00	100	57.14
C6	42.86	33.33	37.50	33.33	33.33	33.33
C7	39.52	33.33	36.16	35.14	33.33	34.21
C8	45.00	50.00	47.37	43.75	50.00	46.67
C9	53.33	50.00	51.61	46.15	50.00	48.00
C10	44.44	33.33	38.09	36.36	33.33	34.78
C11	42.86	33.33	37.50	33.33	33.33	33.33

in Korea with clinical rate. The new clustering method is improved better performance of proposing system than the previous systems. Our proposing system using clustering of disease code with weight based on the survival rate, which is the method using preferred curing clinical preference, is higher 6.99 % in precision, higher 2.27 % in F-measure than the previous system even though it is same to recall. As a result, we could have recommending service with preferred curing clinical rate the for medical treatment plan of cancer patients in Korea. The clustering method of disease code with weight based on the survival rate in an artificial neural network is weak better performance than the previous system using k-means. It is meaningful to present a new clustering method of generating predictive disease pattern, an efficient recommending method for computerized treatment plan of cancer patients in Korea using electronic discharge summary of EMR in HL7 in order to provide recommending service for medical treatment plan by each cluster of disease code with survival rate.

37.5 Conclusion

Recently most hospitals have adopted an artificial neural network form of electric medical record system that computerizes existing medical records which have been written on a paper without any loss of process structure, scope and content of information. The electric medical record system plays a key role in providing information for connection between all of systems managed in a hospital as well as gathering information for clinical research or strategic business. Recently medical treatment recommending service as a application field under ubiquitous computing environment is required by real time accessibility and agility, is in the limelight. We proposed an application of artificial neural network to classify disease pattern with survival rate in order to improve the accuracy of medical treatment recommending service by preferred curing clinical rate of the disease code with survival rate based on successful medical treatment history records. We have described that the performance of the proposal system with an artificial neural network is improved better than previous system (k-means). It could make the medical treatment recommending service for each patient's disease code based on an artificial neural network through successful medical history records in real-time environment. We could simulate the application of an artificial neural network to classify disease pattern with survival rate, generate medical treatment recommending service to be possible to measure the performance of medical treatment recommending service by preferred curing clinical rate of the disease code with survival rate based on successful medical treatment history records. Thus, we could make clusters with the focus of accuracy and efficiency, and validate the system by our results.

Then we could suggest an efficient recommending method for computerized treatment plan of cancer patients in Korea using electronic discharge summary of EMR in HL7. As a result, we could have the disease pattern analysis using electronic discharge summary of EMR in HL7 for computerized treatment plan of cancer patients in Korea. To verify improved better performance of proposing system, we carried out the experiments in the same dataset collected in medical center. It is meaningful to present a new clustering method of generating predictive disease pattern, an efficient recommending method for computerized treatment plan of cancer patients in Korea using electronic discharge summary of EMR in HL7 in order to provide recommending service for medical treatment plan by each cluster of disease code with survival rate. The following research will be looking for ways of a predictive medical treatment recommending service using ART clustering method to increase the efficiency and scalability.

Acknowledgments This work was supported by Basic Science Research Program through the National Research Foundation of Korea (NRF) funded by the Ministry of Science, ICT & Future Planning (No. 2013R1A2A2A01068923) and NRF grant funded by the Korea government (MSIP) (No. 2008-0062611), and by the ITRC (Information Technology Research Center) support program (NIPA-2013-H0301-13-4009) supervised by the NIPA (National IT Industry Promotion Agency).

References

1. Enger E et al (2012) Concepts in biology' 2007, 2007 edn. McGraw-Hill, Boston, p 173. ISBN 978-0-07-126042-8
2. Fayyad UM, Piatetsky-Shapiro G, Smyth P (1996) From data mining to knowledge discovery: an overview. In: Fayyad UM et al (eds) Advances in knowledge discovery and data mining. AAAI/MIT, Menlo Park
3. Hand DJ (1997) Construction and assessment of classification rules. Wiley, Chichester
4. Michie D, Spiegelhalter DJ, Taylor CC (1994) Machine learning, neural and statistical classification. Ellis Horwood, New York

5. Lee HM (2007) Pattern extraction of a disease treatment procedure using sequential pattern mining of discharge summary data. Computer Science in Graduate School of KeiMyung University, Daegu
6. Bekker L-G, Myer L, Catherine Orrell, Lawn S, Wood R (2006) Rapid scale-up of a community-based HIV treatment service: programme performance over 3 consecutive years in Guguletu, South Africa. S Afr Med J 96(4):315
7. Ewer MS, Vooletich MT, Durand J-B, Woods ML, Davis JR, Valero V, Lenihan DJ (2005) Reversibility of trastuzumab-related cardiotoxicity: new insights based on clinical course and response to medical treatment. Am Soc Clin Oncol 23:7820–7826
8. Cios KJ, Moore GM (2002) Uniqueness of medical data mining. Artif Intell Med 26:1–24
9. Hastie T, Tibshirani R, Friedman J (2001) The elements of statistical learning—data mining, inference, and prediction. Springer, Berlin
10. Hand D, Mannila H, Smyth P (2001) Principles of data mining. The MIT Press, Cambridge
11. Collier K, Carey B, Grusy E, Marjaniemi C, Sautter D (1998) A perspective on data mining. Northern Arizona University, Flagstaff
12. Cho YS, Moon SC, Jeong SP, Oh IB, Ryu KH (2014) Clustering method using weighted preference based on RFM score for personalized recommendation system in u-commerce. In: Jeong Y-S et al (eds) Ubiquitous information technologies and applications, LNEE, vol 280. Springer, Heidelberg, pp 131–140
13. Cho YS, Moon SC, Jeong SP, Oh IB, Ryu KH (2014) Efficient purchase pattern clustering based on SOM for recommender system in u-commerce. In: Jeong Y-S et al (eds) Ubiquitous information technologies and applications, LNEE, vol 280. Springer, Heidelberg, pp 617–626
14. Kohonen T (2000) Self-organizing maps. Springer, Berlin
15. Herlocker JL, Kosran JA, Borchers A, Riedl J (1999) An algorithm framework for performing collaborative filtering. In: Proceedings of the 1999 conference on research and development in information research and development in information retrival

Chapter 38
Research on the Network Assisted Teaching System of College Physics Experiments

Yanling Du, Hongxia Bu and Xiaonan Fang

Abstract This paper analyzes the situation of the college physics experiment teaching and discusses the necessity of the research on a network assisted teaching system, which is used in college physics experiments. Combining with our university's condition, the design scheme of a network assisted teaching system is presented. The system includes administration module, teacher module, student module and resource module. All the modules build upon each other to provide students with more efficient guidance, rich content and more autonomy. In addition, students can make full use of the openness and interactions of the system to learn better.

Keywords Network assisted teaching · College physics experiment · Hierarchical modular design

38.1 Introduction

Physics experiments play a very important role in physics teaching. Physics experiments help students not only to master the use of the basic instruments but also to find problems and solve them. At the same time, students could master the knowledge and improve the experimental skill, experimental quality and innovative

Y. Du · H. Bu
School of Science and Technology, Shandong University of Traditional Chinese Medicine, Jinan 250355, China
e-mail: duyanling08@sina.com

H. Bu
e-mail: buhx666@163.com

X. Fang (✉)
School of Information Science and Engineering, Shandong Normal University, Jinan 250014, China
e-mail: franknan@126.com

J.J. (Jong Hyuk) Park et al. (eds.), *Ubiquitous Computing Application and Wireless Sensor*, Lecture Notes in Electrical Engineering 331,
DOI 10.1007/978-94-017-9618-7_38

ability. However, with the expansion of university enrollment scale, the contradiction between the goal of training and the shortage of teaching resources is more and more serious. Education is entering the age of cybersymbosis. This is neither a fad nor a trend; it is education's future [1]. The traditional and single teaching model is unable to adapt to social development. Application of network network assisted teaching system is a useful method to solve the contradiction and improve the teaching quality. With the development of the modern information technology and college physics experiment simulation system, network assisted teaching system of college physics experiments is becoming possible and changing the traditional education mode of the university. How to build a more perfect network assisted teaching system is a subject with practical significance. Combining with our university's condition, we have conducted some active exploration in the research of network assisted teaching system, which is used in college physics experiments.

38.2 Situation Analysis

38.2.1 Traditional Physics Experiments

Traditional physics experiments have advantages in training students' experimental abilities, such as the observational ability, analytical ability, practical problem-solving ability, cooperative ability, and so on. However, with the expansion of the university scales and the limits of education funds, the number and variety of experiment instruments are relatively scarce. And the disadvantages of traditional physics experiments have been emerged, such as the lack of flexibility, the lack of individuality, the lack of research, the lack of design, the lack of innovation, and so on [2, 3]. The traditional and single teaching model is unable to adapt to social development and the trend of the times.

38.2.2 Network Teaching Materials

With the development of multimedia technology and the emergence of supermedia, network teaching materials become nicer and richer. They have provided a large number of more diverse and individual hypertexts to make learning easy and interesting. Some complicated motion can also be simulated by multimedia, such as the interaction between small particles, interference and diffraction of light, collision, and so on. Though there are plenty of network teaching materials, few can meet the need of teaching independently, for these teaching materials are not developed under the uniform standards, instrument types, running environment and the operation way are different from those in the real laboratory. In addition, most of the application codes have been encapsulated, and cannot be further developed to meet the variety of practical teaching requirements [4]. Many schools have to develop the network assisted teaching system on their own.

38.2.3 Virtual Experimental Teaching Software

Virtual experimental teaching software has enriched experiment teaching modes and provided more experiments with less investment. We have introduced a virtual experimental teaching software (College Physics Virtual Experiments V2.0 for Windows) developed by University of Science and Technology of China. The current simulation software of physics experiments can provide students with vivid simulation pictures, dubbing explanations and real-time data. Some abstract things can be presented in three-dimensional animation, such as instrument structure, characteristic, principle of work, and so on. This visualization method can help students to gain more knowledge easily. The operation of simulation physics experiments is safe for students, and can not damage the instruments either. Students can attempt boldly, and the self-exploration ability can be better trained. However, the operational process is so idealistic that some problems frequently occurred in the real operational process are not included in virtual experimental teaching software, such as equipment failure, external interference, and so on. This is not good for students to discover problems, nor to cultivate students' practical problem-solving skills. So, virtual experiments can't completely replace traditional physics experiments. They have to be used as a part of the network assisted teaching system and coexist with traditional experiments.

38.3 System Design

The guiding principle of the design is that the system should take students as main body and strengthen guidance for the students in the course of preview. At the same time, the system should provide convenience for the users and assist experiment teaching effectively [5]. Combining with the actual situation of our university, the system is designed under the campus network environment and opened to all students. The platform of the system includes four modules: administration module, teacher module, student module and resource module.

38.3.1 Administration Module

Considering the safety of users and the work of the network optimization, we design the administration module. Besides, the administrator also pays attention to the link, test, update and maintenance of the system. In order to realize better function of this module, we develop the system with the teacher of computer specialty, who has rich experience and advanced technologies in the aspect of network assisted teaching.

38.3.2 Teacher Module

In order to verify teacher identity, we design the teacher module. The teacher users can be added, modified, and deleted by the system administrator. Teachers can do things such as correcting preview reports, grading experiment reports, distributing assignments or reports, exchanging with students through online communication or message, updating the information, and so on.

38.3.3 Student Module

Student users can be automatically added after selecting the course. And they can be modified, and deleted by the system administrator. Students can do things after login, such as submitting assignments, submitting preview reports, submitting experiment reports, receiving corrected assignments or reports, exchanging with teachers through online communication or message, and so on.

38.3.4 Resource Module

The module includes four parts, they are course introduction, syllabus, calendar and teaching resources.

38.3.4.1 Course Introduction

This part introduces the functions and some learning methods of college physics experiments. Through this part, students can realize the importance of college physics experiments and enhance learning motivation.

38.3.4.2 Syllabus

The syllabus of each major must take into full account the differences among students and should meet the demands of different majors. Our university has 21 undergraduate majors, covering medicine, science, literature, engineering and management disciplines. The base of physics knowledge and experimental ability is different for the students from different majors. Even in the same major, the students' physics knowledge and experimental ability are dissimilar. So the syllabus should vary from major to major. The practice of teaching proves that the modular and hierarchy teaching mode can meet the requirement [6]. Modular teaching mode is that the teaching content of different major is formed by different experiment modules according to the fostering plan. Hierarchy teaching mode is that there are both required and elective experiments in each module, and students have freedom to choose experiments according to their individual interest and knowledge base. The syllabus differs not only in teaching content and methods but also in teaching requirement and aim. By reading the syllabus, students can know the principal tasks and basic requirement of college physics experiments.

38.3.4.3 Calendar

The calendar is also called teaching schedule, and it is different for each class. By reading the calendar, students can know experimental assignments and requirements. They can get everything ready in time, such as previewing experiments, submitting preview reports, submitting experiment reports, and so on.

38.3.4.4 Teaching Resources

Teaching resources include electronic teaching plans, courseware, videos, virtual experimental teaching software, relevant physics knowledge, relevant links, exercises, and so on. Students can click to study them and can do any one of virtual experiments which they are interested in. In order to improve network assisted teaching effects, the module should provide students with richer materials, more diverse expression and more reasonable knowledge structure.

38.4 External Supervision Systems

The network assisted teaching system can reform traditional physics experimental teaching models and improve the effect of experimental teaching. If there are external supervision systems, it can work better.

38.4.1 Teaching Supervision

With the development of the network assisted teaching system, the teachers have to face more challenges and higher requirement. For example, a lot of online assignment and questions place more workload on teachers. However, some teaching work lacks an effective system for monitoring, such as whether or not the teachers have replied their students' questions, whether or not the teachers have corrected their students' reports, and so on. The feedback of students is an alternative for monitoring. Campus network-based students' evaluation of teaching is already running, and has been paid more and more attention. Students' evaluation of teachers' performance is beneficial to improve the teaching quality.

38.4.2 Learning Supervision

Though new comer college students are aged nearly 20, they haven't matured in many aspects, such as self-control and self-autonomous learning. How to efficiently control and promote students' online self-autonomous learning is a problem. In order to assure students understand the principle and steps before doing an experiment, we require students to submit preview reports, and the grade of the preview reports accounts for about 10 % of the final grade. In addition, we have established a network assisted teaching laboratory, and there are 120 computers. In order to help the students without computer to solve the problem of learning online, the laboratory opens free to students regularly.

38.5 System Implementation

This network assisted teaching system is built on B/S structure. Nowadays B/S structure is commonly used in many online learning system because of its validity, safety and ability. On this basis, we develop the background system with C# language under ASP.net platform. As Access 2007 can efficiently manage small to medium datasets, we choose it as the database management system. For web design tools, we use a classical combination of Dreamweaver+Flash+Photoshop. The CSS +DIV method is also selected in the webpage layout stage.

This system utilizes component-ware technology, which can realize a hierarchical system structure, make logic and real data separated. The core of our system is the uniform service interface protocol. Users can interact with the system through web pages, and the system can provide information and services to the users. The basic operations include web browsing, querying, editing and management. At last modular design method is applied in this system, which can place different parts into different modules according to the user requirements.

The network assisted teaching system has been applied in the teaching process of the college physics experiments for 2 years. About 1,500 students and many physics teachers have used the system. The questionnaire surveys show that about 91 % students believe that this system can help them understand the experimental principles and improve efficiency markedly. Most of the teachers point out this system can reduce the pressure of teaching and make the experimental process easier.

38.6 Conclusion

Combining with our university's condition, we have conducted some active exploration in the design of network assisted teaching system. The system provides students with more efficient guidance, rich content and convenient user interface. It stimulates students' learning interest and helps students to preview experiments easily. In addition, students can take full advantage of the openness and interactions of the system to learn better. Thus the system achieves the goal of assisting the college physics experiment teaching. Of course, the system is still incomplete. We should continue and strive to make it more scientific and reasonable.

Acknowledgments This research is supported by the teaching reform projects of Shandong Normal University (2012) and the research projects of Shandong University of Traditional Chinese Medicine (No. ZYDXY1356).

References

1. Gundogan MB, Eby G (2012) A green touch for the future of distance education. Online Submission
2. Zhang XJ, Wang SP (2005) Reforming experimental teaching all-roundly to cultivate students' innovative abilities. Res Explor Lab 1:002
3. Wang J, Zhou Z, Yu L, Wang Y (2002) Distance education and physics experiment in university. Phys Exp 22(5):28–30
4. Guohong G, Ning L, Wenxian X, Wenlong W (2012) The study on the development of internet-based distance education and problems. Energy Procedia 17:1362–1368
5. EY Song Q (2013) A study on the construction of network-assisted teaching platform of college physics. J Lanzhou Jiaotong Univ 32(5):177–179
6. Du Y (2012) Research and practice on the reform of the physics experiment course in comprehensive university. Gaoxiao Shiyanshi Gongzuo Yanjiu 4:24–25

Chapter 39
Scheduling Model and Algorithm of Vertical Replenishment

Shufen Liu and Bin Li

Abstract With the development of vertical replenishment technology, to create a vertical replenishment schedule has become an important way to increase the efficiency of vertical replenishment. In this paper, the vertical replenishment (VERTREP) scheduling model is established. Base on the model, two algorithms are proposed to generate the VERTREP schedules. The Simple Scheduling (SS) algorithm is used to generate the schedule for the VERTREP mission that contains light cargos. The Cargo Capacity Hierarchy Scheduling (CCHS) algorithm can process heavier cargos and increase the cargo weight limit. Empirical results show that the vertical replenishment efficiency of schedules obtained by algorithm CCHS is higher than the vertical replenishment efficiency of schedules obtained by algorithm SS.

Keywords Vertical replenishment · Scheduling algorithm · Cargo capacity · Replenishment efficiency

39.1 Introduction

Underway replenishment or replenishment at sea is a method of transferring fuel, munitions, and stores from one ship to another while under way. One type of performing an underway replenishment is vertical replenishment (VERTREP) [1, 2]. In this method, a helicopter lifts cargo from the supplying ship and lowers it to the receiving ship. The main advantage of this method is that the ships do not need to be close to each other, so there is little risk of collision. However, the maximum load and transfer speeds are both limited by the capacity of the

S. Liu (✉) · B. Li
College of Computer Science and Technology, Jilin University, Changchun, China
e-mail: liusf@jlu.edu.cn

B. Li
e-mail: binli11@mails.jlu.edu.cn

J.J. (Jong Hyuk) Park et al. (eds.), *Ubiquitous Computing Application and Wireless Sensor*, Lecture Notes in Electrical Engineering 331,
DOI 10.1007/978-94-017-9618-7_39

helicopter. Over the past four decades, research on the VERTREP focused on studying the equipments [3–7]. However, the schedule of the VERTREP needs to be standardized and automated. At present the lack of the scheduling algorithm restricts the advance of VERTREP.

The aim of this paper is to discuss a practical VERTREP scheduling problem and provide algorithms for this model. The first step is to establish the VERTREP scheduling model. Base on the model, the Simple Scheduling (SS) algorithm was designed to generate the vertical replenishment schedule. According to the model of the VERTREP that the helicopter will fly back and forth many times during one VERTREP mission, we propose the Cargo Capacity Hierarchy Scheduling (CCHS) algorithm. At last, we use the replenishment efficiency to evaluate the obtained schedule. The efficiency of the proposed algorithm CCHS and SS were verified by computer simulations on scheduling problems.

39.2 VERTREP Scheduling Model

To illustrate the VERTREP scheduling problem, VERTREP mission is defined as follow:

Definition 39.1 (*VERTREP mission*) A VERTREP mission is a 4-tuple VM = (P, S, F, A), where P is a set of cargos, S is a set of receiving ships, F is a set of helicopters, A is a set of operating areas.

We use A ∈ VM to denote the set of operating areas. A = $\{a_1, a_2, a_3 \ldots a_v\}$ is a set which contains V elements. $a_1, a_2, a_3 \ldots a_v$ are operating areas. We use F ∈ VM to denote the set of helicopters. F = $\{f_1, f_2, f_3 \ldots f_K\}$ is a set which contains k elements. $f_1, f_2, f_3 \ldots f_K$ are helicopters. The helicopter should satisfy the following conditions:

$$G_{all}(f_j) = G_h(f_j) + G_l(f_j) + G_f(f_j) + G_p(f_j) + G_e(f_j) \quad (39.1)$$

$$G_t(f_j) \geq G_{all}(f_j) \quad (39.2)$$

$$G_f(f_j) = G_s(f_j) + G_u(f_j) \quad (39.3)$$

$$G_f(f_j) \leq G_f\max(f_j) \quad (39.4)$$

where $G_{all}(f_j)$ is the total weight of helicopter f_j, $G_h(f_j)$ is the operating empty weight (OEW) of helicopter f_j, $G_l(f_j)$ is the cargo weight of helicopter f_j, $G_f(f_j)$ is the usable fuel weight of helicopter f_j, $G_p(f_j)$ is the passengers weight of helicopter f_j, $G_e(f_j)$ is the equipment weight of helicopter f_j, $G_t(f_j)$ is the maximum takeoff weight of helicopter f_j, $G_s(f_j)$ is reserve fuel weight of helicopter f_j, $G_u(f_j)$ is trip fuel weight of helicopter f_j, $G_{fmax}(f_j)$ is the maximum allowed fuel weight of helicopter f_j. Let $G_r(f_j) = G_h(f_j) + G_p(f_j) + G_e(f_j) + G_s(f_j)$ denote the constant weight of helicopter f_j.

Definition 39.2 (*transporting process*) A transporting process contains four steps. First, the helicopter lifts a cargo from the operating area of the supplying ship. Second, the helicopter flies from the place of supplying ship to the place of receiving ship. Third, the helicopter unloads the cargo to the receiving ship. Fourth, the helicopter flies from the place of receiving ship to the place of supplying ship.

We use S ∈ VM to denote the set of receiving ships. S = $\{s_1, s_2, s_3 \ldots s_O\}$ is a set which contains O elements. $s_1, s_2, s_3 \ldots s_O$ are receiving ships. We use $T_v(s_i,f_j)$ to denote the time that helicopter f_j spends on performing a transporting process which supplies receiving ship s_i. $T_v(s_i,f_j)$ is given below.

$$T_v(s_i,f_j) = \frac{2 \times D(s_i)}{V(f_j)} + T_{up}(f_j) + T_{down}(f_j) \tag{39.5}$$

where $D(s_i)$ is the distance between the supplying ship and the receiving ship s_i, V (f_j) is the average speed of helicopter f_j during transporting process, $T_{up}(f_j)$ is the time that helicopter f_j spends on lifting the cargo, $T_{down}(f_j)$ is the time that helicopter f_j spends on unloading the cargo.

We use U_v (s_i, f_j) to denote the fuel that helicopter f_j uses for performing a transporting process which supplies receiving ship s_i. U_v (s_i, f_j) is given below.

$$U_v(s_i,f_j) = U_t(f_jj) \times T_v(s_i,f_j) \tag{39.6}$$

where $U_t(f_j)$ is the fuel consumption rate of helicopter f_j.

We use P ∈ VM to denote the set of cargos. P = $\{p_1, p_2, p_3 \cdots p_n\}$ is a set which contains n elements. $p_1, p_2, p_3 \cdots p_n$ are cargos. In order to more clearly describe the VERTREP scheduling model, we first introduce the concept of VERTREP batch in this paper.

Definition 39.3 (*VERTREP batch*) A VERTREP batch is a 5-tuple VB = {P', s, f, a, λ}, where P' is a cargo set which contains N elements, s is a receiving ship, f is a helicopter, a is an operating area. We use $\lambda(p_n)$ to denote the supplying order of cargo p_n which is in this VERTREP batch.

In many real-life VERTREP, the helicopter did not refuel during one VERTREP batch. According to this, the number of the cargos in one batch, which is denoted by N, need to complies with the following constraint

$$N \leq \left\lfloor \frac{G_u(f_j)}{Uv(s_i,f_j)} \right\rfloor = \left\lfloor \frac{G_u(f)}{U_v(s,f)} \right\rfloor \tag{39.7}$$

where $G_u(f)$ is trip fuel weight of the helicopter in one batch, $U_v(s,f)$ is the fuel that the helicopter uses for performing one transporting process in the batch.

Definition 39.4 (*VERTREP Scheduling problem*) Given a VERTREP mission VM, each cargo of the VERTREP mission has to be assigned to a VERTREP batch. This can be defined as VS(VM) = [VB_1, VB_2, VB_3,…,VB_m]. VB_1, VB_2, VB_3,…,VBm are VERTREP batches.

The objective of VERTREP scheduling problem is to find the optimal schedule. The objective functions considered here are the minimization of completion time. An objective function Obj(VS) has the form:

$$Obj(VS) = \min(T(VS(VM))) = \min\left(\sum_{k=1}^{m}(T(VB_k) + T(R_f(f_j)))\right), \quad T(VB_k) = N \times T_v(s_i, f_j),\ VB_k \in VS(VM), f_j \in VB_k \tag{39.8}$$

where T ($R_f(f_j)$) is the time that helicopter fj spends on refueling once. In this paper, the number of receiving ship is assumed as one. In reality, if there are multiple receiving ships in one VERTREP mission, several schedules can be given.

39.3 Scheduling Algorithm

Base on the model, two algorithms are proposed to generate the VERTREP schedules. The Simple Scheduling (SS) algorithm is used to generate the schedule for the VERTREP mission that contains light cargos.

From Eq. (39.7), the max number of cargos which can be assigned to one VERTREP batch can be formulated as follow:

$$N_{\max} = \min\left(\left\lfloor \frac{G_{f\max} - G_s}{U_v} \right\rfloor, \max(C(a))\right) \tag{39.9}$$

where C(a) is the number of the cargos which placed in operation area a. SS algorithm can be described as follows.

Algorithm 39.1 (*Simple Scheduling algorithm*)

Input: total number of cargos in the VERTREP mission n, the maximum number of cargos in one batch N_{max}.

Output: VERTREP schedule b_{mj}.

Step 1: set m = 0, $N_f = 0$.
Step 2: if $N_f > n$, end of the algorithm, else, m = m + 1, set j = 1, $b_{mj} = 0$.
Step 3: if j <= N_{max}, and N_f <= n, go to step 4, else, turn to step 2.
Step 4: if exist unlabeled p_x, $b_{mj} = p_x$, label the p_x, j = j + 1, $N_f = N_f + 1$, and then turn to step 3.

Base on the VERTREP scheduling model, in the same VERTREP batch, cargo capacity increase with the increasing count of executions of transporting process. Since the changeable of the cargo capacity, the upper weight limit of the cargo can be increase which can improve the replenishment efficiency. We proposed a Cargo Capacity Hierarchy Scheduling (CCHS) algorithm.

Before scheduling, we use the cargo capacity evaluation functions to extract the cargo capacity level of the helicopter. From Eqs. (39.1) to (39.6), we obtain

$$\begin{aligned} Z_L = & \{G_t(f) - G_r(f) - L \times U_v(s_i,f)\}, \\ & L \in [1,Y], Y \leq \frac{G_{f\max}(f) - G_s(f)}{U_v(si,f)} \end{aligned} \tag{39.10}$$

where Z_L is cargo capacity level L. We use $[Z_1, Z_2 \ldots\ldots Z_Y]$ to denote the set of cargo capacity levels which contains Y elements. According to the cargo capacity levels, we sort the cargos. This operation is described as follows:

$$\begin{aligned} Z(p_n) = & \{Z_L | Z_{L+1} \leq W(p_n) < Z_L\}, \\ & p_n \in P,\ L \in [1,\ Y],\ Z_{Y+1} = 0 \end{aligned} \tag{39.11}$$

where $Z(p_n)$ is the cargo capacity level of cargo p_n, $W(p_n)$ is the weight of cargo p_n. CCHS algorithm can be described as follows.

Algorithm 39.2 (*Cargo Capacity Hierarchy Scheduling algorithm*)

Input: total number of cargos in the VERTREP mission n, the maximum number of cargos in one batch N_{max}, cargo capacity levels Y.

Output: VERTREP schedule.

Step 1: set $m = 0$, $N_f = 0$.
Step 2: if $N_f > n$, end of the algorithm, else, $m = m + 1$, $j = 1$, $b_{mj} = 0$, $y = 1$, $k = 1$.
Step 3: if $j <= N_{max}$, $y <= Y$, and $N_f <= n$, go to step 4, else turn to step 2.
Step 4: if $k == 0$, go to step 5, else, go to step 6.
Step 5: set $k = k + 1$, if $y == Y$, go to step 3, else, $y = y + 1$, go to step 3.
Step 6: if exist unlabeled p_x, where $Z(p_x) == Z_y$, then $b_{mj} = p_x$, label the p_x, $j = j + 1$, $k = k - 1$, $N_f = N_f + 1$, and then turn to step 3, else go to step 7.
Step 7: set $y = y + 1$, $k = k + 1$, go to step 3.

39.4 Simulation Results and Analysis

In VERTREP Scheduling problem, the replenishment efficiency is used as the optimization criteria. We use the replenishment efficiency $E = W_{all}/T_{all}$ to evaluate the obtained schedule. Where W_{all} is the total cargo weight of a VERTREP mission, T_{all} is the completion time of a VERTREP mission.

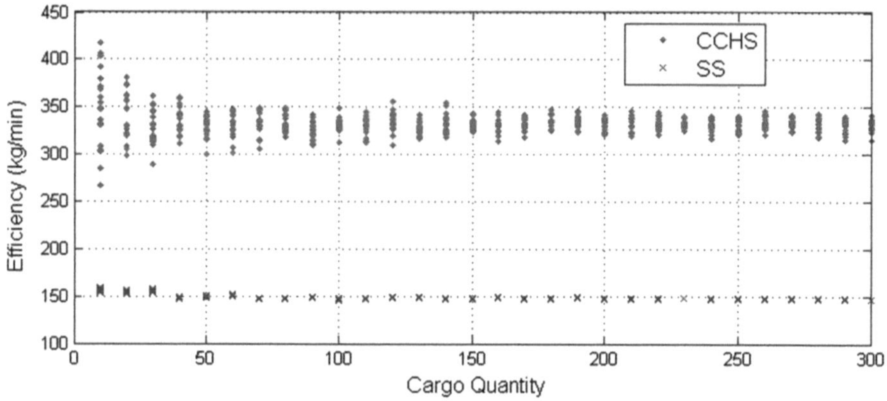

Fig. 39.1 The effect of cargo quantity on replenishment efficiency

The algorithms are implemented in MATLAB. In the following example, the fourteen parameter values are given. Moreover, randomly generate data were used for testing the proposed algorithms by analyzing the VERTREP efficiency.

Figure 39.1 shows the effect of the cargo quantity on replenishment efficiency. Several VERTREP missions with different cargo quantity are selected to test the performances of SS and CCHS. Thirty test problems are selected. The algorithms were run 20 independent times on each test problem. For each VERTREP mission in test problem i, we suppose the cargo quantity $n_i = 10 \times i$, $i = 1, 2, \ldots, 30$, and the cargo weights were random generated. It can be seen that the replenishment efficiency of schedules obtained by both algorithm SS and CCHS tend to be stable with the increasing cargo quantity. Under any cargo quantity, the replenishment efficiency of schedules obtained by algorithm CCHS is higher than the replenishment efficiency of schedules obtained by algorithm SS.

39.5 Conclusion

This paper is a research of a vertical replenishment scheduling model and vertical replenishment scheduling algorithms. The main contributions are as follows: (1) We establish the vertical replenishment scheduling model. (2) Base on the model, the Simple Scheduling algorithm was proposed to generate the vertical replenishment schedule. (3) In order to increase the upper weight limit of the cargo, we proposed a Cargo Capacity Hierarchy Scheduling algorithm which base on the changeable of the cargo capacity. Empirical results show that the Cargo Capacity Hierarchy Scheduling algorithm outperforms the Simple Scheduling algorithm in terms of replenishment efficiency.

References

1. Fu SH, Cheng CC, Yin CY (2004) Nonlinear adaptive tracking control for underway replenishment process—networking, sensing and control. In: 2004 IEEE international conference, vol 2, pp 707–712
2. Erik K, Kristin YP, Michiel W, Henk N (2007) Output synchronization control of ship replenishment operations: theory and experiments. Control Eng Pract 15(6):741–755
3. Bernard M, Kondak K (2009) Generic slung load transportation system using small size helicopters. In: Proceedings of the IEEE international conference on robotics and automation. IEEE Press, Kobe, pp 3258–3264
4. Bernard M, Kondak K, Hommel G (2010) Load transportation system based on autonomous small size helicopters. Aeronaut J 114:191–198
5. Markus B, Konstantin K, Gunter H (2008) A A slung load transportation system based on small size helicopters. In: Autonomous systems-self-organization, Management, and ControlSpringer, Netherlands, pp 49–61
6. Maza I, Kondak K, Bernard M, Ollero A (2010) Multi-UAV cooperation and control for load transportation and deployment. J Intell Rob Syst 57:417–449 (Springer, Netherlands)
7. Konstantin K, Markus B, Fernando C, Ivan M, Anibal O (2009) Cooperative autonomous helicopters for load transportation and environment perception. Advances in robotics research. Springer, Heidelberg, pp 299–310

Chapter 40
Assisted Lung Ventilation Control System as a Human Centered Application: The Project and Its Educational Impact on the Course of Embedded Systems

Diego Cabezas, Alexei Vassiliev and Evgeny Pyshkin

Abstract This paper describes the control modules for assist-control lung ventilation devices and introduces an embedded system application considered in the broader context of human centered computing. We highlight aspects of our current understanding of computer-assisted health care with special attention paid to patient/workstation interaction. The paper explores the possibility of using the technical project as a starting point for creating computer models which can be useful not only to verify technical solutions before they are embedded into the end devices but which can also be used for academic needs, thereby bridging the gap between classroom and enterprise scale projects. We involve students in an end-to-end project, from primary specifications and mathematical models up to the implementation of the end device. This is not usually easy to realize within the context of an academic course, but in so doing we aim to reduce the distance between the learning environment and real life projects of embedded systems and their applications.

Keywords Embedded systems · Human centered computing · IT education · Medical applications · Computer-assisted health care

40.1 Introduction

Human centered computing is usually defined as a field of knowledge and technology aimed to bring gaps between various disciplines with respect to design and implementation of computer systems supporting human activities [7]. Hence they relate strongly to human experience, human knowledge and human abilities for

D. Cabezas · A. Vassiliev · E. Pyshkin (✉)
Institute of Computing and Control, St. Petersburg State Polytechnic University, Politechnicheskaya, 21, 195251 St. Petersburg, Russia
e-mail: pyshkin@icc.spbstu.ru

J.J. (Jong Hyuk) Park et al. (eds.), *Ubiquitous Computing Application and Wireless Sensor*, Lecture Notes in Electrical Engineering 331,
DOI 10.1007/978-94-017-9618-7_40

cognition. Since 1997 (when the term was introduced in [2]) researchers were mostly focused on how humans organize their life around computer technologies and in what way computer technologies can assist human cognitive manifestations. A vast majority of human centered computing applications are in the domains of computer supported collaborative work, computer assisted learning and knowledge discovery, social networking and services.

That's why a traditional discourse on computer centered applications often include research works on supporting new interfaces (like in search systems, see [6] for examples), creating algorithms focusing on human perception peculiarities, etc.

In the complex area of medical applications computer technologies are often considered mostly with regards to big medical data in the areas of patient records keeping, medical image processing, material modeling, and others. An idea to consider computer-assisted health care as a human centered area of computing (not limited to the health-related information retrieval and processing) is relatively recent. The philosophy of such an understanding is wider than simply using computers to support medical processes. The main goal is to support a sort of co-operation between patients and medical equipment or medical information services, all being integral parts of an ambient ecosystem.

The paper is organized as follows. In the Sect. 40.2 we review several related works both in medicine oriented and information oriented environments. Section 40.3 describes shortly our contribution to the design and implementation of the lung ventilation unit. Section 40.4 is focused on educational issues. The computer model of respiratory compliance units is introduced as an element supporting teaching and learning process for the course of embedded systems.

40.2 Related Work

One of obvious computer oriented implementations of the above mentioned patient/workstation interaction philosophy is creating adaptive technologies of medical maneuvers. Instead of simple prescription of the defined therapy such technologies should provide conditions for a kind of adaptive therapy process. The respiratory compliance unit control system serves as a good example of such a technique. The boundary phases corresponding to moments when a patient falls asleep under anesthesia or emerges and tries to draw a breath by oneself are especially important. As it is mentioned by Gilstrap and MacIntyre in [1], assisted modes of mechanical ventilation might have advantages over controlled modes with respect to ventilator muscle function or recovery and patient comfort. At the same time, they are agree that the assisted breaths demand patient/workstation interaction during all three phases of breath delivery: trigger, target, and cycle. Most mechanical ventilation units are based on one of two types of the gas delivery algorithms: flow/volume targeting (volume assist-control ventilation) or pressure targeting with time or flow cycling (pressure assist-control ventilation). It is still open to discussion on what approach is synchronized better with patient breathing efforts (in order to minimize patient sedation needs) [5].

In this article we introduce aspects of a discourse in terms of human centered computing applied to health care together with educational issues: how the real life project affects the university educational process in the domain of embedded intelligent control systems.

40.3 The Implementation

In the previously published works we reported our contribution to the design of the assist-control lung ventilation unit including the following components:

- Built-in control system for the serial manufactured lung ventilation unit [8]
- Medical gas humidifier control system [4]

Both components use the microcontroller module designed and developed in St. Petersburg State Polytechnic University. The module is compatible with different kinds of medical devices and is based on the general-purpose control system architecture described in [8].

Figure 40.1 gives some images of units produced by Krasnogvardeets, Ltd, where the designed control systems were implemented.

In particular, we implemented the controller for an automated medical gas humidifier aimed to maintain the gas mixture temperature and humidity at the desired level specified by a physician (the functional diagram of the unit is described in [4]). Here we only mention that the gas humidifier water pump capacity Q is determined by the following fuzzy logic function $Q = F(T_u, T_m, \dot{V})$, where T_u is the user specified required respiration mixture temperature obtained from the temperature sensors, T_m is the actual respiration mixture temperature, and $\dot{V}$ is the gas mixture average flow rate determined usually by the built-in lung ventilation unit flow meter.

Let us note that in medical control systems decision making based on fuzzy algorithms is subject of continuous discussions in the medical community. On the one part, with using fuzzy approaches we might expect evident shortening the length of a decision chain connecting a medical researcher idea to the final system implementation. The reason for this is that many decision supporting knowledge can be defined by using so called linguistic descriptions. On the other part, many believe that fuzzy based control algorithms might lead to potentially unsafe or even dangerous consequences conditioned by the fact that we don't know exactly what the apparatus will do in a certain moment of time. So it's important to define boundaries indicating in what part the fuzzy logic based solutions can be used for non-critical control operations.

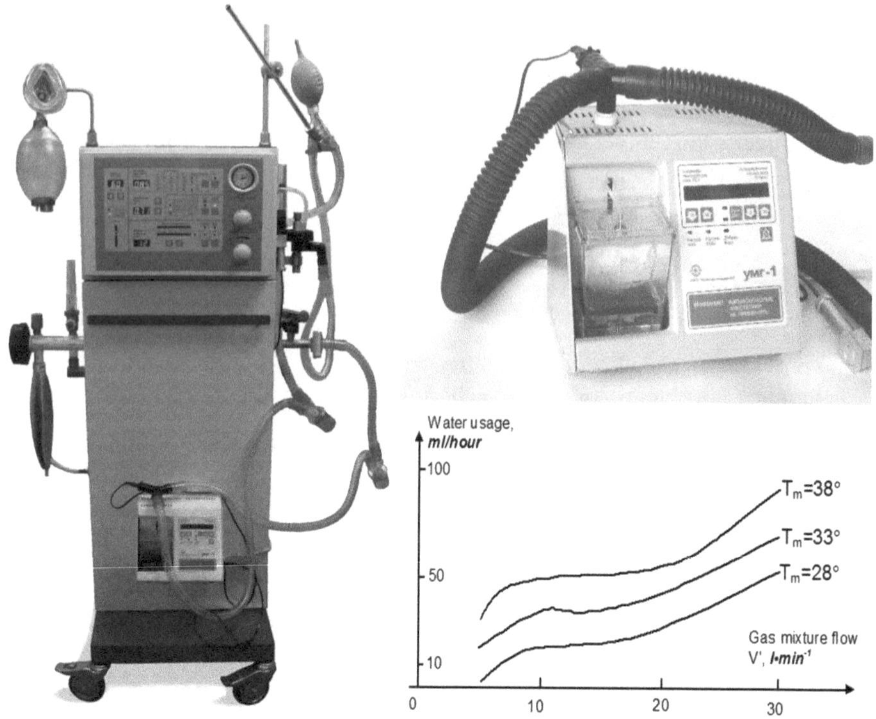

Fig. 40.1 Lung ventilation unit, gas humidifier unit and the set of characteristics for humidifier performance evaluation

40.4 Educational Perspective

Despite evident practical needs of the contribution presented in Sect. 40.3 we also pay attention to improve academic courses of intelligent control systems by using examples of solutions accepted and consumed by the workstation manufacturers. This research is resulted in defining a design and learning methodology of creating CAD-systems for functionally-oriented embedded microcontrollers [9].

This article is another step to include the design of medical control systems into the broader context of human centered computing technologies and to improve the contents and the practical impact of the academic course of embedded systems.

Practically, we used an approach similar to reverse engineering in the software domain. We "ported" the enterprise scale project to the academic course and, as a matter of things, we encouraged students to participate in the end-to-end project, beginning with the control process equation system up to the end device implementation. Here is the list of main stages of design process that students learn while attending the course:

1. Formal model defined as a mathematical description of the physiology processes (the control system project input data), as shown in Fig. 40.2 for the patient breaths state chart.
2. Modeling by using modeling environments (we used MvStudium—a computer modeling software for complex dynamic systems simulation [3]). The example of the modelled prototype operator panel implementation is shown in Fig. 40.3.

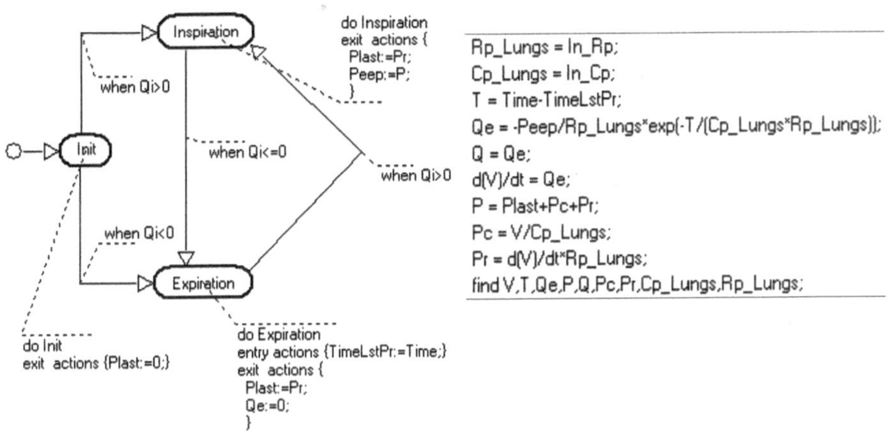

Fig. 40.2 Patient breaths state chart and its algebraic differential description

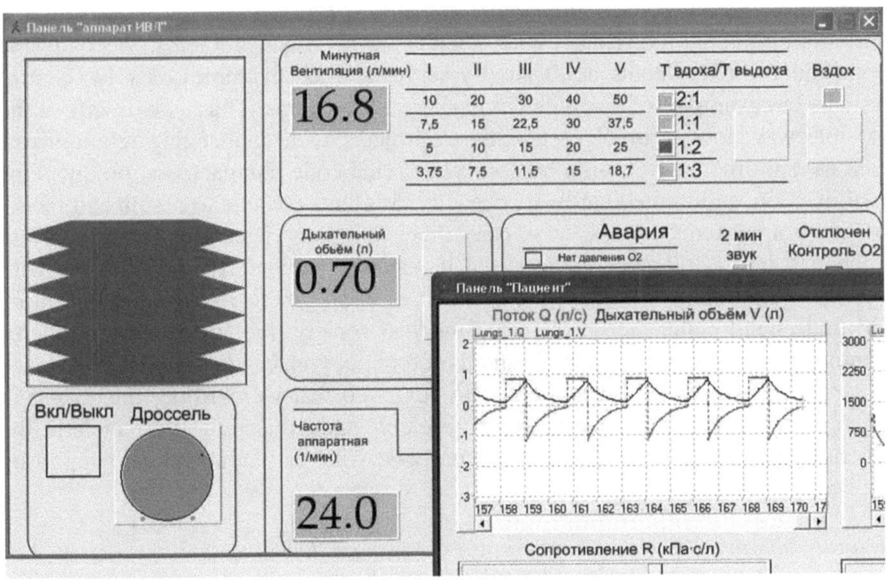

Fig. 40.3 An example of the operator panel implementation

3. Structural synthesis of the control system with taking results obtained after the modeling stage into the consideration.
4. Implementation and integration with the controlled unit.

The visit to a medical company so as to see how the solutions similar to projects students learned and developed in the class can extend the academic course and give students better understanding of how the designed control modules and algorithms are deployed and used in industry.

Now let us give some more consideration to the gas humidifier unit. By using graphical charts shown in Fig. 40.1 students learn the device control laws which, in turn, can be considered as initial data for the traditional fuzzy control unit design task. The example of gas humidifier control system allows student constructing linguistic rules for the fuzzy knowledge base by using graphical description of the input–output relationship. Unlike to the medical practice, in a computer modeling lab we can safely extend boundaries of fuzzy logic applications for wider range of learning experiments. Students are able to create models in order to compare their solutions with the control system embedded into a full-scale device used in medical practice.

40.5 Conclusion and Future Work

Let us conclude with emphasising some important aspects of this contribution. First, we believe that we show how the specific implementation with strongly defined usage requirements affects the educational process and illustrates how the classroom material correlates with society needs and problems. Second, the described control systems don't simply serve as good examples of *some* control systems but examples of *complex* but manageable systems. Third, we described the possible way how to involve students to projects with challenging requirements such as real time control, high reliability, extreme code compactness, product cost minimization, and limited hardware capacity. Students are able to design and model the control systems, and to compare their solutions with the control system embedded into a full-scale device used in medical practice. Finally, dealing with models and equipment used in the real life medical practice students know more and understand better the problems of medical society (including some problems mentioned in Sects. 40.1 and 40.2) and therefore get greater professional experience as systems analysts and developers. The next step that we consider interesting in regards to better connection between research and education is to include the function-oriented chips to the learning process, with the command set supporting the control process directly.

Acknowledgments The authors wish to thank Angelique Antonova from the Herzen State Pedagogical University of Russia for her valuable suggestions.

References

1. Gilstrap D, MacIntyre N (2013) Patient–ventilator interactions. Implications for clinical management. AJRCCM 188(9):1058–1068
2. Kling R, Star SL (1998) Human centered systems in the perspective of organizational and social informatics. ACM SIGCAS Comput Soc 28(1):22–29
3. Kolesov Y, Senichenkov Y (2010) Modeling hybrid systems in MvStudium. SNE 20(1):31–34
4. Leonov G, Filippovskii V, Vasilev A, Vasilev A (2005) The UMG-1 medical gas humidifier. Biomed Eng 39(6):301–305
5. MacIntyre N (2011) Counterpoint: is pressure assist-control preferred over volume assist-control mode for lung protective ventilation in patients with ards? No. CHEST J 140(2): 290–292
6. Pyshkin E, Kuznetsov A (2010) Approaches for web search user interfaces. J Convergence 1 (1):1–8
7. Sebe N (2010) Human-centered computing. In: Nakashima H et al (eds) Handbook of ambient intelligence and smart environments. Springer, New York, pp 349–370
8. Vasil'ev A, Leonov G (2004) A controller for medical devices. Biomed Eng 38(2):96–100
9. Vasiliev A, Do Xuan T, Cabezas D, Sadin Y, Dontsova A (2013) Methodological aspects and CAD tools for functionally-oriented embedded microcontrollers [in Russian]. St Petersburg State Polytech Univ J Comput Sci Telecommun Control Syst 2(169):123–134

Chapter 41
Research of SSO Based on the Fingerprint Key in Cloud Computing

Yong-Sheng Zhang, Jia-Shun Zou and Yan Gao

Abstract To solve the problem of single sign-on and to realize the destination of fast access, an improved single sign-on system is presented, which is based on the combination between biometric key technology and elliptic curve cryptography. The system analyses the mechanism of status register and verification, which is based on the platform of cloud sign-on. The system needs dual authentication of both fingerprint and password. What's more, it adopts a two-way authentication. So it has strong security, and it can resist replay attack and decimal attack effectively.

Keywords Single sign-on · Elliptic curve · Fingerprint verification · Cloud computing

41.1 Introduction

Although the exact definition of cloud computing has not been presented, there are more and more according services and applications, such as cloud killing provided by antivirus software, cloud storage provided by SkyDrive. In order to facilitate the

Y.-S. Zhang · J.-S. Zou (✉) · Y. Gao
School of Information Science and Engineering, Shandong Normal University,
Jinan 250014, China
e-mail: 1010336028@qq.com

Y.-S. Zhang
e-mail: zhangys@sdnu.edu.cn

Y. Gao
e-mail: 1063594899@qq.com

Y.-S. Zhang · J.-S. Zou · Y. Gao
Shandong Provincial Key Laboratory for Novel Distributed Computer
Software Technology, Jinan 250014, China

J.J. (Jong Hyuk) Park et al. (eds.), *Ubiquitous Computing Application and Wireless Sensor*, Lecture Notes in Electrical Engineering 331,
DOI 10.1007/978-94-017-9618-7_41

study of SSO problem, we give the definition of the cloud computing: cloud computing is a kind of data storage and application of the Internet and the remote service center service [1].

41.2 Single Sign-on in Cloud

41.2.1 Occurrence of the Problem

With the arrival of the cloud era, more and more users and enterprises begin to use a variety of application under the cloud environment. As for ordinary users, they may use multiple cloud services provided by different enterprises to complete a transaction. This requires the user to re-enter the account number and password when switching service or switching node, causing unnecessary trouble. This is adverse for the popularity of cloud computing service. So It is imminent to solve the SSO problem.

41.2.2 Conception and Application of Single Sign-on

The meaning of Single sign-on (Single Sign-on, SSO) is that the user only need to make the sign-on system assemble users' management. Different systems carry out identity authentication automatically relying on the relations of mutual trust between each other [2]. Literature [3] introduces the single sign-on technology based on the traditional script and the receipt such as the Windows Passport technology. Because of the use of the Cookie, there exists a certain security risk for systems that require high security. Literature [4] introduces a single sign-on system based on Kerberos. But it requires the existence of third party certification system. There are many ways to realize the current single sign-on, mature schemes consist of three kinds: (1) Integrated design method. The whole system has only one login module and one account management module, (2) Security gateway method. Through the establishment of unified identity authentication platform to accomplish single sign-on, (3) Account mapping method. Develop a separate entrance portal and a new set of independent account and password. Then mapping each user's account in every application system to the portal system [5]. This paper will also adopt this scheme.

41.2.3 Design of Single Sign-on System

This paper accomplishes single sign-on based on account mapping method. The system structure chart of account mapping method is as shown in Fig. 41.1. The system can be divided into four modules: (1) Client End, (2) Login Platform

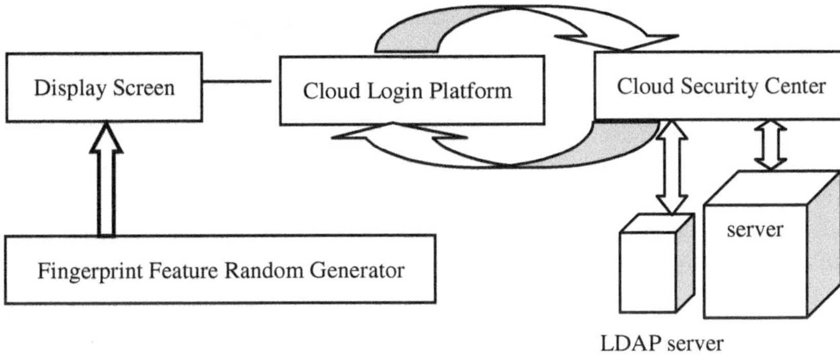

Fig. 41.1 Single sign-on system structure

(single sign-on platform), (3) Cloud Security Center, (4) Cloud Service Center. Cloud sign-on platform is the main part of single sign-on system, adopting the method of account mapping to achieve single sign-on. In the next section, we will discuss four module functions.

41.3 Single Sign-on System Based on the Fingerprint

Single sign-on system consists of four modules:

(1) The client module. The client is composed of fingerprint feature random generator and the user can input the username and password through client manually, and can get the fingerprint key by pressing the fingerprint.
(2) Cloud login module. This module is the core of single sign-on system under cloud environment. It adopts the account mapping technique and the unified identity authentication technology to achieve the goal that visit everywhere with only one login. Cloud login module can interact with the cloud security center so as to finish the identity registration and certification. As once the authentication works the user can visit everywhere, so the validation process of cloud sign-on platform is very important which is the key issue to study in this paper.
(3) Cloud Security Center module. The module mainly stores users' authentication information, and exchange information with the cloud login module. It is a bridge between the cloud login platforms and cloud service center.
(4) Cloud service center module. What stored in the cloud service center is the resources user inclined to access. Users who pass the authentication of the cloud security center can access the resources.

This paper adopts the encryption algorithm based on elliptic curve in the process of identity registration and certification to complete cloud login platform.

41.4 Cloud Login Platform Based on ECC Identity Authentication

The intractability of the discrete logarithm of Elliptic curve is presented independently by Miller and Koblitz. ECC has many advantages such as short key, high security, fast speed, small storage space and low bandwidth requirement. Because of these characteristics, the industry generally believe that ECC will become the most common public key encryption algorithm standard in the next generation [6].

41.4.1 Getting the Fingerprint Key

It is a long research history to use the fingerprint as the users' identity authentication information. But there are few key extracted from the fingerprint. The main reason is that the objective conditions of extraction of fingerprint key are complex. Many factors will affect the accuracy of fingerprint key such as finger sweat, scar, pressure, fade skin and etc. These factors will make the fingerprint's image sharpness insufficient, thus affecting the obtained data.

The paper adopts the fingerprint key extractor presented by literature [8] to produce fingerprint key. The fingerprint extractor introduces the fingerprint data into the fuzzy extractor. The fingerprint extractor consists of two algorithms: the key producing algorithm named KeyGen and the key recovery algorithm named KeyRec. The KeyGen receives the input fingerprint figure named FP and a random figure named i and then it extract help string of a certain length and a key named k. If fingerprint figure FP is extracted from the same finger with figure FP′, the algorithm KeyRec can recover the key k with a high probability with the condition of accepting FP′, i and help string P. In this paper, we assume the fingerprint key that we get is Q, which is used as the private key in the authentication.

41.4.2 Elliptic Curve Cryptosystem and the Fingerprint Key

Definition 1 elliptic curve is a plane curve composed of all points satisfying the following Wells Dreas Weierstrass equation and a infinite point O together:

$$E : y2 + a1xy + a3y = x3 + a2x2 + a4x + a6$$

The coefficient ai (i = 1, 2, …, 6) is defined in a domain Fp. It may be a rational number field, a field of real numbers, a complex domain, or a finite field. (x, y) is the point on the elliptic curve [9].

Definition 2 ECDLP: Given the elliptic curve E(Fp), G is the generator of cyclic subgroups whose order is n. In the formula $mG = G + G + \cdots + G = Q$, it is easy to work out Q when knowing G and m. Whereas it is very difficult to work out the m when Q and K are known, which is called the discrete logarithm problem on point group of elliptic curve. The security of elliptic curve crypto system just depends on the difficulty of elliptic curve discrete logarithm problem.

Many manufacturers have produced fingerprint client to login cloud service system through the fingerprint verification, but in the combination of fingerprint and single sign-on is not ideal. This paper will describe how to combine the two in detail.

41.4.3 Registration and Certification in Cloud Login Platform

The first stage, users' registration:

According to the secure elliptic curve recommended by NIST, we select one of the elliptic curves and fix a point G. We define that ID represents a user's registered account when landing single sign-on platform. PW represents the use of passwords. Si, Ri represents the random number in the ith certification session. We use the random fingerprint key U as the user's private key. Through the ECDLP, we can get the users' public key Q = UG. While the public key KSR and private key KSS of cloud server are produced by cloud security center. The meaning of Eg(m) indicates that the message m is encrypted with key k, while the meaning of Dg(m) is that message m is decrypted with key k. The H() means hash function.

(1) The cloud user inputs ID, password PW and then cloud security center searches the LDAP server. If the ID exists, the user is asked to enter again. Or else, go to the second step and submit data.
(2) The user presses the fingerprint machine, the fingerprint key U is produced. The client allocates two functions to produce random number P() and F(). The client generates a random number Ri = P(ID, PW), Si = F(ID, PW) based on the functions. Then count the function Ci = H(ID Ri, PW Si) and H(Ri), and convey the Ci, ID, H(Ri) through a secure channel to cloud security center. The random number generating function is stored on the client.
(3) When cloud security center receives, the H(Ri), EQ(Ci) are stored in the LDAP server correspond with ID, which are used in the certification. The registration period ends.

The second stage, the authentication:

(1) The cloud user produces an authentication request and input ID1, PW1. Then random number Ri′, Si′ are generated. The calculation of a = EKSR(ID1, Ri′, Si′), Ci′ = H(ID1 ⊕ Ri′, PW1 ⊕ Si′), Request = EQ(Ci′) and the time stamp are sent together to cloud security center, storing Ri′ and Si′.
(2) After receiving Request and a, cloud security center checks the time stamp. if T-Tnow is larger than the normal time, the authentication fails. Or else, ID1, Ri′, Si′ are gotten through decrypt. Search ID1 in the LDAP server. If the user exists, verify if EU(Ci) = EU(Ci′). If the verification is successful,

the cloud security center generate two functions Q() and R() randomly. Random numbers Ri + 1′ = Q(ID1, PW1), Si + 1′ = R(ID1, PW1). Sent B = EU(ID1, Ri, Si, Q(), R()) to the user, and store Ri + 1′, Si + 1′ momentarily.

(3) Cloud users get ID1, Ri, Si, Q(), R() through calculation of DU(B) with their own fingerprint key. If Ri = Ri′, Si = Si′, it can pass the cloud security center verification. And the random functions P(), F() are replaced by Q(), R(). Random numbers Ri + 1, Si + 1 are generated. Calculation of Ci + 1 = H (ID1 ⊕ Ri + 1, PW1 ⊕ Si + 1), d = EKSR(Si + 1) is operated, and EQ(Ci + 1), d are sent to cloud security center.

(4) Cloud security center receives d and gets Ri + 1, Si + 1 by decrypting DKSS (d), If Ri + 1 = Ri + 1′ Si + 1 = Si + 1′, the authentication is successful. The successful information will be sent to users. At the same time, the random number and user password stored in the cloud security center will be updated. Replace Ri, Si, EQ(Ci) with Ri + 1, Si + 1, EQ(Ci + 1) in case of the use of next authentication or otherwise, the access will be denied.

41.5 Safety Analysis

Concerning the above certification process, the decrypted message appears only in the client according to the characteristics of discrete intractability of the elliptic curve password. And what transmitted in the process of authentication are encrypted information. Even if it is acquired by the third, it is difficult to get the decrypted message. Secondly, because of the using of two-way authentication during the certification process, so the attacker cannot impersonate cloud security center, preventing the decimal attack. What is more, the data stored in the cloud security center updates along with authentication. Users have access to cloud service center only when they pass the certification. As a result, it can resist forgery attack and denial of service attack. Finally, it can prevent replay attacks because of the message in time stamp which improve the safety of certification.

At present, some scholars have applied the fingerprint recognition into the single sign-on system. But more of them use it as a login method rather than an encryption mode [10]. Single sign-on system based on fingerprint key has more safety proposed in this paper.

41.6 Conclusion

The number of researches applying fingerprint key into encryption technology are not many. The scheme that combines fingerprint key and the elliptic curve crypto system proposed in this paper has high safety and reliability. The sign-on system improved in this paper can effectively resist all kinds of attacks so that users can be convenient to use single sign-on platform for resource access.

References

1. Ainul ACF, Noraziah A, Tutut H et al (2012) Cloud computing security issues. In: 4th Asian conference on intelligent information and database systems, pp 560–569. Springer
2. Dai FS, He SC, Huang JH (2005) Research of single sign-on technology. J Second Ind Univ 22:25–30
3. Lin MS, Guo HQ (2004) Present situation and development of single sign-on technology. Comput Appl S1:248–251
4. Qiu H, Quan Y (2003) The research and design of single sign-on system based on kerberos. Comput Appl 23:142–145
5. Yan J (2011) Design and implementation of single sign-on system based on bypass technology. Comput Appl S1:95–98
6. Zhao ZJ, Zhang XQ (2013) ECC-based image encryption using code computing. In: 11th American institute of steel construction, pp 859–864. Springer
7. Girgis MR, Sewisy AA, Mansour RF (2009) A robust method for partial deformed fingerprints verification using genetic algorithm. Expert Syst Appl 36:2008–2015
8. Li XM, Yang B (2011) Fingerprint key extractor to construct high performance. Comput Sci 38:107–109
9. Li DW, Wang ZY, Zhao JG (2012) Analysis of elliptic curve cryptosystem security. Comput Technol Dev 22:227–230
10. Wu XP (2012) Single sign-on model based on fingerprint identification and CAS. Appl Res Comput 29:381–1383

Chapter 42
Study of Local Monitoring System Based on SMS Under the Cloud Environment

Yong-sheng Zhang, Jia-shun Zou and Yan Gao

Abstract To enhance the overall safety of cloud service and ensure the data integrity and to ease the burden of monitoring system, cluster analysis is introduced into the paper to divide users into different trust domain. The monitoring system's objects only include special users and users whose trust domain is minimal. The system has function of SMS alarm, and it permits users to recover the data by means of smart mobile phone. The system takes both special users and general users into consideration and ensures the reliability of the service and enhances the experience of users.

Keywords Cloud service · Cluster analysis · Trust · SMS · Data recovery

42.1 Introduction

The advent of cloud computing has changed the traditional commercial mode and service concept, more and more users' data of enterprise is stored in the public cloud. Compared with the traditional pattern, cloud service is more efficient and convenient. But the cloud service brings us with both convenience and safety problems. For example, the availability of services, hacker attacks and other

---

Y. Zhang · J. Zou (✉) · Y. Gao
School of Information Science and Engineering, Shandong Normal University, 250014 Jinan, China
e-mail: 1010336028@qq.com

Y. Zhang
e-mail: zhangys@sdnu.edu.cn

Y. Gao
e-mail: 1063594899@qq.com

Y. Zhang · J. Zou · Y. Gao
Shandong Provincial Key Laboratory for Novel Distributed Computer Software Technology, 250014 Jinan, China

J.J. (Jong Hyuk) Park et al. (eds.), *Ubiquitous Computing Application and Wireless Sensor*, Lecture Notes in Electrical Engineering 331,
DOI 10.1007/978-94-017-9618-7_42

problems. Data security is also one of the most important concerns in the public cloud. In the traditional architecture, user data resource is stored in the client and the user has the absolute control of data. In the cloud service, the user's data is stored in the cloud server, users lose the absolute control, outsourcing the data, which is equal to the outsourcing of control [1]. Whether some sensitive data is leaked, the user always feel no sense of security towards this problem. Especially if the data is uploaded, cloud server will not has any active interaction with the client end, which increased the users' concerns.

42.2 Monitoring System in the Cloud

42.2.1 Status of Traditional Monitoring System

Objects of security monitoring can be divided into two categories: information and operation. The information mainly refers to the text information and documents in system. The operation mainly refers to the operation behavior of user generated artificially [2]. At present, because of the different research directions, there are many kinds of researches in the theory of monitoring system. According to the different research fields, they can be divided into two categories, one is the study of the mechanism of monitoring system and its operation method. As mentioned in the literature [3], the directed network crawler strategy can reduce the search range, achieving directional monitoring. Literature [4] proposes improved scheduling algorithm based on integrity detection through a combination of online and offline monitoring method. The two is the research of realization mechanism of monitoring system. According to its different realization mechanism, monitoring system can be divided into internal control and external control [5]. The typical internal monitoring system to realize the virtual platform are Lares [6] and SIM system.

42.2.2 Monitoring System in Cloud Environment

Although there are many researches about file monitoring system, researches about file monitoring system applied to the cloud services are not too many. File monitoring system in application cloud service is not popular mainly due to the following reasons: Complex operation and Enormous expense and Users' experience is poor.

Even if the file monitoring system is configured in the cloud server, the user also can not experience the advantages of this system. It is the transparent to the users of the system leads to the cloud service provider not making a significant investment in this aspect.

File monitoring system proposed in this paper is to solve the problems above.

42.2.3 Design of Local Monitoring System Based on Short Message System

The local monitoring system proposed in this paper is composed of five modules, as shown in Fig. 42.1.

(1) Client module of monitoring system.
This module is mainly responsible for the collection of user client security information and parameters, and the collected data is transmitted to the user monitor and analysis module. The client module is only a program to collection data, and it will not affect the user experience.
(2) User monitor and analysis module. This module is the main module of local monitoring system. Its main function is to produce behavior vector through the collected data, and to make clustering analysis, so as to obtain the user group who have the minimum trust degree. Then the monitoring system make real-time local monitoring on the system, so as to realize the global optimization.
(3) Special user module. Design of this module is to consider some special users who have paid, and require a higher security level service, such as companies, banks and other institutions.
(4) Mobile phone alarm module.
The users' data is usually stored in the cloud server in the era of cloud computing, and the module's function is to send alarm message to users when

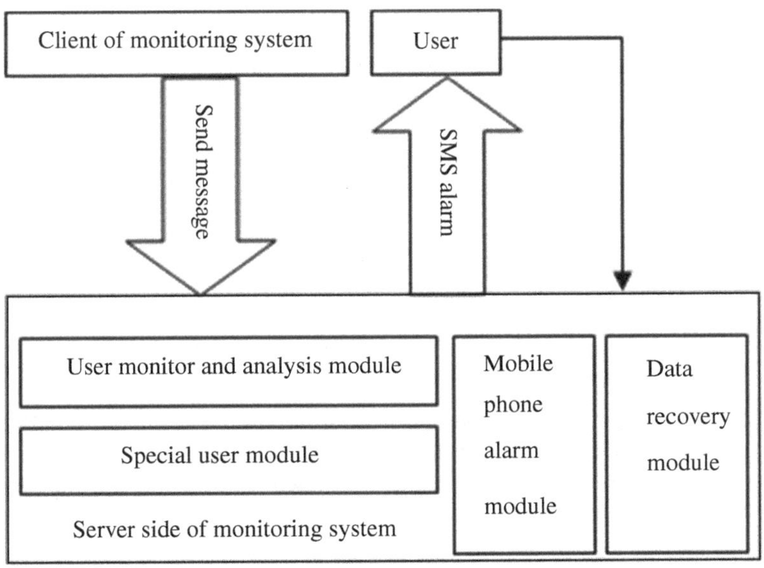

Fig. 42.1 Structure of monitoring system

triggering corresponding operation events (create, modify, delete), At the same time, it will record monitoring information to log in order to recover the data.
(5) Data recovery module.

Users will get a dynamic verification code as soon as they receive a alarm message. Users login via web service system and pass the authentication, then they can recover the damaged data by the module. But the process has a certain time limit.

Ordinary users must complete the operation before deadline after receiving SMS alarm. Otherwise the user data will not be able to recover. But the special users don't have this concern so as to encourage more users to use the more stable service by paying. The monitoring object is dynamic as the local monitoring system. So the system need to be updated at times. There also exists a update time. Then the monitoring system will only provide service for special users not the ordinary users in the period of T. As a result, it gives users enough time for data recovery operation. At the same time, user monitor and analysis module will analyze the data collected from the client, so as to update the monitored user group.

42.3 Algorithm of User Monitor and Analysis Module

User monitor and analysis module is the most important part of the local monitoring system. It determines the monitoring object through the analysis of data from the client. The paper adopts the method of cluster analysis. Cluster analysis can put objects which has higher similarity together. And the similarity between objects of different classes is low.

Minimal fragile user group can be produced by analysis of users' trust vector. Then the system can supervise the group. This module has obvious difference with intrusion detection system. The intrusion detection system is to classify all the users' behaviors, including registered users and intruders. But this module only analyzes trust parameter of registered users.

42.3.1 Algorithm Based on the Improved K-means

The idea of this algorithm is to select two farthest and most optimal cluster center, so as to avoid the problem that the initial clustering center is too close occurring in the choose of initial values. The algorithm finds two farthest object from many data objects at first. Then two initial cluster centers are created. It make sure that objects come from the same class have higher similarity with each other [7].

42.3.2 Determine the Initial Cluster Center

When the data collected from client is sent to user monitor and analysis module, the user parameter flow is formed into w dimension trust vector by ways of free parameter. Then we regard each trust vector as a data object.

Assuming that the database has M data objects, Om = {x1, X2, …, xm}. We select a data object XP randomly. $d_{i,j}$ indicates the distance between the object i and j. It can be calculated as: $d(i,j) = \sqrt{|x_{i1} - x_{j1}|^2 + |x_{i2} - x_{j2}|^2 + \cdots + |x_{iw} - x_{jw}|^2}$.

(1) Find object x_i who has the most distant distance with x_p. $d_{p,i}$ represents the distance between the two.
(2) Find the object x_k who has the most distant distance with x_k. $d_{i,k}$ represents the distance between the two.
(3) If K is equal to P, the x_p and x_i have the most distant distance. Turn to (5). Or else, turn to (4).
(4) If $d_{p,i} < d_{ik}$, x_i and x_k are the two most distant points.
(5) We use c_1 and c_2 to store the two most distant points as the two initial cluster center.

42.3.3 Search the Clustering Objects of C1 and C2

Select a threshold value q. Assuming that the latest clustering center has been found as c_m, as to the other data objects in the database, if the distance between x_h and c_m is less than q, put this data objects into the cluster. Calculate each data object in the database until initial clusters are generated.

42.3.4 Determine the Remaining Cluster Center

In order to determine the number K of initial cluster and the center point and to evaluate the clustering results, the paper introduces DBI index. The DBI index is a cluster evaluation index of non fuzzy type. It based on the close degree and the discrete degree of the same cluster. Make sure that data has higher similarity in the same cluster while having higher difference among different clusters [8]. The calculation method of DBI is:

$DBI = \frac{1}{k}\sum_{i=1}^{k} \max\left\{\frac{S_i+S_j}{d_{i,j}}\right\}$ and $S_i = \frac{1}{|c_i|}\sum_{x \in c_i} ||x - v_i||$ indicates the standard error compared with data center in cluster i. ||x|| indicates the Euclidean distance. x indicates the each data object in the cluster i. Here refers to the behavior vector. C_i indicates the number of data objects in the cluster i. $d_{i,j}$ indicates the Euclidean distance between the class i and class j. The algorithm can determine the remaining initial clustering center.

(1) Find two furthest points x_1, x_2 as the two initial clustering center.
(2) Calculate the distance to x_1 and x_2 as d_{j1} and d_{j2} for the remaining objects x_j in the collection O_m.
(3) $dk = \max\{\min(d_{j1}, d_{j2})\}$, take x_k as the new clustering center. Find the nearest neighbor of x_k which satisfy q. Add it into c_3, and calculate the center of c_3.
(4) Calculate the value of new clustering c_3 according to the formula.
(5) Repeat step (2), calculate the new DBI of the new cluster center. Compared it with the previous DBI. If $DBI_{new} < DBI_{last}$, then a new cluster can be created near x_j; Do it until the DBI value is greater than the last DBI value, then the process to search the initial clustering number K is terminated.
(6) After getting all the cluster, select the cluster which has the lowest trust. Then provide local supervision on its user group.

42.4 Conclusion

The come of cloud computing era provides us with more convenient service and also saves resources for us. Many users query the safety of the cloud service while using it. Many users still concern the sensitive data stored in the cloud server. The local monitoring system presented in this paper maximizes the users' experience degrees. Make sure that the users can be informed when the intrusion occurs, which can reduce the worry from the users.

References

1. Chow R, Golle P, Jakobsson M et al (2009) Controlling data in the cloud: outsourcing computation without outsourcing control. In: ACM workshop on cloud computing security, pp 85–90
2. Yu Y, Yang ZH, Jia PF (2007) Research on key issues of computer security monitor system. Comput Eng 33:146–149
3. Qu ZX, Zhu WC (2013) Research of directed searching and monitoring based on cloud computing. Comput Eng Sci 35:82–88
4. Abdullah ZH, Udzir I, Mahmod R et al (2007) File integrity monitor scheduling based on file security level classification. In: 4th international conference on software engineering and computer science. Springer, Berlin, pp 77–189
5. Xiang GF, Jin H, Zou DQ et al (2012) Virtualization-based security monitoring. J Softw 23:2173–2187
6. Payne B, Carbone M, Shari M, Lee W (2008) An architecture for secure active monitoring using virtualization. In: IEEE computer society, Oakland, pp 233–247
7. Wu SY, Yen E (2009) Data mining-based intrusion detectors. Expert Syst Appl 36:5605–5612
8. Du Q, Sun M (2011) Intrusion detection system based on improved clustering algorithm. Comput Eng Appl 47:106–108

Chapter 43
Effects of Meal Size on the SDA of the Taimen

Guiqiang Yang, Zhanquan Wang, Ding Yuan, Shaogang Xu and Junfeng Ma

Abstract Specific dynamic action (SDA), the metabolic phenomenon resulting from the digestion and assimilation of a meal, is generally influenced by body mass. The effects of ration level on SDA of taimen (*Hucho taimen*, Pallas) was evaluated, by measuring the temporal pattern of the oxygen consumption rates of taimen with meal size, 1/3 satiation, 2/3 satiation and satiation, after feeding, at 17.5 °C. With the approximate body mass but different meal size, both Peak VO_2 and SDA had a tendency to increase with increased meal size. Factorial scope of peak Vo_2 and Duration had a tendency to increase with increased meal size.

Keywords Specific dynamic action · Meal size · Taimen

43.1 Introduction

Feeding causes an increase of metabolic rate, which initially escalates rapidly, reaches a peak value and then gradually declines to the pre-feeding rate. This phenomenon, termed “specific dynamic action”(SDA) [1, 2], reflects the energy requirements of the behavioral, physiological and biochemical processes that constitute feeding, including capture, handling, ingestion, digestion, the assimilation of prey and the increased synthesis of proteins and lipids associated with growth [3–5].

The parameters that are usually used to describe SDA include SMR (Standard metabolic rate), peak Vo_2, factorial scope of peak Vo_2, Duration, SDAE (energy

Supported by the major projects of Agriculture Ministry (201003055-05); Youth Fund of BAAFS(QNJJ201433); Beijing major projects (D121100003712002); major projects of Agriculture Ministry (99124120); Beijing project (SCSYZ201411-4).

G. Yang (✉) · Z. Wang · D. Yuan · S. Xu · J. Ma
Beijing Fisheries Research Institute, National Engineering Technology Research Center for Freshwater Fisheries, Beijing, China
e-mail: ygqbj2013@163.com

J.J. (Jong Hyuk) Park et al. (eds.), *Ubiquitous Computing Application and Wireless Sensor*, Lecture Notes in Electrical Engineering 331,
DOI 10.1007/978-94-017-9618-7_43

expended on SDA, kJ kg^{-1}) and SDA coefficient [3, 6]. These parameters of the SDA depend on the species [6, 7], meal size [7, 8], meal type [6, 8, 9], feeding frequency [10], temperature [11, 12], body mass [13], and composition of the diet [14].

Results from earlier studies suggest that SDA is dependent on meal size in some ectotherms [6]. Only a limited number of studies have evaluated on the effects of body mass on SDA were related to fish [7, 8]. No previous study, to the best of the authors' knowledge, has reported the influence of body mass on SDA response in taimen.

Taimen are the largest specie in the Salmonidae family and the distribution of taimen has been seriously diminished by dams, water diversion, pollution and overfishing [15]. As a result of its decline, the taimen is now listed as a threatened species throughout its native range, and is being considered for listing in the International Union for Conservation of Nature's (IUCN) red list [16]. The relationship of the metabolic rates of this unfed fish with body mass has been studied by Kuang [17]. But no studies have explored the effects of meal size on the SDA responses of taimen. There were two major objectives of the present study: (1) to determine whether the taimen responds to increasing meal size by increasing peak Vo_2 and SDA; (2) to document Duration and SDA coefficient of this taimen.

43.2 Materials and Methods

Taimens were obtained from Yudushan Coldwater Fishery Base of Beijing Fisheries Research Institute and acclimated to the diets (*Oryzias latipes*, body mass 0.45 −0.55 g) in a rearing system for 2 weeks prior to the metabolism experiment. During this period, the fish were fed with their diet twice per day (at 10:00, 16:00). The ingredients of the experimental diets are listed in Table 43.1. The oxygen content was kept above 5 mg L^{-1}, the pH ranged from 7.3 to 8.0 and ammonia-N was kept below 0.025 mg L^{-1} during the experimental period. A 12/12-h (light/dark) photoperiod was used to simulate natural light cycle.

Experimental protocol: The SDA response was measured with meal size (1/3 satiation, 2/3 satiation and satiation) using fifteen fish (five fish per group, three

Table 43.1 Effects of meal size on several variables of SDA in taimen

Variables	1	2	3
Meal size (% of body mass)	2.15 ± 0.15	3.95 ± 0.21	6.36 ± 0.23
Body mass (g)	29.75 ± 1.18	27.00 ± 0.95	27.25 ± 1.23
SMR (mg kg^{-1} h^{-1})	144.58 ± 4.37	148.89 ± 2.36	146.92 ± 5.12
Peak VO_2 (L/min)	175.81 ± 2.89^{ab}	186.67 ± 5.84^{b}	218.42 ± 6.03^{c}
factorial scope of peak Vo_2	1.22 ± 0.06 ^{ab}	1.25 ± 0.06 ^{b}	1.49 ± 0.10 ^{c}
Duration (h)	14.67 ± 0.41	16.00 ± 0.28	17.33 ± 0.35
SDA (kJ/kg)	3.72 ± 0.05 ^{a}	4.37 ± 0.05 ^{b}	9.11 ± 0.08 ^{c}
SDA coefficient (%)	31.55 ± 1.24	22.23 ± 0.98	28.51 ± 1.36

Note: The different superscripts in the same row show significant difference ($P < 0.05$)

repeated groups in each ration level). Five fish were acclimated in perforated plastic cages with a surface area of 1,500 cm^2 and a water volume of 60 L. Each cage was put in a recirculating system, and the temperature was controlled at 17.5 ± 0.5 °C. Then the fish were acclimated for 2 weeks prior to the experiment. The oxygen consumption rate (OR) was measured continually during this period for three repeated groups per temperature level. After 24 h fast, the fish were placed in the chamber and allowed to acclimate for 24 h without food. OR was measured one time at 1 h intervals and used as standard metabolic rate. Then the different ration level (1/3 satiation, 2/3 satiation and satiation) of experiment diet was offered to the experimental fish. After the fish finished the diet, the chambers were closed immediately, and the OR was monitored. The duration was determined by a pilot experiment and guaranteed the postprandial metabolic rate returning to the prior status. Energy content of the diets (*Oryzias latpes*) was determined by bomb calorimetry (CN61 M/1B, Beijing zhongxi taian Co., Ltd. China). Dissolved oxygen concentration was measured at the outlet of the chamber by an oxymeter (YSI 550A, YSI Incorporated, USA).

For each metabolic trial, several variables were quantified, as described by Jobling [3] and Stephen [6]: (1) standard metabolic rate (SMR); (2) peak Vo_2; (3) factorial scope of peak Vo_2; (4) duration; (5) SDA_E, and (6) SDA coefficient (%). The oxygen consumption was converted to energy using a conversion factor of 13.56 J mg O_2^{-1} [18]. A monofactorial variance analysis was performed using SPSS16.0 to compare the variables among different levels. $P < 0.05$ was considered significant. All data are presented as means ± S.E. Non-linear estimation was also used where necessary. Figures were drawn by Microsoft Excel software.

43.3 Results

There was no significant difference in body mass (n = 15, P = 0.394) among different meal size groups (Table 43.1). With each ration level, metabolic rate increased significantly 1 h after feeding, peaked at 6 to 7 h post-feeding, and then decreased gradually to the fasting level (Fig. 43.1). With the approximate body mass but different meal size, both Peak VO_2 and SDA had a tendency to increase with increased meal size (SMR, n = 15, P < 0.001; Peak VO_2, n = 15, P < 0.001) (Table 43.1). The relationship between Peak VO_2 (mg kg^{-1} h^{-1}) and meal size (m, %) was described as: Peak $VO_2 = 155.17e^{0.0523m}$, ($R^2 = 0.9716$, n = 15, $P < 0.001$). The relationships between SDA (kJ/kg) and meal size (m, %) was described as: $SDA = 2.1412e^{0.2178m}$, ($R^2 = 0.9285$, n = 15, $P < 0.001$).

On the contrary, factorial scope of peak Vo_2 and Duration had a tendency to increase with increased meal size (Table 43.1). The relationship between factorial scope of peak Vo_2 and meal size (m, %) was described as: factorial scope of peak $Vo_2 = 1.0731e^{0.0489m}$, ($R^2 = 0.8959$, n = 15, $P = 0.047$). The relationship between Duration (h) and meal size (m, %) was described as: Duration = 0.6274 m + 13.394, ($R^2 = 0.9931$, n = 15, $P < 0.001$).

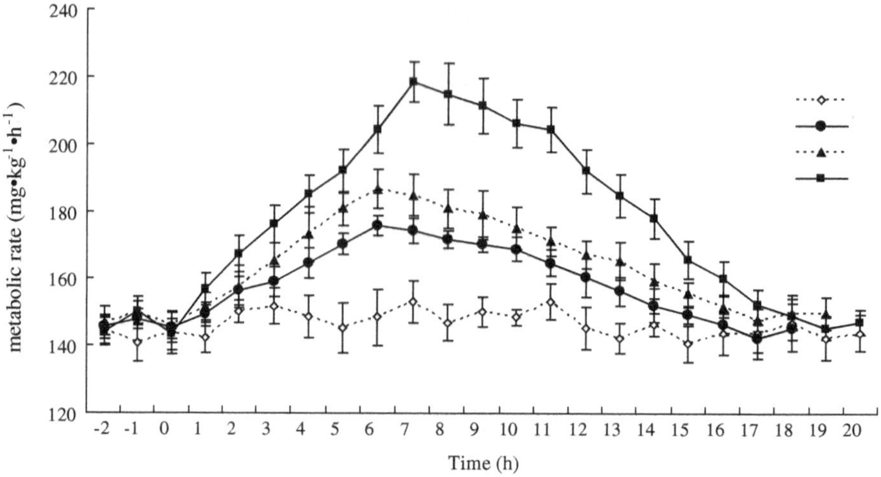

Fig. 43.1 The metabolic rate (mg kg^{-1} h^{-1}) in taimen with different body masses (● meal size 2.15 %; ▲ meal size 3.95 %; ■ meal size 6.36 %; □ unfed)

43.4 Discussion

Similar to most published studies, SDA duration increased with increased meal size [19, 20]. But SDA duration exhibited a slow–fast increase course with a gradual increase of meal size (Table 43.1). It is interesting that in relationship: Duration = a + b m, when meal size (m) goes to zero, the duration goes to a, which could be considered as minimum duration of SDA in this Hucho taimen. In this study the value of "a" was 13.39, which was a little lesser than the duration of 2.15 % body mass group (14.67). So the minimum duration of SDA in this Hucho taimen was about 14 h.

The relationship between meal size and peak Vo_2 was complicated. The research on polar animals such as Antarctic plunderfish (*Harpagifer antarcticus*) and limpet (*Nacella concinna*) showed that the peak Vo_2 was not significantly increased with increased meal size because of the narrow metabolic scope, and they primarily relied on extending the digestive course to meet the energy requirement [21, 22]. Some studies in fish and other animals found peak Vo_2 increased curvilinear with meal size [19, 23]. In this study, peak Vo_2 increased with meal size as meal size increased from 2.15 to 6.36 % body mass.

Factorial scope of peak Vo_2 of this Hucho taimen was 1.22−1.49. The value was much lesser than those of all other documented work on any fish species [21, 24, 25], which usually ranged from 1.5 to 2.5.

Acknowledgments This study was supported by the major projects of Agriculture Ministry (201003055-05); Youth Fund of BAAFS(QNJJ201433); Beijing major projects (D121100003712002); major projects of Agriculture Ministry (99124120); Beijing project (SCSYZ201411-4).

References

1. Xie XJ, Sun RY (1991) Advances of the studies on the specific dynamic action in fish. Acta Hydrobiol Sin 15:82–90
2. McCue MD (2006) Specific dynamic action: a century of investigation. Comp Biochem Physiol A 144:381–394
3. Stephen MS (2009) Specific dynamic action: a review of the postprandial metabolic response. J Comp Physiol B 179:1–56
4. Zeng LQ, Fu SJ, Li XM, Li FJ, Li B, Cao ZD, Zhang YG (2014) Physiological and morphological responses to the first bout of refeeding in southern catfish (*Silurus meridionalis*). J Comp Physiol B 184:329–346
5. Wells MJ, Odor RK, Mangold K, Wells J (1983) Feeding and metabolic rate in *Octopus*. Mar Fresh Behav Physiol 9:305–317
6. Stephen MS, Jessica AW (2007) Effects of meal size, meal type, and body temperature on the specific dynamic action of anurans. J Comp Physiol B 177:165–182
7. Jobling M, Davis PS (1980) Effects of feeding on metabolic rate, and the specific dynamic action in plaice. J Fish Biol 16:629–631
8. Iain JM, Chantelle MP (2014) Effect of meal type on specific dynamic action in the green shore crab, *Carcinus maenas*. J Comp Physiol B 184:425–436
9. Pan ZC, Xiang J, Lu HL, Ma XM (2005) Influence of food type on specific dynamic action of the Chinese skink *Eumeces chinensis*. Comp Biochem Physiol A 140:151–155
10. Guinea J, Fernandez F (1997) Effect of feeding frequency, feeding level and temperature on energy metabolism in *Sparus aurata*. Aquaculture 148:125–142
11. Luo YP, Xie XJ (2009) The effect of temperature on post-feeding ammonia excretion and oxygen consumption in the southern catfish. J Comp Physiol B 179:681–689
12. Robertson RF, El-Haj AJ, Clarke A, Taylor EW (2001) Effects of temperature on specific dynamic action and protein synthesis rates in the Baltic isopod crustacean, *Saduria entomon*. J Exp Mar Biol Ecol 262:113–129
13. Beaupre SJ (2005) Technical comment: ratio representations of specific dynamic action (mass-specific SDA and SDA coefficient) do not standardize for body mass and meal size. Physiol Biochem Zool 78:126–131
14. Brodeur JC, Calvo J, Johnston IA (2003) Proliferation of myogenic progenitor cells following feeding in the sub-antarctic notothenioid fish Harpagifer bispinis. J Exp Biol 206:163–169
15. Holcik J, Hensel K, Nieslanik J (1988) The Eurasian huchen Hucho hucho: largest salmon of the world. Kluwer Academic Publishers, Hingham USA
16. Matveyev AN, Pronin NM, Samusenok VP, Bronte CR (1998) Ecology of Siberian taimen *Hucho taimen* in the Lake Baikal Basin. J Great Lakes Res 24:905–916
17. Kuang YY, Yin JS, Jiang ZF, Xun W, Li YF (2003) The correlation between oxygen consumption of *Hucho taimen* and body weight, water temperature. Chin J Fish 16:23–30
18. Elliott JM, Davison W (1975) Energy equivalents of oxygen consumption in animal energetics. Oecologia 19:195–201
19. Toledo LF, Abe AS, Andrade DV (2003) Temperature and relative meal size effects on the postprandial metabolism and energetics in a boid snake. Physiol Biochem Zool 76:240–246
20. Fu SJ, Cao ZD, Peng JL (2006) Effect of meal size on postprandial metabolic response in Chinese catfish (*Silurus asotus Linnaeus*). Comp Biochem Physiol B 176:489–495
21. Anneli S, Swaantje B, Elettra L, Katja M, Hans OP, Felix CM (2012) Metabolic shifts in the Antarctic fish *Notothenia rossii* in response to rising temperature and PCO_2. Front Zool 9:1–15
22. Peck LS, Veal R (2001) Feeding, metabolism and growth in the Antarctic limpet. Nacella concinna. Mar Biol 138:553–560
23. Fu SJ, Xie XJ, Cao ZD (2005) Effect of meal size on postprandial metabolic response in southern catfish (*Silurus meridionalis*). Comp Biochem Physiol A 140:445–451

24. Hunt von Herbing I., White L (2002) The effects of body mass and feeding on metabolic rate in small juvenile Atlantic cod. J Fish Biol 61:945–958
25. Peck MA, Buckley LJ, Bengtson DA (2003) Energy losses due to routine and feeding metabolism in young-of-year juvenile Atlantic cod (*Gadus morhua*). Can J Fish Aquat Sci 60:929–937

Chapter 44
Management Information Systems

Lidia Ogiela and Marek R. Ogiela

Abstract In this publication authors present the most important aspects of semantic inference dedicated for description and analysis of financial data in cognitive management information systems. Analysis of financial data may be implemented not only to improve the results of the analysis process, i.e. insight analysis, the effectiveness of the proposed solutions, increase the use of analysis, but also to rationalization of data management processes of financial data in enterprises or organizations. The financial information management systems which semantically analyse the economic conditions, may execute their tasks using semantic information available to them. The foundation for the operation of financial and economic cognitive systems, consists in both the semantic analysis of the situation of the enterprise. This situation are described by various economic ratios and the assessment of the future situation of this enterprise.

Keywords Management information systems · UBMLRSS systems · Semantic analysis

44.1 Introduction

The general concepts and ideas of cognitive data analysis systems [5, 6] are dedicated to describe and semantic analyse of financial data started on CFAIS systems (Cognitive Financial Analysis Information Systems)—used to analyse strategic data for enterprises, i.e. to analyse economic and financial ratios.

L. Ogiela (✉) · M.R. Ogiela
Cryptography and Cognitive Informatics Research Group,
AGH University of Science and Technology, Al. Mickiewicza 30,
30-059 Krakow, Poland
e-mail: logiela@agh.edu.pl

M.R. Ogiela
e-mail: mogiela@agh.edu.pl

J.J. (Jong Hyuk) Park et al. (eds.), *Ubiquitous Computing Application and Wireless Sensor*, Lecture Notes in Electrical Engineering 331,
DOI 10.1007/978-94-017-9618-7_44

These systems were developed as four main subclasses of economic information systems, especially defined as [7]:

- Understanding Based Management Liquidity Ratios Support Systems—UBMLRSS;
- Understanding Based Management Financial Leverage Ratios Support Systems;
- Understanding Based Management Profitability Ratios Support;
- Understanding Based Management Activity Ratios Support.

The selection of ratios depends on the class of the cognitive system, which is enable to analyse various financial or economical ratios in the context of global situation of the enterprises. The formalisms proposed for introducing the linguistic description [4, 9] necessary for performing data analysis processes [1], have the form of sequential grammars [8], and were designed for simple single-factor analysis, two-factor analysis and the more complex multi-factor analysis [2, 3]. The results of conducted analyses may show whether the decision taken is right and whether it is weighed with risk, and if so, whether this risk is high. For an entrepreneur, the answer to these questions is fundamental for the correct operation of the enterprise in future. It was also therefore appropriate to establish that the set of financial data had been absolutely correctly selected for its semantic analysis by cognitive management information systems.

44.2 Cognitive Management Systems

The financial information management systems described in this chapter were especially concentrated around a selected group of financial cognitive systems—UBMLRSS systems. Financial data analysis in cognitive UBMLRSS systems is conducted in a different combinations of liquidity ratios (Fig. 44.1). The liquidity ratios were assigned symbols consistent with the order in which the values of individual ratios are analysed, namely:

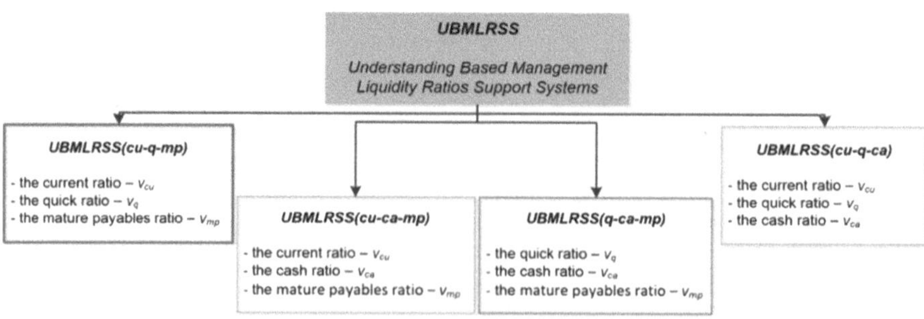

Fig. 44.1 Subclasses of cognitive UBMLRSS systems

- The value of the current ratio is represented by v_{cu};
- The value of the quick ratio is represented by v_q;
- The value of the cash ratio is represented by v_{ca};
- The value of the mature payables ratio is represented by v_{mp}.

The analysis of current, quick and cash ratio was particularly discussed in [7]. In this publication authors present the impact if the analysis of the fourth liquidity ratios, which are the mature payables ratios for the remaining ratios. Due to the fact that the cognitive analysis of liquidity ratios in enterprises is implemented in semantic systems using a minimum of three from the four parameters—ratios, below will be presented three examples of UBMLRSS class systems.

44.2.1 UBMLRSS$_{(cu\text{-}q\text{-}mp)}$ System

The first example of UBMLRSS system was assigned to following values of financial ratios:

- the value of the current ratio—v_{cu};
- the value of the quick ratio—v_q;
- the value of the mature payables ratio—v_{mp}.

A grammatical formalism was defined for the above ratios describing the short-term liquidity of an enterprise. The proposed mathematical formalism has the following form:

$$G_{UBMLRSS(cu-\mathrm{q}-\mathrm{mp})} = \left(V_{N(cu-\mathrm{q}-\mathrm{mp})}, V_{T(cu-\mathrm{q}-\mathrm{mp})}, P_{(cu-\mathrm{q}-\mathrm{mp})}, S_{(cu-\mathrm{q}-\mathrm{mp})}\right) \quad (44.1)$$

where:

$V_{N(cu\text{-}q\text{-}mp)}$ {LIQUIDITY1, EXCESS_LIQUIDITY1, OPTIMAL_LIQUIDITY1, SOLVENCY_PROBLEMS1}—the set of non-terminal symbols

$V_{T(cu\text{-}q\text{-}mp)}$ {a, b, c, d, e},—the set of terminal symbols,

where: a ∈ [0; 1), b ∈ [1; 1,2], c ∈ (1,2; 1,5), d ∈ [1,5; 2], e ∈ (2; +∞)

$$S_{(cu-\mathrm{q}-\mathrm{mp})} \in V_{N(cu-\mathrm{q}-\mathrm{mp})}, S_{(cu-\mathrm{q}-\mathrm{mp})} = \mathrm{LIQUIDITY1}$$

$P_{(cu\text{-}q\text{-}mp)}$—set of productions:

1. LIQUIDITY1 → EXCESS_LIQUIDITY1 | OPTIMAL_LIQUIDITY1 | SOLVENCY_PROBLEMS1
2. **EXCESS_LIQUIDITY1** → EEE | EDE | EED
3. **OPTIMAL_LIQUIDITY1** → DCA | DCB | DCC | DCD | DCE | DBB | DBC | DBA | DBD | DBE | CBA | CBB | CBC | CBD | CBE | CCA | CCB | CCC | CCD | CCE

4. **SOLVENCY_PROBLEMS1** → DEE | |AAA | ABA | AAB | ABB | BAB | BBA | ABC | BAC | ACB | BCA | AAC | ACA | CAA | AAD | ADA | DAA | AAE | AEA | EAA | ACD | ADC | ABD | ADB | DAB | ABE | AEB | BAA | BAD | BAE | BEA | EAB | EBA | CAB | ACC | CAC | BCC | CAD | CDA | CAE | CEA | ACE | ADE | AED | DAE | DEA | EAD | EDA | BBB | DDD | BBC | BDA | BCB | BBD | BDB | BBE | BEB | EBB | CDC | CEC | ECC | DDE | DED | EDD | BCD | BDC | CDB | BCE | BEC | ECB | EBC | CEB | CDE | CED | EDC | ECD | DEC | EEA | EAE | AEE | EEB | EBE | BEE | CEE | ECE | EEC | DDA | DAD | ADD | BDD | DDB | DDC | CDD | BDE | BED | DAC | EAC | EBD | ECA | EDB | AEC | DEB
5. A → a
6. B → b
7. C → c
8. D → d
9. E → e

44.2.2 UBMLRSS$_{(cu\text{-}ca\text{-}mp)}$ System

The second example of UBMLRSS system was assigned to symbols consistent with individual following ratios:

- The value of the current ratio—v_{cu};
- The value of the cash ratio—v_{ca};
- The value of the mature payables ratio—v_{mp}.

A grammatical formalism has the following form:

$$G_{UBMLRSS(cu-\mathrm{ca}-\mathrm{mp})} = \left(V_{N(cu-\mathrm{ca}-\mathrm{mp})}, V_{T(cu-\mathrm{ca}-\mathrm{mp})}, P_{(cu-\mathrm{ca}-\mathrm{mp})}, S_{(cu-\mathrm{ca}-\mathrm{mp})}\right) \tag{44.2}$$

where:

$V_{N(cu\text{-}ca\text{-}mp)}$ {LIQUIDITY2, EXCESS_LIQUIDITY2, OPTIMAL_LIQUIDITY2, SOLVENCY_PROBLEMS2}—the set of non-terminal symbols

$V_{T(cu\text{-}ca\text{-}mp)}$ {a, b, c, d, e}—the set of terminal symbols,

where: a ∈ [0; 1), b ∈ [1; 1,2], c ∈ (1,2; 1,5), d ∈ [1,5; 2], e ∈ (2; +∞)

$$S_{(cu-\mathrm{ca}-\mathrm{mp})} \in V_{N(cu-\mathrm{ca}-\mathrm{mp})}, S_{(cu-\mathrm{ca}-\mathrm{mp})} = \text{LIQUIDITY2},$$

$P_{(cu\text{-}ca\text{-}mp)}$—set of productions:

1. LIQUIDITY2 → EXCESS_LIQUIDITY2 | OPTIMAL_LIQUIDITY2 | SOLVENCY_PROBLEMS2
2. **EXCESS_LIQUIDITY2** → EEE | EDE | EED | EEC

3. **OPTIMAL_LIQUIDITY2** → DCA | DCB | DCC | DCD | DCE | DBA | DBB | DBC | DBD | DBE | CBA | CBB | CBC | CBD | CBE | CCA | CCB | CCC | CCD | CCE
4. **SOLVENCY_PROBLEMS2** → DEE | AAA | ABA | AAB | ABB | BAB | BBA | ABC | BAC | ACB | BCA | AAC | ACA | CAA | AAD | ADA | DAA | AAE | AEA | EAA | ACD | ADC | ABD | ADB | DAB | ABE | AEB | BAA | BAD | BAE | BEA | EAB | EBA | CAB | ACC | CAC | BCC | CAD | CDA | CAE | CEA | ACE | ADE | AED | DAE | DEA | EAD | EDA | BBB | DDD | BBC | BDA | BCB | BBD | BDB | BBE | BEB | EBB | CDC | CEC | ECC | DDE | DED | EDD | BCD | BDC | CDB | BCE | BEC | ECB | EBC | CEB | CDE | CED | EDC | ECD | DEC | EEA | EAE | AEE | EEB | EBE | BEE | CEE | ECE | DDA | DAD | ADD | BDD | DDB | DDC | CDD | BDE | BED | DAC | EAC | EBD | ECA | EDB | AEC | DEB
5. A → a
6. B → b
7. C → c
8. D → d
9. E → e

44.2.3 UBMLRSS$_{(q\text{-}ca\text{-}mp)}$ System

The last example of UBMLRSS system was assigned to following values of individual ratios:

- The value of the quick ratio—v_q;
- The value of the cash ratio—v_{ca};
- The value of the mature payables ratio—v_{mp}.

A mathematical formalism has the following form:

$$G_{UBMLRSS(q-\text{ca}-\text{mp})} = \left(V_{N(q-\text{ca}-\text{mp})}, V_{T(q-\text{ca}-\text{mp})}, P_{(q-\text{ca}-\text{mp})}, S_{(q-\text{ca}-\text{mp})}\right) \quad (44.3)$$

where:

$V_{N(q\text{-}ca\text{-}mp)}$ {LIQUIDITY3, EXCESS_LIQUIDITY3, OPTIMAL_LIQUIDITY3, SOLVENCY_PROBLEMS3}—the set of non-terminal symbols

$V_{T(q\text{-}ca\text{-}mp)}$ {a, b, c, d, e}, where: a ∈ [0; 1), b ∈ [1; 1,2], c ∈ (1,2; 1,5), d ∈ [1,5; 2], e ∈ (2; +∞)—the set of terminal symbols

$$S_{(q-\text{ca}-\text{mp})} \in V_{N(q-\text{ca}-\text{mp})}, S_{(q-\text{ca}-\text{mp})} = \text{LIQUIDITY3},$$

$P_{(q\text{-}ca\text{-}mp)}$—set of productions:

1. LIQUIDITY3 → EXCESS_LIQUIDITY3 | OPTIMAL_LIQUIDITY3 | SOLVENCY_PROBLEMS3

2. **EXCESS_LIQUIDITY3** → EEE | EDE | EED | DDC | DDD | DDE | DED | DEE | EDD
3. **OPTIMAL_LIQUIDITY3** → DCA | DCB | DCC | DCD | DCE | DBA | DBB | DBC | DBD | DBE | CBA | CBB | CBC | CBD | CBE | CCA | CCB | CCC | CCD | CCE | BBA | BBB | BBC | BBD | BBE | BCA | BCB | BCC | BCD | BCE | BDA | BDB | BDC | BDD | BDE | BEA | BEB | BEC | BED | BEE
4. **SOLVENCY_PROBLEMS3** → AAA | ABA | AAB | ABB | BAB | ABC | BAC | ACB | AAC | ACA | CAA | AAD | ADA | DAA | AAE | AEA | EAA | ACD | ADC | ABD | ADB | DAB | ABE | AEB | BAA | BAD | BAE | EAB | EBA | CAB | ACC | CAC | CAD | CDA | CAE | CEA | ACE | ADE | AED | DAE | DEA | EAD | EDA | EBB | CDC | CEC | ECC | CDB | ECB | EBC | CEB | CDE | CED | EDC | ECD | DEC | EEA | EAE | AEE | EEB | EBE | CEE | ECE | DDA | DAD | ADD | DDB | CDD | DAC | EAC | EBD | ECA | EDB | AEC | DEB
5. A → a
6. B → b
7. C → c
8. D → d
9. E → e

The results of the operation of four defined subclasses of UBMLRSS system are presented in Fig. 44.2.

The mathematical linguistic formalisms proposed for cognitive analysing liquidity ratios, used for interpreting and analysing situation of an enterprise, allow the following to be identified the current situation, by analysing the value of the each liquidity ratios—the current ratio, the quick ratio, the cash ratio, the mature

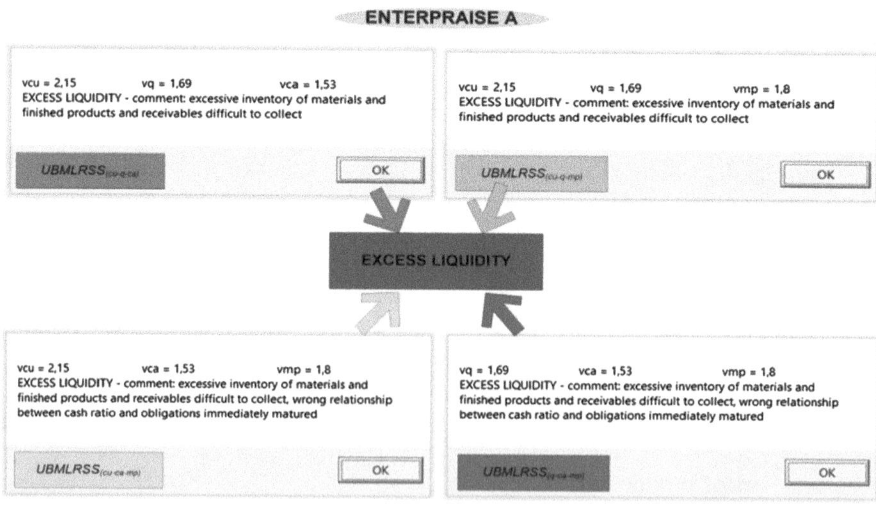

Fig. 44.2 Example of cognitive financial analysis in four subclasses of UBMLRSS systems—excess liquidity for each subsystem

payables ratio. Also the semantic analysis is used to determine the significance of the analysed ratios for the current and future situation of the given organisation.

The examples of the operation of a UBMLRSS economic system, demonstrate how this system assesses the financial situation of an enterprise with regard to debt service. The important information is whether the enterprise has payment problems or not, and what consequences this causes. The cognitive management system can assess whether current liquidity is maintained. Another information which needs to be correctly identified is whether there are problems with collecting amounts receivable and with their timely payment by debtors. Analysed information influence the assessment of enterprise's operations as part of the analysis of its current global situation and form the basis for the cognitive description of its situation in the aspect of their semantic analysis.

44.3 Conclusion

The methodology for cognitive interpreting liquidity ratios in the financial area provides the foundation for developing aspects of management information systems. Understanding the global enterprise situation is the fundamental of cognitive financial systems. The idea of the presented analysis is the semantic interpretation of strategic data of enterprises and company management achieved by conducting a cognitive analysis of the interpreted data which illustrate certain economic phenomena occurring within these enterprises. Analyzed situations may be also connected with fluctuations of the selected financial ratios. The systems for the semantic analysis of strategic data for enterprises will, to a significant extent, help to manage these enterprise better, more efficiently and rationally.

Acknowledgments This work has been supported by the National Science Centre, Republic of Poland, under project number DEC-2012/05/B/HS4/03625.

References

1. Albus JS, Meystel AM (2001) Engineering of mind: an introduction to the science of intelligent systems. Willey, New York
2. Bodzioch S, Ogiela MR (2009) New approach to gallbladder ultrasonic images analysis and lesions recognition. Comput Med Imaging Graph 33(2):154–170
3. Hachaj T, Ogiela MR (2011) A system for detecting and describing pathological changes using dynamic perfusion computer tomography brain maps. Comput Biol Med 41(6):402–410
4. Kornai A (2008) Mathematical linguistics. Springer, Berlin
5. Ogiela L (2008) Syntactic approach to cognitive interpretation of medical patterns. In: Xiong C et al (eds) Intelligent robotics and applications, first international conference, ICIRA 2008, Wuhan, China, 15–17 Oct 2008, LNAI 5314, pp 456–462

6. Ogiela L (2010) Cognitive informatics in automatic pattern understanding and cognitive information systems. In: Wang Y et al (eds) Advances in cognitive informatics and cognitive computing, studies in computational intelligence, vol 323. Springer, Berlin, pp 209–226
7. Ogiela L (2013) Data management in cognitive financial systems. Int J Inf Manage 33(2):263–270
8. Ogiela L, Ogiela MR (2012) Advances in cognitive information systems, COSMOS 17. Springer, Berlin
9. Ogiela MR, Ogiela U (2012) DNA-like linguistic secret sharing for strategic information systems. Int J Inf Manage 32(2):175–181

Chapter 45
Security in Management of Distributed Information

Marek R. Ogiela, Lidia Ogiela and Urszula Ogiela

Abstract In the chapter will be shortly presented some advances in the area of computer methods used for encryption and division of confidential data, as well as modern approaches for management of shared information. Cryptographic techniques for secret information distribution allow to secure strategic data against disclosure to unauthorized persons. The chapter will shortly describe algorithms for information sharing on the basis of personal biometric features. The development of computer techniques for classified information sharing should also be useful in the process of shared information distribution and management.

Keywords Cryptographic protocols · Secret sharing · Secure information management

45.1 Introduction

One of the fastest developing subjects associated with application of modern and advanced information technologies to manage information in commercial organizations, comprises the acquisition, flow and intelligent analysis of the information management process. During several recent years there were developed many different cryptographic procedures for secure information splitting or sharing, which may be applied for secure information distribution in particular organization. Among such algorithms it is possible to find some special examples of splitting

M.R. Ogiela (✉) · L. Ogiela · U. Ogiela
Cryptography and Cognitive Informatics Research Group, AGH University of Science and Technology, Al. Mickiewicza 30, 30-059 Krakow, Poland
e-mail: mogiela@agh.edu.pl

L. Ogiela
e-mail: logiela@agh.edu.pl

U. Ogiela
e-mail: ogiela@agh.edu.pl

J.J. (Jong Hyuk) Park et al. (eds.), *Ubiquitous Computing Application and Wireless Sensor*, Lecture Notes in Electrical Engineering 331,
DOI 10.1007/978-94-017-9618-7_45

procedures, which are based on using personal data in the sharing algorithm. From strategic point of view it may be very interesting, when as input values we can also put biometric or personal information for generation of particular shares. In the next section we'll try to describe some important features of crypto-biometric techniques, which may be used for secret sharing tasks [8, 9].

45.2 Security of Threshold Schemes

Secure information management procedures may be oriented towards developing cryptographic threshold schemes for sharing and secure distribution of information. The ideas of such schemes are hide information and guarantee its confidentiality, but in our research we also made attempts to use such techniques to create new models for intelligent management of strategic information [10, 11]. Especially challenging problem is to perform hierarchical secret splitting and shares management. What characterizes such a split is the possibility of reconstructing information from sets containing various numbers of shares of split secret. For this purpose we proposed a special threshold schemes called linguistic threshold schemes [9] dedicated for information sharing and using them to manage secret data in various hierarchical organisational structures [7].

Such algorithms allow effectively use threshold techniques of information sharing for multilevel management of data in digital form. The proposed general model for sharing information was additionally based on mathematical linguistic formalisms including protocols for information retrieval, and the range of application of such techniques is very broad for various organizational structures.

Linguistic threshold schemes allow to move from purely mathematical models of information sharing, or from using them only in dedicated, specialized information sharing problems, to a broader application of such techniques to manage secret data, designed for broader user groups. Such information can be stored by any commercial organization or state institution, and its meaning can be used only if it is accessed as authorized by appointed, entitled groups of users or employees. This is why we will attempt to define a model structure of the flow and assignment of information shares to individual groups of stakeholders. The proposed model could then be rolled out for its practical use in any commercial organization or state institution based on its legacy information system [1].

The new method of information splitting is called linguistic threshold schemes [8]. Mathematical linguistic techniques have not yet been used in information splitting, so building a new protocol for splitting secret data using these techniques represents a new research element in this field.

45.3 Crypto-Biometric Sharing Schemes

A great number of cryptographic threshold procedures were developed. Some of them may also use individual human information or biometric patterns [2–4]. Nowadays, information frequently needs to be kept back from unauthorized persons, so it is not always enough to just encrypt it with various types of algorithms.

For better supporting the authentication and authorization process, we can also verify biometric features, like fingerprints, voice characteristics or the retina. DNA molecules are also playing an increasing role in cryptography, but it was only in the 21st century that science offered opportunities of using them as information media, and the replication processes taking place in them as information coding techniques. Recent years have seen increasingly frequent reports of further discoveries, while the results of DNA research are becoming significant not just in biology or genetics, but also in the field of cryptography and steganography [11].

People have not realises the computational potential associated with molecules for many years. The first ideas of combining computers with DNA chains appeared in 1973, when Charles Bennett published a paper in which he proposed a model of a programmable molecular computer capable of executing any algorithm [11]. Since then, many new proposals for using DNA sequences as an information medium, have been made. Practically every such method of classifying data boils down, at least at one stage, to storing this data in the appropriate DNA molecules. At this level there are several available possibilities of using these acids as the medium for coded information.

The most obvious one is using the structure of particular nucleotides. As four types of them can be distinguished, one base can store 2 bits of information. We can thus assume that the coding will, for example, be executed as presented in Fig. 45.1. One can also start from the assumption that one pair of nucleotides (a single hydrogen bond irrespective of its polarisation) corresponds to one bit of information (Fig. 45.1).

Such information coding methods are used in biological solutions which have inspired us to development of a new class of algorithms for secret splitting [11]. However, linguistic threshold scheme operates in a more general way and supports coding secret information (to be split) in longer sequences, i.e. containing more than 2 bits of information [9]. The purpose of this algorithm is a threshold split of strategic data managed within hierarchical structures, with varied access capabilities dependent on the rights granted [11].

45.4 Security Features for Strategic Information Management

In presentation of strategic data splitting and sharing algorithms it has become necessary to describe security features of linguistic algorithms used for information splitting and data reconstruction. The essence of this approach, representing an

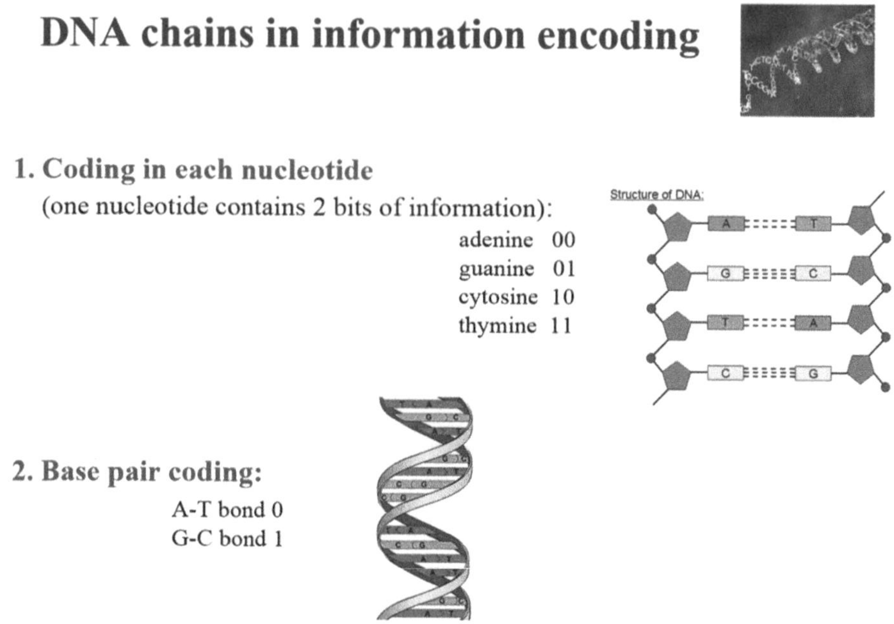

Fig. 45.1 Possible methods of coding information using DNA molecules

interdisciplinary topic straddling the border between the subjects of commercial organisation management and of information theory, is an attempt to use cryptographic methods more commonly applied in engineering and technical fields for purposes for which such solutions have not yet been proposed like economy and management sciences.

The proposed algorithmic solutions for data splitting and sharing have the following important properties and characteristic security features:

- Linguistic cryptographic threshold algorithms are suitable for dividing important strategic data and assigning shares to members from the authorized group;
- The algorithms are based on digital data (texts, images, voice recordings) which needs to be intelligently split among authorized persons, and then its secret reconstruction must be possible;
- There are wide opportunities to combine traditional methods of cryptographic information splitting ((m, n)-threshold schemes) with the presented protocols;
- The ability to present information in the form of its bit recording or sequences of blocks containing n bits;
- Introducing additional safeguards against the unauthorized reconstruction of the information and the possibility of implementing two independent versions of protocols for assigning the created shadows to individual protocol participants: the option with a trusted arbitrator intermediating in assigning and reconstructing the information and the option without an arbitrator (an additional

trusted party), but only with assigning the introduced grammar as an additional part of the secret;
- The ability to introduce restrictions of the length of coded bit blocks in the proposed scheme, as a result of which the defined grammar will not contain a large number of derivation rules;
- The computational complexity of the proposed schemes is polynomial.

The above characteristics of the linguistic algorithms of information division constitute their advantages and show how universal these proposed methods for splitting and sharing secret or strategic information in commercial organisation are.

45.5 Conclusions

In this chapter were described some advances in using biometric information to develop new procedures for secret information sharing called linguistic thresholds schemes. Processes of splitting strategic data are currently used in many fields of life, science and economy. Application of linguistic coding methods in the concealment and analysis processes, offers the full capability of using personal information for such purposes. Concealing biometric or personal data constitutes a very important problem because it is highly probable that personal data will be taken over by unauthorized persons. The individual DNA code and many other standard or non-standard biometrics may be used during sharing procedure [5, 6].

Acknowledgments This work has been supported by the National Science Centre, Republic of Poland, under project number DEC-2013/09/B/HS4/00501.

References

1. Cohen H, Lefebvre C (eds) (2005) Handbook of categorization in cognitive science. Elsevier, Amsterdam
2. Hachaj T, Ogiela MR (2012) Framework for cognitive analysis of dynamic perfusion computed tomography with visualization of large volumetric data. J Electron Imaging 21 (4):043017
3. Ogiela L (2008) Syntactic approach to cognitive interpretation of medical patterns. Lect Notes Artif Intell 5314:456–462
4. Ogiela L (2008) Cognitive systems for medical pattern understanding and diagnosis. Lect Notes Artif Intell 5177:394–400
5. Ogiela L (2009) UBIAS systems for cognitive interpretation and analysis of medical images. Opto-Electron Rev 17(2):166–179
6. Ogiela L (2010) Cognitive informatics in automatic pattern understanding and cognitive information systems. Studies in computational intelligence, vol 323. Springer, Berlin, pp 209–226
7. Ogiela L, Ogiela MR (2009) Cognitive techniques in visual data interpretation. studies in computational intelligence, vol 228. Springer, Berlin

8. Ogiela MR, Ogiela U (2009) Security of linguistic threshold schemes in multimedia systems. In: 2nd international symposium on intelligent interactive multimedia systems and services, Mogliano Veneto, Italy, 16–17 July 2009. New directions in intelligent interactive multimedia systems and services—2, Studies in computational intelligence, vol 226, pp 13–20
9. Ogiela MR, Ogiela U (2010) The use of mathematical linguistic methods in creating secret sharing threshold algorithms. Comput Math Appl 60(2):267–271
10. Ogiela MR, Ogiela U (2012) DNA-like linguistic secret sharing for strategic information systems. Int J Inf Manage 32(2):175–181
11. Ogiela MR, Ogiela U (2014) Secure information management using linguistic threshold approach. Advanced information and knowledge processing. Springer, London

Chapter 46
Research of ABAC Mechanism Based on the Improved Encryption Algorithm Under Cloud Environment

Long-xiang Zhang and Jia-shun Zou

Abstract To solve the data security problem based on the Attribute Based Access Control (ABAC) mechanism, an improved CP-ABE algorithm which is applicable to ABAC environment is proposed. The research is done from two aspects of access system structure and formal definition to study the ABAC mechanism. Then the formal definition of improved algorithm is given. The simulation and performance analysis is also finished. The algorithm has less storage consumption and higher efficiency under ABAC environment through the comparison with traditional CP-ABE algorithm.

Keywords Usage control · Attribute · Obligation · Condition

46.1 Introduction

Access control technology is one of the core of the information system security technology, the technology emerged in the 1970s, the original is in order to solve the mainframe grant access to Shared data management problems. Access matrix Based on the early Access Control policy of Access Control, Discretionary Access Control (DAC) and Mandatory Access Control (MAC), but with the development of computer network, the user node and the increase of the resources, the traditional Access Control policy already cannot satisfy the needs of practical application,

L. Zhang (✉) · J. Zou
School of Information Science and Engineering, Shandong Normal University, Jinan 250014, China
e-mail: 756387259@qq.com

J. Zou
e-mail: 1010336028@qq.com

L. Zhang · J. Zou
Shandong Provincial Key Laboratory for Novel Distributed Computer Software Technology, Jinan 250014, China

J.J. (Jong Hyuk) Park et al. (eds.), *Ubiquitous Computing Application and Wireless Sensor*, Lecture Notes in Electrical Engineering 331,
DOI 10.1007/978-94-017-9618-7_46

therefore the Role Based Access Control model (RBAC) [1]. For the research is still a hot spot of RBAC model, successively appeared ARBAC97 (Administrative RBAC97) [2], ARBAC02 extended RBAC model [3], etc. But under the Web environment, because of the great number of users, resource types, strong dynamic, when make the RBAC model, assign roles and permissions for the user to become a very heavy work [4]. And attribute-based Access Control (ABAC) can solve the complex information system of fine-grained Access Control and the massive user dynamic scaling issues, as an open network environment provides the ideal Access Control scheme, which is ABAC Access Control model to become the main reason for the new research hotspot in [5].

46.2 ABAC Policy Model

Here is no like DAC, MAC, RBAC model has been widely accepted definition ABAC model, so this article will attempt to give formal definition of ABAC. ABAC strategy including the subject, object, environment, operation, permissions, five elements, the system according to the subject, object and environment attributes to decide whether to permit the corresponding permissions.

46.2.1 Strategy Model Related Concepts

Definition 46.1 Entity attribute describes the physical properties of the variables. Can be abstracted into binary group A (name, Ran). This paper respectively with SA, OA, EA said main body, object attribute and environmental attribute.

Definition 46.2 ABAC can be abstracted into a quad U (SA, OA, EA, P). In which P said privilege set. Set {SA, OA, EA} represents the ABAC the collection of all attributes in the model.

Definition 46.3 Attribute predicate ap was defined as a triple (name, α, Ran), among them, the $\alpha \in$ {= indicates, < , or less, > , or greater, “, ”} as the operator, to limit the scope of property.

Definition 46.4 ABAC defined p is defined as the sign please (SAP, OAP, EAP, ACT), SAP, OAP, EAP see Definition 3, ACT = {act1, act2, … A collection of, act} said operations. Sign to permit or deny, values are positive and negative authorization.

46.2.2 Strategy Evaluation Related Concepts

Definition 46.5 Attribute name the specific values of NVP said properties, abstract for binary group (name, val), in which the name and val respectively corresponding to the name of the attribute and value.

Definition 46.6 User requests can be abstracted as quad the Req (SNVP onvp, envp, act) which SNVP, onvp, envp see Definition 5, act represents the user's request.

46.2.3 Based on the XACML Access Control Policy

Language XACML policy language is a kind of distributed strategy, for the same resource by different strategy management (PAP) for the different strategies. In XACML defined four combination algorithm to solve the conflict, and avoid unnecessary operation. Are several:

(1) refused to priority algorithm. Thought of the algorithm was once a rule or policy is applied to get a reject result, it returns the result as a refusal.
(2) the licensed priority algorithm. Thought of the algorithm is as long as there is a rule or policy application results for permission, it returns the results is licensing.
(3) the first application of algorithms. Thought of the algorithm is in the process of application of a set of rules or strategies, if there is a rule or policy can be applied, the algorithm to immediately stop, and return the application result of a rule or policy.
(4) the only application of algorithms. Thought of the algorithm is if there is only one strategy can be applied in the strategy, it returns the strategy application results [7].

46.2.4 The Decision-Making Model Based on XACML

ABAC strategy management basically follows the IETF proposed framework, the framework discussed the basic components and their mutual relations. Model diagram is shown in Fig. 46.1.

Context processor converts information into acceptable language conveyed to strategy information point (PIP). Strategic information point (PIP) from the subject, object and environment function module to obtain related attributes, the processor is returned to the context. After a certain amount of processing the information to the policy decision point (PDP) put forward the corresponding request. Policy decision point (PDP) according to the judgment of relevant policies, and the judgment

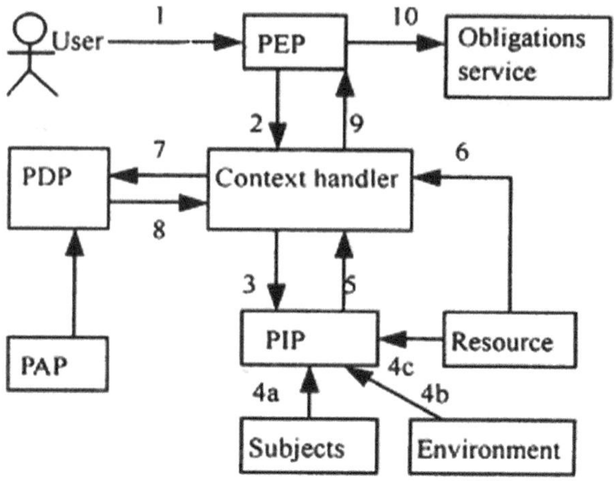

Fig. 46.1 XACML decision-making model figure

results returned to the policy enforcement point (PEP), by the policy enforcement point (PEP) to implement the decision results, submit a task to task service or refused to the user request [8].

46.3 CP-ABE Algorithm

46.3.1 Traditional algorithms of CP-ABE

Traditional CP-ABE algorithm mainly includes three parts:
Properties:

(1) set P = {P1, P2, …, Pn} for the collection of all attributes, is the attribute of each user C P a set of the loophole, C {P1, P2, …, Pn}.
(2) the access structure: access structure is complete {P1, P2, …, Pn} a set of the loophole, T 2 (P1, P2, …, Pn)\ ∅. T represent an attribute judgment conditions.
(3) access to the tree: the access tree is used to describe an access structure, each tree leaf nodes represent an attribute, each internal node represents a relationship between function AND relationship between function can be AND n of (n), OR (1 of n) AND n of m (m > n) threshold, etc. In the process of implementation, node traversal method first sequence traversal [9].

CP-ABE algorithm mainly includes four steps

(1) Setup. Generate the master key MK PK and public parameters.
(2) CT = Encrypt (PK, M, T). T use PK, access structure and data clear M, for CT data encrypted cryptograph.

(3) the SK = the KeyGen (MK, C). Using MK and user attributes to generate the user's private key SK C.
(4) M = Decrypt (CT, SK). Using the private key SK unlock CT plaintext M.

46.3.2 CP-ABE Algorithm in the Research of Access Control

For CP-ABE algorithm was applied to access control mechanism, scholars have done the corresponding research, such as document [10] proposes a cipher text access control mechanism based on CP-ABE algorithm. Song et al. [11] was studied based on the attribute based encryption cipher strategy (CP-ABE) algorithm of cloud storage security mechanism. And this article is proposed to improve the algorithm of CP-ABE, under the environment of ABAC for CP-ABE algorithm to the formal definition. And in several other typical access control under the environment of CP-ABE mechanism is used in the comparison.

46.3.3 Improve CP-ABE Algorithm

CP-ABE algorithm and ABAC strategies have in common is both are based on the properties and operation requires attribute collection was carried out on the subject. This article will try to give suitable for CP ABAC environment—ABE formalized definition of the algorithm.

Definition 46.7 CP-ABE with subject attribute SA (see Definition 1) said that each user's C is no longer a separate attribute set, C H {sap1, sap2, sap3, … The sapn}. Among them, the collection of H is subject attribute predicate the SAP each attribute can be predicate abstraction into a new attribute name.

Definition 46.8 T access structure is a subset of the set H, T H. Represents the authorized collection. T 2 (sap1, sap2, sap3, …, sapn)\ ∅.

Definition 46.9 access can form into a tree: sap1 problem sap2 problem sap3 problem… Because the sapn. Among them, the operator problem ∈ {∨, Sunday afternoon}. For the constraints of SAP.

After the above definition, CP-ABE algorithm can through ABAC original attributes, without having to set up relevant properties, thus save the storage space. Steps are basically the same with the traditional algorithm, the algorithm only use the access structure of T and user C name is replaced with subject attribute predicate name. The improved algorithm decryption diagram as shown in Fig. 46.2.

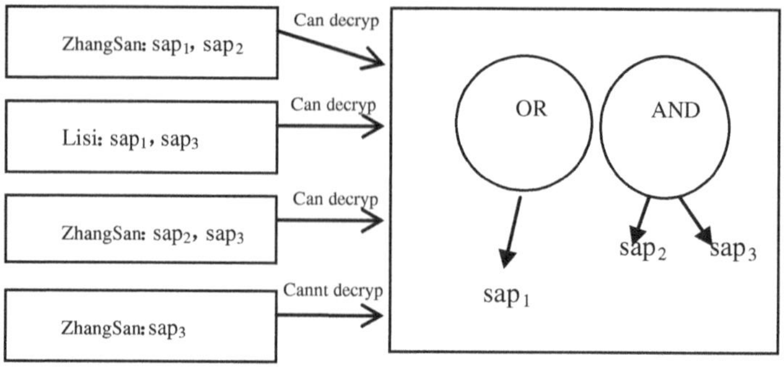

Fig. 46.2 Improve CP-ABE decryption schematic diagram

46.4 Simulation and Performance Analysis

In this paper, the experimental environment for: Intel Core 1.73 GHz CPU, 2 gb of memory, the operating system for Windows Server 2003, in the VMware Workstation tactical fix packs for 6.5.2 installed Ubuntu 10.10, given a 2 gb of memory. For test to improve the efficiency of the algorithm and the traditional algorithm in ABAC environment question has carried on the traditional CP-ABE algorithm and improved algorithm CP-ABE in same ABAC environment test efficiency.

Experimental results show that because do not need additional properties are stored, and strategy information point (PIP) auxiliary reduced CP-ABE time of information collection, so the improved algorithm in ABAC environment with lower storage cost and higher efficiency.

46.5 Conclusion

ABAC access control model is a hotspot of current research, compared with the RBAC model, ABAC model has strong extensibility, the advantages of fine-grained access control, has the stronger vitality. The RBAC to see as a kind of ABAC model based on single attribute. An improved algorithm of CP-ABE proposed in this paper can be well combined with ABAC access control model, improve the efficiency of data encryption and save the storage space.

References

1. Sandhu R, Coyne E, Feinstein H et al (1996) Role-based access control models. IEEE Comput 29(2):38–47
2. Sandhu R, Bhamidipati V, Munawer Q (1999) The ARBAC97 model for role-based administration of roles. ACM Trans Inf Syst Secur 2(1):105–135

3. Oh S, Sandhu R, Zhang XW (2006) An effective role administration model using organization structure. ACM Trans Inf Syst Secur 9(2):113–137
4. Zhang B, Zhang Y (2012) Based on the properties and role access control model. Comput Eng Des 33(10):3807–3813
5. Wang XM, Fu H, Zhang LC (2010) Based on the attributes of the access control research progress. J Electron 38(7):1660–1668
6. Cheng XR, Chen XY, Zhang B, Yang Y (2010) The access control policy model based on attribute. Comput Eng 36(15):131–134
7. Chen WP, Wang NN (2013) Based on the research of XACML strategy evaluation optimization technique. Comput Appl Res 30(3):900–906
8. Gao Y, Zhang GY, Wu M (2006) Based on the XACML and RBAC access control system. J Comput Appl Softw 23(8):65–67
9. Zhang HJ, Fan XF (2013) A trusted third party based on the CP-ABE cloud storage access control. J Wuhan Univ 59(2):153–158
10. Sun GZ, Dong Y, Li Y (2011) Cloud storage based on CP-ABE algorithm data access control. J Commun 32(7):146–153
11. Song KB, Luo J, Sun JT (2012) Based on cloud storage data protection mechanism of CP-ABE algorithm. J Huazhong Univ Sci Technol 40(SI):266–270

Chapter 47
A Novel Design of Education Video Personalized Recommendation System Based on Collaborative Filtering Recommendation Technology

Xiao Jun and Wang Min

Abstract With rapid development of internet, the expansion of information restricts the effect of learning. This paper designs an education video personalized recommendation system based on collaborative filtering recommendation technology. We design this system mainly based on collaborative technology and content analysis technology. This system can increase the effect of teaching, also can increase the learning autonomy of students and make students studying high-efficiency. This system has been applied to Shanghai Lifelong Learning Network and achieves significant influence.

Keywords Personalized recommendation · Collaborative filtering recommendation technology · Teaching resource · U-learning

47.1 Introduction of the Intelligent Retrieval Recommendation Technique

In the 21st century, ubiquitous learning is blooming as a new learning style. Education video resources account for a very large proportion in u-learning resources, however, there are still some problems in the application. For example, there is insufficient support in the switch and adaptive distribution and scheduling of video resources between different devices, education video resources can't realize the seamless and continuous learning between different devices.

X. Jun · W. Min (✉)
Shanghai Engineering Research Centre of Open Distance,
Education Shanghai Open University, Shanghai, China
e-mail: wangm@shtvu.edu.cn

X. Jun
e-mail: ecnuxj2003@163.com

J.J. (Jong Hyuk) Park et al. (eds.), *Ubiquitous Computing Application and Wireless Sensor*, Lecture Notes in Electrical Engineering 331,
DOI 10.1007/978-94-017-9618-7_47

Facing huge amounts of video resources, in addition to realize full-text retrieval service, we must also focus on the intelligent video recommendation. And we also consider the following factors: those resources associated with the contents, the current equipment as well as learners' personal character (hobby information, learners' portfolios (such as learning process, access history)). Based on the above information, the system can intelligently recommend more individualized video resources.

Implicit evaluation is adopted for this paper. Users can watch videos through the recommended video list or by searching. The viewing record of each time can be seen as an evaluation to the system without further evaluation by users, which increases the user experience. Through the viewing records each time by users, the system has done a comparison to obtain the accurate rating mark of the video. On receiving the viewing request of the video by users, recalculation of similarity can be conducted according to the set time, so as to re-search the adjacent users with similar interests and re-recommend videos.

47.2 System Introduction

47.2.1 Data Support Module

Data support module refers to an information database including four data tables, namely, user information table, learning behaviour data table, material information table and resource evaluation data table.

User information table: The personal information of users including the basic information in registration and other relevant information obtained through Web data mining is stored in this table.

Learning behavior data table: It is to preserve the learning behavior record of learners in the study process.

Material information table: It preserves various learning resource information such as courseware, cases, tests, news and literatures.

Material evaluation data table: It preserves the evaluation information of learning resources by learners. This table is the main data support for collaborative filtering algorithm.

47.2.2 Recommendation Engine Module of Combinational Algorithm

This engine is the core module of the recommendation system as well as the main centre for realizing personalized recommendation of learning resources. Its algorithm process can be summarized into the following few steps:

Step 1: retrieve the database and form the evaluation matrix of users and resources;

Step 2: calculate and define the data sparsity;

Step 3: according to the sparsity degree, the method can be selected to correct the collaborative filtering algorithm. Here we set a threshold value "Th-value" as the critical value to select the evaluation prediction or content filtering. When Sparsity <Th-value, it is considered that the system is in the state of "cold boot" or "pre-cold boot". At this time, the content filtering shall be selected as the correction for collaborative filtering algorithm. When Sparsity >Th-value, the evaluation prediction algorithm shall be adopted for correction;

Step 4: form the adjacent users and generate the recommendation courses finally, so as to form the recommendation list in accordance with TOP-K.

47.2.3 New Resource Recommendation Module

This module is mainly designed for the "cold boot" problem in collaborative filtering recommendation technology. Its main function is to analyze the interest, hobby and major (profession) categories of each learners, and recommend the latest resources in the relevant fields for them. However, if a newly added resource hasn't been accessed or evaluated by learners, it will never get the chance to be recommended by this system. By adding such module, the cold boot problem in collaborative filtering can be conquered to certain degrees, so as to improve the click rate of learning resources that have been newly added to the library. In Shanghai Lifelong Learning Network, the effective visit mechanism of the new courses is available.

47.3 System Features and Realization

Intelligent learning recommendation is the main feature for the personalized video resource recommendation system and realizes the personalized recommendation of the entire system based on content and collaborative filtering algorithm. In Shanghai Lifelong Learning Network where the personalized resource recommendation system is mainly applied to, the application of intelligent recommendation technology is especially prominent. Through the resource evaluation by adjacent users, the calculation process of recommendation aims to predict the evaluation on learning resources by targeted users.

47.3.1 Theoretical Modelling

(i) Presentation of the evaluation data
As is shown by Fig. 47.1, the evaluation information of the resource programs by users can be represented by a $m \times n$ order matrix $R(m, n)$, the row "m" of which refers to the number of users in the system, the column "n" means the number of resource programs in the system, and R_{ij} refers to the evaluation value of user i on the program j. $R(m, n)$ is known as the evaluation matrix of users and programs. The binary 0 and 1 can be used to represent the preference or purchase state of specific users for relevant programs as the evaluation value. Otherwise some evaluation value interval can be used to show the preference degree of users for the programs. The target of algorithm is to decide which programs shall be recommended to the target users by predicting the non-evaluated value in the matrix (Table 47.1).

(ii) Construction of the neighbourhood formation
One key step for user-based collaborative filtering algorithm is to calculate the neighbourhood formation of targeted users. As for the target user c, k (the number) users with the highest similarity degree shall be searched to form the neighbourhood formation to that user $N_c = \{c_1 c_2 c_3 \cdots c_K\}, c \notin N_c$. The users c_k in N_c shall be listed in descending order according to its similarity with user c $sim(c_k, c)(1 \leq k \leq K)$. Generally, the value range of $sim(c_k, c)$ is [−1, 1]. When $sim(c_k, c)$ is closer to 1, it means the similarity degree between c_k and c is higher,

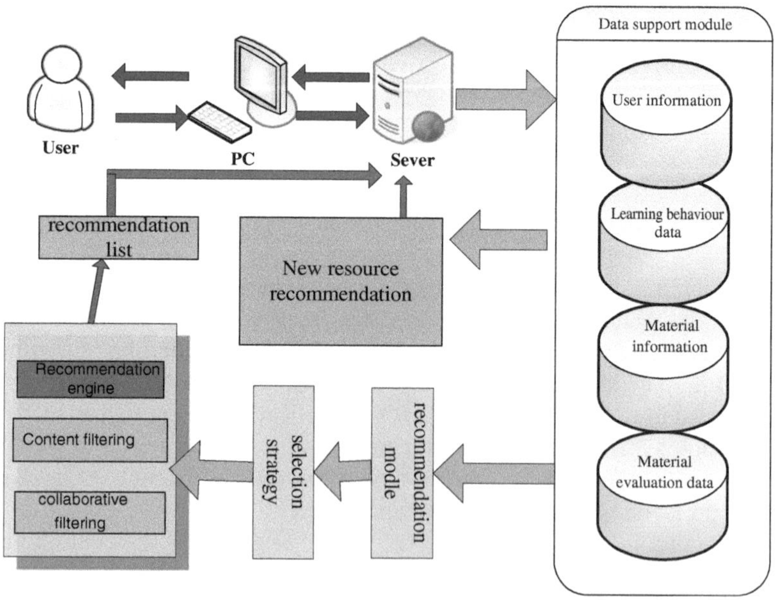

Fig. 47.1 Solutions for the personalized video resource recommendation system

Table 47.1 Evaluation matrix on resources by users

Item / User	I_1	I_2	…	I_j	…	I_n
U_1	R_{11}	R_{12}	…	R_{1j}	…	R_{1n}
U_2	R_{21}	R_{22}	…	R_{2j}	…	R_{2n}
…	…	…	…	…	…	…
U_i	R_{i1}	R_{i2}	…	R_{ij}	…	R_{in}
…	…	…	…	…	…	…
U_m	R_{m1}	R_{m2}	…	R_{mj}	…	R_{mn}

and when $sim(c_k, c)$ is closer to -1, it means that c_k and c owns the contrary interests and hobbies and the similarity degree between c_k and c is 0. In the user-based recommendation algorithm, there are three most common approaches: Pearson correlated coefficient, cosine similarity, and constrained Pearson correlated coefficient. Supposing R_{ui} and R_{vi} represent the evaluation of program i by user u and user v respectively, all three approaches have defined the common evaluation program collection of u and v.

- Pearson correlated coefficient

The Pearson correlated coefficient between user u and user v can be defined as:

$$sim(u,v) = \frac{\sum_{i \in I_{uv}} (R_{ui} - \overline{R}_u)(R_{vi} - \overline{R}_v)}{\sqrt{\sum_{i \in I_{uv}} (R_{ui} - \overline{R}_u)^2 \sum_{i \in I_{uv}} (R_{vi} - \overline{R}_v)}} \tag{47.1}$$

where $\bar{R}_u$ and $\bar{R}_v$ represent the average evaluation value of user u and user v on the joint evaluation program collection I_{uv}, then:

$$\bar{R}_u = \frac{1}{|I_{uv}|} \sum_{i \in I_{uc}} R_{ui}, \quad R_v = \frac{1}{|I_{uv}|} \sum_{i \in I_{uc}} R_{vi} \tag{47.2}$$

- Cosine similarity

In the included angle cosine approach, the evaluation information by users can be seen as the spatial vector on the n-dimensional program. The similarity of users is measured by the cosine included angle among the vectors. The smaller the included angle is, the higher similarity degree it shows. Taking vectors $\vec{u}, \vec{v}$ as the evaluation ones on the joint evaluation program collection I_{uv} by user u and user v, the included angle cosine between u and v can be defined as:

$$sim(u,v) = \cos(\vec{u}, \vec{v}) = \frac{\vec{u} \cdot \vec{v}}{\|u\| \times \|v\|} = \frac{\sum_{i \in I_{uv}} R_{ui} \cdot R_{vi}}{\sqrt{\sum_{i \in I_{uv}} R_{ui}^2 \sum_{i \in I_{uv}} R_{vi}^2}} \tag{47.3}$$

- Restraint Pearson correlated coefficient

The restraint Pearson correlated coefficient between user u and user v can be defined as:

$$sim(u,v) = \frac{\sum_{i \in I_{uv}} (R_{ui} - R_{med})(R_{vi} - R_{med})}{\sqrt{\sum_{i \in I_{uv}} (R_{ui} - R_{med})^2 \sum_{i \in I_{uv}} (R_{vi} - R_{med})^2}} \quad (47.4)$$

where R_{med} refers to the mid-value of the system evaluation. In Shanghai Lifelong Educational Network, the evaluation interval of courses by learners is from 1 to 5, and only the integer can be taken, so the mid-value of R_{med} will be 2.5 point.

The result can be calculated by making use of the user similarity introduced above, so as to generate a m × m order matrix S (m, m) to store the user similarity. The elements in S show the similarity between user u and user v. As the user himself or herself cannot be chosen as the nearest adjacent user, the diagonal elements of matrix S can all be 0. After getting the user similarity matrix S, the neighbourhood formation shall be determined.

In general, there are three ways: The first is to set the similarity threshold value by the system, then select all users whose similarity with the target user is greater than threshold users as its neighborhood formation. The second way is to set the nearest adjacent user number K by the system, and obtain the top k users with highest similarity with the target user to add to its neighborhood formation. The third way is to combine the above two methods, that is to say, the top K users with greatest similarity whose target user similarity is greater than that of threshold users shall be selected to join the neighbourhood formation of the target user. If there are less than k users with greater similarity of target users than threshold users, only this part of users shall be selected to join the nearest adjacent user collection.

(iii) Generation of the recommendation list
As can be seen from (47.2), the neighbourhood formation of the target user u is $N_u = \{u_1 u_2 u_3 \cdots u_K\}$, the two kinds of recommendation results can be calculated: one is the generation of evaluation on specific program i by the target user u, the other is the generation of the top-N recommendation list for target users. The simplest way is to use the average evaluation on program i by all users in the collection directly to represent the predicted value P_{ui}, then:

$$P_{ui} = \frac{1}{K} \sum_{v \in N_u} R_{vi} \quad (47.5)$$

Formula (47.5) treats all users in the neighbourhood formation the same. However, users with different similarity degrees actually have different influence on the evaluation by the target user. Therefore, the prediction result with this method is with great different with the actual evaluation value, and an improvement way is to calculate the weighted average of adjacent users' evaluation, which is the most widely used approach by far:

$$P_{ui} = \frac{\sum_{v \in N_u} sim(u,v) \cdot R_{vi}}{\sum_{v \in N_u} |sim(u,v)|} \tag{47.6}$$

Based on the above features, the combined recommendation way on the basis of content and collaborative filtering is adopted in the design of personalized course recommendation system as the main approach to realize the algorithm.

47.3.2 Concrete Realization

The technological principle of personalized video recommendation system can be explained by the following case:

As is shown in the following course information table, it is assumed that there are four users. 1 represents that the course has been taken, and 0 means hasn't. All five courses are from Shanghai Lifelong Learning Network. They are "Learn to Speak Shanghai Accent", "Commercial Street of the Old Shanghai", "Style Presentation of Changning District", "Expo Pavilions", and "Style Presentation of Putuo District" (simplified as "Learn to Speak", "Commercial Street", "Changning District", "Expo" and "Putuo District"). The course taking behaviours of the four users are shown as Table 47.2.

Taken user 1 of the above table as an example, the simple calculation process of collaborative filtering algorithm of personalized course recommendation can be explained:

- According to the course taking behaviors of the user, the vector of "resource-user" is generated. When transforming each clause of the course division in the table into vector, the clauses along with their correspondent vectors are as follows: "Learn to Speak": Vec1 = <1,1,1,0>; "Commercial Street": Vec2 = <0,1,1,0>; "Changning District": Vec3 = <1,1,0,0>; "Expo": Vec4 = <0,0,0,0>; "Putuo District": Vec5 = <0,1,0,1>.
- The similarity of two resource courses shall be calculated with the adoption of cosine similarity calculation way. User 1 has taken the courses of "Learn to Speak" and "Changning District". Their similarity with other courses is shown below: "Learn to Speak": Sim12 = 0.82, Sim13 = 0.82, Sim14 = 0.82, Sim15 = 0.41; "Changning District": Sim21 = 0.82, Sim23 = 0.5, Sim24 = 0, Sim25 = 0.5.

Table 47.2 Course taking information table of users

	User 1	User 2	User 3	User 4
Learn to speak shanghai accent	1	0	1	0
Commercial street of the old shanghai	0	1	1	0
Style presentation of changning district	1	1	0	0
Expo pavilions	0	0	0	0
Style presentation of putuo district	0	1	0	1

- Through comparison of the generated recommendation result, "Commercial Street" and "Changning District" are with the greatest similarity with "Learn to Speak", and "Learn to Speak", "Commercial Street" and "Putuo District" are with the greatest similarity with "Changning District". Finally, the recommendation system selects the intersection of the two and recommends the "Commercial Street" to user 1.

47.3.3 Check and Evaluation

The check and evaluation of personalized video recommendation system is conducted from the following few aspects:

- Data record: (1) Record on the recommendation site and window site: The user click time for the recommendation course, login users, recommendation course ID, original course ID (specific to FILTER3). (2) Record on the recommendation course: The time summation for the appearance of the recommendation course shall be recorded.
- Check method: (1) Successful rate of recommendation: the proportion of the show-up times and click times of the recommendation course. (2) Successful rate of course selection: the proportion of recommendation course click times and course selection number.

Through the record of original data, we can use the above steps to calculate the property of the personalized course recommendation engine. Based on the systematic tests, it is found that this engine could excavate the interest of users effectively with high successful rate of recommendation and course selection.

Taking Shanghai Lifelong Learning Network as an example, a use has just finished the course of fund transaction, the window will show up the recommended courses for relevant study. In addition, as the job orientation of the registered user is engineer, we recommend the "Engineering Mathematical Method" and "Design Algorithm Introduction" to him or her for further study according to the interest direction that the user stored in the database. Therefore, it can be seen that this personalized course recommendation system could excavate and satisfy the demands of users well.

Additionally, we can also recommend high-quality courses with high evaluation by similar users on the basis of the collaborative filtering algorithm in Shanghai Lifelong Learning Network. For example, when a course of "Nutrition and Diet Therapy for the Elderly" is clicked by a user, the most similar users with that user can be found out through collaborative filtering algorithm based on the user information table and learning behaviour table stored in the database, and then the course with best evaluation by those similar users will be recommended to that user by showing up on the window. By searching with the algorithm, courses are recommend to the user, and the theme of all courses owns satisfactory consistency. Besides, it verifies that good recommendation effect can be achieved by the personalized course recommendation system based on collaborative filtering algorithm.

47.4 Application Effect and Conclusion

The personalized video resource recommendation system is more convenient for users to check the courses, which saves a lot of browsing and course finding time to a large degree and allows more users to be involved in the Lifelong Learning Network. In the operation of its resource alliance platform, over 80 % of the trainees believe the course content on Lifelong Learning Network is abundant and approximately 90 % of users speak highly of it and consider conducting lifelong study on it. The resource alliance platform of Shanghai Lifelong Learning Network has accumulated more than 8,000 video courses, and its click rate has been increased by 12.4 % on year-on-year basis ever since the application of personalized video resource recommendation system.

As for personalized video resource recommendation system, it could provide learning resources to users specifically along with better support for their personalized study. Therefore, these systems are with great user experience and convenience, which could carry out intelligent and efficient management on multimedia resources effectively.

Acknowledgments This research was supported by Engineering Technology Research Centre of Shanghai Science and Technology Research Program (13DZ2252200), and also supported by Innovation Program of Shanghai Municipal Education Commission.

References

1. Zhang Z, Liu H (2007) Research on video retrieval using high-level semantic. Comput Eng Appl 43:168–170
2. Hu S, Li J, Li J (2007) Video retrieval based on latent semantic analysis. Comput Eng 33:216–217
3. Wengang C, De X (2002) Content-based video retrieval using audio and visual clues. In: IEEE Proceeding of 2002 region 10 conference on computers, communications, control and power engineering, Beijing, China, 28–31 Oct 2002, pp 586–589
4. Liu J, Zhou T, Wang B (2009) Research progress of personalized recommendation system. Prog Nat Sci 19:1–15
5. Breese J, Hecherman D, Kadie C (1998) Empirical analysis of predictive algorithms for collaborative filtering. In: Proceedings of the 14th conference on uncertainty in artificial intelligence, Medison, US, 24–26 July 1998. Morgan Kaufmann Publishers Inc., San Francisco, pp 43–52

6. Deng A, Zhu Y, Shi B (2003) A collaborative filtering recommendation algorithm based on item rating prediction. J Softw 14:1621–1627
7. Bong RP, Iacovou N, Suchak M (1994) Group lens: an open architecture for collaborative filtering of Netnews. In: Proceedings of the 1994 ACM conference on computer supported cooperative work, Chapel Hill, NC, 22–26 Oct 1994. ACM, New York, pp 175–186
8. Zeng C, Xing C, Zhou L (2003) A personalized search algorithm by using content-based filtering. J Softw 14:999–1004

Chapter 48
The Design of a Medical Rules Synchronization System

Yi-Hsing Chang, Leng-Kang Chang Chien and Rong-Jyue Fang

Abstract The study incorporates the semantic web to achieve the synchronization for medical rules. Web Ontology Language (OWL) is first used as the knowledge model and a medical rule is viewed as a class of medical statuses representing knowledge of medical rules. A classification is then applied to decide the legality of medical usage. The characteristics of the proposed solution are manifold. OWL supports the syntax of class inherences, so it can develop a cognitive economy and shorten the encoding length of medical rules. OWL syntax can represent the styles of most medical rules, so the automatic synchronization system only need one document and one inference engine to handle all medical rules. OWL simplifies the design of the medical rules of National Health Insurance automatically, and synchronizes the design to hospital information systems. Therefore, doctors' ordering systems rapidly reflects the changes of medical rules.

Keywords Synchronization system · Semantic web · Medical rule · Hospital information system

48.1 Introduction

In Taiwan, after the implementation of National Health Insurance (NHI), abuse of medical resources has become a serious issue. To control abuse the medical resources, the Bureau of NHI has formulated rules and restraints for medical

Y.-H. Chang (✉) · L.-K.C. Chien · R.-J. Fang
Department of Information Management, Southern Taiwan University of Science and Technology, No. 1, Nan-Tai Street, Yongkang, Tainan 710, Taiwan
e-mail: yhchang@mail.stust.edu.tw

L.-K.C. Chien
e-mail: fhopecc@gmail.com

R.-J. Fang
e-mail: rxf26@mail.stut.edu.tw

J.J. (Jong Hyuk) Park et al. (eds.), *Ubiquitous Computing Application and Wireless Sensor*, Lecture Notes in Electrical Engineering 331,
DOI 10.1007/978-94-017-9618-7_48

services provided by medical associations. This forces hospitals to instruct doctors to consider the health insurance rules when they prescribe medicine. It also discourages doctors from prescribing too much, thus, avoiding very big losses for hospitals. When doctors prescribe medicine, they need a good diagnosing and treating system to provide information about rules.

The artificial sequence for renewing medical rule from the Bureau of NHI to the hospital information systems is as follows. When the Bureau of NHI changes medicals rules, it transmits to each coherent unit through the archives to announce the renewed command. After medical administrators of the hospital receive the archives, they will enforce the rule accordingly in the hospital situation, and propose revision formulas in hospital meetings. Then, the hospitals will send the formulas to the hospital information center to apply for revision. After completing system analysis, the programmer revises the formula and deploys it to its system again. At this stage, the new rule is actually carried out in the hospital's information system. Because the Bureau of NHI makes rule changes for newly invented drugs, policies, and new medical research, hospitals have to stay up-to-date on changes to medicine and drug control formulas; moreover, the same formulas must actually be integrated in each hospital system. This process wastes a lot of manpower. Therefore, it is necessary to design a system which allows the Bureau of NHI and hospital information systems to renew medical rules immediately, synchronously and automatically. Hence, developing this type of effective renewal system is the main purpose of this study.

48.2 Literature Review

48.2.1 Rules Deduction Model

A rules control system must depend upon the rule deduction function to apply and control drugs prescription behavior. A rules deduction system may classify each case situation to a suitable rule. After classification, a case should be handled according to whether it is within the rules (legal act), or it is outside the rules (tort). Skalak [1] induced a rules deduction system according to a disaggregated model, dividing the system into three types of models: classical picture, probability type and case models.

Taiwan's legal framework is primarily based on written law and the rules for drugs are quite clear. The classical model can be used to adequately implement a rules deduction system. Therefore, this study's rules deduction module is mainly implemented based on the classical model.

48.2.2 Drug Rules Checking System

At present, the main application of Taiwan's drug rules checking system is to act as a warning signal. That is, the system is triggered to provide doctors with knowledge related to the prescription of medicine immediately and to promote medication security when doctors prescribes the medicine.

In the other hand, Lau et al. [2] informed HIS practice and research by consolidating existing evidence from published systematic reviews on health information system (HIS) evaluation studies. Fifty reviews published during 1994–2008 were selected for meta-level synthesis and covered the five areas: medication management, preventive care, health conditions, data quality, and care process/outcome. Thus, the check of drug rules is an important issue.

48.2.3 Sharing Medical Information

Ouziri and Verdier [3] mentioned that medical data are stored on multiple health information systems which are heterogeneous and non-communicating and then difficult to get a complete and consistent long-life medical record. They proposed a user interface in which the patient's medical records rebuilt by the end-user himself in a simple interface where concepts are linked automatically together. The user can navigate in this space of concepts to obtain information lie needs, as easily as in a web site.

Stanfill et al. [4] evaluated all types of automated coding and classification systems to determine the performance of such systems. Automated coding and classification systems themselves are not generalizable, nor are the results of the studies evaluating them. The research showed that these systems hold promise, but these data must be considered in context, with performance relative to the complexity of the task and the desired outcome.

The Bureau of NHI controlling each hospital's information system is the same case. Its information processing system is essentially composed of each heterogenic, an independent design, a semiautomatic-like hospital system and the Bureau of NHI system. If the Bureau of NHI were to adopt medical information centralism, the conversion and transmission costs would excessively high, for both hospital systems and the Bureau of NHI. Regarding this question, Minsky and Ungureanu [5] proposed a Law-Governed Interaction (LGI) design principle to solve problems related to the centralized sharing of information.

For the proposed system to conform to the LGI principles, the first demand is the effective description of medical service rules, with the majority of the rules described through the sole and consistent methods. Therefore, this study induces the semantic networking model for the effective description of medical service rules.

48.3 The Medical Rules Automatic Synchronization System

48.3.1 Medical Rules Synchronization

Figure 48.1 shows the synchronization sequence for medical rules, where MARS denotes the medical rules synchronization system proposed by this study.

To achieve the purpose of medical rules synchronization, the Bureau of NHI, MRAS, HIS and doctors, contained in the system must carry out specific independent functions respectively. By the mechanism of decomposing and appointing functions to different units, the redundant rule control codes of the hospital applications, as well as the control scope of the Bureau of NHI, can be reduced. In addition, this design conforms to the LGI principle.

48.3.2 MRAS System Architecture

The MRAS system architecture is shown in Fig. 48.2, containing the Jena function module, the inference engine module, the MRASP module and the knowledge ontology of medical rules. All documents are first presented by OWL. The MRAS applies Jena to analyze OWL documents, infers the implicit rules the model by the OWL inference function of Jena and makes them coexist in the memory to increase the efficiency of inference engine. The implicit rule means that it does not clearly appear in the original OWL model but the facts can be obtained by the inference rules defined in the OWL model.

1. OWL document:
   This is the machine storage form for the health insurance rules. The OWL documents express the meanings of health insurance rules and can be easily compiled literally by the program.
2. Jena function module:
   Jena is a group of tools processing RDF and OWL semantic network technology and is adopted as the OWL document parser.

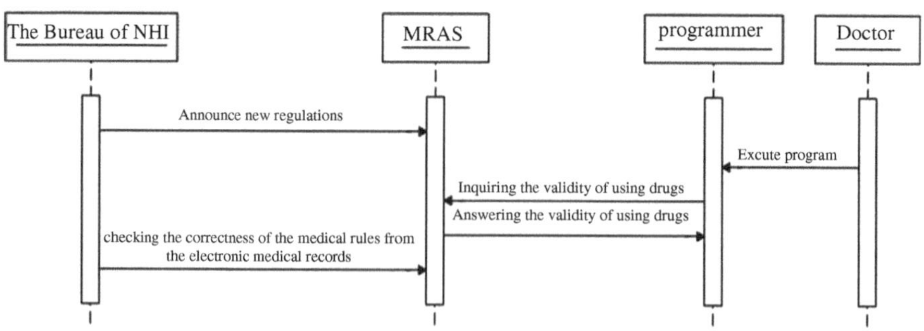

Fig. 48.1 Synchronization sequence for medical rules

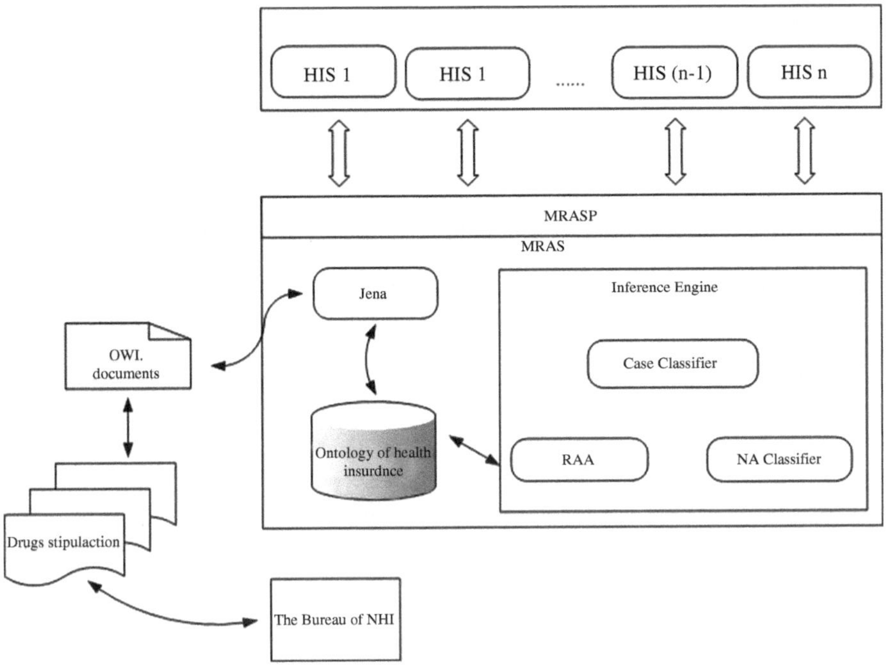

Fig. 48.2 Medical rules automatic synchronization

3. Inference engine module:
   This module distinguishes legitimate functions of medication behavior.
   (a) RAA module: this module calculates and obtains the medical status information of valid medication needs based on the medicals passed from the HIS by the MRASP.
   (b) Case Classifier module: this module executes classification based on the drugs and medical treatment passed by HIS. If a case is categorized as a legitimate medical treatment, then it is legitimate; otherwise, it is an illegal medication treatment.
   (c) NA Classifier module: this module supports numeric sector classification for the Case Classifier module, which originally was not serviced by the OWL.
4. MRASP module: by this agreement, a hospital's HIS can easily obtain the services of medical rules by the MRAS.
   Knowledge ontology of medical rules: It is based on the inference of the facts of health insurance rules; the rules inference engine still inspects prescription legitimacy based on the ontological description.

48.4 System Design

48.4.1 Medical Rules' Knowledge Ontology

In this study, the knowledge domain is limited to health insurance medical payment rules. The essence of medical legislation is based deciding whether or not the medical treatment for the present medical condition is legal. Therefore, the ontology of a health insurance rule is constructed as in Fig. 48.3.

The root category is "medical treatment". The next level which is the subclass of "medical treatment" expresses that all categories in the level belong to the medical domain knowledge.

The "medical condition" shows the center of the entire medical activity and refers to each person when they go to see a doctor. Therefore, many attributes are defined in "medical condition", such as "medication", "dosage", "prescribing doctor", "patients' gender", "age", etc., to precisely describe the medical activity.

"Medical regulation" essentially carries out the classification of "medical condition". Therefore, "medical regulation" is defined as a subclass of "medical condition". In addition, we increased the numeric sector attributes, such as "patient body weight is more than", "patient takes the medicine cycle longer than", etc.

Since "medical regulation" can be applied to describe the "health insurance rules", the "regulation of NHI self coverage" is therefore defined as a subclass of "medical regulation". The subcategories of the "regulation of NHI self coverage" roughly follow the stipulation stratum, as shown in Fig. 48.4.

Fig. 48.3 The ontology of health insurance rule

Fig. 48.4 The ontology of regulation of NHI self coverage

48.4.2 OWL Design

48.4.2.1 Medical Condition

"Medical condition" contains all of the information medical behavior such as the medication, prescribing doctor, patient gender, etc. should record. We use a group of standard diagnosis codes, ICD9, to express the disease conditions.

48.4.2.2 Medical Rules

"Medical rules" is essentially the classification of the "medical condition". Thus, the examining operation of the medication rules might inspect whether a medical case obeys the regulations.

48.4.2.3 Set Attributes Value to Define "Medical Rules"

A class is essentially a group of cases. The following two expressions are used to describe a set of cases.
1. Enumerated type:
For example, defining sector category 1 … 10 as
Class Range (1 … 10) = {1, 2, 3, 4, 5, 6, 7, 8, 9, 10}
2. Attribute value limit:
For example, defining sector category 1 … 10 as
Class Range (1 … 10) = {for t of Range (1 … 10), 1 < = t < = 10}

48.4.2.4 Define the Value Sector Category

The OWL limit is incapable defining the value sector category. That is, we are unable to use the category operation to define the value sector category. For example, the "medical rules" category of "patient age less than 70" cannot be defined by OWL. To carry on value sector category classification, "the medical rules" must define a new attribute, "patient age less than", where its attribute value scope is the same with "patient age", and its semantics surmount the OWL model definitions of classified semantics. Therefore, we carry out extra operations with the inference engine.

1. Implement the semantic logic of the value sector category limitation attribute.
2. Implement arithmetic logic.

48.4.3 Mrasp

This study utilizes the bank counter pattern to obtain the rules service to redundant information transmissions.

48.4.3.1 Bank Counter Pattern

As shown in Fig. 48.5, although the hospital system must carry out requests twice, however, so long as the necessary information of medication validity is uploaded to MRAS, unnecessary network traffic is dramatically reduced and processing efficiency is promoted.

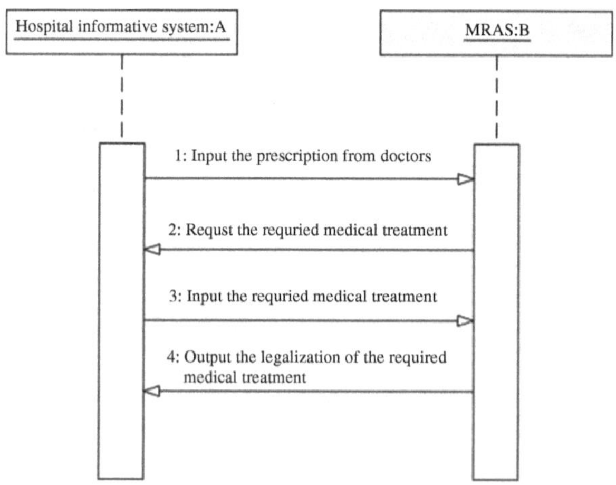

Fig. 48.5 Bank counter pattern flow sequence

48.4.3.2 Message Format of the MRASP

Five messages are defined for the MRASP to realize the bank counter pattern:

1. USEMED message: this is the medical information passed from HIS to the MRAS system to indicate for which medicine medication validity needs to judged. Afterward, the MRAS returns the determined validity of the medical condition.
2. REQ_PROPS message: this is the message for the MRAS to reply to the USEMED message passed by HIS; it indicates the medical condition for which medication validity needs to be determined. Afterward, the hospital system should enter the requested medical record to allow the MRAS complete the assessment of medication validity.
3. MED_STATUS message: let the HIS transmit the medical service condition. Then, the MRAS makes judgment according validity of the medication.
4. LEGALITY message: MRAS replies whether the drug is legitimate.
5. INVALID message: Both sides respond with the wrong message.

48.4.4 Inferences Engine

The inferences engine contains the following four modules.

1. RAA module: this module assesses the desired information of the medical condition to judge medication validity.

To achieve the bank counter pattern, the inference engine must assess the medication attribute of the medical condition to deduce the information which the user should provide to the MRAS. The following three steps show the operation logic of RAA.

It first defines the "medres" to represent the categories of medicine rules which are the same as the medication prescription transmitted by user.

The medres then defines all of its subclass "medcls", to express all limits to the category of this medicine.

All attribute limit category "restrictions" are extracted again and inherited by medcls. By the "on Property" attribute value in the restriction category, all attribute limits of this medication are indicated.

2. Case Classifier module:
This module analyzes the OWL document and follows the ontology model described by the OWL document to create the classification operation for the transmitted medical condition in question and simulate the rules examining the operation.

3. NA Classifier module:
This module focuses on the numeric attribute limitation operation and its rule is:

IF an individual case attribute is classified as a numeric and as a numeric sector THEN

IF the case attribute is located in the numeric sector THEN
It passes as a legitimate attribute. ENDIF
ELSE Call the case classification module to continue another attribute classification operation. END IF

48.5 Conclusion

A medical rules synchronization system based on semantic web techniques is implemented in this study. By the following design strategies, the system enables the Bureau of the NHI to carry out its rules policy and to manage medical resources properly and effectively and enables hospitals to reduce the system costs of information changes. Furthermore, it promotes the government's ability to implement electronic rule controls to a wider range of goals. The OWL is used to encode. However, OWL only inputs the fixed rules category definition while the more implicit rules are deducted after Jena execution. Thus, the machine coding length of the medical rules is reduced and the transmission cost for renewal rules is decreased. The medical rules are regarded as a medical condition category. To evaluate a medication's validity, we simply assess whether the medical condition case transmitted by a doctor belongs to the legitimate medical condition category. Therefore, the inference engine needs only to implement the case classification operation. After that, we just update the category definition information to achieve the purpose of rule renewal.

The OWL grammar is enough to describe the majority of drug rules. The system can use one document file to describe all rules concisely by using the OWL model. This simplifies the design of HIS to renew the health insurance pay stipulation. The inference engine can implement the continuously revised OWL files and actually function to synchronize rules. The data and the inference engine are completely separated in the system design, it easily divides each piece work into different entities to reduce the works.

Acknowledgment This research is supported in part by the National Science Council of Republic of China under the contract number NSC102-5111-S-218-001.

References

1. Skalak DB (1989) Taking advantage of models for legal classification. In Proceedings of the 2nd international conference on artificial intelligence and law, pp 234–241
2. Lau F, Kuziemsky C, Price M, Gardner J (2010) A Review on systematic reviews of health information system studies. J Am Med Inform Assoc 17(6):637–645
3. Ouziri M, Verdier D (2008) Semantic data integration and navigation in Web-based medical records. Inform Health Soc Care 33(3):191–210

4. Stanfill MH, Williams M, Fenton SH, Jenders RA, Hersh WR (2010) A systematic literature review of automated clinical coding and classification systems. J Am Med Inform Assoc 17 (6):646–651
5. Minsky NH, Ungureanu V (2000) Law-governed interaction: a coordination and control mechanism for heterogeneous distributed systems. ACM Trans Softw Eng Methodol 9(3):274–305

Chapter 49
Associative Recommendation of Learning Contents Aided by Eye-Tracking in a Social Media Enhanced Environment

Guangyu Piao, Xiaokang Zhou and Qun Jin

Abstract In this paper, an approach to presenting the learning resources, especially those existing user-generated contents associated with learners' activities, as the recommendation to satisfy their current requirements in a social media enhanced learning system, is proposed. Users' attentions are caught and analyzed from the browsing behaviors of learners on a webpage through an eye-tracking device.

Keywords Browsing behavior · Eye-tracking · Associative recommendation · Social media

49.1 Introduction

Social media enhanced learning systems provide a new learning paradigm with abundant resources and communications. Recommender systems, as a type of personalization service to support users' online searching, have been widely used in recent years. How useful information is going to be recommended to users has become an ever-growing issue. In the collaborative learning environment, learners can use social media to exchange ideas and share information instantly [1]. With the support from social media, learners can easily share information, including their learning contents and progresses. Social media can make learning process effectively [2]. Social media, such as Twitter, is often considered as an e-learning tool [3]. Learners can read tweets from participants such as classmates, instructors, friends, and etc. via the Twitter. When the learners are searching information on

G. Piao · X. Zhou · Q. Jin (✉)
Graduate School of Human Sciences, Waseda University, Tokyo, Japan
e-mail: jin@waseda.jp

G. Piao
e-mail: piao@fuji.waseda.jp

X. Zhou
e-mail: xkzhou@ruri.waseda.jp

J.J. (Jong Hyuk) Park et al. (eds.), *Ubiquitous Computing Application and Wireless Sensor*, Lecture Notes in Electrical Engineering 331,
DOI 10.1007/978-94-017-9618-7_49

timelines, in many cases, learning contents, teachers and students' conversations, questions and answers are mingled, which forces learners to spend a lot of time on searching for the useful contents. Therefore, it is necessary to utilize the analysis of the information seeking behaviors of learners to figure out their current attentions which can facilitate the arrangement of learning resources.

This study focuses on the extraction of attentions of learners while they are browsing the learning contents. With the development of the eye-tracker, it is possible to measure the position and motion of the eye gaze more accurately and precisely [4], which also provides us a way to record and analyze the browsing behaviors of learners to assist the learning contents organization process and further benefit the associative recommendations in the social media enhanced learning environment.

In our previous study, we proposed an integrated approach to analyze and extract gaze patterns in order to capture users' unselfconscious information behavior based on the eye gaze movement data which is collected by an eye-tracker and then processed by our proposed method [5]. We showed a set of experiment results based on the analysis of two patterns in terms of the reading efficiency using eye gazing data, in order to demonstrate how the subjects look for specified keywords in the Twitter timeline, which can further contribute to categorization of viewing patterns [6].

In this study, a new approach that recommends useful learning contents for a specific learning purpose associatively is proposed for a social media enhanced learning system, in which we analyze learners' eye movement on the Twitter pages (e.g., specific keywords, pictures, icons, etc.) to figure out the learners' current attentions. The locations and the learning contents are linked together to extract and analyze the browsing behaviors for learning contents aided by eye-tracking technology.

49.2 Related Works

Eye-tracking has been widely applied in the research field of human-computer interaction, usability study, etc. Buscher et al. used the eye-tracker to detect a user's gaze behaviors, and their results in the experiment indicated that gaze-based feedback is very useful and can greatly improve the quality of Web search [7]. Xu et al. developed a new online content recommender system for documents, images and videos [8]. Yoshitaka et al. described a method of information recommendation based on social filtering, in which the preference of users was implicitly acquired through gaze detection [9]. Giordano et al. presented a proactive content based recommender system that employed web document clustering by using eye gaze data [10]. Chen et al. investigated the efficacy of recommender interface design in affecting users' decision making strategies through the observation of their eye movements and product selection behavior [11]. Goldberg et al. extended several compact visual scan path representations, which can provide additional insight about scanning strategies [12].

49.3 Pre-process for Associative Recommendation

We firstly pre-process the eye-gaze, in which two basic ways to seek for the learning contents are categorized. One is seeking learning contents on the basis of keywords, and another one is seeking learning contents on the basis of icon. When the learner starts to get gaze on the screen, the fixation time of gaze point over a certain threshold time will be recorded. As shown in Fig. 49.1, at the beginning, when a specific learner begins to seek information, the calculation of gazing time of each fixation starts simultaneously. If the gazing time is bigger than the pre-defined threshold, this fixation will be recognized. Continuously, if the identified gaze point is on the icon, we extract the corresponding user ID for the icon, and store it into the User ID Database. If the identified gaze point is on the text area, the corresponding keyword in a given area of the gaze point will be extracted and stored into the Keyword Database. The extraction process will repeated until the learner finish the information seeking process.

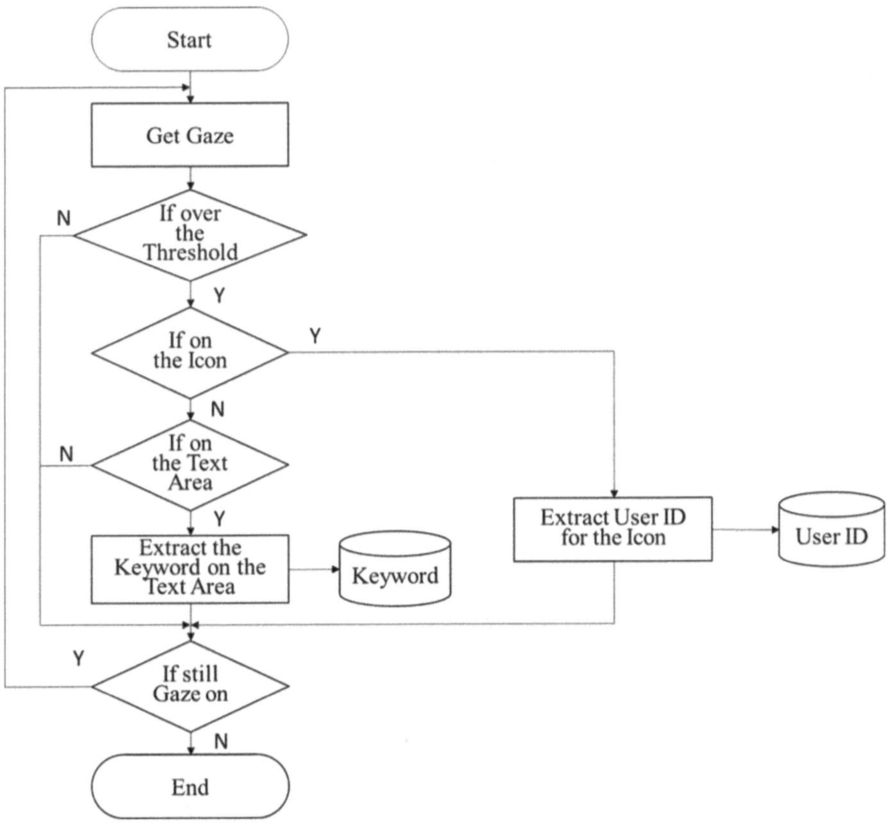

Fig. 49.1 Pre-processing of gaze: seeking by keyword or by user icon

As for seeking learning contents on the basis of keywords, when the learner is gazing on the keyword extracted from the gaze point, he/she may have questions in terms of the extracted keyword. Thus, in this situation, the learning contents related to this specific keyword should be retrieved and provided to this learner.

Specifically, the data in Q&A, FAQ, and key point tweets are employed to compose the related learning contents for the recommendations. As shown in Fig. 49.2, if the extracted keyword is related to the data in the Q&A, which means this specific learner may intend to read some explanation or discussion related to a specific keyword, we match the keyword with the contents stored in the Q&A database and recommend the related answers to him/her. We have the FAQ for the learners' frequently asked questions. Thus, following the answers or explanations given by others from the Q&A, and if the extracted keyword is related to the data in the FAQ, which shows that the learner may be interested in the FAQ, we match the keyword with the contents stored in the FAQ database and recommend the related answers to him/her. In this case, the related learning contents frequently asked from the history data will have the higher priority to be recommended. During some discussions, the instructor may summarize some key points of the important knowledge. And in some time, the TA may arrange the main points on some issues after the discussions. Meanwhile, the classmates who have been involved into the discussions will also post sort of contents related to the key points. All these will be

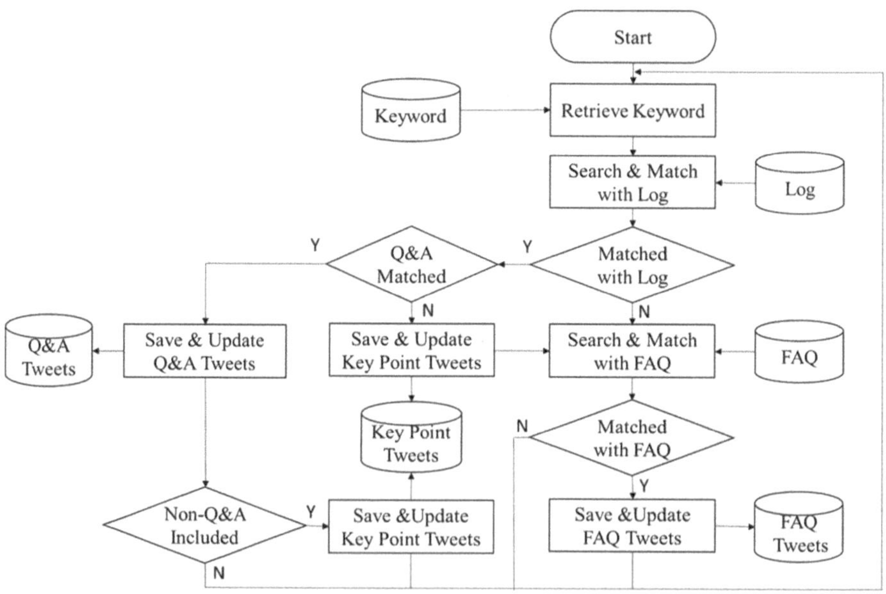

Fig. 49.2 Flowchart to match the content based on the keyword

organized and stored in the key point database. Thus, if the extracted keyword is related to the data in the key point tweets, we match the keyword with the contents stored in the key point database and recommend the related contents to him/her.

As for seeking learning contents on the basis of icon, when the specific learner is seeking the information through the timeline of Twitter, some other learners may post lots of useful information in the timeline. Therefore, he/she may search for these persons to read all their posts during a specific time period.

In details, as shown in Fig. 49.3, when a specific user icon is identified, we will retrieve the corresponding user ID by analyzing the link between the user ID and icon we defined in the database. If the user ID is matched, we will search and match the tweets from the log to find the related learning contents. If the user ID is identified as the instructor, we will store the corresponding contents into the instructor tweets database. If the user ID is the TA, the corresponding contents will be stored into the TA tweets database. And if the user ID is the classmate, the extracted learning contents will be stored into the classmate tweets database for recommendation.

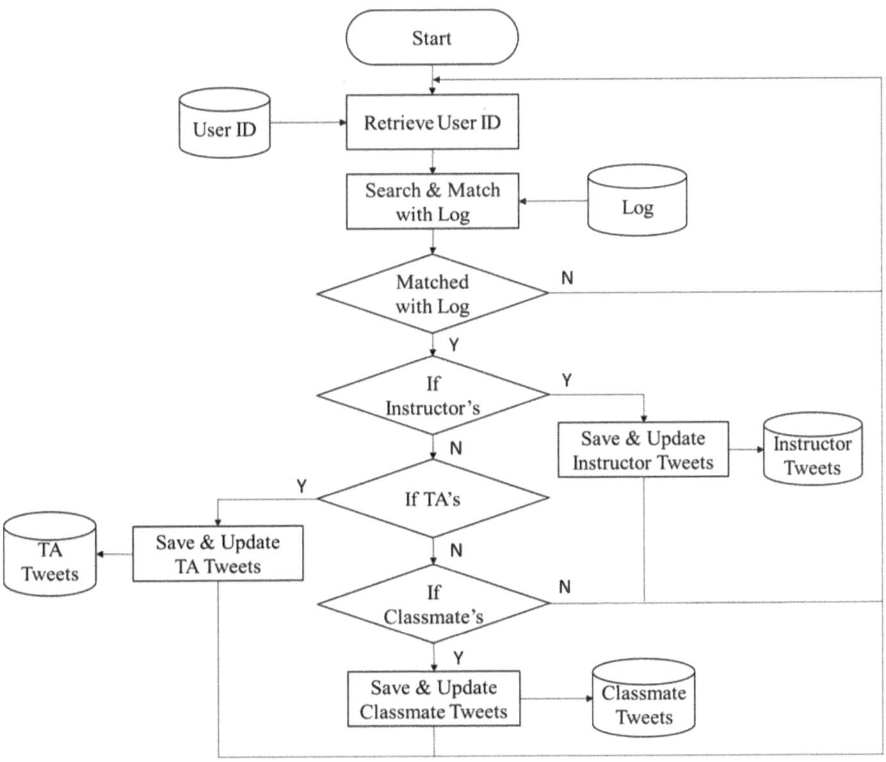

Fig. 49.3 Flowchart to match the content based on the user ID

49.4 Associative Recommendation Procedure

Based on the discussion above, six basic databases are generated based on learners' different browsing behaviors. That is, three keyword-based databases, the related contents in Q&A, the related contents in FAQ, and the related contents in key points, are extracted and organized in accordance with the keyword-based seeking behaviors, while three icon-based databases, instructor tweets database, the TA tweets database, and the classmate tweets database, are extracted and organized in accordance with the icon-based seeking behaviors. All these are utilized to associatively provide the learners with the more suitable information to satisfy his/her requirements. The set relations of the contents to be recommended are shown in Fig. 49.4. The details of the recommendation procedure are described as follows.

For a specific learner, with his/her identified target (keyword or icon),

Step 1: if he/she has sought the learning contents based on both keyword and icon, and the extracted keyword is related to the contents in Q&A,

- 1.1: if the identified user ID is the instructor, select the related contents in both Q&A and instructor tweets database into the recommendation result list, which are marked as ① in Fig. 49.4.
- 1.2: if the identified user ID is the TA, select the related contents in both Q&A and TA tweets database into the recommendation result list, which are marked as ② in Fig. 49.4.
- 1.3: if the identified user ID is a classmate, select the related contents in both Q&A and classmate tweets database into the recommendation result list, which are marked as ③ in Fig. 49.4.

Step 2: if he/she has sought the learning contents based on both keyword and icon, and the extracted keyword is related to the contents in key points,

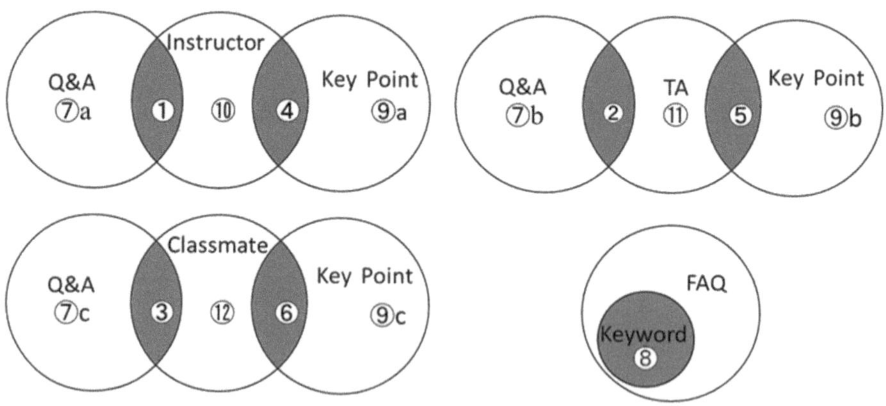

Fig. 49.4 The set relations of contents to be recommended

2.1: if the identified user ID is the instructor, select the related contents in both key points and instructor tweets database into the recommendation result list, which are marked as ④ in Fig. 49.4.

2.2: if the identified user ID is the TA, select the related contents in both key points and TA tweets database into the recommendation result list, which are marked as ⑤ in Fig. 49.4.

2.3: if the identified user ID is a classmate, select the related contents in both key points and classmate tweets database into the recommendation result list, which are marked as ⑥ in Fig. 49.4.

Step 3: if he/she has sought the learning contents based on keyword,

3.1: if the extracted keyword is related to the contents in Q&A, select the related contents that have not been in the recommendation results according to the user rank order: instructor, TA, and then classmate, into the recommendation result list, which are marked as ⑦a, ⑦b, and ⑦c in Fig. 49.4.

3.2: if the extracted keyword is related to the contents in FAQ, select the related contents in FAQ into the recommendation result list, which are marked as ⑧ in Fig. 49.4.

3.3: if the extracted keyword is related to the contents in key points, select the related contents that have not been in the recommendation results according to the user rank order: instructor, TA, and then classmate, into the recommendation result list, which are marked as ⑨a, ⑨b, and ⑨c in Fig. 49.4.

Step 4: if he/she has sought the learning contents based on icon,

4.1: if the identified user ID is the instructor, select the related contents that have not been in the recommendation results from the instructor tweets database into the recommendation result list, which are marked as ⑩ in Fig. 49.4.

4.2: if the identified user ID is the TA, select the related contents that have not been in the recommendation results from the TA tweets database into the recommendation result list, which are marked as ⑪ in Fig. 49.4.

4.3: if the identified user ID is a classmate, select the related contents that have not been in the recommendation results from the classmate tweets database into the recommendation result list, which are marked as ⑫ in Fig. 49.4.

Step 5: return and present the recommendation results in the list following the order of steps conducted above to the specific learner.

49.5 Conclusions

In this study, a new approach to associatively recommending the appropriate learning contents to the learners is proposed in a social media enhanced learning system. Firstly, we developed a method to identify the learners' attentions during the browsing process of learning contents based on the keyword and/or the icon, which were extracted from their browsing behavior sequences according to the eye-tracking technology. An associative recommendation mechanism was developed to provide the learners with more suitable and related learning contents among the categorized learning resources (such as Q&A, FAQ, and key points based on the contents, or instructor, TA, and classmate based on the publisher), in order to improve the learning efficiency in the social media-enhanced learning system.

As for future work, we will continuously enhance the proposed approach and add more related factors to improve the performance.

Acknowledgments The work has been partly supported by 2013 and 2014 Waseda University Grants for Special Research Projects No. 2013A-6395, No. 2013B-207, and No. 2014 K-6214.

References

1. Wang Q (2009) Design and evaluation of a collaborative learning environment. Comput Educ 53(4):1138–1146
2. Rienties B, Tempelaar D, Bossche P, Gijselaers W, Segers M (2009) The role of academic motivation in computer-supported collaborative learning. Comput Hum Behav 25:1195–1206
3. Grosseck G, Holotescu C (2008) Can we use Twitter for educational activities? In: Proceedings of the 4th international scientific conference: eLearning and software for education, Bucharest, Romania
4. Andrew D (2007) Eye tracking methodology, 2nd edn. Springer, Berlin
5. Piao G, Zhou X, Jin Q, Nishimura S (2011) Analyzing and extracting gaze patterns to capture unselfconscious information behavior via eye tracking experiments. In: Proceedings of the ICISS11 (2011 IEEE international conference on intelligent computing and integrated systems), Guilin, China
6. Piao G, Zhou X, Jin Q, Nishimura S, Wattanachote K, Shih T, Yen N (2013) Eye-tracking experiment design for extraction of viewing patterns in social media. In: Proceedings of the UIC2013 (10th IEEE international conference on ubiquitous intelligence and computing), Sorrento Peninsula, Italy
7. Buscher G, Dengel A, Biedert R, Elst L (2012) Attentive documents: eye tracking as implicit feedback for information retrieval and beyond. ACM Trans Interact Intell Syst 1(2):30 (Article 9)
8. Xu S, Jiang H, Lau F (2008) Personalized online document, image and video recommendation via commodity eye-tracking. In: Proceedings of the 2008 ACM conference on recommender systems, Lausanne, Switzerland, pp 83–90
9. Yoshitaka A, Wakiyama K, Hirashima T (2006) Recommendation of visual information by gaze-based implicit preference acquisition, vol 4351. LNCS, Berlin, pp 126–137
10. Giordano D, Kavasidis I, Pino C, Spampinato C (2012) Content based recommender system by using eye gaze data. In: Proceedings of the ETRA 2012 (Symposium on eye tracking research and applications), Santa Barbara, California, USA, pp 369–372

11. Chen L, Pu P (2010) Eye-tracking study of user behavior in recommender interfaces, UMAP, pp 375–380
12. Goldberg J, Helfman J (2010) Visual scanpath representation. In: Proceedings of the ETRA 2010 (Symposium on eye-tracking research and applications), Austin, TX, USA, pp 203–210

Chapter 50
A Novel Strategy for Colonies Recognition and Classification in Super-Resolution Images

Qi Zhang, Xueqing Li and Xianlun Dong

Abstract This paper demonstrates a method for colonies recognition and classification in super resolution images. According to the features of the colony images, we propose a novel strategy for colonies recognition and classification using image processing, binarization partitioning curved surface (BPCS), adhesion segmentation and support vector machines (SVM). The experimental results show that the new colonies algorithm is better than usual algorithm in reducing recognition classification time, and it also can display all colonies in different conditions, such as max length, sphericity, color feature, etc.

Keywords Partitioning curved surface (BPCS) · Adhesion segmentation · Support vector machines (SVM) · Edge detection

50.1 Introduction

In microbiology research, the colonies recognition and classification is the most important step and a key processing in colonies analysis [1]. So we propose a series of methods which can be used in colonies recognition and classification. The algorithms use gauss filter, partitioning curved surface (BPCS), adhesion segmentation and support vector machines (SVM) [2, 3] to recognize and classify the different colonies in super resolution images. The gauss filter will remove noise in the images and the BPCS aims to segment colonies from the images, then the adhesion segmentation algorithm will separate the adherent colonies for all the possibility in order to improve the classification, finally the improved Euclidean

Q. Zhang (✉) · X. Li · X. Dong
School of Computer Science and Technology, Shandong University, Jinan, Shandong, China
e-mail: D.Steven@sdu.edu.cn

J.J. (Jong Hyuk) Park et al. (eds.), *Ubiquitous Computing Application and Wireless Sensor*, Lecture Notes in Electrical Engineering 331,
DOI 10.1007/978-94-017-9618-7_50

distance-based support vector machine decision tree will classify all the colonies in super resolution images. And we have proved the correctness of the algorithm through the special experiments.

50.2 The Image Pre-processing

50.2.1 The Image Denoising

We capture the bacterial colony images using the EOS 5D mark ii camera, and the images resolution are 3,648*2,736. The images contain a lot of sharp noises caused by discrete pulse, zero mean gauss noise and other reasons. Under this situation, we adopt the gauss filter to reduce the sharp noises, and the Eq. 50.1 shows the gauss function.

$$f[x,y] = \frac{1}{2\pi\sigma^2} e^{\frac{-x^2-y^2}{2\sigma^2}} \tag{50.1}$$

We adopt 7 × 7 template to remove the sharp noises because of the separability of gauss function, and the Eq. 50.2 shows the way to calculate the standardized coefficient.

$$c = \frac{1}{\sum_{0}^{m-l} \sum_{0}^{n-l} f[x,y]} \tag{50.2}$$

The experiments show that this algorithm can effectively reduce sharp noises and recover some details of the colonies images caused by white gauss noise.

50.2.2 BPCS

There are lots of methods for colonies recognizing from images. In this paper we use binarization partitioning curved surface (BPCS) [4] to separate all the colonies from images. The single edge detection algorithm will cost longer time because of the super resolution, so we adopt the BPCS to reduce the calculate time. The BPCS process is as follows:

(a) we define the super resolution colonies images as SRCI, and the SRCI resolution is ($M \times N$). Then the SRCI will be divided into ($K \times L$) pieces and the size of ($K \times L$) depends on the size of ($m \times n$) and the size of Eq. 50.2.

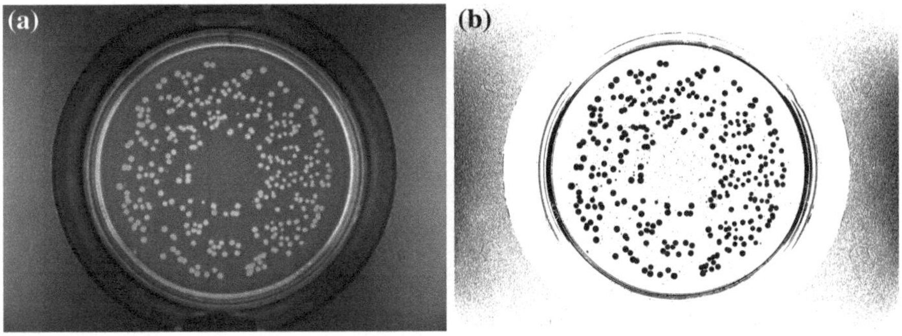

Fig. 50.1 The results of the BPCS. **a** Is the pre-processing image, **b** is the corresponding processed image

(b) Calculate the local minimum and maximum threshold, and the Eq. 50.3 shows the way. The m, n is the number of the set of $I \in (K \times L)$, the lE is the set of low point in I and the hE is opposite.
(c) According to the set of I establish the quadratic surface $f(x, y)$.
(d) Apply binary for each pixel in the set of I.

$$\begin{cases} LT = \dfrac{1}{m} \sum\limits_{(i,j) \in lE} I(i,j) \\ HT = \dfrac{1}{n} \sum\limits_{(i,j) \in hE} I(i,j) \end{cases} \tag{50.3}$$

Experiments show that the BPCS can distinguish the colonies and the background effectively and improve the binarization processing speed greatly. The Fig. 50.1. shows the results of BPCS.

50.2.3 Adhesion Segmentation

The segmented images usually contain lots of adhesive and defective colonies because of the complexity of colonies images. So we need to separate the adhesive colonies firstly. In this paper, we use chain code sum (CCS) [5] algorithm to solve this problem. The CCS algorithm process is as follows:

(a) Calculate the CCS and the curve feature.
(b) Calculate the angle point, smooth curve and estimate central angle which the boundary segment point to. The angle point's judgment is based on the Eq. 50.4, the smooth curve judgment is based on the Eq. 50.5 and the estimating is based on Eq. 50.6.

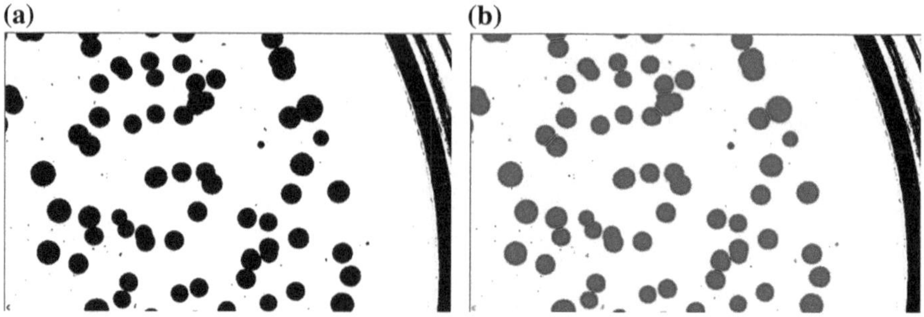

Fig. 50.2 The results of the CCS. **a** Is the BPCS image, **b** is the corresponding processed image

$$D_{iff}(i) < -1 \quad \text{or} \quad D_{iff}(i) > 2 \tag{50.4}$$

the m and n are the ends of the smooth curve line.

$$\begin{cases} (m-n) > 15 \\ [\sum(m) - \sum(n)] > 3 \\ -2 < D_{iff}(i) < 6 \end{cases} \tag{50.5}$$

$$\theta = 15 \times [\sum(m) - \sum(n)] \tag{50.6}$$

(c) Find all the angle points and separate colonies.

The experiment results show that it can separate the colonies effectively. The Fig. 50.2. shows the results of CCS.

50.3 The SVM for Colonies Recognition and Classification

50.3.1 Extract the Characteristics of All the Colonies

Although the colonies have diffident characteristics, we can put the special characteristics into the SVM according to the length, width, sphericity, colour and other characteristics. In this paper, the colonies characteristics we defined are as follows: the colony I, the boundary of the colony I_l, area S, circumference L, sphericity C, rectangularity R, maximum diameter Φ_{max}, minimum diameter Φ_{min}, colour O [6–8].

50.3.2 Construction of SVM Decision Tree for Colonies Classification

According to the characters of diffident colonies, we need to design the optimized SVM decision tree classifier which aims to improve the classification effect. First of all, we define set S [9] including all the characteristics and categories, and then we put one of characteristics into a decision node because of the diffident characteristics. The set S is divided into set S_1 and set S_2, and $S = S_1 \cup S_2$. And the iteration process above is repeated. The model partitioning strategy is as follows [10–12]:

(1) Obtain the center of each set. U is the behalf of the specific colony classes.

$$c_i = \frac{1}{n_k} \sum_{u_i \in U_i} u_i \tag{50.7}$$

(2) Calculate the center distance between each of classes.

$$d_{ij} = \left\| c_i - c_j \right\| \quad c_i, c_j \in C \tag{50.8}$$

(3) Obtain the average distance of D_{s1} and D_{s2} which is between any two classes within S_1 and S_2 [4].

$$\begin{cases} D_{s1} = \dfrac{2}{N_{s1} \times (N_{s1} - 1)} \times \displaystyle\sum_{i \in s1, j \in s1, i \neq k} d_{ij} \\ D_{s2} = \dfrac{2}{N_{s2} \times (N_{s2} - 1)} \times \displaystyle\sum_{i \in s2, j \in s2, i \neq k} d_{ij} \end{cases} \tag{50.9}$$

Assuming that $D_{s1} > D_{s2}$ and calculate the mean value.

$$d_m = \frac{D_{s1} \times N_{s1} + D_{s2} \times N_{s2}}{N_{s1} + N_{s2}} \tag{50.10}$$

(4) Cluster: if $\|c_i - d_m\| \geq 0 \; i \cup S_1$ else $i \cup S_2$.
(5) Repeat the above iteration process.

The super resolution complex colonies images can be classified effectively according to the SVM decision tree which includes the characteristics in Chap. 3. The better classification depends on the selected priority of the characteristics. The experiment shows that the classification accuracy of the klebsiella is 72.67 %, escherichia coli is 83.73 %, the pseudomonas aeruginosa is 82.44 %, the actinomycetes is 94.57 %, and the staphylococcus aureus is 98.36 %. The Table 50.1 shows all the characteristics of the colonies images.

Table 50.1 (**a**) Shows the result of the SVM. (**b**) Is the classify parameters

Area statistics	Colonies in area	The number of unit area	The total number in sample
693.644	373	0.538	373,000

(a)

Class name	*S*	*L*	*C* (%)	*O*
A	3–8	1–12	89–100	RGB(210,18,32)–RGB (255,28,64)
B	9–12	9–22	64–89	RGB(123,89,54)–RGB (187,97,64)
C	10–15	12–32	45–78	RGB(1,180,4)–RGB (10,235,12)
D	15–36	33–71	78–100	RGB(203,250,238)–RGB (255,255,255)
E	24–77	32–98	34–87	RGB(38,176,99)–RGB (34,195,186)
F	45–100	67–183	21–56	RGB(100,128,231)–RGB (111,188,245)
G	76–210	132–480	54–72	RGB(65,83,152)–RGB (70,92,180)

(b)

50.4 Conclusions and Prospect

Colonies reorganization and classification is one of the import links of microorganism research. In this paper, we has proposed the algorithm for the colonies recognition and classification in super resolution images which involves the gauss filter, BPCS, CSS and SVM, and the decision tree classifier is the key point in this solution. We propose the improved construction of decision tree and the better way to work out the ideal result. And the most needed to be done is the improvement for both segmentation algorithm and overlapped colonies handling to obtain better result in the future.

References

1. Shen W, Ya-chun W, Lie Z, Hui Z (2010) Experimental study for automatic colony counting system based on image processing. In: 2010 international conference on computer application and system modeling, V6-612–V6-615
2. Sergio V, Frederic P, Laura L, Mario C, Noemí C, Javier Herrero J, González Ballester MA (2013) Automated annotation removal in agar plates. In: 35th annual international conference of the IEEE EMBS, pp 3016–3019
3. Arenas-Garcia J, Ptrez-Cruz F (2003) Multi–class support vector machines-a new approach. In: Proceedings of the IEEE, pp 781–784

4. Zhang Q, Wang Y, Jiang X, Feng Z (2009) Development of a analytical software for dynamic microscopic granule based on VS2005. J Jinnan Univ Jinan (Sci. and Tech.) 23(1):64–67
5. Lu Z, Tong T (2002) The application of chain code sum in the edge form analysis. J Image Graphics 7(12):1323–1328
6. Osowski S, Markiewicz T (2007) Support vector machine for recognition of white blood cells in leukemia, In: Camps-Valls G, Rojo-Alvarez JL, Martinez-Ramon M (eds) Kernel methods in bioengineering, signal and image processing. Idea Group, London, pp 93–123
7. Amari S, Wu S (1999) Improving support vector machine classifiers by modifying kernel functions. Neural Netw 12:783–789
8. Garcia J, Barbedo A (2012) Method for counting microorganisms and colonies in microscopic images. In: 12th international conference on computational science and its applications, pp 84–87
9. Hsu CW, Lin CJ (2002) A comparison of methods for multi-class support vector machines. IEEE Trans Neural Netw 13(2):415–425
10. Shih FY, Wong WT (1994) An improved fast algorithm for the restoration of images based on chain codes description. In: Proceedings of computer vision and graphic image processing conference, pp 348–351
11. David C et al (2000) Advances in real time oil analysis. Pract Oil Anal Mag 11:28–34
12. Tang GY (1998) Region filling with the use of the discrete green theorem. In: Proceedings of computer vision and graphics

Chapter 51
Study on Intelligent Course Scheduling System

Teng Li and Xuqing Li

Abstract Course scheduling is a challenge task in the teaching process, which highly related with the quality of university. Given the number of the class rooms and time range, the aim of the algorithm is to arrange proper time and place to make sure there is no conflict. Usually, this kind of algorithms is NP-complete problem, and takes a lot of time to find the exact solution. To improve the speed and the quality, we propose a time searching algorithm consisting of three steps: initial, bottom-up and up- bottom. In order to speed the nearest capacity classroom searching, we also incorporate an equivalence class division algorithm. Besides, we provide a well-designed interactive interface to make the course scheduling more simple and flexible.

Keywords Intelligent course scheduling · Time searching algorithm · Equivalence class division algorithm · Interactive interface

51.1 Introduction

Course scheduling is one of the important works in college academic activities. It is the premise and foundation in ensuring the normal teaching process. Now, manual course-arranging methods are used in many colleges. But with the increase of students and teacher, the shortage of the manual methods become more and more obvious such as wide range, multi-constraints. Therefore, a solution using computer is considered.

T. Li · X. Li (✉)
The Department of Computer Science and Technology, Shandong University,
Jinan 250101, Shandong, China
e-mail: xqli@sdu.edu.cn

T. Li
e-mail: gogoahead@126.com

J.J. (Jong Hyuk) Park et al. (eds.), *Ubiquitous Computing Application and Wireless Sensor*, Lecture Notes in Electrical Engineering 331,
DOI 10.1007/978-94-017-9618-7_51

Time First searching and place first searching are two main methods of course scheduling. We use time first searching because normally classroom is enough but we are short for time resource.

Now many course scheduling algorithm use random time searching methods. It is an easy way to find time, but the time is often unreasonable. We create a time searching algorithm to solve this problem so that we can find the best time we need.

The greedy algorithm is always used to find the nearest capacity classroom. It will find the most reasonable classroom but it will take a lot of time. We will introduce an equivalence class division algorithm to find the same place costing less time.

The different kinds of courses will increase the difficulty of course scheduling for its different rules such as we can't go to the physical class after breakfast and we also can't put the professional course of computer science into the classroom belongs to other major and so on. We provide a well-designed interactive interface to divide the courses to make the course scheduling more simple and flexible.

51.2 Description of Course Scheduling

To realize the course scheduling algorithm, we need describe the teaching resource which participates in this activity in a reasonable way. In this paper, we use math symbol to describe the resource including teaching task (course to arrange), teacher, class, time and place in the first part.

Our time searching algorithm and place searching algorithm rely a resource management container. This paper will introduce the data structure which is used to realize that container in the second part.

To find the right time and place, we should set the priority of them. We will introduce how to determine priority in the third part.

The algorithms used in course scheduling are the kernel of our paper. We will use the whole next chapter to introduce them.

51.2.1 The Definition of Course Scheduling Using Math Symbol

We use $\{p_1, p_2 \ldots p_n\}$ to describe the teaching task to arrange. Let $T_i = \{t_1, t_2\}$ describe teachers belong to p_i, $C_i = \{c_1, c_2 \ldots c_n\}$ describe classes belong to p_i.

If $(T_i \cap T_j \neq \varnothing) \| (C_i \cap C_j \neq \varnothing)$, then $p_i \oplus p_j$. That means p_i and p_j can't share the same time and classroom; We use a 3-dimensional vector (d, s, k) to describe the time. d means the day of a week. d $\in \{1, 2, 3, 4, 5\}$; s means the start of each section, we suppose there are four sections of each day $s \in (0, 2, 4,6)$, k means the continue time $k \in (2,3)$ If $p_i \oplus p_j$ then $(d_i, s_i, k_i) \neq (d_j, s_j, k_j)$, If $p_i \oplus p_j$ then $r_i \neq r_j$.

The simple r is on the behalf of classroom. We will solve the problem if we can design an algorithm to make sure that for each 'p_i', we can find the appropriate combination of (d, s, k) and the classroom without conflict.

51.2.2 Resource Management Container Based on Map

Suppose we use the time first searching algorithm. We will first find the right time, and then we find the right unused classroom. We use map to store the time and place resource. The time is the key of the map. The value is the list of the classroom. Since the size of week is five and the size of start time is four and the K is fixable for a special teaching task, we will have 20 different keys. The List of the classroom will decrease when it is assigned. And with the length of the list decrease, the priority of the time section will decrease too.

This resource management container has many advantages. Firstly, we find all the resource we need and put them into the map. Actually, we put them into the memory state. Then we will not access the database any more. That will greatly improve the speed. Secondly, the resources in the container will decrease during the course scheduling process. Moreover the time section will be removed from the container when the list is empty. Thirdly, we can easily deduce the time utilization depending on the length of the list which can help us find the best time. And the time section we find will be more reasonable.

51.2.3 Priority Setting

We first set the priority of teaching tasks grouped by the interactive surface. We use the course attributes, course name, college and other information to divide the teaching tasks. The priority of the teaching task is determined by D(g).

$$D(g) = J(g) * C1 + T(g) * C2 + P(g) * C3$$

J(g)	means course attribute;
T(g)	means weekly hours;
P(g)	means class size;
C1, C2, C3	is the weights of the parameters.

Secondly we set the priority of the time section. The lower the utilization of time is, the higher the priority will be. At last we will set the priority of the classroom. The bigger the classroom capacity is, the higher the priority will be.

51.3 Algorithms We Design During the Process of Arranging Course

We will first find the right place using time searching algorithms. Then we will use place searching algorithm to find the available classroom. If we can't find the answer in the end, we will use call-back algorithm.

51.3.1 Time Searching Algorithms

The searching time algorithm includes unavailable time search algorithm, initial optimal time searching algorithm, bottom-up time searching algorithm and up-bottom time searching algorithm.

51.3.1.1 Unavailable Time Search Algorithm U_t (Avoid Conflict)

p_i has the collection of $T_i = \{t_1, t_2\}$ and $C_i = \{c_1, c_2 \ldots c_n\}$.

P_{t_1} means the collection of teaching tasks which have the primary teacher belongs to T_i, $P_{t_1} = \bigcup_{p_j \supseteq t_1} p_j$, P_{t_2} means the collection of teaching tasks which have the assistant belongs to teacher T_i, $P_{t_2} = \bigcup_{p_j \supseteq t_2} p_j, P_t = P_{t_1} \cup P_{t_2} \cdot P_c$ is the collection of teaching tasks whose class belongs to C_i. $P_{c_i} = \bigcup_{p_j \supseteq c_i} p_j$, $P_t = P_{t_1} \cup P_{t_2} \cdot P_c$ is the collection of teaching tasks which have conflict with p_i. $P' = P_t \cup P_c$, p_f is the collection of teaching tasks which are organized. Finally, we get $U_t = \bigcup_{time \in p_i, p_i \in P' \wedge p_i \in P_f} time.$

51.3.1.2 Initial Time

The best time is initialized as (1, 0, k). When one of the teaching tasks is organized, the list of classroom belongs to its time section will decrease, which will increase the priority of the time section.

51.3.1.3 Bottom-Up

Input: $(d, s, k) \notin U_t$

if (d, s, k) is (5, 6, 2) or (5, 4, 3), that means we have reached the end of time section, the algorithm returns −1 to show that no time section available. Otherwise we search the time section after this, s = s + 2, d = d. if c > 2, we can't use these time sections which start at 2 or 6. When we meet this situation, s = s + 2, When s > 6, we need carry the digit d, d = d + 1, s = 0. Finally, return (d, s, k).

51.3.1.4 Up-Bottom Time Search Algorithm

Input: $(d, s, k) \notin U_t$, if d = 0, s = 1, that means we have reached the end of time section, the algorithm returns −1 to show that no time section available. Otherwise we search the time section before this, s = s −2; d = d; if c > 2, s = s −2. If s < 0, we need borrow the digit d, d = d −1, s = 6. Finally, return (d, s, k).

51.3.1.5 Search Rules

If d >= 3, we first use the up—bottom search algorithm, then use the bottom-up one. If d < 3, we first use the bottom–up search algorithm, then use the up—bottom one.

51.3.2 Place Search Algorithm

The place search algorithms in other paper are based on the greedy algorithm. Suppose we have the classroom with the capability of 130, 115, 100, 85, 65, 45, 45, 35, 25, 20 in descending order.

If the class size is 112, the program will execute twice, but if the class size is 23, we will find it for nine times. If we sort the classroom in increasing order, it will take us more time to arrange large class.

51.3.2.1 Fast Search Algorithm Based on Equivalence Classes

We divide the classrooms into five categories. Category A: 0–30, Category B: 31–60, Category C: 61–90, Category D: 91–120, Category E: more than 120.

If we find 23, we just look once. We use the head of the list to store the address of Category A–E.

51.4 Experiment and Result

In this chapter, we first describe our result in three aspect, speed, rationality and successful rate. Then we will introduce our interactive surface.

51.4.1 About the Speed

As there are six campus of Shandong University, the teaching tasks can only use the classrooms belong to its own campus. We just choose about 200 teaching tasks belong to the software campus. In order to be more convincing, we put another 200 tasks we make into the list.

It takes about 20 s to finish the arranging. It is a good result.

51.4.2 About the Rationality

Although the random time searching algorithm can give good result sometimes. As we have divided the tasks into many groups, when the result of one group is not good, we need to arrange them again which may affect other groups. Look at the chart Fig. 51.1, the number of tasks on each week day is between 340 and 390, which are distributed evenly. That achieves the desired effect.

51.4.3 About the Successful Rate

The interactive surface can allow us use different rules to face different conditions. So the successful rate is improved. Look at the chart Fig. 51.1, the unassigned part is very low.

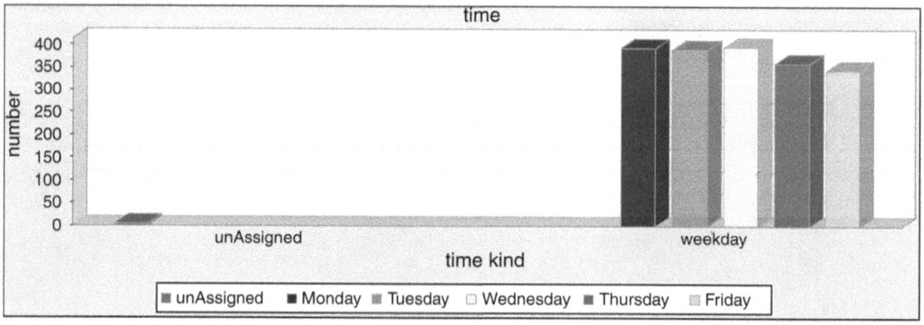

Fig. 51.1 The result of course scheduling

51.5 Conclusion

Course scheduling need uniform transfer teaching resource. Good way to coordinate limited time and place resources will make the arrangement more reasonable which will increase the quality of the algorithm. Shandong University has been using the course scheduling supported by Tsinghua University. But with the development of the Shandong University, we have more needs for our education activities. The design and implement of course scheduling is absolutely basic to the undergraduate educational system.

References

1. Chen H, Li X (2010) Research and implementation of intelligent scheduling algorithm
2. Santos HG, Och IH S, Souza MJF (2004) An efficient tabu search heuristic for the school timetabling problem, Berlin, Springer, pp 468–481
3. Abdullah S (2006) Heuristic approaches for university timetabling problems. University of Nottingham, Nottingham
4. Schaerf A (1999) Local search techniques for large high school timetabling problems. IEEE Trans Syst Man Cybern Part A Syst Hum 29(4):368–377
5. Wang L (2001) Intelligent optimization algorithm and its application. Tsinghua University, Beijing

Chapter 52
The Study and Research on Chinese Sports Network Marketing Strategy

Jiang Yong

Abstract Taking the network marketing of sports products in China as the research object, through statistics to the application situation and effects of the Internet in the marketing of Chinese sports equipment, combined with the history of the Internet application in domestic and international sports events and some actual cases, using a variety of research methods of documents literature searching, comparative study, logic analysis and case study, this paper put forward a relatively complete theoretical reference to strategies and technique of the present sports network marketing in our country.

Keywords Sports marketing strategy · Network marketing · Chinese sports brand

52.1 Introduction

Since the 2008 Olympic Games, each city in China has undertaken more and more a various of world-class sports competition, which shows charm and infinite potential of Chinese sports industry development for the world. Entering the 21st century, the impact of the Internet on sports has been developing with the accelerated speed, and the Internet has become an important promotional media and distribution channel for many sports goods companies. Sports network or "network sports" will be the distinct symbol of modern sports.

The network marketing is an activity that carrying on transactions and approaching profit through the Internet. The sports network marketing is an advanced marketing method that building sports on the basis of the Internet. However, China's sports product marketing way is still confined to the traditional mode.

J. Yong (✉)
Shandong Jiaotong University, No. 5 Jiaoxiao Road, Jinan, Shandong, China
e-mail: Xiaojiang_531@163.com

J.J. (Jong Hyuk) Park et al. (eds.), *Ubiquitous Computing Application and Wireless Sensor*, Lecture Notes in Electrical Engineering 331,
DOI 10.1007/978-94-017-9618-7_52

52.2 Concept Statement and Status Research

52.2.1 Studies on Sports Network Marketing Concept

52.2.1.1 Marketing

Modern marketing concepts are divided into two parts—marketing and social marketing. The marketing concept regards the direct customers as the center. The definition of Kotler in "Introduction to Marketing" is that "marketing is a social and managerial process by which individuals and organizations create products and value and then exchange with others to meet needs and desires". Market sales personnel's mission is to find and understand the needs of consumers, and sports marketers certainly have to pay attention to different consumer demands to different sports events.

52.2.1.2 Network Marketing

Network marketing (also known as the Internet Marketing) originated in the late 90s. There have been different versions of definition for the Internet so far and more representative among them are as follows:

A definition holds that: "Internet marketing is a part of the whole marketing strategies of enterprises, and are various activities that the enterprise uses the Internet as the primary measure to create the environment of online business and then achieves the total objectives of the business".[1]

Another definition thinks that: "network marketing is a marketing mode based on the online network, computer communication and digital interactive media to achieve marketing goals".[2]

Foreign scholars believe that: "the Internet marketing or the marketing based on the network marketing refers to the activity that uses the Internet as a medium to carry out transactions and profits. This kind of marketing mode is also known as the "e-marketing".[3]

52.2.1.3 Sports Marketing

Sports marketing is social and managerial process by which human creates and exchanges sports goods and value to let individuals and organizations meet the desires and needs of doing physical exercise and enjoying sports.

---

[1] Feng [1].

[2] Guo [2].

[3] Pittsburgh and Stotlar [3].

52.2.2 The Current Research Situation of Sports Marketing Based on Internet

At present, foreign research about sports network marketing has already been mature, involving more extensive sides. For example, American Delpy and Bosetti (1998), Pope and Forrest (1997), Mainardi (1997) all specifically described characteristics that successful webs should have in their monographs. The books "*Sports Marketing Theory and Practice*", and "*Sports Marketing Case Analysis*" written by American scholars Brenda G. Pitts and David K. Stotlar, the book "Marketing e-technique" written by the British scholar Ian Chastain and the book "The Network Marketing" written by the American scholar Ai Lusi Coupet all show detailed studies on the networking of sports industry and its application.

In summary, the current foreign research about sports network marketing has already been mature, involving more extensive sides. In contrast, the domestic study on the sports network marketing is still in the initial stage, and all aspects of research are relatively weak, therefore there is a huge research and development space.

52.3 Analysis of the Current Condition and Prospect of Domestic Sports Network Marketing

52.3.1 The Current Sports Network Marketing Condition in China

With respect to the current marketing condition of our domestic sports goods, there is the situation where product categories and the ways to promote those products of many private enterprises are similar with each other because of their imitation and reference mutually. As a result, it is very easy to duplicate the experience to success of other enterprises to your own business. This imitation, formed by geographical relationship—neighbor effect, makes the marketing ways of most sports goods enterprises be the same, such as investing large sums of money on television advertising, keening on celebrity endorsement etc.

In this marketing environment where the homogeneous marketing model is widely used, only a handful of sports goods firms attempt to choose such marketing way that is much more convenient and more international, which is called network marketing. Taking the domestic sports goods business leader Lining as example, we can see that the company has been developing external cooperation vigorously and taking differentiated marketing strategies based on strengthening its own independent research and development. And in 2005, Lining cooperated with NETEASE company and officially established "Lining-NETEASE sports channel". In addition, the "ANTA-Sohu sports channel", launched jointly by ANTA company and Sohu, the leading Chinese media Internet Service Corporation, has been full operation. The efforts to promote network marketing made by these domestic sports equipment companies including Lining and ANTA have been reflected remarkably by the annual financial statements in 2012.

The present situation where sports goods businesses develop network marketing in our country is that enterprise websites the primary place for businesses to carry out network marketing. So far, many sports goods companies in China have set up their own enterprise network. However, some of these companies didn't register in the portal site, and they didn't publish Internet advertising. According to the statistics, up to now, there are about 160–210 sports goods businesses who have established business website and registered in the portal site in China.

52.3.2 The Prospect of the Domestic Sports Network Marketing

The number of the users who communicate with each other, carry on commercial transaction and transmit information through the Internet has made a breakthrough of 600,000,000 in 2012. The popularity of the Internet, the sharp increase of the Internet users and the huge business potential of the Internet are all transmitting a message that the advanced e-commerce will certainly replace the traditional business model and the trend of online shopping will become more and more popular. The data shows that the Internet users who visit shopping websites frequently are more than 48.9 % in 2013, which is up by more than 6 % in 2012. The Chinese personal e-commerce market scale is in an unprecedented growth.

52.4 The Characteristics and the Competitive Advantage of Sports Network Marketing

52.4.1 The Characteristics of Sports Network Marketing

With the in-depth development of the Internet, its marketing characteristics are being excavated by operators gradually:

(1) Across time and space. The Internet has broken through the limit of time and space, as a result, marketers can provide marketing services to global customers and consumers every day a week and every hour a day, which can increase market share as far as possible.

(2) The using of the multimedia. Both parties involved in the transaction can process various forms of media information such as text, sound and image through the Internet. Thereby, the information exchange and transaction achievement of marketing activities can become colorful and fantastic. Besides, this characteristic has provided a stage to marketers to express their creativity and initiative.
(3) Interactivity. Through the Internet, enterprises can display products catalogue and information to customers, sell some products that have network characteristics directly, interact and communicate with customers, collect market information and set about testing those products and investigating customer satisfaction.
(4) Humanity. The marketing activities carried on the Internet reflect some features such as one-to-one, rationalization and customer management and so forth, which can effectively prevent the occurrence of strong sales in traditional marketing in order to make customers be more satisfied and establish a long-term good relationship of mutual trust with customers.

52.4.2 The Competitive Advantage of Sports Network Marketing

Based on the above characteristics and compared with traditional business model, Internet marketing has gradually formed its unique competitive advantage.[4]

(1) Most widely spread. The spread of network advertisement it is not affected by time and space, and the ad information can be spread every hour a day uninterruptedly spread around the world through the internet.
(2) Strong interaction. The Internet media is not one-way information dissemination as traditional media, but the interactive communication of information, in which way the users can get information what they think is useful and manufacturers can get valuable feedback information form users at all times.
(3) Highly targeted. According to the analysis results, the network advertising audience is such a group of people who is the youngest, the most vigorous, the highest educated, and they have the strongest purchasing power. Online advertising can help you find the most likely potential users directly.

[4] CNNIC [4].

52.5 The Discussion About the Study Results and Analysis

52.5.1 The Current Network Marketing Condition and Analysis of ANTA Sports Goods

52.5.1.1 The Network Marketing Strategy of ANTA Sports Goods

In August 2004, the "ANTA-Sohu sports channel", launched jointly by ANTA company and Sohu, the leading Chinese media Internet Service Corporation, has been full operation. The official website of ANTA contains five columns; there are ANTA Spot News; ANTA Hot Survey; ANTA Star; ANTA Club; ANTA Cool Show, respectively. Among them, ANTA Club is the main column that ANTA Company publicizes, and also be the platform for consumers' communication. The promotion activity of ANTA—"ANTA brings you to Athens", is carried out on the network platform, and it is the first domestic network sports marketing case.

52.5.1.2 The Status and Analysis of ANTA Products Network Marketing

The results of "2012 Famous Sports Shoes Brand Value Study" (From the cooperation between Horizon Research and Horizon Index Data Network) suggests: the brand loyalty of domestic brand LiNing (53.4 %) is higher than foreign brand ADIDAS (39.8 %) and NIKE (39.1 %), however, the brand loyalty of two domestic brands Doublestar (13.4 %) and ANTA (15.1 %) is relatively low. From the perspective of brand competitiveness, among Chinese enterprises, only LiNing has the ability to compete with other international brands, such as NIKE and ADIDAS, and now it has been the main opponent of NIKE and ADIDAS in China's market.

The 16th China international sports goods fair reflects two problems, on the one hand, China's sporting goods industry has the rapid development momentum and the market share is growing; on the other hand, there is huge gap between quality and quantity, many domestic manufacturers don't have their own international famous brands.

52.5.2 361° and Tencent

52.5.2.1 The Marketing Strategy of 361°

In 2006, Tencent and 361° company formally signed a strategic cooperation partnership agreement in Nanjing Olympic Sports Center. They would implement a wide variety of interactive advertising forms, such as four championships broadcast and "Dare playing community" 020 activities.

52.5.2.2 The Marketing Situation and Analysis of 361°

From the cooperation between Tencent and 361°, we could find the joint point that links sports product to network media. The platform advantage of Tencent has been a vital interactive bridge that companies link up to users. As China's biggest Internet Service Provider, Tencent has the number 1 instant messaging software—QQ, and it is the most important interactive media to communicate with users. Otherwise, the platform of Tencent has the prominent advantages of diversified marketing. Strategic alliance stress "Win-Win", 361° also would provide a series of patronal competitions and display opportunities, report and advertise with Tencent together.

52.5.3 The Strategical Analysis of Sports Network Marketing

Traditional competition marketing strategy has developed into 4C strategy mainly based on consumers' needs, but 4P (product, price, place, promotion) marketing mix is still the basis of marketing activities. Now, China's sports network marketing activity is still in its infancy phase, better execution of 4P strategy will be beneficial to the continuous improvements of the our country's sports network marketing theory and practice.

52.5.3.1 The Product Strategy of Sports Network Marketing

The aim of sports network marketing is as the same as traditional marketing, namely providing consumers with satisfied products and services, and at the same time getting the interests of marketers. As the bridge between the interest of enterprise and consumer, product could be simply summarized tangible product and intangible service. New product exploitation strategy mainly includes: differentiation exploitation strategy and product update and improvement strategy.

52.5.3.2 The Price Strategy of Sports Network Marketing

Resulting from the implementing of network marketing activity, HR cost and store rents are saving, general goods are often taking low-price strategy. But the marketing campaigns, as sports event the core, make pricing strategy more complex on account of different degree of popularity and rarity of sports event resources.

(1) Free pricing strategy. In order to occupy the market and promote, generalize products, general products always take different forms of free strategy, including permanent free, short-term or periodic free and partial free forms. Besides, free products often have the characteristics of digitization, formlessness, growth and indirect benefit.

(2) Differentiation pricing strategy. In virtue of the influence of match and the competition environment of network, it always takes differentiation pricing strategy in sports event marketing process.

52.5.3.3 The Place Strategy of Sports Network Marketing

The place strategy is also a significant part of sports network marketing. In marketing activities, sports events operators hope to present high-quality games with high efficiency and low distribution channels cost; official network media also wish that their operational matches could be familiar to Internet users and advertisers quickly, and begin to find ideal sites to conduct text links and advertising. However, enterprises also want to put their ads on any appropriate games websites and have the ideal publicity effect to achieve the purpose of conveying their brands effectively.

52.5.3.4 The Promotion Strategy of Sports Network Marketing

Traditional promotions of sport market mainly include four forms, advertising, sales promotion, public relations and personal selling, respectively. However, network marketing is a kind of promotion activity that mainly conducted in the platform of internet. New strategy mainly includes: internet advertising strategy, site publicity strategy, sales promotion strategy and relationship marketing strategy.

52.6 Conclusion and Suggestions

52.6.1 Conclusion

(1) The cyber marketing of domestic sport goods is in the initial period, currently only a few sports product enterprises try to build network sales platform. But restricted by technical level and inherent consumption concept, there is still a certain resistance in the implementing of network media marketing of domestic sports goods.
(2) Domestic large-scale sports product enterprises could basically integrate with the international level in the network marketing. Based on strong market value, the marketing of sport goods could being transferred a part of the market share to the network, and gradually accomplish a sales pattern of specialization, humanization, generalization and a certain degree of nationalization.
(3) China's sports network marketing activities are lack of systematicness and integrality.

(4) Although the market competitiveness of domestic sports goods is a little low, it could help to create a bigger development space for domestic sports goods through strengthening the construction of own websites, building WIN-WIN partnership and weak-strong alliance platform, with reference to the network marketing model of international famous brands.

52.6.2 Suggestions

(1) Aiming at the immature present situation of network marketing, I suggest that public should advocate for the increase popularity rate of computers and for the training of network elite. Through media publicity, we should change people's inherent traditional consumption concept, increase awareness for network marketing, strengthen integrity, completely eradicate internet fraud and boost consumers' credibility of internet marketing.
(2) The large-scale sports goods enterprises should reinforce and establish websites with strong national characteristics in network marketing. Meanwhile, avoid going into the erroneous zone, and attach great importance to the right positioning of brand.
(3) Small and medium-sized sports product corporations should find ways in alliance, apply to the strategy of "Relying on the union, Clear division of labor, Perform their duties" and make their brands famous.
(4) Aiming at the present situation that domestic sports goods network marketing is at low levels of development, I consider that the breakthrough of domestic sports goods network marketing strategy is website construction. By being more innovative, attracting more PV (Page View), improving its product technology content, enhancing the competitiveness of domestic sports goods, China's sports goods enterprises could gradually occupy the international market share and integrate with the world trade way of sports goods as soon as possible.

References

1. Feng Y (2002) The network marketing theories and practices. Tsinghua University Press, Beijing, p 1
2. Guo X, Pei Y, Cao H (2006) The network marketing. Machinery Industry Press, Beijing
3. Pittsburgh BG, Stotlar DK (2005) Marketing theories and practices. Liaoning Science and Technology Press, Shenyang
4. CNNIC (2014) The statistical report of the thirty-third Internet development in China. China Internet Network Information Center, Beijing
5. Liu Q (2007) Sports operating case select. People's Sport Publishing House, Beijing, p 3

6. Ma J, Qu Y, Fan W (2007) The vane of sports marketing. The Chinese and foreign management, p 10
7. Guo B, Xu X (2007) Contemporary sports present situation and development trend of media operation. J Sport Cult Tribune 9:25
8. Li N, Yuan G (2006) Sports economics. FuDan University Press, Shanghai, p 8

Chapter 53
Interactive Visualization of Enrollment Data Using Parallel Coordinates

Wei Xinglei and Li Xueqing

Abstract As one of the main tasks in colleges and universities, enrollment has a vital role in higher education. Visualization of enrollment data serves as an auxiliary to analyze and extract the potential and value of it. Due to the limitations of spatial imagination ability on multivariate data set, people can't gain an intuitive sense of it. What is more, it's difficult to display the variables and characteristics globally and simply using the common methods such as histogram, line chart and so on. It's proposed that interactive visualization on multivariate enrollment data using parallel coordinates should be a simple and flexible approach. The comprehensive utilization of data modelling, data analysis and parallel coordinates aid to interpretation or mining on model, characteristics and rules hidden in the interior of enrollment data. Results of the application on the multivariate enrollment data set prove that, the method of parallel coordinates can implement visual analysis to multivariate Enrollment data flexibly and effectively.

Keywords Parallel coordinates · Enrollment · Data analysis · Data visualization

53.1 Introduction

With the constant development of economy and progress of technology, the demand of high-level talents is increasing. Higher education is the main way of high-level talent training, and the high quality students are the congenital condition to ensure the quality of high-level talent training. The quality of students directly affects the quality of higher education and is also the lifeblood for the survival and development of the colleges and universities. At present, within the informatization

W. Xinglei · L. Xueqing (✉)
The Department of Computer Science and Technology,
Shandong University, Jinan 250101, Shandong, China
e-mail: xqli@sdu.edu.cn

W. Xinglei
e-mail: weixingleisdu@gmail.com

J.J. (Jong Hyuk) Park et al. (eds.), *Ubiquitous Computing Application and Wireless Sensor*, Lecture Notes in Electrical Engineering 331,
DOI 10.1007/978-94-017-9618-7_53

construction of the enrollment in colleges and universities, the development and application of business operation information system has been relatively successful. Furthermore, enrollment work has accumulated abundant data resources over the years. With the enrollment data increasing, the problem ahead of the Enrollment management of the colleges and universities is how to manage data resources effectively, explore the potential and value of information contained in them and improve cultivation level of higher education in colleges and universities. Some data mining techniques and algorithms for decision makers can't be understood and used immediately, but the visualization technology can make the results easier to understand and allow the comparison and test of results.

Currently, the common data visualization methods include histogram, line chart, scatter plot and so on, and the data sets are merely represented as two-dimensions in the Euclidean space by these methods. Beyond third dimension, the dynamic elements like color, shape or movement will be additionally used to enhance the visual effects of the data [1]. However, the enrollment data sets which are ordinarily much more complex and multivariate can't be visualized by using the above methods. Therefore, the utilization of parallel coordinates to display multivariate enrollment data and transform from the research of the relationship among variables to identification of the two-dimensional model, makes it easier to associate a large amount of variables and visualize enrollment data [2].

Multivariate data can be expressed by parallel coordinates which have great mathematical basis. And projective geometry interpretation and duality properties make it very suitable for data visualization and analysis. At present, the parallel coordinates are applied for many visualization techniques, such as stock chart, weather forecast, NM graph and Andrew's graph.

In this paper, through modeling and analysis of enrollment data and the utilization of parallel coordinates which's an interactive visualization method, we can assist colleges and universities in the efficient and flexible analysis of characteristics and rules contained in the enrollment data and the improvement of the cultivation level of higher education in colleges and universities.

53.2 Enrollment Data Analysis and Visualization

53.2.1 Data Model

By means of data filter, data integration, data transformation on the data sets and the abstract of characteristics from the enrollment data, we model and analyze it to gain the concepts and definitions of the enrollment.

53.2.1.1 Data Filter

The goal of data filter is to analyze and filter the data sets which are objective and relevant. Through the analysis of fields in the data sets, we can determine whether it's necessary to choose it. For example, nearly 99 % of the values on national field are Han. Obviously, the result of the analysis is of no significance. Besides, the consideration and analysis about correlation between fields is essential. There are identity card number field, native place field (included province, city and county) and other fields which are highly correlated evidently in the same personal information table. In this case, we select only one of them, such as the native place field. Moreover, some unique fields which are abandoned exist but don't make sense [3].

53.2.1.2 Data Integration

Data integration is to integrate the data from multiple data sources. For instance, we expect to acquaint the ranking of the universities according to some rules defined by us (comprehensive score, colleges attribute and so on). However, the universities' information resulting from enrollment data contains universities' code and name only. Consequently, the integration of multiple data sources about universities' information is crucial.

Additionally, it is to consider the matter of entity recognition on data integration. Entity recognition is to deal with matching entities from the multiple information sources in the real world. For the universities, we can adopt the general standards or interfaces (such as college code) to match, construct the relationships between multiple data sources and eventually merge data from multiple data sources into a consistent data source, so as to achieve the purpose of data integration.

53.2.1.3 Data Transformation

Data transformation is aiming at converting the data into a form which's suitable for data visualization through data generalization, data standardization and variables structure.

1. Data generalization is to replace the low layer or raw data with high layer concept using concept hierarchy. For example, categorical variables, such as native place and major, can be generalized for higher level concepts, such as provinces and discipline.
2. Data standardization is to establish the mapping between variables and data and to preprocess the measure of variables. In addition, we retain 9 significant variables and the amount of data is more than 18,000. Due to the diverseness of the data's dimension in all variables, it's inconvenience for the visualization and analysis of multivariate data. Thus, we must standardize and preprocess all variables of original data matrix X resulting in standardized data matrix Z [4]:

$$X = \begin{pmatrix} x_{11} & \cdots & x_{1m} \\ \vdots & \ddots & \vdots \\ x_{n1} & \cdots & x_{nm} \end{pmatrix} \quad Z_{i,j} = \frac{x_{i,j} - \overline{x_1}}{s_i^2}, \quad \overline{x_1} = \frac{1}{n}\sum_{j=1}^{n} x_{i,j}, \quad s_i^2 = \frac{1}{n-1}\sum_{j=1}^{n}\left(x_{i,j} - \overline{x_1}\right)^2 \tag{53.1}$$

3. Variables structure is to construct new variables which added to the variables' set. For example, the values of university variable are calculated through additional weighted variables (such as comprehensive evaluation, 985 Project (985) or not, 211 Project (211) or not). The formula is as follows:

$$U = a_1 \times 70\,\% + [100 \times (3a_2 + 2a_3)/5] \times 30\,\% \tag{53.2}$$

Among the formula (53.2), a_1 is the comprehensive evaluation, a_2 is 985 or not, a_3 is 211 or not.

53.2.2 Visualization Using Parallel Coordinates

By means of variables discretization, dimension reorder and custom dynamic filter, we expect to avoid some problems, such as the information clutter, overlap of lines, the limitation of display and so on.

53.2.2.1 Variables Discretization

When the value points of variables on several neighbouring axes are reduced, the number of lines between the axes displayed on the graph owing to overlap is less obviously, leading to be unable to demonstrate the convergence characteristics and comparison. Furthermore, this's to discount the value of parallel coordinates plot.

In order to avoid this, variables discretization on the less value points of variables has a beneficial effect (Fig. 53.1). The range of the value point can be represented by the 10^{-2} orders of magnitude of parallel axis scale.

53.2.2.2 Dimension Reorder

There are data points between neighbouring dimensions that don't belong to any cluster are called outliers. Due to the fact that outliers often obscure structure and thus confuse the user, clutter in parallel coordinates can be defined as the proportion

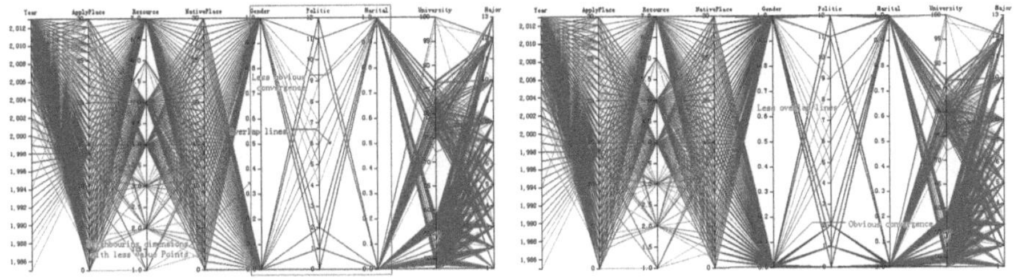

Fig. 53.1 The comparison of effect before and after variables discretization

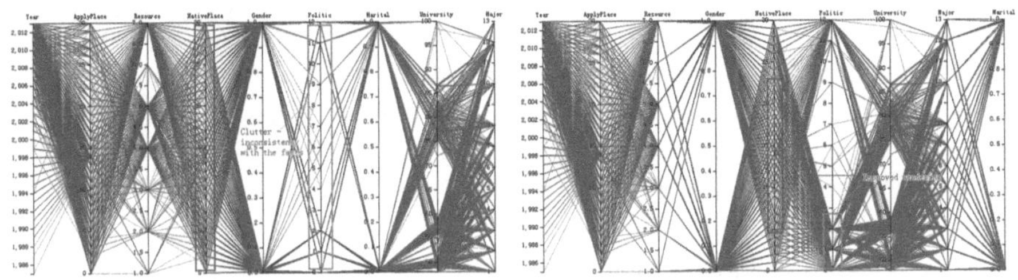

Fig. 53.2 The comparison of effect before and after dimension reorder

of outliers against the total number of data points. To reduce clutter in this technique, our task is to reorder the dimensions to minimize the outliers between neighbouring dimensions (Fig. 53.2).

To calculate the score for a given dimension order, we count the total number of outliers between neighboring dimensions, $S_{outlier}$. If there are n dimensions, the number of neighbouring pairs for a given order is n−1. The average outlier number between dimensions is defined to be $S_{avg} = S_{outlier}/((n-1)$. Let S_{total} denote the total number of data points. The clutter C, defined as the proportion of outliers, can then be calculated as follows: [5]

$$C = S_{avg}/S_{total} = S_{outlier}/((n-1) \times S_{total}) \quad (53.3)$$

53.2.2.3 Custom Dynamic Filter

The feature and tendency of the whole data set can be visualized by parallel coordinates, but it has some limitations on the display of details or custom conditions. Therefore, the data visualization should be effective and flexible with the method of custom dynamic filter.

Parallel coordinate plot varies with the modification of each dimension selected range when custom ranges in any dimension are picked [6] (Fig. 53.3). Meanwhile,

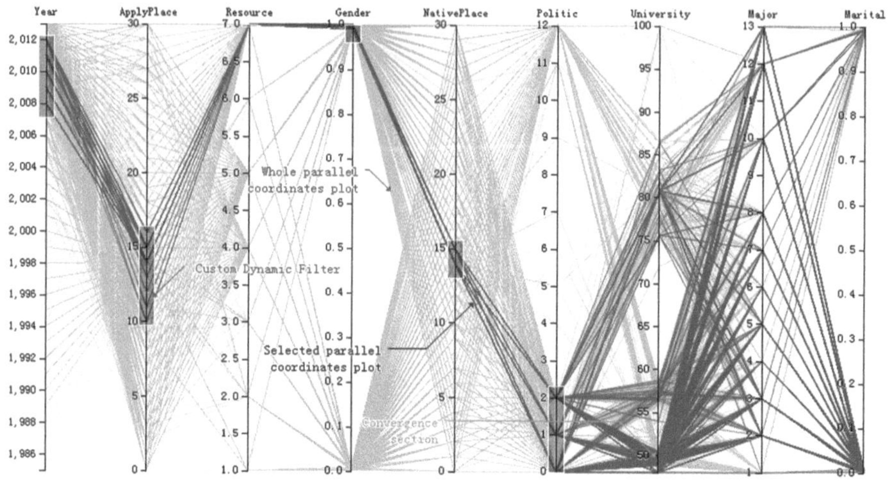

Fig. 53.3 The application of custom dynamic filter

the whole parallel coordinate plot is displayed in the background. As a result, it can't only shows the local characteristics and trends, but also reflects the relationship between the local and the whole.

53.3 Conclusion

In this paper, we present a new approach to analyze and visualize enrollment data using parallel coordinates and good results have been achieved. We make the most of the advantage of visualization technique—parallel coordinates, to reveal the multivariate data from different perspectives. On the other side, we provide a flexible and interactive method for users to analyze and understand enrollment data conveniently. The focus of future research is how to provide better methods to implement collaborative visualization and analysis of enrollment data using parallel coordinates.

References

1. Wegman EJ (1990) Hyperdimensional data analysis using parallel coordinates. J Am Stat Assoc 85(411):664–675
2. Inselberg A (1985) The plane with parallel coordinates. Visual Comput 1(2):69–91
3. Inselberg A (2008) Parallel coordinates: visualization, exploration and classification of high-dimensional data. In: Handbook of data visualization. Springer, Berlin, pp 643–680
4. Keim DA (2001) Visual exploration of large data sets. Commun ACM 44(8):38–44

5. Peng W, Ward MO, Rundensteiner EA (2004) Clutter reduction in multi-dimensional data visualization using dimension reordering. In: IEEE symposium on information visualization, INFOVIS 2004, pp 89–96
6. Xu Y, Hong W, Chen N, Li X, Liu W, Zhang T (2007) Parallel filter: a visual classifier based on parallel coordinates and multivariate data analysis. In: Advanced intelligent computing theories and applications. With aspects of artificial intelligence. Springer, Berlin, pp 1172–1183
7. Fua YH, Ward MO, Rundensteiner EA (1999) Hierarchical parallel coordinates for exploration of large datasets. In: Proceedings of the conference on Visualization'99: celebrating ten years. IEEE Computer Society Press, pp 43–50
8. Zhao K, Liu B, Tirpak TM, Schaller A (2003) Detecting patterns of change using enhanced parallel coordinates visualization. In: Third IEEE international conference on data mining, ICDM 2003, pp 747–750
9. Siirtola H, Räihä KJ (2006) Interacting with parallel coordinates. Interact Comput 18 (6):1278–1309
10. Bendix F, Kosara R, Hauser H (2005) Parallel sets: visual analysis of categorical data. In: IEEE symposium on information visualization, INFOVIS 2005, pp 133–140
11. Kosara R, Bendix F, Hauser H (2006) Parallel sets: interactive exploration and visual analysis of categorical data. IEEE Trans on Visual Comput Graphics 12(4):558–568
12. Huh MH, Park DY (2008) Enhancing parallel coordinate plots. J Korean Stat Soc 37 (2):129–133

Chapter 54
The Effect of Peer's Progress on Learning Achievement in e-Learning: A Social Facilitation Perspective

Po-Sheng Chiu, Ting-Ting Wu, Yueh-Ming Huang and Hong-Leok Ho

Abstract Electronic learning (e-Learning) is the next generation of learning. It provides comprehensive supports to students and can overcome geographical and time limitations compared to traditional classroom learning. There is a dramatic difference between traditional classroom learning and e-Learning. One of the significant discrepancy is the nature of the social context, which is a main research issue in this study. The lack of in-person contacts in e-Learning does not isolate students from the social thoroughly, yet with the assistance of technology, there is still room for much more creative learning activities that facilitates interactivity between students. Therefore, in this study, social presence theory and social facilitation theory were reviewed and designed to examine the influence of information about peers' progress on students' learning effect. Forty-four students of an elective course in a senior high school at Tainan City, Tainan participated in the experiment. The result shows that students' learning effect is affected by information about peers' progress, and results were discussed with social presence theory and social facilitates theory. Displaying of information about peers on e-Learning system can have influence on students. Consequently, students' learning can more or less be influenced by controlling the display of peer information.

Keywords Social facilitation · Social presence · Peer progress

P.-S. Chiu · Y.-M. Huang (✉) · H.-L. Ho
National Cheng Kung University, Tainan, Taiwan
e-mail: huang@mail.ncku.edu.tw

T.-T. Wu
Chia Nan University of Pharmacy and Science, Tainan, Taiwan

J.J. (Jong Hyuk) Park et al. (eds.), *Ubiquitous Computing Application and Wireless Sensor*, Lecture Notes in Electrical Engineering 331,
DOI 10.1007/978-94-017-9618-7_54

54.1 Introduction

Electronic learning (e-Learning) is the next generation of learning. It provides comprehensive supports to students and can overcome geographical and time limitations compared to traditional classroom learning. Welsh et al. [1] defined e-Learning, under the sense of using it in organizational training, as the use of computer network to deliver information and instruction to individuals. Eklund et al. [2] suggested that e-Learning includes computer-based learning, virtual classrooms, web-based learning, digital collaboration and it is flexible and comprise of a wide set of applications and processes, delivering education and training through electronic media. The advance of computer technology and the power of the Internet and web-based applications can change the way people learn and the role of teachers [3–7]. With the application of a suitable pedagogy, technology is able to assist the learning process and enhance the learning outcome [8–10].

There is a dramatic difference between traditional classroom learning and e-Learning. One of the significant discrepancies is the nature of the social context, which is a main research issue in this study. Unlike traditional classroom learning, face-to-face contacts with peers and teacher are missing in e-learning. Students have fewer chances to communicate directly with classmates, whereas computers and Internet may serve as the communication medium. The influence of such social context distinction is often studied in e-learning [11] and in computer-mediated communication research [12].

However, conscious social interaction is not a necessary condition in order to elicit social influence. In light of social presence theory [13], only a feeling of the presence of others, in the digital environment, is able to influence the performance, and satisfaction of e-learners [14, 15]. Online status, displayed pictures (avatars), webcam video, and etc., is able to increase the level of one's social presence. In those cases, it could be no interactions at all, yet it is enough to produce social influence.

Social facilitation theory [16] is another social psychological theory suggesting a social influence effect that, like social presence theory, one's performance is affected when others are presence. Unlike social presence theory, social facilitation theory predicts performance, whereas social presence theory predicts mainly satisfaction. There are hardly any studies of applying social facilitation in education context or human learning. Therefore, in this study, social facilitation theory was taken into account to investigate the effect of social influence in e-learning environment.

The objective of this study is to reveal the social influence of the mere presence of others (i.e. unconscious social interaction) in a web-based learning system and its effect on learning. An experiment approach was taken to investigate the problem. A class of senior high school students participated in the experiment. A learning system which displays peers' learning progress during learning is developed. The learning progress is a kind of social information that is delivered over the digital environment, and is believed to be producing certain effect on students learning performance.

54.2 Method

54.2.1 Participants

Forty-four senior high school students of a high school in Tainan City, Taiwan, Republic of China participated in the experiment. They were first year or second year senior high school students, enrolled in an elective course named "computer concepts". The mean age of participants was 15.5 years with a standard deviation of 0.3 year.

Taking part in the experiment was assigned to the students as part of the course. No compensation or reward was given to them. Their performance in the experiment was not counted in their course grades. Participants were instructed to give their best to learn and perform in the experiment.

The experiment carried out in the computer laboratory of participants' school. Each participant was provided with one desktop personal computer running the learning system used in this study. Participants sit side-by-side. They could see each other in person, yet they were not allowed to talk with others.

54.2.2 Learning System

A web-based learning system was developed to present the learning material to the participants with the pseudo peer learning progress displayed. The system was composed of a login page, instruction screen, the learning section, and the learning assessment.

Figure 54.1 respectively show the experiment group and control group during learning. Before the experiment started, an experiment instruction was conducted to briefly introduce the learning system, the experiment objective, the experiment procedure, and participants' privacy. The serial number, namely 1–44, was unique for each participant for identifying and group assigning. The serial numbers were assigned according to the sequence of participants' seats in the computer laboratory.

Fig. 54.1 The experiment groups during learning (*left*) and the control group during learning (*right*)

Table 54.1 The experiment design

Course / Group	1 History (easy)	2 Psychology (medium)	3 Metaphysics (difficult)
1 Leading	Leading positions		
2 Average	Average positions		
3 Trailing	Trailing positions		
4 Control	No pseudo progress displayed		

54.2.3 Procedure

The experiment has one independent variable, pseudo peer progress display. In the experiment, participants learnt three short courses of subject matter history, psychology, and metaphysics with the learning system. The system logged the time instances of turning the pages and the course assessments measured participants' learning effects.

Participants were assigned randomly into four groups; three experiment groups (leading, average, and trailing) and one control group (pseudo progress not displayed). The experiment design is illustrated in Table 54.1. Each group had 11 participants. Participants of all groups learn together in the same computer laboratory with the learning system, in the sequence; course 1 (history), course 2 (psychology), and course 3 (metaphysics). Their corresponding experiment controls (pseudo progress) were delivery with the learning system while learning the materials.

54.3 Results

Learning performance is operationally defined by outcome of the three course assessments. Scores of the participants were calculated. The score means, by groups, are shown in Table 54.2 shows the line graph of scores in the three courses. In terms of their scores, trailing and average groups showed a greater fluctuation at different levels of course difficulties whereas leading and control groups show a relatively lower fluctuation at different levels of course difficulties.

The average scores of course 1 were higher than that of course 2 and course 3. These results confirm our predefined difficulties of the courses. Moreover, there is an interesting interaction effect in the leading group, average group, and trailing group with respect to course difficulty. The leading group had an average score lower than the control group, but the average group and the trailing group had average scores higher than the control group in the easy course. However, the leading group has an average score higher than the control group, but the average group and the trailing group had average scores lower than the control group in the difficult course.

Table 54.2 Learning performance score

Group	Assessment score			Average
	Course 1	Course 2	Course 3	
1(Leading)	0.788	0.803	0.742	0.778
2(Average)	0.900	0.817	0.617	0.778
3(Trailing)	0.879	0.773	0.617	0.756
4(Control)	0.828	0.818	0.697	0.781
Average	0.849	0.803	0.668	0.773

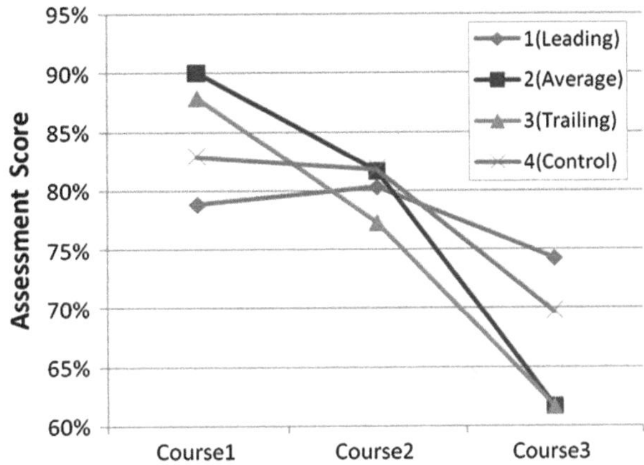

Fig. 54.2 Assessment scores of the four groups

Further analysis on the interview feedback indicates that, learning effect was found to be positively correlated with ability in the easy course for all of the participant combinations but not for the medium course and the difficult course; Generally, Agreeableness is negatively correlated with learning effect in most of the cases; Openness to experiences shows an interaction effect to score between leading group and other groups; there were strong negative correlations in leading group but it tends to be positive correlations in the other three groups (Fig. 54.2).

54.4 Conclusion and Discussion

In this study, we have carried out an experiment to examine the effect of displaying peers' progress to learning effect in web-based learning environment, and discussed the results in terms of social facilitation and social presence. The results may provide supports to existing theories and to future theories development. The result shows that students' course performance is affected by information about peers' progress. Displaying of information about peers on e-Learning system can have influence on students. Consequently, students' learning can more or less be influenced by controlling the display of peer information. More potential determining factors, such as gender, are still to be discovered. Meanwhile, an interaction effect of learning effect are yet to be analyzed in more detail. On the other hand, other kinds of peer information, such as performance of doing course exercise, are potential socially influencing factors that may worth studying.

Acknowledgments This work was supported in part by the NSC, under Grant NSC 102-2511-S-041-003.

References

1. Welsh ET, Wanberg CR, Brown KG, Simmering MJ (2003) E-learning: emerging uses, empirical results and future directions. Int J Train Dev 7(4):245–258
2. Eklund J, Kay M, Lynch HM (2003) E-learning: emerging issues and key trends: a discussion paper
3. Huang Y-M, Chiu P-S, Liu T-C, Chen T-S (2011) The design and implementation of a meaningful learning-based evaluation method for ubiquitous learning. Comput Educ 57(4):2291–2302
4. Chen T-S, Chiu P-S, Huang Y-M, Chang C-S (2011) A study of learners' attitudes using TAM in a context-aware mobile learning environment. Int J Mobile Learn Organ 5(2):144–158
5. Hsieh W-J, Chiu P-S, Chen T-S, Huang Y-M (2010) The effect of situated mobile learning on Chinese rhetoric ability of elementary school students. In: IEEE, Proceedings of WMUTE'10, pp 177–181
6. Wu T-T, Huang Y-M, Chao H-C, Park JH (2011) Personlized English reading sequencing based on learning portfolio analysis. Inf Sci 257:248–263
7. Huang Y-M, Wu T-T (2011) A systematic approach for learner group composition utilizing u-learning portfolio. Educ Technol Soc 14(3):102–117
8. Chiu P-S, Kuo Y-H, Huang Y-M, Chen T-S (2008) A meaningful learning based u-learning evaluation model. In: Proceedings of the eighth IEEE international conference on advanced learning technologies, pp 77–81
9. Wu T-T, Wang C-S, Huang S-H, Chung M-Y (2012) Using Google+ to conduct high interaction and interdependence learning environments based on group investigation strategy. J Internet Technol 13(6):997–1004
10. Wu T-T, Sung T-W, Huang Y-M, Yang C-S, Yang J-T (2011) Ubiquitous english learning system with dynamic personalized guidance of learning portfolio. Educ Technol Soc 14(4):164–180
11. Cochrane T (2005) Interactive quicktime: developing and evaluating multimedia learning objects to enhance both face-to-face and distance e-learning environments. Interdisc J E-Learn Learn Objects 1(1):33–54
12. Walther JB, Anderson JF, Park DW (1994) Interpersonal effects in computer-mediated interaction a meta-analysis of social and antisocial communication. Commun Res 21(4):460–487
13. Zajonc RB, Heingartner A, Herman EM (1969) Social enhancement and impairment of performance in the cockroach. J Pers Soc Psychol 13(2):83
14. Lai C-F, Chiu P-S, Huang Y-M, Chen T-S, Huang T-C (2014) An evaluation model for digital libraries' user interfaces using fuzzy AHP. Electr Libr 32(1):83–95
15. Chen C-C, Chiu P-S, Huang Y-M, Wang Y-H (2013) A power-saving mechanism of body sensor network for long-term care. Asia Life Sci 23(1):221–229
16. Zajonc RB (1965) Social facilitation. Research Center for Group Dynamics, Institute for Social Research, University of Michigan

Chapter 55
An Integration Framework for Clinical Decision Support Applications

Xiang Zheng, Yin-sheng Zhang, Zhen-zhen Huang, Zheng Jia, Hui-long Duan and Hao-min Li

Abstract Clinical Decision Support (CDS) applications have been widely recognized for improving health care quality and medical knowledge translation. However, its adoption and utilization in a real clinical environment is still subject to many factors, an important one of which is the lack of mechanism to easily integrate these applications to clinical workflow. This paper designs and develops an extensible CDS application integration and management system framework. This framework achieves interoperability between Clinical Information System (CIS) and CDS applications at two levels under predefined integration protocols; and through a mechanism of registration, management and delivery, various CDS applications can be delivered to targeted scenarios and executed in a context-aware way. This framework was validated in real clinical environment, and proved to be effective in both integration and interoperability. Meanwhile, targeted CDS application delivery further facilitates CDS in clinical promotion.

Keywords Clinical decision support · Service integration · Extensible framework · Service for application delivery

X. Zheng · Y. Zhang · Z. Huang · Z. Jia · H. Duan · H. Li (✉)
Department of Biomedical Engineering, Zhejiang University, Hangzhou, China
e-mail: lhm@vico-lab.com

X. Zheng
e-mail: zhengx@vico-lab.com

Y. Zhang
e-mail: zys@vico-lab.com

Z. Huang
e-mail: hzz@vico-lab.com

Z. Jia
e-mail: jiaz@vico-lab.com

H. Duan
e-mail: dhl@vico-lab.com

J.J. (Jong Hyuk) Park et al. (eds.), *Ubiquitous Computing Application and Wireless Sensor*, Lecture Notes in Electrical Engineering 331,
DOI 10.1007/978-94-017-9618-7_55

55.1 Introduction

Reducing medical errors and improving health care quality by obtaining CDS has become the strong driving force to promote clinical informatization [1]. Demand for high-quality CDS applications is getting urgent in medical care [2]. However, the CDS applications which can impact clinical practice widely and really are still very few [3]. The reasons are manifold [4]. A very important one is that most of the existing CIS do not provide an extensible mechanism for CDS applications which are of variety in both of function and source, resulting in the majority of CDS applications can not effectively be merged into the workflow. Burdensome and repetitive data entry working is one of the adverse effects. What is worse, it also caused that the results produced can by no means affect clinical practice completely and accurately.

Even if such integration barriers existing, due to the extensive needs for daily use, specific CDS tools have been clinically accepted, including various calculators (e.g. Calculator for creatinine clearance rates, BMI and body surface area) and clinical modeling evaluator (e.g. Grace, TIMI score for Cardiology and DAS-28 score for Rheumatology etc.). However, because lack of effective mechanism for CDS tool dissemination, update and management [5], a number of users do not know the existence of these tools and way to access them, while others may still use outdated ones. In short, the current CIS lack the management mechanism of CDS applications, which fails to provide the right people the right CDS tools at the right time [6].

To solve the above two issues, this paper proposes and implements an extensible CDS mechanism. By designing and implementing interoperable data integration interface and using scalable plug-in technology, the mechanism makes all types of CDS applications possible to be embedded in CIS, with the ability to allow interaction between CIS and CDS applications, reduces the obstacle in CDS applications data acquisition and makes CDS interventions produced submitted to CIS completely and accurately. Meanwhile, through the design and implementation of a management platform for CDS applications, all types of CDS applications are managed, delivered and updated properly, which makes CIS users can easily discover, access and update these applications. In this paper, the design and implementation of the mechanism are described, and its validity is verified by the integration of different types of real CDS applications in CIS.

55.2 Method

55.2.1 *Extensible CDS Application Mechanism*

Providing publicly available standardized and interoperable data integration interfaces is an effective way of scalable system implementation [7], where plug-in is one of these and has been widely used in web browser, media player, etc. However,

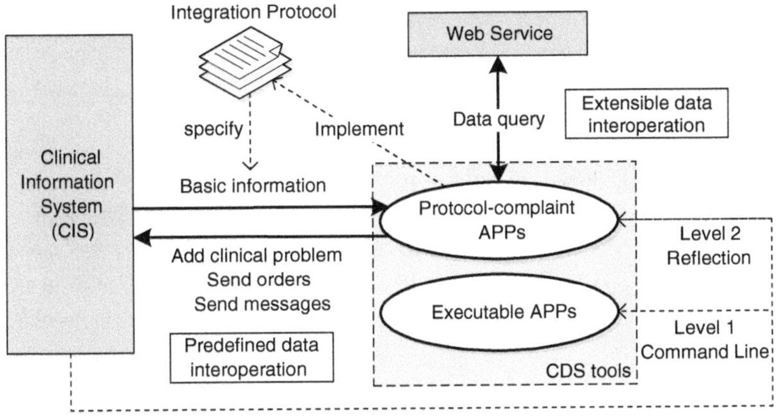

Fig. 55.1 Framework of extensible CDS application integration

health information technology is quite different, which did not form a widely used and accepted scalable interface protocol, so that most existing CDS applications are not developed orienting plug-in environment. This paper proposed mechanism to achieve the two levels of interoperability to meet the need of expansion (Fig. 55.1).

55.2.1.1 The First Level of Interoperability

This level enables CIS to invoke CDS applications in appropriate situations. The plug-in mechanism is able to call the applications developed for a specific protocol, but a number of third-party and stand-alone CDS applications always does not implement interface, such as the previously mentioned various clinical calculators, evaluators and intervention planner. In the clinical environment, such tools are often independent of the electronic medical records (EMR) or other CIS. The required information and produced recommendations or intervention usually requires manual translation and entry to interact with the CIS. Therefore, on this basis, this paper extends the plug-in mechanism to enable it to call these applications: They can be registered into CIS by the method described in Sect. 55.2.2. In CIS, users can call related applications directly through the registration information, thus CDS applications are integrated in CIS on the first level.

55.2.1.2 The Second Level of Interoperability

This level ensures data between CIS and CDS applications are exchangeable. As mentioned earlier, the interoperation of data among CIS and CDS applications and the ability of intervention produced by CDS applications directly entering CIS is a key factor to enhance clinical acceptance of CDS applications. However, because

Table 55.1 Protocol of extensible CDS application integration

Interface name	Type	Parameter	Implementation
SetPatientInfo	Input	Patient ID	Protocol
SetUserInfo	Input	Patient ID	Protocol
GetContextItemId	Input	Context item name	Web service
GetFact	Input	Patient ID	Web service
SendOrders	Output	Order list	Protocol
SendMessage	Output	Target ID, message	Web service
SetFact	Output	Patient ID, context item name, unit	Web service
AddClinicalProblem	Output	Problem name, user ID	Protocol

the different CDS applications' data requirements are various, defining a specific fixed data integration interface usually cannot meet the constantly emerging changing needs. Therefore, this paper designs a hybrid data acquisition mechanism, which on one hand provides the most basic data support for applications by a fixed predefined plug-in interface such as basic patient ID and user ID, on the other hand, opens data access service through WebService to provide an expanding mechanism. While CDS intervention outputs are mainly in the following three categories: new-generated clinical problems and proper clinical interventions (orders) and alarm or reminder message. For all of the above, this paper designs a CDS application integration protocol (Table 55.1).

55.2.2 Mechanism for CDS Application Management, Delivery and Update

With in-depth clinical information technology, CDS applications will be more comprehensive and various. While the majority of CDS applications are only for specific clinical scenarios or departments, how intelligently to push CDS applications targeted to target clinical scenarios is another issue concerned in this paper. Moreover, management and update of the CDS application version should also be taken into account.

This paper designs a CDS application management platform to collaborate with CIS (Fig. 55.2). All the CDS applications which are integrated in CIS should be registered on the management platform with its application package uploading. During the registration process, besides some basic information (e.g. Type, Icon, Brief description) should be set, what is more important is to establish the combination between the CDS application and clinical conception or event, such as set them for a specific clinical problem (e.g. Disease diagnosis, Surgery, Processes), or for a specific order execution process. And users who subscribe this application will be recorded so that the tool would be found in one's toolbox. In CIS, a proxy module of CDS application is developed to push the specific CDS applications to

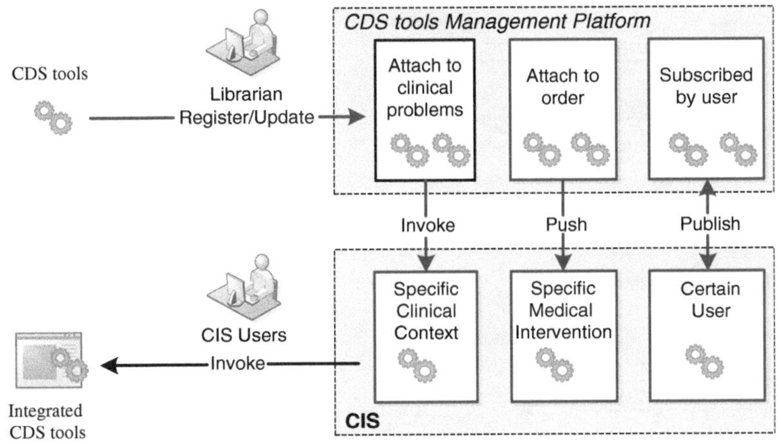

Fig. 55.2 Mechanism of management, delivery, launch and subscribing of CDS tools

specific clinical problem, specific order execution process or certain subscriber's toolbox. Additionally, by registering, the CDS application version can be effectively managed, so that the latest applications can be pushed to various CIS clients, which can facilitate CDS applications spread.

55.2.3 System Implementation

The CDS application management platform is realized through knowledge management platform website. Moreover, a CDS service proxy module is extended and implemented in the CIS, which is currently being implemented in a hospital. This proxy can be used to maintain the patient's current clinical problem list, and is able to obtain the current patient and user ID. Because the plug-in technology is an implementation alternative of scalable personalized software architecture [8], integration mechanism of CDS applications are achieved based on the plug-in technology.

55.3 System Assessment

In order to assess whether this mechanism and system implementation could meet various CDS application integration needs, this paper collected and developed different types of CDS tools and integrated them into an actual CIS.

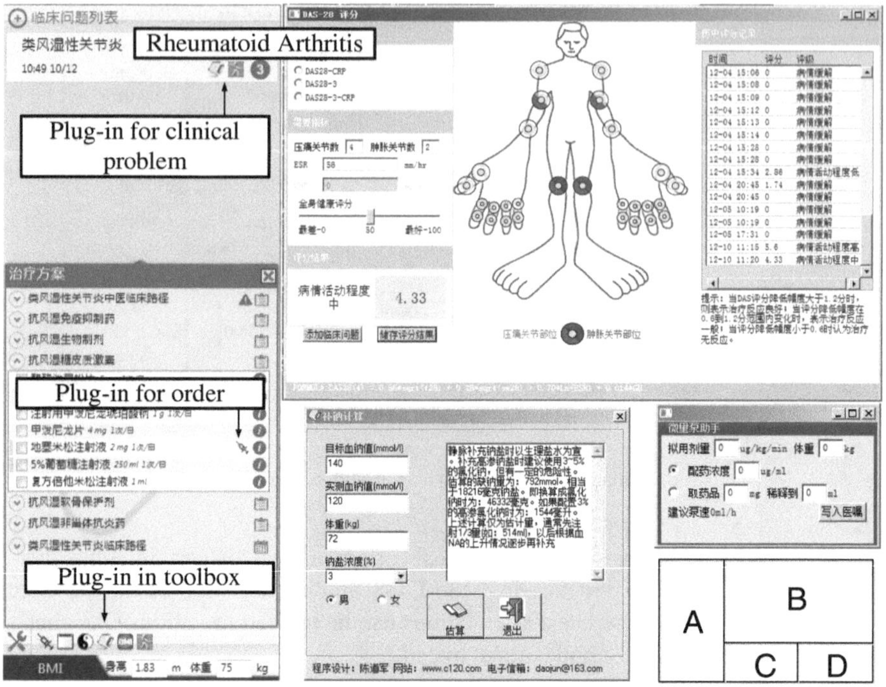

Fig. 55.3 Instances for CDS tools: **a** CDS service proxy module and docked BMI calculator, **b** DAS-28 evaluator, **c** micropump helper and **d** sodium supplement calculator

A variety of CDS applications are delivered to CIS systems through a proxy module. Various-functioned Applications were pushed to the target order, clinical problem or user's toolbar (Fig. 55.3a).

55.3.1 CDS Tools for Specific Clinical Problem

DAS-28 is a quantitative measure used to monitor the treatment of rheumatoid arthritis proposed by the Radboud University Nijmegen Medical Centre (RUNMC) of Netherland [9]. It helps clinicians to measure the disease activity in patients with Rheumatoid Arthritis (RA), access information to adjust the treatment plan. This assessment method is also widely accepted by Rheumatology specialist. This paper develops a DAS-28 score tool based on the integration protocol (Fig. 55.3b). After associating with the clinical problem of RA at the time of registration, the icon of this tool will display on clinical problem RA when RA occurs in the patient clinical problem list, which enables clinicians to launch the tool. In the context, CIS will pass the patient and user ID to the plug-in, which can be used to get essential data such as erythrocyte sedimentation rate (ESR) and high-sensitive c-reactive protein (hs-CRP) via the interface GetFact of WebService provided by CIS, without having to transcribe related data. The plug-in provides a visualized evaluation service to help clinical staff to quickly complete the relevant assessments. The result produced can be submitted to CIS via interfaces easily.

55.3.2 CDS Tools for Specific Treatment Protocol

For some medication, personalized settings (e.g. Dosage, Infusion speed) are in need according to patient condition (e.g. Weight, Body surface area). Clinicians always need to manually calculate related results in order to obtain specific parameters in performing process. This process is prone to miscalculate, or in most cases, clinicians take experience value to avoid troubles, so patients fail to get personalized treatment.

This paper has developed and implemented micropump helper which is compliant to integration protocol (Fig. 55.3d). This tool should be associated with a specific drug using micropump, such as "dopamine". When this order is included in clinical protocol, the tool can be launched. After the system automatically obtaining patient data (Weight) and clinician manual supplementing data (Planned dose and concentration), the required parameter of pump speed can be calculated based on the scientific formulas. Clicking on the "Write into order" can write the personalized information into CIS, with nothing to be transcribed.

55.3.3 Daily Tools Which Can Be Docked in the Layout

Some gadgets serving the daily work will be occasionally used. Repeated launching may be tiresome. This paper provides a docking mechanism for this kind of gadget, which allows gadgets to be docked in the layout of CIS. A good example is BMI (Body Mass Index) calculator. For doctors of endocrinology, BMI calculating is necessary when every patient received. This paper realized a BMI calculator based on the integration protocol as shown at the bottom of Fig. 55.3a. This gadget can get patient data automatically and display the BMI directly when data available. Meanwhile, it also allows the clinical staff to manually input the relevant values to calculate BMI. The mode of docking can improve the working efficiency.

55.3.4 Third-Party Developed CDS Tools

As mentioned in the introduction, there are now a lot of CDS applications developed by third parties, which may be good supplement for CIS. The integration mechanism can satisfy the registration and calling for these applications. For them, integration is supported to be associated with any forms of specific order, clinical problem and user's toolbar. Several third-party applications are collected from the Internet, including diabetic diet calculator, rehydration calculator, body surface area calculator and creatinine clearance measurement tool. Sodium supplement calculator is one of them (Fig. 55.3c). After registered by the user, the icon will emerge and one can use it to generate the treatment plan for filling sodium, according to planned and actual serum sodium and patient weight. Due to absence of implementation of integration protocol, only the first level of interoperability is supported and the result cannot be input into the CIS directly, where manual transcription is necessary.

55.4 Discussion

Achieving a scalable CDS service framework in CIS enables existing and coming CDS applications to be integrated into the workflow of current CIS. Effective integration and actionable output can greatly save the time consumed on data entry. High efficiency promotes busy clinicians to be more willing to try and accept all types of CDS applications, in turn, demand of clinicians stimulates the development of CDS applications, thus facilitating the knowledge translation [2]. Therefore, this method not only effectively enhances the significance of CIS applications, but also greatly improves the clinician acceptance of CDS applications.

However, some limitations still exist. Clinical data are diverse in types and complex in contexts. And this relation information is usually very important for certain CDS, while the current data integration is providing isolated data element query. To solve this problem, creating more complex models to describe clinical data is essential. HL7's Virtual Medical Record (vMR) [10] in this regard is a good attempt. However, it relies on standardization of data elements, data structures and clinical documentations, which is very hard to realize on current level of clinical informatization.

Finally, the profound and comprehensive standardization makes it possible to solve this problem and many research teams have made a lot of good work. With the development of scalable and sharable CDS service concepts, more and more CDS services are achieved in a web-based service form. To access these online CDS services in a standardized way is also a future development direction. HL7's InfoButton-related standards provide standardized interfaces to access online knowledge-base in a context-specific way [11], but only in a URL-based way. Thus, building more profound integration standard for CDS has become the goal of some research teams [12]. But given the complexity of the clinical data itself, the development of this standard will be a very long process.

55.5 Conclusion

This paper provides a solution of extensible CDS service framework. It achieves CDS application integration and interoperability, and embeds all kinds of CDS applications into an existing CIS workflow. Registration, management, delivery and update of CDS applications are realized through a management platform, while a proxy module is used in CIS to deliver and execute these tools automatically in proper clinical context in the latest version. The solution is verified by integrating various existing CDS applications in the clinical environment, achieving at a certain level of scalability and interoperability under standardized conditions. This kind of CIS will greatly enhance the clinical user acceptance for all types of CDS applications.

Acknowledgments This research was financially supported by the National High-tech R&D Program (2012AA02A601) and National Natural Science Foundation of China (30900329).

References

1. Kawamoto K, Houlihan CA, Balas EA, Lobach DF (2005) Improving clinical practice using clinical decision support systems: a systematic review of trials to identify features critical to success. BMJ 330:765
2. Sittig DF, Wright A, Osheroff JA, Middleton B, Teich JM, Ash JS, Campbell E, Bates DW (2008) Grand challenges in clinical decision support. J Biomed Inform 41:387–392
3. Wu S, Chaudhry B, Wang J, Maglione M, Mojica W, Roth E, Morton SC, Shekelle PG (2006) Systematic review: impact of health information technology on quality, efficiency, and costs of medical care. Ann Intern Med 144:742–752
4. Osheroff JA, Teich JM, Middleton B, Steen EB, Wright A, Detmer DE (2007) A roadmap for national action on clinical decision support. J Am Med Inform Assoc 14:141–145
5. Liu J, Wyatt JC, Altman DG (2006) Decision tools in health care: focus on the problem, not the solution. BMC Med Inform Decis Mak 6:4
6. Kawamoto K, Lobach DF (2007) Proposal for fulfilling strategic objectives of the US roadmap for national action on decision support through a service-oriented architecture leveraging HL7 services. J Am Med Inform Assoc 14:146–155
7. Wright A, Sittig DF (2008) SANDS: a service-oriented architecture for clinical decision support in a national health information network. J Biomed Inform 41:962–981
8. Wolfinger R, Dhungana D, Prähofer H, Mössenböck H (2006) A component plug-in architecture for the. NET platform, in modular programming languages. Springer, Berlin, pp 287–305
9. Fransen J, Creemers M, Van Riel P (2004) Remission in rheumatoid arthritis: agreement of the disease activity score (DAS28) with the ARA preliminary remission criteria. Rheumatology 43:1252–1255
10. Kawamoto K, Del Fiol G, Strasberg HR, Hulse N, Curtis C, Cimino JJ, Rocha BH, Maviglia S, Fry E, Scherpbier HJ (2010) Multi-national, multi-institutional analysis of clinical decision support data needs to inform development of the HL7 virtual medical record standard. In: AMIA annual symposium proceedings, vol 2010. American Medical Informatics Association, p 377

11. Del Fiol G, Huser V, Strasberg HR, Maviglia SM, Curtis C, Cimino JJ (2012) Implementations of the HL7 context-aware knowledge retrieval ("infobutton") standard: challenges, strengths, limitations, and uptake. J Biomed Inform 45:726–735
12. Zhou L, Hongsermeier T, Boxwala A, Lewis J, Kawamoto K, Maviglia S, Gentile D, Teich JM, Rocha R, Bell D (2013) Structured representation for core elements of common clinical decision support interventions to facilitate knowledge sharing. In: MEDINFO 2013: proceedings of the 14th world congress on medical and health informatics, vol 192. IOS Press, p 195

Chapter 56
Method of Generating Intelligent Group Animation by Fusing Motion Capture Data

Jie Song, Xiang-wei Zheng and Gui-juan Zhang

Abstract To achieve the realistic simulation of intelligent group animation, a method of generating intelligent group animation by fusing intelligent algorithms and motion capture data is proposed. This method mainly proposed a novel algorithm that can effectively fuse the animation path extracted from experiments about group simulation with particles based on the Artificial Fish-Swarm Algorithm (AFSA) and the original motion capture data, so as to synthesize one new motion data. As a result, we can achieve multi-roles path animation with reality through binding the synthesized motion data to specific roles. Depending on the experimental results, it demonstrates that the method of generating intelligent group animation by fusing motion capture data and intelligent group path planning can generate group animation with higher reality.

Keywords AFSA · Motion capture · C3D file · Character binding · Group animation

56.1 Introduction

Group animation [1, 2] has grown up to be a hotspot in the fields of artificial intelligence applications and computer graphics. However, most researches on group animation simulation only implement the effect of particle motion simulation and therefore they have a lack of reality and cannot guarantee the realistic visualizations

J. Song · X. Zheng (✉) · G. Zhang
School of Information Science and Engineering,
Shandong Normal University, Jinan 250014, China
e-mail: xwzhengcn@gmail.com

J. Song · X. Zheng · G. Zhang
Shandong Provincial Key Laboratory for Distributed
Computer Software Novel Technology, Water Conservancy
Research Institute of Shandong Province, Jinan 250014, China

J.J. (Jong Hyuk) Park et al. (eds.), *Ubiquitous Computing Application and Wireless Sensor*, Lecture Notes in Electrical Engineering 331,
DOI 10.1007/978-94-017-9618-7_56

of group motion. So how to improve the reality of the group animation by combining motion capture [3, 4] with swarm intelligence has been a meaningful research.

On one hand, path planning [5, 6] technology is part of the core content of the researches on intelligent group animation. Since the AFSA has been submitted, many scholars have conducted a lot of optimized algorithms [7, 8]. But the path planning and path editing are not combined. On the other hand, motion capture [9, 10] has been widely accepted and used in computer animation. However, it cannot change and rebuild its motion path information. To solve those problems, this paper presents a method of generating intelligent group animation by fusing motion capture data and AFSA, which can achieve realistic behavior simulation of the intelligent group animation in place of the particle simulation.

56.2 Method of Generating Intelligent Group Animation Based on Intelligent Algorithms and Motion Capture Data

Flow chart of the proposed method that generates intelligent group animation based on intelligent algorithms and motion capture data is shown in Fig. 56.1.

56.2.1 Generation of Group Animation Path Based on AFSA

The primary design task of the group animation is to determine the path for the group motion. AFSA is under a flexible architecture that can well solve the path planning problems in swarm intelligence applications. The following brings an example of the aggregation behavior. Algorithm is presented as follows.

Step 1: Defining the symbolic variables. Number of fish that the number of test particles is denoted as N (1,2…N), state of fish is denoted as Xi (X1,X2… XN), function f (Xj) is the objective function that defines the degree of the fish itself adaptation whose value can be referred to judge and determine the direction of individual fish movement. Sensing range of fish is

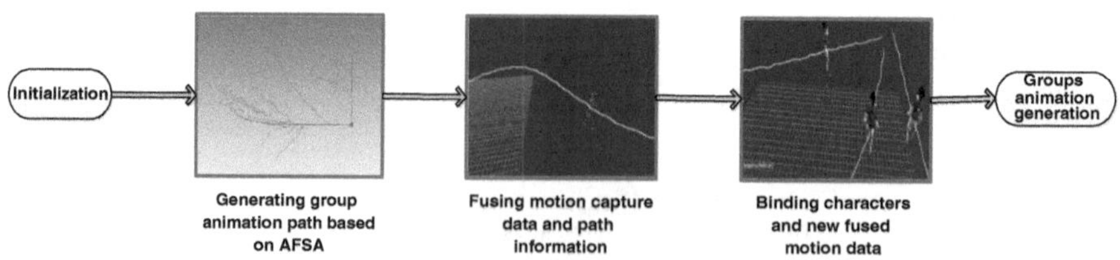

Fig. 56.1 Flow chart of method

denoted as visual, Step is denoted as step, crowding factor is denoted as δ and times for routing attempts is denoted as try-times.

Step 2: Initialization. N = 15, step = 9 (as an integer and its value closer to the step of human movement from our sample C3D files), visual = 10 (decides the field of view for each step), δ = 10, try-times = 3, gathering point status as D = (a, b).

Step 3: There supposed a total of N fish that is N = (1,2,…N), each fish mobile steps from the start position to the target position is M as j = (1,2,…M), starting from the first step (j = 1) we should set the status of individual fish as $X_j = Random(N(X_i, visual))$, the adjacent particles under the perception range of observed individual fish as $n_f = |N(X_j, visual)|$, we find the next state according to the formula as $X_{j+1} = f(N(X_j, visual))$, determine the direction of movement must meet the condition that $n_f/N < \delta$ and $f(X_j) < f(X_{j+1})$, determine the moving direction for next step of each fish individual base on the condition that $X_{j\backslash next} = X_{j+1}$, if the conditions are not met to Step 4.

Step 4: When meeting the routing attempts $try-times \leq 3$, repeat Step 3, when $try-times \geq 3$ to Step 5.

Step 5: When routing attempts is greater than try-times, randomly selected the next moving state as $X_{j\backslash next} = Random(N(X_j, visual))$.

Step 6: Traversing the aggregation swimming direction of a total of N fish, repeat Step 3, 4, 5.

Step 7: Determine the termination of the algorithm under the condition as $|(\mathrm{d}(\mathrm{X_j}, \mathrm{D})/N)| < \varepsilon$, where in ε is a threshold to judge the fish whether have reached the final destination to stop swimming.

As is shown from Fig. 56.2, we can get the path renderings generated by the path automatically planning platform based on AFSA illustrated by 15 fish, then we save it as the path information needed for the next task.

56.2.2 Fusion of Motion Capture Data and Path Data

C3D is a common notion capture data file, the sample C3D motion data used in our experiment come from human motion capture database called mocap which is constructed by CarnegieMellon University [11]. C3D file contains a minimum of three sections of information [12]: header section, a parameter section and 3D data section.

Basic processes of the fusion of motion capture data and intelligent algorithm as follows:

Step 1: C3D file needs to be edited, mainly contains three sections: excluding the original path information, cropping frames of human motion and synthesis of loop animation.

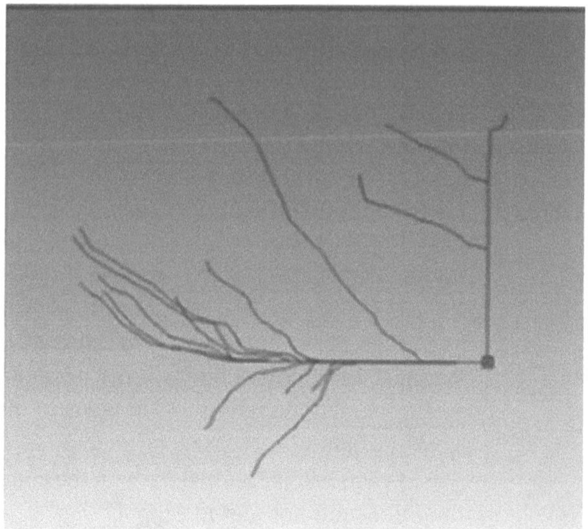

Fig. 56.2 Roadmap of the fish aggregation simulating

Step 2: Initialization of path information. C3D file after editing and initialization was retained the orientation of initial human movement, stride, as well as the total number of the cycle animation frames. In the experiment implemented by the platform that automatically generating group animation path based on AFSA, we set step (best consists with the step of the original C3D file), set number of particle (15 particles in the aggregation experiment), set the crowding factor and distance weight and other parameters to automatically generate group animation path information. Then we select one path data from one of the 15 particles, and intercept the number of points needed which makes up of the number of cycle animation frames in Step 1.

Step 3: Define relevant parameters. Oriented angle θ for each human movement frame need to start seeking from the second frame, Fig. 56.3 shows the human orientation vector as a, b is the vector toward to the next frame, the angle between a and b is just the rotation angle we calculated from each frame movement.

Step 4: Initialization. Number of new motion data frames as N (1…N), number of path points as N (1…N), number of sample points of human movements as M(1…M) (M = 41).

Step 5: Calculating the oriented angle of human movement for each frame relevantly, refer to the formula $a \cdot b = |a| \cdot |b| \cdot Cos{<}a, b{>} = |a| \cdot |b| \cdot Cos{<}\theta{>}$,we can work out the orientated angle for each human movement frame as $\theta = arccos\,\theta = (a \cdot b)/(|a| \cdot |b|)$. d(dpf) stands for distance per frame, total distance in from the path information as D, and $d = D/N$.

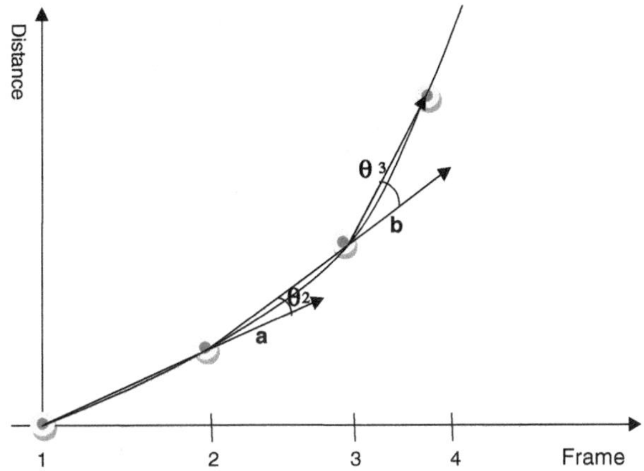

Fig. 56.3 Diagram of the calculation for the orientation angle of the sample points from the top 4 frames

Step 6: Direction for the first frame is a default that is the value of the orientation from the original C3D file, so we need to calculate the rotation information of each sample points from the second frame.

Step 7: Define the number of frames as the variable i, variable j is the number of sample points, k is the number of the calculated current frames.

```
for (i=2 … N)//Calculating the rotation angle θ i from the second frame
{
 for (j=1… M)//Traversing a total number of M sample points in Frame i
{
   for (k=2… i)
   {
     ①Rotating the current sample point j, centering on the root point
(LBWT) in Frame k.
     ②Moving forward one unit distance d based on the calculated rotation θ i
in Frame k
    }
}
}
```

The experiment as follows is role animation along the curve like sin a (a = 20), via fusing human walking motion capture data to synthesize a new motion data. As is seen in Fig. 56.4, the new motion data has larger space and higher simulation and realization for the actual motion path.

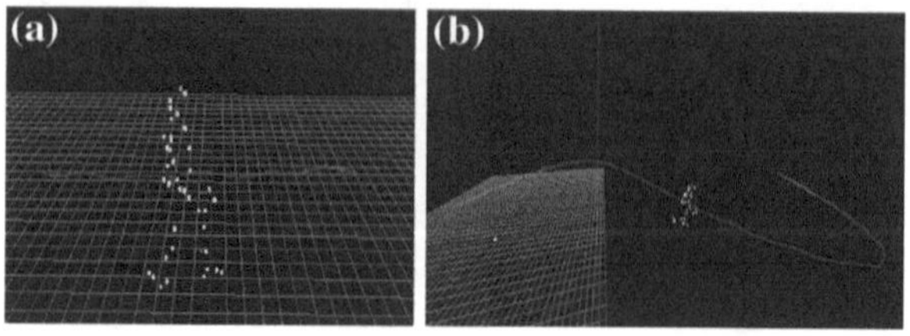

Fig. 56.4 **a** The role path from the original C3D file, **b** The role path from the new synthesized C3D file

56.3 Experimental Results and Analysis

There is performed out two experiments such as the single-role path animation and the multi-roles path animation during the animation designing task.

First, in the experiment of single-role path animation, through merging in the well rigged role, we change the role mode into the mode of baking to the skeleton, then merging in the motion data from the saved previously file named mocap (saving the well-bound role animation moving along sin curve as the form FBX), Fig. 56.5 below shows the final results of the experimental single-role path animation.

Next, in the experiment of multi-roles path animation that is based on AFSA path planning design. Figure 56.6 shows the three typical paths separately marked out by a, b, c, which are extracted from the paths generated in the aggregation simulation by a number of 15 fish, then we separately fuse them by the motion capture data C3D file into three new motion data file, last, we bind the three new synthesized motion data to three individual roles that are merged in the same 3-dimension space as to complete each role path animation. The relevant position of the role animation paths as is shown in Fig. 56.6(c) well consists of the three paths shown in Fig. 56.6(b) which are extracted from the aggregation experiment shown in Fig. 56.6(a) by comparison, as a result, we can raise the authenticity of the intelligent group animation through using the vivid roles to take place of the

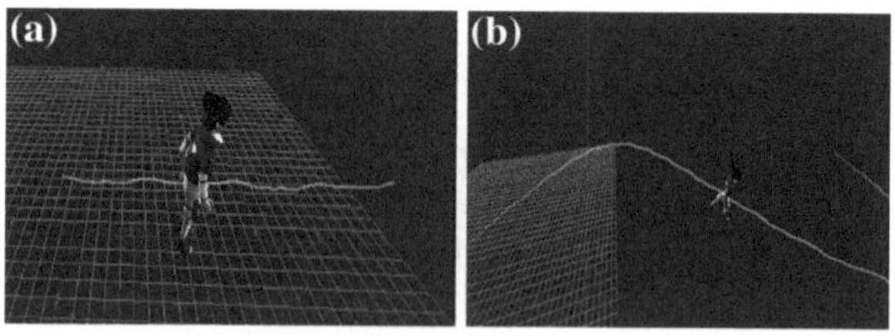

Fig. 56.5 **a** The comparison renderings between role animation from the original C3D file, **b** The role animation from the new synthesized motion file

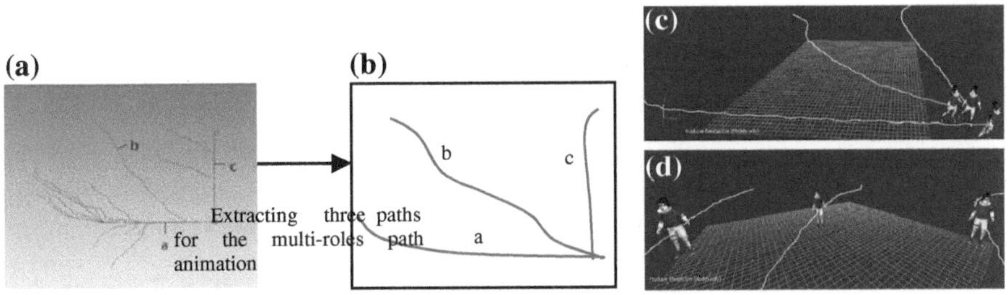

Fig. 56.6 **a** Path generated by AFSA, **b** Three path extracted from (**a**), **c** The overlook of the final experimental renderings, **d** A closer look of the multi-roles path animation

experimental particles, Fig. 56.6(d) just shows the enlarged renderings of the multi-roles path animation, we can see that each role has its own specific movements, so, better reflecting the authenticity of the group animation.

56.4 Conclusion

This paper mainly studied a method of generating intelligent group animation by fusing motion capture data. First, the group paths are generated based on AFSA and raw motion capture data is edited. Second, group paths data and edited motion data are synthesized with the proposed algorithm. At last, roles and above data are bound and group animation is generated automatically. The advantages of the method perform that: avoiding tedious manual operation, enhancing the efficiency of animation, saving cost, and at the same time completing the path animation simulation of the complex groups, as a result to improve the authenticity and value of application of intelligent group animation.

Acknowledgments We are grateful for the support of the National Natural Science Foundation of China (61373149), the Promotive Research Fund for Excellent Young and Middle-aged Scientists of Shandong Province (BS2010DX033) and a Project of Shandong Province Higher Educational Science and Technology Program (J10LG08).

References

1. Zhan B, Monekosso DN, Remagnino P, Velastin SA, Li-Qun Xu (2008) Crowd analysis: a survey. Mach Vis Appl 19(5/6):345–357
2. Silverman BG, Badler NI, elechano NP, O'Brien K (2008) Crowd simulation incorporating agent psychological models, roles and communication[C]. In: Proceedings of the 1st international workshop on crowd simulation, St. Philadelphia, PA, [s. n.], pp 21–30
3. Shen Y, Wu X, Lü C, Cheng H (2011) National dances protection based on motion capture technology [C]. In: Proceedings of 2011 international conference on computer science and information technology (ICCSIT), pp 491–493

4. Long J, Jones MD (2012) 3D Tree modeling using motion capture. In: Proceedings of 2012 IEEE 4th international symposium on plant growth modeling, simulation, visualization and applications (PMA 2012), pp 242–249
5. Kala R, Shukla A, Tiwari R (2011) Robotic path planning using evolutionary momentum-based exploration. J Exp Theor Artif Intell 23(4):469–495
6. Chen Z, Ma L, Li Z, Wu X, Gao Y (2006) Editing human motion path. J Comput Aided Des Comput Graph 18(5):651–655
7. Zheng X, Wang X, Huang Y (2006) A modified artificial fish-swarm Algorithm. In: Proceedings of intelligent control and automation (WCICA), IEEE, pp 3456–3460
8. Wang L, Hong Y, Shi Q (2009) Global edition artificial fish swarm algorithm. J Syst Simul 21 (23):7483–7502
9. Cai M, Zou B, Xin G (2012) Extraction of key-frame from motion capture data based on pre-selection and reconstruction error optimization. J Comput Aided Des Comput Graph 24 (11):1485–1492
10. Li J, Ya Y (2011) Animation system of virtual human using motion capture devices [C]. In: Proceedings of 2011 IEEE international conference on computer science and automation engineering, pp 544–547
11. [DB/OL], http://mocap.cs.cmu.edu (2011)
12. The C3D file format user guide. printed in the united states of America motion lab systems, Inc pp 17–65 (2008)

Chapter 57
Histogram-Based Masking Technique for Retinal Fundus Images

Rachel M. Chong and Jeziel C. Suniel

Abstract The number of eye disease cases, especially those that will result to blindness, is increasing. Retinal fundus images are usually used for disease detection and monitoring, which are essential for proper treatment and care. Image processing algorithms can be used to help increase the speed of analyzing these images. Since analyses require more computational effort and time, operations should be focused only on the object pixels. This paper will present a new technique that can identify the object pixels in a simple and fast manner. Unlike existing methods, this uses only a few assumptions and computations. Its performance is tested by experiments that used various publicly available datasets. Results will show the accuracy, effectiveness, and robustness of the proposed method.

Keywords Retinal imaging · Image histogram · Adaptive threshold · Mask generation

57.1 Introduction

The World Health Organization (WHO) estimates that 285 million people are visually impaired worldwide and that 90 % of which live in developing countries [1]. Potentially blinding eye conditions such as age-related macular degeneration (AMD), diabetic retinopathy, and glaucoma are increasing [2]. These require long-term care that includes constant monitoring commonly accomplished through the

R.M. Chong · J.C. Suniel (✉)
Cebu Institute of Technology—University, N. Bacalso Ave, Cebu City, Philippines
e-mail: jezielsuniel@yahoo.com

R.M. Chong
e-mail: rmchong.ece.citu@gmail.com

J.J. (Jong Hyuk) Park et al. (eds.), *Ubiquitous Computing Application and Wireless Sensor*, Lecture Notes in Electrical Engineering 331,
DOI 10.1007/978-94-017-9618-7_57

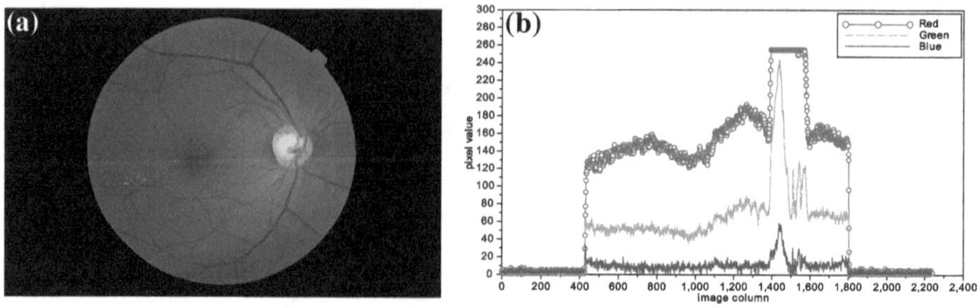

Fig. 57.1 Example of a retinal fundus image (image 19 in [11]). **a** Colored or RGB image **b** pixel values of a selected row

use of retinal fundus images or sometimes simply called fundus. An example of this is shown in Fig. 57.1 where Fig. 57.1a is the actual fundus and Fig. 57.1a is the plot of the pixel values marked by the white line in Fig. 57.1a.

The use of image processing algorithms provides an additional tool for faster screening of eye diseases. In order to detect the disease, accurate segmentation and measurement of anatomical structures and/or lesions are essential. An obvious prerequisite for these is the fundus image quality that can be evaluated by various techniques proposed in [3–7]. Analysis of images is usually computationally demanding thus, segmentation is initially done to separate the object from the background. Applying mathematical operations only to the object pixels can increase the speed of processing.

Fundus image processing sometimes do not employ segmentation since masks are already provided beforehand. However, this is not true for most of the images and as a result various computational techniques are proposed to produce the mask or to accomplish segmentation. Current approaches can be grouped according to is basic operations, which are thresholding, boxing, and scaling. Thresholds that require empirical data are used in [6–9]. In [6, 7], the threshold is determined by a statistical study of 361 images from various datasets. Using this value, the resulting binary image is then processed using two morphological openings. This is followed by the determination of the left, right, upper, and lower edges to reduce the size of the image. In [8], the mask is generated by calculating statistical values of each color band followed by a four-sigma thresholding with a free parameter that is empirically chosen. This resulted to mask sizes that is not always the same for each band due to different responses. A technique that uses only the red channel uses an empirically determined threshold of 35 [9]. The thresholded image is then morphologically processed (opening, closing, and erosion) using a 3×3 kernel. Although one channel is used, it still requires a lot of processing due to the morphological processes.

The rectangular boxes used in [4, 5] are pre-defined. The assessment method of [4] uses the RGB values inside a 121×121 box that is placed at the center of the image. In contrast, the work of [5] uses seven regions then for each region features are computed using the RGB and CIELab values. These methods require that the placement and size of the boxes must coincide with the object. Thus, the object's approximate location must be known beforehand and is assumed to be almost the same for all images.

The scaling technique in [3, 10] uses only the green channel. It is scaled down to a 160 × 120 image then region growing is performed from its four corners. Once the mask is formed, it is scaled back to its original size.

The existing methods that have been discussed mostly use priori information and are not adaptive for fundus images that are acquired by different cameras and in various situations. Additionally, several processes are required to produce the final mask. In this paper, a new technique for generating a fundus mask will be presented. It uses the red channel and an adaptive threshold that is computed based on the current image histogram. It requires only a few number of processes and is relatively simple. The next sections will discuss the proposed algorithm and the experiments performed to test its performance.

57.2 Fundus Mask Generation

Fundus processing involves operations for the image's object that is obviously lighter than the background. In Fig. 57.1, it can be seen that the background is dark but not exactly zero. This is why if thresholding is used, the value cannot be simply set to zero for the three color channels. In image processing, the green channel is usually used since it shows most of the image's details. It can be seen in Fig. 57.1b that green pixel values for the object is higher than the background. However, this difference is most pronounced in the red channel. It can be shown later that this characteristic is very useful for accurate segmentation results. In view of these, the proposed method can be outlined as follows:

1. Extract the red channel of the fundus.
2. Create a histogram of the pixel values. The bins are assigned for the values 0–255 since most fundus images are encoded using 8-bits.
3. Determine which of the bins (0–127) has the maximum value or peak. Note that this is only half of the total histogram width.
4. Starting from the bin in the previous step and using a window size of eight bins, compute the average value.
5. With a step size of two, shift the window to the right and compute the next average value.
6. Compare the previous and new values. Repeat step 4 until the new average is greater than the previous.
7. Determine the threshold by selecting the bin with the minimum value from the last window.
8. By thresholding the red channel, a binary image will result. This can now be used as a mask.

9. In some cases when the mask will have an irregular shape, a complete region can be formed by using the ellipse equation centered at (x_0, y_0):

$$\frac{(x - x_0)^2}{a^2} + \frac{(y - y_0)^2}{b^2} = 1 \quad (57.1)$$

where a and b are the major and minor radii, respectively. The variables x and y are locations in the rectangular coordinate. In some literature, the shape is assumed to be circular but this is not applicable for all fundus images. In this paper, we use the ellipse equation to provide more flexibility. Furthermore, we assume that the major axis will always lie on the x axis.

57.3 Experiment Descriptions and Results

To test the proposed method, we performed two experiments. These use several datasets, which are publicly available. Their specific names and descriptions are in Table 57.1. It is worthy to note that the images have different types and consequently, compression methods. Additionally, image sizes vary as well as the fundus images. Only DRIVE and MESSIDOR have images that form a complete retinal fundus (may be circular or elliptical) while the rest are sectors since the top and bottom segments are removed. STARE originally contains 400 plus images however, we only selected those that have sectors. If the top, bottom, left, and right segments are removed then these are not included in the experiment.

The first experiment will determine the accuracy in classifying the pixels (object or background). The result right after thresholding is used for this. Accuracy assessment is similar to that used in [6], which involves the computation of the following:

Table 57.1 Datasets of test images with their descriptions

Data set	No. of images	File type	Size (pixels)
DIARETDB0 [12]	130	png	1,500 × 1,152
DIARETDB1 [13]	89	png	1,500 × 1,152
DRIVE [14]	40	tif	565 × 584
HRF [15]	45	jpg	3,504 × 2,336
MESSIDOR [10]	1,200	tif	1,440 × 960
			2,240 × 1,488
			2,304 × 1,536
STARE [16]	390	ppm	700 × 605

Table 57.2 Sensitivity and specificity after thresholding

Data set	No. of images	Sensitivity			Specificity		
		Min.	Max.	Ave.	Min.	Max.	Ave.
DRIVE [14]	40	0.99585	1.00000	0.99857	0.97814	1.00000	0.99577
HRF [15]							
DR	15	0.99961	1.00000	0.99991	0.99785	1.00000	0.99921
Glaucoma	15	0.99970	1.00000	0.99989	0.99844	1.00000	0.99930
Healthy	15	0.99977	1.00000	0.99995	0.99587	0.99999	0.99861

$$\text{sensitivity} = \frac{t_p}{t_p + f_n} \tag{57.2}$$

$$\text{specificity} = \frac{t_n}{t_n + f_p} \tag{57.3}$$

where:

t_p = number of true positives or the number of retina pixels correctly identified

f_n = number of false negatives or the number of retina pixels incorrectly identified

t_n = number of true negatives or the number of non-retina pixels correctly identified

f_p = number of false positives or the number of non-retina pixels incorrectly identified.

Only DRIVE and HRF are used since these come with masks, which are considered as the ground truth. HRF dataset is further divided into three groups: Diabetic retinopathy (DR), glaucoma, and healthy. It can be observed that the results shown in Table 57.2 have values that are very close to one. These indicate that majority of the object and background pixels are correctly classified based on proposed adaptive thresholding technique.

In the second experiment, all the nine steps are performed to ensure that a complete region will result. All images in Table 57.1 are used resulting to segmented images whose background pixels are set to white. These are then visually inspected whether the whole fundus is completely segmented or not. In most datasets, the proposed method effectively segmented the image as shown in Table 57.3. It can be observed that only STARE is not 100 % due to grossly uneven lighting in some images.

Table 57.3 Segmentation or masking performance using various datasets

Data set	No. of images	Accuracy (%)
DIARETDB0 [12]	130	100.00
DIARETDB1 [13]	89	100.00
DRIVE [14]	40	100.00
HRF [15]	45	100.00
MESSIDOR [10]	1,200	100.00
STARE [16]	390	97.44
	Average	99.57

57.4 Conclusion

This paper presented a new technique that can create a mask or segment fundus images based on adaptive thresholding of the image's red channel. An experiment involving comparisons with the provided mask of 85 images yielded sensitivity and specificity values that are close to one. These translate to high accuracy in classifying the object and background pixels. Another experiment applied the proposed method to several datasets for a total of 1,894 images. Visual inspection after ellipse fitting resulted to an average accuracy of 99.57 %. Majority have the whole retinal fundus segmented while a few have not due to grossly uneven illumination across the image. The consistently high values illustrate the proposed method's effectiveness and robustness. Additionally, it requires less computations and processing time thus, this is very applicable for pre-processing purposes.

References

1. World Health Organization (2013) Visual impairment and blindness. http://www.who.int/mediacentre/factsheets/fs282/en/
2. World Health Organization (2014) Global trends in the magnitude of blindness and visual impairment. http://www.who.int/blindness/causes/trends/en/
3. Giancardo L, Abramoff MD, Chaum E, Karnowski TP, Meriaudeau F, Tobin KW (2008) Elliptical local vessel density: a fast and robust quality metric for retinal images. In: 30th annual international conference of the IEEE engineering in medicine and biology society (IEMBS), pp 3534–3537
4. Pires R, Jelinek HF, Wainer J, Rocha A (2012) Retinal image quality analysis for automatic diabetic retinopathy detection. In: 25th conference on graphics, patterns and images (SIBGRAPI), pp 229–236
5. Davis H, Russell SR, Barriga ES, Abramoff MD, Soliz, P (2009) Vision-based, real-time retinal image quality assessment. In: 22nd IEEE international symposium on computer-based medical systems (CBMS), pp 1–6
6. Dias JMP, Oliveira CM, da Silva Cruz LA (2012) Retinal image quality assessment using generic image quality indicators. Inf Fusion 13:1–18
7. Dias JMP, Oliveira CM, da Silva Cruz LA (2012) Evaluation of retinal image gradability by image features classification. Procedia Technol 5:865–875

8. Gagnon L, Lalonde M, Beaulieu M, Boucher MC (2001) Procedure to detect anatomical structures in optical fundus images. http://www.cs.rpi.edu/research/groups/vision/ken/fundusstructureid.pdf
9. Ter Haar F (2005) Automatic localization of the optic disc in digital colour images of the human retina. Utrecht University, Utrecht
10. Giancardo L, Meriaudeau F, Karnowski TP, Chaum E, Tobin K (2010) Quality assessment of retinal fundus images using elliptical local vessel density. In: Campolo D (ed) New developments in biomedical engineering. InTech, Croatiapp, pp 201–224
11. MESSIDOR: Digital retinal images. http://messidor.crihan.fr/
12. Kauppi T, Kalesnykiene V, Kamarainen J-K, Lensu L, Sorri I, Uusitalo H, Kalviainen H, Pietila J (2006) DIARETDB0: evaluation database and methodology for diabetic retinopathy algorithms. ImageRet Project
13. Kauppi T, Kalesnykiene V, Sorri I, Raninen A, Voutilainen R, Kamarainen J-K, Lensu L, Uusitalo H (2009) DiaRetDB1 V2.1—diabetic retinopathy database and evaluation protocol. Machine Vision and Pattern Recognition Laboratory, Lappeenranta University of Technology. http://www.it.lut.fi/project/imageret/diaretdb1_v2_1/
14. Staal JJ, Abramoff MD, Niemeijir M, Viergever MA, van Ginneken B (2004) Ridge-based vessel segmentation in color images of the retina. IEEE Trans Med Imaging 23(4):501–509
15. Kohler T, Budai A, Kraus M, Odstrcilik J, Michelson G, Hornegger J (2013) Automatic no-reference quality assessment for retinal fundus images using vessel segmentation. In: 26th IEEE international symposium on computer-based medical systems, pp 95–100
16. Hoover A, Kouznetsova V, Goldbaum M (2000) Locating blood vessels in retinal images by piece-wise threshold probing of a matched filter response. IEEE Trans Med Imaging 19(3):203–210

Chapter 58
Effect of Electronic Scoring System for Scenario Group Tutorial Implementation for Supporting Medical Student Studies

Piyapong Khumrin and Volaluck Supajatura

Abstract Small group discussion is a part of the medical program for second and third year medical students at the Faculty of Medicine, Chiang Mai University. In the past, the students were scored manually by facilitators. The scores were checked and transferred to a spread sheet by at least three administrative assistants. This massive workload delayed the process of finalizing scores in each study block which meant students with low scores were not identified promptly. Consequently some of these students did not pass their study. An electronic scoring system was implemented to improve the scoring process. The system facilitates the administrative assistant to create a scoring form, manage scenario details, store data, perform backups, and export data. One year analysis showed the major effect of the new system was that it significantly reduced the time taken for the scoring process. Finally, the improvement of the scoring system decreased the number of repeat student from 10 (on average between 2008 and 2011) to 2 students in 2012.

Keywords Electronic scoring system · Electronic data management · Medical education

P. Khumrin (✉)
Department of Physiology, Faculty of Medicine, Chiang Mai University,
Chiang Mai 50200, Thailand
e-mail: u4507075@hotmail.com

V. Supajatura
Department of Microbiology, Faculty of Medicine, Chiang Mai University,
Chiang Mai 50200, Thailand
e-mail: volaluck.supajatura@gmail.com

J.J. (Jong Hyuk) Park et al. (eds.), *Ubiquitous Computing Application and Wireless Sensor*, Lecture Notes in Electrical Engineering 331,
DOI 10.1007/978-94-017-9618-7_58

58.1 Introduction

Small group discussion is an important part of the medical educational program for second and third year medical students at the Faculty of Medicine, Chiang Mai University, Chiang Mai, Thailand. The group discussion runs regularly every Monday and Friday. Lecturers within the Faculty of Medicine are assigned to facilitate the group discussion as a "facilitator". The knowledge content is organized using a block system. The lecture committees for each block prepare scenario cases for group discussion and the study schedules are prepared by the Medical Curriculum Section. Medical students are divided into 100 groups, each with 9–10 students and each group is supervised by one facilitator. Facilitators observe group process, and score student performances.

In the previous work flow system, the administrative assistant printed scoring forms for facilitators. After each group discussion finished, facilitators filled the scores into the forms and handed them back to the administrative assistant. Annually, over 23,000 forms were collected and data transferred into a spread sheet. At least three administrative assistants were needed to check the score transfer for diminishing human errors. This process produced a massive workload for calculating final scores. Consequently, the process caused a delay for announcing final grades and students with low scores were not able to be detected.

Nowadays, the electronic data system plays an important role for modern data management. Previous research consistently shows that electronic data management increases the effectiveness and efficiency of systems [1–3]. In this research, we aimed to implement an electronic scoring system to reduce redundant human work, reduce work flow, reduce natural resources, increase the effectiveness of grading, and early detection and intervention for low-scoring students.

58.2 Materials and Methods

First, we explored the core problems underlying the delay. We found that the main problem was the manual score transfer process. Therefore, we concluded that an electronic system should replace the existing process.

The scoring system is divided into three parts:

1. Input form generation
2. Scoring program
3. Recording and backup system

Input form generation includes the process of generating a scoring paper form for using in scenario rooms. The administrative assistant enters scenario details and the facilitators' name into the MySQL database. After the details are entered into the database, the administrative assistant runs input form generation software to generate pdf forms. The forms are automatically generated by pulling data from the

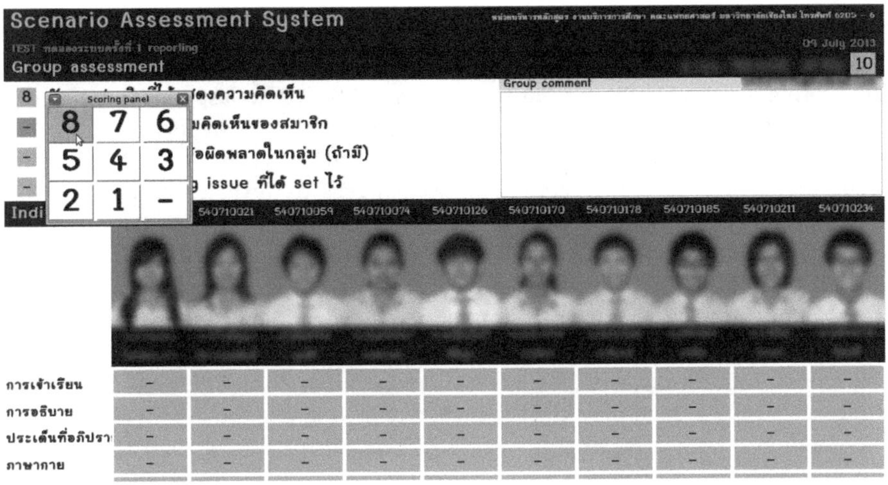

Fig. 58.1 Scoring program

database. The details consist of scenario details, date, facilitator's name, topics of assessment, students' pictures, and students' details. The forms are printed out and handed to the facilitators before attending the class. After finishing discussion in the scenario room, facilitators fill scores into the form and come to the facilitator's room. Within the room, facilitators use computers to transfer scores into the database using medical scoring software. The software provides graphical user interfaces (GUI) to fill scores. After the facilitator pushes a submit button, the data are transferred to store in the database via local area network and the Transmission Control Protocol/Internet Protocol (TCP/IP) was used for passing messages across the network. Additionally, the data are also created to an html file and sent via emails to facilitators and administrative assistants. Regarding the security policy, every facilitator was provided with a secure username and password for logging into the system. The password is required twice: the first time, when the facilitator performs login, and the second time, before the facilitator submits the scores.

Figure 58.1 shows the main panel of the scoring program. The electronic form was designed similarly to the original paper sheet. The scoring software provides a digit panel for numbering input. Student's pictures are presented at the head of each scoring column for helping the facilitator to identify students. After the data were submitted, facilitators can recheck the scores from their email. Normally, one block of study period is around 1–2 months. At the end of each block, the administrative assistant exports data from the database into a spread sheet and analyzes data for final grades.

All software which is mentioned above are written in Java and run with JDK version 7. Apart from the Java common libraries, the software used additional libraries: MySQL JDBC Driver [4], and Commons email [5]. MySQL JDBC Driver is a library for connecting and transferring data between a client computer and database. Commons email is a library for creating and sending email.

58.3 Results

The scoring system has been used since January 2012. The last annual report was presented at the faculty meeting on 22 February 2013. The report included data from December 2011 to February 2013. There were 500 students and 131 facilitators participating in the system. The system records revealed that overall the system was used 2,163 times. Statistical analysis showed that the electronic system reduced the use of paper by 21,630 sheets. The analysis of system efficiency showed that the system decreased pre-class scoring sheet preparation time from 30 to 5 min for each scenario. In summary, the preparation time was reduced by 2,150 min (or 5 days and 3 h calculated based on regular working hours) in a year. The scoring transfer process took approximately 3 min during the scoring process. Interestingly, the system played a big impact on shortening the final grade process. The duration of preparing final grades with the paper-based system took around 420 min whereas the duration was reduced to 5 min using the new system. For 1 year, we saved 35,690 min (or 85 days calculated based on regular working hours) for final grades (Table 58.1). For second year medical students, around 41 % of total facilitators supervised the scenario group discussion for 90–120 min and 38 % of facilitators supervised for 60–90 min. For third year medical students, around 37 % of total facilitators supervised the scenario group discussion for 90–120 and 27 % supervised for 60–90 min.

In August 2013, we sent a survey to 104 facilitators' email using Google form application. We received 30 responses back. Seventy six percent were female with an average age of 41 years. The results of the survey are shown in two tables. Table 58.2 shows general details of system usage and Table 58.3 shows user satisfaction of the scoring system.

58.4 Discussion

The scoring system was designed as simply as possible and the most important consideration was based on user needs. The system can be used by the faculty administrative assistant and does not require advanced skills in computer science. Currently, the information is filled into the database, and the scoring sheets are generated by the administrative assistant. Importantly, these two processes are independently done by the administrative assistant because the goal of the system implementation is the complexity of the system must be designed simply enough to

Table 58.1 The summary of the effects of using electronic scoring system in 2012

Major effects	Result
Paper-use reduction	21,630 sheets
Paper preparation time reduction	Five working days
Scoring time reduction	Eighty five working days

Table 58.2 General detail of system usage

Topics	Results (%)
How long have you used the electronic scoring system?	
Less than 6 months	13
6–12 months	40
More than 12 months	47
How long do you spend using the electronic scoring program when entering data?	
5 min	70
5–10 min	27
More than 10 min	3

Table 58.3 User satisfaction of the scoring system

User satisfaction (Grading form one to five, five is the most satisfaction)	Results
The system increases the flexibility of scoring process	4.57
The system is user friendly, easy to understand	4.68
The data are stored correctly and accurately	4.54
The system is convenient and reduced scoring time	4.68
The summary of score which is sent to the facilitator's email is necessary	3.32
The overall of system satisfaction	4.54

be suitable for use by local workers and with limited facilities and they can work without computer experts. The structures of the database are simple and easy to reproduce. Within the process of form generation, the details in the forms are retrieved by the software from the database. This automatic process reduces human errors and reduces manual document preparation works for 5 days from a whole year.

Concerning data security issues, the facilitators have to fill in their username and password twice during the scoring process. Also, the scoring system was enabled only during the period of group discussion. The duration of the scoring process varied depending on facilitators. The facilitators who finished filling the scores in paper before transferring data into the scoring software used the program for less than 1 min whereas the facilitators who filled in the paper while they were using the program took more than 5 min.

After the data are submitted, the data are sent to three different places; MySQL database, facilitator's email, and administrative assistant's email. Primarily, the MySQL database is a main storage location. Afterward, the data are arranged into a human readable html format, and sent to the facilitator's email for confirming that data were stored completely and correctly. This mailing system provides function that allows facilitators to revise the scores. Finally, the data are sent to the administrative assistant's email as backup data. At the end of the year, all scoring data in the administrative assistant's emails are exported and written onto a DVD. The backup system helps to protect against data loss.

The cost of system implementation including computers and networks was 336,156 baht. Also, the maintenance and depreciation cost have to be included every year. Overall, if we consider only the visible infrastructure cost, the electronic system costs more than the paper system. However, we have gained the effectiveness of data management. Based on a minimum salary of 15,000 baht per month, the new system saves approximately 45,000 baht per year (5 days of preparation time plus 85 days of data entry). Therefore the system would recoup its initial set-up costs within 7.47 years.

At the end of the study block, the data are exported from the MySQL database as a spread sheet file. After that, the administrative assistant calculates final scores. The system massively reduces the exporting process. The final scores reached the head of the course much quicker and finally students with low scores could be quickly identified and these students could enter the faculty supporting system in a timely way. From 2008 to 2011, we had 9, 12, 5, and 17 repeat students respectively. On average, around 10 students failed the study program every year. After the scoring system was implemented, the system helped to decrease repeat students from ten to two students in 2012. Although the early detection is not the only one factor involved, we believe that the early detection of students with low scores is a major contributing factor to fixing this problem. In general, the survey showed the user satisfaction of the electronic scoring system was optimal. The summary of score via email was not valued by participants as a means of data backup but it played more of a role in rechecking data. Further research should follow up these results for the long-term outcomes of the scoring system.

58.5 Conclusion

The electronic scoring system was implemented for managing data for scenario discussion. The result showed the system gained rich benefits in the effectiveness of data management. The scoring time process was significantly reduced and this effect helped the administrative assistants to quickly detect students with low scores. This early detection helped the students to obtain prompt support and help from the faculty.

Acknowledgments We would like to thank the Tilley family who always give me advice, our team from the IT department who are working very hard to help me implement the system. Most importantly, thank you to the Medical Curriculum unit, Department of Physiology, Faculty of Medicine, and Chiang Mai University for facilitating this project.

References

1. Minshall S (2013) A review of healthcare information system usability and safety. Stud Health Technol Inform 183:151–156
2. Shekelle P, Morton SC, Keeler EB (2006) Costs and benefits of health information technology (Cited 2013). http://www.ncbi.nlm.nih.gov/books/NBK37988/
3. Kinn JW et al (2002) Effectiveness of the electronic medical record in improving the management of hypertension. J Clin Hypertens 4(6):415–419
4. Oracle (2013) Overview of MySQL connector (Cited 1 Sept 2013). http://dev.mysql.com/doc/refman/5.1/en/connector-j-overview.html
5. Foundation TAS (2012) Commons email (Cited 1 Sept 2013) 1.3.1. http://commons.apache.org/proper/commons-email/. Accessed 3 March 2013

Chapter 59
Perceived Risk of Anti-corruption e-Learning, Email Phishing Literacy, and Anomia

Juneman Abraham and Sharron

Abstract In order to prevent corruption in Indonesia, the use of anti-corruption e-learning needs to be optimized and to reach young generation whose life nowadays mostly inseparable from the internet connections. The aim of this research is to identify predictor variables of perceived risk of anti-corruption e-learning. When risk perception is low, attitude toward e-learning is expected to be more positive. This psychotechnological research employs predictive correlational design, with the predictors (1) email phishing literacy and (2) anomia. The participants of this research were 71 students of faculty of psychology at a private university in Jakarta, Indonesia (20 males, 51 females; Mean of age = 19.93 years old; Standard Deviation of age = 1.397 years old). The data were analyzed by using simple linear regression analyses. Results of this research show that both variables are able to predict perceived risk of anti-corruption e-learning, consecutively in negative and positive ways.

Keywords Psychology of e-Learning · Anti-corruption · Phishing literacy · Anomia

59.1 Introduction

Andersen et al., through their empirical research, indicated their support for a proposition stating that the internet is a powerful anti-corruption technology [1]. It is driven by the characteristic of the internet which emphasizes at disseminating transparent information, facilitating the public to identify or to detect corrupt behavior (and hence it can generate 'deterrent effect' to do corruption), and cutting the bureaucracy between government and public. Zinnbauer, however, stated carefully that the effectiveness of the internet to fight corruption is not conclusive

---

J. Abraham (✉) · Sharron
Psychology Department, Bina Nusantara University, Jakarta, Indonesia
e-mail: juneman@binus.edu

J.J. (Jong Hyuk) Park et al. (eds.), *Ubiquitous Computing Application and Wireless Sensor*, Lecture Notes in Electrical Engineering 331,
DOI 10.1007/978-94-017-9618-7_59

yet, and it needs further researches [2]. In addition, he also mentioned the importance of accompanying factor which is society empowerment. This factor will enhance civil engagement to guard the implementation of technology, so that technology will not only be a 'make up' without actual substances of accountability and integrity enforcement. Finnegan specifically highlighted the role of online collaborative learning networks–either utilized by activists or organizations which are the stakeholders of government–in enhancing governance accountability and preventing corruption [3]. One of the manifestations of online collaborative learning networks is Anti-corruption e-Learning (ACL, for instance: http://thefightagainstcorruption.unodc.org).

With the growing expectations for ACL, it is beneficial to identify models of relationship between variables which explain the attitude of actual or potential ACL users. No matter how sophisticated the ACL system is, it will be 'useless' (in terms of expected competence and learning outcomes are not achieved) if the attitude of its users is less positive which has an implication to less optimal utilization. In this research, negative attitude toward ACL is operationalized as perceived risk of ACL, based on the understanding that it has a contribution to the negative attitude [4].

Some researchers have identified that attitude toward e-learning is contributed by the following variables: (1) privacy concern, (2) perception on e-learning quality (including the instructor), (3) perception on cheating level of e-learning participants in online assessment for the sake of obtaining certificate, (4) level of the need of 'blended' learning sessions (e-learning combined with classroom learning facilitated by subject matter expert), (5) level of client involvement in e-learning production and evaluations phases, (6) level of accommodation for varied learning styles, as well as (7) aspects of aesthetics, error prevention, redundancy reduction, memorability, and clearly defined learning outcomes and paths [5–7]. Nevertheless, there is a crucial factor that has not been much studied in terms of attitude toward e-learning, i.e. email phishing literacy. The definition of phishing is as follows: "A phishing attack occurs when a user receives a fraudulent or 'spoofed' email representing a trusted source (e.g., bank, retailer or credit card company). This e-mail leads them to an equally fraudulent web site that maliciously collects personal information, including account information, passwords and PINs" [8].

Tembe et al., through his empirical research, found that the consequences of being phished experience can be personal, technological, economical, social, and psychological [9]. They identified seven experiences coming afterward, i.e. (1) Providing private information to an unauthorized person, (2) Experiencing identity theft as a result of stolen personal information, (3) Lost money or property as a result of stolen personal information, (4) Loss of use of a service, such as an email account, (5) Unwillingness to use a service in the future, (6) Reduced trust in technology, and (7) Reduced trust in people. The fifth to the seventh consequences, in our opinion, can further predict positive or negative individual attitude toward ACL. Thus this hypothesis needs empirical testing. It is in line with the 'stimulus generalization' concept of behavioristic psychology. Conceptual definition of the terminology is as follows [10]: "The generalization of the conditioned response to stimuli that is similar to the initial stimulus causing the response. For example,

Pavlov's infamous dog may have started to salivate on hearing a sound similar to the original sound that was used to elicit the conditioned response. Other examples include 'Little Albert', who was conditioned to fear rats. A few days later, the child not only feared rats but also a rabbit, a dog, and even a Santa Claus mask. As not many stimuli in real life are exactly the same, stimulus generalization serves an adaptive purpose. As a general rule, the closer the stimulus is to the initial stimulus causing the response, the greater the likelihood of *stimulus generalization*" (p. 493).

We assume that there is a similar characteristic between email phishing and e-learning participation invitation or offering. Both of them appear as authoritative and trustworthy parties. They also employ persuasion for their prospective users participating further to enter and to involve in the system and the material offered. In addition, Chien found that lack of computer knowledge contributes on computer anxiety and further decreasing the effectiveness of e-learning [11]. This research is different from Chien's since predictor used in this research is more specific and directly related to the online world, that is, email phishing literacy, not computer knowledge in general. Thus, **the first hypothesis** of this research is: "The lower an individual's email phishing literacy, the higher his/her risk perception of ACL will be".

Furthermore, in relation to anti-corruption materials, the fundamental question arising is: "Why should one believe on ACL?" This is a sociopsychotechnological question which is related to field of psychology of trust. Cheshire specifically investigated factors contributing on online trust [12]. She stated that there is an urgency to differentiate trust in human and trust in the system, although it is also necessary not to get trapped in oversimplification and pedantic distinction since equalization between interpersonal trust and trust in the computer/internet system on several facets indeed has an analogical likeness. Meanwhile, if they have to be differentiated, the assumption would be as follows [12]: "When a human betrays a friend's trust, the friend knows who is culpable, and the consequences are often clear for both parties. When a system 'betrays' a human's trust, assigning blame can have enormous repercussions" (p. 56).

We note that the previous studies still lack for addressing one of essential contextual-socio-psychological factors in influencing an individual's negative attitude and risk perception toward online system and e-learning system, and more specifically ACL. The factor is *anomia*, or feeling of *anomie*. Anomia is "subjectively perceived anomie", "perceived unpredictability of outcomes upon given moral beliefs" (Olsen 1969, p. 290, as cited in Lytkina [13], Seeman 1959, Srole 1956, as cited in Lytkina [14]), while *anomie* is "a state of society characterized by lack of norms and regulations" (Durkheim 1897, as cited in Lytkina [13]). Further question arising is "Will the 'new norm' offer from ACL become a significant moral-ethical guidance? Or will it just become another source of norm confusion, incoherence view of social life, and source of uncertain consequences in the individual psychological reality?" Since that kind of anomia can lead an individual to disoriented and anxious feelings; thus **the second hypothesis** is: "The higher an individual's experience of anomia, the higher his/her risk perception of ACL will be". Both of these hypotheses are visualized in a hypothetical diagram as shown in Fig. 59.1.

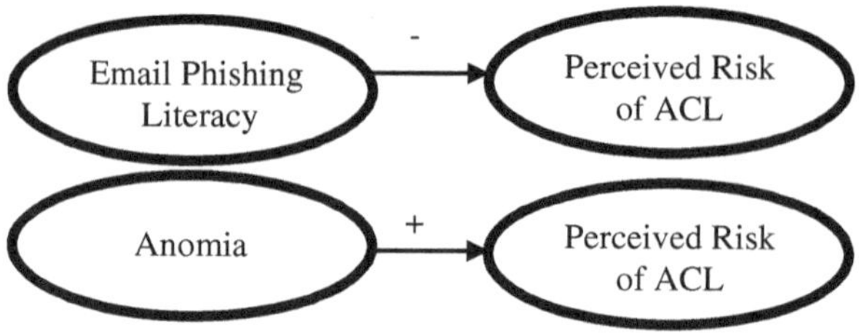

Fig. 59.1 Hypothetical model. *Note* (+) positive predictive correlation. (−) Negative predictive correlation

59.2 Methods

This research employs quantitative, predictive correlational design. As shown in Fig. 59.1, the independent variables of this research are email phishing literacy and anomia while the dependent variable of this research is perceived risk of anti-corruption e-learning (ACL). The statistical analyses used in testing the prediction of independent variables toward dependent variable are simple linear regression analyses.

Participants of this research were 71 students of faculty of psychology at a private university in Jakarta, the capital city of Indonesia, who were taken by using convenience sampling technique. Here are the descriptions of the participants: mean of age 19.93 years old, standard deviation of age 1.397 years old; gender: 20 males, 51 females; majority of the participants are Javanese (30 %) and Chinese (25 %). In Indonesia, Psychology is indeed a study program dominated by female students.

Measurement instrument for Risk Perception of ACL variable is adapted from Perceived Risk in Online Education Questionnaire [15]. This questionnaire is adapted by employing the ACL as the object of risk perception. The dimensions of this instrument are perceived performance risk, perceived time-demand risk, perceived social risk, perceived psychological risk, and perceived source risk. The response scale of this questionnaire ranged from *Strongly Disagree* (score of 1) to *Strongly Agree* (score of 6). It begins by the following instruction: "*Nowadays, Indonesia is active in preventing corruption, including through anticorruption e-learning. Below is the display taken from the example of the e-learning site* [the picture is taken from http://e-integrity.net/elearning/; this website was established by a collaboration of TIRI-Integrity Action Indonesia, United States Agency for International Development (USAID), the Royal Kingdom of Netherlands, and Partnership for Governance Reform.]. *You are kindly requested to carefully read the home page display. To give response to statements in scales, please imagine that you are offered to attend online classes of this e-learning*".

Sample items of perceived performance risk: "I doubt the instructor will be able to make this type of class work for all of the students", "I do not believe the instructor will be highly accessible by e-mail". Sample items of perceived time-demand risk:

"I'm not sure I'll have the time needed to successfully complete online courses", "If I take an online course, I'll have less free time". Sample items of perceived social risk: "In general, people who earn their degrees through online programs are held in higher esteem than are traditional students" (*unfavorable item, reversed score*). Sample items of perceived psychological risk: "Just thinking about taking an online class makes me feel stressed". Sample items of perceived source risk: "It is difficult to determine the credibility of some universities offering anti-corruption e-learning". The result of validity and reliability testing shows that the instrument containing the risk perceptions dimensions of ACL is reliable (Cronbach's Alpha index of internal consistency = 0.762) after aborting 17 items, with corrected item-total correlations ranging from 0.252 to 0.507.

Measurement instrument for Email Phishing Literacy variable is adapted from Ability to Identify Phishing Emails Questionnaire [16]. This questionnaire consists of 8 items and is preceded by the following question: "*Which of the following generally indicates that an email may be phishing or fraudulent?*" The example of the items is as follows: "Asking you to enter information about your account", "Conveying sense of urgency and surprise", "Containing attachment, notifying you that it might contain viruses that could harm your computer", "Directing you to the web site with URL starting with https". The options to response this questionnaire are "*Phishing*" ("*Penipuan*" in Indonesian) and "*Legitimate*" ("*Bukan Penipuan*" in Indonesian). The answer of the participants will be valued by score 1 (True) or 0 (False) based on the answer keys from Al-Hamar et al. [16]. The maximum score of the participants for this sale is 8, and the minimum is 0. This questionnaire has been tested to 2,000 respondents consisting of 1,000 Qatari citizens and 1,000 United Kingdom citizens, by a margin of error of 3.1 % with a confidence level of 95 %.

Measurement instrument for anomia variable is adapted from Multidimensional Scale of Feeling of Anomie [17]. This questionnaire consists of three dimensions, i.e. (1) meaninglessness and distrust, (2) powerlessness, and (3) fetishism of money. Sample items of meaningless and distrust: "Everything is relative, and there just are not any definite rules to live by"; "There is little use writing to public officials because often they are not genuinely interested in the problems of the average man". Sample items of powerlessness: "The world is changing so fast that it is hard for me to understand what is going on"; "I lead a trapped or frustrated life". Sample items of fetishism of money: "A person is justified in doing almost anything if the reward is high enough". The response scale of this questionnaire ranged from *Strongly Disagree* (score of 1) to *Strongly Agree* (score of 6). The result of validity and reliability instrument testing shows that the instrument is reliable (Cronbach's Alpha = 0.836) after aborting 6 items, with corrected item-total correlations ranging from 0.272 to 0.717. It means the instrument is reliable ($\alpha > 0.6$) and valid (corrected $r_{it} > 0.250$).

59.3 Results and Discussion

Data analysis with *SPSS 21 for Windows* gives results as shown in Table 63.1. Email phishing literacy is able to predict risk perception of ACL in a negative way while anomia is able to predict it in a positive way. They show that both hypotheses are supported by the data. This research brings out the new insight related to individual perspective toward e-learning, especially anti-corruption e-learning.

Perceived risk of ACL can be predicted by email phishing literacy. This is understandable since there is parallelism between them. For instance, phishing email can waste an individual's time to restore the infected computer condition to its condition before being exposed to phishing. ACL can also be perceived as time-consuming, lowering efficacy in utilizing it. For example, the users will feel the loss of leisure time since they should complete a series of online courses which may also be complicated with potential conflicts [18]. Perceived performance risk also can be felt in email phishing and e-learning system. Email phishing usually has a good "face validity" as it involves complex social engineering. Actually, however, behind surface appearance, there is a trap or a trick which harm the users either economically or psychosocially. Generally, e-learning system does not trap, although it does have technical and security risks [19] similar to that of phishing (integrity and confidentiality attack, unauthorized use, etc.). Nevertheless, the potential discrepancies between credible display (e.g. with the inclusion of nationally or internationally-wide instructors' names and reputations, the inclusion of educational institution and authoritative government or NGO logos) and the actual quality of its contents and instructors can make users feel deceived, have a sense of risk for "consuming" or being involved in the e-learning system and give negative attitude to the e-learning system. All of this explanation can be bridged by the *stimulus generalization* concept in psychological science as described in the Introduction.

This research also finds that anomia can predict perceived risk of ACL. Anomie effect on the learning process has actually been studied by Hibbert et al. [20]. They stated: "*Anomie*... It is also possible for *learning to be limited* because of disconnection, a lack of a sense of collective purpose, or even uncertainty about the nature or reality of the collective and a resultant lack of effective social norms" (p. 462).

Table 63.1 Simple linear regression analyses predicting perceived risk of anti-corruption e-learning ($n = 71$)

Independent variable 1	*B*	*SE B*	*ß*	*p*
Email phishing literacy	−1.685	0.592	−0.324	0.006
Note $F(1,70) = 8.097$, $R^2 = 0.105$, $p < 0.01$; *SE* standard error				
Independent variable 2	*B*	*SE B*	*ß*	*p*
Anomia	0.222	0.087	0.293	0.013
Note $F(1,70) = 6.463$, $R^2 = 0.086$, $p < 0.05$				

One of the characteristics of anomia is feeling of normless, meaningless, isolated, helpless, powerless, and materialistic. Such individuals tend to avoid their ethical obligation because of disabled moral thinking and moral disengagement as the effect of anomic situation, and to behave less positive toward social changes. In fact, anti-corruption education requires simultaneously sustainable effort and collective work which are supported by the state as well as stakeholders having creativity and obvious stance against corruption [21, 22]. Based on that argument, it is clear that anomie becomes a barrier for e-learning participants to coordinate in collective learning and action of which, in this research, is shown by high perceived risk of ACL when anomia is high.

59.4 Conclusion

This research concludes that email phishing literacy is able to predict perceived risk of ACL in a negative way, and anomia is able to predict it in a positive way.

In the context of enhancing effective participation of the young generation in anti-corruption e-learning environment, the result of this research implicate that personal and socio-contextual factors are necessary to be seriously managed. Personal factor in this research is email phishing literacy while contextual factor is anomia. Anti-corruption e-learning providers and organizers need to do networking with other parties to overcome the barrier either in a personal level or socio-contextual level in order to make the users' involvement in the e-learning system optimal.

This research has added the body of knowledge in psychotechnology field by successfully extending the Yamakami's propositions that: "Human beings are social beings. Therefore, the hook, retention … phases have direct links to social contexts" [23] (p. 114).

References

1. Andersen TB, Bentzen J, Dalgaard C-J, Selaya P (2010) Does the internet reduce corruption? Evidence from U.S. states and across countries. World Bank Econ Rev 25(3):387–417. http://www.econ.ku.dk/okotand/images/InternetCorruptionWBER.pdf
2. Zinnbauer D (2012) Governments using ICTs for integrity and accountability: some thoughts on an emergent research and advocacy agenda. In: Finlay A (ed) Global information society watch: the internet and corruption: transparency and accountability online. APC and Hivos, Uruguay
3. Finnegan S (2012) Using technology for collaborative transparency: risks and opportunities. In: Finlay A (ed) Global information society watch: the internet and corruption: transparency and accountability online. APC and Hivos, Uruguay
4. van Osch SMC, Stiggelbout AM (2007) The development of the health-risk attitude scale. Universiteit Leiden. https://openaccess.leidenuniv.nl/bitstream/handle/1887/12363/07.pdf;jsessionid=30366083D50E9A3EC5568D4C257C830E?sequence=10

5. Borcea-Pfitzmann K, Stange A-K (2007) Privacy: an issue for eLearning? A trend analysis reflecting the attitude of european elearning users. TUD-FI07-01, Technische berichte Technical Reports, Institut für Systemarchitektur, Technische Universität Dresden
6. Gross A (2012) Thirsty learners: why is your eLearning training failing? Training Dev 39(5):14
7. Oztekin A, Kong ZJ, Uysal O (2010) UseLearn: a novel checklist and usability evaluation method for elearning systems by criticality metric analysis. Int J Ind Ergonom 40:455–469
8. Entrust (2014) Securing digital identities and information. http://www.entrust.com/anti-phishing/
9. Tembe R, Hong KW, Murphy-Hill E, Mayhorn CB, Kelley CM (2013) American and Indian conceptualizations of phishing. In: Proceedings of the 3rd workshop on socio-technical aspects in security and trust (STAST). http://www4.ncsu.edu/~khong/papers/rt_etal_stats_13.pdf
10. Sullivan LE (2012) The SAGE glossary of the social and behavioral sciences. SAGE Publications, Thousand Oaks
11. Chien T-C (2008) Factors influencing computer anxiety and its impact on e-Learning effectiveness: a review of literature. http://files.eric.ed.gov/fulltext/ED501623.pdf
12. Cheshire C (2011) Online trust, trustworthiness, or assurance? Daedalus 140(4):49–58
13. Lytkina E (2012) Anomie and anomia: an approach towards the measurement of social well-being and deviance. Laboratory for Comparative Social Research, National Research University, Higher School of Economics, Moscow, Russia. http://www.hse.ru/data/2012/11/10/1249461488/Lytkina%20%D0%94%D0%BE%D0%BA%D0%BB%D0%B0%D0%B4_%D0%9D%D0%BE%D1%8F%D0%B1%D1%80%D1%8C_2012%20(Progress_Report_November_2012).pdf
14. Lytkina E (2013) Anomie: social reality or a unique measure of all the possible negative phenomena? http://opec.ru/data/2013/03/21/1233151969/%D0%90%D0%BD%D0%BE%D0%BC%D0%B8%D1%8F.pdf
15. Mohamed FA, Hassan AM, Spencer B (2011) Conceptualization and measurement of perceived risk of online education. AELJ 15(4):1–16
16. Al-Hamar M, Dawson R, Al-Hamar J (2011) The need for education on phishing: a survey comparison of the UK and Qatar. Campus-Wide Inf Syst 28(5):308–319
17. Heydari A, Davoudi I, Teymoori A (2011) Revising the assessment of feeling of anomie: presenting a multidimensional scale. Procedia Soc Behav Sci 30:1086–1090
18. Whitworth A (2005) The politics of virtual learning environments: environmental change, conflict, and e-Learning. Brit J Educ Tehnol 36(4):685–691
19. Barik N, Karforma S (2012) Risks and remedies in e-learning system. IJNSA 4(1):51–59
20. Hibbert P, Huxham C, Sydow J, Lerch F (2010) Barriers to process learning: authority and anomie in regional clusters. Manage Learn 41(4):453–471
21. World Bank Institute (2008) Fighting corruption through collective action: a guide for business. http://actoolkit.unprme.org/wp-content/resourcepdf/Collective%20action%20guide.pdf
22. Clara RD (2012) Pendidikan antikorupsi butuh kerja kolektif bangsa. http://berita.upi.edu/2012/02/19/pendidikan-antikorupsi-butuh-kerja-kolektif-bangsa/
23. Yamakami T (2012) Taxonomy of emotion engineering: lessons from mobile social game business. In: Park JJ, Jin Q, Yeo MS, Hu B (eds) Human centric technology and service in smart space (HumanCom 2012), LNEESpringer, vol 182. Heidelberg, pp 113–120

Chapter 60
A Model of Business Intelligence Systems for School: A Case Study (Perception Applications and Benefits)

Adhi Nugroho Chandra and Yohannes Kurniawan

Abstract Usually the purpose with business intelligence is to increase the profit, through capturing, storing, sharing, and utilizing knowledge in an innovative way. Are the concepts of business intelligence applicable to schools? If that is the case, then the primary and secondary education sectors should be replete with examples of institutions that leverage business intelligence to achieve operational excellence. This paper discusses the need for business intelligence from the perspectives of teachers in academic services, perceptions of application and benefits. However, although some examples exist, they are the exception rather than the rule. We believe there is tremendous value to school institutions that develop initiatives to achieve business objectives. The business intelligence can be classified as support for teaching staff decision making. The conclusion is that business intelligence could be used in schools, facilitating the business intelligence within the organization for academic services.

Keywords Business intelligence · School · Application · Benefit

60.1 Introduction

Usually the data warehouse and data mining concepts and methods will be applied in various profit sectors, like marketing, banking, health care, telecommunication, supply chain management, customer relationship management, etc. With advances in the business intelligence sector, there is an increasing interest for the introduction of business intelligence model into organizations, especially for education sectors. The perceived benefits from business intelligence systems, in terms of better information quality or achievement of information quality improvement goals, are

---

A.N. Chandra · Y. Kurniawan (✉)
Information Systems Department, Bina Nusantara University, 11480 Jakarta, Indonesia
e-mail: Kurniawan_yohannes@yahoo.com

J.J. (Jong Hyuk) Park et al. (eds.), *Ubiquitous Computing Application and Wireless Sensor*, Lecture Notes in Electrical Engineering 331,
DOI 10.1007/978-94-017-9618-7_60

far from being neglected, these are only indirect business benefits or the business value of such systems.

The true business benefit of Business Intelligence (BI) is to improve business processes and thus in improved business performance. The purpose of this paper is to propose a conceptual model of business intelligence systems that was developed on literature review, interview, and case study analysis for researching business intelligence systems.

Currently, BI has an important role in the creation of current information for operational, managerial and strategic business decision making. BI as tools to help specifically decision processes at the analytical level. According to a research by IT Strategies [1], BI have one of the greatest potentials to achieving of asymmetry information and differentiation from the competitors respectively and thus achieve competitive advantage with information technology [2]. Regardless of this, we perceived that when organizations think about introducing BI, the key factor is improvement of information processes as a different way for providing the right information. Information quality improvement goals, such as increased self service access to data, data integration from different sources, and interactive and convenient access to operational data are important.

Business Intelligence requires the historical data or data collected from various sources of operational database. The solution is found in data warehouses, which are the main technology used for ETL process (extract, transform, and load). The development cycle of a data warehouse involves a lots of resources, time, high costs and above all, it is built only for some specific tasks. The paper explores the concepts of BI, benefits of BI, conceptual designing business intelligence, and various BI techniques. The paper would be useful for budding researchers in the field of BI to understand the basic concepts of BI in case study approach.

60.2 Paper Preparation

Stackowiak et al. [3] define Business intelligence as the process of taking large amounts of operational data, analyzing that data, and presenting a high level set of the reports that data into the basis of business actions, enabling management to make daily business decisions.

And Zeng et al. [4] define BI as "The process of collection, treatment and diffusion of information that has an objective, the reduction of uncertainty in the making of all strategic decisions." Zeng et al. [4] categorized the BI technology based on the method of information delivery, like reporting, statistical analysis, ad hoc report analysis and predicative analysis.

BI in a management term, which refers to the systems and technologies that are used to gather, provide access, and analyze data and information about organization operations. BI systems can support the organizations have a more comprehensive knowledge of the factors affecting their business, such as metrics on marketing, internal operations, and they can support organizations to make a better operation decisions.

Arnott and Pervan [5] argue "BI is a poorly defined term and its industry origin means that different software vendors and consulting organizations have defined it to suit their products; some even use 'BI' for the entire range of decision support approaches." The BI environment integrate all of the development, information processing, and support activities required to deliver reliable and highly relevant business information and analytical capabilities to the business decisions [6]. Within their BI environments, organizations look for analyzing, designing and implementing successful BI systems. It can be defined as information systems providing the quality of data and information for operational and analytical decision making as a source for supporting the business towards achieving organizational targets.

In the table below it can categorized into a series of data mining techniques, which are classified and described in Table 60.1. BI use amount of collected data during the daily operational processes, and transforms the data into information and knowledge [7].

The main characteristics of a BI system are: the capability of providing information to the management, to support strategic processes such as target setting, prediction and forecasting, to gather, analyze, and integrate internal and external data into indicators. Based on each decision maker information needs, BI can access historical and real time data through queries.

Table 60.1 Business intelligence techniques [8]

Techniques	Description
Predictive modeling	Predict value for a specific data item attribute
Characterization and descriptive data mining	Data distribution, dispersion and exception
Association, correlation, causality analysis (link analysis)	Identify relationships between attributes
Classification	Determine to which class a data item belongs
Clustering and outlier analysis	Partition a set into classes, whereby items with similar characteristics are grouped together
Temporal and sequential patterns analysis	Trend and deviation, sequential patterns, periodicity
OLAP (online analytical processing)	OLAP tools enable users to analyze different dimensions of multidimensional data. For example, it provides time series and trend analysis views
Model visualization	Making discovered knowledge easily understood using charts, plots, histograms, and other visual means
Exploratory data analysis (EDA)	Explores a data set without a strong dependence on assumptions or models; goal is to identify patterns in an exploratory manner

BI helps the links in the new form organization, bringing the real time data and information to centralized databases to create precisely targeted analytics that can be exploited at every level within and outside the organization [9, 10].

60.3 Method

Writing method for this paper is qualitative. The method used in data collection was reviews from the literature and direct observation and interview at BINUS International School. Reference sources used are a variety of books, journals, and articles obtained from the library. Another source of internet is includes electronic book and other supporting sites. Retrieving information or data by quoting the contents of the books or from the internet and using the available data to be used as supporting evidence the authors put forward of a statement. The nature and form of paper to be presented in descriptive format.

60.4 Result and Discussion

BINUS international school has a great amount of operational data which can be analyzed and extracted for the data mining system. School has also important detail data regarding courses and modules which are in document form. This section will discuss each operational data source in BINUS International School:

- Admission System is developed for the purpose of marketing usage, which is used to manage admission data. We can acquire more admission information from this system.
- Student Activities System holds information about student profiles for example student background, examination results and course enrolment.
- Teaching and Learning System (e-learning) allows students to access course materials for a particular course to extend their classroom teaching using more interactive techniques (included discussion forum).
- Library System provides information data which could be utilized for student academic integration. It can help the school to track how often the students borrow books.
- Question Bank System enables all to take online skills test. It can help the school to understand the student academic performance (real time).
- Registration, Administration, and Scheduling. Student services for students and teachers/staff at the school so that they are well informed to service the students. Information could include procedures related to financial services, registration, request document/legalizing and other services.

- Student Performance System. Pedagogy and assessment techniques, including, lesson objectives and learning outcomes. Analyzed student evaluations results each semester for lessons learned and learning outcome.
- Clinic System included all record related health consultation and monitoring. It can help us to understand the student health performance.

Figure 60.1 shows the system model of business intelligence for schools. Data sources cover student enrolment (time table), student results, course/module data, learning capabilities, and student activities. Data sources are then integrated and transformed into data warehouse. Data warehouse then generates appropriate data to the data mining machine. As SQL Server 2005 is the most commonly used database in our data sources, Ms. SQL Server 2005 is chosen as a platform.

Based on the previous discussion and analysis, we propose a conceptual model for researching business value and key factors of BI (Fig. 60.1). It starts by looking at BI capability as a source of true benefits of BI initiatives as suggested by many authors [11, 12]. In connection with the BI system stage, an important issue on the path of achieving business value and benefits such a system provides for stakeholders decision making. In order to be of value, quality information must be used within processes to improve school stakeholders decision making, to improve business process [13], and to fulfill stakeholder needs [14].

Providing that organizations strive after better quality of information for decision making, an important issue is about the use of such information in school business processes for the purpose of their improvement or change. We are developing a webbased application to monitor student performance and behavior, module, and course information. When a student logs in, they can check the detail personal information, they report card, whether they are at risk.

As for teachers and staff, they can login to find detail related course and information module, which students are at risk (potential failed). They can also compare the information of related modules in different semesters. Similarly, subject heads and the head of department or principal can also login to find the information related to their program. Different reports with different formats and templates will be sent out automatically by the system as well. For example, students are divided into different year level, class, pathway and different rules will be applied to different levels and groups. Students will access their detail of their progress report in an PDF report.

At different times of the semester, email notification will also be sent automatically to provide students potential failed, such as at the beginning of semester, before exams etc. Triggered emails will be sent once students need help are beyond the threshold, and explain how the students could make improvement, and who should be contacted. Teachers not get excel reports of all their students at scheduled times, but they also receive automatic email each students that may need teachers attention.

Business intelligence refers to a system based set of processes business users use to capture and analyze the data related to school operations and academic area. A business intelligence model comprises the specific systems that can help transform

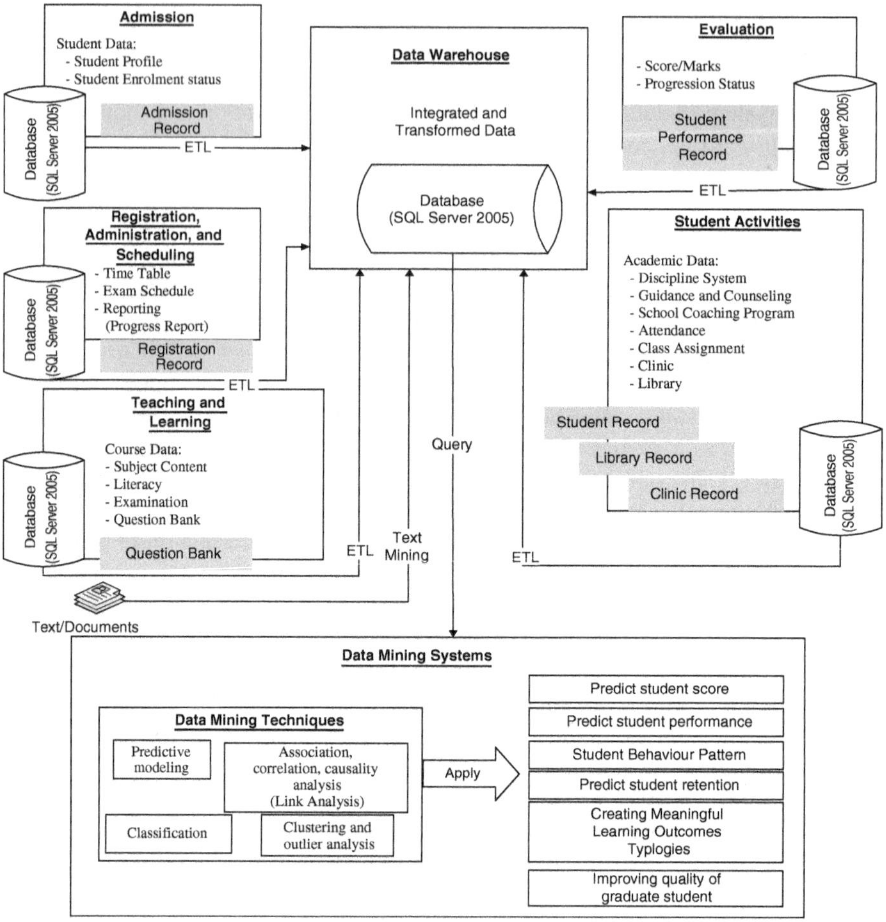

Fig. 60.1 Business intelligence models

business data into understandable and relevant information and knowledge in order to support business decisions making relating to operational and academic improvements. An business intelligence model support school spend less time ordering through information and allows them to develop systems that are easily repeatable for future business decisions. These models are related with some performance analysis techniques.

From the above statement we can conclude the main purpose of BI is developed performance via business decisions based on operational information and analysis of the business. Now therefore, important reports (business information) that must be generated from BI. Reports generated not as regular reports, such as reports generated OLTP (Online Transaction Processing) system, but more analytical report.

The analytical report is a report that can be viewed from many parameters, so it can help the analysis of information and knowledge in various perspectives. Which will it be associated with the cube view in BI. From Fig. 60.1, it seems simply how the position of a data warehouse on a BI Platform. There was described that a few things. The first, the data source is derived from the line of operational and academic process in school. And from sources of data, data will be transformed into a data warehouse. The process of data transformation is better known as ETL process (Extract Transform and Loading). From the ETL described the three main processes, namely the process of extracting data from data sources, data transformation and loading of data into the data warehouse. The data from the data source selected, cleaned (cleansing), and then loaded (load) to the data warehouse. One of the tools that we can use for these processes is Microsoft SQL Server 2005/2008 Integration Services.

As described above, BI should be able to generate reports and analytical ad hoc queries. Therefore, we must make the process of analysis from the data warehouse. In the data warehouse is defined of data storage in the form of fact tables and dimension. Fact table contains the values that will be reported, while the dimensions would be the value of facts parameters (Perfective).

And based on the fact tables and dimension, we build a cube. In our definition of this cube measures, target monitoring (Key Performance Indicators) and others. We can use Microsoft SQL Server 2005/2008 Analysis Services to help realize this analysis process. And if needed also the aspects discovered patterns or trend based on data on the data warehouse also provides the features associated with the Data Mining techniques.

The last process, we stayed deliver the reports and query results into various media analysis required. We can use Microsoft SQL Server 2005/2008 Reporting Services or Microsoft Office Excel. BI is the delivery of accurate, useful information to the appropriate decision makers within the necessary timeline to support effective decision making.

Business intelligence provides organizational data in such a way that the organizational knowledge filters can easily associate with this data and turn it into information for the school stakeholders. These can then be categorized into a series of data-mining techniques for schools, which are classified and illustrated in Table 60.2.

Data mining in education sectors are used to research and build models in several areas that can influence school operation and academic services. At the simplest level, the school stakeholders can detect when a student in a problems. Educational data mining have the potential to make visible data that unseen, unnoticed, and therefore unactionable. To help further the fields and gain value from their practical applications.

Table 60.2 Analysis BI technique and benefits for school

No.	Objective	Technique	Knowledge	Educational process
1	Use of data mining in score prediction	Prediction (regression)	The patterns of previous student score associated with their previous semester score	Evaluation
2	Creating meaningful learning outcomes typologies	Cluster analysis	The patterns of previous students learning outcome per subject	Evaluation
3	Predicting student performance (progression)	Classification	Classified pattern of previous students based on their grade, discipline system, and attendance	Evaluation
4	Improving quality of graduate student using data mining (curriculum)	Association and classification	Characteristic patterns of previous students who took a particular subject and the patterns of previous students which were likely to be good in given subject	Student activities
5	Student behavior patterns	Classification	Classified pattern of previous student behavior based on their discipline system and guidance and counselling	Evaluation
6	Predicting student retention	Prediction (regression)	The patterns of previous students score per subject	Evaluation

60.5 Conclusion

BI systems have a powerful impact on strategic decisions quality to reduce the time for making decisions and thus these systems must have the ability to allow stakeholders to view data in different perspective, to drill-down to aggregate levels, to navigate and online query data sets in order to discover new indicators that affect business process and also to anticipate and prediction changes inside and outside the school. BI system improve the quality of management in school through new type of techniques for extracting, transforming, loading and presenting data in order to provide strategic information.

The proposed conceptual model enables researching BI system absorbability capabilities or key factors facilitating the usage of quality information in school provided by BI system to generate business benefit from business processes. In the future research, we will investigate the model in the direction of the analysis of the impact of quality of access and content on school performance. An empirical research will be carried out to confirm the proposed conceptual model (links from the model between schools).

Based on the empirical research, our future work will focus on defining the key factors facilitating the usage of quality information provided by BI system to extract business value from current business processes. We expect the analysis of the measurement data to provide an understanding of the relationship between stakeholder perceptions and the use of information within school. The data collection and analysis together should also provide empirical evidence regarding the validity of our proposed conceptual model.

References

1. Fact gap fuels ERP investments and invites printer vendors to ride the wave. http://www.techexchange.com/thelibrary/RideTheWave.html
2. Marchand DA, Kettinger WJ, Rollins JD (2002) Information orientation: the link to business performance. Oxford University Press, Oxford
3. Stackowiak R, Rayman J, Greenwald R (2007) Oracle data warehousing and business intelligence solutions. Wiley Publishing, Indianapolis
4. Zeng L, Xu L, Shi Z, Wang M, Wu W (2007) Techniques, process, and enterprise solutions of business intelligence. In: IEEE Conference on systems, man, and cybernetics, vol 6, p 4722, Taipei, Taiwan, 8–11 Oct 2006
5. Arnott D, Pervan G (2005) A critical analysis of decision support systems research. J Inf Technol 20(2):67–87
6. Williams S, Williams N (2007) The profit impact of business intelligence. Morgan Kaufmann, San Francisco
7. Wang Z (2005) Business intelligence. DrMaster Culture Limited Company, Taipei
8. Goebel M, Le G (1999) A survey of data mining and knowledge discovery software tools, vol 1, issue 1. ACM, New York
9. Albert S (1998) Knowledge management: living up to the hope. Midrange Syst 11(13):52
10. Malhotra Y (2000) From information management to knowledge management: beyond hi-tech hidebound systems. In: Srikantaiah TK, Koenig MED (eds) knowledge management. Information Today, Medford, pp 37–62
11. Chamoni P, Gluchowski P (2004) Integration trends in business intelligence systems—an empirical study based on the business intelligence maturity model. Wirtschaftsinformatik 46 (2):119–128
12. Williams S (2004) Delivering strategic business value. Strateg Finance 86(2):40–48
13. Najjar L (2002) The impact of information quality and ergonomics on service quality in the banking industry. University of Nebraska, Lincoln
14. Salaun Y, Flores K (2001) Information quality: meeting the needs of the consumer. Int J Inf Manage 21(1):21–37

Chapter 61
A Surveillance Platform of Antimicrobial Use in Hospital Based on Defined Daily Dose

Guowei Liang, Yinsheng Zhang, Haomin Li, Weihong Chen and Huilong Duan

Abstract Antimicrobial abuse is very serious and is increasing threat China. It is important to make antimicrobials' better service for patients through the information technology. The surveillance platform which provides the public health authorities the patterns of consumption of antimicrobial drugs is necessary for a constructive approach to many problems that arise from the multiplicity of antibiotics now available, their high cost and the ecological sequelae of their use. In this study, DDD (Defined Daily Dose) which is an international general concept to investigate the drug consumption is used to build such a surveillance platform in a Chinese 3A hospital. First, we will explain how to classify and grade each antimicrobial drug and obtain their corresponding DDD. Then the surveillance platform was described. In the end, this paper shows how we analyzed the consumption pattern of antimicrobial drugs through this platform.

Keywords Surveillance of antimicrobial use · Defined daily dose (DDD) · Hospital management

---

G. Liang · Y. Zhang · H. Duan
College of Biomedical Engineering and Instruments Science,
Zhejiang University, Hangzhou, China
e-mail: lgw@vico-lab.com

Y. Zhang
e-mail: zys@vico-lab.com

H. Duan
e-mail: dhl@vico-lab.com

H. Li (✉)
School of Medicine, Zhejiang University, Hangzhou, China
e-mail: lhm@vico-lab.com

W. Chen
Shanxi Dayi Hospital, Taiyuan, China
e-mail: whhchen@sina.cn

J.J. (Jong Hyuk) Park et al. (eds.), *Ubiquitous Computing Application and Wireless Sensor*, Lecture Notes in Electrical Engineering 331,
DOI 10.1007/978-94-017-9618-7_61

61.1 Introduction

Medication is currently one of the most common clinical treatments. Among all approved drugs, antimicrobial is the most widely used clinically. Meanwhile, antimicrobial abuse problem is rather serious, especially in China [1]. However, the surveillance of antimicrobial is insufficient [2]. In order to strengthen the management of antimicrobial use, it is important to offer public health authorities and hospital managers the overall pattern of antimicrobial drugs' consumption. Presentation and comparison of drug consumption statistics is one of the best approaches.

The challenge faced by investigators when comparing drugs' consumption is that different medications can be of different potencies, a highly potent drug (e.g. morphine) evokes a larger response at low concentrations, while a drug of lower potency (e.g. ibuprofen) evokes a small response at low concentrations. In order to standardize the comparison of drug consumption among different medications or among different health care environments, the WHO adopted the ATC/DDD (Anatomical Therapeutic Chemical (ATC) Classification and Defined Daily Dose) in 1981. The DDD is the assumed average maintenance dose per day for a drug used for its main indication in adults. DDD provides an objective measurement for various drugs consumption, and it helps to compare single drug consumption in different hospitals/regions [3].

Antimicrobial resistance, a global concern, is particularly pressing in developing countries, including China [4]. As Kass has pointed out [5], information about patterns of usage of antimicrobial drugs is necessary for a constructive approach to the many problems that arise from the multiplicity of antibiotics now available, their high cost and the ecological sequelae of their use. In this study, a surveillance platform of antimicrobial use was designed and developed. Through this platform, user can classify antimicrobials and statistical visualize the pattern of antimicrobial drugs' consumption in hospital.

61.2 Method

61.2.1 Construction of Knowledge Base

In our previous study, a Knowledge Translation Platform (KTP) aiming to manage and disseminate different kinds of medical knowledge to accelerate knowledge translation was developed [6]. A medication knowledgebase which is part of the diverse knowledge bases was maintained on the KTP. In this medication knowledgebase, basic knowledge of each drug such as dosage, daily dosage, administration route, contra-indications and drug-drug interaction was maintained through an authoring web portal. In this study, we extend this knowledge base with DDD and grouping attributes.

DDD acquisition is a very important part of this surveillance platform, the accuracy of DDD directly affects the statistics result. However, there is no a comprehensive DDD knowledge base which could cover all the drugs used in practice. In this study, 3 DDD knowledge sources as listed below were used to construct the knowledgebase.

1. DDD provided by WHO Collaborating Centre for Drug Statistics Methodology. The DDD value could be queried on its web pages (http://www.whocc.no).
2. DDD provided by MOH (The Ministry of Health of the People's Republic of China) Center for Antibacterial Surveillance in its "Drug Dictionary and DDD of MOH Center for Antibacterial Surveillance" (2011.08).
3. DDD calculated from drug fact sheet. Acquire the DDD by analyzing the dosage and administration instructions of the drug. Table 61.1 lists the common rules.

The three DDD sources' priority declines in turn. In other words, if a drug's DDD can be obtained from all three sources, we choose the higher priority one. In this way, we maintained 129 antibiotic drugs which have at least one DDD value.

In order to manage the antimicrobials more efficiently, we designed and develop a web page to help grouping the antimicrobials in the knowledge base. And classification categories can be added according to demand. On the other hand, one drug can be classified into several groups according to different standards. In this study, antibiotics grouping mainly depends on "The Guiding Principles of Antibacterial Drugs' Clinical Usage" compiled by the Chinese Medical Association. In this way, we maintain 19 antimicrobial groups in the knowledge base.

61.2.2 Statistical Indicator Been Used

Defined Daily Doses (DDDs) is an indicator of drug consumption. DDDs can reflect hospital's overall drug utilization [7], as well as a tool for comparison over nationwide drug consumption. The formula of calculating DDDs:

Table 61.1 Rules to obtain DDD from drug fact sheet

Rule	Description
Average	Averaging when the dosage in drug fact sheet is range value; e.g. '0.1–0.2 g once, 3 times daily', thus the DDD = 0.15*3(g)
Unit	Conform to the unit of the specifications
Interval	Some drugs' administration has interval (such as "interval of 4 days"), thus the DDD = average/4
Single standard	Some drugs' dosage are based on patient's weight or height, we choose the unified adult standard (60 kg, 1.70 m^2)
Main indication	If a drug is valid on various indications, choose the main indication's DDD

$$DDDs = \frac{the\ Sum\ of\ used\ in\ a\ certain\ period\ of\ time}{Defined\ Daily\ Dose}$$

If a drug has high DDDs value in certain time, it means the intensity of that drug use is strong and clinician tends to choose that drug. Furthermore, different drugs' DDDs can be added. So we can add DDDs of the drugs which belong to the same drug group to obtain the DDDs of a drug group. DDDs of drug group make it possible to compare different type of antimicrobial drugs' consumption in the hospital. DDDs from different hospitals could also be aggregated to makes the region and the nation's drug consumption data comparable. And it facilitates monitoring and management of long-term drug utilization.

61.2.3 Integration with Clinical Information System

This surveillance platform will monitoring pattern of drugs' consumption in hospital in which a Computerized Physician Order Entry (CPOE) system had been implemented. In the CPOE system, clinicians place the electrical medical orders first. Then medication orders will be sent to pharmacy. When the medications were dispensed and delivered back to nurse station, nurses could cross-check the drugs and then administrate the medication. Information of each step was recorded and stored in the CPOE database. The data table which records medication administration information was used to calculate the real dosage of drug in this study. That table not only provides information about the medication administration, including dosage, administration route, and administration time, but also could link to which patient who from which department. All those information will support the surveillance platform to give different level drugs utilization pattern.

61.2.4 Building the Surveillance Platform

Based on the knowledge base and clinical data, a surveillance platform was designed and developed as shown in Fig. 61.1. There are three main parts in the platform technically: edit drug knowledge, query clinical data and statistics application. Our platform adopts the Microsoft's ADO.NET Entity Framework [8] technique to operate the data in the database. Both knowledge edit interface and statistical result are present to the healthcare workers in the form of web pages. We build website using the MVC (Model_View_Controller) technique.

The surveillance platform could be customized to monitoring drugs' consumption in a period of time. Its basic steps can be summarized as follows:

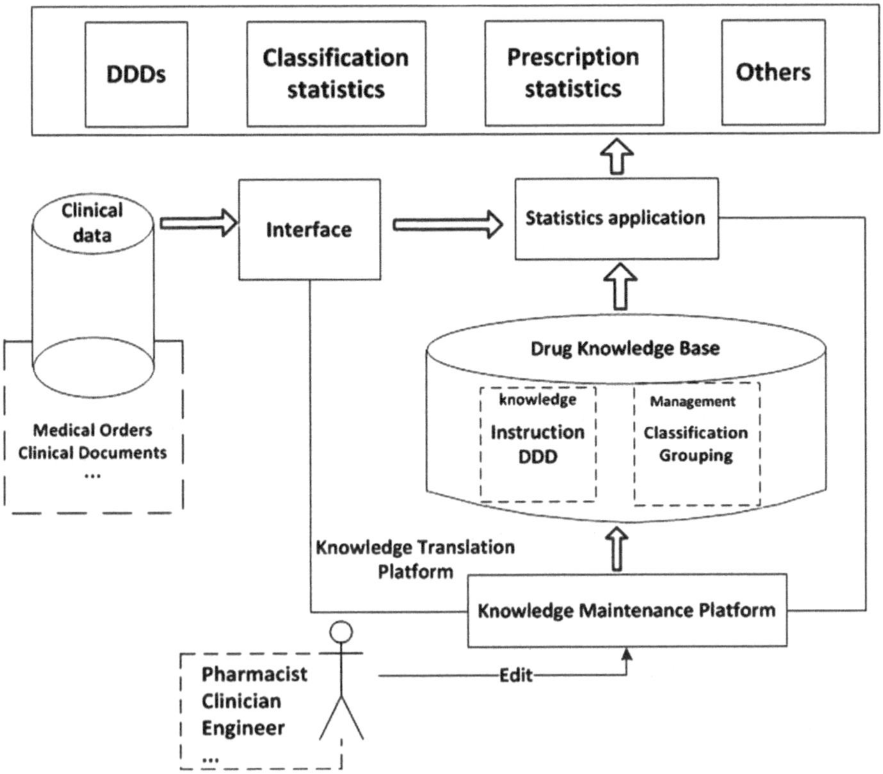

Fig. 61.1 System framework of surveillance platform based on KTP

1. Determine the drug consumption period to be monitored.
2. Check whether the DDD value of each drug was maintained in the knowledge base. The drug without DDD available will be ignored.
3. Query the corresponding medication administration information in a specific period from CPOE database. As is well-known, the same drug may have different DDD values because of its various specifications or dosage form. So additional information (e.g. Dosage unit in the record) was used to validate whether the DDD value is feasible in each record. Those inappropriate records whose dosage unit is not the same as the DDD's will be ignored.
4. Sum the dosage in the drug's every medication order record, and then divided by the drug's DDD value. In that way, we can obtain every drug's DDDs value.
5. As DDDs is a ratio, it is possible to compare various medications of different potencies. As a result, DDDs sorting can clearly show the hospital's overall antimicrobial consumption.
6. To compare different kinds of antimicrobials' consumption, we just need to put the DDDs of drugs belong to the same group together, and then sum them. In that way we can get the DDDs of the whole group.

61.3 Results

Figure 61.2 shows the surveillance platform that embedded in the KTP. There are two views, "Common Drug Investigation" and "Antimicrobial Investigation", in the surveillance platform's statistics web page.

In the default dashboard of the "Common Drug Investigation" view, the bar chart in the top left corner shows the top 10 drugs (depending on drug utilization frequency, not on DDDs) used in hospital. And there are pie charts displaying the distribution of departments which consumed each top 3 used drug.

The second view, "Antimicrobial Investigation", mainly focus on monitoring antimicrobials consumption. In the DDDs sorting part, it provides a web form to let user determine the antimicrobials consumption period they are interested in. Once the period of time is confirmed, the corresponding statistical result will be calculated and shown on the page. Similarly, we can obtain DDDs of antimicrobial which belongs to the same group.

A clinical dataset which contains 8 months data from a 3A hospital was used to demonstrate this platform. Two antimicrobial surveillance outcomes of the applications mentioned in "Antimicrobial Investigation" view were shown below.

Figure 61.3 shows the result of antibiotics' DDDs within a certain period of time (2013/08/15–2014/4/4). As can be seen from the table, though the antimicrobial may have different DDD values or units, DDDs still measures those drugs' consumption in single standard. In that way, we can compare the consumption of different drugs.

Figure 61.4 displays the specific antibiotics group's consumption in a period, the bar chart shows the DDDs of each antimicrobial which belongs to the group. The drop-down list provides all the antibiotic drug groups, those groups can be maintained through KTP authoring web portal based on the demand.

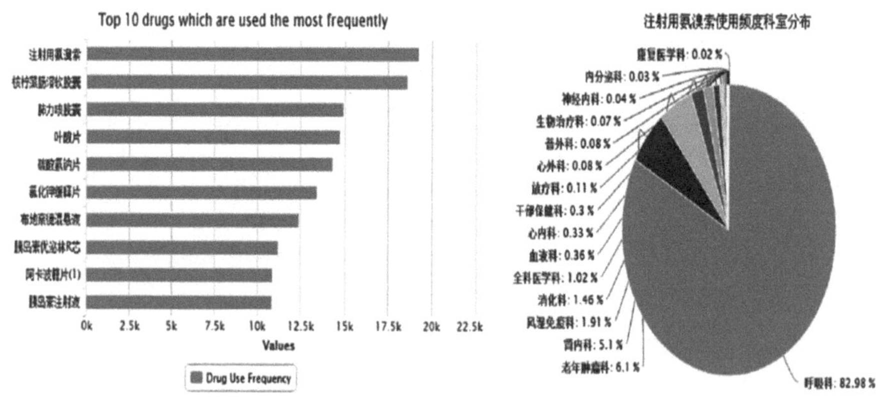

Fig. 61.2 Screen shot of the surveillance platform

显示 10 项结果查询全院药物DDD排行

药物	DDD	DDDs
叶酸片	0.4mg	342189.00
碳酸氢钠片	2.25g	26045.33
肝素钠注射液	3.5万IU	24296.69
制霉素片	75万IU	13138.67
布地奈德混悬液	1.5mg	11957.00
左旋氨氯地平片	2.5mg	10082.20
桉柠蒎肠溶软胶囊	0.825g	6898.55
注射用头孢哌酮舒巴坦	1.5g	5733.53
硝苯地平缓释片(II)	30mg	5572.50
氯化钾缓释片	3g	4512.50

显示第 1 至 10 项结果，共 85 项

Fig. 61.3 DDDs statistics table (List the top 10 DDDs)

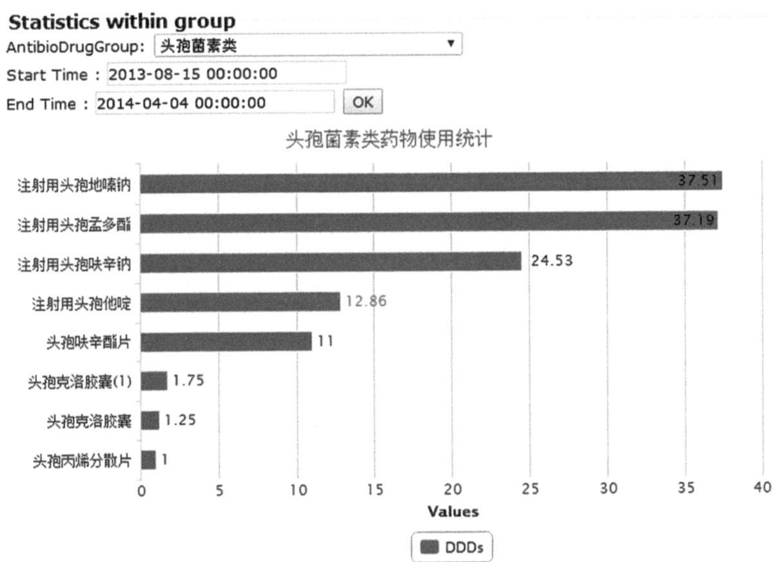

Fig. 61.4 DDDs bar chart

61.4 Discussion

This paper introduces a surveillance platform based on DDD. We can compare the consumption of different drugs through DDD objectively. There are still a few improvements we can do. First, there are many other statistics indicators based on DDD [9], such as DUI (Drug Utilization Index), DDC (Daily Dose Cost), we could get a more comprehensive pattern of antimicrobial consumption. Second, the microbiology laboratory could be a key partner on the issue of antimicrobial resistance. Third, our platform needs to spread the scope of statistics, which is helpful to assess the regional antimicrobial consumption.

On the other hand, the accurate value of DDD's is rather significant to the surveillance platform. Wang et al. suggest building a DDD database which is fit to the local population [10]. This database localizes the DDD data from WHO, so that it is more suitable for the local population.

61.5 Conclusion

In this study an antimicrobial surveillance platform based on Defined Daily Dose is designed and implemented. It helps to evaluate the different regions' drug consumption more objectively.

Acknowledgments This research was financially supported by the National High-tech R&D Program (2012AA02A601) and National Natural Science Foundation of China (30900329).

References

1. Lan-gui C, Lei Z, Tie-liang Z, Zhi-heng W, Rong-hui Z, Jie Z (2009) Abuse of antibacterials and infection management in hospital. Chin J Nosocomiol 19:464–471
2. Yuan W, Wenju X, Fei P, Yujie L, Qingping W (2007) Development and application of the hospital drug use audit and analysis system. China Pharm 18:513–515
3. Polk RE, Fox C, Maboney A, Letcavage J, MacDougall C (2007) Measurement of adult antibacterial drug use in 130 US hospitals: comparison of defined daily dose and days of therapy. Clin Infect Dis 44:664–670
4. Deli S, Zhongyi H, Ying F, Huili S, Yonghong Z (2005) Relevant analysis of antimicrobial consumption and MSSA resistance. Chin J Pharmacoepidemiol 14:244–246
5. Saboor S, Chimiak-Opoka J, Ammenwerth E (2007) Supporting the systematic assessment of clinical processes: the MedFlow method. Methods Inf Med 46:586–593
6. Yin-Sheng Z, Zhen-Ye LI, Hao-Min L, Xiang Z, Xu-Dong L, Hui-Long D (2014) A knowledge representation and translation method for clinical protocols. Chin J Biomed Eng 33: 23–28
7. Rui Z, Zhongshan X, Zhiyong J (2012) Application of DDDs in the hospital management of antimicrobial. China Med Pharm 101(2):178–184

8. Adya A, Blakeley JA, Muralidhar S, Team AN (2007) Anatomy of the ADO.NET entity framework. In: Proceedings of the 2007 ACM SIGMOD international conference on management of data, pp 1070–1072
9. Teng L, Xin H, Blix HS, Tsutani K (2012) Review of the use of defined daily dose concept in drug utilisation research in China. Pharmacoepidemiol Drug Saf 21:1118–1124
10. Wang J, Ma J, Wu, N, Chen Q, Peng Y (2008) Study of establishing CPPN drugs defined daily dose database, evaluation and analysis of drug-use in hospitals of China 8:879–880

Chapter 62
Identification of Adverse Drug Events in Chinese Clinical Narrative Text

Caixia Ge, Yinsheng Zhang, Huilong Duan and Haomin Li

Abstract Drug is an effective measure of alleviating pain and treating diseases. Whereas medication-related harms due to both adverse drug effects and drug errors have become the leading iatrogenic injury. However, such medication-related harms often remain unrecognized and unreported. The purpose of this study is to automatically identify adverse drug events (ADEs) in routine clinical documents. Firstly, ADE related Chinese lexical resource was collected and maintained. Then, a natural language processing (NLP) application which could automatically extract ADE symptom from drug manuals was developed and applied for building an ADE knowledge base for 3,733 drugs. Finally, based on these resources, an ADE detection algorithm was proposed to identify ADEs in the clinical free-text. Results revealed that the precision of the ADE detection algorithm was 80.8 %.

Keywords Adverse drug event · Natural language processing · Knowledge base

C. Ge · Y. Zhang · H. Duan
College of Biomedical Engineering and Instrument Science, Zhejiang University, Hangzhou, China
e-mail: gecx@vico-lab.com

Y. Zhang
e-mail: zys@vico-lab.com

H. Duan
e-mail: dhl@vico-lab.com

H. Li (✉)
Institute for Translational Medicine School of Medicine, Zhejiang University, Hangzhou, China
e-mail: haomin_li@yahoo.com

J.J. (Jong Hyuk) Park et al. (eds.), *Ubiquitous Computing Application and Wireless Sensor*, Lecture Notes in Electrical Engineering 331,
DOI 10.1007/978-94-017-9618-7_62

62.1 Introduction

For each approved drug, adverse effects are investigated through multiple phases of clinical trials. However, as clinical trials usually target only one drug, it is quite difficult to capture detailed effects resulting from multiple drug administration [1]. It is estimated that there are 1.5 million cases of adverse events (AEs) as a result of improper medication, with related cost coming up to $177.4 billion per year [2–5]. In China, according to 2012 Annual Report of Drug Reaction Monitoring issued by the State Food and Drug Administration, a total number of 1.2 million adverse events were reported in the whole year, among which the events that have never occurred and those have led to serious effects add up to 240,000. Furthermore, as a large amount of AEs were unreported due to several reasons such as ADE bias and work load [6]. These data may only be the tip of iceberg.

Chart review can detect adverse events in research settings, but fails in routine use for its high expense [7]. Information technology, such as NLP which is capable of extracting related information from documents, provides a potential way to automatically detect some adverse events in a timely and cost-effective way [8], thus reducing medical harms to patients in the early stage.

Detection of ADEs by informatics techniques has been widely studied. Kilbridge et al. [9] constructed an expert system with a rule-based computer program to monitor pediatric patients to monitor pediatric patients exploring the demographic, encounter, laboratory, and pharmacy data to identify possible ADEs. Wang et al. [10] parsed discharge summaries using MedLEE to identify drug–potential ADE relationships through filters. Honigman et al. [11] utilized computerized data including diagnosis codes, allergy rules, event monitoring rules, and text searching to detect ADEs in outpatients.

Although a significant of relevant ADE studies are conducted and several methods have been proposed, most of them are available conditional on the English lexical resources. In fact, few studies offer methods to detect ADEs in Chinese medical documents. This paper presents the first attempt to explore the ADE related detection in Chinese clinical narrative text.

In this study, we construct drug and ADE related lexical resource and knowledgebase in Chinese culture. Then an NLP based ADE detection algorithm is proposed and evaluated.

62.2 Method

62.2.1 Drug and ADE Related Lexical Resource and Knowledgebase

The performance of NLP technology in identifying and extracting named entity such as drug and symptom, relies very heavily on a large comprehensive lexical resource. Most relevant researches which are focused on western language are based on standard terminology such as RxNorm, SNOMED CT, WHO Adverse Reactions Terminology (WHO-ART) and Unified Medical Language System (UMLS). While in Chinese environment, such lexical resources are not available.

The medication lexicon adopted in this study was constructed based on the medicine dictionary in a hospital which contains the general name,product name and chemical name of drugs used in practice. Total 3,733 drugs with 5,127 terms were collected in our medication lexicon.

The Chinese version WHO-ART provides a basic to construct the AE symptom lexicon. Nevertheless, it's not enough to support NLP since there are many diverse terms that can describe the same symptom. The AE section of drug manual provided a potential resource to supplement the lexicon.

Total 3,733 drug manuals clawed from the internet were used to extract AE symptoms. First, each of HTML formatted drug manuals was segmented based on its semi-structured sections. Then, a general purpose lexicon maintained by our group in previous studies [12] was used to detect symptoms in the AE section of the drug manual by way of reverse maximum matching. Finally, the words and phrases annotated as Symptom and Diagnosis type were extracted to add to the AE symptom lexicon.

In a summary 3,065 AE terms and 5,127 drug terms were enrolled in the lexicon used in this study. Total 39,325 relationships between AE terms and drugs obtained during this step were also being used to build an ADE knowledgebase.

62.2.2 ADE Detection in Clinical Narrative Text

We designed a workflow for ADE detection. First of all, we selected three keywords to localize the corresponding AE location. Next, a negation detective method was applied to rule out the AE with negative status. Then, drug terms or AE terms were annotated based on the lexicon. Finally, drug-AE pairs were extracted by method of pattern matching.

Step1: General sentence segmentation and keywords localization

A hierarchical clause parser based on punctuation marks was adopted in this step. Although the period in Chinese (。) does not denote a decimal point as in English, there are still many misused period (.) at the end of sentence. Therefore, the hierarchical punctuation marks were designed as "{new char{'.', '。', '.', '\n'}, new char{',', ';',', ', '; ',':',': '}}" in c sharp language. First level represented a complete sentence. Second level stood for the clauses in the whole sentence. Segmented clauses were organized in an XML format to facilitate further processing.

We decided to use keywords to localize ADE related information. The chosen keywords were "不良反应 (adverse event)", "副作用 (side effect)", "副反应 (side reaction)". Once keywords were located, the sentence identification (ID) which was created as an attribute of each node in the XML file was collected for further detection.

Example：昨日应用环磷酰胺0.2g静点后无恶心、呕吐等消化道**不良反应**。

(Example: After intravenous cyclophosphamide 0.2g yesterday there is no nausea, vomiting and other digestive **adverse reactions**.)

Fig. 62.1 Negation filter example

Step2: Negation detection filter and Named Entity Recognition (NER)

A negation detection algorithm maintained by our group [12] which checked whether the status of keywords was affirmed or not was added to filter out the negative status of AE. An example shows in Fig. 62.1. The status of keyword "adverse reactions" in the example sentence was negated for the reason that it is modified by the character "no".

After negation detection filtering, the whole sentence where keywords existed and the status of keywords were affirmed was used to tag drug and AE based on the ADE terminology we created before. Similarly, the named entity recognition method was reverse maximum matching.

However it was noticed that drug names were usually missed by the method. The reason was that the suffix or prefix of a drug name was usually omitted by the physicians when recorded in the medical narratives. For instance, "阿莫西林胶囊 (Amoxicillin Capsule)" is always called "阿莫西林 (Amoxicillin)", "注射用香菇多糖 (Lentinan for injection)" is called "香菇多糖" (Lentinan). To address this, each drug was prepared a term by removing the common suffix and prefix (Table 62.1). While these terms led to ambiguity when several drugs shared the same shortname. For example, after tagging the drug "左氧氟沙星 (levofloxacin)" in the text, we could not be sure which one it specifically referred to, "左氧氟沙星片 (Levofloxacin tablets)", "左氧氟沙星胶囊 (Levofloxacin capsule)" or "左氧氟沙星滴眼液 (Levofloxacin eye drops)". In order to complete the judgment we turned to the patients' medication administration record. Finally, we tagged the drug concept as "d", adverse event as "a".

Table 62.1 Common suffix and prefix of drug name

Prefix	Prefix	Suffix	Suffix
注射用 (for injection)	醋酸(acetate)	注射液 (injection)	片 (tablet)
医用 (medical)	硫酸 (sulfate)	口服液 (oral solution)	乳膏 (cream)
复方 (compound)		溶液 (solution)	糖浆 (syrup)
重组 (recombinant)		滴丸 (dripping pill)	凝胶 (gel)
口服 (oral)		颗粒 (granule)	软膏 (ointment)
小儿 (pediatric)		滴眼液 (eye drop)	胶囊 (capsule)
盐酸 (hydrochloride)		栓 (suppository)	滴剂 (drop)
硝酸 (nitrate)		酊 (tincture)	丸 (pill)

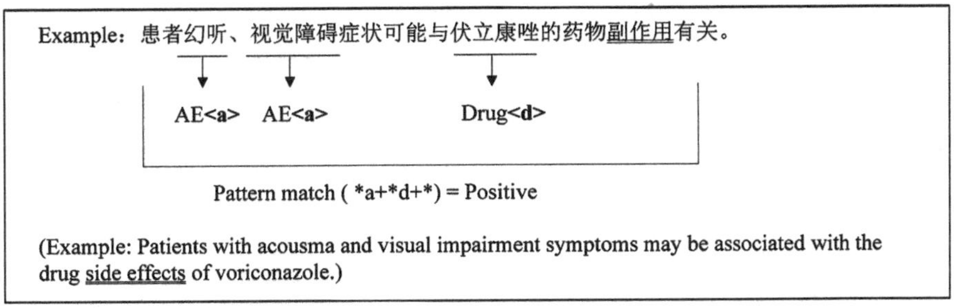

Fig. 62.2 A named entity tag and relation extraction example

Step3: Drug-adverse event pair relation extraction

After step2, the sentence contained only two types of symbols except some other characters. A regular expression was created to match the drug-AE pattern. There are two kinds of patterns, namely "*d+*a+*" and "*a+*d+*", in which "*" denotes n Chinese characters ($n \geq 0$) and "+" represents that the number of term is one or more. The matched drug-AE pair would be extracted as the result, which would be fed into the knowledgebase if the clinical specialist or pharmacist confirmed them as real ADE. A named entity tag and relation extraction example is shown in Fig. 62.2.

62.3 Results

The performance of our method was assessed on a corpus consisting of 17,430 progress notes acquired from the clinical pilot hospital with 3,000-bed. The corpus was collected about 1 year in an EHR system.

Of all the progress notes, 2,572 contained the keywords. A total of 125 ADEs were extracted. To validate the precision of results two annotators (first and second author) independently annotated the progress notes. Then all the ADEs were classified into three categories. One was the false positive results which contained 24, the second was the drug-adverse event found in the progress notes and also presented in the knowledgebase which contained 87, the last one was that only found in the progress notes while not found in the knowledgebase which contained 14 (Table 62.2). The precision of method is 80.8 % (101/125 = 80.8 %). The 14 results in the last category were pending conclusion of casual drug relationship. We regarded them as the possible drug-AE. Then the possible drug-AE would be submitted to clinical experts or pharmacist to confirm their accuracy. From Table 62.2 we can find that a majority of drugs are for injection. At the same time, statistics manually showed that 57 drugs in the 87 true positive results were injection medicine.

Table 62.2 Some ADE results that are not found in the knowledgebase

Drug	Adverse event
氟尿嘧啶注射液 (fluorouracil injection)	骨髓抑制 (myelosuppression)
注射用香菇多糖 (lentinan for injection)	体温升高 (elevated body temperature)
氟尿嘧啶注射液 (fluorouracil injection)	胃肠道反应 (gastrointestinal reaction)
注射用三氧化二砷 (arsenic trioxide for injection)	低血钾 (hypokalemia)
注射用三氧化二砷 (arsenic trioxide for injection)	骨髓抑制 (myelosuppression)
注射用丙戊酸钠 (sodium valproate for injection)	不适 (discomfort)
环孢素注射液 (ciclosporin injection)	水钠储留 (sodium and water retention)
注射用伏立康唑 (voriconazole for injection)	幻听 (acousma)
注射用多索茶碱 (doxofylline for injection)	烦躁 (irritable)
注射用硼替佐米 (bortezomib for injection)	神经炎 (neuritis)
注射用硼替佐米 (bortezomib for injection)	骨髓抑制 (myelosuppression)
榄香烯注射液 (elemene injection)	腹痛 (abdominal pain)
紫杉醇注射液 (aclitaxel injection)	神经炎 (neuritis)
利妥昔单抗注射液 (rituximab injection)	肺浸润 (pulmonary infiltration)

The major reason leading to the false positive results is that the description is about treating the symptom of adverse event not about narrating adverse drug event. For example, the sentence "针对CTX所致的不良反应为出血性膀胱炎予美司钠治疗" (Aiming at the adverse reaction hemorrhagic cystitis of CTX mesna is given for treating) is talking about to cure the hemorrhagic cystitis by mesna. Indeed 23 of all the 24 false positive results are caused by this reason.

62.4 Discussion

Comprehensive medical terminology dictionary is the basis and premise of drug related knowledge discovery and mining, while there is no one including both drug and adverse event in China. To achieve the extraction goal, we established the ADE terminology and knowledgebase based on the drug manuals.

We designed and developed methods to automatically extract drug-adverse event in clinical narratives, which showed great ability to identify ADEs. Meanwhile, the results could be validated by our knowledgebase. If the relationship was not found in the knowledgebase, it could be submitted to the clinical experts, and after confirmation they may be added into the knowledgebase as supplement. At the same time, the automated ADE reporting can provide services for decision support systems or as the basis of further data-driven research into causal relationship between drug and AE.

Result shows that the precision of our method is 80.8 %. We can't compare this result with recently proposed models, because it is the first time to construct an ADE knowledgebase and to achieve automated ADE detection in clinical text in

Chinese. The result is considerable encouraging. However, the frequency of the adverse information is quite low (125/17,430 = 0.7 %). What should do later to improve the proportion is to expand the keywords and maybe some machine learning method is desired.

What has to be mentioned is that the ADE terminology remains limited because of the flexible and irregularity of natural language. As mentioned in the Method section, we get rid of the prefix or the suffix of the drug name. Nevertheless, some names cannot be processed like this. An example is "甲状腺片" (Thyroid tablet). If "片" (tablet) is removed, "甲状腺" (thyroid) cannot be treated as a shorthand of Thyroid tablet. Hence, it is necessary to artificially tidy the shorthand of drug names later. The construction of the lexicon costs nearly 3 months, more efforts are continually needed to put based on the present one.

62.5 Conclusion

In this study a method to automatically extract ADE in clinical narratives was designed and implemented. We also provide a more comprehensive ADE lexicon and knowledgebase as the main dictionary of the algorithm. Based on evaluation results, we believe that the ADE information in clinical narratives could be identified automatically. Furthermore, these results could support further ADE studies.

To our knowledge, it is important to emphasize that this is the first time to construct an ADE knowledgebase and to achieve automated ADE detection in clinical text in a language other than English. We believe that our knowledgebase and method can provide a solid foundation for further studies and research on ADE especially in China.

Acknowledgments This research was financially supported by the National High-tech R&D Program (2012AA02A601) and National Natural Science Foundation of China (30900329).

References

1. Ratanawijitrasin S, Wondemagegnehu E, Organization WH, Drugs E, Policy M (2002) Effective drug regulation: a multicountry study. World Health Organization, Geneva
2. Classen DC, Phansalkar S, Bates DW (2011) Critical drug-drug interactions for use in electronic health records systems with computerized physician order entry: review of leading approaches. J Patient Saf 7:61–65
3. Ernst FR, Grizzle AJ (2001) Drug-related morbidity and mortality: updating the cost-of-illness model. J Am Pharm Assoc (Washington, DC: 1996) 41:192
4. Lazarou J, Pomeranz BH, Corey PN (1998) Incidence of adverse drug reactions in hospitalized patients: a meta-analysis of prospective studies. JAMA 279:1200–1205
5. Gandhi TK, Weingart SN, Borus J, Seger AC, Peterson J, Burdick E, Seger DL, Shu K, Federico F, Leape LL (2003) Adverse drug events in ambulatory care. N Engl J Med 348:1556–1564

6. Hazell L, Shakir SA (2006) Under-reporting of adverse drug reactions. Drug Saf 29:385–396
7. Brennan TA, Localio AR, Leape LL, Laird NM, Peterson L, Hiatt HH, Barnes BA (1990) Identification of adverse events occurring during hospitalization. A cross-sectional study of litigation, quality assurance, and medical records at two teaching hospitals. Ann Intern Med 112:221–226
8. Bates DW, Evans RS, Murff H, Stetson PD, Pizziferri L, Hripcsak G (2003) Detecting adverse events using information technology. J Am Med Inform Assoc 10:115–128
9. Kilbridge PM, Noirot LA, Reichley RM, Berchelmann KM, Schneider C, Heard KM, Nelson M, Bailey TC (2009) Computerized surveillance for adverse drug events in a pediatric hospital. J Am Med Inform Assoc 16:607–612
10. Wang X, Chase H, Markatou M, Hripcsak G, Friedman C (2010) Selecting information in electronic health records for knowledge acquisition. J Biomed Inform 43:595–601
11. Honigman B, Lee J, Rothschild J, Light P, Pulling RM, Yu T, Bates DW (2001) Using computerized data to identify adverse drug events in outpatients. J Am Med Inform Assoc 8:254–266
12. Li HM, Li Y, Duan HL, Lv XD (2008) Term extraction and negation detection method in chinese clinical document. Chin J Biomed Eng 27:716–721

Chapter 63
Control System and Control Method for Automatic Adjustment of Outdoor LED Display Brightness

Feng Yang, Xu-fei Qin and Lin-bo Zhai

Abstract The paper involved a control system and control method for automatically adjusting the outdoor LED display screen brightness. The control system includes outdoor LED Display, test module, the optical system which was collect emitted light and reflect Light, a light sensor, a signal amplifier, analog to digital conversion control, LED display brightness automatic control system. The outdoor LED display controlled by LED display; the optical system which was collect emitted light includes a hood and fixed focus lens, fixed focus lens; the test modules controlled, scanned and drived by LED display; the hood is used to shield unwanted ambient light, fixed focus lens video test module emit light to the light sensor, the light sensor convert an optical signal into an electrical signal and amplified by the signal amplifier and the analog signal which amplified by the signal amplifier was sent to the digital converter and turned into the corresponding digital signals and was sent to the LED display brightness automatically control system and then achieves automatic brightness adjustment.

Keywords Automatic adjustment of outdoor LED display · Reflected light · Gray scale

63.1 Introduction

The clarity of content for the outdoor LED display is own to the screen brightness of their own, and also related with the environment. Exactly, the clarity of LED display screen observed by people is effected by the light in the eyes which was

F. Yang (✉) · X. Qin · L. Zhai
School of Information Science and Engineering, Shandong Normal University,
Jinan 250014, China
e-mail: 2405022641@qq.com

F. Yang · X. Qin · L. Zhai
Shandong Provincial Key Laboratory for Distributed Computer
Software Novel Technology, Jinan 250014, China

J.J. (Jong Hyuk) Park et al. (eds.), *Ubiquitous Computing Application and Wireless Sensor*, Lecture Notes in Electrical Engineering 331,
DOI 10.1007/978-94-017-9618-7_63

reflected from the display by the ambient light, the stronger of the reflected light, the more it will reduce the clarity of display screen [1]. So we have to adjust the screen brightness according to the light intensity reflected on the display to ensure its effective adjustment. The paper touches on a control system and control method for automatic adjustment of outdoor LED display brightness.

63.2 Control of Method

The clarity of content for the outdoor LED display is own to the screen brightness of their own, and also related with the environment, such as the sun shines. If the sun light is stronger, the display content is not easy to be seen, then the brightness of the display need to be improved. If the sun light is weaker, of course it is weakest in the night, then the brightness of the screen can be reduced, the benefits for the reduction as follows: 1, to make the display not dazzling; 2, to reduce the power consumption of the screen; 3, to improve the service life of the LED lamp display.

Broadly speaking, the stronger ambient light will decrease more clarity of display screen. Exactly, the clarity of LED display screen observed by people is not directly affected by the ambient light, actually effected by the light in the eyes which was reflected from the display by the ambient light, the stronger of the reflected light, the more it will reduce the clarity of display screen. The present light metering methods mainly measures the intensity of ambient light, rather than the intensity of the reflected light. The ambient light of the LED display and the reflected light are not a linear relationship, and in some cases are not related. Case 1: in order to reduce the effect of sunlight on the outdoor LED display, LED light emitting pixel has a brim hat shading, light emitting pixels have a part of the dark area to increase the display contrast, if the hat brim is too short, the shading effect is not good; but if the hat brim is too long, it will shield other light emitting pixel light, when the sun is nearly perpendicular to the floor radiation screen, such as at noon, when the sun is very strong, the ambient light is also very strong, because of the brim of the occlusion, the reflected light will not be too strong; when the sun is nearly parallel to the ground radiation screen, such as sunrise and sunset, even though the sun is weak, the ambient light is also weak, due to the direct sunlight to the LED light emitting pixels, then the reflected light will not be too weak. When it is cloudy weather, the sunlight through the refraction by the clouds is changed to be multiple direction light, and cloudy weather also makes the reflected light and environmental light not establish the corresponding relationship. Case 2: in addition to the angle of the sunlight, the reflected light is also related to the structure and material of display: peak structure of LED light pixels is different, then the reflected light intensity is different; reflective performance of the display's surface materials also affect the reflected light intensity, when the display screen is blow insolation and its surface is covered with dust, the reflected light will be increased. So we have to adjust the screen brightness according to the light intensity reflected on the display to ensure its effective adjustment.

Gray scale of the LED display screen: to show the contents of a sense of hierarchy, the design of LED display screen has a gray level, generally it is realized by the emission ratio [2]. For example, the 256 gray level, the gray value of 0/255, 1/255, 2/255, … 255/255. One of them, which is implemented as of 1/255 shades of gray: 1 unit of time LED lights glow, 254 units LED lights do not light, repeat the above process frequency exceeds 50 Hz, the human eye can't see flashing. As the average duty cycle and brightness nonlinear, you can increase the number of grayscale levels for nonlinear correction. LED display for each pixel can have different gray values.

LED display brightness level: LED display design has brightness level, and can also be achieved by the light duty cycle [3]. For example, 16 levels of brightness, the brightness value is 0/15, 1/15, 2/15, …, 15/15. Brightness level has a nonlinear characteristic, generally it does not need nonlinear correction. The LED display wholly has a brightness level.

Attenuation of LED lamp: if the brightness of the LED lamp is not attenuated, brightness level of the LED display screen can be completely determined by the reflected light intensity of LED display screen. However, the brightness of the LED lamp is in use after a period of time, usually a year or so, there will be a more obvious attenuation. In the same reflected light, the emission light which original could meet the visible luminance level, but through the attenuation the brightness of the LED lamp, then its brightness does not necessarily meet the LED display clearly visible effects.

The purpose of this paper is to solve the problem of automatic adjustment outdoor LED display brightness, provide an automatical adjustment control system for the outdoor LED display screen brightness. The method of the present paper determine the level of brightness of LED display based on the ratio of the real time measured emitted light and the reflected light, the signal-to-noise ratio.

In order to achieve the above purpose, the paper takes the following technical scheme:

An automatical adjustment control system of outdoor LED display brightness, including outdoor LED display, test module, collection system for emitting light and reflected light optical systems, optical sensors, signal amplifier, AD converter, LED display screen brightness automatic control system; the above outdoor LED display is controlled by the LED screen controller; the test module is arranged in the outdoor LED display screen, or being independent set, and in the same plane as the outdoor LED display screen; the collection system for emitting light and reflected light optical systems includes the hood and fixed focus lens, the hood and fixed focus lens is installed on the opposite below of the outdoor LED display screen; the fixed-focus lens and test modules are controlled, scanned and drived by the described LED display controller; the described hood shield the useless ambient light, prime lens imaging test module described the launch of light to the light sensor, then light sensor converts light signals into electrical signals, and through the signal amplifier amplification, signal amplifier amplification process the analog signal to the AD converter into the corresponding digital signal, sent to the LED display brightness, brightness adjusted automatically.

The described is used for the control system of automatic adjustment of outdoor LED display brightness control method [4, 5], the steps are as follows:

Step 1: the test module is not luminous controlled by LED display controller, metering system measured data is the numerical data from reflection of light, no noise, assumed to be N.

Step 2: LED display controller controls the display brightness level test module, according to the light, the measured data from the metering system for are the sum by transmission and reflection light, minus N, get the light, for the effective signal, assuming S.

Step 3: calculate according to formula one to get the signal to noise ratio SNR;

$$\mathrm{SNR} = \mathrm{S/N}. \tag{63.1}$$

Step 4: according to SNR determine the brightness level of outdoor LED display, when SNR is less than the threshold SNRmin, LED screen brightness automatic control system control of the outdoor LED display the current brightness level of incremental, maximum level until the current brightness level reaches the brightness level, or until the brightness level can satisfy the SNR is greater than SNRmin and the brightness level minus 1 class can not meet the SNR is greater than SNRmin, LED display with the current level of brightness; when SNR is greater than the threshold SNRmin, automatic control system to control the outdoor LED display the current brightness level gradually reduced brightness of the LED display screen, the minimum level until the current brightness level reaches the brightness level, or until the brightness level can meet below conditions: satisfy the SNR is greater than SNRmin and the current brightness level minus 1 level. Even it can not meet that the SNR is greater than SNRmin, LED screen displays with the current level of brightness; when SNR is equal to SNRmin, LED screen displays with the current level of brightness.

Among them, method for determination of SNRmin as follows: in the different weather conditions, setting the outdoor LED display for various brightness levels, repeating the first step to the third step from the described method way, test the signal noise ratio SNR, get the minimum signal-to-noise ratio of SNRmin to satisfy people to clearly see the contents of the outdoor LED display, especially to test the signal-to-noise ratio SNRmin when the sun's rays are the strongest and mostly close to the light in the direction parallel to the ground conditions.

Beneficial effects of the present paper are as follows: compared with the prior art, the invention, using the signal-to-noise ratio of SNR evaluation index, the real-time measured emission and reflection light ratio, in order to meet the people to see the outdoor LED display content, can more accurately control the outdoor LED screen brightness, the outdoor LED display with a display effect, better energy-saving effect, reduce the attenuation of LED lamp, to extend the life of outdoor LED display screen.

63.3 Application

Below are the further details for this paper combined with the attached drawings and examples.

Example: Figs. 63.1 and 63.2 for a set of control system used to automatically adjust the outdoor LED display brightness [6], including outdoor LED display, test module, collection and emitting light and reflected light optical systems, optical sensors, signal amplifier, AD converter, LED display brightness automatic control system; the described outdoor LED display is controlled by the LED display controller; the described test module is set on the outdoor LED display, or the test module is set independently in the same plane with outdoor LED display; Acquisition of emitting light optical system including the hood, the prime lens, and prime lens hood installed on the outdoor LED display on the opposite side of below, will be subject to not keeping out the LED display screen; Prime lens and the test module described by scanning LED display controller to control and drive; Described hood shielding useless ambient light, described prime lens imaging test module launch the light to the light sensor, light sensor, converts light signals into electrical signals, and through the signal amplifier amplification, after signal amplifier amplification of analog signal to the AD converter, into the corresponding digital signal, sent to the LED display brightness automatic control system for its brightness automatically adjust.

As shown in Fig. 63.3, used in the control system of automatic adjustment of outdoor LED display brightness control method [7]. Simplified steps as follows:

The first step: the test module is not luminous controlled by LED display controller, metering system measured data is the numerical data from reflection of light, no noise, assumed to be N.

The second step: by LED display controller test control module, in accordance with the display brightness level current light-emitting, metering system measured data for transmission and reflection light and

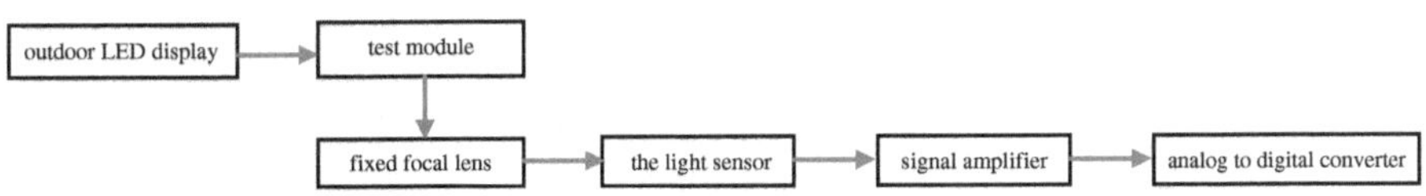

Fig. 63.1 The control system of the present paper

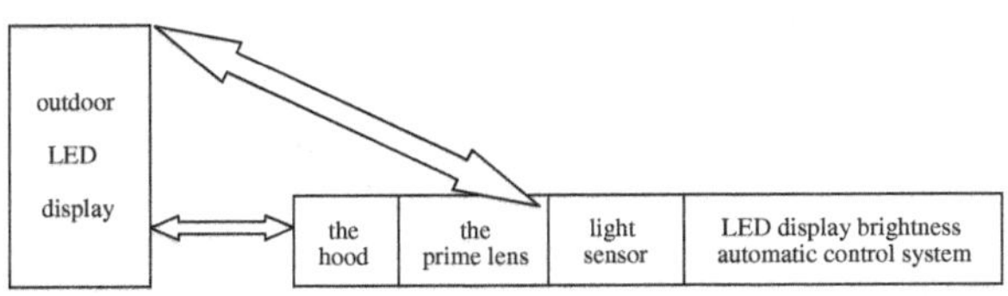

Fig. 63.2 A schematic diagram of a flat outdoor LED display control system

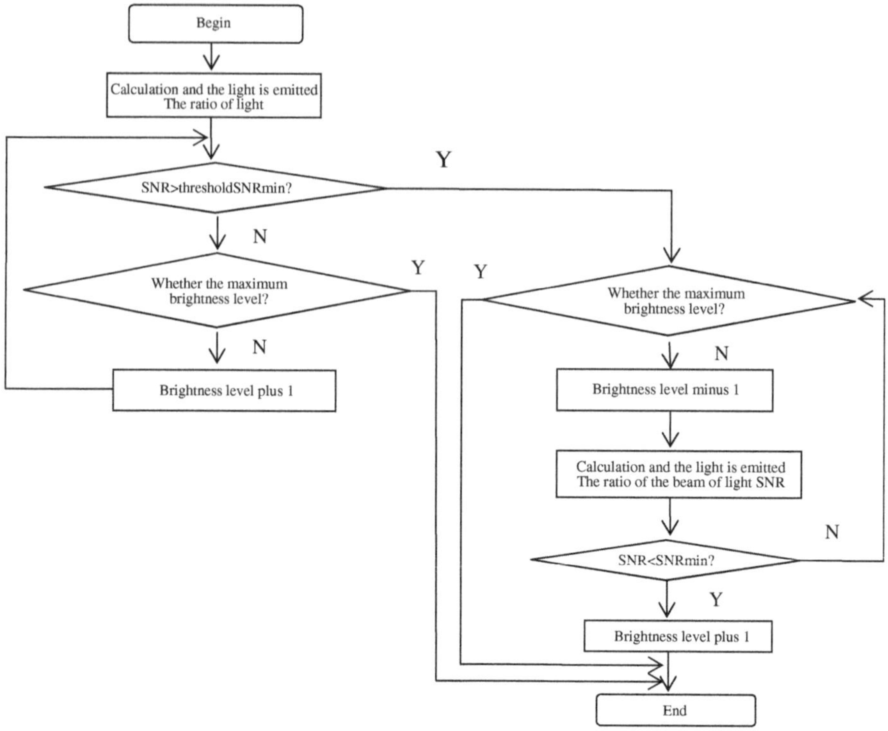

Fig. 63.3 The paper metering step flow char

reflection light, minus N, get the light, for the effective signal is assumed to be S.

The third step: calculated according to formula one, get the signal to noise ratio SNR.

The fourth step: according to the SNR determines outdoor LED display 101 of brightness grade, when SNR is less than the threshold SNRmin, automatic control system of the control of outdoor LED display, LED display brightness of the brightness level is gradually increased, the most high level until the current brightness level reaches the brightness level, or until the brightness level can satisfy the SNR is greater than SNRmin and the current brightness level minus 1 level can not meet the SNR is greater than SNRmin, outdoor LED display to the brightness level of display; when SNR is greater than the threshold SNRmin, LED screen brightness automatic control system of the control of outdoor LED display the brightness level of the minimum level gradually reduced, until the current brightness level reaches the brightness level, or until the current level of brightness to meet the SNR is greater than SNRmin and the current level of brightness by 1 level can not meet the SNR which is greater than SNRmin, outdoor LED display screen on the display brightness level; when SNR equals SNRmin, outdoor LED display on the display brightness level.

According to the change of the sun light, the first step to step four need to implemented timingly, such as to perform once every 30 min.

63.4 Summary

The paper designs a control system and control method for automatically adjusting the outdoor LED display screen brightness. Compared with the prior art, the paper, using the signal-to-noise ratio of SNR evaluation index, the real-time measured emission and reflection light ratio, in order to meet the people to see the outdoor LED display content, can more accurately control the outdoor LED screen brightness, the outdoor LED display with a display effect, better energy-saving effect, reduce the attenuation of LED lamp, to extend the life of outdoor LED display screen. The experimental results show that the control system and control method for automatically adjusting the outdoor LED display screen brightness is suitable for the large amount of cars in light when meeting each other, and in a company has carried out the actual technology production and application, and has achieved good economic benefit.

References

1. Ahn JH (2013) Implementation of an LED tile controller for high-quality image display. Displays 34(1):17–26
2. Marks P (2013) LED carpet turns the floor into a screen. New Sci 220(2945)
3. Svilainis L (2008) LED brightness control for video display application. Displays 29(5):506–511
4. Anonymous (2013) New distributor of Foreground LED screens. The Stage (46):41
5. Wang Q, Sun X (2004) Structural optimization design of outdoor LED display. Adv Display
6. Song Y, Feng Y, Juanli MA, Zhang X (2011) Design of LED display control system based on AT89C52 single chip microcomputer. J Comput 6(4):718–724
7. Puettjer D, Praemassing F, Buss R, Jaeger D (2002) LED-display for an intraocular microoptic system. Biomed Tech 47(1):164–166

Chapter 64
A Novel Quantitative Evaluation Metric of 3D Mesh Segmentation

Xiao-peng Sun, Lu Wang, Xingyue Wang and Xiaona Zhao

Abstract This paper presents a novel quantitative metric for comparison of 3D mesh segmentations, Ultimate Measurement Accuracy, basing on the screening data sets, to support our quantitative comparisons of seven recently published mesh segmentation algorithms; and the experiment results suggest that, our metrics is robust to degenerative segmentation and hierarchical refinement, stable to the imprecision of cut boundaries.

Keywords 3D segmentation evaluation · Quantitative evaluation · Ultimate measurement accuracy · Misclassified error

64.1 Introduction

3D mesh segmentation benefits various algorithms in Digital Geometry Processing, such as modeling, simplification, texture synthesis, collision detection, skeleton extraction, compression, and 3D shape retrieval and the establishment of the corresponding relations in morphing and deformation [1].

In the last several years, a number of segmentation algorithms have been proposed to cut 3D mesh into simple sub-meshes, including a few components with higher level shape semantic, or lower level disk-like patches, and many of them aim at a meaningful segmentation, but little work on the qualitative evaluation of quality, even little on the quantitative evaluation. [2–4].

The main contributions of this work is that, we investigate a novel evaluation metrics for quantitative comparing mesh segmentations—UMA, and discuss its properties; To improve the accuracy of our three metrics, we establish a subset of 3D mesh segmentation Benchmark as our experimental dataset.

X. Sun (✉) · L. Wang · X. Wang · X. Zhao
Computer Systems Institute, Liaoning Normal University, Dalian 116029, China
e-mail: cadcg2008@gmail.com

J.J. (Jong Hyuk) Park et al. (eds.), *Ubiquitous Computing Application and Wireless Sensor*, Lecture Notes in Electrical Engineering 331,
DOI 10.1007/978-94-017-9618-7_64

64.2 Related Work

Most work evaluated their segmentation only by the vision; little work has been done on the qualitative and quantitative quail evaluation of segmentation.

The qualitative criteria frequently used to evaluate the automatic segmentation algorithms in a qualitative way: exhibit the segmentation image of the same set of meshes side by side, and point out which results look good or not. And the criteria include: the robustness and complexity, the number of control parameters, the hierarchical and multi scale properties, whether to produce over-segmentation, the smoothness of the segmentation boundary, and whether the segmentation results have a clear visual geometry semantics (meaningful), whether the segments boundaries lie on concave seams and valleys, as predicted by the minima rule.

Quantitative evaluation of 3D segmentation first appeared in 2009. To address the key problem of quantitative evaluation, Benhabiles et al. [2] proposed a ground truth corpus: two objective dissimilarity measures, which provide a quantitative comparison between the segmentation algorithms. The limitations of this work are the number of model and category, and the meaningful of manual segmentations.

Comparing with the work of Benhabiles et al. [2], this data set of Chen et al. [3] has a greater size and a higher sampled distribution. To provide a basis for quantitative comparisons between the human-generated segmentations and automatic segmentation [5–10] for the same mesh, they propose four quantitative evaluation metrics. In 2010, Benhabiles et al. [4] present an extensive experimental comparison of existing similarity metrics addressing the quality assessment problem of mesh segmentation and introduce a new metric, and basing on their own corpus and Chen et al. corpus [3], they get a better performs, but their work has a drawback with the accuracy of probability of two vertices belonging to the same segment in the ground-truth set.

All of these works provide a benchmark basing on a ground-truth corpus of human segmented 3D meshes. Using a set of similarity metrics, they evaluate the segmentation algorithm by quantifying the consistency between the reference segmentations of the ground-truth corpus and those obtained by this algorithm on the same models.

Some problems with the benchmark as follow: First, the manual segmentation results produced by different volunteers show significant subjectivity and randomness. Second, although Consistency Error evaluation metric proposed independently by Chen et al. and Benhabiles et al. are similar, they produce a significant different evaluation metrics for the same segmentation algorithm. Third, the four kinds of evaluation criteria proposed by Chen et al. still produce significant different evaluation metrics for the same segmentation algorithm with others. Finally, all the six quantitative metrics produce very differences evaluation to which by human visual perception.

In this paper, we first analyze the serious inconsistencies, errors and other problems of the data set produced by manual interaction segmentation in [2–4], and build a subset of Chen's dataset with higher consistency and accurate. Finally, we perform quantitative comparison of seven automatic mesh segmentation algorithms according to the new metric UMA, and our other two metrics, FME and AME [11].

64.3 Ultimate Measurement Accuracy

The ultimate goal of 3D mesh segmentation is to obtain the precise measurement of target feature, so we can evaluate the quality of segmentation by means of measuring and calculating the target feature. The common features frequently-used to measure the target feature precisely include target area, perimeter, circular, eccentricity, shape parameters, Ultimate Measurement Accuracy (UMA) et al.

64.3.1 Shape Feature Form Factor

First, we define the sub-grid shape feature Form Factor (FF) as follows: Let S be a 3D boundary mesh, and S1 the set of sub-grid generated by manual interactive segmentation, i.e., $S1 = \{S_1^1, S_1^2, \ldots, S_1^m\}$, and S2 the automatic segmentation produce by the given algorithm, i.e., $S2 = \{S_2^1, S_2^2, \ldots, S_2^n\}$, m and n are the number of segments. Here we require m equals to n. For each sub-grid S_1^i in S1, note its match position in S2 is sub-grid S_2^j, i.e., there is a correspondence $j = p(i)$. We define the Form Factor of arbitrary sub-grid S_1^i in S1 as follows:

$$FF(S_1^i) = \sum_{a=1}^{N_e} L(S_1^i, e_a) \Big/ \sum_{b=1}^{N_f} A(S_1^i, f_b) \tag{64.1}$$

where e_a and f_b are the arbitrary boundary edge and face in sub-grid S_1^i respectively, and $L(S_1^i, e_a)$ is the length of edge e_a, $A(S_1^i, f_b)$ is the area of f_b, N_e and N_f are the number of boundary edges and faces of S_1^i respectively. Similarly, the Form Factor of arbitrary sub-grid S_2^j in S_2 is defined as follows:

$$FF(S_2^j) = \sum_{a=1}^{N_e} L(S_2^j, e_a) \Big/ \sum_{b=1}^{N_f} A(S_2^j, f_b) \tag{64.2}$$

where the e_a, f_b, L, A, N_e and N_f are defined similarly with that as before. Then, we obtain the Form Factors of all the sub-grid produced by manual interactive segmentation and automatic algorithm segmentation.

64.3.2 Ultimate Measurement Accuracy

With the manual interactive segmentation S_1 as criterion, the UMA of the segmentation S_2 produce by the given algorithm is defined as follows:

$$UMA(S_1, S_2) = \frac{1}{m} \sum_{i=1, j=p(i)}^{m} \frac{FF(S_1^i) - FF(S_2^j)}{\left\| S_1^i \right\|} \quad (64.3)$$

where $\left\| S_1^i \right\|$ is the area of S_1^i. The smaller the value of $UMA(S_1, S_2)$, the closer the automatic algorithm segmentation results with manual interactive segmentation results, the better the performance of segmentation algorithm, or vice versa.

To a given automatic segmentation algorithm, for each 3D boundary-mesh S in the data set of benchmark, one of its manual interactive segmentation is S_1^i and its segmentation produced by the given automatic algorithm is $S_2 = \{S_2^1, S_2^2, \ldots, S_2^n\}$, note the Ultimate Measurement Accuracy as $UMA(S_1^i, S_2)$, then for the set of manual interactive segmentation $\{S_i\}$, the $UMAs$ of this automatic segmentation algorithm is defined as $\sum_i UMA_S(S_1^i, S_2)/n$, where n is the number of manual interactive segmentation models.

For the manual segmentation results of a data set with N different meshes, we define its quantitative evaluation metrics using Ultimate Measurement Accuracy as $\mathrm{UMA} = \sum_\mathrm{j} \mathrm{UMAS}/\mathrm{N}$. The smaller the UMA is, the closer the automatic algorithm segmentation results with manual interactive segmentation results, the better the performance of segmentation algorithm.

64.4 Results and Analysis

64.4.1 Experimental Environment

Our experiments are performed on a subset of the benchmark data set proposed by Chen et al. [3], but reject 3 kinds degenerative segmentation results as following: First, the segmentation with the sub-grid number less than or equal to 3; Second, and the ratio of that the face number in sub-grid to the total face number of the original mesh less than or equal to 0.01; Third, the segmentation results without a clear meaningful shape. The subset contains 3,697 manually generated segmentations for 380 meshes models spread evenly amongst 19 different object categories, and has stronger consistency and higher quality.

Our quantitative evaluations are performed on 7 automatic segmentation algorithms frequently-used recently [5–10]: Randomized Cuts, Shape Diameter Function, Normalized Cuts, Core Extraction, Random Walks, Fitting Primitives and K-Means.

64.4.2 Properties Comparison of UMA, FME and AME

In this section, we perform three experiments to illustrate some properties of UMA metrics, comparing with FME and AME, and the results are shown in Figs. 64.1, 64.2 and 64.3.

The first property is the robust when deal with some segmentation in degenerative case. We provide a k-mean segmentation with random cluster center, which includes some degenerative segmentation with only 2 or 3 segments or more than 50 segments, and some random segmentation with a number similar to that of ground-truths. The **score** of three new metric are computed for this segmentation and averaged over the entire data set, Fig. 64.1 presents the experiment results. Since this segmentations is random or degenerative, so the data set is totally different from the manually ground-truth, the scores of three metrics are low, but stable for the segmentation.

The second property is the sensitivity to refinement. We compared 12 versions to the ground-truths, Fig. 64.2 illustrates that all of the three novel metrics present a slight variation (not full invariant to refinement), but still be stable.

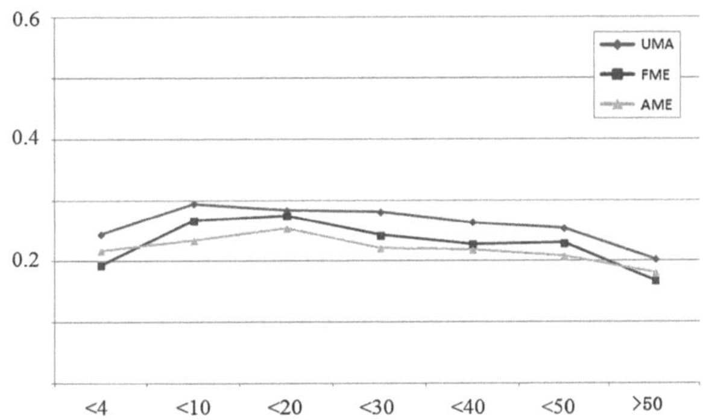

Fig. 64.1 Robust to random and degenerative cases

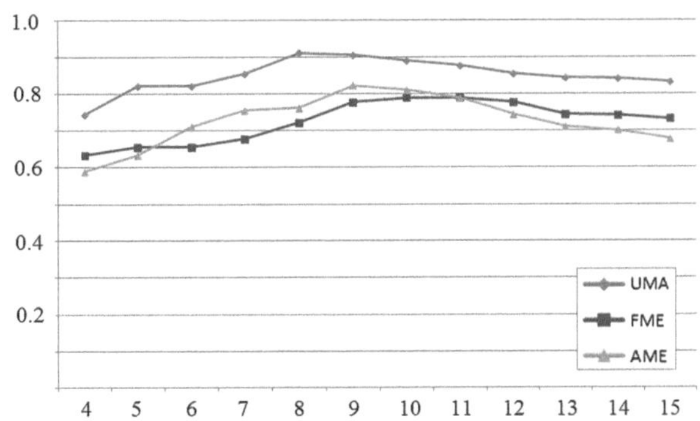

Fig. 64.2 Sensitivity to hierarchical refinement

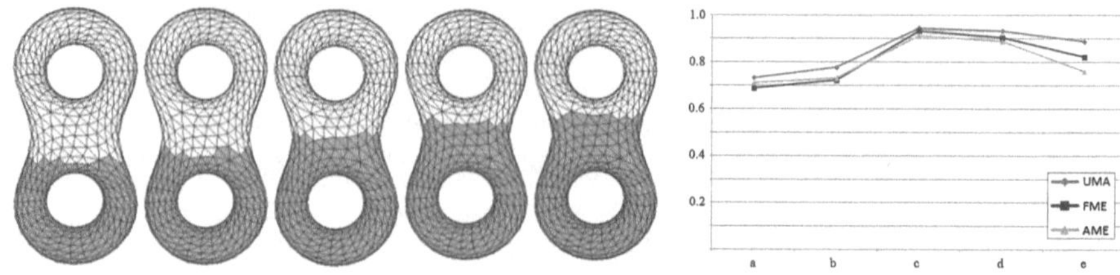

Fig. 64.3 Tolerance to the imprecision of cut boundaries

We proposed five versions of segmentation by manually segmented the model bi-torus into two segments, and each has a slight difference to others with the boundary position. And the results in Fig. 64.3 shows that, segmentation (a) and (b) are more imprecision than the segmentation (c–e), the scores illustrate this difference indeed, and all the three novel metrics present slight variation and a good tolerance to the imprecision of cut boundaries.

64.4.3 Comparisons of Our Metric and Rand Index

Basing on the subset with 3,697 manually generated segmentations and according to the metrics of UMA, FME, AME and Rand Index [3], we perform other experiments also to compare the rank of the seven algorithms for each object category, the results in Table 64.1 shows that, according to the rank of UMA, the algorithm Randomized Cuts, Shape Diameter function, Normalized Cuts are on the top three, basically identical with that of Rand Index metrics, the rank of other four algorithms are similar too, this indicates that the performance of UMA metric and Rand Index metric are almost same, but FME and AME show a slight difference to them. And there is little difference between the ranks according to the metric of UMA and Rand Index too.

From the chart of UMA, FME, AME and Rand Index in Fig. 64.4, we see that each of the four bars of all four evaluation metrics is remarkably consistent with others. Although the UMA metric is boundary-based, and other three metrics focus on region dissimilarities (FME, AME and Rand Index), all variants of all four

Table 64.1 Comparison of UMA, FME, AME and Rand Index

Metrics	Rand Cuts	Shape Diam	Norm Cuts	Core Extra	Rand Walks	Fit Prim	K Means
UMA	2	3	1	4	5	7	6
FME	1	3	2	5	4	6	7
AME	1	2	3	6	4	5	7
Rand Index	1	3	2	5	6	4	7

Fig. 64.4 Comparison of FME, AME, UMA and Random Index metrics

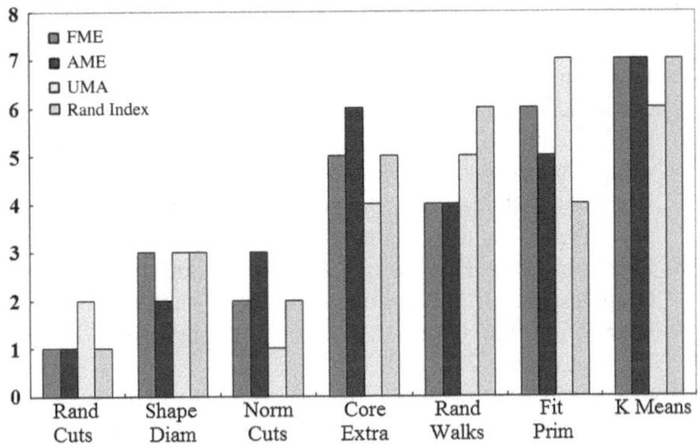

metrics suggest almost the same relative performance of the seven algorithms. The ranks of the four kinds of metrics are basically same. Because our experiments are basing a selection of the Chen's manual interactive segmentation data set, it has higher consistency, and leads to the rank of algorithms are significantly different with that in Chen et al. [3].

64.5 Conclusion

This paper presents a novel metric UMA to evaluate the algorithm of 3D mesh segmentation quantitatively, and experiments show that UMA is robust to degenerative segmentation and hierarchical refinement, stable to the imprecision of cut boundaries. Because we have excluded some manual segmentation with obvious error and serious inconsistencies, and use a data set with larger size, the UMA metric gives a more precise quantitative evaluation than others.

Acknowledgments This work is supported by National Natural Science Foundation of China (No. 61472170, No. 61170143, No. 60873110), and Beijing Key Laboratory of Intelligent Telecommunications Software and Multimedia, Beijing University of Posts and Telecommunications (No. ITSM201301).

References

1. Shamir A (2008) A survey on mesh segmentation techniques. Comput Graph Forum 27(6):1539–1556
2. Benhabiles H, Vandeborre J, Lavoue G et al (2009) A framework for the objective evaluation of segmentation algorithms using a ground-truth of human segmented 3d-models. In: Proceedings of IEEE international conference on shape modeling and applications, Beijing
3. Chen X, Golovinskiy A, Funkhouser T (2009) A benchmark for 3D mesh segmentation. ACM Trans Graph 28(3):73:1–73:12

4. Benhabiles H, Vandeborre J, Lavoue G et al (2010) A comparative study of existing metrics for 3D-mesh segmentation evaluation. Visual Comput Int J Comput Graph 26(12):1451–1466
5. Shlafman S, Talm A, Katz S (2002) Metamorphosis of polyhedral surfaces using decomposition. Comput Graph Forum 21(3):219–228
6. Lai YK, Hu SM, Martin RR, Rosin PL (2009) Rapid and effective segmentation of 3D models using random walks. Comput Aided Geom D 26(6):665–679
7. Attene M, Falcidieno B, Spagnuolo M (2006) Hierarchical mesh segmentation based on fitting primitives. Visual Comput 3(22):181–193
8. Golovinskiy A, Funkhouser T (2008) Randomized cuts for 3D mesh analysis. ACM Trans Graphics (TOG) 5(27). doi:10.1145/1409060.1409098
9. Katz S, Leifman G, Tal A (2005) Mesh segmentation using feature point and core extraction. Visual Comput 21(8–10):865–875
10. Shapira L, Shamir A, Cohen-Or D (2008) Consistent mesh partitioning and skeletonisation using the shape diameter function. Visual Comput Int J Comput Graph 4(24):249–259
11. Sun XP, Zhao XN, Li CF et al (2012) Misclassified evaluation metrics for 3D mesh segmentation. J Chin Comput Syst 33(8):1811–1815

Chapter 65
A Survey Analysis of Chinese Virtues Questionnaire in Medical Postgraduates

Miao Yu, Wei Wang, Zhilei Hu, Huibin Ji and Wei Xing

Abstract Objective: To probe the applicability conditions of Chinese Virtues Questionnaire (CVQ-96) in medical postgraduates, and to provide theoretical guidance for the establishment of corresponding policies. Method: Carrying out CVQ-96 on-site testing in elective class, using SPSS17.0 for the data analysis of the questionnaire. Results: According to the principal component analysis (PCA) results, four dimensions that gregariousness, flexibility, adaptability and principle are extracted in this paper, the scale and its dimensions presented good internal consistency reliability and construct validity during the validation among medical postgraduate. There is only gender difference of medical postgraduates in the adaptation dimension but little demographic difference in other dimensions. And there is no obvious difference in other demographic aspects. In addition, the strengths that have demographic differences among medical postgraduates are also not much. Conclusion: The results of the questionnaire can be used to analyze the distribution of character strengths among the medical postgraduates, but it still need to be revised. The relevant guidance of the aspects that have difference should be carried out to make the medical graduate students receive better education.

Keywords Medical postgraduates · Chinese virtues questionnaire · Positive psychology · Character strengths

M. Yu · W. Wang · Z. Hu · H. Ji
College of Humanities and Social Science, China Medical University, Shenyang, Liaoning, China

W. Xing (✉)
College of Sciences, Northeastern University, Shenyang 110004, Liaoning, China
e-mail: awxing@mail.neu.edu.cn

J.J. (Jong Hyuk) Park et al. (eds.), *Ubiquitous Computing Application and Wireless Sensor*, Lecture Notes in Electrical Engineering 331,
DOI 10.1007/978-94-017-9618-7_65

65.1 Introduction

Seligman put forward positive psychology in 1997, which is a way to use the relatively perfect and effective experimental method and the measuring method in psychology to study the positive aspects of human power and virtue [1]. Peterson put forward the Values in Action Classification of Strengths system, 24 character strengths were divided into six virtues, and to cultivate the 24 kinds of character strengths is the way to obtain virtue [2]. Duan created the Chinese Virtues Questionnaire (CVQ-96), according to this classification system, combining with Chinese culture. Three virtue subscales based on Chinese culture were established: Interpersonal (32 items), vitality (40 items), and cautiousness (24 items) [3].

However, the questionnaire mainly on the basis of workers and undergraduates, studies still needed to probe the potential replication [4]. Medical postgraduates is a specific group that most of them will go to the departments of hospital after graduation, their psychological status is related to the doctor-patient relationship and social stability. To investigate their strengths can help to provide theoretical guidance for the establishment of corresponding policies.

65.2 Object and Methods

65.2.1 Object

A total of 92 questionnaires were distributed randomly to the medical postgraduates of China Medical University and 86 valid questionnaires were recycled (valid return rate is 94.48 %). Among the 86 subjects, there are 37 boys and 49 girls; 49 from one-child family and 37 from non-one child family; 23 from countryside, 27 from town and 36 from city; 38 of basic research direction, 15 of humanistic research direction and 33 of clinical research direction.

65.2.2 Methods and Tools

The Chinese Virtues Questionnaire (CVQ-96), created by Duan is used for this study, it has 96 items (4-item per strength), measuring 24 strengths and 3 virtues [3]. This questionnaire uses Likert 5 points. On-site testing is given in elective class, with a unified instructions that asking participants read sentence by sentence and choose the most appropriate one from “very inconsistent” to “very consistent” five levels according to their own reality. Removing the invalid questionnaire and using SPSS17.0 for the data analysis.

65.3 Data Analysis and Results

65.3.1 The Factor Analysis of the CVQ-96

The principal component analysis (PCA) was used to extract the factors that characteristic roots greater than 1 in the 24 strengths. Four factors are extracted according to the gravel figure results and the cumulative variance contribution rate is 65.22 %. Therefore, according to the loading and adjusting with the actual situation, the 5 strengths that kindness, teamwork, love, leadership and gratitude are classified into the first dimension, named gregariousness; the 7 strengths that humor, curiosity, zest, creativity, hope, beauty and learning are classified into the second dimension, named flexibility; the 7 strengths that perspective, social, bravery, belief, regulation, modesty and perseverance are classified into the third dimension, named adaptation; the 5 strengths that fairness, authenticity, forgiveness, judgment and prudence are classified into the fourth dimension, named principle. The factor loading of the 24 strengths are shown in Table 65.1.

Table 65.1 The factor loading of the 24 strengths

	Dimension 1	Dimension 2	Dimension 3	Dimension 4
Kindness	0.069	−0.070	−0.048	−0.230
Teamwork	0.072	−0.134	−0.139	0.003
Fairness	0.069	−0.155	−0.175	0.145
Love	0.064	−0.168	−0.037	−0.107
Authenticity	0.060	−0.120	−0.041	0.331
Leadership	0.066	−0.151	−0.146	0.043
Forgiveness	0.056	−0.061	−0.069	0.272
Gratitude	0.072	−0.128	−0.091	−0.180
Humor	0.057	0.146	−0.153	0.148
Curiosity	0.063	0.120	−0.243	−0.094
Zest	0.062	0.135	−0.062	0.051
Creativity	0.056	0.224	−0.172	0.069
Perspective	0.070	0.115	0.239	−0.046
Hope	0.064	0.099	−0.085	0.041
Social	0.059	0.079	0.342	−0.223
Beauty	0.060	0.133	−0.260	−0.117
Bravery	0.077	−0.051	0.094	−0.144
Belief	0.059	−0.085	0.109	−0.283
Judgment	0.062	0.182	0.110	0.251
Prudence	0.064	0.046	0.290	0.228
Regulation	0.057	0.053	0.263	0.181
Modesty	0.059	−0.184	0.129	−0.071
Perseverance	0.061	−0.089	0.184	0.080
Learning	0.044	0.228	−0.017	−0.319

65.3.2 The Construct Validity of the CVQ-96

The author inspected the construct validity at the same time of the principal component analysis, the result of KMO is 0.88 and the significance of Bartlett is 0.00, indicated that the high structure validity in the division of four dimensions.

65.3.3 The Internal Consistency Reliability of the CVQ-96

The Cronbach's α coefficient of the whole questionnaire is 0.94; gregariousness 0.89; flexibility 0.86; adaptation 0.87 and principle 0.81. It seems that there is high reliability of the CVQ-96 in medical postgraduates, and the divide of the dimensions can stably reflect the strengths of subjects.

65.3.4 The Demographic Differences of the CVQ-96

As is shown in Table 65.2, there is gender difference in the adaptation dimension of medical postgraduates ($t = 2.57$, $p < 0.05$). However, there is little demographic difference in other dimensions.

The author made further examination of the demographic differences in the 24 strengths, and found that there are gender differences in the 5 strengths of humor,

Table 65.2 The demographic differences of the CVQ-96 in the four dimensions

		Gregariousness	Flexibility	Adaptation	Principle
Gender	Boy	80.41 ± 8.93	101.81 ± 11.58	116.30 ± 13.36	63.24 ± 6.50
	Girl	78.49 ± 9.24	96.94 ± 14.34	108.44 ± 14.40	60.86 ± 6.39
t		0.97	1.69	2.57*	1.70
One-child	Yes	79.94 ± 8.83	99.69 ± 13.95	113.08 ± 13.17	62.31 ± 6.15
	No	78.49 ± 9.51	98.16 ± 12.70	110.27 ± 15.94	61.32 ± 7.00
t		0.73	0.52	0.89	0.69
Come from	Countryside	82.04 ± 9.22	101.13 ± 11.44	115.17 ± 13.82	62.61 ± 7.31
	Town	78.33 ± 9.19	95.96 ± 17.25	108.00 ± 17.26	61.00 ± 6.42
	City	78.31 ± 8.85	100.00 ± 10.96	112.66 ± 11.92	62.08 ± 6.14
F		1.43	1.09	1.65	0.40
Major	Basic	81.05 ± 9.86	98.68 ± 14.57	112.71 ± 16.09	62.87 ± 7.03
	Humanistic	77.67 ± 10.97	102.07 ± 14.13	108.80 ± 14.15	60.93 ± 6.88
	Clinical	78.06 ± 6.96	98.06 ± 11.69	112.28 ± 12.58	61.18 ± 5.70
F		1.26	0.48	0.41	0.78
Evenly distributed		79.31 ± 9.10	99.03 ± 13.37	111.86 ± 14.42	61.88 ± 6.51

*$p < 0.05$

Table 65.3 The strengths that have gender differences in CVQ-96

		Humor	Zest	Judgment	Prudence	Regulation
Gender	Boy	15.00 ± 2.65	14.73 ± 1.91	15.22 ± 1.95	15.62 ± 2.28	13.62 ± 2.13
	Girl	13.59 ± 2.73	13.59 ± 2.22	14.18 ± 2.37	14.04 ± 2.52	12.27 ± 2.26
t		2.40*	2.50**	2.16*	3.00**	2.83**

*p < 0.05, **p < 0.01

Table 65.4 The strengths that have differences whether one-child or not in CVQ-96

		Humor	Social
One-child	Yes	14.73 ± 3.11	13.88 ± 2.59
	No	13.49 ± 2.08	12.65 ± 2.51
t		2.23*	2.21*

*p < 0.05

Table 65.5 The strengths that have differences among countryside, town and city in CVQ-96

		Regulation	Learning
Come from	Countryside	13.70 ± 2.42	14.52 ± 2.64
	Town	11.48 ± 1.85	12.00 ± 3.46
	City	13.33 ± 2.10	13.72 ± 2.55
Levene		0.866	1.395
F		8.42**	5.13**

**p < 0.01

zest, judgment, prudence and regulation. Humor and judgment have significant gender differences; and zest, prudence and regulation have very significant gender differences. The results are shown in Table 65.3.

There is no significant difference between one-child and non-one child in the four dimensions, but a significant difference in humor and social strengths, and the scores in the two strengths are both one-child higher than non-one child. The results are shown in Table 65.4.

There is no significant difference among those come from countryside, from town and from city in the four dimensions, but very significant differences in regulation and learning strengths. The LSD test is used for the back testing, and the results show that the differences lie in those from towns and those from countryside or cities in the regulation strength. It is as the case of the LSD test results in learning strength. The results are shown in Table 65.5.

There is no significant difference among the direction in the four dimensions, but significant differences in fairness and leadership strengths. The LSD test is used for the back testing of the fairness strength, and the results show that the differences lie in those of basic direction and those of clinical direction. The Tamhane test is used for the back testing of the leadership strength, and the results show that the differences also lie in those of basic direction and those of clinical direction. The results are shown in Table 65.6.

Table 65.6 The strengths that have differences among directions in CVQ-96

		Fairness	Leadership
Direction	Basic	16.55 ± 2.02	16.13 ± 2.28
	Humanistic	15.40 ± 2.59	14.87 ± 2.80
	Clinical	15.45 ± 1.91	14.85 ± 1.72
Levene		1.929	3.142*
F		3.03*	3.62*

*$p < 0.05$

65.4 Discussion

The analysis results of this study show that CVQ-96 has good reliability and validity in medical postgraduates. However, the structure still need to be further discussed. The PCA results show that factors for medical postgraduates have certain discrepancy with the standard CVQ-96 dimensions [5]. This may be due to the unique group characteristics of medical postgraduates.

There is only gender difference of medical postgraduates in the adaptation dimension, but little demographic difference in other dimensions. And there is no obvious difference in other demographic aspects. This may be due to that the medical postgraduates have certain heterogeneity, or that the compiling of CVQ-96 mainly depends on workers and undergraduates and the distinction degree of the questions in medical postgraduates is insufficient.

65.5 Conclusions

The scale and the division of the four dimensions present good internal consistency reliability and construct validity among medical postgraduate. However, it still needs to be revised. According to the PCA results, four dimensions that gregariousness, flexibility, adaptability and principle are extracted in this paper, and this result is consistent with Badar, Kashdan and others scholars [6].

The scores of girls are lower than that of boys in the adaptability dimension, and it can be a theoretical basis for relevant departments to carry out the adaptive ability cultivating for medical postgraduate girls. Gender difference is also reflected in humor, zest, judgment, prudence, regulation, etc.; The scores of non-one child are lower than the one-child in humor and social intelligence dimension; The scores of those from towns are lower than those from countryside or cities in regulation and learning dimension; The scores of those of clinical direction are lower than those of basic direction in fairness and leadership dimension. The relevant guidance on those aspects should be carried out to make the medical postgraduates receive better education.

Acknowledgments This paper was supported in part by the Shenyang Special Research Project (SDSZ2014-047) and the Project of CMU (YDJK2012004).

References

1. Sheldon M, King L (2001) Why positive psychology is necessary. Am Psychol 56(3):216–2171
2. Peterson C, Seligman MEP (2004) Character strengths and virtues: a handbook and classification. Oxford University Press, American Psychological Association, New York, Washington
3. Duan WJ, Ho SMY, Bai Y, Tang XQ, Zhang YH, Li TT, Yuen T (2013) Psychometric evaluation of the Chinese virtues questionnaire. Res Soc Work Pract 00:1–10
4. Duan WJ, Ho SMY, Bai Y, Tang XQ, Zhang YH, Li TT, Yuen T (2012) Factor structure of the Chinese virtues questionnaire. Res Soc Work Pract 22(6):680–688
5. Duan WJ, Li TT, Zhang YH (2011) Research progress of values in action the inventory of strength questionnaire and its application. Chin J Clin Psychol 19(2):205–208
6. Badar I, Kashdan TB (2010) Character strengths and well-being in Croatia: an empirical investigation of structure and correlates. J Res Pers 44(1):151–154

Chapter 66
Practice and Exploration of All-in-English Teaching of Compiler Principles

Xinxin Liu and Hongyun Xu

Abstract All-in-English teaching (AET) is a pedagogical reform performed by many universities of China. In this paper, we explore all-in-English teaching of compiler principles, which is a disciplinary basic course of the computer science curriculum. Teaching goals of compiler principles are analyzed and an action learning approach to all-in-English teaching of compiler principles is explored. We analyze the effect of all-in-English teaching of compiler principles through our 3 years of teaching experience.

Keywords All-in-English teaching · Compiler principles · Action learning

66.1 Introduction

All-in-English teaching (AET) is a pedagogical reform performed by many universities of China. In AET model, all teaching materials are given in English and English is used as the teaching language for non-language courses. The purpose of AET is to lead the way to the cultivation of students with international view and international communication ability. In 2011, our university began the construction of all-in-English courses as well as all-in-English specialities. Computer science and technology is one of all-in-English specialities and compiler principles is one of the disciplinary basic courses of the computer science curriculum.

The course of compiler principles introduces the fundamental principles of compiler construction and the basic techniques of realization. It covers lexical analysis, syntax analysis, semantic analysis, runtime environments, code generation,

---

X. Liu (✉) · H. Xu
School of Computer Science and Engineering, South China University of Technology, Guangzhou 510640, China
e-mail: csliuxx@scut.edu.cn

H. Xu
e-mail: hongyun@scut.edu.cn

J.J. (Jong Hyuk) Park et al. (eds.), *Ubiquitous Computing Application and Wireless Sensor*, Lecture Notes in Electrical Engineering 331,
DOI 10.1007/978-94-017-9618-7_66

and optimization. Upon completion of this course, students should be able to understand the theoretical foundation of compiler principles and know the basic organization and operations of a compiler. They will as well have the experience of implementing a full compiler for a simplified programming language in this course.

We took all-in-English teaching project of our university in 2011 and began to perform AET of compiler principles from then on. In the following 3 years, we keep exploring the better ways of teaching this speciality course in total English. Compiler principles is a course of rigid logic and abstraction. Even teaching and learning it in the native language, students always find this course particularly challenging.

There are two major goals of learning compiler principles. On the one hand, students are expected to grasp the basic techniques of compiler construction. On the other hand, and the more important for learning this course is to use these techniques to solve many other problems in computing. Compiler construction involves the basic ideas and methods of problems abstracting and solving in computing. How to help students gain the ability of computational thinking is a big challenge, especially when teaching is not given in the native languages. With the experience of teaching compiler principles for many years, we develop an action learning approach to all-in-English teaching of this course, and have received certain effect.

The rest of this paper is organized as follows. We begin with a discussion of the teaching of compiler principles in top universities in Sect. 66.2. Action learning approach to the teaching of compiler principles is proposed in Sect. 66.3, and the cultivation of computational thinking in the teaching of compiler principles is presented in Sect. 66.4. The evaluation of all-in-English teaching is summarized in Sect. 66.5. We conclude the paper in Sect. 66.6.

66.2 Teaching of Compiler Principles in Top Universities

In this section, we survey the teaching of compiler principles in top universities like MIT and Stanford through their open course projects. The teaching content, teaching approaches and tools are investigated.

Both Stanford online course and MIT open courseware provide courses on compiler principles. In Stanford, the course is called compiler offered by Prof. Alex Aiken [2] and in MIT the course is called computer language engineering offered by Prof. Saman Amarasinghe and Prof. Martin Rinard [1].

The course computer language engineering of MIT analyzes issues associated with the implementation of higher-level programming languages. Topics covered include: fundamental concepts, functions, and structures of compilers, the interaction of theory and practice, and using tools in building software. The course includes a multi-person project on compiler design and implementation. Students in this course write a compiler for the Decaf language, a simple imperative language similar to C or Pascal. Teaching process includes lectures, recitations, projects and quizzes.

The course compiler of Stanford discusses the major ideas used today in the implementation of programming language compilers, including lexical analysis, parsing, syntax-directed translation, abstract syntax trees, types and type checking, intermediate languages, dataflow analysis, program optimization, code generation, and runtime systems. An optional course project is to write a complete compiler for COOL, the Classroom Object Oriented Language. Teaching process includes lectures, quizzes, exams and projects.

The biggest difference between the courses on compilers offered by MIT and Stanford is that the Stanford's course is in the self-service mode while the MIT's course is in the teaching model. Therefore in the compiler course of Stanford, there are in-lecture questions to answer and exercises done and checked on-line. An on-line discussion forum is set up in which students can ask questions and receive answers too.

66.3 Action Learning of Compiler Principles

Action learning is an educational process whereby people work and learn together by tackling real issues and reflecting on their actions [4]. Learners in action learning acquire knowledge through actual actions and practice rather than through traditional instruction.

Compiler principles is deemed by students to be a hard course because of its abstraction. Even teaching in the native language, seldom students can really grasp the essence. The tough problem of all-in-English teaching is to evolve a suitable approach to teaching this course.

The traditional teaching model of compiler principles is that theory should be first taught with lectures and then verified in laboratory exercises. Chinese philosopher Confucius (551 BC to 479 BC) stated "I hear and I forget. I see and I remember. I do and I understand". In order to help students gaining a better understanding of the compiling theories and make good use of English as a tool, we adopt the action learning approach where theory is regarded as something which must be built by each individual. In our approach, teaching materials are structured around the construction of a real compiler for a small language. Students begin the course with a compiler writing project. During the process of compiler construction, the theories and techniques needed are illustrated as supporting activities. In this learning process, knowledge is created within the students' minds rather than transferred from the teacher.

The framework of action learning of compiler principles is shown in Fig. 66.1. In this framework that goes from the concrete to the abstract, students reflect on and review the action they have taken and the learning points arising.

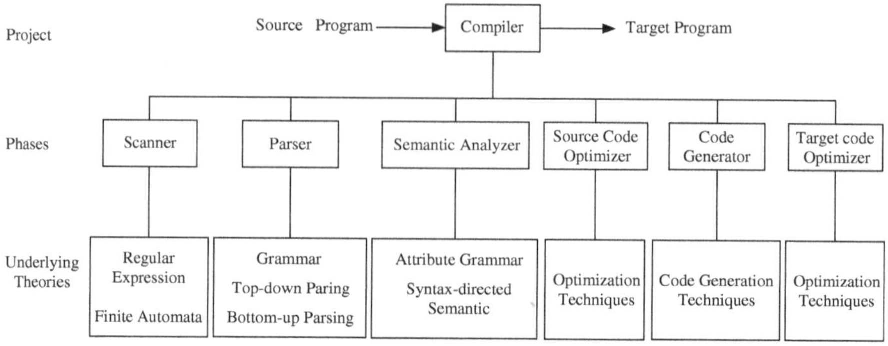

Fig. 66.1 Action learning of compiler principles

66.4 The Cultivation of Computational Thinking

Because many disciplines of computer science are employed in a compiler, learning this course will improve the ability of students on analyzing and solving problems systematically. In the compiler textbook "Compilers: Principles, Techniques, and Tools" [3], Aho et al. write that: "the principles and techniques of compiler writing are so pervasive that the ideas found in this book will be used many times in the career of a computer scientists." In all-in-English teaching of compiler principles, we try to reveal not only the techniques of compiler construction themselves but also the ideas underlying the techniques. Students are expected to know both what it is and why it is. For example, it is less useful to know the sequence of logical steps needed to derive a specific result of the parsing method than lead students going through the way in which computer scientists develop the idea and design the method.

The basic computational methods used in compiler construction are given in Table 66.1. Learning these methods will spur students' computation thinking and improve their innovation ability.

66.5 The Evaluation of All-in-English Teaching

We have conducted all-in-English teaching of compiler principles in the last 3 years. We find the application of the action learning approach is effective even education happened not in the native language. Performance improves on attendance, on engagement and on learning. In Fig. 66.2, we compare the teaching effect of all-in-English teaching (AET) and the traditional teaching model (TTM) in 3 years. It turns out that AET is effective in boosting students' scores. AET dramatically reduces the percentage of the total number of students who fail the exam and increase the percentage of outstanding students. What's more, students have a

Table 66.1 Computational methods used in compiler construction

Compiler phases	Computational methods
Scanner	Formal description, formal methods
Parser	
Semantic analyzer	
Scanner	Pattern recognition
Parser	Top-down method, bottom-up method, iteration and recursion, tree manipulation
Code optimizer	Methods of optimization
Putting it all together	Language recognition
	The ability to learn new programming languages quickly
	Learn to design and use some useful tools
	Build a large project in a team

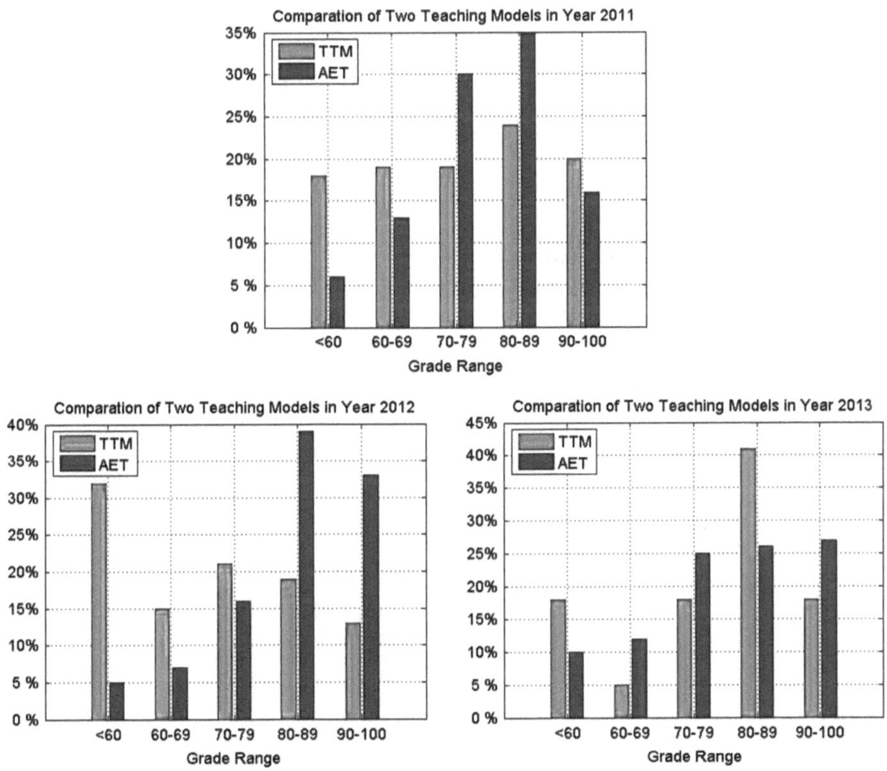

Fig. 66.2 Comparison of all-in-English teaching and traditional teaching

better understanding of compiler and more satisfaction with course. Investigation reports that this course get 4.786 points in 2011, 4.765 points in 2012, and 4.8 points in 2013 with the full mark is 5. This course ranks the forth in 2011, the sixth in 2012, and the third in 2013 among more than thirty courses of our school.

66.6 The Conclusion

In this paper, we showed our practice and exploration of all-in-English teaching of compiler principles. We analyzed the teaching goals of this course that is to provide students with not only the basic techniques of compiler construction but also the appropriate problem solving skills and computational thinking. To implement these high level goals in an actual curriculum, we explored an action learning approach to all-in-English teaching. The teaching effect is evaluated through our teaching experiences.

Acknowledgments This work is partially supported by the Quality and Educational Reform Project of Guangdong Institutions of Higher Education (Grant No. N912059a).

References

1. http://ocw.mit.edu/courses/electrical-engineering-and-computer-science/6-035-computer-language-engineering-spring-2010/index.htm
2. https://www.coursera.org/course/compilers
3. Aho AV, Lam MS, Sethi R, Ullman JD (2007) Compilers: principles, techniques, and tools, vol 1009. Pearson/Addison Wesley, Boston
4. Boshyk Y, Dilworth RL (2010) Action learning: history and evolution. Palgrave Macmillan, Basingstoke

Chapter 67
An Application of Ant Colony Optimization Clustering Approach for Primary Headache Diagnosis

Wu Yanping and Duan Huilong

Abstract In this study, an Ant Colony Optimization (ACO) clustering approach is proposed for medical diagnosis to assign patients into different primary headache groups with Visual C++ program. By experiments, 375 patients were classified into Migraine, Tension-Type Headache (TTH) and Trigeminal Autonomic Cephalalgias (TACs). Results show that the clustering algorithm with ant colony can be important supportive for the medical experts in diagnostic.

Keywords Ant colony algorithm · Clustering · Diagnosis · Primary headache

67.1 Introduction

Clustering, a method aiming to discover sensible organization of objects in a given dataset by identifying and quantifying similarities or dissimilarities between the objects [1], is applied generally in medical diagnosis. Due to its strategic importance, several algorithms have been proposed in literature to solve clustering problems. Ant Colony Optimization (ACO) clustering algorithm [2], firstly proposed by Shelokar in 2004, is an approach to solve a clustering problem based on minimizing the sum of squared Euclidean distance between each object and cluster center. It is a fast suboptimal heuristic algorithm and achieve intelligent outcome

This Article is part of the issue entitled: Zhejiang medical and health science and technology project (No. 2013KYA050).

W. Yanping (✉) · D. Huilong
College of Biomedical Engineering and Instrument Science, Zhejiang University, Hangzhou 310027, China
e-mail: wyp@zjmc.net.cn; w_yp@zju.edu.cn

W. Yanping
Public Health Department, Zhejiang Medical College, Hangzhou 310053, China

J.J. (Jong Hyuk) Park et al. (eds.), *Ubiquitous Computing Application and Wireless Sensor*, Lecture Notes in Electrical Engineering 331,
DOI 10.1007/978-94-017-9618-7_67

through ant pheromone communication iteratively and the pheromone-meditated cooperative search process.

Primary headache, including three main categories: migraine, tension-type headache (TTH) and trigeminal autonomic cephalalgias (TACs) according to The International Classification of Headache Disorders, 3rd edition (beta version) (ICHD, 3rd), is one of the most common disorder for neurological consultation. While the diagnostic criteria developed by the International Headache Society (IHS) have been widely validated, there are only few studies [3, 4] with expert systems or intelligent technologies that would help physicians make diagnoses.

In this work, ACO clustering method was performed to classified primary headache patients into three clusters with program in Visual C++ tool. As a result of clustering algorithms, patients' statues are identified as Migraine, TTH and TACs that demonstrated the ACO clustering algorithm can increase identification sensitivity in medical diagnosis.

67.2 Ant Colony Optimization Clustering Algorithm

Ant Colony Optimization-Sacc Clustering (ACO clustering) algorithm [2], as a classic for ant colony clustering approach, is proposed for medical diagnosis in this work. The approach considers a given dataset of N patients $\{x_1, x_2, \ldots, x_N\}$ in $\mathcal{R}^n$-dimensional symptoms to be partitioned into a number of K disease groups. Here the process of ant searching food is the progress to classify diseases and mainly includes three stages. On the first stage, R number of solution strings S is constructed for the same number of ant. Each solution string S is in length of N where each element value corresponds to each patient disease cluster number. To construct the solution, a pheromone matrix of size $N \times K$ is used. Trail value τ_{ij} at location (i, j) in matrix represents the pheromone concentration of patient x_i associated to the jth $(j = 1, \ldots, K)$ disease cluster. After generating a population of R trial solutions of R number of ant agent, a local optimal search procedure based on same process is performed to further improve fitness of the best selected L solutions at the second stage. Finally, the pheromone trail matrix is updated through the best L solutions. With the progress of iteration, the pheromone matrix is modified and solutions improved in terms of the value of objective function continually until a certain number T of iterations and solution having lowest function values which mean the optimal partitioning of patients into several disease groups. Figure 67.1 showed the flow chart of ACO clustering algorithm-Sacc with main steps as following.

Step 1. Define integer $t(t = 1, \ldots, T)$ for iteration, integer $r(r = 1, \ldots, R)$ for agent, string S of length N in number R for solution and pheromone matrix τ. Initialize iteration parameter $t = 1$ and pheromone matrix value $\tau(t) = \tau(0)$ where $\tau(0)$ is generated randomly from uniform distribution in the range between 0 and 1.

Step 2. Send solution string S with empty of length N and $r = 1$ at the start.

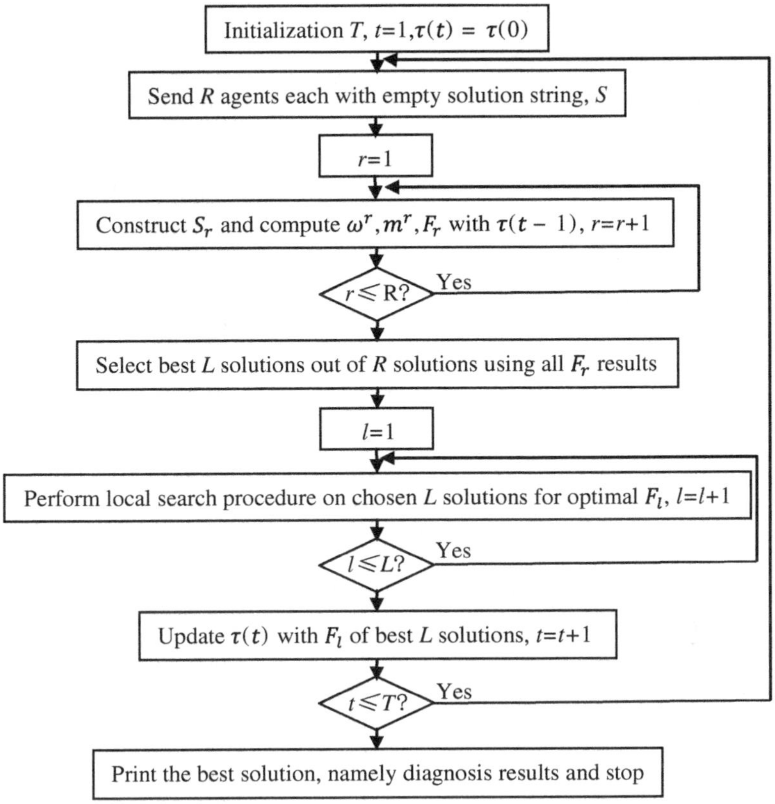

Fig. 67.1 The flow chart of ACO clustering algorithm

Step 3. Construct the rth ($r = 1, 2, \ldots, R$) solution S_r. Each element value as possible cluster number in the solution string is generated by one of the following ways according to the pheromone matrix updated at the end of iteration $t - 1 (t = 1, \ldots, T)$.

- Cluster having the maximum $\tau(t-1)$ is chosen when $\tau_{ij}(t-1) \leq q_0$ (q_0 being a priori defined number).
- Cluster having the maximum $p_{ij}(t-1)$ is chosen when $\tau_{ij}(t-1) > q_0$. $p_{ij}(t-1)$ is a normalized pheromone probability value given by Eq. (67.1).

$$p_{ij}(t-1) = \frac{\tau_{ij}(t-1)}{\sum_{k=1}^{K} \tau_{ik}(t-1)}, j = 1, 2, \ldots, K \tag{67.1}$$

Step 4. Compute objective function F_r of solution S_r using following equations.

$$MinF_r(\omega^r, m^r) = \sum_{j=1}^{K}\sum_{i=1}^{N}\sum_{v=1}^{n} \omega_{ij}^r \left\| x_{iv} - m_{jv}^r \right\|^2 \quad (67.2)$$

where, x_{iv} is a value of vth symptom of ith patient; m a cluster center matrix of size $K \times n$; m_{jv}^r an average attribute value of the vth symptom in all patients in the cluster j; ω a weight matrix of size $N \times K$; ω_{ij}^r an associated weight of patient x_i with disease j which can be assigned as Eq. (67.3) based on element values of solution S_r.

$$\omega_{ij}^r = \begin{cases} 1, \textit{if } i \textit{ belongs to cluster } j \\ 0, \textit{otherwise} \end{cases}, i = 1, \ldots, N, j = 1, \ldots, K \quad (67.3)$$

$$m_{jv}^r = \frac{\sum_{i=1}^{N} \omega_{ij}^r x_{iv}}{\sum_{i=1}^{N} \omega_{ij}^r}, j = 1, \ldots, K, v = 1, \ldots, n \quad (67.4)$$

Step 5. Make $r = r + 1$ and judge iterations for solution string S construction. If $r \le R$ go to step 3, else stop.

Step 6. Choose the best L solution $S_l (l = 1, \ldots, L)$. Define a probability threshold p_{ls} in [0,1], a temporary solution string S_{Tem} of length N, integer parameter $l(l = 1, \ldots, L)$ and give $l = 1$.

Step 7. Assign $S_{Tem}(i) = S_l(i), i = 1, \ldots, N$. For each $S_{Tem}(i)$, draw a random number r_{Tem} in (0,1). If $r_{Tem} \le p_{ls}$, an integer j in the range $(1,K)$, such that $S_l(i) \ne j$ is randomly selected and let $S_l(i) = j$. Calculate cluster centers, weights and its objective function value F_{Tem} associated with solution string S_{Tem} with above Eqs. (67.2–67.4). If $F_{Tem} < F_l$, then $S_l = S_{Tem}$ and $F_l = F_{Tem}$. Redefine $l = l + 1$. If $l \le L$ continue this step 7, else stop.

Step 8. Update the pheromone matrix $\tau(t)$ with the best L solution on the local search operation results using Eq. (67.5).

$$\tau_{ij}(t) = (1 - \rho)\tau_{ij}(t - 1) + \sum_{l=1}^{L} \Delta\tau_{ij}^l, i = 1, \ldots, N, j = 1, \ldots, K \quad (67.5)$$

Where, $(1 - \rho)$ denotes evaporation rate. The amount $\Delta\tau_{ij}^l$ is equal to $1/F_l$, if cluster j is assigned to ith data point of the solution built by ant l and zero otherwise.

Step 9. Let $t = t + 1$. If $t \le T$ continue to step 2, else stop. Transfer the best solution string S to be the medical diagnosis results.

67.3 Primary Headache Diagnosis with ACO Clustering Approach

Headaches, a general chronic disease, are broadly divided into primary and secondary. And primary headache disorders constitute the vast majority of headache disorders, with migraine and TTH being the most prevalent that affects 75–95 % of the headache population totally while TACs headache is uncommon, but often misdiagnosed and mismanaged [5].

Since the features of headache are more than twenty, an important first step in the diagnostic process is to distinguish primary headaches on the basis of their important symptoms. According to research [6], 13 headache features are chosen as diagnostic variables due to their influence on primary headache. These features and symptoms include headache history, monthly days, attack duration, frequency, sensory pain, pain intensity, pain site, whether or not a new headache, headache attack season, attack daily time, movement worsen, headache aura and accompany symptoms. 375 patients from certain hospital in China are chosen for diagnosis. Class distribution or number of instance for migraine is 213, for TTH is 127 and for TACs is 35.

ACO clustering algorithm is used to assign the patients to three different groups of primary headache as migraine, TTH and TACs. The program was developed by Visual C++ tool based on ACO clustering algorithm (Fig. 67.1). Experiments have demonstrated parameters in ACO clustering would be assigned as values in Table 67.1. The application results is in Table 67.2 and its final results of headache patients in distribution is given in Fig. 67.2.

Table 67.1 Parameters' values in ACO clustering

Parameter	N	n	K	R	T	L	ρ	q_0	p_{ls}
Value	375	13	3	2	20,000	100	0.1	0.99	0.01

Table 67.2 ACO clustering algorithm results

Primary headache	Sum	Correct classified results	Correct rate (%)
Migraine	213	190	89.2
TTH	127	107	84.3
TACs headache	35	30	85.7

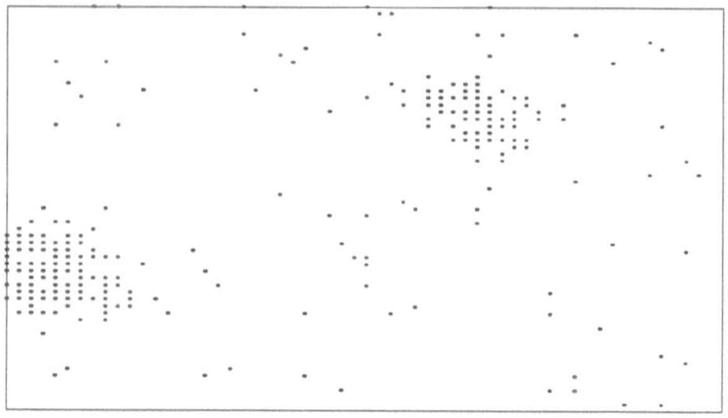

Fig. 67.2 The distribution of headache patient dataset in clustering

67.4 Conclusion

Headache is common in working people and represents a significant health as well as social and economic problem. By the application of IHS criteria, physicians' experience was used to normalize headache features. The study presents ACO clustering algorithm to classify primary headache types. The research demonstrated ACO clustering approach is a promising support in diagnostic decision making in medicine.

References

1. Han J, Kamber M (2012) Data mining concepts and techniques. Bumaby: Morgan Kaufmann Publishers Inc., 3rd ed, pp 89–103
2. Shelokar PS, Jayaraman VK, Kulkarni BD (2004) An ant colony approach for clustering. Anal Chim Acta 509:187–195
3. Svetlana S, Dragan S, Petar S (2008) Computer-assisted diagnosis of primary headache. HAIS, pp 314–321
4. Jeong Y, Kyung SH, Sun YO (2011) An application of interval-valued intuitionistic fuzzy sets for medical diagnosis of headache. Int J Innovative Comput Inf Control 5(7):2755–2762
5. Fayyaz A (2012) Headache disorders: differentiating and managing the common subtypes. The British Pain Society 6(3):124–132
6. Lora A, Hasse P, Neal R, Robert S (2012) Quantifying headache symptoms and other headache features from chart notes. Headache 44:873–884

Chapter 68
Comparisons of Iterative Closest Point Algorithms

Lu Wang and Xiaopeng Sun

Abstract With the huge advancement in shape scanning and acquisition technology, 3D shape matching becomes one of hotspots in machine vision. This paper introduces the classical ICP together with Sparse ICP, EM-ICP and ICNP algorithms, which respectively refined ICP with weighted factor, closest normal vector as well as sparsity inducing norms, discusses their improvements and insufficient, and gives the direction of future research.

Keywords Registration · ICP · Closest normal vector · Sparsity · EM algorithm

68.1 Introduction

Registration of 3D point cloud data directly affects the subsequent geometric modeling and related algorithm robustness [1]. 3D registration is generally divided into initial coarse registration and fine registration. The most often used fine registration method is ICP (Iterative Closest Point) algorithm, which is used to address lots of data form, e.g. 3D point set, curves, implicit surfaces and parametric surfaces, etc. [2]. And many algorithms to improve ICP are presented successively [3].

This paper deeply analyzed and compared ICP algorithm, EM-ICP (Expectation-Maximization) algorithm [4], Sparse ICP (Sparse Iterative Closest Point) algorithm [5] and ICNP (Iterative of Closest Normal Point) algorithm [6], and investigated weighting factor, closest normal vector, sparsity and other factors on ICP algorithm, then summarizes the advantages and disadvantages of these four algorithms.

L. Wang · X. Sun (✉)
Computer Systems Institute, Liaoning Normal University, Dalian 116029, China
e-mail: cadcg2008@gmail.com

J.J. (Jong Hyuk) Park et al. (eds.), *Ubiquitous Computing Application and Wireless Sensor*, Lecture Notes in Electrical Engineering 331,
DOI 10.1007/978-94-017-9618-7_68

68.2 Classical ICP Algorithm

In 1992, in order to find space transform of two 3D data points sets and achieve data point set spatial registration, Besl et al. [7] proposed a precision registration algorithm which based on geometric transformation and free-form surface approximation-iterative closest point (ICP) algorithm. This algorithm was fast becoming the mainstream of the registration algorithm. ICP algorithm is based on the least squares method, which is a precise rigid optimal registration algorithm.

Given two sets of 3D point cloud models X and Y, which $X = \{x_i, i = 1, 2, \ldots, N_x\}$ and $Y = \{y_j, i = 1, 2, \ldots, N_y\}$, let x_i and y_j represent the position vector of points in three-dimensional space, N_x and N_y represent the number of reference model and target model, respectively. Let R represent the rotation transformation in 3D space, and let t represent the translation transformation in 3D space, then the error function between model X and Y can be written:

$$E = \sum_{i=1}^{N_x} \sum_{j=1}^{N_y} \left\| Rx_i + t - y_j \right\|_2^2 \tag{68.1}$$

Let corresponding pairs (x_i, y_j) represent the point y_j which on the target model with the smallest distance to the point x_i which on the reference model. Calculating iteratively all corresponding pairs between X and Y, using SVD method to solve the rotation transformation R and translation transformation t between corresponding pairs, until the error function E is less than a specified threshold τ, in order to get the best registration of the data set X and Y.

ICP Algorithm can obtain accurate registration, handle these surfaces which is expressed by forms including 3D point, parametric surfaces, etc. scilicet, and with a good initial transformation, or it will converges slowly, only converge to a local minimum, produces a certain amount of false corresponding points. The corresponding points search is a high computational cost. And it is a rigid transformation, so it is not sensitive to expression faces and etc.

68.3 Improved ICP Algorithms

Sparse ICP, EM-ICP and INCP introduce sparsity, weighting factor and the closest normal vector to improve ICP algorithm.

68.3.1 Sparse ICP

In 2013, Bouaziz et al. [5] proposed the Sparse ICP algorithm, which omits artificial interacting tedious steps and improves the robustness.

Let $\varphi(x, y) = \phi(\|x - y\|_2)$, $\phi(r) = |r|^p$, $p \in [0, 1]$, and $z_i = Rx_i + t - y_i$, then the l_p norm alignment error between model X and Y can be written as:

$$E = \sum_{i=1}^{N_y} \|z_i\|_2^p, \quad \delta_i = 0 \tag{68.2}$$

where $\delta_i = Rx_i + t - y_i - z_i$. Iteratively optimize the registration using the Lagrange method until the minimizing error function E is less than threshold τ, and then l_p can be written as:

$$L_A(R, t, Z, \Lambda) = \sum_{i=1}^{n} \|z_i\|_2^p + \lambda_i^T \delta_i + \frac{\mu}{2} \|\delta_i\|_2^2 \tag{68.3}$$

where $\Lambda = \{\lambda_i \in R, i = 1, \ldots, n\}$ is a set of Lagrange multipliers, and $\mu > 0$ is the penalty weight. Then using the alternating direction method of multipliers, the Eq. (68.3) is effectively decomposed into three simple sub-problems:

(a) $\underset{Z}{\operatorname{argmin}} \sum_i \|z_i\|_2^p + \frac{\mu}{2} \|z_i - h_i\|_2^2$

(b) $\underset{R,t}{\arg\min} \sum_i \|Rx_i + t - c_i\|_2^2$

(c) $\lambda_i = \lambda_i + \mu\delta_i$

where $c_i = y_i + z_i - \frac{\lambda_i}{\mu}$, and $h_i = Rx_i + t - y_i + \frac{\lambda_i}{\mu}$. The shrinkage operator is used to solve the above three sub-problems.

The l_p norm, instead of the Euclidean distance, maximizes the number of zero distance between correspondences, and then avoid local alignment and other problems. The algorithm uses the Lagrange method to redefine the error function, solving the optimization problem caused by the alignment of the model is non- smooth and non-convex, improve the accuracy of the algorithm, and the algorithm is more reliable.

68.3.2 EM-ICP

In 2002, Granger et al. [8] proposed EM-ICP algorithm to directly register the large transformational model without the initial alignment. Basing on GPU, Tamaki et al. [4] divided the EM-ICP algorithm into vector and matrix operations in 2010, and accelerate the speed with CUDA kernel function. EM-ICP algorithm can obtain the best 3D transformation R and t to get the minimum of error function E:

$$E = \sum_{i=1}^{N_x} \sum_{j=1}^{N_y} \alpha_{ij} d_{ij}^2 \tag{68.4}$$

where $d_{ij} = \left\| Rx_i + t - y_j \right\|_2$, and $\alpha_{ij} = \frac{1}{C_i} \exp\left(\frac{-d_{ij}^2}{\sigma_p^2} \right)$ is the probability of x_i matches to y_j, and $C_i = \exp\left(\frac{-d_0^2}{\sigma_p^2} \right) + \sum_{k=1}^{N_x+1} \exp\left(-\frac{-d_{ik}^2}{\sigma_p^2} \right)$, σ_p and d_0 are constants.

EM-ICP algorithm applies EM algorithm to ICP using multiple matches weighted by normalized Gaussian weights and the match probability on each point in models to prevent local minimum. And it omits the initial registration step and improves the robustness of the algorithm. Comparing with ICP, it's more suitable for large-scale data in rigid registration and more accurate.

68.3.3 ICNP

In 2013, Mohammadzade [6] proposed ICNP algorithm, to address the challenge of 3D registration of local change.

Let $\{\rho_i\}$ and $\{\pi_i\}$ represent the normal vectors of the reference model X and target model Y, and point set $\{x_i^c\}$ of X with the smallest distance to each point on Y. Then find the point set $\{x_i^{cn}\}$ in $\{x_i^c\}$ that their normal vectors have the smallest angle to that of the target point, the approximate smoothed closest normal point is noted as $x_i^{cn'}$. Finally, select the closest normal point $x_i^{cn''}$ with the smallest distance to each point $x_i^{cn'}$ on X:

$$x_i^{cn''} = x_k, \quad k = \underset{1 \le j \le N_x}{\arg\min} \left(\left\| x_i^{cn'} - x_j \right\|_2 \right) \tag{68.5}$$

Updating $X = RX + t$ until the sum of the squared differences between the current rotation matrix and the identity matrix is less than a threshold, otherwise select the point on the reference model X with the smallest distance to each point on the target model Y again.

Iterative closest normal point (ICNP) algorithm is another improved ICP algorithm, whose essence is the sample of reference model according to the feature of target model point set. ICNP algorithm can effectively help identify corresponding point set that having the same characteristics in all reference models which are called the closest normal points (CNPs) and also help avoid some rigid registration algorithms only applied almost invariant regions, such as the regions under expression variation can't achieve the accurate registration results. However, using ICNP algorithm obtained CNPs cannot judge the quality of registration between models. It needs to be applied to the DA method, and classifying the CNPs of all reference models by minimizing the within-class variability of the model while maximizing the between-class variability.

68.4 Results and Analysis

According to the analysis of ICP algorithm, EM-ICP algorithm, Sparse ICP algorithm and ICNP algorithm, we apply the four ICP methods to registration of 3D ear clouds. The ear cloud data are scanned with degree of 45°, 60°, 90° and 135°, we registry the ear data of 45° to that of 45°, 60°, 90° and 135° respectively, the computing time, registry error and results are shown in Table 68.1. And some comparisons of these algorithms on their stages and technologies are summarized in Table 68.2.

According to Table 68.2, ICP algorithms and Sparse ICP algorithms all use the Euclidean distance to calculate the closest point of reference model at any point on the target model. However, in solving the error function, Sparse ICP algorithms using sparsity inducing norms replacing the Euclidean distance. EM-ICP algorithms and ICNP algorithms respectively use the weighting factor and Euclidean distance combining with the normal vector angle to change the method by which ICP algorithms finds corresponding points. Comparing with other three algorithms, ICNP algorithms can output the normal point set. The advantages and disadvantages of four ICP algorithms are shown in Table 68.3 as following.

Table 68.1 Registration comparison of time, error and results

		45°–45°	45°–60°	45°–90°	45°–135°
Time (s)	ICP	1473.002	1780.146	1036.520	1315.633
	SICP	543.788	843.293	862.006	911.362
	ICNP	2489.621	2408.834	6482.670	6879.124
	EM-ICP	1.07643	1.08359	1.03587	0.65621
Error (mm^2)	ICP	45657.243	5293.221	19519.271	726493.074
	SICP	35251.470	1964.168	14832.862	565261.402
	ICNP	51571.221	13586.843	25599.155	745948.639
	EM-ICP	51775.881	77216.344	114863.948	945725.893
Results	ICP				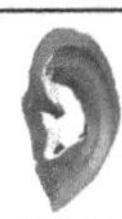
	SICP				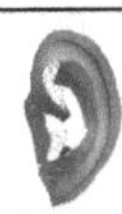
	ICNP	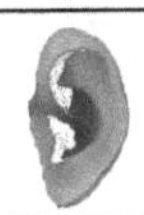			
	EM-ICP				

Table 68.2 Comparison of four kinds of ICP algorithm

	Determine the corresponding point	Error function
ICP	Euclidean distance	$E = \sum_{i=1}^{N_x} \sum_{j=1}^{N_y} \|\|Rx_i + t - y_j\|\|_2^2$
Sparse ICP	Euclidean distance	$E = \sum_{i=1}^{N_y} \|\|z_i\|\|_2^p$
EM-ICP	Weighting factor	$E = \sum_{i=1}^{N_x} \lambda^2 \|\|Rx_i + t - y_i'\|\|_2$
ICNP	Euclidean distance and normal vector angle	$E = \sum_{i=1}^{N_x} \sum_{j=1}^{N_y} \|\|Rx_i + t - y_j\|\|_2^2, \arccos(\rho_i \cdot \pi_i)$

Table 68.3 Comparison of advantages and disadvantages

Algorithm	Advantage	Disadvantage
ICP	✓ Accurate registration results	✓ Need a good initiation
	✓ Can handle curved surface	✓ Sensitive to outliers
	✓ A fast convergence	✓ Insensitive to local region
Sparse ICP	✓ Deal with noise and outliers	✓ Need a good initiation
	✓ Solve non-convex, non-smooth phenomenon	✓ Insensitive to local region
	✓ Decompose error function into sub-problem	
EM-ICP	✓ Deal with noise and outliers	✓ Complex corresponding criterion
	✓ Without a initiation	✓ Large computational cost
	✓ Reduce the local minimum of model surface	
ICNP	✓ Processing local features	✓ Need a good initiation
		✓ Large computational cost

68.5 Conclusion

This paper reviews that ICP, Sparse ICP, EM-ICP and ICNP algorithms, and discusses the way of weighting factor, the closest normal and the sparsity to improve ICP. And the next improvement should be the time complexity, the quick search on the high-performance of GPU parallel computing should be used to speed up the searching of the corresponding points.

Acknowledgments This work is supported by National Natural Science Foundation of China (No. 61472170, No. 61170143, No. 60873110), and Beijing Key Laboratory of Intelligent Telecommunications Software and Multimedia, Beijing University of Posts and Telecommunications (No. ITSM201301).

References

1. Chang W (2011) Global registration of dynamic range scans for articulated model reconstruction. ACM TOG 30(3):26:1–26:15
2. Jain V, Li X (2004) Point matching methods: survey and comparison. CMPT 888
3. Jost T, Hügli H (2002) A multi-resolution scheme ICP algorithm for fast shape registration. In: 3D data processing visualization and transmission, first international symposium, Padova, Italy, pp 540–543
4. Tamaki T, Abe M, Raytchev B et al (2010) Softassign and EM-ICP on GPU. In: First international conference on networking and computing, ICNC 2010, Higashi Hiroshima, Japan, pp 179–183
5. Bouaziz S, Tagliasacchi A, Pauly M (2013) Sparse iterative closest point. Comput Graph Forum 32(5):113–123
6. Mohammadzade H, Hatzinakos D (2013) Iterative closest normal point for 3D face recognition. IEEE Trans Pattern Anal Mach Intell 35(2):381–397
7. Besl PJ, Mckay HD (1992) A method for registration of 3-D shapes. IEEE Trans Pattern Anal Mach Intell 14(2):239–256
8. Granger S, Pennec X (2002) Multi-scale EM-ICP: a fast and robust approach for surface registration. In: 7th European conference on computer vision (ECCV2002), Copenhagen, Denmark, pp 69–73

Chapter 69
Study on the Effectiveness of Distance Education in the Judicial Examination Training

Rongxia Zhang and Weisheng Wang

Abstract Distance education for judicial examination training is different from face-to-face teaching. The difference is mainly reflected in the authority and the pertinence of the teaching content. In addition, distance education can save the cost of hiring well-known experts as it in the field of face to face. However the actual effect of distance teaching method has achieved in the judicial examination training is poor. The main reason lies in the lack of teaching process among many reasons. Some measures must be taken to improve the teaching effect of distance judicial examination training: firstly we must solve the problem of short-term behavior of educational institutions, also should we strengthen the communication and interaction between teachers and students in the teaching process.

Keywords Distance education · Judicial examination training · Effectiveness

69.1 Introduction

Judicial examination is a qualification examination that people engaged in the legal occupation activities must pass. The pass rate of judicial examination is very low in Chinese numerous qualification examination in the present situation. Therefore, judicial examination is referred to as the first test of the Chinese. How to pass the judicial examination is a practical problem that the legal professionals facing. For

R. Zhang (✉)
School of Law, Beihua University, Jilin 132013, China
e-mail: zhangrx55@163.com

W. Wang
Jilin Petrochemical Company of PetroChina Co Ltd, Jilin 132021, China
e-mail: jh_wweish@petrochina.com.cn

J.J. (Jong Hyuk) Park et al. (eds.), *Ubiquitous Computing Application and Wireless Sensor*, Lecture Notes in Electrical Engineering 331,
DOI 10.1007/978-94-017-9618-7_69

which to participate in judicial examination training has become the main choice. In recent years, modern distance teaching mode gradually becomes a very important way in the judicial examination training. Distance teaching is "the teacher teaches students the knowledge and train students to master certain skills course by the modern education technology under the guidance of modern distance education theory and education theory" [1].

69.2 Application of Distance Teaching Method in the Judicial Examination Training

At present, the judicial examination training organizations are numerous in the nationwide. There are all training, New Oriental and Sanxiaomingshi training with high reputation in this industry, and so on. These training institutions all use the distance teaching method in the judicial examination training. The method of distance education is different from face-to-face teaching mode in the judicial examination training. There are deep-seated reasons that distance teaching mode for judicial examination training can be widely used. Firstly, the reason is the authority of the distance teaching teachers and the teaching content. The most attractive factor of distance teaching mode being applied in the judicial examination training is that the teachers of distance education usually come from authority law institutions of Beijing. As the judicial examination training institutions market their distance learning products, the teachers' authority, influence and the proposition background usually are regarded as a selling point under the circumstances of judicial examination training education market. The value that law students and relative personnel pursuit is the teacher's background in the distance teaching. They think that the content of these teachers teach is either about the judicial examination question or has a high hit rate. Secondly, the reason is the pertinence of the distance teaching content. The teaching content of judicial examination training class is more pertinent compared with the traditional teaching. For example, the judicial examination training class will summarize the high-frequency test centers over the years and reduce the pressure for review. Thirdly, the reason is the operation of distance teaching mode. Educational institutions do not need to employ outstanding legal teachers if the distance teaching mode is used in the judicial examination training. The only work they need to do is cooperating with the training organization of Beijing and open network. Finally, the reason is the low cost of distance teaching mode. The distance teaching mode does not need to employ well-known teachers to teach face-to-face, which can eliminates the high cost.

69.3 Analysis on the Effect and Reasons About Application of Distance Teaching Mode in the Judicial Examination Training

69.3.1 Application Effect of Distance Teaching Mode in the Judicial Examination Training

The actual application effect of distance education in the judicial examination training was not ideal. Now take the actual situation of distance education training in the judicial examination of our school as an example. In order to improve the pass rate of the judicial examination, law school authorities cooperated with the judicial examination training institutions since 2011. They used the distance teaching mode to train the students in judicial examination training. More than 100 students of the third grade for law professional participated in the training. The training was conducted by playing remote video teaching manner. And the students were organized watching online video every Saturday and Sunday and the usual night time. But only a dozen students passed the judicial examination in the end.

69.3.2 The Reason of Poor Distance Teaching Effect in the Judicial Examination Training

Among the reasons of distance teaching method in the judicial examination training's poor effect, the main reason is the absence of distance teaching process. The teaching process is an important link to ensure the quality of teaching, and it is also the embodiment of educational law. Although distance education is different from the traditional teaching, "distance education is education. Since it is the education, we should follow the law of education" [2]. The teaching process is the combination of the teachers and students in the teaching activities of the school, and it is a process of communication and interaction that the teacher to teach students the knowledge and the students to accept the knowledge. The teaching process is the most important link to guarantee the teaching quality. The teaching process is the key to ensure the quality of teaching in both the traditional teaching and modern distance education. The most outstanding advantage of traditional teaching lies in the process of teaching in which teachers and students teach face to face. So teachers can master the situation that the students accept the knowledge and students can also question promptly to make learning problems in teaching resolve in a timely manner. The specialized distance leaning system actually pays attention to the exchange of teachers and students, which is equipped with some corresponding measures, such as a certain hours of face-to-face teaching, question answering online, homework online, and so on. The distance education teaching quality is guaranteed with these arrangements.

The key problems affecting the quality of teaching is the lack of interaction and communication during the teaching process in distance teaching of the judicial examination training. Legal professional is abstract and very practical, and simple mechanical memory can not achieve better learning results, so it needs more interaction between teachers and students in the teaching process. It can help students improve the ability to solve the problem with legal professional knowledge through the exchange of teachers and students. Therefore, the interaction between teachers and students in the teaching process is an important aspect of improving teaching quality in the judicial examination training. However there is no teacher teaches face to face, no online exercises, and there is no more question and answer online in distance teaching of the judicial examination training. The common practice of educational institutions is to let the students watch the video network. Communion part does not exist in teachers and students. In fact the distance teaching mode is used wrongly.

69.4 Measures to Improve the Distance Teaching Method Application in the Judicial Examination Training

The purpose that the students participate in the distance teaching of the judicial examination training is very clear, which is to pass the national judicial examination. What the students pay attention to is the effectiveness of the teaching. The students want to grasp the content and skill of judicial examination, and to grasp how to improve their test ability through special training for short-term. Therefore, the teaching effectiveness should be improved in the distance teaching of judicial examination training. Specifically, we can take the following measures to solve the problem.

69.4.1 To Solve the Problem of Short-Term Behavior of Educational Institutions

The aim that training institutions organize judicial examination training is to make money. Almost all of the training institutions approach to distance teaching method in judicial examination training, because the cost of distance teaching method is much lower compared to face-to-face training. The training institution only plays a distance network video and teaching and answering face-to-face is not arranged in order to reduce costs. Obviously this is a short-term behavior of running. The short-term behavior can be timely profit, but it can not guarantee the quality of teaching and can not survive long time. Therefore, to improve the application effect of distance teaching mode in the judicial examination training, we must solve the philosophy of educational institutions firstly. We should change the short-term education thought based on long-term development.

69.4.2 To Strengthen the Communication and Interaction Between Teachers and Students in the Teaching Process

Communication and interaction between teachers and students is the basic embodiment of the educational process, and it is also the effective measures to improve the teaching effect. The communication and interaction between teachers and students should be taken as the basic teaching quality guarantee measure in the distance training of the judicial examination. In view of this, we should adopt the following measure: Firstly, Adding distance answering and practice link. The training institutions and the source of unit of distance teaching information determine to answers to students' specific problems by the teachers from the source organization of network teaching. Or they may adopt the other way of answering. A certain number of simulation tests should be arranged in order to test the students grasping situation on the teaching content. Secondly, employing outstanding legal teachers by educational institutions in their own place to teach students face-to-face, coach and answer their questions [3]. Thirdly, they can set up a specialized learning on Internet discussion area to provide a platform for exchanges between teachers and students. Thus it can make up for the lack of teaching process in distance education and improve the teaching quality of distance teaching mode in the judicial examination training.

69.5 Conclusion

To solve the problem of short-term behavior of teaching institutions and the teaching process deficiency in distance training of the judicial examination in time, we can guarantee the teaching quality of judicial examination distance training and ultimately promote its far-reaching development under the circumstance that distance education mode widely used in judicial examination training.

Acknowledgment The thesis is a result of staged research of education and scientific research project of Jilin Province "Research on interactive path selection of university legal education and judicial examination" (Project No.ZC13034).

References

1. Liu R, Liu E (2005) Lectures on modern distance education, vol 1. Heilongjiang people's Publishing Press, Heilongjiang
2. Liu R (2009) Thinking of the distance practice education teaching reform. Mod Distance Educ 6:1
3. Sun C (2010) Analysis of modern distance education social value and its application—The effectiveness of distance teaching mode in the judicial examination training. Mod Distance Educ 6:27

Chapter 70
Greedy Strategy Based Self-adaption Ant Colony Algorithm for 0/1 Knapsack Problem

De-peng Du and Yue-ran Zu

Abstract The notion of using a meta-heuristic approach to solve the Knapsack Problem has been intensively studied in recent years. By comparing and analyzing the research of Ant Colony Algorithm (AVA) for 0/1 Knapsack Problem the authors propose an improved ACA based on greedy strategy and normal distribution. Experimental results show that the proposed approach performs better than the basic ACA.

Keywords 0/1 knapsack problem · Ant colony algorithm · Greedy strategy · Normal distribution

70.1 Introduction

The 0-1, or Binary, Knapsack Problem (KP) is defined as follows: given a set of n items and a knapsack, with

p_j profit of item j,
w_j weight of item j,
c capacity of the knapsack,

Select a subset of the items so as to maximize

$$z = \sum_{j=1}^{n} p_j x_j \tag{70.1}$$

D. Du (✉) · Y. Zu
School of Information Science and Engineering, Shandong Normal University, Jinan 250014, China
e-mail: woxinfei_xiang@126.com

D. Du · Y. Zu
Shandong Provincial Key Laboratory for Distributed Computer Software Novel Technology, Jinan 250014, China

J.J. (Jong Hyuk) Park et al. (eds.), *Ubiquitous Computing Application and Wireless Sensor*, Lecture Notes in Electrical Engineering 331,
DOI 10.1007/978-94-017-9618-7_70

$$\text{subject to}\begin{cases}\sum\limits_{j=1}^{n} w_j x_j \leq c \\ x_j \in \{0,1\}, (i = 1, 2, \ldots n)\end{cases}$$

where $x_j = \begin{cases} 1 & \text{If item j is selected} \\ 0 & \text{otherwise} \end{cases}$

Many industrial problems can be formulated as KP, such as: cutting stock, cargo loading, project selection, and budget control to mention a few examples. Lots of combinatorial problems can be reduced to KP, and KP arises also as a subproblem in several algorithms of integer linear programming [1].

KP is a classical NP-hard problem in combinatorial optimization [2], there are many traditional exact algorithms for solving the problem such as: dynamic programming, backtrace, branch and bound, etc. Generally, the time complexity is $o(2^n)$, so they are not adaptable to solve large scale problems. As a result, many approximate algorithms have been put forward. They are greedy algorithm, genetic algorithm, ant colony algorithm, etc.

70.2 Ant Colony Algorithm

Ant Colony Algorithm (ACA) was firstly proposed by Dorigo M in 1991 as an analog algorithm for simulating real ant colony behavior. Ants communicate with each other through a substance called pheromone, in the process of moving they leave this kind of substance and can perceive the existence and density of it. Furthermore, ants guide themselves according to the pheromone. Therefore, a large number of ants collective actions have a positive feedback for information. The phenomenon is described as follow: a path with more ants passing by, more pheromone will be left and the probability for newcomers to choose the path will be greater. Ant individuals communicate through this kind of information and search for a shortest path from the ant nest to the food source. ACA has been applied extensively in TSP [3], quadratic assignment problem, sorting problems, multi-objective combinatorial optimization problems and so on.

70.3 ACA for 0/1 Knapsack Problem

The 0/1 Knapsack Problem and TSP are essentially different. Therefore, the ACA should be extended to solve the KP. The main idea to solve TSP is to choose the path with more pheromone [4], here is an analogy, item with more pheromone would be more likely to be put into the knapsack.

70.3.1 The Basic ACA

Every ant chooses next item according to the selection probability formula (70.2) which denotes the probability for the kth ant to choose item i not selected before.

$$p_i^k(t) = \begin{cases} \frac{\tau_i^{\alpha}(t)\eta_i^{\beta}(t)}{\sum_{k \in allowed_k} \tau_k^{\alpha}(t)\eta_k^{\beta}(t)} & \mathrm{I} \in \mathrm{allowed_k} \\ 0 & \text{otherwise} \end{cases} \quad (70.2)$$

$\tau_i(t)$ is the amount of pheromone on item i in the tth iteration. It is often initialized as a constant $\tau_i(0) = c$. $\eta_i(t)$ stands for the heuristic factor where

$$\eta_i = \frac{p_i}{w_i} \quad (70.3)$$

p_i is the profit of item i, w_i is the weight of item i. α and β represent the influence of the pheromone and heuristic factor respectively. Each ant k calculate the selection probability of every surplus item j, and find the item with the maximum value as the next one to be selected. Then add the item j to the taboo table. In consideration of the constraint of the knapsack, after putting the item into table taboo, the weight sum of items ant k selected should be calculated to see whether exceed the constraint and a Boolean table flag(k) is introduced. In this way, when all the ants completed a cycle, update each item according to formula (70.4).

$$\tau_i(t+1) = (1-\rho)\cdot\tau_i(t) + \sum_{k=1}^{m} \Delta\tau_i^k \quad (70.4)$$

ρ is the pheromone volatilization coefficients in the domain of $0 \sim 1$. $\Delta\tau_i^k$ denotes the pheromone left by ant k on item i.

$$\Delta\tau_i^k = \begin{cases} Q * L_k & \text{the item is seleted by ant k} \\ 0 & \text{otherwise} \end{cases} \quad (70.5)$$

where Q is a constant, the positive feedback is more obvious with bigger Q but the global search ability become worse. Thus, different Q should be used with different instances.

$$L_k = p_i / P^k \quad (70.6)$$

p_i is the profit of item i and P^k is the total profit of selected items by ant k. A better rule is imposed in the following.

70.3.2 Comparison and Analysis ACA for KP

There has been much research about ACA application in KP. In literature [3], the authors make state of the item selected or not as the target of ant. The difference of objective functions is the expected value of choosing item or not. Ants transfer in the two states according to the probability. In this way there is only one feasible solution emerged and without making use of the group intelligence. As a result the outcome is not very good. Literature [5] chose items as ants target and left pheromone on the item which had been passed by ants. After each iteration every ant will have a new solution. And the residual item element information can guide the next iteration of the ants. In 2005, literature [4] modified the ant algorithm to adapt the KP and took fully consideration to the capacity restriction of the KP. A flag table was introduced to insure the constraint. A new type of ant colony algorithm based on the dissimilarity of the solutions was presented to solve the 0-1 Knapsack Problem in literature [6].

Paper [7] imposed a quick ant colony algorithm for solving the 0-1 Knapsack Problem in 2007, changed the probability sum from 1 to u and then used roulette-wheel to increase the randomness of the search. In 2011, literature [8] presented an improved ant colony algorithm based on normal distribution for Knapsack Problem, improved the robustness of the algorithm and over come the initial parameters sensitivity.

The main algorithm is described as the flow chart below (Fig. 70.1).

70.3.3 Improved ACA

Here, an improved ant colony algorithm is put forward to combine the normal distribution and greedy strategy.

In step ①: set the values of α, β, ρ, Q, the number of ants m, the maximum number of iteration times N_{max}, the amount of pheromone on the item $\Delta\tau_i$.

In step ②: To sum up formulas (70.2), (70.3), (70.4), in the tth iteration, denominator of formula (70.2) is the same. For the kth ant's selection, as the $allowed_k$ table is certain, the denominator is the same. So the relative value is needed only. Here define $p^*(t) = \tau_i^\alpha(t)\eta_i^\beta(t)$. Before each iteration, calculate the p*(t) of each item previously. There is no need for ant to recompute. Therefore, the time complexity of this step will reduce from O(mn2) to O(n2). The efficiency is greatly raised.

Formulas (70.2) (70.3) (70.4) and (70.5) demonstrate that the values of (α, β, ρ, τ, Q) influence the p*(t). If improper values are chosen, the p*(t) of different items may be different in countless times. As a result, some items have so small p*(t) that the probability to be selected nearly to be zero. And thus it cannot get the optimal solution. To avoid this kind of situation, many solutions were put forward, such as set bound for τ, adjust the value of ρ dynamically. They have obtained the certain effect, but not completely solve the problem. Here, a new train of thought is proposed. Through

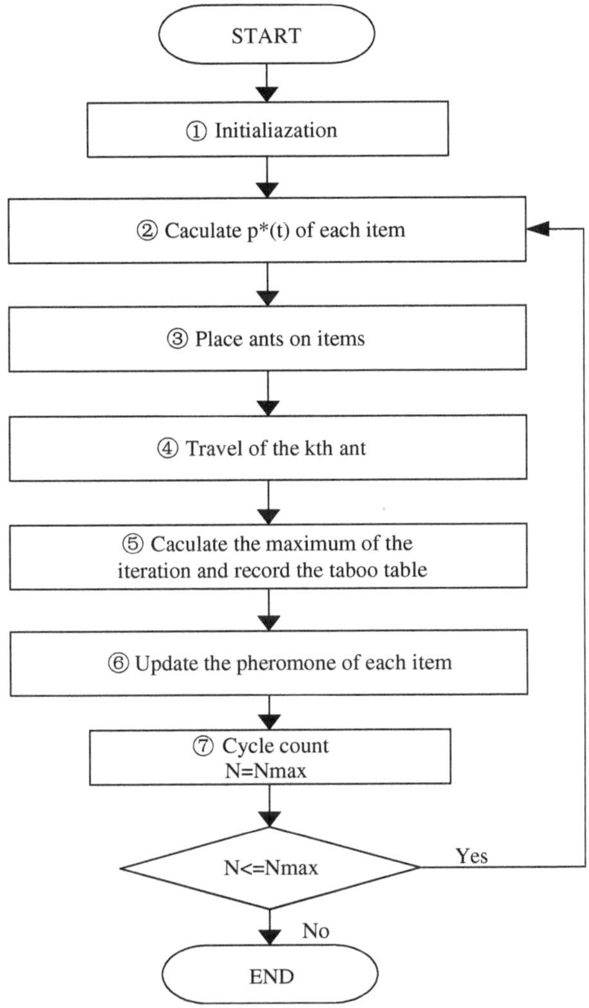

Fig. 70.1 Flow chart of ACA

adjustment the value of p directly, control the distribution of p in the macroscopic level. Specific methods are as follows:

(a) Introduce distribution coefficient λ (λ > 0), select n spots in [0, λ] uniformly (n is the number of items).

$$x_k = k * \frac{\lambda}{n}, k \in \{1, 2, \ldots, n\}$$

According to the standard distribution density function

$$\varphi(x) = \frac{1}{\sqrt{2\pi}} e^{-x^2/2}$$

Compute the probability distribution density φ at every point;

(b) Before the tth iteration, calculate p(t) of each item, and order them in descending order;
(c) Make $p * (t) = \varphi$;
(d) Every ant does the roulette according to the new p*(t).

Through such conversion, the value of p*(t) is limited in a section and the basic ant colony algorithm originally in the control of five parameters, now only adjust the distribution coefficient will change the distribution of p*(t).

On the other hand, in the formula (70.3), the strategy is prone to get a local best answer. In this paper, a better way is present. It is based on the greedy strategy. When the result of successive iteration are the same, it may be has been trapped into local optimum. With making the

$$j = \begin{cases} \arg \max_{i \in allowed_k} (p_i / w_i) \\ formula\,(70.2) \end{cases}$$
$$\text{If} \quad wi < \frac{\sum_{i \in n} w_i}{n} \;\&\&\; pi > \frac{\sum_{i \in n} p_i}{n} \tag{70.7}$$

Otherwise, use formula (70.2) to select the next item. In step ⑥ modify the update rule as $L_k = p_k / p_{\max}$ where $p_{\max}$ is the best profit of this iteration. In order to make convergence of the algorithm faster, the way to update the pheromone is: only performance of ants with better performance than the average profit producing pheromone, the final formula (70.8) is:

$$\Delta\tau_i^k = \begin{cases} Q * \frac{p_k}{p_{\max}} & \text{if } p_k > \text{pavg} \;\&\; i \in \text{tabuk} \\ 0 & \text{otherwise} \end{cases} \tag{70.8}$$

The proposed method can effectively avoid being trapped into local optimum and speed up the convergency.

70.4 Experimental Results and Analysis

Use the algorithm of this paper to solve the problem in literature [3] as case 1 and case 2 and [9] as case 3. The experimental environment is Visual C++ 6.0 on a computer with Intel Core(TM) i5 CPU 2.67 GHz, 4 GB memory and 32 bit Win7 system.

After running ten times of the basic ACA and ACA in this paper, the result is as the Table 70.1. It is clearly that the improved ACA performs better.

Table 70.1 The experimental results contrast

		n=10	n = 20	n = 50
Basic	Iteration times	150	200	300
ACA	Best value	295	1024	3064.6
ACA in	Iteration times	1.1	1.5	10
This paper	Best values	295	1024	3013

The charts below express the convergence and robustness of the algorithm respectively for case 1, case 2 and case 3. The curve blow is for the basic ACA and the upper one is for the improved ACA (Figs. 70.2, 70.3, 70.4).

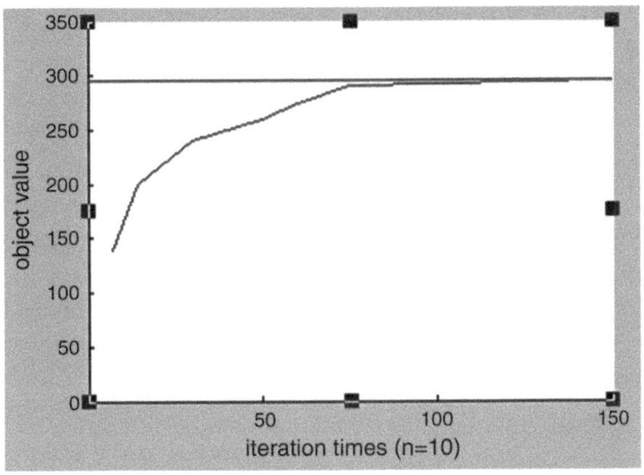

Fig. 70.2 Convergence state of case 1

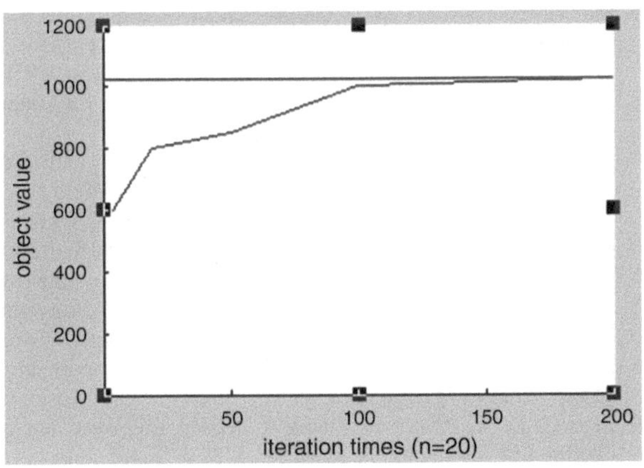

Fig. 70.3 Convergence state of case 2

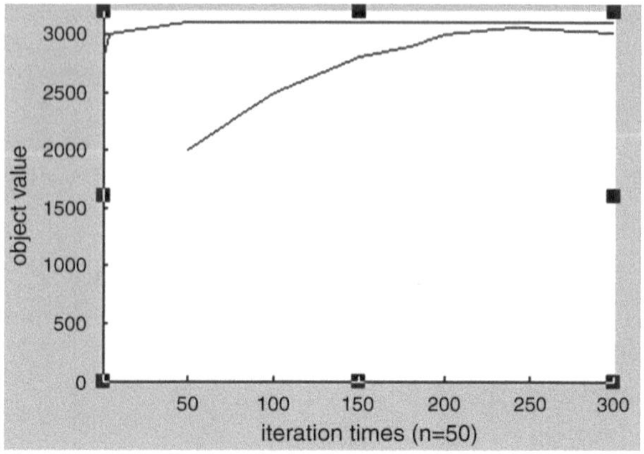

Fig. 70.4 Convergence state of case 3

70.5 Conclusion

Applying ACA to 0/1 Knapsack Problem is one of the ever developing approaches to this traditional problem explored in recent years. This paper proposes an improved algorithm r based on normal distribution and greedy strategy. The experimental results show that the robustness and astringency of the proposed approach are both better than the basic ACA. This research is at its very preliminary stage, and the property of ACA and combining it with other algorithms will be our next bid for 0/1 Knapsack Problem.

References

1. Martello S, Toth P (1990) Knapsack problem: algorithms and computer implementations. DEIS, University of Bologna, Biddles Ltd, Guildford, Great Britain, p 13
2. Pisinger D (1995) Algorithms for knapsack problems. PhD thesis, p 33
3. Ma L, Wang L (2001) Ant optimization algorithm for knapsack problem. Comput Appl 21(8). doi:1001-9081(2001)08-0004-02
4. Liu H, Liu Y, Liu J (2005) Solution to the 0/1 knapsack problem based on ant algorithm. J Daqing Petrol Inst 29(3):59–62. doi:1000-1891(2005)03-0059-04
5. Luo X, Zhao L (2004) Ant colony algorithm for 0/1 knapsack problem. J Soochow Univ 24 (1):41–44 (engineering science edition). doi:1000-1999(2004)01-0041-04
6. Qin L, Bai Y, Zhang C, Chen L (2004) Ant colony algorithm for 0/1 knapsack problem
7. Wang H, Jia R, Zhang Y, Qi P (2007) A quick ant colony algorithm of solving the 0-1 knapsack problem. Comput Technol Dev 17(1):104–107. doi:1673-629X(2007)01-0104-04
8. Xiong Y (2012) Application of self-adaption ant colony algorithm based on greedy strategy to TSP. Comput Digit Eng 1(37)
9. Liao C, Li X, Zhang P, Zhang Y (2011) Improved ant colony algorithm based on normal distribution for knapsack problem. J Syst Simul 23(6):1156–1160. doi:1004-731X(2011)06-1156-05

Chapter 71
Non-Chinese Students Speak: Sectional and Clinical Anatomy Learning in a Chinese Medical School

Xu He, Fu-Xiang Liu and Aihua Pan

Abstract Non-Chinese students in Xiangya medical School had their reflection on the learning difficulty of sectional and clinical anatomy. Their suggestions of these two disciplines in various aspects may be helpful to the improvement of teaching.

Keywords Teaching model · Anatomy

71.1 Introduction

Xiangya Medical School of Central South University has been one of the most prestigious medical schools in China since 1914. Each year, it enrolls numbers of foreign students in undergraduate and postgraduate courses. Among those disciplines, sectional and clinical anatomy is considered to be the toughest.

Anatomy is one of the most fundamental medical sciences. On the one hand, it is essential for clinical medicine; on the other hand, it provides an organizational framework for other medical disciplines. Sectional anatomy is the study of the relationship of the structures of the body by examination of cross sections of tissues and organs while clinical anatomy is the practical application of anatomical knowledge to diagnosis and treatment, some times termed applied anatomy or clinical anatomy [1]. For many students and teachers, the representation and integration of spatial and symbolic information pertaining to anatomical entities presents considerable challenges, which largely account for the difficulties in teaching and learning.

X. He · F.-X. Liu · A. Pan
Teaching and Research Section of Human Anatomy, YiYang Medical College, Yiyang 413000, Hunan, China

A. Pan (✉)
Department of Anatomy and Neurobiology of Xiangya School of Medicine, Central South University, Changsha 410013, China
e-mail: panaihua@csu.edu.cn

J.J. (Jong Hyuk) Park et al. (eds.), *Ubiquitous Computing Application and Wireless Sensor*, Lecture Notes in Electrical Engineering 331,
DOI 10.1007/978-94-017-9618-7_71

This year, foreign students who were master in general surgery had their reflection on the learning difficulties and suggestions of these two disciplines in various aspects, which may be helpful to the improvement of teaching.

71.2 Difficulties

Difficulties faced in the anatomy class are various. First of all, compression of the two disciplines into 10–11 months means less content in class and more self-learning after class, which demands more innovative integrated teaching skills and facilities for the studying in the class as well as by oneself. Then the assessment and feedback system doesn't fulfill its duty fully. Apart from that, language barrier and other factors affect the teaching and learning.

Major problems are illustrated as follows.

71.2.1 Language Barrier and Cultural Differences

The number of non-Chinese speaking students is increasing, when language barrier becomes a key problem affecting non-Chinese speakers in learning courses in this school. Generally speaking, since the school primarily trains Chinese students based on Chinese medium, there are teachers who are well endowed with the right skills and teaching mostly in Chinese medium. Though there are quite a good number of teachers and lecturers that can lecture in English. The foreign students attended class with the Chinese students sharing the same materials and resources in traditional method. Usage of foreign languages materials and mobilization of resources which also suits the non-Chinese speakers were insufficient, as they reflected which hindered the efficiency of learning the courses. Their common unsatisfying academic results prompts the use of the language, teaching style and teaching aid that can promote the understanding of the course by more foreign students.

Cultural differences might limit interaction and communication which might increase the difficulties. Though it is not a major factor since the Chinese community is friendly, it still needs mention for those unsocial students.

71.2.2 Inappropriate Learning Pattern

Students have to complete mastering the anatomy course within just 10–11 months. The arduous tasks for teachers include lecturing specific subtopics for emphasis and guiding students on ways of reading by giving most of the contents as self-taught. Students may primarily expect mastery upon exiting the lecture hall. During lectures, students often compulsively take notes to the exclusion of thought and comprehension, thinking they will write it down now and learn it later. Common experience is that anatomy classes often induces sleep in the students as they don't follow what is taught. Consequent poor performance in assessments causes frustration and dislike of the subject.

In the dissection laboratory, thorough dissection is essential yet is one of several means of learning that should be engaged in. But the goal of laboratory study is not clear enough to the students. Students are actually spending their time and effort preparing the dissection. There is a tendency for them to feel that when they have completed the days' assignment that they are done. All too often, the laboratory period is the most wasted portion of the student's day. Stress from dissecting influences the efficiency emotionally. Testing method paled into the assessment of effectiveness of the laboratory experience.

Most students believe their study problems are unique; however, some typical pitfalls are encountered as they "learn how to learn anatomy". These include:

- Place too much emphasis on memorization and use of the textbook alone in the place of dissection.
- Excessive re-reading.
- Lack of confidence in learning from visual experience.
- Studying in the state of fatigue or in isolation.
- Using study techniques that are consumptive rather than productive.

71.2.3 Deficiency of Teaching Materials and Methods

One question that crops in lessening the burden of anatomy is "Can the teachers make the subject palatable to facilitate self-learning by the students?" Since there are no standard methodologies for this, every teacher has to evolve newer and newer ideas to impart instruction, relevant to the theme-anatomical thinking for solving clinical problems. Most of the reading is required to complete out of the class, which is not fully done by the students. Lack of the mobilization of resources and contact between teachers and students may account for this partly.

Additionally, problems of materials influence the teaching as follows:

- Difficulty in obtaining simplified anatomy illustrated textbooks. Recommended textbooks are mostly complicated to cope with alone, thus resulting in lack of interest.
- Materials of limited usefulness. For example, the embedded body sections are only marginally useful.
- Labels awkward in atlases. Students found the "porcupine" approach to labeling images in most cross-sectional atlases awkward and overwhelming. The extent of a structure often is not apparent, especially on CTs and MRIs. Students also found it difficult to orient to cross-sectional images.

- Lack of correlated CTs and MRIs. The lack of exact correspondence between clinical images and body sections in atlases made comparison cumbersome and inefficient.

71.2.4 Problems of Assessment and Feedback

Practices and assignments are considered to better reveal the understanding of the course, which facilitates the teacher to make further teaching plan.

Current assessments include norm-referenced and criterion-referenced one. The former is successful in a large and diverse class with performances compared against each other, while the latter occurs when students are compared against some predetermined set of criteria. Objective-type tests are selection of pre-determined responses, which include multiple choice questions, true-or-false, fill-in-the-blank, short answer and matching questions. The advantages of this method are that it is efficient, easy and the scoring is reliable. Its disadvantages include much memorization and lack of application of the learning.

As they see, tests were monotonous which failed to assess how students are progressing.

71.2.5 Interdepartmental Co-ordination in Selecting Student

From their point of view, there is a lack of interdepartmental co-ordination in the university to select foreign students. Some students are admitted to department by foreign students' affairs without prior arrangement regarding the courses and language proficiency.

71.3 Suggestions

Principles of good teaching are discussed: (1) knowledge of the subject and teaching resources; (2) Critical thinking and problem solving skills; (3) Knowledge of students and their learning; (4) Teaching and communication skills. Effective instruction is the most important in the process of teaching. Skills considered to make up a good teacher includes the ability to: (1) motivate the students; (2) merge the classroom; (3) assess prior knowledge of the students; (4) communicate ideas effectively; (5) take into account the characteristics of the learners; (6) assess learners' outcomes; (7) review information. Besides, the teachers are preferred to start in English medium classes and guide students to gain more outside the class as self-taught.

As Facinet (one of the non-Chinese students) sees, effective medical education should be viewed as a continuum. Integration of the basic sciences and clinical medicine should occur throughout the curriculum and self-directed learning should be emphasized. Curricular revision is appropriate if these fundamental concepts are absent. The principles of curricular models are generalized as traditional, problem-based and systems-oriented. The ideal curriculum may draw from the models: A truly integrated curriculum. However, the curricular model chosen must meet the needs of the institution and its students. Flexibility and innovation are expected to be incorporated into the educational approaches when the curricular design is proposed.

The students gave us numerous suggestions in the previewing, teaching, reviewing, testing and other aspects.

71.3.1 Teaching and Learning Skills

First of all, teachers should establish a positive attitude of readiness in students at the beginning of a lesson, which may raise more interest in cognition of the learning later. Then, review of the prerequisites is necessary. Teachers need to ensure that students have mastered the skills and knowledge needed beforehand in order to link to the new information that they are about to receive. Just a few quick questions before the new lesson will remind students of what they already acquired and give them the framework to incorporate the new information. Lectures should be approached with the idea that the lecture represents parts of the general topic. Within the constraints of lecture period, teachers usually select specific subtopics for emphasis, or to supplement information neglected by those references. The lecture is the opportunity to find out what teachers consider to be important. During lectures, students should maximize their listening and observing, and minimize their note-taking. Notes should be limited to the amount sufficient to keep students actively involved. Participation must be active rather than a passive process.

On one hand, lesson structure should be clear and well-ordered, while the emphasis should be clearly indicated. Repeat important information which may bring students back into the lesson whenever possible. On the other hand, methods should be taken to ensure efficiency in the class: (1) Proper variety, activity or humor in the lesson may live it up and maintain students' attention, yet too much variation can be counterproductive- balance must be achieved. (2) Teachers must be constantly aware of the effects of their instruction by regularly probing their students' understanding of the material being presented. Learning probes can be various. Teachers can ask for brief responses to the content of the lesson. Or brief written. (3) Wait time and calling order are concerns in classroom questioning. Wait time is the amount of time that the teacher will wait for a student to answer before going to another student. Less wait time appears to tell the student that the teacher expects little from him/her. As for calling order, teachers often call on volunteers which allow avoidance of participation in the class. This can be solved by asking a randomly selected student. (4) Choral response is favorable when there is only one possible response to the question. (5) In-class seatwork and homework are important to get better understanding for students as well as to provide feedback for teachers.

For presenting new material, recommended processes are listed below:

Direct instruction occurs when teaching well-defined anatomical information. Explanation is effective with explanatory words (i.e., because, consequently). Worked examples are used for certain kinds of problem solving techniques. When presenting a problem, a teacher can model the strategies which expert may use to solve the problem by working through it and explaining the thinking process. A better way to present new concepts may be explanation- rule- example- rule-example. Visual and manual experience is also important throughout the learning. Visual representations are considered to be retained in the long-term memory more than when the information is only heard.

In laboratory learning, dissection of the human body is usually the first in a long series of students' encounters with death and such experiences often are seen as unpleasant or frightening [2]. Many students go through a process of psychological accommodation [3]. So, respect and compassion should be reinforced in the students before their dissection. First, respectful language is encouraged. The term "donor" is more appropriate than "cadaver" or "corpse" in referring to the donated body, promotes appreciation for the students' first "patient". Second, provide the students with the actual name, age, history and likely cause of death of the donor in honor of the donor as having once been a living human being. Third, prompt students to explore feelings and discuss topics stimulated by the intense experience of dissection. Suggested topics include the feelings about dissection, the difficulty in deciding to donate one's body, the importance of anatomy to a medical practitioner's role, and the historical development of the anatomy study.

The laboratory should be the site where the most anatomical learning occurs. Teachers are responsible for ensuring the efficiency in the laboratory. Students must be taught how to be efficient. Students are advised to spend a few minutes before each lab, previewing the assignment. As proceeding to the laboratory, students should look at their dissections firstly. Direct observation is the first experience. It is much more important to observe the structures than to duplicate information that is already written out in the text. Because of great differences in the dissected specimens due to anatomical variance, sex and dissection skills, it is imperative for students to observe each other's dissections. Secondly, for the goal of the dissection exercise which is to gain a clear mental image of the dissected specimen, structures must be cleaned sufficiently so that borders, attachments as well as direction of muscle fibers can be observed clearly. Even if there is no time for perfection, the structures must be distinguishable in order for a clear formed mental image. An atlas is necessary for the constant comparison with the dissection, which also promotes recalling mentally after the laboratory session. Thirdly, specialized terms should be used in the discussion with the dissecting session. Thus students become proficient in the anatomy language with application. Fourthly, review after the dissecting visual aid is finished. With the prepared visual aid, the demonstrating student shows the structures and what he/she obtained in the dissection, while others listen in courtesy to organize their thoughts and correct the errors. Finally, self-testing should be taken with the help of dissecting manual and atlases.

71.3.2 Assessment and Feedback

Assessment is essential to check how students have mastered the objectives that are set for the anatomy lesson. Informal test can be done with the teacher asking questions, or with independent work as well as a quiz in class to assess understanding. Tests should be given frequently and short. The directions should be clear. The material that a test should cover includes the information from both published sources such as the text and lectures. The feedback may help teachers to decide the pace and the emphasis of their lectures.

Diverse modalities of assessments are expected. Essay test may assess the mastery of knowledge in a greater depth. The advantages include that it reveals how well the students recall, organize and communicate the obtained information, and that it assesses higher-level abilities of analysis, synthesis, evaluation and critical thinking. Its disadvantages are that consistency in grading is difficult and only limited questions can be asked. Another type is the performance test which requires students to apply knowledge and/or skills in a realistic way. It is a way to measure what students can do, rather than what they know, by using portfolios of the students' work, exhibitions and demonstrations.

There are general guidelines advised for grading and reporting student's performance. Grades should be based on multiple sources. Feedbacks should be taken into consideration. Teachers should explain their system of grading clearly to the students. As well, grades must be made available to the students.

71.4 Benefit

Reflection from non-Chinese students will promote more flexibility and innovation in the educational approaches and improvement of teaching skills and methods. It will help reduce the stress and unnecessary psychological burden in students who feel that they can not understand and benefit from the lectures offered in Chinese. Better understanding of the knowledge equips the students for a good clinical practice. Finally, it will benefit better training of Chinese and non-Chinese students.

References

1. Cahill DR (1997) Lachman's case studies in anatomy, 4th edn. Oxford University Press, New York
2. Finkelstein P, Mathers L (1990) Post-traumatic stress among medical students in the anatomy dissection laboratory. Clin Anat 3:219–226
3. Bertman SL, Marks SC Jr (1989) The dissection experience as a laboratory for self-discovery about death and dying: another side of clinical anatomy. Clin Anat 2:103–113

Chapter 72
Dynamic and Efficient Search System for Digital Encyclopedia of Intangible Cultural Heritage: The Case Study of ICHPEDIA

Jung Song Lee, Soon Cheol Park and Han Heeh Hahm

Abstract In this paper, we have presented the three ways of search functions in Ichpedia, the web-based intangible cultural heritage encyclopedia database and archives system. The search system consists of simple search, semantic search and map search. All functions can promote dynamism and efficiency in the achievement of best results from the user's point of view. First, the simple search provides not only exact results associated with keywords but also providing two statistical graphs. The graphs show the ratios of the search results according to classification and regional distribution of an individual ICH element. Second, map search provides graphical interface so that users are easy to follow the search result. It is constructed extensively by interconnecting two other searches, simple and semantic search functions. Third, semantic search is the most difficult and meaningful function but not perfect yet. However, it already shows some important implications in terms of that an ICH element is associated with other elements and its characteristics can be understood within the wider networks of ICH resources. The proficiency of our search system highlights in finding the connected elements and by doing so, in expanding users' knowledge and safeguarding awareness of intangible cultural heritage.

Keywords Intangible cultural heritage · Ichpedia · Digital archives · Semantic search · LAMP

J.S. Lee · S.C. Park
Division of Electronics and Information Engineering,
Chonbuk National University, Jeonju, South Korea
e-mail: ei200411147@jbnu.ac.kr

S.C. Park
e-mail: scpark@jbnu.ac.kr

H.H. Hahm (✉)
Department of Archeology and Cultural Anthropology,
Chonbuk National University, Jeonju, South Korea
e-mail: hanheeh@jbnu.ac.kr

J.J. (Jong Hyuk) Park et al. (eds.), *Ubiquitous Computing Application and Wireless Sensor*, Lecture Notes in Electrical Engineering 331,
DOI 10.1007/978-94-017-9618-7_72

72.1 Introduction

The authors are presently involved in the management of Ichpedia, a web-based encyclopedia and archives of intangible cultural heritage (ICH). We established its basic structures of database and archives in 2010 and have been collecting ICH data for the last three years. Ichpedia [1] has currently approximately 30,000 ICH data and it is one of the biggest ICH data-base with multimedia archives not only in Korea but also in the world. The digital encyclopedia, Ichpedia, is still expanding its data quantitatively and qualitatively. In addition, we have been endeavoring to make an advanced retrieval system by generating diverse search functions including keywords, semantics and maps. The reason why the function of search is significant in the area of digital encyclopedia of ICH is that the size of data and the depth of contents are increasing so that users face serious difficulties in finding the right and exact information among a large amount of ICH data. In this paper we present three levels of the search system that are working together as an integrated system for making the best results. The digital technology and philosophy employed in our Ichepedia project is innovative in light of searching ICH data as well as ICH data recording. The web-based inventory has become a key tool for promoting general awareness of ICH and for safeguarding ICH worldwide. In particular, the web can offer various advantages in the process of inventory and search, such as utilizing interactive mode of communications among multiple voices, bringing together of scattered resources, and efficient sharing of information [2, 3]. Ichpedia has similar features as the aforementioned digital platforms but it offers more advanced features which allows for dynamism and efficiency in the process of inventorying, excavating and sharing of resources. In this paper we focus on the functional features in excavating and sharing ICH information, that is, the search function in general in Ichpedia.

The second chapter demonstrates main of Ichpedia three search functions with respective analytical results. Next, Ichpedia development environment which is characterized as an integrated model is described. In conclusion the use of digital technology is an efficient way to increase of people's knowledge and awareness of the importance of ICH resources.

72.2 Search System in Ichpedia

72.2.1 Simple Search

Ichpedia offers various ways to search ICH data in the Ichpedia system while the results of them have the same output format. Figure 72.1 is also an example of a search result with the search keyword, 'Samulnori' which is a traditional folk Korean music with four instruments. In a simple search, if user inputs a query keyword in the window, A, the result of the query in shown again as like in Fig. 72.1.

Fig. 72.1 Result of keyword search of "Samulnori"

The result of the search lists indicated with B in the Fig. 72.1 consist of the alphabetical order list, lately inserted list and most frequently searched list. Users can choose one of the lists in convenient way in order to accomplish the further related search. The terms which are the same as the search keyword, 'Samulnori' are highlighted as indicated with C in the figure. Two statistical graphs are shown as indicated with D. The left graph shows the ratio of the search results according to classification defined in Ichpedia. The right shows the regional distribution ratio about the search results. These two statistical graphs are instrumental in finding new facts and clues for a new analysis of Samulnori for researchers concerned.

72.2.2 Map Search

The mapping of cultural elements has long been considered a meaningful device of a cultural configuration. We, however, tried to go further in using the mapping method as a research device. For this, the use of map search is extensively utilized. Since Ichpedia is keeping records in real time so that we can keep track of the current situation of ICH [4]. Researchers, administrators and other outsiders who are interested in the safeguarding of ICH can find out in which 'areas' or which 'villages' ICH are viable. The map presents a more detailed visual of the current situation of ICH when one of the circles on the map is touched.

The map in Fig. 72.2 shows the results searching the keyword 'arirang' in the Ichpedia system. It has the list of 479 contents about 'arirang' distributed all over the Korean peninsula. The data distributions of Gyeonggi and Gangweon are denser than those of other areas relatively. After clicking one of the circles in the map, a user can see an ICH data list or an Ichpedia content page through the popup windows. The left in the figure is the popup window of the Ichpedia content page of the 'arirang' ICH data collected in the area of Milyang, located in the south-eastern Korea.

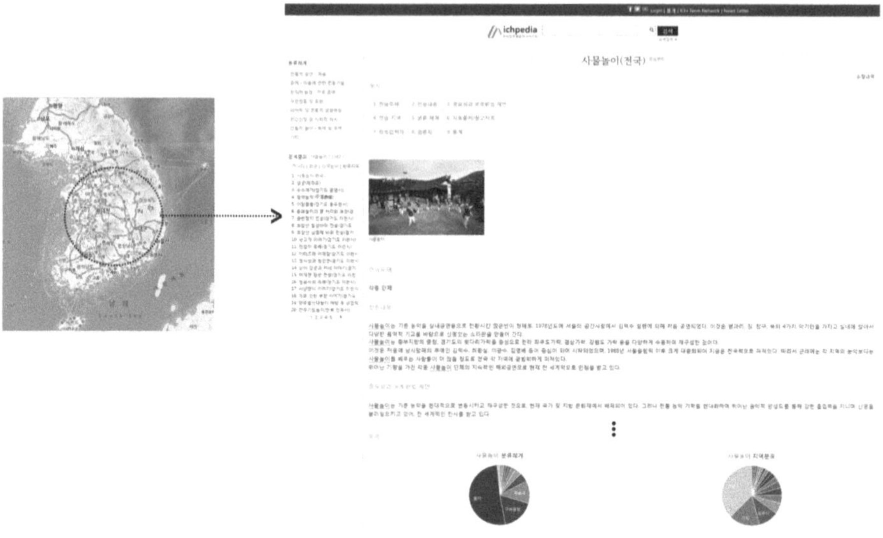

Fig. 72.2 Ichpedia system using location information

72.2.3 Semantic Search

The semantic search is one of the main functions of Ichpedia. It is not perfect yet but its implications are quite meaningful. The ICH ontology which is one of the necessary components for semantic search has been being built by the scholars and specialists who are engaged in Ichpedia. We operate the software to construct the ontology through web and general users can check and build the ICH ontology restrictedly. Semantic search is significant in respect that an individual ICH element is associated with other elements and its characteristics can be understood within the wider networks of a specific domain. Searching the ICH data in this system semantically, the user query may be expanded to get the more precise results [5–7]. The expanded query, Q′, can be defined as follows;

$$Q' = t_0^{\wedge} w_0 + \left\{ \sum\nolimits_{i=1\ldots n} t_i^{\wedge} w_i \right\} \tag{72.1}$$

where $\boldsymbol{t}_0$ is the clicked node (or term) and $\boldsymbol{w}_0$ is the weight of $\boldsymbol{t}_0$. $\boldsymbol{t}_i$ and $\boldsymbol{w}_i$ are the expanded term and its weight. The weight, $\boldsymbol{w}_i$, is a floating-point numbers calculated using the term to term relationships. The operation ^ followed by $\boldsymbol{w}_i$ is the boost factor for the preceding term, $\boldsymbol{t}_i$. The expanded terms may be the first and second level terms from the clicked node.

Taking the handicraft domain as an example, 'geomungo (Korean traditional musical instrument),' one of the ontology terms, is selected. The term,'geomungo,' is related to 'bokjori (fortune rice picker),' 'but (traditional writing pen),' 'buchae (traditional fan)' through the source material of 'taenamu (bamboo)' as shown in

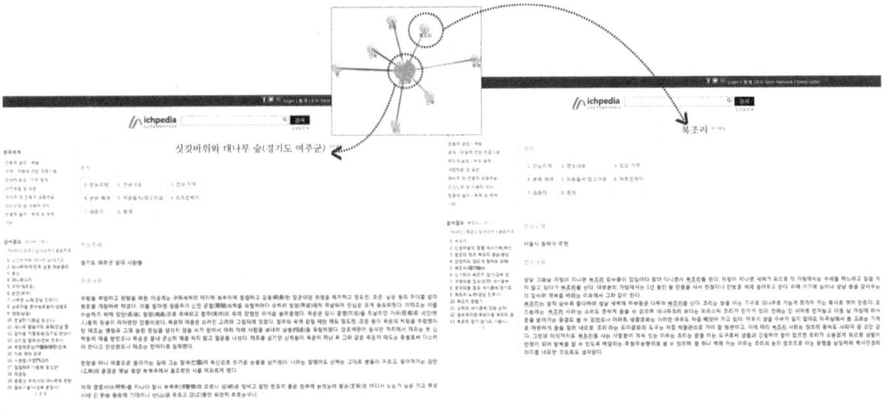

Fig. 72.3 Search "bokjori" using term-network

Fig. 72.3. At first glance, 'geomungo'and 'bokjori' seem to have nothing to do with each other. However, 'taenamu' is essential for the manufacture of both 'geomungo' and 'bokjori'. Because of that reason, both craftsmen of the traditional musical instrument and manufacturing communities of bamboo rice pickers were found near (in map search) where good quality bamboo trees were produced.

One of the benefits of semantic search is that even if no exact query information is known, a user can find what he/she wants through the visualized term network. Users may click one of the nodes in the tree to look for interconnected terms and contents. The ability to find the connected elements is significant in encouraging an integrated search in Ichpeida. Ichpedia's semantic search thus shows that an individual ICH element is associated with other elements and its associated structure makes it possible to understand how one ICH element fits within the wider networks of the domain.

72.3 Ichpedia Development Environment

The Ichpedia system has been designed and coded under the LAMP environment. LAMP is a combination of free, open source software. The acronym LAMP refers to the first letters of Linux (operating system), Apache HTTP Server, MySQL (database software), and PHP, Perl or Python. When used together, they support web application servers. Therefore, countries or groups can share this system without any economic burden to manage and have easy access to intangible cultural resources through Web.

The structure of the Ichpedia system has 3-layer architecture, as shown in Fig. 72.4. The data layer is implemented through MySQL and Lucene [8] on a Linux server and contains all the system's data structure and information. The application layer is built on the MVC framework written on PHP and Python which

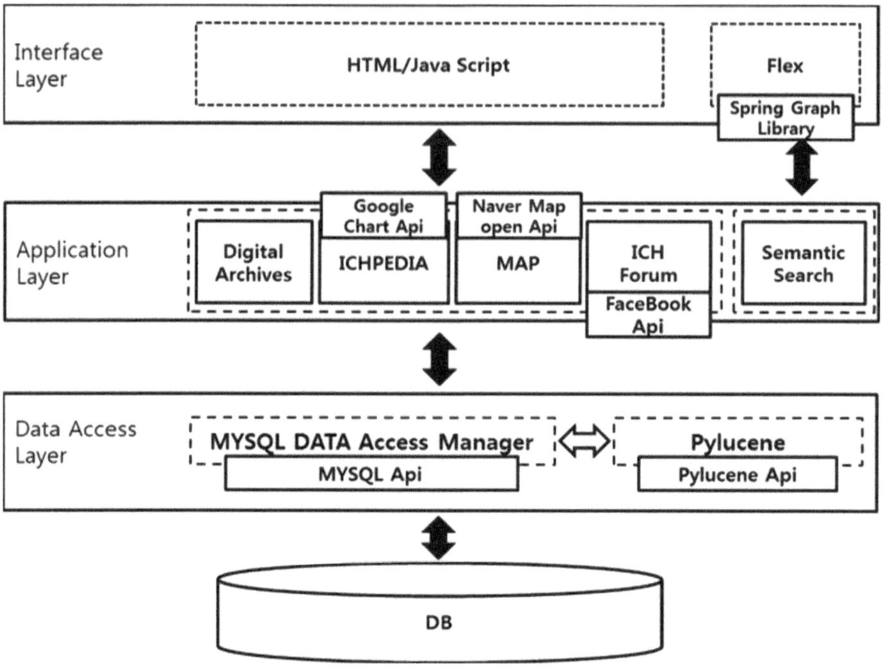

Fig. 72.4 Software structure of Ichpedia

implement all the functionalities such as digital archives, Ichpedia, culture map and etc. The interface layer is based on HTML, CSS and Java Scripts to allow a smooth user interaction with the applications, and Flex to manage a high geographical data set like the term networks.

72.4 Conclusion

In this paper, we have presented the three ways of search functions in the web-based ICH encyclopedia database and archives system, Ichpedia. The search system consisting of simple search, semantic search and map search can promote dynamism and efficiency in the achievement of best results from the user's point of view. The characteristics of ICH data are complicate, unfixed and holistic since many ICH data are closely related to everyday life style, practices and ideas. Ichpedia provides users who want to get satisfactory search results with various advanced functions. First, even the simple keyword search provides two statistical graphs which show the ratios of the search results according to classification and regional distribution of an individual ICH element. Second, map search is crucial to find out the current situation of a particular element of ICH by keeping records in real time and showing a graphic user interface in Ichpedia. Map search is also interconnected with simple and sematic searches in order to expand the search result. Third, semantic search, the most difficult and meaningful function in the search system, has been developed with strenuous efforts by the Ichpedia team and not perfect yet. However, it has already produced important results in terms of an ICH element is associated with other elements and its characteristics can be understood within the wider networks of ICH resources. We have developed various functions of digitalized resources including the co-related structure of search function. Lastly, this system has been developed in the LAMP (Linux, Arpache, MySQL and Python) environments. Therefore, our Ichpedia system has the advantage of the least expensive for development and maintenance of the system. Furthermore, the system can be run in various language environments in high portability with a little effort. We hope that Ichpedia will make a contribution to pave the way for digital innovation in the area of digital encyclopedia and library as well.

Acknowledgments This research was supported both by Basic Science Research Program through the National Research Foundation of Korea (NRF) funded by the Ministry of Education, Science and Technology (NRF-2013R1A1A2063572) and by the National Research Foundation of Korea (NRF) grant funded by the Korean Government (NRF-2013S1A5B8A01072201), and the Brain Korea 21 PLUS Project, National Research Foundation of Korea.

References

1. Intangible Cultural Heritage Online EncycloPEDIA, http://www.ichpedia.org/
2. Park SC (2008) A Study of The digital archives of 20th century people's life research center. J Korean Yougnam Cul Inst 14:35–61
3. Hahm HH, Park SC (2006) Digital archives of cultural archetype contents: its problems and direction. J Korean BIBLIA Soc Libr Inf Sci 17:23–42
4. Davis CA, Gerais CM, Fonseca FT (2007) Assessing the certainty of locations produced by an address geocoding system. GeoInformatica 11(1):103–129
5. Voorhees EM (1994) Query expansion using lexical-semantic relations. In: Proceedings of the 17th annual international ACM SIGIR Conference on Research and development in information retrieval. Dublin, Ireland, pp 61–69
6. Ogilvie P, Voorhees EM, Callan J (2009) On the number of terms used in automatic query expansion. J Inf Retrieval 12:666–679
7. Choi LC, Choi KU, Park SC (2008) An automatic semantic term-network construction system. In: Proceedings of the international conference on computing, engineering and information. Fullerton, CA, pp 217–220
8. Hatcher E, Gospodnetic O (2004) Lucene in action. Manning, Shelter Island

Chapter 73
On the Performance of Quasi-orthogonal Space Time Block Coded Massive MIMO with up to 16 Antennas

Khin Zar Chi Winn, Phyu Phyu Han, Kasun Bandara and Yeon-Ho Chung

Abstract Massive multiple-input multiple-output (MIMO) using a large number of antennas at both transmitter and receiver sides based on quasi-orthogonal space time block code (QOSTBC) is presented. Space-time block code (STBC) is a MIMO transmit strategy that applies transmit diversity and high reliability. QOSTBC is attractive because it achieves higher code rate than orthogonal STBC and lower decoding complexity than non-orthogonal STBC. We present the performance of massive MIMO systems using the QOSTBC with multiple antennas up to the 16 × 16 configuration. The performances of 2 × 2, 4 × 4, 8 × 8 and 16 × 16 massive MIMO systems have been presented. Simulation results show that the massive MIMO systems with QOSTBC give significant performance improvement with full rate and full diversity, compared with previously considered massive MIMO systems.

Keywords Massive MIMO · Quasi-orthogonal STBC · Full rate

73.1 Introduction

The demand for mobile communication systems with high data rates has significantly increased in recent years. Many technical challenges remain in designing robust wireless networks that deliver the performance necessary to support

K.Z.C. Winn · P.P. Han · K. Bandara · Y.-H. Chung (✉)
Department of Information and Communications Engineering,
Pukyong National University, Busan, Republic of Korea
e-mail: yhchung@pknu.ac.kr

K.Z.C. Winn
e-mail: khinzar111@gmail.com

P.P. Han
e-mail: phyu2han@gmail.com

K. Bandara
e-mail: kassae6@gmail.com

J.J. (Jong Hyuk) Park et al. (eds.), *Ubiquitous Computing Application and Wireless Sensor*, Lecture Notes in Electrical Engineering 331,
DOI 10.1007/978-94-017-9618-7_73

emerging applications [1]. Multiple-input multiple-output (MIMO) systems offer numerous benefits over conventional single-input single-output (SISO) systems, such as the potential to facilitate data rates considerably higher or to significantly improve the reliability of a wireless link. In fact, MIMO technology is envisioned to be a core transmission technology for high speed wireless communications, due to increased data rate and performance obtainable largely from transmit diversity and spatial multiplexing [2].

MIMO scheme can be split into two categories: space time coding (STC) and spatial multiplexing (SM). STC improves the reliability while SM increases data rate [3]. STC, introduced by Tarokh [4], is promising method where the transmitted code symbols per time slot are equal to the number of transmit antennas. The class of linear STBC is the major category of STC and can be divided into subclasses, such as orthogonal STBC (OSTBC) [5] and quasi-orthogonal STBC (QOSTBC) [6] that is typically designed for more than two antenna systems with increased decoding complexity.

The massive MIMO is considered a candidate for a next generation wireless communication system [7]. For full benefits from the massive MIMO, the number of antennas is required to be at least more than 10 antennas at each side of the transmitter and receiver [8]. However, conventional performance analyses have been presented with only up to 8 antennas in the literature [8]. Therefore, a thorough report on a larger scale massive MIMO system still needs to be presented.

In this paper, we examine 2×2, 4×4, 8×8 and up to 16×16 massive MIMO systems. Comparative studies are undertaken for these different number of antenna configurations.

The rest of this paper is organized as follows. The details of massive MIMO system are described in Sect. 73.2. Section 73.3 describes STBC. In Sect. 73.4, the orthogonal designs (ODs) are explained. Section 73.5 describes quasi-orthogonal design. Section 73.6 present simulation results and conclusions are drawn in Sect. 73.7.

73.2 Massive MIMO System

A potential candidate for future mobile communication systems is massive MIMO, also known as Full-dimension MIMO or Hyper-MIMO [7]. This is a form of multi-user multiple antenna wireless communication which promises improvements in spectral efficiency and energy efficiency over 4G technology. A recognized feature of massive MIMO is that a large number of antennas can function for a significantly smaller number of active autonomous terminals. Large antenna arrays can potentially reduce uplink (UL) and downlink (DL) transmit power through coherent combining with increased antenna aperture [9]. Most recently, massive MIMO has been investigated as new cellular network architecture with several attractive characteristics [10].

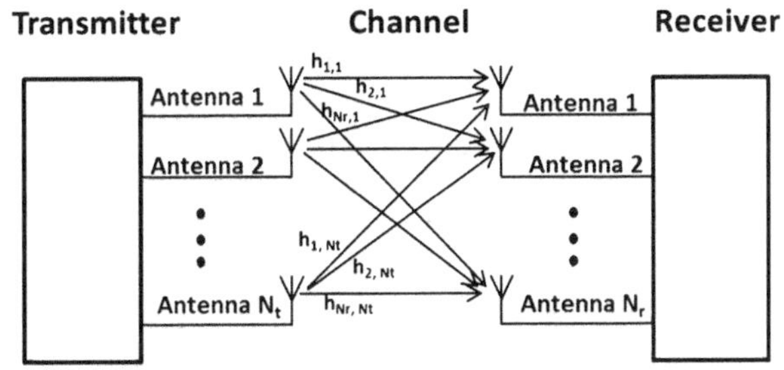

Fig. 73.1 Block diagram of massive MIMO system

Let N_t and N_r denote the number of transmit and receive antennas, respectively. The channel with N_r outputs and N_t inputs is denoted as a $N_r \times N_t$ matrix. A massive MIMO usually employs a large number of transmit and receive antennas, at least larger than 10 antennas. Figure 73.1 depicts a massive MIMO system block diagram. We consider a transmission scheme with multiple transmit and receive antennas. The MIMO system model can be represented as

$$\mathrm{x} = \mathrm{G}z + w, \tag{73.1}$$

where $x \in C^{(Nr\times 1)}$ and $z \in C^{(Nt\times 1)}$ denote the received and transmitted complex vector, respectively. $G \in C^{(Nr\times\ Nt)}$ is the complex channel matrix and $w \in C^{(Nr\times 1)}$ is noise vector.

73.3 Space-Time Block Code

Space-time block codes (STBC) are designed to obtain the maximum diversity order for the provided number of transmit and receive antennas. These codes are orthogonal and can achieve full transmit diversity specified by the number of transmit antennas. The STBC matrix has columns equal to the number of the transmit antennas and rows equal to the number of the time slots required to transmit the data. The matrix representation of OSTBC, D, is expressed in (73.2).

$$\mathrm{D} = \overset{\text{Transmit antennas}\ \longrightarrow}{\begin{bmatrix} a_{11} & a_{12} & \cdots & a_{1N_t} \\ a_{21} & a_{22} & \cdots & a_{2N_t} \\ \vdots & \vdots & \ddots & \vdots \\ a_{T1} & a_{T2} & \cdots & a_{TN_t} \end{bmatrix}} \Big\downarrow \text{Time slots} \tag{73.2}$$

where a_{ij} denotes information data symbol transmitted at time slot i and from jth transmit antenna. T represents the number of the time slots and N_t is the number of the transmit antennas. If the number of symbols transmitted per time slot is k, the code rate is defined as k/T [11].

73.4 Orthogonal Design

There are two types of orthogonal design theory: one for real numbers and other for complex numbers. For real or complex numbers, matrices satisfying the following equation are called orthogonal designs [12].

$$D^H D = \left(|a_1|^2 + |a_2|^2 + \cdots + |a_i|^2\right) I, \tag{73.3}$$

where D^H is the Hermitian of the matrix D, if D has complex numbers. D^H may also represent the transpose of the matrix D, if D has real numbers. In the real orthogonal designs, an $N_t \times N_t$ matrix is with real entries $\pm a_i$, $i = 1, \ldots, N_t$. In the complex orthogonal designs, an $N_t \times N_t$ matrix is with complex entries $\pm a_i$, $\pm_{ai}*$, $\pm\, ja_i$, and $\pm ja_i*$, $i = 1, \ldots, N_t$.

In complex orthogonal design, not only do the information symbols appear but also their conjugates. The ith column of the code matrix can contain either the information symbols $a_1, \ldots, a_T$, or their conjugates $a_1*, \ldots, a_T*$ only.

The most popular OSTBC is the Alamouti code [5]. This code uses a complex orthogonal design with full diversity at full rate. Alamouti code matrix can be shown as

$$\mathrm{D} = \begin{bmatrix} a_1 & a_2 \\ -a_2^* & a_1^* \end{bmatrix} \tag{73.4}$$

where * denotes complex conjugate. Here, two time slots are required to transmit two symbols. Thus, $k = 2$ and $T = 2$. Hence, the code rate of Alamouti Code is 1. However, when three or more transmit antennas are employed, the maximum transmission rate of the complex valued OSTBC with linear processing reduces to 3/4 [5].

r_1 and r_2 are the received signals at time t and $t + T$. Conjugating the signal r_2, the received signal can be expressed as (73.5)

$$\begin{aligned} r_1 &= h_1 a_1 + h_2 a_2 + \tilde{n}_1 \\ r_2^* &= -h_1^* a_2 + h_2^* a_1 + \tilde{n}_2. \end{aligned} \tag{73.5}$$

Therefore, the Eq. (73.5) can be described as

$$\begin{bmatrix} r_1 \\ r_2^* \end{bmatrix} = \begin{bmatrix} h_1 & h_2 \\ h_2^* & -h_1^* \end{bmatrix} \begin{bmatrix} a_1 \\ a_2 \end{bmatrix} + \begin{bmatrix} \tilde{n}_1 \\ \tilde{n}_2 \end{bmatrix} \tag{73.6}$$

or in short notation:

$$y = Ha + \tilde{n}, \tag{73.7}$$

where $y = [r_1, r_2^*]^T$ represents the receive vector. For MIMO channel matrix, the rows and columns of the virtual channel matrix H of Alamouti STBC are orthogonal:

$$HH^H = H^H H = \left(|h_1|^2 + |h_2|^2\right) I_2 = |h|^2 I_2, \tag{73.8}$$

where I_2 is the (2×2) identity matrix and h^2 is the power gain. It is certain that the equivalent virtual channel matrix [12] depends on the structure of the code and the channel coefficients. The channel coefficients h_1 and h_2 can be used by the decoder as channel state information (CSI) if they can be absolutely estimated at the receiver.

73.5 Quasi-orthogonal Design

Orthogonal space-time block code (OSTBC) can provide full transmit diversity with simple linear maximum-likelihood (ML) decoding complexity. In spite of these advantages, OSTBC has a code rate that is less than one when the number of transmit antennas more than two [4]. Therefore, the quasi-orthogonal space-time block codes are considered at the expense of slight orthogonality [12]. Each row and column contains all elements of a as shown in (73.9). Any element in a code word may occur with a positive or negative sign. The conjugate complex operation of symbols is only allowed on entire rows of the block matrix. This is required for holding quasi-orthogonality and low decoding complexity. Quasi-orthogonal STBC coding matrix is similar to the orthogonal space-time codes. However, any column vector in encoding matrix of the quasi-orthogonal space-time block codes is related to only one of the columns. In this quasi-orthogonal code design, the columns of the transmission matrix are divided into groups. While the columns within each group are not orthogonal to each other, different groups are orthogonal to each other. The most famous quasi-orthogonal space-time block code is proposed by Jafarkhani [6]. For 4 transmit antennas, a code with symbol transmission rate 1 was constructed from the Alamouti scheme as shown in (73.9):

$$D_4 = \begin{bmatrix} a_1 & a_2 & a_3 & a_4 \\ -a_2^* & a_1^* & -a_4^* & a_3^* \\ -a_3^* & -a_4^* & a_1^* & a_2^* \\ a_4 & -a_3 & -a_2 & a_1 \end{bmatrix} \tag{73.9}$$

OSTBC is decoded by decoding symbol one by one, and quasi-orthogonal STBC is decoded by decoding two pairs of symbols. The full transmission rate is useful and important for lower signal to noise ratios (SNRs) and higher bit error rates (BERs). Full diversity is efficient for higher SNRs and lower BERs. Since lower SNR range is of practical interest, QOSTBCs have attracted much attention recently.

73.6 Simulation Results

In order to evaluate a large massive MIMO, we have conducted performance evaluation of the QOSTBC based massive MIMO under additive white Gaussian noise (AWGN) channel and with ideal channel state information. We analyze the bit error rate (BER) performance of the massive MIMO system. It is assumed that all antennas in the system use the same energy level. The BER comparison has been undertaken according to the number of antennas. Note that it is possible to create different matrices that obey the quasi-orthogonal property. QOSTBC can achieve the full transmit diversity of $N_t = 2, 4, 8$ and 16 and code rate is 1.

The BERs of QOSTBC for 2 and 4 transmit and receive antennas are plotted in Fig. 73.2a and 8 and 16 transmit and receive antennas are plotted in Fig. 73.2b. When the number of antennas increases, the BER performances of the system dramatically improve. For a large antenna configuration such as 16 × 16, the

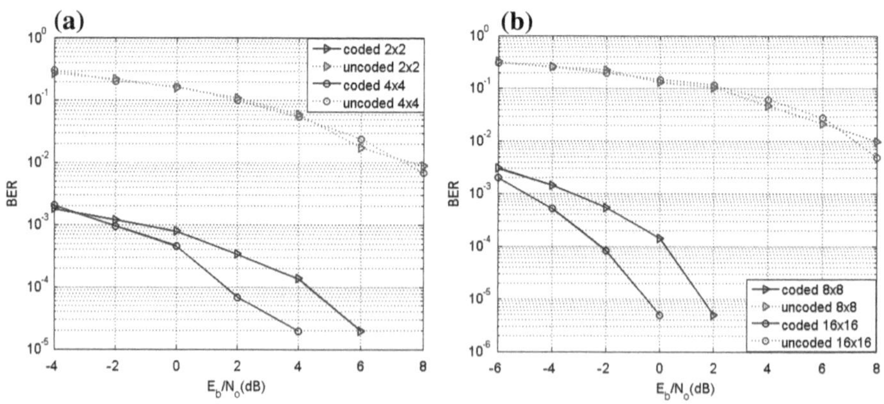

Fig. 73.2 Performances of massive MIMO systems: **a** 2 × 2 and 4 × 4 antennas **b** 8 × 8 and 16 × 16 antennas

performance gain increases further. This is due to the fact that increasing antennas in massive MIMO provide significant increase in the BER performance through a diversity effect.

73.7 Conclusions

A large massive MIMO system with up to 16 × 16 antenna configuration has been evaluated. For a massive MIMO, performance analysis as well as energy efficiency must be conducted in a large antenna configuration. By making use of quasi-orthogonal space-time block code (QOSTBC) with full diversity and full rate, the simulation results show that the 16 × 16 configuration provides the best performance among other considered configurations. A more thorough analysis may need to be undertaken with realistic channel conditions.

References

1. Foschini G (1996) Layered space-time architecture for wireless communication in a fading environment when using multi-element antennas. Bell Labs Tech J 1(2):41–59
2. Bliss DW, Amanda MC (2005) MIMO wireless communication. Lincoln 15(1):97–126
3. Foschini GJ, Gans MJ (1998) On limits of wireless communications in a fading environment when using multiple antennas. Wireless Pers Commun 6:311–335. Kluwer Academic, Netherlands
4. Tarokh V, Calderbank AR (1999) Space-time block codes from orthogonal designs. IEEE Trans Inf Theory, 45(5):1456–1467. IEEE Press
5. Alamouti SM (1998) A simple transmit diversity technique for wireless communications. IEEE J Sel Areas Commun, 16(8):1451–1458. IEEE Press
6. Jafarkhani H (2001) A quasi-orthogonal space time block code. IEEE Trans Commun, 49 (1):1–4. IEEE Press
7. Cornell University Library, http://arxiv.org/abs/1304.6690
8. Lee G, Park J, Sung Y, Seo J (2012) A new approach to beamformer design for massive MIMO systems based on k-regularity. In: IEEE Globecom Workshops, IEEE Press, Anaheim, pp 686–690
9. Ngo HQ, Larsson EG, Marzetta TL (2013) Energy and spectral efficiency of very large multiuser MIMO systems. IEEE Trans Commun, 61(4):1436–1449. IEEE Press
10. Marzetta TL (2010) Noncooperative cellular wireless with unlimited numbers of base station antennas. IEEE Trans Wirel Commun, 9(11):3590–3600. IEEE Press
11. Su W, Xia XG (2004) A systematic design of high-rate complex orthogonal space-time block codes. IEEE Commun Lett, 8:380–382. IEEE Press
12. Jafarkhani H (2005) Space-time codes: theory and practice. Cambridge University Press, Cambridg

Chapter 74
Encrypted Data Group Authentication for Outsourced Databases

Miyoung Jang, Ara Jo and Jae-Woo Chang

Abstract Cloud computing has been spotlighted as a new paradigm of database management system. However, privacy needs to be preserved for databases that are valuable and sensitive against unauthorized accesses. For this, two issues of data security, including data confidentiality and query result integrity, become major concerns for users. Existing bucket-based data authentication methods have problems of data disclosure and transmission overhead, due to the unsophisticated data grouping strategy. In this paper, we propose a privacy-aware query authentication index which guarantees data confidentiality and query result integrity for users. We privately partition a spatial database into small groups by using periodic function and generate a signature of each group. The group signature is used to check the correctness and completeness of outsourced data when answering a range query to users. Through performance evaluation, it is shown that proposed method outperforms the existing method in terms of range query processing time up to 3 times.

Keywords Database outsourcing · Encrypted database · Data authentication index · Query result integrity

74.1 Introduction

Due to the advancement in cloud computing technologies, the research on the outsourced database has been spotlighted as a new paradigm of database management system. Small-size Location-based Service (LBS) businesses outsource

M. Jang · A. Jo · J.-W. Chang (✉)
Department of Computer Engineering, Chonbuk National University, Jeonju-si, Republic of Korea
e-mail: jwchang@jbnu.ac.kr

M. Jang
e-mail: brilliant@jbnu.ac.kr

A. Jo
e-mail: ara10626@jbnu.ac.kr

J.J. (Jong Hyuk) Park et al. (eds.), *Ubiquitous Computing Application and Wireless Sensor*, Lecture Notes in Electrical Engineering 331,
DOI 10.1007/978-94-017-9618-7_74

their spatial database to a service provider in order to reduce cost for data storage and management. The services provider maintains the data and answers queries to users. However, privacy needs to be preserved for sensitive databases (e.g., personal location data, financial and medical record) that are valuable and sensitive to unauthorized accesses. In addition, because a service provider is not the owner of outsourced database, it must prove the correctness and completeness of query result to users. In this context, two issues of data security including data confidentiality and data integrity become major concerns for users. Data confidentiality refers to protecting data access and disclosure from unauthorized parties. Data integrity refers to the trustworthiness of data, especially for query users. For example, since the service provider is not fully trusted, the service provider may inject fake data in the query result, or may delete some genuine result data before sending them to users. The straightforward solution for protecting data confidentiality is database encryption. Many researches have been studied on database encryption, such as AES [1], OPES [2], schemes. By adopting those encryption schemes, data confidentiality for users can be solved.

On the other hand, in order to guarantee data integrity, verification information is sent to users with query result so that the result can be verified by using data owner's signature. Previous data authentication researches [3–18] can be categorized into three classes: signature-based approaches, authenticated data structures and bucket-based authentication. However, authentication data structure does not guarantee the data confidentiality, since it cannot be built on the encrypted data. In addition, tree-based index suffer from inflexible data update and verification object transmission costs. To address these problems, J. Wang et al. [4] proposed a bucket-based authentication where a bucket contains a bucket id, data range (upper-lower bound), number of tuples in a bucket and a checksum. A checksum is similar concept to the data signature and generated by using a Hash function. The limitations of existing bucket-based authentication are as followings. First, because it generates a bucket with equal width of the data range, the original data distribution can be disclosed. Moreover bucket id is assigned as the ascending order of the data range. The existing bucket-based authentication cannot fully provide the data security. Secondly, existing bucket-based authentication methods only consider conventional relational database. The distribution of database may not be the objects to be protected. However, in terms of spatial database, the distribution of them could involve meaningful knowledge. To provide full-range of privacy protection coverage for spatial database, any additional information, including data distribution, of database can be revealed.

To solve the problems, we propose a privacy-aware query authentication which supports data correctness and completeness for users. To support data confidentiality, it is assumed that databases are encrypted before outsourced to a service provider by using AES algorithm. In order to process range queries over encrypted database, a privacy-aware query authentication index is proposed. The authentication scheme includes a periodic function-based data grouping scheme, which privately divides a database into small groups. The prime goal of using periodic function for database partitioning is to minimize the disclosure risk of spatial data distribution while reducing the number of false-positive results bounded within the cyclic ranges. To provide query result integrity, a group signature generated by condensed-RSA function is stored in the index. When answering a range query, the signatures of retrieved data group are sent with results. Then, the users checks the correctness and completeness of the query result by comparing the received signatures with genuine signatures.

The rest of this paper is organized as follows. Section 74.2 presents the related Work. In Sect. 74.3, our proposed data integrity scheme is introduced. Section 74.4 provides an experimental evaluation on existing and proposed methods. Section 74.5 concludes this paper with brief summary.

74.2 Related Work

Existing data authentication researches can be categorized into signature-based approaches, authenticated data structures and bucket-based authentication. In this section, we introduce them briefly.

First, three signature-based approaches were proposed: tuple level signature [5], aggregated signature [5, 6] and signature chaining [6]. In tuple level signature, each tuple is signed by a data owner and the integrity of tuples in query result is verified based on the signature. Secondly, in terms of aggregated signature, a signature is calculated by a concatenation of individual signatures. Compared to the tuple level signature, aggregated signature has faster verification time than that of the tuple level signature since the signature condensing scheme is faster than signature verification of multiple tuples. Finally, a signature chain is introduced in [6] where a signature is signed on three neighboring tuples, i.e. $s_i = \text{sign}(t_{i-1}|t_i|t_{i+1})$ Note that t_1, ...,t_N are sorted tuples in the database based on the data owner's preference or frequently performed query. The tuple lever signature and aggregated signature only support data correctness. The signature chain approach guarantee both data correctness and completeness but the query processing cost is drastically increased with the large dataset.

Second, in authenticated data structures, Merkle Hash Tree (MH-Tree) is first studied in the field of Cyptography [7]. Tuples are organized into a tree so that one signature to the root node can guarantee the data integrity of other nodes in the tree. Merkel Hash Tree (MHT) is a main-memory binary tree, where each leaf node contains the hash of a tuple, and each internal node contains the hash of the concatenation of its child nodes. To authenticate range queries, the records are sorted on the query attribute and indexed by a MH-tree. The verification process contains following steps. The service provider determines the boundary records whose are neighboring the query range. Then, the paths from the root to the boundary nodes are stored. The verification object contains the paths of the boundary nodes and all visiting nodes' hash information for searching the query result nodes. This verification object is sent to the query user so that the user can rebuild the root's signature. It the reconstructed signature matches the original signature, the result is sound. Since one MHT is built on one attribute, to support authentication of multi-dimensional range queries, multiple MHTs are required to be constructed. The tree-based authentication methods suffer from data update cost, since all the hash values are needed to be updated even for a single tuple insertion. Thus, high update cost for the data owner.

Finally, Bucket-based index was proposed by Hacigumus [16] and extensively studied in [4, 17, 18]. A bucket is d-dimensional rectangle and bucket-based index contains a bucket id, data range (upper-lower bound), number of tuples in the bucket and a checksum. A bucket is generated by partitioning databases with the equivalent data range (values) or with the equal number of data (count) in a bucket. A bucket checksum is a hash digest such that unique and efficient in calculation. Upon receiving the result with checksums, the user re-generates the checksum with results and compares them with received ones. If the reconstructed checksum matches the original, it guarantees the query result integrity. In bucket-based index, security lies in the anonymity of tuples within the same bucket. The larger the bucket size, the less the information disclosure and more secure the bucket index is. In bucket-based authentication, the authentication index contains lower and upper range of each bucket. Thus, applying existing methods to the spatial database is valuable to the data distribution disclosure attack.

74.3 Periodic Function-Based Data Group Index

Figure 74.1 depicts overall system architecture and data flow of the authenticated query processing. System models assumed that there are three main components in this architecture: data owner, service provider and trusted user. The scenario of data flow in Fig. 74.3 can be explained as follows: In the pre-processing phase, data owner performs data encryption for outsourcing spatial data, and authentication bucket index generation for data integrity guaranteeing query processing. Then, the data owner outsources the E(P) and the authentication index to the service provider (1-a). At the same time, the owner forwards the database encryption information and the auditing data to trusted users so that they can utilize the range query processing with the service provider as well as the query result verification (1-b). At query processing time, a trusted user transforms its query (Q) to a transformed query (E(Q)) by applying the same encryption of outsourced database. Then, the trusted user issues the query to the service provider (2-a). When answering a query, the service retrieves the authenticated data index (2-b) and forwards the encrypted query result data with the verification data (i.e., signature) (2-c). Upon receiving query results, the authenticated user generates group signature for query results by using the key from the data owner. If generated signature is identical to the signature that is sent with the query results, the user can verify the correctness and completeness of query results.

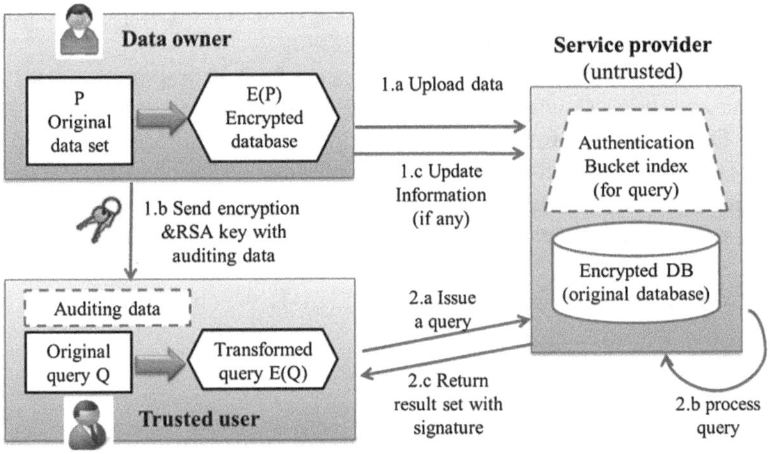

Fig. 74.1 System architecture and data flow

74.3.1 Periodic Function Based Private Data Authentication Index

In this part, we introduce a periodic function based database grouping scheme and a private data authentication index for processing queries at service provider. The periodic function based database grouping scheme partitions a database into small-groups. Then, each small-group is signed by the data owner using Condensed-RSA [19] with POI ids within the group. By this means, the private authentication index is generated without revealing the partitioning information to unauthorized accesses.

Our system model assumed tuple-level encryption for processing range queries. For query processing efficiency, existing work generally partition the database into groups with varying methods. Wang et al. [4] generate a bucket-based authentication index by partitioning databases with the equivalent data range (values) or with the equal number of data (count) in a bucket. However, their authentication index contains lower and upper range of each bucket. Hence, applying existing methods to the spatial database suffers from the data distribution problems. To resolve this problem, in this paper, a periodic function based database partitioning scheme is proposed. In this scheme, a spatial database is divided into Group $G = \{G_1, G_2, \ldots, G_n\}$ and then the periodic function is applied on each G_i to divide them into small-group $g = \{g_1, g_2, \ldots, g_k\}$. Here, a function f is said to be periodic with period P (P being a non-zero constant) if we have $f(x + P) = f(x)$ for all values of x. For example, the sine function is periodic with period 2π, since $f(x + 2\pi) = \sin x$ for all values of x. This function repeats on intervals of length 2π. By using this property, the database is divided where the key values are multiple of periods. Then, to generate small groups g in each G_i, we compute the upper and lower peaks of the periodic graph where the differentiation results $f'(x)$ of the function $f(x)$ are zero. These peaks become the partitioning points for generating small-group g. Figure 74.2 illustrates the example of small group generation with the periodic function $f(x) = \{\sin 2x + \cos x/2\}/2$.

In Fig. 74.2 (b), there are 8 upper and lower peaks for $f(x) = \{\sin 2x + \cos x/2\}/2$. So the dataset within a G_i will be divided into 9 small-groups $g = \{g_1, g_2, \ldots, g_9\}$ as shown in Table 74.1. To protect the privacy of partitioning information, group ids are randomly assigned to small-groups. To support range query in spatial database, we perform a multi-dimensional partition for x and y axes. After constructing a data partition, a signature for each small-group is generated by using Condensed-RSA. An RSA signature is computed on the hash of the input message. Let $h()$ denote a cryptographically strong hash function (such as SHA-1) which takes a variable-length

(a)

$$f(x) = \frac{\left(\sin(2x) + \cos\left(\frac{x}{2}\right)\right)}{2} \; \textit{is a periodic function}$$

$$\textit{with period } 4\pi \; (12 < 4\pi < 13)$$

$$\begin{aligned} f'(x) &= \left(\frac{1}{2} + \sin 2x'\right) + \left(\frac{1}{2} + \cos\left(\frac{x}{2}\right)'\right) \\ &= \left(\frac{1}{2} \times \sin'(2x) \times 2x'\right) + \left(\frac{1}{2} \times \cos'\left(\frac{x}{2}\right) \times \frac{x}{2}'\right) \\ &= \left(\frac{1}{2} \times \cos 2x \times 2\right) - \left(\frac{1}{2} \times \sin\frac{x}{2} \times \frac{1}{2}\right) \\ &= \cos 2x - \frac{1}{4} \times \sin\frac{x}{2} \end{aligned}$$

(b)

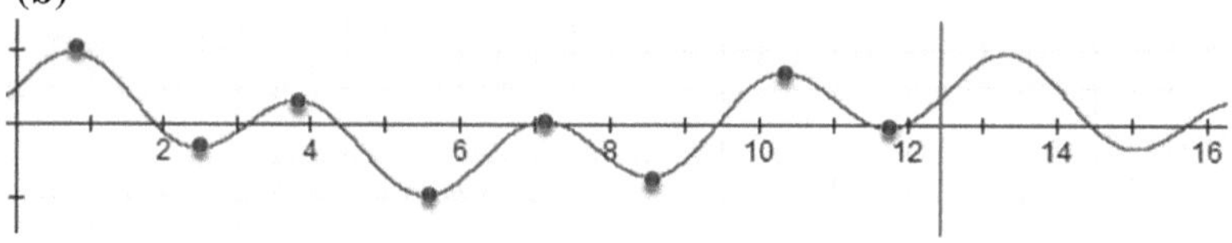

Fig. 74.2 Private data partition with periodic function **a** Sub-group calculation based on cyclic block **b** Small-group division with peaks

Table 74.1 Database divided into 9 groups

Order	Group id	Range
sub_1	2	$0 \le x \le 0.740$
sub_2	7	$0.740 \le x \le 2.475$
sub_3	1	$2.475 \le x \le 3.808$
sub_4	3	$3.808 \le x \le 5.543$
sub_5	5	$5.543 \le x \le 7.119$
sub_6	9	$7.119 \le x \le 8.526$
sub_7	4	$8.526 \le x \le 10.324$
sub_8	8	$10.324 \le x \le 11.730$
sub_9	6	$11.730 \le x \le 4\pi$

input and produces a fixed-length output. For an input message m, $h(m)$ denotes the hash of m and a standard RSA signature on message m, and computed by using a given formulation from [19]. In this scheme, each tuple signature is generated by using RSA with its id (e.g., POI id).

$$\sigma = \mathrm{h(m)}^{d} (mod\ n) \tag{74.1}$$

Mykletun et al. [6] defined the Condensed-RSA as a product of individual signatures in each group (Eq. 74.2). In the private authentication index, we generate a small group signature as a Condensed-RSA of each data signature.

$$\sigma_{1,t} = \prod_{t=1}^{t} \sigma_i (mod\ n) \tag{74.2}$$

By using the authentication index, the authentication of range query result is performed as illustrated in Algorithm 1. The query processing with query result integrity auditing is performed between two parties: a query user and a service provider. First, the query user encrypts a query range with the periodic function, so that the query range is converted to a set of small-group ids where their covering area overlaps the query range (line 1). Then, the user issues a query to the server (line 2). Upon receiving a query, the service provider retrieves private authentication index to return an encrypted original data set for the query (line 3–4) and their signatures (line 5–8). Finally, the user generates signatures of query results by using the RSA key sent from the data owner, and compares the signature with result signature (line 9–13). If generated signature is identical to the signature with the query results, the user can verify the correctness and completeness of query results.

Algorithm 1: Range query()

Input: Query range [low, upper), Encryption Key CK, RSA
Output: Query result set R, signature set S
/*query user*/
1: encrypt query range with CK;
2: request the server for all data whose sub-group overlaps the query region;
/*service provider*/
3: for each sub_group range ∊ query **do**
4: query result += all data in the sub_group
5: for each requested sub_group id **do**
6: search signature index
7: S += signature
8: return R and S,to the user
/*query user*/
9: generate signature RSA(R)
10: compare RSA(R) and S
11: if matches,
12: select the result R
13: else delete R

74.4 Experimental Evaluation

We compare the performance of the proposed range query processing algorithm with the bucket-based approach [4] under different settings. The algorithms are implemented by using Windows 7 Ultimate K operating system with Microsoft Visual Studio 2010 running on an Intel(R) Core(TM) i5-3470 CPU 2.30 GHz with 4 GB of RAM. We used two different spatial datasets: the real data (Northern East America (NE)), uniform data (UNI). The synthetic datasets containing 100,000 point of interests (POIs) are generated by using Generate Spatio-temporal Data (GSTD) [20], while the NE dataset contains 119,898 POIs. For experiments, we varied the query window extent from 0.01, 0.02, 0.04, 0.06 to 0.08 % and the bucket size from 25*25, 50*50, 75*75 to 100*100. The bucket size refers to the number of buckets when partitioning a database into small groups. The range query window extent mean the area of range query rectangle. We performed 100 queries with different parameters and an average value over 100 query results is used.

Figure 74.3 illustrates the construction cost for proposed authentication index and existing bucket-based authentication as a function of the bucket size. This cost includes both the time for partitioning a database into small groups (buckets) and signature creation time. In all cases, proposed algorithm requires less time than the bucket-based approach. When inserting the Point of interest (POI) to the index, our method decides the bucket id for each POIs by simple calculation of polynomial expression. In case of bucket-based approach, the bucket id is assigned by comparing its (x, y) coordinates with bucket partition values. This is the reason why our index has fast construction time compared to the existing approach.

We measured the query processing time with query result verification time on both approaches: bucket-based approach and proposed algorithm. The verification time includes signature re-construction time and the time required for comparing the reconstructed signatures with original ones. In this experiments, we varied the query window extent from 0.01 to 0.08 % and the bucket size from 25*25 to 100*100. The experimental results are shown in Fig. 74.4. This shows that the proposed approach achieves the improved performance compared to the existing

Fig. 74.3 Index construction time

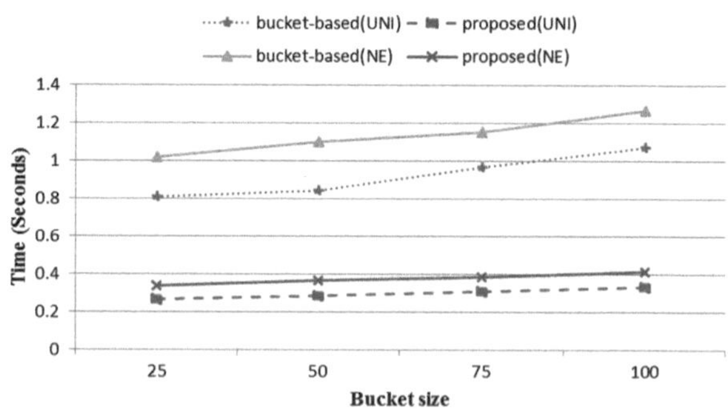

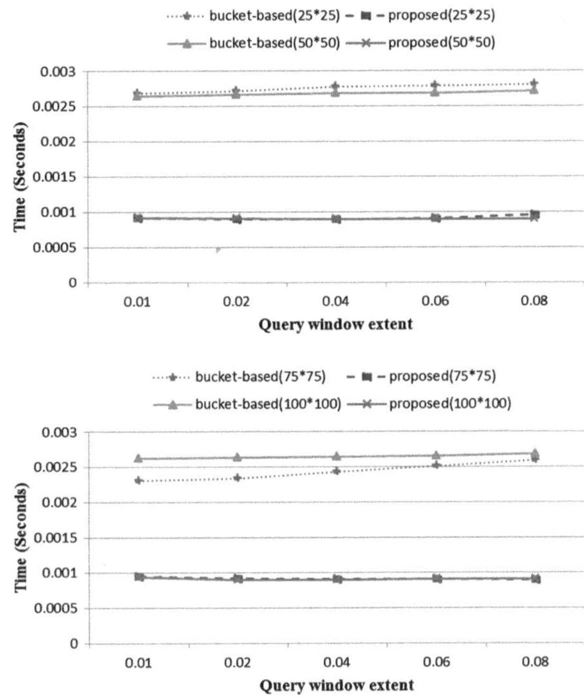

Fig. 74.4 Query processing time with verification time (NE data)

work. For all cases, the retrieval time is almost constant as the query window size is increased. For example, when the bucket size 50∗50, our range query processing algorithm requires only 0.0075 s whereas the bucket-based approach takes 0.00272 s for processing query ranged 0.08 % of the whole area.

We measured the size of false positive data in query result on both approaches Since the primary goal of bucketization is to minimize the data disclosure while providing a query result integrity to users, it is important to consider the false-positive data size in query result. The size of false-positive data is shown in Fig. 74.5. For the simplicity, we only delivered the result from NE data. The number of false positive data is calculated by subtracting the genuine query result size from the query result set size acquired by query processing. For all cases, the size of the false positive data is almost same for two approaches. Based on this result, it is proved that our method takes less time in query processing while returning reasonable number of false-positive data compared to the existing method. In addition, since our private authentication index generates data partition by using periodic function, we provide enhanced data privacy compared to the bucket-based authentication which partitions the database with equal-range or equal-number of data.

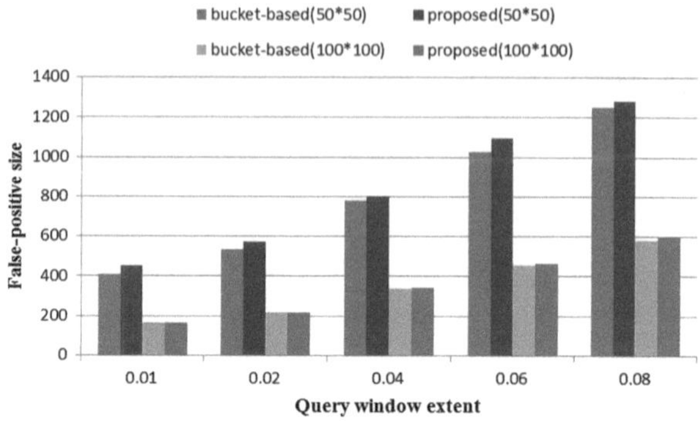

Fig. 74.5 Size of false positive data in query results (NE data)

74.5 Conclusion

In this paper, we propose a privacy-aware query authentication which supports data correctness and completeness for users. A periodic function-based data grouping scheme is designed to partition a database into small groups for generating a signature of each group. The group signature is used to check the correctness and completeness of outsourced data when answering a range query to users. Through performance evaluation, it is shown that proposed method outperforms the existing method in terms of range query processing time up to 3 times while providing similar performance in returning number of false positive data.

As a future work, we will extend the proposed algorithm to support variety of query types (e.g., k-NN, skyline queries) in spatial database.

Acknowledgments This research was supported by Basic Science Research program through the National Research Foundation of Korea (NRF) funded by the Ministry of Education, Science and Technology (grant number 2013010099)

References

1. FIP Standard (2001) Advanced encryption standard (AES). National Institute of Standards and Technology (NIST)
2. Chow CY, Mokbel MF, Liu X (2011) Spatial cloaking for anonymous location-based services in mobile peer-to-peer environments. GeoInformatica 15(2):351–380
3. Kerr S, Krkpatrick MS, Bertino E (2010) PEAR: a hardware based protocol authentication system. In: Proceedings of the 3rd ACM SIGSPATIAL international workshop on security and privacy in GIS and LBS, pp 18–25
4. Wang J et al (2010) Bucket-based authentication for outsourced databases. Concurrency Comput Pract Experience 22(9):1160–1180
5. Mykletun E, Narasimha M, Tsudik G (2004) Signature bouquets: immutability for aggregated/condensed signatures. In: European symposium on research in computer security (ESORICS), pp 160–176

6. Mykletun E, Narasimha M, Tsudik G (2006) Authentication and integrity in outsourced databases. J ACM Trans Storage (TOS) 2(2):107–138
7. Merkle RC (1990) A certified digital signature. Advances in cryptology—CRYPTO'89 proceedings. Springer, New York, pp 218–238
8. Narasimha M, Tsudik G (2005) DSAC: integrity for outsourced databases with signature aggregation and chaining. In: Proceedings of the 14th ACM international conference on information and knowledge management, ACM, New york, pp 235–236
9. Sacharidis D, Mouratidis K, Papadias D (2010) k-Anonymity in the presence of external databases. IEEE Trans Knowl Data Eng 22(3):392–403
10. Yang Y, Papadias D, Papadopoulos S, Kalnis P (2009) Authenticated join processing in outsourced databases. In: ACM SIGMOD international conference on management of data, ACM, New york, pp 5–18
11. Liu D, Wang S, (2012) Query encrypted databases practically. In: Proceedings of the ACM conference on computer and communications security, ACM, New York, pp 1049–1051
12. Hore B et al (2012) Secure multidimensional range queries over outsourced data. Int J Very Large Data Bases 21(3):333–358
13. Balpande S et al (2012) Data integrity and confidentiality in outsourced database. In: International conference and workshop on recent trends in technology, (TCET)
14. Devanbu P, Gertz M, Martel C, Stubblebine S (2003) Authentic data publication over the internet. J Comput Secur 11(3):291–314
15. Hacigumus H, Iyer B, Li C, Mehrotra S, (2002) Executing SQL over encrypted data in the database service provider model. ACM SIGMOD, New York, pp 216–227
16. Hore B, Mehrotra S, Tsudik G (2004) A privacy-preserving index for range queries. In: Proceedings of the 30th international conference on very large data bases-Volume 30, VLDB Endowment, pp 720–731
17. Wang J, Du X (2008) A secure multi-dimensional partition based index in DAS. Progress in WWW research and development. Springer, Heidelberg, pp 319–330
18. Wang C, Ku W, (2012) Efficient evaluation of skyline queries in wireless data broadcast environments. In: Proceedings of the 20th international conference on advances in geographic information systems, ACM, New York, pp 442–445
19. Rivest R, Shamir A, Adleman L (1978) A method for obtaining digital signatures and public-key cryptosystems. Commun ACM 21(2):120–126
20. Theodoridis Y, Silva J, Nascimento M (1999) On the generation of spatiotemporal datasets. Adv spat databases. Springer. Heidelberg, pp 147–164

Chapter 75
De-Word Classification Algorithm Based on the Electric Power of Large Data Library Retrieval

Xiaoli Guo, Huiyu Sun, Ling Wang, Zhaoyang Qu and Wei Ding

Abstract In order to improve the performance of text classification and information retrieval in big data of electric power domain, we propose a novel Chinese language classification algorithm—De-word classification algorithm. Focusing on the key role played by the De-word in modern Chinese language, this algorithm examines Chinese text classification method from a unique angle. Besides, on the basis of traditional weighted algorithm, it designs a novel relevance weighting model—De-TFIDF, and achieves a higher correlation in text information retrieval. Experiments show that, De-word classification algorithm significantly improves the efficiency of text classification, significantly improved information retrieval performance.

Keywords Big data · Classification algorithm · De-word · Relevance

75.1 Introduction

The development of information technology and computer promote the progress of human civilization. In recent years, the number of all kinds of electronic documents is growing at an unprecedented speed. The global data volume has reached 2.8 ZB

X. Guo · H. Sun · L. Wang (✉) · Z. Qu · W. Ding
Northeast Dianli University, Jilin 132012, China
e-mail: smile2867ling@gmail.com

X. Guo
e-mail: 243589657@qq.com

H. Sun
e-mail: 690011768@qq.com

Z. Qu
e-mail: qzywww@mail.nedu.edu.cn

W. Ding
e-mail: 649993290@qq.com

J.J. (Jong Hyuk) Park et al. (eds.), *Ubiquitous Computing Application and Wireless Sensor*, Lecture Notes in Electrical Engineering 331,
DOI 10.1007/978-94-017-9618-7_75

at 2012, and is expected to reach 40ZB at 2020. With the continuous expansion of power grid, power informatization construction is gradually advancing, a large number of text type data has been produced [1]. Face of such vast ocean of information, it is particularly important of how to quickly retrieve relevant information. It may take longer time to understand information without effective classification and retrieval method [2]. In addition, the reliability management of massive amounts of information is also a basic requirement of the smart grid construction [3]. In the process of power grid construction and maintenance, all kinds of text information such as all kinds of hidden information, job security, patrol inspection information, etc., all these information in the form of text, on the information classification of efficient and effective management, to improve the level of information management of the construction of the smart grid. Therefore, how to dig out the useful information from the power big data resource has become the hotspot in the research of data mining and semantics research area.

There are many kinds of classification methods, which are through the text and been easily to be converted into representation to realize classification or mechanical text matching, and neglected the practical significance of the word itself on the semantic, caused the classification effect is not ideal, such as inaccurate retrieval problem. In order to solve this problem, this paper proposed De-word classification algorithm based on the power of big data library retrieval. Based on the weight of the existing algorithm, the article used the word "De", which often being ignored and the most common words, dig deeper into the correlation between the text information and increase the correlation text retrieval accuracy.

75.2 Related Work

Typical text categorization algorithm mainly include decision trees, K nearest neighbor (KNN), association rules, support vector machine (SVM) [4], Bayesian algorithm (Bayes theorem), rough set and neural network, etc. Now is generally agreed that K nearest neighbor (KNN) method and support vector machine (SVM) method is the best way to effect of text categorization. The main text categorization algorithm accuracy is shown in Fig. 75.1.

In recent years, the mainstream of text classification research is basically around the text representation method, the characteristics of the reduction, the classifier fusion and classifiers to improve the direction. An improved KNN text classification algorithm based on clustering center was proposed in Yong Z, et al. To the traditional KNN text classification algorithm didn't reflect the different importance of different samples [5]. A Sectile algorithm based on mutual dependence and the classification of the equivalent radius was proposed in Wang J, et al., which had good extension performance, suited for the occasion of large amount of data [6]. But the distance coefficient of this method is randomly selected, the classification result is instability.

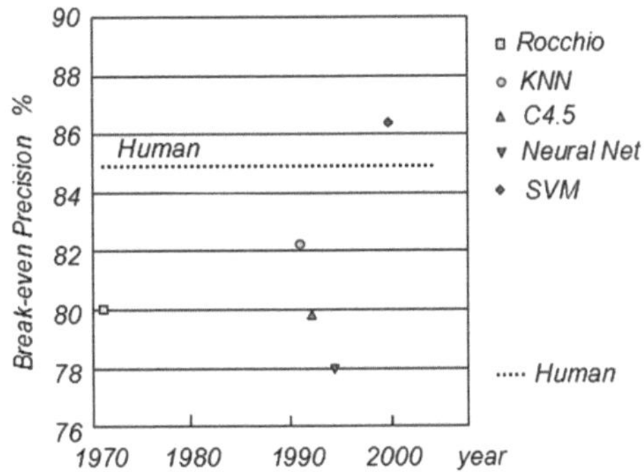

Fig. 75.1 Precision of the main text classification algorithm

For text classification technology applied in different fields encountering various problems, many scholars also give the corresponding solutions. A single label text classification method was proposed in Colace F, et al. that performed better than baseline methods when the number of labeled examples is small [7]. A multi-subjects text classification algorithm based on hyper-sphere support vector machine was proposed in Ai Q et al. to solve multi-subjects text classification can not be solved by standard SVM multi-class classification algorithms [8]. But when the sample size are not balanced, the methods of training the classification error prone to small sample size categories, so it is not applicable to the sample number of unbalanced data set.

The above described classification algorithms improve the classification performance, but also a common problem existence: these methods are starting from the angle of how to improve the efficiency of the traditional text classification method, but did not find the characteristics of the language itself; there is no unique role as special word statement. These special words are usually in the statement to the language logic function, which makes the performance limits. Therefore, by the integrated use of traditional text classification algorithm and special words, this article puts forward De-word classification algorithm, the algorithm of text classification constraints, greatly improves the accuracy and the precision of text categorization.

75.3 De-Word Classification Algorithm

75.3.1 Classifications Idea

Study found that the modern Chinese language habits the word “De” is the most commonly in modern Chinese word, especially in the journal title, there is a common feature of some items, that these terms can be divided into two parts, with “De” as the boundary, This paper define it as the prefix and suffix word. Prefix words are almost the same circumstances, suffix words although different, but because of the existence of a certain relationship, these suffixes and prefixes are therefore, there are some inherent links between these suffixes, such as containment, parallel relationship, as shown in Table 75.1.

The prefix words are almost the same, the suffix words are not the same, but the suffix words such as relay protection and static voltage stability of a correlation. Between the static voltages stability—will affect the relay protection. Therefore, the unique role can use "De" word in Chinese grammar in the classification of the text, the discovery of association between the text implied, so as to improve the text classification and retrieval efficiency, solve the low retrieval performance.

75.3.2 *Classification Process*

Based on the research of the current mainstream method, considering the characteristics of electric power big data library, De-word classification algorithm mainly includes three steps:

Step 1: text preprocessing, including the removal of stop words, assign ID, segmentation.

Step 2: semantic relevancy algorithm is used to calculate and store entries between search terms related degrees, and according to the correlation and cluster size rearrange entries.

Step 3: Design relevance weighting model De-TFIDF, based on correlation between vocabularies for entry sequence table Step 2 was subjected to re-adjust the weight indicators, to get the final retrieval results.

Table 75.1 Sample data table

(a)
火电的低碳生活
当前燃煤火电的效益审视
改进火电的燃烧方式亟待进行
燃煤火电的技术开发状况
(b)
Thermal power ***of*** the low carbon life
The current efficiency ***of*** coal-fired thermal power
To improve the way ***of*** coal combustion
Development status ***of*** coal-fired thermal power technology

75.4 Relevance Weighting Model de-TFIDF

75.4.1 Relevancy Ranking Table

First, Classifying data set, and setting the ID for text entry, and word processing. Second, remove stop words and some operations had no effect on the classification of vocabulary.

The proposed relevance weighting model De-TFIDF is the text right TF-IDF is based on TF-IDF algorithm design of text weight calculation method. TF-IDF is a statistical method for assessing a term for a set of files or a corpus of a document in which the degree of importance. By TF-IDF algorithm to calculate each class text vocabulary weights sort these words, get the weight sort table.

Select the maximum weighting value K words before the word is characterized, such as text feature vector represents. Using these feature vectors can search words with the training set is represented as a vector of entry, easy to calculate semantic relatedness between them.

Then it is calculated using vector cosine correlation between the terms, obtaining relevance ranking table between a search term and the training set.

75.4.2 Correlation Weighting Correction

After the above calculated relevancy ranking table only calculate the correlation between the traditional words, no more words between digging deeper relevance, and therefore of relevance to sort the table accordingly amended.

Definition 75.1 Correlation correction factor α

Assuming the number of entries in the dataset direct number entry contains the search term t is N_t, indirect association with t exist through other vocabulary, max {relevancy ranking table} relevancy ranking table indicates the maximum correlation to the relevant entry degree value, then the correlation correction coefficient α is defined as follows:

$$\alpha = \frac{N_t}{N_t + N_t'} \cdot \max\{ relevancy \quad ranking \quad table \} \tag{75.1}$$

Correlation value reflects the correction coefficient α is associated with the search term correlation between words, the paper used as the standard correction relevance ranking tables to meet different conditions, by the value of α is adjusted to a reasonable within the scope of the amendment, in order to achieve a high correlation to the text retrieval.

Definition 75.2 Intra-class correlation correction value W_{St}

Class S is said on behalf of (the data set according to different categories are set to S1, S2, S3 ……), if the total number of entries in the category S is m, then $term_l$ for Class S W_{St} relevance correction value can be expressed as:

$$W_{St} = \alpha \frac{m_t}{m} \tag{75.2}$$

where: m_t of $term_l$ number of S in the category.

W_{St} correlation correction value is reflected in the value of each class among the respective association with a search word associated with each class of cluster analysis are obtained in order to achieve a high degree of correlation between the information retrieval based on their relevance.

De-word classification algorithm with the previous classification algorithm biggest difference is that the algorithm is weighted by relevance weighting model De-TFIDF sort of text-related degrees, so that when the text retrieval, easy to find the greatest correlation with the search term text.

75.5 Experimental Analyses

In the experiment, using the electricity library paper names from China National Knowledge Infrastructure as the experimental data, a total of 100,000 paper titles, choosing 80,000 titles, including 70,000 paper titles of training sets, are divided into five categories as Parts 1, 2, 3, 4, and 5 and test sets include 10,000 paper titles.

75.5.1 Evaluation Method

Five kind of data in the process of experiment, to test their accuracy and recall rate respectively and F test values, at the same time, will be compared with the classification algorithm based on support vector machine (SVM). After that, the keywords retrieval experiment was carried out.

75.5.2 The Result of the Experiment

For five types of data classification based on SVM and based on the power of big data library retrieval De-word classification algorithm testing their precision, recall, and F-measure, the results are shown in Figs. 75.2, 75.3, and 75.4.

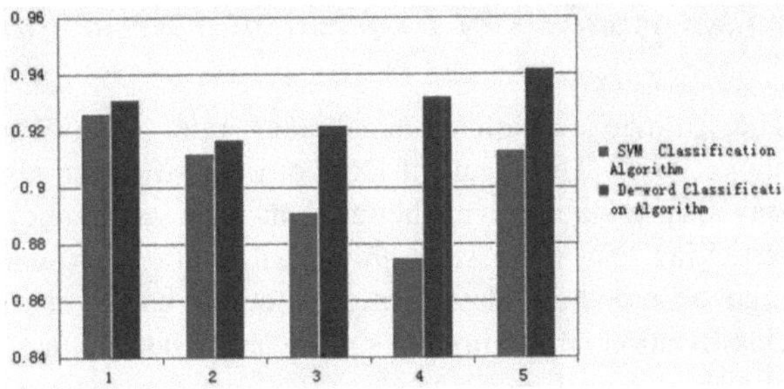

Fig. 75.2 Categorization precision

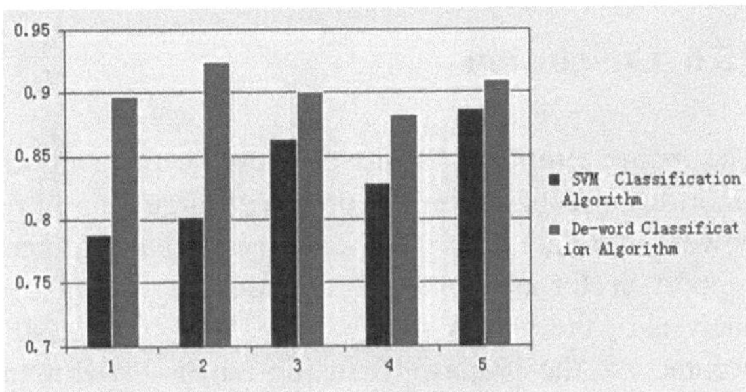

Fig. 75.3 Recall

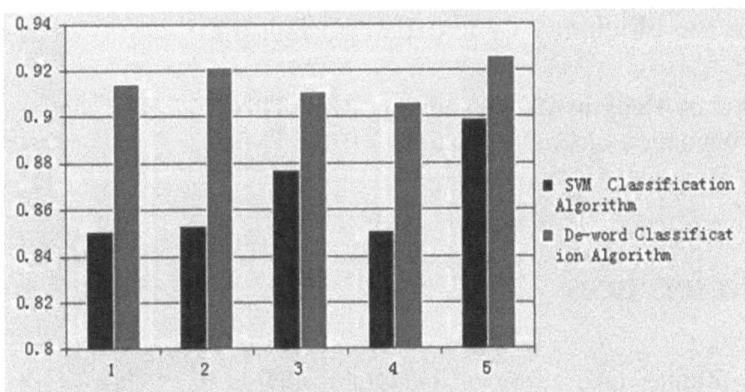

Fig. 75.4 F-measure

When retrieving the word "simulation" is, by calculation, will be as shown in the table sorted according to relevance ranking search results in Table 75.2, just to name a part of it.

Table 75.2 Examples retrieval

Journal title	Relevance weights
Online monitoring of substation equipment	0.00362
Corona experiment of apparatus in the UHV AC power system	0.00331
Islanding algorithm of distribution networks with distributed generators	0.00068

75.5.3 Analysis of Experimental Results and Thinking

The average classification accuracy rate of De-word classification algorithm is above 90 %. The recall of De-word classification algorithm is above 85 %, and it stayed at a stable level, better than the classification method based on SVM. By retrieving test, the results showed that after verification, the retrieved entries had high accuracy in classification method based on SVM. In summary, De-word classification algorithm is significantly better than traditional text classification algorithm based on SVM.

75.6 Conclusion

The paper proposed a novel classification algorithm—De-word classification algorithm that based on the power of large data library retrieval, its application in power large data library retrieval by calculating precision and recall, and with the support vector machine classification algorithm based on doing a comparative analysis of the results showed that, De-word classification algorithm improve the accuracy of the premise, focusing on the text found between implied correlation significantly improve the recall of text categorization, making comprehensive test value F has increased dramatically. Secondly, the algorithm can effectively reduce the dimensions of data, reducing the amount of computation, a substantial increase in the efficiency of text categorization.

Acknowledgments Foundation item: This work was supported by the National Natural Science Foundation of China (No.51077010).

References

1. Zhijian QU, Liang GUO, Qiulin CHEN et al (2013) Intelligent dispatching lossless cluster compression technology based on hadoop cloud framework. Autom Electr Power Syst 18:93–98
2. Xiaoli Guo, Xiao Han (2014) Research on grid knowledge collaborative discovery strategies. J Northeast Dianli Univ 34(1):94–98
3. He L, Jia Y, Han WH, Tan S, Chen ZK (2012) Research and development of large scale hierarchical classification problem. Jisuanji Xuebao (Chinese J Comput) 35(10):2101–2115
4. Kumar MA, Gopal M (2010) A comparison study on multiple binary-class SVM methods for unilabel text categorization. Pattern Recogn Lett 31(11):1437–1444
5. Yong Z, Youwen L, Shixiong X (2009) An improved KNN text classification algorithm based on clustering. J comput 4(3):230–237

6. Wang J, Wang H, Shen Z, Hu Y (2005) An simple and efficient algorithm to classify a large scale of texts. Jisuanji Yanjiu yu Fazhan (Comput Res Dev) 42(1):85–93
7. Colace F, De Santo M, Greco L, Napoletano P (2014) Text classification using a few labeled examples. Comput Hum Behav 30:689–697
8. Ai Q, Qin YP, Li YC (2010) Multi-subjects text classification algorithm based on hyper-sphere support vector machines. Comput Eng Des 31(10):2272–2275

Author Index

A
Abraham, Juneman, 577
Ahmad, Rashidi, 377
Azhar, Muhaimin Noor, 377

B
Bandara, Kasun, 687
Bang, Hyo-Chan, 211
Bu, Hongxia, 405
Bustam, Aida, 377

C
Cabezas, Diego, 421
Cao, Ning, 197
Chan, Jun-Fu, 239
Chandra, Adhi Nugroho, 585
Chang, Jae-Woo, 175, 695
Chang, Yi-Hsing, 481
Chen, Hui-I, 273
Chen, Sih-Yang, 297
Chen, Weihong, 595
Chen, Yi-Ming, 259
Chien, Leng-Kang Chang, 481
Chiu, Hsiao-Hsien, 31
Chiu, Po-Sheng, 537
Cho, Jungho, 1
Cho, Kyungeun, 333
Cho, Seung-il, 323
Cho, Young Sung, 219, 393
Choi, Dong-You, 123
Choi, Min, 1
Choi, Seung-Hyun, 67
Chong, Rachel M., 561
Chung, Yeon-Ho, 687

D
Ding, Wei, 707
Dong, Xianlun, 503
Doo, Seokjoo, 361
Du, De-peng, 663
Du, Yanling, 405
Duan, Huilong, 543, 595, 605

F
Fang, Rong-Jyue, 481
Fang, Xiaonan, 405

G
Gao, Yan, 429, 437
Ge, Caixia, 605
Guo, Xiaoli, 707

H
Hahm, Han Heeh, 679
Halim, Shamimi A., 377
Han, Jong-Wook, 89, 105
Han, Phyu Phyu, 687
He, Xu, 671
Higgs, Russell, 197
Ho, Hong-Leok, 537
Hsieh, Pei-hua, 283
Hsu, Li-Min, 273
Hu, Zhilei, 629
Huang, Hsin-Mao, 305
Huang, Teng-Yao, 305
Huang, Yueh-Ming, 537
Huang, Zhen-zhen, 543
Huilong, Duan, 643
Hung, Jason C., 297
Hwang, Sang Yun, 143

J
Jang, Miyoung, 695
Jang, Uijin, 53
Jeong, Young-Sik, 333
Ji, Huibin, 629
Jia, Zheng, 543

J.J. (Jong Hyuk) Park et al. (eds.), *Ubiquitous Computing Application and Wireless Sensor*, Lecture Notes in Electrical Engineering 331,
DOI 10.1007/978-94-017-9618-7

Jin, Qun, 493
Jo, Ara, 695
Jun, Xiao, 471
Jung, Min-Soo, 133

K

Kao, Jui-Hung, 273, 283
Khan, Muhammad Sajjad, 77
Khumrin, Piyapong, 569
Kim, ChongGun, 349
Kim, Hyungjoo, 53
Kim, Il Kon, 151
Kim, Jang Hyun, 143
Kim, Jeong-Nye, 341
Kim, Marie, 211
Kim, Seong-Kweon, 323
Kim, Tae-Eun, 115
Kim, Yong Kyun, 89
Ko, Chih-Hsiang, 385
Kong, Ning, 229
Koo, Insoo, 77
Kurniawan, Yohannes, 585

L

Lai, Feipei, 273, 283
Lee, Deok-Gyu, 105, 341
Lee, Do Yun, 151
Lee, Hoojin, 11
Lee, Hyunjo, 175
Lee, Jung Song, 679
Lee, Kyung-Ryang, 323
Lee, Seong-Won, 67
Lee, Seongjo, 333
Lee, Seoung-Hyeon, 105
Lee, Soojin, 361
Lee, Xiaodong, 229
Lei, Yanle, 23
Li, Bin, 413
Li, Changle, 23
Li, Hao-min, 543
Li, Haomin, 595, 605
Li, Teng, 511
Li, Xueqing, 503
Li, Xuqing, 511
Liang, Guowei, 595
Liaw, Horng-Twu, 239, 247, 273, 283
Lim, Hyungmin, 53
Lim, Jungmin, 361
Lim, Kyung-Soo, 341
Lin, Kuan-Cheng, 297
Lin, Yen-Wen, 163
Liu, Fu-Xiang, 671
Liu, Peng, 229
Liu, Shufen, 413
Liu, Xinxin, 637

M

Ma, Junfeng, 443
Manikam, Rishya, 377
Min, Wang, 471
Moon, Song Chul, 219, 393

O

O'Hare, Gregory M.P., 197
Ogiela, Lidia, 449, 457
Ogiela, Marek R., 449, 457
Ogiela, Urszula, 457
Oh, Seokho, 45
Omar, Alghanmi Ali, 349

P

Pan, Aihua, 671
Park, Hyodal, 89
Park, Namje, 211
Park, Neungsoo, 67
Park, Soon Cheol, 679
Peng, Chun-Lung, 305
Piao, Guangyu, 493
Prasain, Prakash, 123
Pyshkin, Evgeny, 421

Q

Qin, Xu-fei, 613
Qu, Zhaoyang, 707

R

Ryu, Keun Ho, 219, 393
Ryu, Kwang Sun, 393
Ryu, Yeonseung, 45

S

Sharron, 577
Sim, Sohyun, 333
Sin, Jae Woo, 151
Song, Jie, 553
Song, Joon Hyun, 151
Song, Sihoo, 361
Song, Yong Ho, 67
Song, Yueyang, 23
Suh, Hyo-Joong, 143
Sun, Huiyu, 707
Sun, Xiao-peng, 621, 649
Sung, Junghwan, 143
Suniel, Jeziel C., 561
Supajatura, Volaluck, 569

T
Tang, Yi-Jiun, 31
Tian, Ye, 229

U
Um, Kyhyun, 333

V
Vassiliev, Alexei, 421

W
Wang, Hai-Hui, 305
Wang, Ling, 707
Wang, Lu, 621, 649
Wang, Ming-Shi, 31
Wang, Po-Hsiang, 163
Wang, Wei, 629
Wang, Weisheng, 657
Wang, Xingyue, 621
Wang, Zhanquan, 443
Winn, Khin Zar Chi, 687
Wu, Ting-Ting, 537
Wu, Wei-Chen, 247, 259

X
Xing, Wei, 629
Xinglei, Wei, 529
Xu, Hongyun, 637
Xu, Shaogang, 443
Xueqing, Li, 529

Y
Yan, Baoping, 229
Yang, Feng, 613
Yang, Guiqiang, 443
Yang, Li, 23
Yang, Ta-Chih, 239
Yanping, Wu, 643
Yeo, Sung-Dae, 323
Yong, Jiang, 519
Yoon, Hyunsoo, 361
Yoon, Min, 175
Yu, ChunGun, 349
Yu, Miao, 629
Yuan, Ding, 443
Yuan, Xiaoming, 23

Z
Zhai, Lin-bo, 613
Zhang, Gui-juan, 553
Zhang, Long-xiang, 463
Zhang, Qi, 503
Zhang, Rongxia, 657
Zhang, Yin-sheng, 543, 595, 605
Zhang, Yong-Sheng, 429, 437
Zhao, Xiaona, 621
Zheng, Xiang, 543
Zheng, Xiang-wei, 553
Zhou, Xiaokang, 493
Zou, Jia-Shun, 429, 437, 463
Zu, Yue-ran, 663

Subject Index

0–9
0/1 knapsack problem (KP), 663
ACA for, 664
basic ACA, 665
comparison and analysis ACA for, 666
convergence state, 669*f*, 670*f*
NP-hard problem, 664
361° company
marketing situation and analysis of, 525
marketing strategy, 524
3D segmentation evaluation, 622

A
Access control policy, 463
based on XACML, 465
of DS, 231
Access control protocols, for VANETs, 261. *See also* VANETs (Vehicular ad hoc networks)
differentiated service, 260
PAACP, 261
cryptanalysis of, 261–263
Access Point (AP), 33
Action learning, 639, 640*f*
Adaptability, WSN
network lifetime, 168–169
adapative $I_{\min}$, 169*f*
adaptive *k*, 169*f*
scalability, 169–172
performance comparison, 170*f*, 171*f*
Adaptive cluster transformation (ACT), 362, 464
advanced (AACT), 363
Adaptive intrusion tolerance systems, 363, 364
Adaptive threshold, 563, 565, 566
sensitivity after thresholding, 565, 565*t*
specificity after thresholding, 565, 565*t*
Adaptive trickle scheme, 163–164, 165, 168. *See also* Wireless sensor networks (WSNs), adaptive trickle scheme for
Additive white Gaussian noise (AWGN), 80, 125
channel, 128, 692
Adhesion segmentation, 503, 505
chain code sum (CCS) algorithm, 505
results of, 506*f*
Adiabatic operation, 324–326
current LED controller using, 326–330, 327*f*
power consumption, 328*f*, 330*f*
structure, 329*f*
pull-up network, 325*f*
total energy consumption of, 326
Adverse drug events (ADEs), 606
detection in clinical narrative text, 607–608
common suffix and prefix of drug name, 608*t*
named entity tag and relation extraction example, 609*f*
negation filter example, 608*f*
related lexical resource and knowledge base, 606–607
results, 610–611
not found in knowledgebase, 610*t*
Adverse events (AEs), 606
AES (Advanced encryption standard) algorithm, 696
AES-256, 62
Aggregated signature, 697
Air pressure database, 101*f*
Air Traffic Control (ATC) system, 105
Air traffic management (ATM) system, 89
Air traffic service chart, 98*f*
Aircraft Reconnaissance, 106

J.J. (Jong Hyuk) Park et al. (eds.), *Ubiquitous Computing Application and Wireless Sensor*, Lecture Notes in Electrical Engineering 331,
DOI 10.1007/978-94-017-9618-7

Aircraft Target Ghost Inject, 106
AirDroid, 309, 309*t*
Airline procedure model, 91*f*
Airspeed indicator reading (ASIR), 92
Airspeed theory, 92–95
  Mach number, 93
  Coriolis parameter, 95
Alamouti code, 690
All-in-English teaching (AET), 637. *See also* Compiler principles, of AET
  comparison with traditional thinking, 641*f*
AMD-V (hardware), 362
American Telephone and Telegraph Company (AT & T), 308
Analysis module algorithm, 440
Anatomical Therapeutic Chemical (ATC) Classification, 596
Anchor Node (AN), 33
AND
  in CP-ABE algorithm, 466, 468*f*
  fusion rule, 78
Android, 308, 311
  API (Application Programming Interface), 312
  audio architecture, 312*f*
  bitmap, 311
  connection
    on PC's end, 313*f*
    on phone's end, 314*f*
  HTC One V smartphone with, 309
  Processing Touch Event and Keyboard Event in, 312
  RAW to RGB, 312*t*
  version 4.0.3, 306, 307
Android, memory management, 144–145
  Out of Memory (OOM) killer, 144
  policy based on periodic habits of user, 145
    implemented codes for data collection, 147*f*
    with priority control, 146*f*
Anomia (Anomie), 579, 581, 582
  barrier for e-learning participants, 583
  perceived risk of ACL, 582
Ant colony algorithm (ACA), 664, 670
  for 0/1 knapsack problem, 664
    basic ACA, 665
  comparison and analysis for BP, 666
  flow chart of, 667*f*
  improved, 666–667
    standard distribution density function, 668
Ant colony optimization (ACO) clustering algorithm, 643, 644–646
  flow chart of, 645*f*
ANTA company, network marketing
  status and analysis of, 524
  strategy of, 524
Anti-Corruption e-Learning (ACL), 578
Anycast (network), 230, 232
  improving performance of node of, 236, 237
  IP address, 233, 234, 237
Apache HTTP Server, 683
Appell hypergeometric function, 15
Application specific instruction set processors (ASIPs), 68. *See also* ASiPEC
APTEEN protocol, 351
Arithmetic coding, 68, 74
ARM series, 362
Artificial Fish Swarm Algorithm (AFSA), 298, 556
  experimental results of, 303*t*
  feature selection, 298–299
    algorithm stops, 300
    feature set representation, 299*f*
    fitness evaluation, 299
    follow, 299
    initialization, 299
    prey, 300
    of SVM using, 301*f*
    swarm, 299–300
  generation of group animation path based on, 554–555
  modified version of (*see* Modified Artificial Fish Swarm Algorithm (MAFSA))
  path generated, 559*f*
  roadmap of fish aggregation simulating, 556*f*
Artificial neural network, 298, 394, 396
  clustering of disease code with weight, 398
    learning algorithm of, 398*t*
  predictive pattern analysis, 397
ASiPEC
  application specific instructions of, 73*t*
  architecture, 68–69
    undescribed features of, 71
  datapath of, 69*f*
  performance of, 74*t*
  variable length decoding, 69–72
  variable length encoding, 72
Assisted GPS (AGPS), 188
Association, correlation, causality analysis (BI technique), 587*t*
Association rules, 708
Associative recommendation of learning contents, 493–494. *See also* Eye-tracking, in social media enhanced environment

FAQ, 496, 498, 499, 500
key points, 496, 498, 499, 500
pre-process for, 495
flowchart for matching keywords, 496*f*
flowchart for matching user ID, 497*f*
Keyword Database, 495
User ID Database, 495
procedure, 497–499
set relations of contents to, 498*f*
Q&A, 496, 498, 499, 500
Attribute Based Access Control (ABAC) mechanism, 464
policy model, 464
based on XACML access control policy, 465
decision-making model based on XACML, 465–466, 466*f*
strategy evaluation related concepts, 465
strategy model related concepts, 464
Audio streaming, 311–312
Augmented reality, 116, 120, 121
augmentation of pen and proposed sketch system, 121*f*
Authenticated data structures, 696, 697–698
Authentication, 254, 256. *See also* Authentication System (AS); Third-generation mobile system Universal Mobile Telecommunication System (3G/UMTS)
full authentication protocol, 250
latency of, 255
Keyed-Hash Message Authentication Code (Keyed-HMAC)
message authentication codes (MAS) function, 255
mutual, 255
re-authentication protocol
of AS, 253
of F-AAA, 252–253
of H-AAA, 253
Authentication scheme, in VANETs. *See also* VANETs (Vehicular ad hoc networks)
authentication phase, 264
mutual, 260
security analysis, 270
Authentication System (AS), 240, 245, 250, 251, 252
fast re-authentication protocol of, 253
Security Setup process, 57, 59
Authorized Credential (AC), 261
Auto regressive moving average (ARMA), 188
Automatic Dependent Surveillance-Broadcast (ADS-B), 105, 106
diagram, 107*f*
overview, 106
security framework for data protection, 111–112
communication process at, 112*f*
security threats, 107
sensor authentication, 108
creation and transmission of SPKI certificate, 111
signed XML document creation and verification for, 108–109, 109*f*
Autonomous component architecture (ACA), 200

B

Base metal (BM) zone, 5, 6, 8
Base stations (BS), 352, 355
Basic probability assignment (BPA), 78, 80, 81, 82
Baum-Welch algorithm, 336–337
Bayesian algorithm (Bayes theorem), 708
Bessel function, zeroth-order modified, 12
BGP (Border Gateway Protocol), 232
Big data, 298, 708, 710
Binarization partitioning curved surface (BPCS), 503, 504–505
results of, 505*f*
single edge detection algorithm, 504
Binary knapsack problem, 663
BINUS International School, 588
Microsoft SQL Server, 589, 591
operational data source in, 588–589
Bit error probability (BEP), 12
exact and approximated, 20*f*
Bit error rates (BERs), 692
of QOSTBC, 693
Bit reverse operation, 70–71
Bit stream engine (BSE), 69
datapath of, 69*f*
Bluetooth, 32, 188, 240, 306, 315
Boolean table flag, 665
Boxplot method, 126, 128, 129, 130
Browsing behavior, 494, 498, 500
Bucket-based authentication, 696, 698
Buffer-overflow attack income, 368
Bureau of NHI, 481, 482, 483, 484
in synchronization sequence for medical rules, 484*f*, 485*f*
Business intelligence (BI), 586
characteristics of, 587
definition, 586

Business intelligence (BI) (*cont.*)
for schools
benefits for, 592*t*
system model, 589, 590*f*
techniques, 587*t*
Business-to-business (B2B) commerce, 212

C
C3D file (motion capture data file), 555, 556, 557
and motion file, comparison of rendering, 558*f*
role path from, 558*f*
Calibrated Air Speed (CAS), 91, 92, 93
Care record summary (CRS)
document validator, 153–154
documentation service, 155
and HIE, 152
Schematron transformer, 153
specification service, 155
tutorial service, 155
validation history repository, 154
validation tool (VT), 153*f*, 154, 160
language compatibility, 159
social networks, comparison of, 158*t*
socialized service structure of, 155*f*
Cargo Capacity Hierarchy Scheduling (CCHS), 414
algorithm, 417–418, 418*f*
Cartesian coordinate system, 33
CAVLC decoder, 68
performance of ASiPEC on, 74*t*
CCTV video records, 342
CDS tools
for docking, 549
instances for, 548*f*
for specific clinical problem, 548–549
for specific treatment protocol, 549
third-party developed, 549–550
Cell phone service, 247
Centroid-GPS model, 187–188
experimental results, 192–194, 193*f*
autonomous vehicle design, 192*f*
coordinate position values, 194*f*, 194*t*, 195*t*
specifications of receivers, 193*t*
physical layout of, 189*f*
proposed approach, 188–190
accurate calculation of reference point, 190
distance threshold and centroid calculation of receivers, 191–192
invalid data check, 190
position improvement, 191
Character strengths, CVQ-96
differences among
countryside, town and city in, 633*t*
directions in, 634*t*
factor loading of the 24 strengths, 631*t*
of gender differences in, 633*t*
of one child, 633*t*
positive psychology, 630
Characterization and descriptive data mining (BI technique), 587*t*
China National Knowledge Infrastructure, 712
Chinese sports brand, 519. *See also* Sports network marketing
current sports network marketing condition, 521
PV (Page View), 527
sports network marketing concept, 520
Chinese Virtues Questionnaire (CVQ-96), 630
construct validity of, 632
demographic differences of, 632, 632*t*
factor analysis of, 631
factor loading of 24 strengths, 631*t*
internal consistency reliability of, 632
CIELab values, 562
Classification (BI technique), 587*t*
Classification algorithm. *See* De-word classification algorithm
Clear channel assessment (CCA), 25
Client-side script, 399
Clinical anatomy learning, 671
anatomy, 671
difficulties faced in, 672
deficiency of teaching materials and methods, 673–674
inappropriate learning pattern, 672–673
interdepartmental co-ordination in selecting student, 674
language barrier and cultural differences, 672
problems of assessment and feedback, 674
suggestions, 674–675
assessment and feedback, 677
teaching and learning skills, 675–677
Clinical Contents Model (CCM), 152
Clinical data repository (CDR), 395
Clinical Decision Support (CDS) applications, 544
extensible mechanism, 544–545
first level of interoperability, 545
integration, 545*f*
protocol of integration, 546*t*

second level of interoperability, 545–546
mechanism for management, delivery and update, 546–547, 547*f*
system delivery, 547–548
CDS tools (*see* CDS tools)
system implementation, 547
Clinical Document Repository (CDR), 152
Clinical Information Support (CIS) applications, 544, 545. *See also* Clinical Decision Support (CDS) applications
first level of interoperability, 545
mechanism for management, delivery and update, 546–547, 547f
protocol of integration, 546*t*
second level of interoperability, 545–546
system delivery, 547–548
CDS tools (*see* CDS tools)
system implementation, 547
Closest normal points (CNPs), 652
Closest normal vector, 649
Cloud computing, 239, 437
problem, 430
single sign-on (*see* Single sign-on (SSO) system)
Cloud login platform
authentication, 433–434
ECC identity authentication, 432
registration, 433
safety analysis, 434
Cloud service, 437–438
monitoring system in
based on SMS, 439
in cloud environment, 438
traditional monitoring system in, 438
Cluster analysis, 440, 592*t*
Clustering algorithm, 220, 394
using SOM to classify profitable customer, 222
Clustering analysis (BI technique), 587*t*
Cognitive Financial Analysis Information Systems (CFAIS), 449
Cognitive management systems, 450
subclass of, 450*f*
Cognitive radio. *See* Cooperative spectrum sensing, in CR
Collaborative filtering recommendation technology, 471–472
College physics experiment, 406, 411
Columnar crystal, 5, 6, 8
in FZ, 7, 7*f*, 8
Compiler principles, of AET
action learning of, 639, 640*f*
computational methods used in compiler construction, 641*t*
computational thinking, 640
evaluation of, 640–641
teaching of, in top universities, 638–639
Complementary extended Kalman filter (CEKF), 38–40
process diagram, 39*f*
Complementary metal oxide semiconductor (CMOS) circuit, 324
Computational thinking, 640
Computer-assisted health care, 422
Computer-based learning, 538
Computerized treatment plan, of cancer patients in Korea, 399
experiment and evaluation, 400
experimental data for evaluation, 399–400
mean absolute error (MAE), 400, 400*f*
precision, recall, F-measure, 401, 401*f*
Condensed-RSA, 699, 700
Consistency Error evaluation, 622
Constant current, 324, 330
Content Delivery Network (CDN), 232
Controller working position (CWP), 106
COOL, Classroom Object Oriented Language, 639
Cooperative spectrum sensing, in CR, 78, 123–124
for combination rule, 80–83
proposed sequential, 81*f*
simulation results, 83–85, 84*f*, 85*f*, 86*f*
system model, 79, 79*f*
outlier detection techniques, 126–127
future works, 130
simulation results, 83–85, 128–129, 130*f*
system model, 124–125
Core Extraction algorithm, 624
Coriolis parameter, 95
Correlation correction factor, 711
Cosine similarity, 475
Cost Saving, 284
Course scheduling, 511
algorithm designed during course arrangements, 514
place searching algorithm, 515
time searching algorithms, 514–515
description, 512
definition using math symbol, 512–513
priority setting, 513
resource management container based on map, 513
results, 515, 516*f*
about rationality, 516

about speed, 516
about successful rate, 516
CP-ABE algorithm
improved algorithm, 467
decryption diagram, 468*f*
in research of access control, 467
traditional algorithms of, 466–467
Cross cloud authorization mechanism
license management mechanism, 245–246
security analysis, 244
counterfeiting attacks, 245
middle attack, 245
replay attacks, 244
system architecture, 241–244, 242*f*
system introduction, 241
Cryptanalysis, of PAACP, 261
cryptanalysis 1, 262
cryptanalysis 2, 262–263
Crypto-biometric sharing schemes, 459
Cryptographic threshold schemes, 458, 459, 460
CSIM 20 simulator, 368
CSMA/CA, 25, 29–30
comparison, traffic distribution
average packet delay, 29*f*
normalized throughput, 28*f*
simulation parameters, 26*t*
Customer Lifetime Value (CLV), 284
Customer value analysis, 289
F1 indicator, 292
GP indicator, 292
lost customers, 294
new customers, 294
potential customers, 292–293
in losing/project customers:, 293–294
proportion of customers, 294*f*
R indicator, 289, 291
result, 295
RFGP model
customer type in, 293*t*
distribution of various customer indicators, 291*t*
matrix, 292*f*
model results, 291*t*
scoring standard, 290*t*
stable customers, 292
in losing, 293
strategic customers, 293
thin margin customers, 294
Cyclical scheduling method, 378

D
DAG (Directly Acyclic Graph), 163
metric container, 166*f*
DAS-28 score tool, 548
Data authentication index, 699–700
Data confidentiality, 696
Data integrity, 696
Data mining concepts, 585
in schools, 591
Data mining, 274–275, 297, 394
data pre-processing, 276
decision tree, 275–276
pruning of, 279*f*
two-step
clustering, 275
monitor output, 279*f*
Data recovery module, 439*f*, 440
Data warehouse, 585
Database correlation, 32
Database encryption, 696
Database outsourcing, 695–698
periodic function-based data group index (*see* Periodic function-based data group index)
DDC (Daily Dose Cost), 602
Dead-reckoning (DR), 33
future works, 42
test environment, 34, 35*f*
Decision trees, 708
Defined daily dose (DDD), 596. *See also* Surveillance of antimicrobial use
bar chart, 601*f*
rules to obtain, 597*t*
statistics table, 601*f*
Delaunay triangulation, 1
Dempster combination rule, 81
Dempster-Shafer (D-S) theory, 78
Detailed Clinical Model (DCM), 152
De-word classification algorithm
analysis and thinking, 714
categorization precision, 713*f*
classification process, 710
classifications data, 709–710
F-measure, 713*f*
recall, 713*f*
Differential GPS (DGPS), 188
Differentially coherent phase shift-keying (DPSK), 13
M-ary DPSK, 13–14, 17
Digital archives, 684. *See also* Ichpedia

Digital collaboration, 538
Digital evidence, 342, 347. *See also* Network video surveillance
Digital Geometry Processing, 621
Digital signal processors (DSP), 68
DIO (DAG Information Object) messages, 163
Direction averaging technique, 188, 193, 195*t*
Directional source aware routing protocol (DSAP), 351
Discovery service (DS), 230
 analysis, 236
  improving performance of node of, 236
  ISP-unfriendly policies, avoiding influence of, 237
  shortening response time, 236–237
  transparency to users, 237
 architecture of, 231*f*
 BRIDGE DS, 231, 231*f*
 performance improvement scheme for, 232–233, 233*f*
  data update and distribution process, 234–235, 234*f*
  user's lookup process, 235–236, 235*f*
Discretionary Access Control (DAC), 463
Distance calculation
 flight plan message format, 95*f*
 simulation results, 99*f*
 two point information, 98*t*
  at Arctic regions, 99*f*
 between two points, 95–97
Distance education, effectiveness, 658
 distance teaching
  application effect of teaching mode, 659
  reason of poor effect, 659–660
 improving, 660
  communication and interaction between teachers and students, 661
  short-term behavior of educational institutions, 660
 in judicial examination training, 658–659
Distributed denial of service (DDoS) attacks, 362
 resistance through cluster substitution, 363*f*
Distributed grid scheme, 178–179
 data structures, 178–179
 structure of, 179*f*
Distributed Hash Table (DHT) network, 230
Dixon's test, 127, 128, 129, 130
DNA chains, in information encoding, 460*f*
DNS systems, 362
 of Korea Advanced Institute of Science and Technology (KAIST), 369
DODAG (Destination-Oriented DAG), 164
Domain Name System (DNS), 230
 Anycast, 232
Dominating set approach, 353
 based hierarchical clusters of WSN, 359
 SCH selection, 353, 354, 354*f*, 357
Downlink (DL) transmit power, 688
DRAM (dynamic random-access memory), 46
 hybrid main memory
  with NVRAM, 48
  with PRAM, 47, 48
Droid VNC Server, 309
DUI (Drug Utilization Index), 602
Dynamic Separation of Duty (DSD), 241
 DSoD, 243

E

EAP-AKA, 248, 255
EAP-SIM, 248, 255
Economic information systems, 450
Edge detection algorithm, 504
Education video personalized recommendation system, 471–472
 application effect, 479
 evaluation matrix on, 475*f*
 solution for, 474*f*
 system features and realization, 473
  theoretical modelling , 474–477. *See also* Personalized recommendation, theoretical modelling
  check and evaluation, 478–479
  concrete realization, 477–478
 system introduction
  data support module, 472
  new resource recommendation module, 473
  recommendation engine module, 472–473
Electronic data system, 570
Electronic Health Record (EHR) project, 152
Electronic learning (e-Learning), 538
Electronic Medical Record (EMR) in HL7, 395
 computerized treatment plan, of cancer patients in Korea, 399
Electronic Product Code (EPC), 212, 213
Electronic scoring system, 570
 effects of using, 572*t*
 scoring program, 571*f*
 system usage, 573*t*
 user satisfaction of, 573*t*
Electronic-money technology. *See* Near field communication (NFC)
Elliptic curve cryptography (ECC), 432
 identity authentication, 432
  and fingerprint key, 432–433

getting fingerprint key, 432
EM (Expectation-Maximization) algorithm, 649
EM-ICP, 651
advantages and disadvantages, 654*t*
Embedded processor, 143
Embedded systems, 421–423
educational perspective, 424–425
algebraic differential description, 425*f*
patient breaths state chart, 425*f*
future work, 426
implementation, 423
Emergency department (ED). *See* University of Malaya Medical Centre Emergency Department (UMMC-ED)
Encrypted database, 695–698
periodic function-based data group index (*see* Periodic function-based data group index)
Energy detection techniques, 124, 130. *See also* Outlier detection techniques
Energy detection, 78, 80
Energy-efficient communication methods. *See* Hierarchical clustering WSNs
Enhanced cooperative spectrum sensing scheme, 78–79
Enrollment data analysis and visualization
data model, 530
data filter, 530–531
data integration, 531
data transformation, 531
visualization using parallel coordinates, 532
Entropy coding technique, 68. *See also* ASiPEC
EPCglobal, 212
Equivalence class division algorithm, 512
Error probability
coherent phase shift-keying (PSK), 15–16
differentially coherent phase shift-keying (DPSK), 13–14
high-SNR approximate expressions for
M-ary coherent PSK, 18
M-ary DPSK, 17
M-ary square QAM, 19
noncoherent correlated binary signals, 18
quadrature amplitude modulation (QAM), 16
ETL (extract, transform, and loading) process, 586, 591
Euclidean distance, 176, 221, 396, 441–442, 503–504, 643, 651, 654*t*
Euclidean norm, 396
Euclidean space, 530
Evaluation model. *See also* LEACH Protocol; Nearest Closer (NC) Protocol; Single-hop Protocol
coverage, 202
density, 203, 204*f*, 205*f*, 206*f*, 207
lifetime, 202, 203, 204*f*, 205*f*, 206*f*, 207, 208
reliability, 202, 203, 204*f*, 205*f*, 206*f*, 207, 208
Exploratory data analysis (EDA), 587*t*
Exposure time adjustment, 362, 364, 366–368, 367*f*
Extended Kalman Filter (EKF), 38
eXtensible Stylesheet Language (XSL) file, 153
eXtensible Stylesheet Language Transformation (XSLT) engine, 153
Eye-tracking, in social media enhanced environment, 493–494
human-computer interaction, 494
pre-processing of gaze, 495*f*

F

Facebook, 156, 156*t*, 158
Feature selection algorithms, 298
Federal Communication Commission (FCC), 123
spectrum task report, 78
Fingerprint verification. *See also* Cloud computing; Cloud login platform
authentication, 433–434
in cloud computing, 429
getting fingerprint key, 432
registration, 433
safety analysis, 434
sign-on system based on, 431
Fingerprinting, 32
Firewall systems, 362
Fitting Primitives algorithm, 624
Flight plan message
format, 95, 95*f*
original message, 100*f*
Foreign AAA (F-AAA) Server, 250, 251, 252
fast re-authentication protocol of, 252–253
Forensic artifacts, 342–343
metadata (*see* Forensic metadata)
network camera, 343, 343*t*
in network video surveillance system, 345
Forensic metadata, 345–347
in network camera, 343*t*, 346*f*
in network video analytic system, 345*t*
in network video recorder, 344*t*, 347*f*
Form Factor (FF), 623

Full-dimension MIMO. *See* Massive MIMO
Function optimization, 298
Fusion center (FC), 78
Fusion zone (FZ), 5, 6, 7, 8

G
GARMIN GPS 19x HVS receivers, 192
Gas humidifier unit, 424*f*
Gauss hypergeometric function, 17
Gaussian random variable, 80
General cluster heads (GCH), 352, 355
General purpose processors, 68, 74
General purpose register (GPR), 69
Generating Spatio-Temporal Data (GSTD), 702
Genetic Algorithm (GA), 298, 378
Geographical information system (GIS), 187
Gestures, 334. *See also* Hand gesture recognition
  kinds of, 337*f*
  performance assessment, 338–339
    table, 338*t*
  in smart interior, 338*f*, 339
  tasks corresponding to, 339*t*
Global Positioning System (GPS), 32, 37–38
  accuracy, 188, 191
    standard receiver, 193, 193*t*, 195
  future works, 42
  test environment, 34, 35*f*
  tracking path by, 37*f*
Golomb coding, 68
Google Groups, 156, 156*t*
  CRS validation board for, 157*f*
Grain size, in metal component microstructure analysis, 2, 5, 8
Graphical user interface (GUI), 387, 388, 389, 390, 571
  by J-Sim, 201
Graphics tablets
  Cintiq 21 UX (C), 387
  Intuos 3 (I), 387
Grass-fire labeling method, 117–118
Greedy strategy, 668, 670
  in improved ACA, 666
Ground speed (GS), 91, 92
  calculation, 98*f*
Ground Station Flood Denial, 106
Group animation, 553–554
  generating intelligent group animation, 554–555
Grubb's test, 126, 128, 129, 130
GUTI (Global Unique Temporary Identifier), 54*t*
  initial attach for LTE, 60
    MME changed, 61, 62*f*, 63*f*
    MME unchanged, 60–61, 60*f*

H
HAMC (Hybrid memory-Aware Multi-Core) scheduling
  algorithm, 48–49
    architecture, 48–49
    bandwidth requirement prediction, 49–50
  scheme, 46
Hand gesture recognition, 333–334
  identification of 8-directional vector chain, 335–336
    encoding code, 336*f*
    flow chart, 335*f*
  performance assessment (*see under* Gestures)
  using HMM, 336–337
Hash function, 58, 62, 63, 139, 433, 696
  algorithms, 134, 135, 139
  one-way, 263
  strong Hash function, 700
Hasoo method, 135
HCI (Human Computer Interface) systems, 115
Health information exchange (HIE), 151, 154, 158
  and CRS validator, in Korea, 152, 159, 160
  based on KS standard, 157
  Establishing Medical-IT convergence industry upbringing infra, project, 152
Health Infoway, Canada, 151
Health Level Seven (HL7), 403. *See also* Computerized treatment plan, of cancer patients in Korea
  Clinical Document Architecture (CDA), 152
  standard, 213
  Virtual Medical Record (vMR), 550
Heat affected zone (HAZ), 5, 6, 7, 8
Hidden Markov model (HMM), 334
  hand gesture recognition using, 336–337
  state transition diagram in, 337*f*
Hierarchical clustering WSNs, 349–351
  proposed models, 352
    Hierarchical Clustered Routing, 354
    routing procedure, 353
    Selection Dominating Algorithm, 353, 354*f*
    system model description, 352
  simulation results, 355, 355*f*, 356*f*

Hierarchical modular design
administration module, 407–408
research module, 408
calendar, 409
course introduction, 408
syllabus, 408–409
teaching resources, 409
student module, 408
teacher module, 408
High-bandwidth communication, 143
Home AAA (H-AAA) Server, 250, 251, 252
fast re-authentication protocol of, 253
Hospital information system (HIS), 483
Law-Governed Interaction (LGI) design principle, 483
sharing medical information, 483
Hospital management. *See* Surveillance of antimicrobial use
Hoyt statistical model, 11. *See also* Nakagami-*q* statistical model
Human centered computing, 421
implementation, 423
Humidifier performance evaluation, 424*f*
Hybrid main memory, 46. *See also* HAMC (Hybrid memory-Aware Multi-Core) scheduling
DRAM/PRAM, 48. *See also* PRAM (Phase change RAM)
NVRAM, 46, 51. *See also* NVRAM (Non-volatile RAM) designs
Hyper-MIMO. *See* Massive MIMO

I

Ichpedia
development environment, 683
search system in
map search, 681
semantic search, 682, 683*f*
simple search, 680–681, 681*f*
software structure of, 684*f*
system using location information, 682*f*
ICNP (Iterative of Closest Normal Point) algorithm, 649, 652
advantages and disadvantages, 654*t*
IEEE 802.11, 200
IEEE 802.15.6 MAC, 23, 24
beacon mode with superframes, 25, 25*f*
comparison to IEEE 802.15.4, 23, 24
performance evaluation, 25–26
network topology, 26*f*
simulation results, 27
simulation scenario, 26
IHIS validator, 159
iKP (i-Key-Protocol [i-1,2,3], 135
Image histogram, 563. *See also* Retinal fundus images
Image processing
algorithms, 562
pre-processing
adhesion segmentation, 505
BPCS, 504–505
image denoising, 504
scaling technique, 563
IMSI (International Mobile Subscriber Identity), 54*t*
initial attach for LTE, 58–60
proposed protocol, 59*f*
Incremental Cell Expansion (ICE) algorithm, 180
example, 181*f*
Index construction time, 702*f*
Indicated airspeed (IAS), 92
Indoor Google Maps, 33
Indoor positioning system (IPS), 32
Inference engine module, 485
Information Services (IS) server, 212
Information technology. *See also* De-word classification algorithm; IT education
BI, 586
CDS applications, 546–547
health, 545
location-based services (LBSs), 187
network assisted teaching system of physics experiments, 406
NLP, 606
in security in mobile RFID terminals, 213
Initial attach
for LTE, proposed security scheme for, 58
with GUTI, 60. *See also* GUTI (Global Unique Temporary Identifier)
with IMSI, 58–60, 59*f*
for UE, 55–56
Instantaneous vertical speed indicator (IVSI), 93
Intangible cultural heritage (ICH), 680
Intelligent algorithms, 554–555
character binding. *See also* C3D file (motion capture data file)
human movement frame, 556–557
orientation angle, 557*f*
Intelligent course scheduling. *See* Course scheduling
Intelligent retrieval recommendation technique, 471–472. *See also* Education video personalized recommendation system

Inter Cell Interference Control (ICIC), 55
Interactive drawing, 115–116
  background subtraction, 116–117
  labeling, 118–119, 119*f*
  skin region detection, 117–118, 118*f*
  synthesis of hand tracking and virtual objects, 119–120, 120*f*
Interactive interface, 512
International Classification of Headache Disorders, 3rd edition (ICHD, 3rd), 644
International Headache Society (IHS), 644
International Standard Atmosphere (ISA), 93
International Union for Conservation of Nature's (IUCN) red list, 444
Internet of Things (IoT), 31, 33, 45
  Intelligent Commuter, 32
Internet Protocol (IP), 232
Internet Service Provider (ISP) networks, 230
  -unfriendly policies, avoiding influence of, 237
Interworking, WLAN, 248
  network architecture, 249–250
Intra-class correlation correction value, 712
Intrusion detection, 298
Investigated weighting factor, 649
IT education, 421–423
  educational perspective, 424–425
    algebraic differential description, 425*f*
    patient breaths state chart, 425*f*
  future work, 426
  implementation, 423
    operator panel implementation, 425*f*
Iterative Closest Point (ICP) algorithms, 649
  classical, 650
  comparison
    advantages and disadvantages, 654*t*
    four kinds of ICP algorithm, 654*t*
    of time, error and results, 653t
  improved, 650
    EM-ICP algorithm, 651–652
    ICNP algorithm, 652
    sparse ICP algorithm, 651

J
Java environment, 399
Java language, 158
Jena function module, 484
J-Sim (JavaSim), 200–201
Judicial examination training, 657–658
  application of distance teaching method in, 658
  distance teaching
    application effect of teaching mode, 659
    reason of poor effect, 659–660
  improving, 660. *See also under* Distance education, effectiveness

K
Kalman Filter (KF), 38
Kalman Gain, 39
Key agreement, in NFC, 133, 137
  in crypto algorithm, 139
  hash function algorithms, 134
  proposed method, agreement phase, 137
  XOR operations, 134
Keyed-hash functions, 135
Keyed-Hash Message Authentication Code (Keyed-HMAC), 248, 256
Kinect sensor (Microsoft), 334
  in gesture recognition, 335, 337, 339
k-means
  algorithm, 624
  clustering, 274, 334, 395–396
  clustering algorithm, 224, 274, 396
  improved algorithm, 440
  segmentation, 625
k-nearest neighbor (KNN/k-NN) query processing algorithms, 176, 708
  incremental cell expansion algorithm for processing, 180
  performance analysis, 182–184
    query processing time, 183*f*, 184*f*
    retrieval time, 182*f*, 184
  queries in SNDB (Spatial Network Databases), 177
  by using outer expansion, 180*f*
Knowledge Discovery in Database (KDD) process, 275, 285
Knowledge Translation Platform (KTP), 596
  Antimicrobial Investigation view, 600
  Common Drug Investigation view, 600
  surveillance platform, 599*f*
    screen shot of, 600*f*
    system framework based on, 599*f*
Knowledgebase, 611. *See also* Adverse drug events (ADEs)
  ADE related lexical resource and, 606–607
  comprehensive medical terminology dictionary, 610
  DDD knowledge sources, 597
  medication, 596
Korean Industrial Standard (KS), 152, 155, 156, 157. *See also* Care record summary (CRS)
Kungpisdan method, 135

L
Lagrange method, 651
LAMP (open source software), 683
 Linux, Arpache, MySQL and Python environments, 685
Lantana validator, 158*t*, 159
Lauricella hypergeometric function, 14
LEACH Protocol, 201
 evaluation model and analysis, 207
 simulation results, 203–205
  density, lifetime and reliability relationship for, 205*f*
LED display brightness level, 615
 flow chart, 618*f*
 schematic control, 617*f*
 steps to control, 617–619
Licensed users (LU), 78
Light emitting diode (LED) system, 324
Linguistic threshold schemes, 458
Link analysis (BI technique), 587*t*
Linux (operating system), 683, 685
 kernel, 144, 313
Liquid crystal tablet, 385
LLN (Low Power and Lossy Networks), 163
Load capacitance
 power dissipation at, 326
 voltage, 325
Location-based services (LBSs), 32, 175, 187, 695
Location-based system. *See* Centroid-GPS model
Long Term Evolution (LTE), 53
 initial attach
  proposed security scheme for (*see under* Initial attach)
  for UE, 55–56
 network structure, 55, 55*f*
 performance analysis, 64–65, 64*t*
  summary of, 65*t*
 Release 9, 54
 Release 12, 54
 security, 56–57, 57*f*
  analysis, 61. *See also* Security analysis, LTE
 threats, 578
Long-term averaging technique, 190
Low Energy Adaptive Clustering Hierarchy (LEACH), 351, 355, 357
 average energy dissipation, 356*f*
Low memory killer (LMK), 144
 processing sequence, 145
Low power LED
 adiabatic operation (*see* Adiabatic operation)
 circuit design, 324
LRU (Least Recently Used) algorithm, 144, 145, 147, 148
Lung cancer, data, 274–275
 comorbidities of, 280*f*
 death rate of, 280*f*
 decision tree, 275–276
  pruning of, 279*f*, 281*f*
 hospitalized patients, 280*f*
 significant comorbidities of, 277*t*
Lung ventilation unit, 423, 424*f*

M
Mach number, 91
 in airspeed theory, 93
Magnetic Tunnel Junction (MTJ), 47
Mahalanobis equation, 38
Malaysian Government General Order on Labour Law, 379
Management information systems, 449–450
 cognitive management systems, 450
 UBMLRSS (*see* Understanding Based Management Liquidity Ratios Support Systems (UBMLRSS))
Mandatory Access Control (MAC), 463
M-ary modulation schemes
 coherent phase shift-keying (PSK), 15–16
 differentially coherent phase shift-keying (DPSK), 13–14
 quadrature amplitude modulation (QAM), 16
Mask generation. *See under* Retinal fundus images
Massive MIMO, 688–689
 block diagram of, 689*f*
 performances of, 692*f*
 system model, 689
Mathematical programming, 378
MATLAB, 199, 355, 418
Maven dependencies, 158
Meal size. *See* Taimen, meal size on SDA
Mean absolute error (MAE), 223, 225
 comparison of SOM
  and k-means, 224*f*
  and KCA, 226*t*
 by RFM score level, 225*f*, 226*t*
Medical applications computer technologies, 422
Medical postgraduates, CVQ-96. *See* Chinese Virtues Questionnaire (CVQ-96)
Medical rules
 knowledge ontology, 486
  of health insurance rule, 486*f*

regulation of NHI self coverage, 487*f*
synchronization system (*see* Medical rules synchronization system, Taiwan)
Medical rules synchronization system, Taiwan, 481–482
automatic synchronization system, 485*f*
medical rules synchronization, 484
MRAS system architecture, 484–485
drug rules checking system, 483
rules deduction model, 482
sharing medical information, 483
synchronization sequence for, 484*f*
MedLEE, 606
Memory-aware multi-core scheduling schemes, 48
Merkle Hash Tree (MH-Tree), 697
Meta(-)heuristic algorithms, 298, 378
Metal component microstructure analysis, 2
example Voronoi diagram
coarsened microstructure of metal, 6*f*
fine microstructure of metal, 6*f*
from random seed points, 4, 4*f*
Metal oxide semiconductor (MOS) circuit, 324
I/V characteristics of, 328*f*
pull-up network, 325*f*
R_{ON} characteristics of, 328*f*
Meteorological database
aerodrome, 101*f*
upper airspace, 100*f*
Microsoft's ADO.NET Entity Framework, 598
Microstructure of weldment. *See* Weldment
MME (Mobility Management Entity), 54*t*
in initial attach for LTE
with GUTI, 60–61. *See also under* GUTI (Global Unique Temporary Identifier)
with IMSI, 58–60
in LTE
network structure, 55
security, 56
threats, 57
Mobile Agent for RFID Privacy Protection (MARP), 212
Mobile communication, 53, 247, 687, 688
Mobile device, remote desktop approach
framework, 308
functions of related development, 309*t*
future works, 320–321
remote control (*see* Remote control mobile device)
research problem, 306
research purpose, 306–308
sequence diagram, 307*f*
screenshot transfer, 311
superuser
obtained on, 310*f*
permission of, on device, 309–310
synchronization of audio streaming, 311–312
system structure, 311*f*
touch event and keyboard event simulation, 312–313
event argument, 313*t*
Mobile Node (MN), 248, 250, 251, 252
Mobile payment. *See* Near field communication (NFC)
Mobile to Mobile function, 133
Mobile wallet, 134
Model visualization (BI technique), 587*t*
Modified Artificial Fish Swarm Algorithm (MAFSA), 298
dynamical vision, 301, 302*f*
experimental results of, 303*t*
feature selection using, 300–302
LIBSVM, 302
Microsoft Window 7 operate system, 302
UCI dataset, 302, 303*t*
Monte Carlo simulations, 130
Motion capture data, 554–555
and path data, fusion of, 555–558
Moving objects, 175, 176, 178, 179, 184
MRASP, 488
bank counter pattern, 488
flow sequence, 488*f*
message format of, 489
module, 485
Multi-core scheduling, 45–46
Multi-hop protocol, 197
Multilevel cluster based Energy efficient Routing Protocol (MERP), 355, 357
average energy dissipation, 356*f*
lifetime-awareness, 355, 356*f*
performance study of, 355*t*
Multimedia content, recognition of gestures, 334
Multiple-input multiple-output (MIMO) systems, 688
massive MIMO, 688–689
MVC (Model_View_Controller) technique, 598
MySQL database, 570, 573, 683
JDBC Driver, 571

N

Nakagami-*m* statistical model, 11
Nakagami-*n* statistical model, 11
Nakagami-*q* fading channels, 11–12

Nakagami-*q* fading channels (*cont.*)
average error probability in (*see* Error probability; M-ary modulation schemes)
high-SNR approximate expressions for (*see under* Error probability)
channel model, 12–13
numerical results, 19–20
Nakagami-*q* statistical model, 11
National Coordinator for Health Information Technology, United States, 151
National Electronic Health Transition Authority (NEHTA), Australia, 151
National Health Insurance (NHI), Taiwan, 481
National Institute of Standards and Technology (NIST)
in cloud login platform, 433
RBAC, redefinition, 241
validator, 158, 158*t*, 159
Natural language processing (NLP), 606, 607
Naver Webtoon application, 147, 147*t*
hit rate, 148*f*
Near field communication (NFC), 133–134, 239–240
efficiency analysis, 139–140
environments, 134*f*
proposed method, 135–136
authentication phase, 137
key agreement phase, 137
payment phase, 137–138
registration phase, 136
safety analysis, 138
replay attack, 138
user impersonation attack, 138139
verification, 138
-SEC method, 133, 135
Nearest Closer (NC) Protocol, 202
evaluation model and analysis, 207–208
simulation results, 206–207
density, lifetime and reliability relationship for, 206*f*
Network assisted teaching, 407
external supervision systems, 409
learning supervision, 410
teaching supervision, 409–410
system design, 407
administration module, 407–408
calendar, 409
course introduction, 408
student module, 408
syllabus, 408–409
teacher module, 408
teaching resources, 409
system implementation, 410
Network marketing, 519, 520
domestic sports network marketing, 519
prospect of, 522
Network video surveillance
forensic artifacts, 342–343
network camera, 343, 343*t*
network video analytic system, 345, 345*t*
network video recorder, 344, 344*t*
Network Voronoi diagram, 177
Neural network, 221, 708
SOM results for application of, 224–225
New Oriental training, 658
New rule of combination, 78, 80, 83, 86
NMEA (National Marine Electronics Association), 190
Non-cyclical scheduling method, 378
Non-radar control, 103
Non-volatile memories, 47–48
Non-zero RGB color images, 117
Normal distribution, 670
for Knapsack Problem, 666
Normalized Cuts algorithm, 624
NVR (Network Video Recorder), 344
NVRAM (Non-volatile RAM) designs, 46, 47–48

O

OFs (Objective Functions), 164
OLAP (online analytical processing), 587*t*, 590
One-Pass run length labeling algorithm, 117
Optimisation strategy, in emergency department. *See* University of Malaya Medical Centre Emergency Department (UMMC-ED)
OR
in CP-ABE algorithm, 466, 468*f*
fusion rule, 78
operation, 58, 117, 120
Orthogonal STBC (OSTBC), 688, 689
design, 691–692
full diversity at full rate, 690
Out of Memory (OOM) killer, 144
processing sequence, 145
Outdoor LED display
automatic adjustment of, 613–614
control method, 614–616
attenuation of lamp, 615
display brightness level, 615. *See also* LED display brightness level
gray scale of display screen, 615
steps, 616
Outlier analysis (BI technique), 587*t*
Outlier detection techniques, 126
Boxplot method, 126

Dixon's test, 127
Grubb's test, 126
OWL
design
inferences engine, 489–490
medical condition, 487
medical rules, 487
MRASP, 488–489
value sector category, 488
document, 484
grammar, 490

P

Parallel coordinates, 529–530
visualization using, 532
custom dynamic filter, 533–534, 534*f*
dimension reorder, 532–533, 533*f*
variables discretization, 532, 533*f*
Particle swarm optimization (PSO), 298, 378
Partitioning curved surface. *See* Binarization partitioning curved surface (BPCS)
Partnership for Governance Reform, 580
Pattern recognition, 32, 394, 641*t*
Pearson correlated coefficient, 475
restraint Pearson correlated coefficient, 476
Pedestrian Dead Reckoning (PDR), 33, 36–37
tracking path of, 37*f*
Pen based computing, 386
Pen based input devices, 387, 390
Pen gesture commands, 385, 386–387
Brushes Adjustment, 389
File Editing, 387, 388
Graphics Toolbar, 389
Integrated Applications, 389–389
Layer Control, 388
System Usability Scale (SUS), 390
Pen gestures with rubber grip pen (PGr), 387
and keyboard (PGk), 387
Periodic function-based data group index, 698
private data
authentication index, 699–700
partition with, 700*f*
system architecture and data flow, 699*f*
Perl, 683
Personal health records, 395
Personalized recommendation, theoretical modelling
construction of neighbourhood formation, 474–475
cosine similarity, 475
Pearson correlated coefficient, 475
restraint Pearson correlated coefficient, 476
generation of recommendation list, 476–477
presentation of evaluation data, 474
Personalized wellness services
implementation, 215–216
customized ubiquitous hospital model, 216*f*
M-RPS based customized service, 214
default privacy weight level, 214*t*
electronic signature and authentication, 215*f*
medical examination with proposed system, 215*f*
patient care at hospital, privacy policy, 213
Phase shift-keying (PSK), 15
M-ary coherent PSK, 15–16, 18
Phishing
definition, 578
literacy, 581, 583
Ability to Identify Phishing Emails Questionnaire, 581
hypothetical method, 580*f*
simple linear regression analysis, 582*t*
PHP (program language), 683
Physics experiments, 405–406
network teaching materials, 407
traditional, 406
virtual experimental teaching software, 407
PINE (Progressive Incremental Network Expansion), 177
Place searching algorithm, 515
fast search algorithm, 515
place first searching, 512
Pochhammer symbol, 14
Point of interest (POI) id, 176, 699, 702
Policy enforcement point (PEP), 466
Portable privacy-preserving Authentication and Access Control Protocol (PAACP), 261
cryptanalysis of, 261–263
Positive psychology, 630
PRAM (Phase change RAM), 46, 47–48
Predict Covariance, 39
Predictive modeling (BI technique), 587*t*
Primary headache, data, 644
diagnosis with ACO clustering approach, 647–648
distribution of headache patient dataset in, 648*f*
parameters' values in, 647*t*
results, 647*t*
Primary users (PUs), 123
Principal component analysis (PCA), 631

Privacy preservation, of VANETs, 260, 263, 264, 271
context privacy, 260, 266*t*
PAACP, 261
cryptanalysis of, 261–263
SafeAnon, 260
Privacy-aware query authentication index, 696
Profile-based privacy protection, 214
Pulse width modulation (PWM), 324
Python (program language), 683

Q
Quadrature amplitude modulation (QAM), 16
M-ary square QAM, 16, 19, 22
Quantitative evaluation, 621
metric of 3D mesh segmentation, 621–623
ultimate measurement accuracy (UMA), 623, 624. *See also* Ultimate measurement accuracy (UMA)
Quasi-orthogonal STBC (QOSTBC), 688
BERs of, 693
design, 691–692
full diversity at full rate, 693
Query processing algorithm. *See* k-nearest neighbor (KNN/k-NN) query processing algorithms
Query processing time with verification time, 703*f*

R
Radboud University Nijmegen Medical Centre (RUNMC), Netherland, 548
Radio frequency identification (RFID) technology, 211, 230
Enhancer Proxy (REP), 212
Guardian system, 212
improved privacy, 212–213
Mobile Agent for RFID Privacy (MARP) protection, 212
personalized wellness services (*see* Personalized wellness services)
RFID Enhancer Proxy (REP), 212
Radio mapping, 32
Random Walks algorithm, 624
Randomized Cuts algorithm, 624
Rate of climb (RoC), 93
Raya and Hubaux's communication scheme, 260
Rayleigh statistical model, 11
Recommender system in u-commerce, 220–221
clustering method using SOM to classify profitable customer, 222
experimental data
for evaluation, 223–224
for implementation, 223
SOM results for application of neural network, 224–225
procedural algorithm for recommendation, 222
Reflected light, LED, 614, 615
optical systems, 615, 617
Registration
access control protocol for VANETs, 263–264
in BINUS International School study, 588, 590*f*
in cloud login platform, 433–434
comparison of time, error and results, 653
in cross cloud mechanism, 242
mobile payment system, 136
in scheme for DS, 233*f*
of 3D point cloud data, 649
Relevance weighting model de-TFIDF
correlation weighting correction, 711
relevancy ranking table, 711
Remote control mobile device
file transfer, 319–320
from PC to device, 320*f*
GPS via, 315–319
application on PC's end, 318*f*
initialization and touch simulation, 313–314
actual phone image after connection, 316f
actual screen taken after connection, 316f
drag view on PC's end, 317*f*
phone call via, 318*f*
remote phone call, 314–315
Remote Resource Management (RRM), 55
Replenishment efficiency, 414, 417
effect of cargo quantity on, 418, 418*f*
Retinal fundus images, 561–563
datasets of test images, 564*t*
segmentation or masking performance, 566*f*
sensitivity after thresholding, 565, 565*t*
specificity after thresholding, 565, 565*t*
examples, 562*f*
fundus mask generation, 563–564
Retinal imaging. *See* Retinal fundus images
RFM (Recency, Frequency, and Monetary) score, 220–221

SOM and k-means, comparison, 224*f*
RFM model, 284, 286–287
5 equal intervals of, 287*f*
analysis sheet, 287*f*
customer value, 284–285
analysis (*see* Customer value analysis)
matrix quadrants, 292*f*
two-dimensional matrix, 285*f*
data mining, 285–286
RFGP model and, 287–289
distribution of various customer indicators, 291*t*
group results, 291*t*
scoring standard, 290*t*
RGB values, 562
Rice statistical model, 11. *See also* Nakagami-*n* statistical model
RNTI (Radio Network Temporary Identities), 54*t*
Road networks, 175, 176
Role-Based Access Control (RBAC) technology, 240
authorization mechanisms, 240
National Institute of Standards and Technology (NIST) definition, 241
third-part (T-RBAC) database, 243, 244
token-based architecture, 244*f*
Rough set, 708
Routing protocols. *See also* LEACH Protocol; Nearest Closer (NC) Protocol; Single-hop Protocol
for hierarchical clustering (*see* Hierarchical clustering WSNs)
MERP, 355, 355*t*, 356*f*, 357
TEEN protocol, 351, 356*f*, 357, 358*f*
RPL (IPv6 Routing Protocol for Low Power and Lossy Networks), 163, 164, 166
Rules deduction model, 482–483
RxNorm, 607

S
SafeAnon scheme, 260
Sanxiaomingshi training, 658
Schedule Management, 298
Scheduling algorithm, 416
access phase, 265–266
authentication phase, 264
Cargo Capacity Hierarchy Scheduling algorithm, 417–418, 418*f*
performance, 267–269
average waiting time, 268*f*, 269*f*
efficiency comparison, 267*t*
registration phase, 263–264
Scheme for VANETs, 263
security analysis
authentication, 270
authorization, 270
backward secrecy, 270
forward secrecy, 269–270
replay attack, 270
security features, comparison, 266
Simple Scheduling algorithm, 416–417
Secondary users (SUs), 124
Secret sharing tasks, 458
SECSPP (secure and efficient communication scheme), 260, 261
comparison with other schemes, 271
efficiency, 267, 267*t*
security features, 266*t*
Secure information management, 457–458
crypto-biometric sharing schemes, 459
security of threshold schemes, 458
strategic information management, 459–461
Secure payments, 135
Security analysis, LTE, 61
comparative analysis, 63*t*
decryption, 62
encryption, 62
error detection and verification, 63–64
key definition, 62
Sekhar method, 135
Self Organizing Map, 396. *See also* Artificial neural network
Self-Cleansing Intrusion Tolerance (SCIT), 362
cleansing overhead, 365
virtualization-based, 362–363
Self-selected scheduling, 378
Semantic analysis, 450, 455, 637
Semantic search, 682, 683*f*
Semantic web techniques, 490. *See also* Medical rules synchronization system, Taiwan
Sensor nodes (SNs), 45–46, 352, 355
architecture, 47*f*
Seoul National University Bundang Hospital (SNUBH), 152, 158
Server-side script, 399
Service integration, 545*f*. *See also* Clinical Decision Support (CDS) applications
protocol of integration, 546*t*
Service Provider (SP), 260
Service right list (SRL), 261
SET method, 135, 139
S-GRID (Scalable Grid), 176, 177, 178, 180

Shape Diameter Function algorithm, 624
Shiftwork. *See also* Staff scheduling
demand, 382
scheduling strategy
initialisation mode, 380
insufficient mode, 382–383
sufficient mode, 382
THOR, 381*f*
workforce, 381
Short Message System (SMS), 306. *See also* Cloud service; User monitor algorithm
design of local monitoring system based on, 439–440
structure of, 439*f*
local monitoring system based on, 437–438
Shortcut keys (SK), 387
Signal to noise ratios (SNRs), 692
M-ary coherent PSK, 18
M-ary DPSK, 17
M-ary square QAM, 19
moment-generating function (MGF), 12, 13
noncoherent correlated binary signals, 18
probability density function (PDF), 12
Signature chaining, 697
Signature-based approaches, 696, 697
Signed XML document, 112
creation of, 110*f*
verification for authentication, 108–109, 109*f*, 110*f*
Simple Scheduling (SS) algorithm, 414
Simplified rotation process, 362, 364, 365–366
probability density function of, 366, 367*f*
Single sign-on (SSO) system, 430
based on fingerprint, 431
conception and application of, 430
design of, 430–431
system structure, 431*f*
Single-hop Protocol, 201
evaluation model and analysis, 207
simulation results, 203
density, lifetime and reliability relationship for, 204*f*
Single-input single-output (SISO) systems, 688
Six degree of freedom (6DOF) inertial navigation, 33
Size of false positive data, 704*f*
SkyDrive, 429
Skyhook, 33
Slotted Aloha scheme, 25
Smartphones, 34, 36, 40, 133, 143, 144
HTC One V, 309
user responsiveness, 148
SMR (Standard metabolic rate), 443
SNDB (Spatial Network Databases), 177
SNOMED CT, 607
Social facilitation, 538
experimental design, 540*t*
learning system, 539, 539*f*
procedure, 540
pseudo peer progress display, 540
results, 540–541
assessment score, 541*t*
learning performance score, 541*t*
study, participants, 539
Social facilitation theory, 538
Social media enhanced environment. *See* Associative recommendation of learning contents; Eye-tracking, in social media enhanced environment
Social Network Service (SNS), 152
social media comparison, 156*t*
Social presence theory, 538
Socialization, 153, 155, 158, 159
CRS validation board for, 157*f*
SOM (self-organizing map), 219–220
for application of neural network, 224–225
clustering method, to classify profitable customer, 222
learning algorithm, 221*t*
for clustering of user's information, 223*t*
Source optimizing, 284
Space-time block code (STBC), 689–690
orthogonal STBC, 688, 689
design, 690–691
quasi-orthogonal STBC, 688
design, 691–692
Space time coding (STC), 688
Spare server insertion, 362, 364, 366–368
Sparse ICP (Sparse Iterative Closest Point) algorithm, 649, 651
advantages and disadvantages, 654*t*
Sparsity, 649
Spatial multiplexing (SM), 688
Specific dynamic action (SDA), 443
Spectrum sensing data falsification (SSDF) attack, 124
SPKI (Simple Public Key Infrastructure), 107–108
composition and operation of certificate requesting unit, 111*f*
creation and transmission of certificate, 111
Sports marketing, 520. *See also* Sports network marketing
based on Internet, 521
Sports network marketing
characteristics, 522–523

competitive advantage of, 523
marketing, 520
network marketing, 520
sports marketing, 520
strategical analysis of, 525
place strategy of, 526
price strategy of, 525–526
product strategy of, 525
promotion strategy of, 526
SPSS17.0, software, 630
Staff scheduling. *See also* Shiftwork
leave allocation, 380
mandatory rest period, 380
number of workforce, 379
shift pattern in, 380, 380*t*
workforce allocation, 379
Static Separation of Duty (SSD), 241
SSoD, 243
Stimulus generalization concept, 582
of behavioristic psychology, 578, 579, 580
Strategic information management, security features for, 459–461
STT-MRAM (Spin-Torque Transfer Magnetic RAM), 46, 47–48
Subordinate cluster head (SCH), 350, 352, 355
Support vector machines (SVM), 298, 503, 708
for colonies recognition and classification
characteristics of all colonies, 506
construction of decision tree for colonies classification, 507
Euclidean-distance based, 504
feature selection using AFSA, 301*f*
results of, 508*t*
Surveillance of antimicrobial use, 596
building surveillance platform, 598
integration with clinical information system, 598
knowledge base construction, 596–597. *See also* Knowledge Translation Platform (KTP)
monitoring drugs' consumption, 599–600
statistical indicator, 597–598
Swarm intelligence. *See also* Artificial Fish Swarm Algorithm (AFSA)
follow, 299
prey, 300
swarm, 299–300
Symbol error probability (SEP), 12
exact and approximated, 20*f*, 21*f*

T

Table offset formatter (TOF), 69, 70
circuit diagram of, 70*f*
for VLC decoding, 71
Tag to Mobile function, 133
Taimen, meal size on SDA, 443–444
metabolic rate, 446*f*
variables of, 444*t*
Teaching model
anatomical learning, 676
of compiler principles, 639
efficiency in class, 675–676
MIT's course, 639
traditional (TTM), 640
traditional physics experiments, 406, 409
TeamViewer:QuickSupport, 309
TEEN protocol, 351, 357
average energy dissipation, 356*f*
residual energy of different nodes, 358*f*
Temporal and sequential patterns analysis (BI technique), 587*t*
Temporary Mobile Subscriber Identities (TMSI), 248
Text classification algorithm, 709*f*
Thermal wind, 94
Third-generation mobile system Universal Mobile Telecommunication System (3G/UMTS), 247–248
future works, 256
network architecture, 249–250
performance analysis, 255–256
protocols, 250
fast re-authentication protocol of AS, 253
fast re-authentication protocol of F-AAA, 252–253
fast re-authentication protocol of H-AAA, 253
full authentication protocol, 250–252
security analysis, 254–255
Brute-Force attack, 254
Byzantine insiders, 254
Secret key Pre-shared *k*, 254
Threshold schemes, security of, 458
Time searching algorithms, 514
bottom-up, 514–515
initial time, 514
search rules, 515
Time First searching, 512
unavailable time search algorithm, 514
up-bottom, 515
TIRI-Integrity Action Indonesia, 580
Trajectory prediction techniques and theory
aircraft's position in compliance with wind parameter, 97–98
airspeed theory, 92–95

Trajectory prediction techniques and theory (*cont.*)
distance calculation between two points, 95–97
trajectory prediction theory, 90–92
Trajectory prediction, 102*f*
Trajectory prediction theory, 90
airline procedure model, 91*f*
elements for trajectory prediction, 91*f*
key elements for, 90–92
Transport Layer Security (TLS), 233
Trauma Hospital Online Roster (THOR), 378
implementation, 383
scheduling strategy, 381*f*
Trickle algorithm, 164
True airspeed (TAS), 91, 92, 93
Trust factor, 124, 439, 440, 441
Tuple level signature, 697
Twitter, 156, 156*t*
Two-dimension (2D) hierarchical network topology, 352

U
Ubiquitous computing, 333
Ubiquitous learning (U-learning), 471
Ubiquitous positioning
complementary extended Kalman filter, 38–40
future works, 42
GPS, 37–38
issue description, 36
pedestrian dead reckoning, 36–37
positioning methods, comparison of, 41*f*
proposed algorithm, 40–41
integrated hybrid, 40
tracking path by, 41*f*
test environment, 34, 35*f*
WiFi fingerprinting, 37–38
Ultimate measurement accuracy (UMA), 623, 624
comparisons of metric and Rand index, 626–627, 626*f*, 627*f*
properties comparison of FME, AME and, 625
robust to random and degenerative cases, 625*f*
sensitivity to hierarchical refinement, 625*f*
tolerance to imprecision of cut boundaries, 626*f*
shape feature form factor, 623
Ultra-high frequency (UHF) tags, 212
Understanding Based Management Liquidity Ratios Support Systems (UBMLRSS), 450
subclasses of cognitive systems, 450*f*
cognitive financial analysis in, 454*f*
$\text{UBMLRSS}_{(cu\text{-}ca\text{-}mp)}$ system, 452–453
$\text{UBMLRSS}_{(cu\text{-}q\text{-}mp)}$ system, 451–452
$\text{UBMLRSS}_{(q\text{-}ca\text{-}mp)}$ system, 453–455
Unified Medical Language System (UMLS), 607
United States Agency for International Development (USAID), 580
University of Malaya Medical Centre Emergency Department (UMMC-ED), 378
scheduling constraints, 379
leave allocation, 380
Malaysian Government General Order on Labour Law, 379
mandatory rest period, 380
number of workforce, 379
shift pattern in, 380, 380*t*
workforce allocation, 379
shiftwork scheduling strategy (*see* Shiftwork)
THOR (*see* Trauma Hospital Online Roster)
Unknown Node (UN), 33
Uplink (UL) transmit power, 688
User Equipment (UE), 54*t*
Initial Attach for, 55–56
process, 56*f*
User monitor algorithm, 440
based on improved k-means, 440
determining, 441
initial cluster center, 441
remaining cluster center, 441–442
searching clustering objects, 441–442
User periodic habits
memory management policy, 145
with priority control, 146*f*
implemented codes for data collection, 147*f*
User priorities (UPs), 24, 25
comparison
of average packet delay, 27*f*
of normalized throughput, 27*f*

V
Validation, CRS, 154. *See also* Care record summary (CRS)
board for socialization, 157, 157*f*
socialized service structure of, 155*f*
system of, 153

effects of I_{min}, 167, 167*f*
effects of *k*, 167–168, 168*f*
test bed, 166–167
Wireless sensor networks (WSNs), routing protocols in, 188, 197–198, 349–350. *See also* Hierarchical clustering WSNs
coverage and lifetime, 199
energy, density, latency, accuracy trade-offs, 200
evaluation model and analysis, 207–209
evaluation parameters, 202
coverage, 202
density, 203
lifetime, 202
reliability, 202
hybrid routing protocol for, 351
J-Sim, 200–201
power efficient topologies for, 351
radius and number of sensors, 199
routing protocols, 201
LEACH Protocol, 201
Nearest Closer Protocol, 202
Single-hop Protocol, 201
simulation results, 203–207
Wireless sensor nodes, 45–46. *See also* HAMC (Hybrid memory-Aware Multi-Core) scheduling
experiment, 50–51

X

XACML (eXtensible Access Control Markup Language), 214
access control policy, 465
decision-making model based on, 465–466, 466*f*
Xen (virtual machine software), 362
Xiangya Medical School of Central South University, 671
benefits, 677
difficulties faced in, 672
deficiency of teaching materials and methods, 673–674
inappropriate learning pattern, 672–673
interdepartmental co-ordination in selecting student, 674
language barrier and cultural differences, 672
problems of assessment and feedback, 674
suggestions, 674–675
assessment and feedback, 677
teaching and learning skills, 675–677
XML document for forensic artifacts
digital evidence markup language (DEML), 347
digital forensics XML (DFXML), 347
in network camera, 346*f*
in network video recorder, 347*f*
XML file, for validation, 153
XOR operation, 139

Y

YCbCr color, 334
image, 117, 118

Z

Z-axis diagram, acceleration of, 36, 37*f*
Zero count operation, 70

VT on KS, 152, 156, 160. *See also under* Care record summary (CRS)
VANETs (Vehicular ad hoc networks)
access control protocol for, 261–263
comparison, 266
performance (*see under* Scheme for VANETs)
scheme, 263–266. *See also* Scheme for VANETs
security analysis, 266–270. *See also under* Scheme for VANETs
non–safety application for, 260
Variable length code table (VLCT), 69
Variable length coding (VLC), 68
instruction count comparisons of, 73*f*
TOF for decoding, 71
Vertical replenishment (VERTREP), 413, 414
scheduling model, 414
transporting process, 415
VERTREP batch, 415
VERTREP mission, 414
VERTREP scheduling problem, 416
simulation results, 417–418
Vertical speed indicator (VSI), 93
Video codec, 68
Video evidence, 343, 347. *See also* Network video surveillance
retrieval in network surveillance systems, 342
Video surveillance system. *See also* Network video surveillance
Vincenty's Formula, 96
Virtual classrooms, 538
Virtual Medical Record (vMR), 550
VirtualBox (virtual machine software), 362
Virtualization-based server system
exposure time adjustment, 362, 364, 366–368, 367*f*, 374*f*, 374*t*
artificially-generated packet rate, 373*f*
effectiveness of, 369–371
performance comparison between SCIT, 372*f*
performance with fixed cluster size, 370*f*, 371*f*
simplified rotation process, 362, 364, 365–366
spare server insertion, 362, 364, 366–368, 374*f*, 374*t*
effectiveness of, 371–372
Visual C++ 6.0, 668
VMS (Video Management Software/System), 343, 344
VMWare (virtual machine software), 362
VN3 (Voronoi-based Network Nearest Neighbor), 177
VNC (Virtual Network Computing) protocol, 308
structure of, 308*f*
Voltage and clock scaling, 144
Voronoi diagram, 1
building, 3–4
component, designation of, 4*f*
example of
coarsened microstructure of metal, 6*f*
fine microstructure of metal, 6*f*
from random seed points, 4, 4*f*
in finite element simulation field, 2
methodology, 3*f*
Voronoi tessellation, 1

W

Wards method, 274
Web server, 134, 223, 399
Web service systems, 362
Web-based learning, 538, 539, 541
Weldment
expression of, 4–5
Voronoi diagram, to typical laser weld, 7*f*
WEP (Wired Equivalent Privacy), 248
WHO Adverse Reactions Terminology (WHO-ART), 607
Wide area augmentation system (WAAS), 188
WiFi
fingerprinting, 37–38, 188
future works, 42
reference points for, 38*f*
tracking path by, 38*f*
positioning, 32–33
Windows Passport technology, 430
Wireless Body Area Network (WBAN), 23
Wireless local area network (WLAN), 32, 188, 248. *See also* Third-generation mobile system Universal Mobile Telecommunication System (3G/UMTS)
standard (*see* IEEE 802.11)
Wireless Personal Area Network (WPAN), 23
Wireless sensor networks (WSNs), adaptive trickle scheme for, 163–164
basic operations, 164–165
convergence time, 165
energy model, 165
node residual energy estimation, 165–166
simulations, 166
effects of adaptability, 168–170. *See also* Adaptability, WSN

FSC
www.fsc.org
MIX
Papier | Fördert
gute Waldnutzung
FSC® C083411

Kolibri 360 GmbH
Ferdinand-Jühlke-Straße 7
99095 Erfurt, Deutschland
produktsicherheit@kolibri360.de